LINEAR ALGEBRA

THIRD EDITION

JOHN B. FRALEIGH

RAYMOND A. BEAUREGARD

University of Rhode Island

Historical Notes by Victor J. Katz

University of the District of Columbia

ADDISON-WESLEY PUBLISHING COMPANY

Reading, Massachusetts • Menlo Park, California • New York
Don Mills, Ontario • Wokingham, England • Amsterdam • Bonn
Sydney • Singapore • Tokyo • Madrid • San Juan • Milan • Paris

Sponsoring Editor: Laurie Rosatone
Production Supervisor: Peggy McMahon
Marketing Manager: Andrew Fisher
Senior Manufacturing Manager: Roy E. Logan
Cover Designer: Leslie Haimes
Composition: Black Dot Graphics

Library of Congress Cataloging-in-Publication Data
Fraleigh, John B.
        Linear algebra / John B. Fraleigh, Raymond A. Beauregard ;
historical notes by Victor Katz. -- 3rd ed.

        p.   cm.
        Includes index.
        ISBN 0-201-52675-1
        1. Algebras, Linear.   I. Beauregard, Raymond A.   II. Katz, Victor J.
III. Title.
        QA184.F73   1995
        512'.5--dc20   93-49722
                                                                                              CIP

MATLAB is a registered trademark of The MathWorks, Inc.

Many of the designations used by manufacturers and sellers to distinguish their products are claimed as trademarks. Where those designations appear in this book, and Addison-Wesley was aware of a trademark claim, the designations have been printed in initial cap or all caps.

Reprinted with corrections, November 1995.

6 7 8 9 10-MA-97

PREFACE

Our text is designed for use in a first undergraduate course in linear algebra. Because linear algebra provides the tools for dealing with problems in fields ranging from forestry to nuclear physics, it is desirable to make the subject accessible to students from a variety of disciplines. For the mathematics major, a course in linear algebra often serves as a bridge from the typical intuitive treatment of calculus to more rigorous courses such as abstract algebra and analysis. Recognizing this, we have attempted to achieve an appropriate blend of intuition and rigor in our presentation.

NEW FEATURES IN THIS EDITION

- **Evenly Paced Development:** In our previous edition, as in many other texts, fundamental notions of linear algebra—including linear combinations, subspaces, independence, bases, rank, and dimension—were first introduced in the context of axiomatically defined vector spaces. Having easily mastered matrix algebra and techniques for solving linear systems, many students were unable to adjust to the discontinuity in difficulty. In an attempt to eradicate this abrupt jump that has plagued instructors for years, we have extensively revised the first portion of our text to introduce these fundamental ideas gradually, in the context of $\mathbb{R}^n$. Thus linear combinations and spans of vectors are discussed in Section 1.1 and subspaces and bases are introduced where they first naturally occur in the study of the solution space of a homogeneous linear system. Independence, rank, dimension, and linear transformations are all introduced before axiomatic vector spaces are defined. The chapter dealing with vector spaces (Chapter 3) is only half its length in the previous edition, because the definitions, theorems, and proofs already given can usually be extended to general vector spaces by replacing "vectors in $\mathbb{R}^n$" by "vectors in a vector space V." Most of the reorganization of our text was driven by our desire to tackle this problem.

- **Early Geometry:** Vector geometry (geometric addition, the dot product, length, the angle between vectors), which is familiar to many students from calculus, is now presented for vectors in $\mathbb{R}^n$ in the first two sections of Chapter 1. This provides a geometric foundation for notions of linear combinations and subspaces. Instructors may feel that students will be uncomfortable working immediately in $\mathbb{R}^n$. It is our experience that this causes no difficulty; students can compute a dot product of vectors with five components as easily as with two components.

- **Application to Coding:** An application of matrix algebra to binary linear codes now appears at the end of Chapter 1.

- **MATLAB:** The professional PC software MATLAB is widely used for computations in linear algebra. Throughout the text, we have included optional exercises to be done using the Student Edition of MATLAB. Each exercise set includes an explanation of the procedures and commands in MATLAB needed for the exercises. When performed sequentially throughout the text, these exercises give an elementary tutorial on MATLAB. Appendix D summarizes for easy reference procedures and commands used in the text exercises, but it is not necessary to study Appendix D before plunging in with the MATLAB exercises in Section 1.1. We have not written MATLAB .M-files combining MATLAB's commands for student use. Rather, we explain the MATLAB commands and ask students to type a single line, just as shown in the text, that combines the necessary commands, and then to edit and access that line as necessary for working a sequence of problems. We hope that students will grasp the commands, and proceed to write their own lines of commands to solve problems in this or other courses, referring if necessary to the MATLAB manual, which is the best reference. Once students have had some practice entering data, we do supply files containing matrix data for exercises to save time and avoid typos in data entry. For example, the data file for the MATLAB exercises for Section 1.4 is FBC1S4.M, standing for Fraleigh Beauregard Chapter 1 Section 4. The data files are on the disk containing LINTEK, as explained in the next item.

- **LINTEK:** The PC software LINTEK by Fraleigh, designed explicitly for this text, has been revised and upgraded and is free to students using our text. LINTEK is not designed for professional use. In particular, matrices can have no more than 10 rows or columns. LINTEK does provide some educational reinforcements, such as step-by-step execution and quizzes, that are not available in MATLAB. All information needed for LINTEK is supplied on screen. No manual is necessary. The matrix data files for MATLAB referred to in the preceding item also work with LINTEK, and are supplied on the LINTEK disk. Many optional exercise sets include problems to be done using LINTEK.

FEATURES RETAINED FROM THE PREVIOUS EDITION

- **Linear Transformations:** In the previous edition, we presented material on linear transformations throughout the text, rather than placing the topic toward the end of a one-semester course. This worked well. Our students gained much more understanding of linear transformations by working with them over a longer period of time. We have continued to do this in the present edition.

- **Eigenvalues and Eigenvectors:** This topic is introduced, with applications, in Chapter 5. Eigenvalues and eigenvectors recur in Chapters 7, 8, and 9, so students have the opportunity to continue to work with them.

- **Applications:** We believe that if applications are to be presented, it is best to give each one as soon as the requisite linear algebra has been developed, rather than deferring them all to the end of the text. Prompt work with an application reinforces the algebraic idea. Accordingly, we have placed applications at the ends of the chapters where the pertinent algebra is developed, unless the applications are so extensive that they merit a chapter by themselves. For example, Chapter 1 concludes with applications to population distribution (Markov chains) and to binary linear codes.

- **Summaries:** The summaries at the ends of sections have proved very convenient for both students and instructors, so we continue to provide them.

- **Exercises:** There are abundant pencil-and-paper exercises as well as computer exercises. Most exercise sets include a ten-part true-false problem. That exercise gives students valuable practice in deciding whether a mathematical statement is true, as opposed to asking for a proof of a given true statement. Answers to odd-numbered exercises having numerical answers are given at the back of the text. Usually, a requested proof or explanation is not given in the answers, because having it too readily available does not seem pedagogically sound. Computer-related exercises are included at the end of most exercise sets. Their appearance is signaled by a disk logo.

- **Complex Numbers:** We use complex numbers when discussing eigenvalues and diagonalization in Chapter 5. Chapter 9 is devoted to linear algebra in $\mathbb{C}^n$. Some instructors bemoan a restriction to real numbers throughout most of the course, and they certainly have a point. We experimented when teaching from the previous edition, making the first section of the chapter on complex numbers the first lesson of the course. Then, as we developed real linear algebra, we always assigned an appropriate problem or two from the parallel development in subsequent sections on $\mathbb{C}^n$. Except for discussing the complex inner product and conjugate transpose, very little extra time was necessary. This technique proved to be feasible, but our students were not enamored with pencil-and-paper computations involving complex numbers.

- **Dependence Chart:** A dependence chart immediately follows this preface, and is a valuable aid in constructing a syllabus for the course.

SUPPLEMENTS

- **Instructor's Solutions Manual:** This manual, prepared by the authors, is available to the instructor from the publisher. It contains complete solutions, including proofs, for all of the exercises.

- **Student's Solutions Manual:** Prepared by the authors, this manual contains the complete solutions, including proofs, from the Instructor's Solutions Manual for every third problem (1, 4, 7, etc.) in each exercise set.

- **LINTEK:** This PC software, discussed above, is included with each copy of the text.

- **Testbank:** The authors have created a substantial test bank. It is available to the instructor from the publisher at no cost. Note that each multiple-choice problem in the bank can also be requested as open-ended—that is, with the choices omitted—as long as the problem still makes sense. Problems are easily selected and printed using software by Fraleigh. We use this bank extensively, saving ourselves much time—but of course, we made up the problems!

ACKNOWLEDGMENTS

Reviewers of text manuscripts perform a vital function by keeping authors in touch with reality. We wish to express our appreciation to all the reviewers of the manuscript for this edition, including: Michael Ecker, Penn State University; Steve Pennell, University of Massachusetts, Lowell; Pauline Chow, Harrisburg Area Community College; Murray Eisenberg, University of Massachusetts, Amherst; Richard Blecksmith, Northern Illinois University; Michael Tangredi, College of St. Benedict; Ronald D. Whittekin, Metropolitan State College; Marvin Zeman, Southern Illinois University at Carbondale; Ken Brown, Cornell University; and Michal Gass, College of St. Benedict.

In addition, we wish to acknowledge the contributions of reviewers of the previous editions: Ross A. Beaumont, University of Washington; Paul Blanchard, Boston University; Lawrence O. Cannon, Utah State University; Henry Cohen, University of Pittsburgh; Sam Councilman, California State University, Long Beach; Daniel Drucker, Wayne State University; Bruce Edwards, University of Florida; Murray Eisenberg, University of Massachusetts, Amherst; Christopher Ennis, University of Minnesota; Mohammed Kazemi, University of North Carolina, Charlotte; Robert Maynard, Tidewarter Community College; Robert McFadden, Northern Illinois University; W. A. McWorter, Jr., Ohio State University; David Meredith, San Francisco State University; John Morrill, DePauw University; Daniel Sweet, University of Maryland; James M. Sobota, University of Wisconsin—LaCrosse; Marvin Zeman, Southern Illinois University.

We are very grateful to Victor Katz for providing the excellent historical notes. His notes are not just biographical information about the contributors to the subject; he actually offers insight into its development.

Finally, we wish to thank Laurie Rosatone, Cristina Malinn, and Peggy McMahon of Addison-Wesley, and copy editor Craig Kirkpatrick for their help in the preparation of this edition.

DEPENDENCE CHART

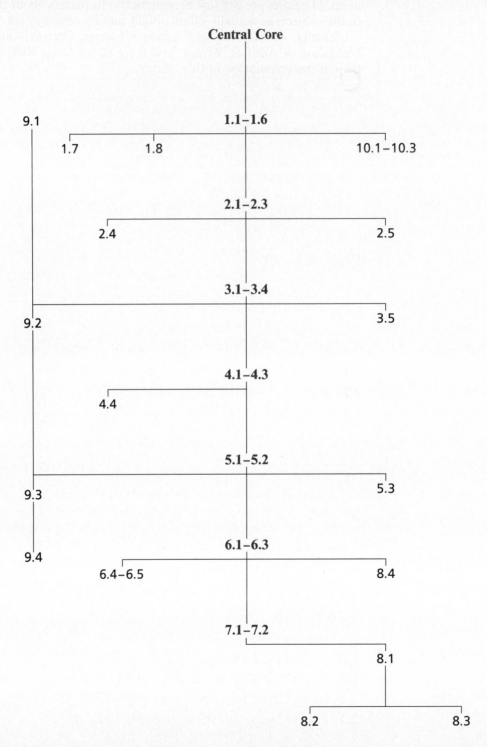

CONTENTS

VECTORS, MATRICES, AND LINEAR SYSTEMS

We have all solved simultaneous linear equations—for example,

$$2x + y = 4$$
$$x - 2y = -3.$$

We shall call any such collection of simultaneous linear equations a *linear system*. Finding all solutions of a linear system is fundamental to the study of linear algebra. Indeed, the great practical importance of linear algebra stems from the fact that *linear systems can be solved by algebraic methods*. For example, a *linear* equation in one unknown, such as $3x = 8$, is easy to solve. But the nonlinear equations $x^5 + 3x = 1$, $x^x = 100$, and $x - \sin x = 1$ are all difficult to solve algebraically.

One often-used technique for dealing with a nonlinear problem consists of *linearizing the problem*—that is, approximating the problem with a linear one that can be solved more easily. Linearization techniques often involve calculus. If you have studied calculus, you may be familiar with Newton's method for approximating a solution to an equation of the form $f(x) = 0$; an example would be $x - 1 - \sin x = 0$. An approximate solution is found by solving sequentially several linear equations of the form $ax = b$, which are obtained by approximating the graph of f with lines. Finding an approximate numerical solution of a partial differential equation may involve solving a linear system consisting of thousands of equations in thousands of unknowns. With the advent of the computer, solving such systems is now possible. The feasibility of solving huge linear problems makes linear algebra currently one of the most useful mathematical tools in both the physical and the social sciences.

The study of linear systems and their solutions is phrased in terms of *vectors* and *matrices*. Sections 1.1 and 1.2 introduce vectors in the Euclidean spaces (the plane, 3-space, etc.) and provide a geometric foundation for our work. Sections 1.3–1.6 introduce matrices and methods for solving linear systems and study solution sets of linear systems.

1.1   VECTORS IN EUCLIDEAN SPACES

We all know the practicality of two basic arithmetic operations—namely, adding two numbers and multiplying one number by another. We can regard the real numbers as forming a line which is a one-dimensional space. In this section, we will describe a useful way of adding two points in a plane, which is a two-dimensional space, or two points in three-dimensional space. We will even describe what is meant by n-dimensional space and define addition of two points there. We will also describe how to multiply a point in two-, three-, and n-dimensional space by a real number. These extended notions of addition and of multiplication by a real number are as useful in n-dimensional space for $n > 1$ as they are for the one-dimensional real number line. When these operations are performed in spaces of dimension greater than one, it is conventional to call the elements of the space *vectors* as well as *points*. In this section, we describe a physical model that suggests the term *vector* and that motivates addition of vectors and multiplication of a vector by a number. We then formally define these operations and list their properties.

Euclidean Spaces

Let $\mathbb{R}$ be the set of all real numbers. We can regard $\mathbb{R}$ geometrically as the Euclidean line—that is, as **Euclidean 1-space**. We are familiar with rectangular x,y-coordinates in the Euclidean plane. We consider each ordered pair (a, b) of real numbers to represent a point in the plane, as illustrated in Figure 1.1. The set of all such ordered pairs of real numbers is **Euclidean 2-space**, which we denote by $\mathbb{R}^2$, and often call *the plane*.

To coordinatize space, we choose three mutually perpendicular lines as coordinate axes through a point that we call the *origin* and label **0**, as shown in Figure 1.2. Note that we represent only half of each coordinate axis for clarity. The coordinate system in this figure is called a *right-hand* system because, when the fingers of the right hand are curved in the direction required to rotate the positive x-axis toward the positive y-axis, the right thumb points up the z-axis, as shown in Figure 1.2. The set of all ordered triples (a, b, c) of real numbers is **Euclidean 3-space**, denoted $\mathbb{R}^3$, and often simply referred to as *space*.

Although a Euclidean space of dimension four or more may be difficult for us to visualize geometrically, we have no trouble writing down an ordered quadruple of real numbers such as $(2, -3, 7, \pi)$ or an ordered quintuple such as $(-0.3, 3, 2, -5, 21.3)$, etc. Indeed, it can be useful to do this. A household budget might contain nine categories, and the expenses allowed per week in each category could be represented by an ordered 9-tuple of real numbers. Generalizing, the set $\mathbb{R}^n$ of all ordered n-tuples $(x_1, x_2, \ldots, x_n)$ of real numbers is **Euclidean n-space**. Note the use of just one letter with consecutive integer subscripts in this n-tuple, rather than different letters. We will often denote an element of $\mathbb{R}^2$ by (x_1, x_2) and an element of $\mathbb{R}^3$ by (x_1, x_2, x_3).

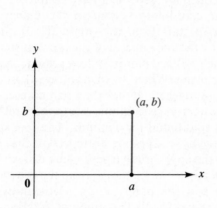

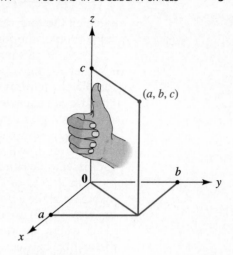

FIGURE 1.1
Rectangular coordinates in the plane.

FIGURE 1.2
Rectangular coordinates in space.

The Physical Notion of a Vector

We are accustomed to visualizing an ordered pair or triple as a point in the plane or in space and denoting it geometrically by a dot, as shown in Figures 1.1 and 1.2. Physicists have found another very useful geometric interpretation of such pairs and triples in their consideration of forces acting on a body. The motion in response to a force depends on the *direction* in which the force is applied and on the *magnitude* of the force—that is, on how hard the force is exerted. It is natural to represent a force by an arrow, pointing in the direction

---

HISTORICAL NOTE THE IDEA OF AN n-DIMENSIONAL SPACE FOR $n > 3$ reached acceptance gradually during the nineteenth century; it is thus difficult to pinpoint a first "invention" of this concept. Among the various early uses of this notion are its appearances in a work on the divergence theorem by the Russian mathematician Mikhail Ostrogradskii (1801–1862) in 1836, in the geometrical tracts of Hermann Grassmann (1809–1877) in the early 1840s, and in a brief paper of Arthur Cayley (1821–1895) in 1846. Unfortunately, the first two authors were virtually ignored in their lifetimes. In particular, the work of Grassmann was quite philosophical and extremely difficult to read. Cayley's note merely stated that one can generalize certain results to dimensions greater than three "without recourse to any metaphysical notion with regard to the possibility of a space of four dimensions." Sir William Rowan Hamilton (1805–1865), in an 1841 letter, also noted that "it *must* be possible, in *some* way or other, to introduce not only triplets but *polyplets*, so as in some sense to satisfy the symbolical equation

$$a = (a_1, a_2, \ldots, a_n);$$

a being here one symbol, as indicative of one (complex) thought; and $a_1, a_2, \ldots, a_n$ denoting n real numbers, positive or negative."

Hamilton, whose work on quaternions will be mentioned later, and who spent much of his professional life as the Royal Astronomer of Ireland, is most famous for his work in dynamics. As Erwin Shrödinger wrote, "the Hamiltonian principle has become the cornerstone of modern physics, the thing with which a physicist expects *every* physical phenomenon to be in conformity."

in which the force is acting, and with the length of the arrow representing the magnitude of the force. Such an arrow is a *force vector*.

Using a rectangular coordinate system in the plane, note that if we consider a force vector to start from the origin $(0, 0)$, then the vector is completely determined by the coordinates of the point at the tip of the arrow. Thus we can consider each ordered pair in $\mathbb{R}^2$ to represent a vector in the plane as well as a point in the plane. When we wish to regard an ordered pair as a vector, we will use square brackets, rather than parentheses, to indicate this. Also, we often will write vectors as columns of numbers rather than as rows, and bracket notation is traditional for columns. Thus we speak of the *point* $(1, 2)$ in $\mathbb{R}^2$ and of the *vector* $[1, 2]$ in $\mathbb{R}^2$. To represent the *point* $(1, 2)$ in the plane, we make a dot at the appropriate place, whereas if we wish to represent the *vector* $[1, 2]$, we draw an arrow emanating from the origin with its tip at the place where we would plot the point $(1, 2)$. Mathematically, there is no distinction between $(1, 2)$ and $[1, 2]$. The different notations merely indicate different views of the same member of $\mathbb{R}^2$. This is illustrated in Figure 1.3. A similar observation holds for 3-space. Generalizing, each *n*-tuple of real numbers can be viewed both as a point $(x_1, x_2, \ldots, x_n)$ and as a vector $[x_1, x_2, \ldots, x_n]$ in $\mathbb{R}^n$. We use boldface letters such as $\mathbf{a} = [a_1, a_2]$, $\mathbf{v} = [v_1, v_2, v_3]$, and $\mathbf{x} = [x_1, x_2, \ldots, x_n]$ to denote vectors. In written work, it is customary to place an arrow over a letter to denote a vector, as in $\vec{a}$, $\vec{v}$, and $\vec{x}$. The *i*th entry x_i in such a vector is the **ith component** of the vector. Even the real numbers in $\mathbb{R}$ can be regarded both as points and as vectors. When we are not regarding a real number as either a point or a vector, we refer to it as a *scalar*.

Two vectors $\mathbf{v} = [v_1, v_2, \ldots, v_n]$ and $\mathbf{w} = [w_1, w_2, \ldots, w_m]$ are **equal** if $n = m$ and $v_i = w_i$ for each i.

A vector containing only zeros as components is called a **zero vector** and is denoted by $\mathbf{0}$. Thus, in $\mathbb{R}^2$ we have $\mathbf{0} = [0, 0]$ whereas in $\mathbb{R}^4$ we have $\mathbf{0} = [0, 0, 0, 0]$.

When denoting a vector $\mathbf{v}$ in $\mathbb{R}^n$ geometrically by an arrow in a figure, we say that the vector is in *standard position* if it starts at the origin. If we draw an

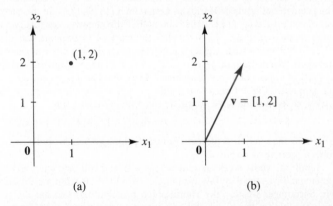

(a)   (b)

FIGURE 1.3
Two views of the same member of $\mathbb{R}^2$: (a) the point (1, 2); (b) the vector v = [1, 2].

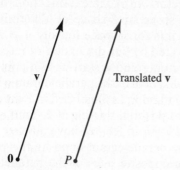

FIGURE 1.4
v translated to *P*.

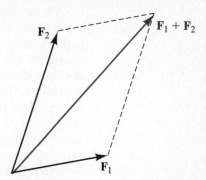

FIGURE 1.5
The vector sum F₁ + F₂.

arrow having the same length and parallel to the arrow representing **v** but starting at a point *P* other than the origin, we refer to the arrow as **v** *translated to P*. This is illustrated in Figure 1.4. Note that we did not draw any coordinate axes; we only marked the origin **0** and drew the two arrows. Thus we can consider Figure 1.4 to represent a vector **v** in $\mathbb{R}^2$, $\mathbb{R}^3$, or indeed in $\mathbb{R}^n$ for $n \geq 2$. We will often leave out axes when they are not necessary for our understanding. This makes our figures both less cluttered and more general.

Vector Algebra

Physicists tell us that if two forces corresponding to force vectors $\mathbf{F}_1$ and $\mathbf{F}_2$ act on a body at the same time, then the two forces can be replaced by a single force, the *resultant force*, which has the same effect as the original two forces. The force vector for this resultant force is the diagonal of the parallelogram having the force vectors $\mathbf{F}_1$ and $\mathbf{F}_2$ as edges, as illustrated in Figure 1.5. It is natural to consider this resultant force vector to be the *sum* $\mathbf{F}_1 + \mathbf{F}_2$ of the two original force vectors, and it is so labeled in Figure 1.5.

HISTORICAL NOTE   THE CONCEPT OF A VECTOR in its earliest manifestation comes from physical considerations. In particular, there is evidence of velocity being thought of as a vector—a quantity with magnitude and direction—in Greek times. For example, in the treatise *Mechanica* by an unknown author in the fourth century B.C. is written: "When a body is moved in a certain ratio (i.e., has two linear movements in a constant ratio to one another), the body must move in a straight line, and this straight line is the diagonal of the parallelogram formed from the straight lines which have the given ratio." Heron of Alexandria (first century A.D.) gave a proof of this result when the directions were perpendicular. He showed that if a point *A* moves with constant velocity over a line *AB* while at the same time the line *AB* moves with constant velocity along the parallel lines *AC* and *BD* so that it always remains parallel to its original position, and that if the time *A* takes to reach *B* is the same as the time *AB* takes to reach *CD*, then in fact the point *A* moves along the diagonal *AD*.

   This basic idea of adding two motions vectorially was generalized from velocities to physical forces in the sixteenth and seventeenth centuries. One example of this practice is found as Corollary 1 to the Laws of Motion in Isaac Newton's *Principia*, where he shows that "a body acted on by two forces simultaneously will describe the diagonal of a parallelogram in the same time as it would describe the sides by those forces separately."

We can visualize two vectors with different directions and emanating from a point P in Euclidean 2-space or 3-space as determining a plane. It is pedagogically useful to do this for n-space for any $n \geq 2$ and show helpful figures on our pages. Motivated by our discussion of force vectors above, we consider the *sum* of two vectors **v** and **w** starting at a point P to be the vector starting at P that forms the diagonal of the parallelogram with a vertex at P and having edges represented by **v** and **w**, as illustrated in Figure 1.6, where we take the vectors in $\mathbb{R}^n$ in standard position starting at **0**. Thus we have a geometric understanding of *vector addition* in $\mathbb{R}^n$. We have labeled as *translated* **v** and *translated* **w** the sides of the parallelogram opposite the vectors **v** and **w**.

Note that arrows along opposite sides of the parallelogram point in the same direction and have the same length. Thus, as a force vector, the translation of **v** is considered to be equivalent to the vector **v**, and the same is true for **w** and its translation. We can think of obtaining the vector **v** + **w** by drawing the arrow **v** from **0** and then drawing the arrow **w** translated to start from the tip of **v** as shown in Figure 1.6. The vector from **0** to the tip of the translated **w** is then **v** + **w**. This is often a useful way to regard **v** + **w**. To add three vectors **u**, **v**, and **w** geometrically, we translate **v** to start at the tip of **u** and then translate **w** to start at the tip of the translated **v**. The sum **u** + **v** + **w** then begins at the origin where **u** starts, and ends at the tip of the translated **w**, as indicated in Figure 1.7.

The difference **v** − **w** of two vectors in $\mathbb{R}^n$ is represented geometrically by the arrow from the tip of **w** to the tip of **v**, as shown in Figure 1.8. Here **v** − **w** is the vector that, when added to **w**, yields **v**. The dashed arrow in Figure 1.8 shows **v** − **w** in standard position.

If we are pushing a body with a force vector **F** and we wish to "double the force"—that is, we want to push in the same direction but twice as hard—then it is natural to denote the doubled force vector by 2**F**. If instead we want to push the body in the opposite direction with one-third the force, we denote the new force vector by $-\frac{1}{3}$**F**. Generalizing, we consider the product r**v** of a scalar r times a vector **v** in $\mathbb{R}^n$ to be represented by the arrow whose length is $|r|$ times the length of **v** and which has the same direction as **v** if $r > 0$ but the opposite direction if $r < 0$. (See Figure 1.9 for an illustration.) Thus we have a geometric interpretation of *scalar multiplication* in $\mathbb{R}^n$—that is, of multiplication of a vector in $\mathbb{R}^n$ by a scalar.

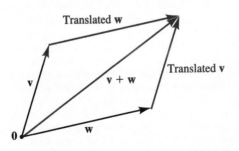

FIGURE 1.6
Representation of v + w in $\mathbb{R}^n$.

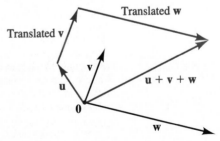

FIGURE 1.7
Representation of u + v + w in $\mathbb{R}^n$.

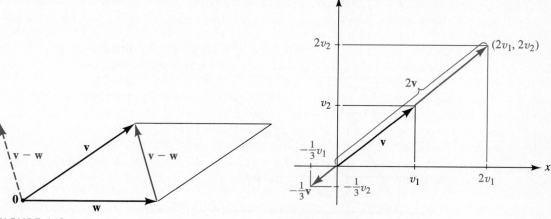

FIGURE 1.8
The vector v − w.

FIGURE 1.9
Computation of rv in $\mathbb{R}^2$.

Taking a vector $\mathbf{v} = [v_1, v_2]$ in $\mathbb{R}^2$ and any scalar r, we would like to be able to compute $r\mathbf{v}$ algebraically as an element (ordered pair) in $\mathbb{R}^2$, and not just represent it geometrically by an arrow. Figure 1.9 shows the vector $2\mathbf{v}$ which points in the same direction as $\mathbf{v}$ but is twice as long, and shows that we have $2\mathbf{v} = [2v_1, 2v_2]$. It also indicates that if we multiply all components of $\mathbf{v}$ by $-\frac{1}{3}$, the resulting vector has direction opposite to the direction of $\mathbf{v}$ and length equal to $\frac{1}{3}$ the length of $\mathbf{v}$. Similarly, if we take two vectors $\mathbf{v} = [v_1, v_2]$ and $\mathbf{w} = [w_1, w_2]$ in $\mathbb{R}^2$, we would like to be able to compute $\mathbf{v} + \mathbf{w}$ algebraically as an element (ordered pair) in $\mathbb{R}^2$. Figure 1.10 indicates that we have $\mathbf{v} + \mathbf{w} = [v_1 + w_1, v_2 + w_2]$—that is, we can simply add corresponding components. With these figures to guide us, we formally define some algebraic operations with vectors in $\mathbb{R}^n$.

DEFINITION 1.1   Vector Algebra in $\mathbb{R}^n$

Let $\mathbf{v} = [v_1, v_2, \ldots, v_n]$ and $\mathbf{w} = [w_1, w_2, \ldots, w_n]$ be vectors in $\mathbb{R}^n$. The vectors are added and subtracted as follows:

Vector addition: $\mathbf{v} + \mathbf{w} = [v_1 + w_1, v_2 + w_2, \ldots, v_n + w_n]$

Vector subtraction: $\mathbf{v} - \mathbf{w} = [v_1 - w_1, v_2 - w_2, \ldots, v_n - w_n]$

If r is any scalar, the vector $\mathbf{v}$ is multiplied by r as follows:

Scalar multiplication: $r\mathbf{v} = [rv_1, rv_2, \ldots, rv_n]$

As a natural extension of Definition 1.1, we can combine three or more vectors in $\mathbb{R}^n$ using addition or subtraction by simply adding or subtracting their corresponding components. When a scalar in such a combination is negative, as in $4\mathbf{u} + (-7)\mathbf{v} + 2\mathbf{w}$, we usually abbreviate by subtraction, writing $4\mathbf{u} - 7\mathbf{v} + 2\mathbf{w}$. We write $-\mathbf{v}$ for $(-1)\mathbf{v}$.

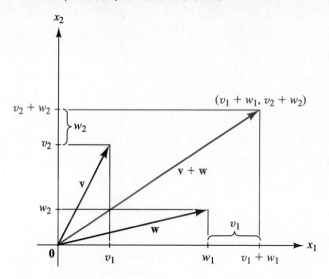

FIGURE 1.10
Computation of v + w in $\mathbb{R}^2$.

EXAMPLE 1   Let $\mathbf{v} = [-3, 5, -1]$ and $\mathbf{w} = [4, 10, -7]$ in $\mathbb{R}^3$. Compute $5\mathbf{v} - 3\mathbf{w}$.

SOLUTION   We compute

$$5\mathbf{v} - 3\mathbf{w} = 5[-3, 5, -1] - 3[4, 10, -7]$$
$$= [-15, 25, -5] - [12, 30, -21]$$
$$= [-27, -5, 16].$$ ∎

EXAMPLE 2   For vectors $\mathbf{v}$ and $\mathbf{w}$ in $\mathbb{R}^n$ pointing in different directions from the origin, represent geometrically $5\mathbf{v} - 3\mathbf{w}$.

SOLUTION   This is done in Figure 1.11.   ∎

The analogues of many familiar algebraic laws for addition and multiplication of scalars also hold for vector addition and scalar multiplication. For convenience, we gather them in a theorem.

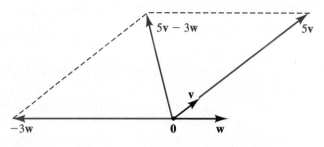

FIGURE 1.11
$5\mathbf{v} - 3\mathbf{w}$ in $\mathbb{R}^n$.

THEOREM 1.1   Properties of Vector Algebra in $\mathbb{R}^n$

Let **u**, **v**, and **w** be any vectors in $\mathbb{R}^n$, and let r and s be any scalars in $\mathbb{R}$.

Properties of Vector Addition

A1	$(\mathbf{u} + \mathbf{v}) + \mathbf{w} = \mathbf{u} + (\mathbf{v} + \mathbf{w})$	An associative law
A2	$\mathbf{v} + \mathbf{w} = \mathbf{w} + \mathbf{v}$	A commutative law
A3	$\mathbf{0} + \mathbf{v} = \mathbf{v}$	**0** as additive identity
A4	$\mathbf{v} + (-\mathbf{v}) = \mathbf{0}$	$-\mathbf{v}$ as additive inverse of **v**

Properties Involving Scalar Multiplication

S1	$r(\mathbf{v} + \mathbf{w}) = r\mathbf{v} + r\mathbf{w}$	A distributive law
S2	$(r + s)\mathbf{v} = r\mathbf{v} + s\mathbf{v}$	A distributive law
S3	$r(s\mathbf{v}) = (rs)\mathbf{v}$	An associative law
S4	$1\mathbf{v} = \mathbf{v}$	Preservation of scale

The eight properties given in Theorem 1.1 are quite easy to prove, and we leave most of them as exercises. The proofs in Examples 3 and 4 are typical.

EXAMPLE 3   Prove property A2 of Theorem 1.1.

SOLUTION   Writing

$$\mathbf{v} = [v_1, v_2, \ldots, v_n] \quad \text{and} \quad \mathbf{w} = [w_1, w_2, \ldots, w_n],$$

we have

$$\mathbf{v} + \mathbf{w} = [v_1 + w_1, v_2 + w_2, \ldots, v_n + w_n]$$

and

$$\mathbf{w} + \mathbf{v} = [w_1 + v_1, w_2 + v_2, \ldots, w_n + v_n].$$

These two vectors are equal because $v_i + w_i = w_i + v_i$ for each i. Thus, the commutative law of vector addition follows directly from the commutative law of addition of numbers.   ∎

EXAMPLE 4   Prove property S2 of Theorem 1.1.

SOLUTION   Writing $\mathbf{v} = [v_1, v_2, \ldots, v_n]$, we have

$$
\begin{aligned}
(r + s)\mathbf{v} &= (r + s)[v_1, v_2, \ldots, v_n] \\
&= [(r + s)v_1, (r + s)v_2, \ldots, (r + s)v_n] \\
&= [rv_1 + sv_1, rv_2 + sv_2, \ldots, rv_n + sv_n] \\
&= [rv_1, rv_2, \ldots, rv_n] + [sv_1, sv_2, \ldots, sv_n] \\
&= r\mathbf{v} + s\mathbf{v}.
\end{aligned}
$$

Thus the property $(r + s)\mathbf{v} = r\mathbf{v} + s\mathbf{v}$ involving vectors follows from the analogous property $(r + s)a_i = ra_i + sa_i$ for numbers.   ∎

Parallel Vectors

The geometric significance of multiplication of a vector by a scalar, as illustrated in Figure 1.9, leads us to this characterization of parallel vectors.

DEFINITION 1.2 Parallel Vectors

Two nonzero vectors $\mathbf{v}$ and $\mathbf{w}$ in $\mathbb{R}^n$ are **parallel**, and we write $\mathbf{v} \parallel \mathbf{w}$, if one is a scalar multiple of the other. If $\mathbf{v} = r\mathbf{w}$ with $r > 0$, then $\mathbf{v}$ and $\mathbf{w}$ have the **same direction**; if $r < 0$, then $\mathbf{v}$ and $\mathbf{w}$ have **opposite directions**.

EXAMPLE 5   Determine whether the vectors $\mathbf{v} = [2, 1, 3, -4]$ and $\mathbf{w} = [6, 3, 9, -12]$ are parallel.

SOLUTION   We put $\mathbf{v} = r\mathbf{w}$ and try to solve for r. This gives rise to four component equations:

$$2 = 6r, \qquad 1 = 3r, \qquad 3 = 9r, \qquad -4 = -12r.$$

Because $r = \frac{1}{3} > 0$ is a common solution to the four equations, we conclude that $\mathbf{v}$ and $\mathbf{w}$ are parallel and have the same direction.   ∎

Linear Combinations of Vectors

Definition 1.1 describes how to add or subtract two vectors, but as we remarked following the definition, we can use these operations to combine three or more vectors also. We give a formal extension of that definition.

DEFINITION 1.3 Linear Combination

Given vectors $\mathbf{v}_1, \mathbf{v}_2, \ldots, \mathbf{v}_k$ in $\mathbb{R}^n$ and scalars $r_1, r_2, \ldots, r_k$ in $\mathbb{R}$, the vector

$$r_1\mathbf{v}_1 + r_2\mathbf{v}_2 + \cdots + r_k\mathbf{v}_k$$

is a **linear combination** of the vectors $\mathbf{v}_1, \mathbf{v}_2, \ldots, \mathbf{v}_k$ with **scalar coefficients** $r_1, r_2, \ldots, r_k$.

The vectors $[1, 0]$ and $[0, 1]$ play a very important role in $\mathbb{R}^2$. Every vector $\mathbf{b}$ in $\mathbb{R}^2$ can be expressed as a linear combination of these two vectors in a *unique* way—namely, $\mathbf{b} = [b_1, b_2] = r_1[1, 0] + r_2[0, 1]$ if and only if $r_1 = b_1$ and $r_2 = b_2$. We call $[1, 0]$ and $[0, 1]$ the **standard basis vectors in** $\mathbb{R}^2$. They are often denoted by $\mathbf{i} = [1, 0]$ and $\mathbf{j} = [0, 1]$, as shown in Figure 1.12(a). Thus in $\mathbb{R}^2$, we may write the vector $[b_1, b_2]$ as $b_1\mathbf{i} + b_2\mathbf{j}$. Similarly, we have three **standard basis vectors in** $\mathbb{R}^3$—namely,

$$\mathbf{i} = [1, 0, 0], \quad \mathbf{j} = [0, 1, 0], \quad \text{and} \quad \mathbf{k} = [0, 0, 1],$$

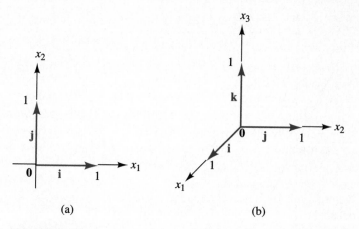

FIGURE 1.12
(a) Standard basis vectors in $\mathbb{R}^2$; (b) standard basis vectors in $\mathbb{R}^3$.

as shown in Figure 1.12(b). Every vector in $\mathbb{R}^3$ can be expressed uniquely as a linear combination of **i**, **j**, and **k**. For example, we have $[3, -2, 6] = 3\mathbf{i} - 2\mathbf{j} + 6\mathbf{k}$. For $n > 3$, we denote the **rth standard basis vector**, having 1 as the rth component and zeros elsewhere, by

$$\mathbf{e}_r = [0, 0, \ldots, 0, 1, 0, \ldots, 0].$$
$$\uparrow$$
$$r\text{th component}$$

We then have

$$\mathbf{b} = [b_1, b_2, \ldots, b_n] = b_1\mathbf{e}_1 + b_2\mathbf{e}_2 + \cdots + b_n\mathbf{e}_n.$$

We see that every vector in $\mathbb{R}^n$ appears as a *unique* linear combination of the standard basis vector in $\mathbb{R}^n$.

The Span of Vectors

Let **v** be a vector in $\mathbb{R}^n$. All possible linear combinations of this single vector **v** are simply all possible scalar multiples $r\mathbf{v}$ for all scalars r. If $\mathbf{v} \neq \mathbf{0}$, all scalar multiples of **v** fill a line which we shall call the **line along v**. Figure 1.13(a) shows the line along the vector $[-1, 2]$ in $\mathbb{R}^2$ while Figure 1.13(b) indicates the line along a nonzero vector **v** in $\mathbb{R}^n$.

Note that the line along **v** always contains the origin (the zero vector) because one scalar multiple of **v** is $0\mathbf{v} = \mathbf{0}$.

Now let **v** and **w** be two nonzero and nonparallel vectors in $\mathbb{R}^n$. All possible linear combinations of **v** and **w** are all vectors of the form $r\mathbf{v} + s\mathbf{w}$ for all scalars r and s. As indicated in Figure 1.14, all these linear combinations fill a plane which we call the **plane spanned by v and w**.

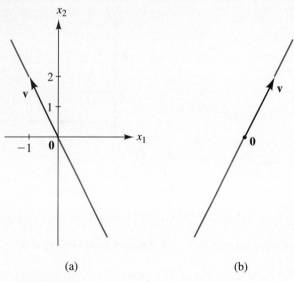

(a)   (b)

FIGURE 1.13
(a) The line along v in $\mathbb{R}^2$; (b) The line along v in $\mathbb{R}^n$.

EXAMPLE 6   Referring to Figure 1.15(a), estimate scalars r and s such that $r\mathbf{v} + s\mathbf{w} = \mathbf{b}$ for the vectors $\mathbf{v}$, $\mathbf{w}$, and $\mathbf{b}$ all lying in the plane of the paper.

SOLUTION   We draw the line along $\mathbf{v}$, the line along $\mathbf{w}$, and parallels to these lines through the tip of the vector $\mathbf{b}$, as shown in Figure 1.15(b). From Figure 1.15(b), we estimate that $\mathbf{b} = 1.5\mathbf{v} - 2.5\mathbf{w}$.   ■

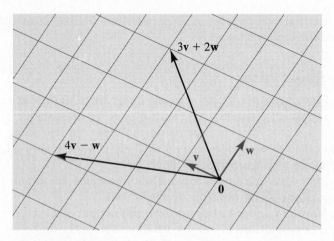

FIGURE 1.14
The plane spanned by v and w.

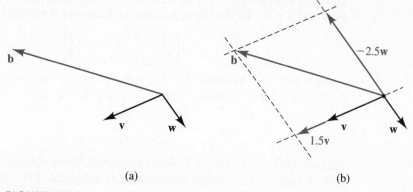

(a)   (b)

FIGURE 1.15
(a) Vectors v, w, and b; (b) finding r and s so that b = rv + sw.

We now give an analytic analogue of Example 6 for two vectors in $\mathbb{R}^2$.

EXAMPLE 7   Let **v** = [1, 3] and **w** = [−2, 5] in $\mathbb{R}^2$. Find scalars r and s such that $r\mathbf{v} + s\mathbf{w} =$ [−1, 19].

SOLUTION   Because $r\mathbf{v} + s\mathbf{w} = r[1, 3] + s[-2, 5] = [r - 2s, 3r + 5s]$, we see that $r\mathbf{v} + s\mathbf{w} =$ [−1, 19] if and only if both equations

$$r - 2s = -1$$
$$3r + 5s = 19$$

are satisfied. Multiplying the first equation by −3 and adding the result to the second equation, we obtain

$$0 + 11s = 22,$$

so $s = 2$. Substituting in the equation $r - 2s = -1$, we find that $r = 3$.   ∎

We note that the components −1 and 19 of the vector [−1, 19] appear on the right-hand side of the system of two linear equations in Example 7. If we replace −1 by b_1 and 19 by b_2, the same operations on the equations will enable us to solve for the scalars r and s in terms of b_1 and b_2 (see Exercise 42). This shows that all linear combinations of **v** and **w** do indeed fill the plane $\mathbb{R}^2$.

Example 7 indicates that an attempt to express a vector **b** as a linear combination of given vectors corresponds to an attempt to find a solution of a system of linear equations. This parallel is even more striking if we write our vectors as *columns* of numbers rather than as ordered *rows* of numbers—that is, as **column vectors** rather than as **row vectors**. For example, if we write the vectors **v** and **w** in Example 7 as columns so that

$$\mathbf{v} = \begin{bmatrix} 1 \\ 3 \end{bmatrix} \quad \text{and} \quad \mathbf{w} = \begin{bmatrix} -2 \\ 5 \end{bmatrix},$$

and also rewrite $[-1, 19]$ as a column vector, then the row-vector equation $r\mathbf{v} + s\mathbf{w} = [-1, 19]$ in the statement of Example 7 becomes

$$r\begin{bmatrix} 1 \\ 3 \end{bmatrix} + s\begin{bmatrix} -2 \\ 5 \end{bmatrix} = \begin{bmatrix} -1 \\ 19 \end{bmatrix}.$$

Notice that the numbers in this column-vector equation are in the same positions relative to each other as they are in the *system of linear equations*

$$r - 2s = -1$$
$$3r + 5s = 19$$

that we solved in Example 7. Every system of linear equations can be rewritten in this fashion as a single column-vector equation. Exercises 35–38 provide practice in this. Finding scalars $r_1, r_2, \ldots, r_k$ such that $r_1\mathbf{v}_1 + r_2\mathbf{v}_2 + \cdots + r_k\mathbf{v}_k = \mathbf{b}$ for given vectors $\mathbf{v}_1, \mathbf{v}_2, \ldots, \mathbf{v}_k$ and $\mathbf{b}$ in $\mathbb{R}^n$ is a fundamental computation in linear algebra. Section 1.4 describes an algorithm for finding all possible such scalars $r_1, r_2, \ldots, r_k$.

The preceding paragraph indicates that often it will be natural for us to think of vectors in $\mathbb{R}^n$ as column vectors rather than as row vectors.

The **transpose** of a row vector $\mathbf{v}$ is defined to be the corresponding column vector, and is denoted by $\mathbf{v}^T$. Similarly, the transpose of a column vector is the corresponding row vector. For example,

$$[-1, 4, 15, -7]^T = \begin{bmatrix} -1 \\ 4 \\ 15 \\ -7 \end{bmatrix} \quad \text{and} \quad \begin{bmatrix} 2 \\ -30 \\ 45 \end{bmatrix}^T = [2, -30, 45].$$

Note that for all vectors $\mathbf{v}$ we have $(\mathbf{v}^T)^T = \mathbf{v}$. As illustrated following Example 7, column vectors are often useful. In fact, some authors always regard every vector $\mathbf{v}$ in $\mathbb{R}^n$ as a column vector. Because it takes so much page space to write column vectors, these authors may describe $\mathbf{v}$ by giving the row vector $\mathbf{v}^T$. We do not follow this practice; we will write vectors in $\mathbb{R}^n$ as either row or column vectors depending on the context.

Continuing our geometric discussion, we expect that if $\mathbf{u}$, $\mathbf{v}$, and $\mathbf{w}$ are three nonzero vectors in $\mathbb{R}^n$ such that $\mathbf{u}$ and $\mathbf{v}$ are not parallel and also $\mathbf{w}$ is not a vector in the plane spanned by $\mathbf{u}$ and $\mathbf{v}$, then the set of all linear combinations of $\mathbf{u}$, $\mathbf{v}$, and $\mathbf{w}$ will fill a three-dimensional portion of $\mathbb{R}^n$—that is, a portion of $\mathbb{R}^n$ that looks just like $\mathbb{R}^3$. We consider the set of these linear combinations to be *spanned* by $\mathbf{u}$, $\mathbf{v}$, and $\mathbf{w}$. We make the following definition.

DEFINITION 1.4   Span of $\mathbf{v}_1, \mathbf{v}_2, \ldots, \mathbf{v}_k$

Let $\mathbf{v}_1, \mathbf{v}_2, \ldots, \mathbf{v}_k$ be vectors in $\mathbb{R}^n$. The **span** of these vectors is the set of all linear combinations of them and is denoted by $\text{sp}(\mathbf{v}_1, \mathbf{v}_2, \ldots, \mathbf{v}_k)$. In set notation,

$$\text{sp}(\mathbf{v}_1, \mathbf{v}_2, \ldots, \mathbf{v}_k) = \{r_1\mathbf{v}_1 + r_2\mathbf{v}_2 + \cdots + r_k\mathbf{v}_k \mid r_1, r_2, \ldots, r_k \in \mathbb{R}\}.$$

It is important to note that $sp(\mathbf{v}_1, \mathbf{v}_2, \ldots, \mathbf{v}_k)$ in $\mathbb{R}^n$ may not fill what we intuitively consider to be a k-dimensional portion of $\mathbb{R}^n$. For example, in $\mathbb{R}^2$ we see that $sp([1, -2], [-3, 6])$ is just the one-dimensional line along $[1, -2]$ because $[-3, 6] = -3[1, -2]$ already lies in $sp([1, -2])$. Similarly, if $\mathbf{v}_3$ is a vector in $sp(\mathbf{v}_1, \mathbf{v}_2)$, then $sp(\mathbf{v}_1, \mathbf{v}_2, \mathbf{v}_3) = sp(\mathbf{v}_1, \mathbf{v}_2)$ and so $sp(\mathbf{v}_1, \mathbf{v}_2, \mathbf{v}_3)$ is not three-dimensional. Section 2.1 will deal with this kind of *dependency* among vectors. As a result of our work there, we will be able to define dimensionality.

SUMMARY

1. *Euclidean n-space* $\mathbb{R}^n$ consists of all ordered n-tuples of real numbers. Each n-tuple $\mathbf{x}$ can be regarded as a *point* $(x_1, x_2, \ldots, x_n)$ and represented graphically as a dot, or regarded as a vector $[x_1, x_2, \ldots, x_n]$ and represented by an arrow. The n-tuple $\mathbf{0} = [0, 0, \ldots, 0]$ is the *zero vector*. A real number $r \in \mathbb{R}$ is called a *scalar*.

2. Vectors $\mathbf{v}$ and $\mathbf{w}$ in $\mathbb{R}^n$ can be added and subtracted, and each can be multiplied by a scalar $r \in \mathbb{R}$. In each case, the operation is performed on the components, and the resulting vector is again in $\mathbb{R}^n$. Properties of these operations are summarized in Theorem 1.1. Graphic interpretations are shown in Figures 1.6, 1.8, and 1.9.

3. Two nonzero vectors in $\mathbb{R}^n$ are *parallel* if one is a scalar multiple of the other.

4. A *linear combination* of vectors $\mathbf{v}_1, \mathbf{v}_2, \ldots, \mathbf{v}_k$ in $\mathbb{R}^n$ is a vector of the form $r_1\mathbf{v}_1 + r_2\mathbf{v}_2 + \cdots + r_k\mathbf{v}_k$, where each r_i is a scalar. The set of all such linear combinations is the *span* of the vectors $\mathbf{v}_1, \mathbf{v}_2, \ldots, \mathbf{v}_k$ and is denoted by $sp(\mathbf{v}_1, \mathbf{v}_2, \ldots, \mathbf{v}_k)$.

5. Every vector in $\mathbb{R}^n$ can be expressed uniquely as a linear combination of the *standard basis vectors* $\mathbf{e}_1, \mathbf{e}_2, \ldots, \mathbf{e}_n$, where $\mathbf{e}_i$ has 1 as its ith component and zeros for all other components.

EXERCISES

In Exercises 1–4, compute $\mathbf{v} + \mathbf{w}$ and $\mathbf{v} - \mathbf{w}$ for the given vectors $\mathbf{v}$ and $\mathbf{w}$. Then draw coordinate axes and sketch, using your answers, the vectors $\mathbf{v}$, $\mathbf{w}$, $\mathbf{v} + \mathbf{w}$, and $\mathbf{v} - \mathbf{w}$.

1. $\mathbf{v} = [2, -1]$, $\mathbf{w} = [-3, -2]$
2. $\mathbf{v} = [1, 3]$, $\mathbf{w} = [-2, 5]$
3. $\mathbf{v} = \mathbf{i} + 3\mathbf{j} + 2\mathbf{k}$, $\mathbf{w} = \mathbf{i} + 2\mathbf{j} + 4\mathbf{k}$
4. $\mathbf{v} = 2\mathbf{i} - \mathbf{j} + 3\mathbf{k}$, $\mathbf{w} = 3\mathbf{i} + 5\mathbf{j} + 4\mathbf{k}$

In Exercises 5–8, let $\mathbf{u} = [-1, 3, -2]$, $\mathbf{v} = [4, 0, -1]$, and $\mathbf{w} = [-3, -1, 2]$. Compute the indicated vector.

5. $3\mathbf{u} - 2\mathbf{v}$
6. $\mathbf{u} + 2(\mathbf{v} - 4\mathbf{w})$
7. $\mathbf{u} + \mathbf{v} - \mathbf{w}$
8. $4(3\mathbf{u} + 2\mathbf{v} - 5\mathbf{w})$

In Exercises 9–12, compute the given linear combination of $\mathbf{u} = [1, 2, 1, 0]$, $\mathbf{v} = [-2, 0, 1, 6]$, *and* $\mathbf{w} = [3, -5, 1, -2]$.

9. $\mathbf{u} - 2\mathbf{v} + 4\mathbf{w}$

10. $3\mathbf{u} + \mathbf{v} - \mathbf{w}$

11. $4\mathbf{u} - 2\mathbf{v} + 4\mathbf{w}$

12. $-\mathbf{u} + 5\mathbf{v} + 3\mathbf{w}$

In Exercises 13–16, reproduce on your paper those vectors in Figure 1.16 that appear in the exercise, and then draw an arrow representing each of the following linear combinations. All of the vectors are assumed to lie in the same plane. Use the technique illustrated in Figure 1.7 when all three vectors are involved.

13. $2\mathbf{u} + 3\mathbf{v}$

14. $-3\mathbf{u} + 2\mathbf{w}$

15. $\mathbf{u} + \mathbf{v} + \mathbf{w}$

16. $2\mathbf{u} - \mathbf{v} + \frac{1}{2}\mathbf{w}$

In Exercises 17–20, reproduce on your paper those vectors in Figure 1.17 that appear in the exercise, and then use the technique illustrated in Example 6 to estimate scalars r and s such that the given equation is true. All of the vectors are assumed to lie in the same plane.

17. $\mathbf{x} = r\mathbf{u} + s\mathbf{v}$

18. $\mathbf{y} = r\mathbf{u} + s\mathbf{v}$

19. $\mathbf{u} = r\mathbf{x} + s\mathbf{v}$

20. $\mathbf{y} = r\mathbf{u} + s\mathbf{x}$

In Exercises 21–30, find all scalars c, if any exist, such that the given statement is true. Try to do some of these problems without using pencil and paper.

21. The vector $[2, 6]$ is parallel to the vector $[c, -3]$.

22. The vector $[c^2, -4]$ is parallel to the vector $[1, -2]$.

23. The vector $[c, -c, 4]$ is parallel to the vector $[-2, 2, 20]$.

24. The vector $[c^2, c^3, c^4]$ is parallel to the vector $[1, -2, 4]$ with the same direction.

25. The vector $[13, -15]$ is a linear combination of the vectors $[1, 5]$ and $[3, c]$.

26. The vector $[-1, c]$ is a linear combination of the vectors $[-3, 5]$ and $[6, -11]$.

27. $\mathbf{i} + c\mathbf{j} - 3\mathbf{k}$ is a linear combination of $\mathbf{i} + \mathbf{j}$ and $\mathbf{j} + 3\mathbf{k}$.

28. $\mathbf{i} + c\mathbf{j} + (c - 1)\mathbf{k}$ is in the span of $\mathbf{i} + 2\mathbf{j} + \mathbf{k}$ and $3\mathbf{i} + 6\mathbf{j} + 3\mathbf{k}$.

29. The vector $3\mathbf{i} - 2\mathbf{j} + c\mathbf{k}$ is in the span of $\mathbf{i} + 2\mathbf{j} - \mathbf{k}$ and $\mathbf{j} + 3\mathbf{k}$.

30. The vector $[c, -2c, c]$ is in the span of $[1, -1, 1]$, $[0, 1, -3]$, and $[0, 0, 1]$.

In Exercises 31–34, find the vector which, when translated, represents geometrically an arrow reaching from the first point to the second.

31. From $(-1, 3)$ to $(4, 2)$ in $\mathbb{R}^2$

32. From $(-3, 2, 5)$ to $(4, -2, -6)$ in $\mathbb{R}^3$

33. From $(2, 1, 5, -6)$ to $(3, -2, 1, 7)$ in $\mathbb{R}^4$

FIGURE 1.16

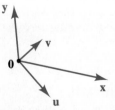

FIGURE 1.17

34. From $(1, 2, 3, 4, 5)$ to $(-5, -4, -3, -2, -1)$ in $\mathbb{R}^5$

35. Write the linear system

$$3x - 2y + 4z = 10$$
$$x - y - 3z = 0$$
$$2x + y - 5z = -3$$

as a column-vector equation.

36. Write the linear system

$$x_1 - 3x_2 + 2x_3 \qquad = -6$$
$$3x_1 \qquad - 4x_3 + 5x_4 = 12$$

as a column-vector equation.

37. Write the row-vector equation

$$p[-3, 4, 6] + q[0, -2, 5] - r[4, -3, 2] + s[6, 0, 7] = [8, -3, 1]$$

as

a. a linear system,
b. a column-vector equation.

38. Write the column-vector equation

$$r_1 \begin{bmatrix} -2 \\ 3 \\ 0 \end{bmatrix} + r_2 \begin{bmatrix} 5 \\ 13 \\ -4 \end{bmatrix} + r_3 \begin{bmatrix} 16 \\ 0 \\ -9 \end{bmatrix} = \begin{bmatrix} 5 \\ -8 \\ 11 \end{bmatrix}$$

as a linear system.

39. Mark each of the following True or False.
___ **a.** The notion of a vector in $\mathbb{R}^n$ is useful only if $n = 1, 2,$ or 3.
___ **b.** Every ordered n-tuple in $\mathbb{R}^n$ can be viewed both as a point and as a vector.
___ **c.** It would be impossible to define addition of points in $\mathbb{R}^n$ because we only add vectors.
___ **d.** If $\mathbf{a}$ and $\mathbf{b}$ are two vectors in standard position in $\mathbb{R}^n$, then the arrow from the tip of $\mathbf{a}$ to the tip of $\mathbf{b}$ is a translated representation of the vector $\mathbf{a} - \mathbf{b}$.

___ **e.** If $\mathbf{a}$ and $\mathbf{b}$ are two vectors in standard position in $\mathbb{R}^n$, then the arrow from the tip of $\mathbf{a}$ to the tip of $\mathbf{b}$ is a translated representation of the vector $\mathbf{b} - \mathbf{a}$.
___ **f.** The span of any two nonzero vectors in $\mathbb{R}^2$ is all of $\mathbb{R}^2$.
___ **g.** The span of any two nonzero, nonparallel vectors in $\mathbb{R}^2$ is all of $\mathbb{R}^2$.
___ **h.** The span of any three nonzero, nonparallel vectors in $\mathbb{R}^3$ is all of $\mathbb{R}^3$.
___ **i.** If $\mathbf{v}_1, \mathbf{v}_2, \ldots, \mathbf{v}_k$ are vectors in $\mathbb{R}^2$ such that $\mathrm{sp}(\mathbf{v}_1, \mathbf{v}_2, \ldots, \mathbf{v}_k) = \mathbb{R}^2$, then $k = 2$.
___ **j.** If $\mathbf{v}_1, \mathbf{v}_2, \ldots, \mathbf{v}_k$ are vectors in $\mathbb{R}^3$ such that $\mathrm{sp}(\mathbf{v}_1, \mathbf{v}_2, \ldots, \mathbf{v}_k) = \mathbb{R}^3$, then $k \geq 3$.

40. Prove the indicated property of vector addition in $\mathbb{R}^n$, stated in Theorem 1.1.
a. Property A1
b. Property A3
c. Property A4

41. Prove the indicated property of scalar multiplication in $\mathbb{R}^n$, stated in Theorem 1.1.
42. a. Property S1
b. Property S3
c. Property S4

42. Prove algebraically that the linear system

$$r - 2s = b_1$$
$$3r + 5s = b_2$$

has a solution for all numbers $b_1, b_2 \in \mathbb{R}$, as asserted in the text.

43. Option 1 of the routine VECTGRPH in the software package LINTEK gives graphic quizzes on addition and subtraction of vectors in $\mathbb{R}^2$. Work with this Option 1 until you consistently achieve a score of 80% or better on the quizzes.

44. Repeat Exercise 43 using Option 2 of the routine VECTGRPH. The quizzes this time are on linear combinations in $\mathbb{R}^2$, and are quite similar to Exercises 17–20.

MATLAB

The MATLAB exercises are designed to build some familiarity with this widely used software as you work your way through the text. Complete information can be obtained from the manual that accompanies MATLAB. Some information is available on the screen by typing **help** followed by a space and then the word or symbol about which you desire information.

The software LINTEK designed explicitly for this text does not require a manual, because all information is automatically given on the screen. However, MATLAB is a *professionally* designed program that is much more powerful than LINTEK. Although LINTEK adequately illustrates material in the text, the prospective scientist would be well advised to invest the extra time necessary to acquire some facility with MATLAB.

Access MATLAB according to the directions for your installation. The MATLAB prompt, which is its request for instructions, looks like ≫. Vectors can be entered by entering the components in square brackets, separated by spaces; do not use commas. Enter vectors **a, b, c, u, v,** *and* **w** *by typing the lines displayed below. After you type each line, press the Enter key. The vector components will be displayed, without brackets, using decimal notation. Proofread each vector after you enter it. If you have made an error, strike the up-arrow key and edit in the usual fashion to correct your error. (If you do not want a vector displayed for proofreading after entry, you can achieve this by typing* ; *after the closing square bracket.) If you ever need to continue on the next line to type data, enter at least two periods* .. *and immediately press the Enter key and continue the data.*

$$a = [2 \quad -4 \quad 5 \quad 7]$$
$$b = [-1 \quad 6 \quad 7 \quad 3]$$
$$c = [13 \quad -21 \quad 5 \quad 39]$$
$$u = [2/3 \quad 3/5 \quad 1/7]$$
$$v = [3/2 \quad -5/6 \quad 11/3]$$
$$w = [5/7 \quad 3/4 \quad -2/3]$$

Now enter **u** *(that is, type* **u** *and press the Enter key) to see* **u** *displayed again. Then enter* **format long** *and then* **u** *to see the components of* **u** *displayed with more decimal places. Enter* **format short** *and then* **u** *to return to the display with fewer decimal places. Enter* **rat(u, 's')**, *which displays rational approximations that are accurate when the numbers involved are fractions with sufficiently small denominators, to see* **u** *displayed again in fraction (rational) format. Addition and subtraction of vectors can be performed using* **+** *and* **−**, *and scalar multiplication using* ∗. *Enter* **a + b** *to see this sum displayed. Then enter* **−3∗a** *to see this scalar product displayed. Using what you have discovered about MATLAB, work the following exercises. Entering* **who** *at any time displays a list of variables to which you have assigned numerical, vector, or matrix values. When you have finished, enter* **quit** *or* **exit** *to leave MATLAB.*

M1. Compute 2**a** − 3**b** − 5**c**.

M2. Compute 3**c** − 4(2**a** − **b**).

M3. Attempt to compute **a** + **u**. What happens, and why?

M4. Compute **u** + **v** in
   a. short format
   b. long format
   c. rational (fraction) format.

M5. Repeat Exercise M4 for 2**u** − 3**v** + **w** by first entering **x** = 2∗**u** − 3∗**v** + **w** and then looking at **x** in the different formats.

M6. Repeat Exercise M4 for $\frac{1}{3}\mathbf{a} - \frac{3}{7}\mathbf{b}$, entering the fractions as (1/3) and (3/7), and using the time-saving technique of Exercise M5.

M7. Repeat Exercise M4 for $0.3\mathbf{u} - 0.23\mathbf{w}$, using the time-saving technique of Exercise M5.

M8. The *transpose* $\mathbf{v}^T$ of a vector $\mathbf{v}$ is denoted in MATLAB by $\mathbf{v}'$. Compute $\mathbf{a}^T - 3\mathbf{c}^T$ and $(\mathbf{a} - 3\mathbf{c})^T$. How do they compare?

M9. Try to compute $\mathbf{u} + \mathbf{u}^T$. What happens, and why?

M10. Enter **help** : to see some of the things that can be done using the colon. Use the colon in a statement starting $\mathbf{v1} =$ to generate the vector $\mathbf{v1} = [-3 \ -2 \ -1 \ 0 \ 1 \ 2]$, and then generate $\mathbf{v2} = [1 \ 4 \ 7 \ 10 \ 13 \ 16]$ similarly. Compute $\frac{1}{3}\mathbf{v1} + \frac{5}{6}\mathbf{v2}$ in rational format. (We use $\mathbf{v1}$ and $\mathbf{v2}$ in MATLAB where we would use $\mathbf{v}_1$ and $\mathbf{v}_2$ in the text.)

M11. MATLAB can provide graphic illustrations of the component values of a vector. Enter **plot(a)** and note how the figure drawn reflects the components of the vector $\mathbf{a} = [2 \ -4 \ 5 \ 7]$. Press any key to clear the screen, and repeat the experiment with **bar(a)** and finally with **stairs(a)** to see two other ways to illustrate the component values of the vector $\mathbf{a}$.

M12. Using the **plot** command, we can plot graphs of functions in MATLAB. The command **plot(x, y)** will plot the $\mathbf{x}$ vector against the $\mathbf{y}$ vector. Enter $\mathbf{x} = -1: .5: 1$. Note that, using the colons, we have generated a vector of x-coordinates starting at -1 and stopping at 1, with increments of 0.5. Now enter $\mathbf{y} = \mathbf{x} .* \mathbf{x}$. (A period before an operator, such as .+ or .*, will cause that operation to be performed on each component of a vector. Enter **help** . to see MATLAB explain this.) You will see that the $\mathbf{y}$ vector contains the squares of the x-coordinates in the $\mathbf{x}$ vector. Enter **plot(x, y)** to see a crude plot of the graph of $y = x^2$ for $-1 \le x \le 1$.

    a. Use the up-arrow key to return to the colon statement and make the increment .2 rather than .5 to get a better graph. Use the up-arrow key to get to the $\mathbf{y} = \mathbf{x} .* \mathbf{x}$ command, and press the Enter key to generate the new $\mathbf{y}$ vector with more entries. Then get to the **plot(x, y)** command and press Enter to see the improved graph of $y = x^2$ for $-1 \le x \le 1$.

    b. Proceed as in part (a) to graph $y = x^2$ for $-3 \le x \le 3$ with increments of 0.2. This time, put a semicolon after the command that defines the vector $\mathbf{x}$ before pressing the Enter key, so that you don't see the x-coordinates printed out. Similarly, put a semicolon after the command defining the vector $\mathbf{y}$.

    c. Plot the graph of $y = \sin(x)$ for $-4\pi \le x \le 4\pi$. The number π can be entered as **pi** in MATLAB. Remember to use * for multiplication.

    d. Plot the graph of $y = 3 \cos(2x) - 2 \sin(x)$ for $-4\pi \le x \le 4\pi$. Remember to use * for multiplication.

1.2   THE NORM AND THE DOT PRODUCT

The Magnitude of a Vector

The *magnitude* $\|\mathbf{v}\|$ of $\mathbf{v} = [v_1, v_2]$ is considered to be the length of the arrow in Figure 1.18. Using the Pythagorean theorem, we have

$$\|\mathbf{v}\| = \sqrt{v_1^2 + v_2^2}. \tag{1}$$

EXAMPLE 1   Represent the vector $\mathbf{v} = [3, -4]$ geometrically, and find its magnitude.

SOLUTION   The vector $[3, -4]$ has magnitude

$$\|\mathbf{v}\| = \sqrt{3^2 + (-4)^2} = \sqrt{25} = 5$$

and is shown in Figure 1.19.   ∎

In Figure 1.20, the *magnitude* $\|\mathbf{v}\|$ of a vector $\mathbf{v} = [v_1, v_2, v_3]$ in $\mathbb{R}^3$ appears as the length of the hypotenuse of a right triangle whose altitude is $|v_3|$ and whose base in the x_1, x_2-plane has length $\sqrt{v_1^2 + v_2^2}$. Using the Pythagorean theorem, we obtain

$$\|\mathbf{v}\| = \sqrt{v_1^2 + v_2^2 + v_3^2}. \tag{2}$$

EXAMPLE 2   Represent the vector $\mathbf{v} = [2, 3, 4]$ geometrically, and find its magnitude.

SOLUTION   The vector $\mathbf{v} = [2, 3, 4]$ has magnitude $\|\mathbf{v}\| = \sqrt{2^2 + 3^2 + 4^2} = \sqrt{29}$ and is represented in Figure 1.21.   ∎

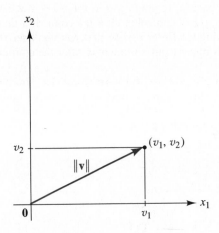

FIGURE 1.18
The magnitude of v in $\mathbb{R}^2$.

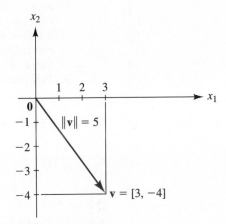

FIGURE 1.19
The magnitude of [3, −4].

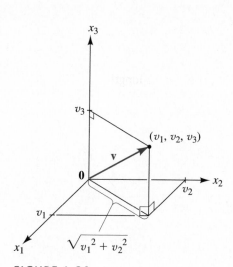

FIGURE 1.20
The magnitude of v in $\mathbb{R}^3$.

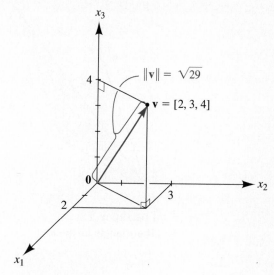

FIGURE 1.21
The magnitude of [2, 3, 4].

The magnitude of a vector is also called the *norm* or the *length* of the vector. As suggested by Eqs. (1) and (2), we define the norm $\|\mathbf{v}\|$ of a vector $\mathbf{v}$ in $\mathbb{R}^n$ as follows.

DEFINITION 1.5   Norm or Magnitude of a Vector in $\mathbb{R}^n$

> Let $\mathbf{v} = [v_1, v_2, \ldots, v_n]$ be a vector in $\mathbb{R}^n$. The **norm** or **magnitude** of $\mathbf{v}$ is
> $$\|\mathbf{v}\| = \sqrt{v_1^2 + v_2^2 + \cdots + v_n^2}. \tag{3}$$

EXAMPLE 3   Find the magnitude of the vector $\mathbf{v} = [-2, 1, 3, -1, 4, 2, 1]$.

SOLUTION   We have
$$\|\mathbf{v}\| = \sqrt{(-2)^2 + 1^2 + 3^2 + (-1)^2 + 4^2 + 2^2 + 1^2} = \sqrt{36} = 6. \qquad \blacksquare$$

Here are some properties of this norm operation.

THEOREM 1.2   Properties of the Norm in $\mathbb{R}^n$

> For all vectors $\mathbf{v}$ and $\mathbf{w}$ in $\mathbb{R}^n$ and for all scalars r, we have
> 1. $\|\mathbf{v}\| \geq 0$ and $\|\mathbf{v}\| = 0$ if and only if $\mathbf{v} = \mathbf{0}$ **Positivity**
> 2. $\|r\mathbf{v}\| = |r|\,\|\mathbf{v}\|$ **Homogeneity**
> 3. $\|\mathbf{v} + \mathbf{w}\| \leq \|\mathbf{v}\| + \|\mathbf{w}\|$ **Triangle inequality**

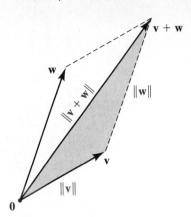

FIGURE 1.22
The triangle inequality.

Proofs of Properties 1 and 2 follow immediately from Definition 1.5 and appear as exercises at the end of this section. Figure 1.22 shows why Property 3 is called the triangle inequality; geometrically, it states that the length of a side of a triangle is less than or equal to the sum of the lengths of the other two sides. Although this seems obvious to us from Figure 1.22, we really should prove it—at least for $n > 3$, where we simply extended our definition of $\|\mathbf{v}\|$ for $\mathbf{v}$ in $\mathbb{R}^2$ or $\mathbb{R}^3$ without any further geometric justification. A proof of the triangle inequality is given at the close of this section.

Unit Vectors

A vector in $\mathbb{R}^n$ is a **unit vector** if it has magnitude 1. Given any nonzero vector $\mathbf{v}$ in $\mathbb{R}^n$, a unit vector having the same direction as $\mathbf{v}$ is given by $(1/\|\mathbf{v}\|)\mathbf{v}$.

EXAMPLE 4   Find a unit vector having the same direction as $\mathbf{v} = [2, 1, -3]$, and find a vector of magnitude 3 having direction opposite to $\mathbf{v}$.

SOLUTION   Because $\|\mathbf{v}\| = \sqrt{2^2 + 1^2 + (-3)^2} = \sqrt{14}$, we see that $\mathbf{u} = (1/\sqrt{14})[2, 1, -3]$ is the unit vector with the same direction as $\mathbf{v}$, and $-3\mathbf{u} = (-3/\sqrt{14})[2, 1, -3]$ is the other required vector.   ■

The two-component unit vectors are precisely the vectors that extend from the origin to the unit circle $x^2 + y^2 = 1$ with center $(0, 0)$ and radius 1 in $\mathbb{R}^2$. (See Figure 1.23a.) The three-component unit vectors extend from $(0, 0, 0)$ to the unit sphere in $\mathbb{R}^3$, as illustrated in Figure 1.23(b).

Note that the standard basis vectors $\mathbf{i}$ and $\mathbf{j}$ in $\mathbb{R}^2$, as well as $\mathbf{i}$, $\mathbf{j}$, and $\mathbf{k}$ in $\mathbb{R}^3$, are unit vectors. In fact, the standard basis vectors $\mathbf{e}_1, \mathbf{e}_2, \ldots, \mathbf{e}_n$ for $\mathbb{R}^n$ are unit vectors, because each has zeros in all components except for one component of 1. For this reason, these standard basis vectors are also called **unit coordinate vectors**.

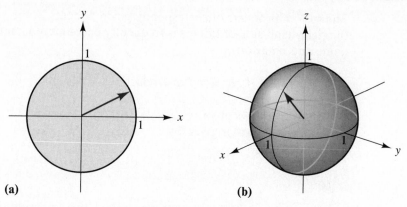

(a)   **(b)**

FIGURE 1.23
(a) Typical unit vector in $\mathbb{R}^2$; (b) typical unit vector in $\mathbb{R}^3$.

The Dot Product

The dot product of two vectors is a scalar that we will encounter as we now try to define the angle θ between two vectors $\mathbf{v} = [v_1, v_2, \ldots, v_n]$ and $\mathbf{w} = [w_1, w_2, \ldots, w_n]$ in $\mathbb{R}^n$, shown symbolically in Figure 1.24. To motivate the definition of θ, we will use the law of cosines for the triangle symbolized in Figure 1.24. Using our definition of the norm of a vector in $\mathbb{R}^n$ to compute the lengths of the sides of the triangle, the law of cosines yields

$$\|\mathbf{v}\|^2 + \|\mathbf{w}\|^2 = \|\mathbf{v} - \mathbf{w}\|^2 + 2\|\mathbf{v}\| \, \|\mathbf{w}\| \, (\cos \theta)$$

or

$$v_1^2 + \cdots + v_n^2 + w_1^2 + \cdots + w_n^2$$
$$= (v_1 - w_1)^2 + \cdots + (v_n - w_n)^2 + 2\|\mathbf{v}\| \, \|\mathbf{w}\| \, (\cos \theta). \quad \textbf{(4)}$$

After computing the squares on the right-hand side of Eq. (4) and simplifying, we obtain

$$\|\mathbf{v}\| \, \|\mathbf{w}\| \, (\cos \theta) = v_1 w_1 + \cdots + v_n w_n. \quad \textbf{(5)}$$

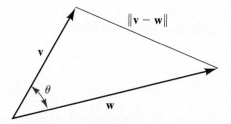

FIGURE 1.24
The angle between v and w.

The sum of products of corresponding components in the vectors **v** and **w** on the right-hand side of Eq. (5) is frequently encountered, and is given a special name and notation.

DEFINITION 1.6 The Dot Product

> The *dot product* of vectors $\mathbf{v} = [v_1, v_2, \ldots, v_n]$ and $\mathbf{w} = [w_1, w_2, \ldots, w_n]$ in $\mathbb{R}^n$ is the scalar given by
>
> $$\mathbf{v} \cdot \mathbf{w} = v_1 w_1 + v_2 w_2 + \cdots + v_n w_n. \tag{6}$$

The dot product is sometimes called the *inner product* or the *scalar product*. To avoid possible confusion with scalar multiplication, we shall never use the latter term.

In view of Definition 1.6, we can write Eq. (5) as

$$\mathbf{v} \cdot \mathbf{w} = \|\mathbf{v}\| \, \|\mathbf{w}\| \, (\cos \theta). \tag{7}$$

Equation (7) suggests the following definition of the angle θ between two vectors **v** and **w** in $\mathbb{R}^n$.

> The **angle** between nonzero vectors **v** and **w** is $\arccos \left(\dfrac{\mathbf{v} \cdot \mathbf{w}}{\|\mathbf{v}\| \, \|\mathbf{w}\|} \right)$.   (8)

Expression (8) makes sense, provided that

$$-1 \leq \frac{\mathbf{v} \cdot \mathbf{w}}{\|\mathbf{v}\| \, \|\mathbf{w}\|} \leq 1, \tag{9}$$

so that we can indeed compute the arccosine of $(\mathbf{v} \cdot \mathbf{w})/(\|\mathbf{v}\| \, \|\mathbf{w}\|)$. This inequality (9) is usually rewritten in the form

$$|\mathbf{v} \cdot \mathbf{w}| \leq \|\mathbf{v}\| \, \|\mathbf{w}\|. \quad \textbf{Schwarz inequality} \tag{10}$$

We obtained it by assuming that Figure 1.24 is an appropriate representation for vectors **v** and **w** in $\mathbb{R}^n$. We give a purely algebraic proof of it at the end of this section to validate the definition in expression (8).

EXAMPLE 5   Find the angle θ between the vectors $[1, 2, 0, 2]$ and $[-3, 1, 1, 5]$ in $\mathbb{R}^4$.

SOLUTION   We have

$$\cos \theta = \frac{[1, 2, 0, 2] \cdot [-3, 1, 1, 5]}{\sqrt{1^2 + 2^2 + 0^2 + 2^2} \, \sqrt{(-3)^2 + 1^2 + 1^2 + 5^2}} = \frac{9}{(3)(6)} = \frac{1}{2}.$$

Thus, $\theta = 60°$. ∎

Equation 7 gives a geometric meaning for the dot product.

> The dot product of two vectors is equal to the product of their magnitudes with the cosine of the angle between them.

THEOREM 1.3 Properties of the Dot Product in $\mathbb{R}^n$

Let $\mathbf{u}$, $\mathbf{v}$, and $\mathbf{w}$ be vectors in $\mathbb{R}^n$ and let r be any scalar in $\mathbb{R}$. The following properties hold:

D1 $\mathbf{v} \cdot \mathbf{w} = \mathbf{w} \cdot \mathbf{v}$,   **Commutative law**

D2 $\mathbf{u} \cdot (\mathbf{v} + \mathbf{w}) = \mathbf{u} \cdot \mathbf{v} + \mathbf{u} \cdot \mathbf{w}$,   **Distributive law**

D3 $r(\mathbf{v} \cdot \mathbf{w}) = (r\mathbf{v}) \cdot \mathbf{w} = \mathbf{v} \cdot (r\mathbf{w})$,   **Homogeneity**

D4 $\mathbf{v} \cdot \mathbf{v} \geq 0$, and $\mathbf{v} \cdot \mathbf{v} = 0$ if and only if $\mathbf{v} = \mathbf{0}$. **Positivity**

Verification of all of the properties in Theorem 1.3 is straightforward, as illustrated in the following example.

HISTORICAL NOTE   THE SCHWARZ INEQUALITY is due independently to Augustin-Louis Cauchy (1789–1857) (see note on page 3), Hermann Amandus Schwarz (1843–1921), and Viktor Yakovlevich Bunyakovsky (1804–1889).

It was first stated as a theorem about coordinates in an appendix to Cauchy's 1821 text for his course on analysis at the Ecole Polytechnique, as follows:

$$\left| a\alpha + a'\alpha' + a''\alpha'' + \cdots \right| \leq \sqrt{a^2 + a'^2 + a''^2 + \cdots} \; \sqrt{\alpha^2 + \alpha'^2 + \alpha''^2 + \cdots}.$$

Cauchy's proof follows from the algebraic identity

$$(a\alpha + a'\alpha' + a''\alpha'' + \cdots)^2 + (a\alpha' - a'\alpha)^2 + (a\alpha'' - a''\alpha)^2 + \cdots + (a'\alpha'' - a''\alpha')^2 + \cdots$$
$$= (a^2 + a'^2 + a''^2 + \cdots)(\alpha^2 + \alpha'^2 + \alpha''^2 + \cdots).$$

Bunyakovsky proved the inequality for functions in 1859; that is, he stated the result

$$\left[\int_a^b f(x)g(x) \, dx \right]^2 \leq \int_a^b f^2(x) \, dx \cdot \int_a^b g^2(x) \, dx,$$

where we can consider $\int_a^b f(x)g(x) \, dx$ to be the inner product of the functions $f(x)$, $g(x)$ in the vector space of continuous functions on $[a, b]$. Bunyakovsky served as vice-president of the St. Petersburg Academy of Sciences from 1864 until his death. In 1875, the Academy established a mathematics prize in his name in recognition of his 50 years of teaching and research.

Schwarz stated the inequality in 1884. In his case, the vectors were functions ϕ, χ of two variables in a region T in the plane, and the inner product of these functions was given by $\iint_T \phi\chi \, dx \, dy$, where this integral is assumed to exist. The inequality then states that

$$\left| \iint_T \phi\chi \, dx \, dy \right| \leq \sqrt{\iint_T \phi^2 \, dx \, dy} \cdot \sqrt{\iint_T \chi^2 \, dx \, dy}.$$

Schwarz's proof is similar to the one given in the text (page 29). Schwarz was the leading mathematician in Berlin around the turn of the century; the work in which the inequality appears is devoted to a question about minimal surfaces.

EXAMPLE 6 Verify the positivity property D4 of Theorem 1.3.

SOLUTION We let $\mathbf{v} = [v_1, v_2, \ldots, v_n]$, and we find that

$$\mathbf{v} \cdot \mathbf{v} = v_1^2 + v_2^2 + \cdots + v_n^2.$$

A sum of squares is nonnegative and can be zero if and only if each summand is zero. But a summand v_i^2 is itself a square, and will be zero if and only if $v_i = 0$. This completes the demonstration. ■

It is important to observe that the norm of a vector can be expressed in terms of its dot product with itself. Namely, for a vector $\mathbf{v}$ in $\mathbb{R}^n$ we have

$$\|\mathbf{v}\|^2 = \mathbf{v} \cdot \mathbf{v}. \tag{11}$$

Letting $\mathbf{v} = [v_1, v_2, \cdots, v_n]$, we have

$$\mathbf{v} \cdot \mathbf{v} = v_1 v_1 + v_2 v_2 + \cdots + v_n v_n = \|\mathbf{v}\|^2.$$

Equation 11 enables us to use the algebraic properties of the dot product in Theorem 1.3 to prove things about the norm. This technique is illustrated in the proof of the Schwarz and triangle inequalities at the end of this section. Here is another illustration.

EXAMPLE 7 Show that the sum of the squares of the lengths of the diagonals of a parallelogram in $\mathbb{R}^n$ is equal to the sum of the squares of the lengths of the sides. (This is the *parallelogram relation*).

SOLUTION We take our parallelogram with vertex at the origin and with vectors $\mathbf{v}$ and $\mathbf{w}$ emanating from the origin to form two sides, as shown in Figure 1.25. The lengths of the diagonals are then $\|\mathbf{v} + \mathbf{w}\|$ and $\|\mathbf{v} - \mathbf{w}\|$. Using Eq. (11) and properties of the dot product, we have

$$\begin{aligned}
\|\mathbf{v} + \mathbf{w}\|^2 &+ \|\mathbf{v} - \mathbf{w}\|^2 \\
&= (\mathbf{v} + \mathbf{w}) \cdot (\mathbf{v} + \mathbf{w}) + (\mathbf{v} - \mathbf{w}) \cdot (\mathbf{v} - \mathbf{w}) \\
&= (\mathbf{v} \cdot \mathbf{v}) + 2(\mathbf{v} \cdot \mathbf{w}) + (\mathbf{w} \cdot \mathbf{w}) + (\mathbf{v} \cdot \mathbf{v}) - 2(\mathbf{v} \cdot \mathbf{w}) + (\mathbf{w} \cdot \mathbf{w}) \\
&= 2(\mathbf{v} \cdot \mathbf{v}) + 2(\mathbf{w} \cdot \mathbf{w}) \\
&= 2\|\mathbf{v}\|^2 + 2\|\mathbf{w}\|^2,
\end{aligned}$$

which is what we wished to prove. ■

The definition of the angle θ between two vectors $\mathbf{v}$ and $\mathbf{w}$ in $\mathbb{R}^n$ leads naturally to this definition of perpendicular vectors, or *orthogonal* vectors as they are usually called in linear algebra.

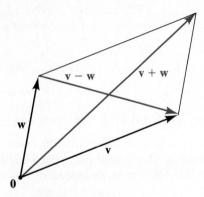

FIGURE 1.25
The parallelogram has v + w and
v − w as vector diagonals.

DEFINITION 1.7 Perpendicular or Orthogonal Vectors

> Two vectors **v** and **w** in $\mathbb{R}^n$ are **perpendicular** or **orthogonal**, and we write **v** $\perp$ **w**, if **v** $\cdot$ **w** = 0.

EXAMPLE 8   Determine whether the vectors **v** = [4, 1, −2, 1] and **w** = [3, −4, 2, −4] are perpendicular.

SOLUTION   We have

$$\mathbf{v} \cdot \mathbf{w} = (4)(3) + (1)(-4) + (-2)(2) + (1)(-4) = 0.$$

Thus, **v** $\perp$ **w**. ∎

Application to Velocity Vectors and Navigation

The next two examples are concerned with another important physical vector model. A vector is the **velocity vector** of a moving object at an instant if it points in the direction of the motion and if its magnitude is the speed of the object at that instant. Physicists tell us that if a boat cruising with a heading and speed that would give it a still-water velocity vector **s** is also subject to a current that has velocity vector **c**, then the actual velocity vector of the boat is **v** = **s** + **c**.

EXAMPLE 9   Suppose that a ketch is sailing at 8 knots, following a course of 010° (that is, 10° east of north), on a bay that has a 2-knot current setting in the direction 070° (that is, 70° east of north). Find the course and speed made good. (The expression *made good* is standard navigation terminology for the actual course and speed of a vessel over the bottom.)

SOLUTION   The velocity vectors **s** for the ketch and **c** for the current are shown in Figure 1.26, in which the vertical axis points due north. We find **s** and **c** by using a calculator and computing

$$\mathbf{s} = [8 \cos 80°, 8 \sin 80°] \approx [1.39, 7.88]$$

and

$$\mathbf{c} = [2 \cos 20°, 2 \sin 20°] \approx [1.88, 0.684].$$

By adding **s** and **c**, we find the vector **v** representing the course and speed of the ketch over the bottom—that is, the course and speed made good. Thus we have **v** = **s** + **c** ≈ [3.27, 8.56]. Therefore, the speed of the ketch is

$$\|\mathbf{v}\| \approx \sqrt{(3.27)^2 + (8.56)^2} \approx 9.16 \text{ knots,}$$

and the course made good is given by

$$90° - \arctan\left(\frac{8.56}{3.27}\right) \approx 90° - 69° = 21°.$$

That is, the course is 021°.   ■

EXAMPLE 10   Suppose the captain of our ketch realizes the importance of keeping track of the current. He wishes to sail in 5 hours to a harbor that bears 120° and is 35 nautical miles away. That is, he wishes to make good the course 120° and the speed 7 knots. He knows from a tide and current table that the current is setting due south at 3 knots. What should be his course and speed through the water?

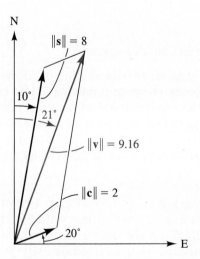

FIGURE 1.26
The vector v = s + c.

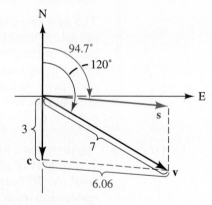

FIGURE 1.27
The vector s = v − c.

SOLUTION   In a vector diagram (see Figure 1.27), we again represent the course and speed to be made good by a vector **v** and the velocity of the current by **c**. The correct course and speed to follow are represented by the vector **s**, which is obtained by computing

$$\mathbf{s} = \mathbf{v} - \mathbf{c}$$
$$= [7 \cos 30°, -7 \sin 30°] - [0, -3]$$
$$\approx [6.06, -3.5] - [0, -3] = [6.06, -0.5].$$

Thus the captain should steer course $90° - \arctan(-0.5/6.06) \approx 90° + 4.7° = 94.7°$ and should proceed at

$$\|\mathbf{s}\| \approx \sqrt{(6.06)^2 + (-0.5)^2} \approx 6.08 \text{ knots.} \qquad \blacksquare$$

Proofs of the Schwarz and Triangle Inequalities

The proofs of the Schwarz and triangle inequalities illustrate the use of algebraic properties of the dot product in proving properties of the norm. Recall Eq. (11): for a vector **v** in $\mathbb{R}^n$, we have

$$\|\mathbf{v}\|^2 = \mathbf{v} \cdot \mathbf{v}.$$

THEOREM 1.4   Schwarz Inequality

Let **v** and **w** be vectors in $\mathbb{R}^n$. Then $|\mathbf{v} \cdot \mathbf{w}| \leq \|\mathbf{v}\| \, \|\mathbf{w}\|$.

PROOF   Because the norm of a vector is a real number and the square of a real number is nonnegative, for any scalars r and s we have

$$\|r\mathbf{v} + s\mathbf{w}\|^2 \geq 0. \tag{12}$$

Using relation (11), we find that

$$\|r\mathbf{v} + s\mathbf{w}\|^2 = (r\mathbf{v} + s\mathbf{w}) \cdot (r\mathbf{v} + s\mathbf{w})$$
$$= r^2(\mathbf{v} \cdot \mathbf{v}) + 2rs(\mathbf{v} \cdot \mathbf{w}) + s^2(\mathbf{w} \cdot \mathbf{w}) \geq 0$$

for all choices of scalars r and s. Setting $r = \mathbf{w} \cdot \mathbf{w}$ and $s = -(\mathbf{v} \cdot \mathbf{w})$, the preceding inequality becomes

$$(\mathbf{w} \cdot \mathbf{w})^2(\mathbf{v} \cdot \mathbf{v}) - 2(\mathbf{w} \cdot \mathbf{w})(\mathbf{v} \cdot \mathbf{w})^2 + (\mathbf{v} \cdot \mathbf{w})^2(\mathbf{w} \cdot \mathbf{w})$$
$$= (\mathbf{w} \cdot \mathbf{w})^2(\mathbf{v} \cdot \mathbf{v}) - (\mathbf{w} \cdot \mathbf{w})(\mathbf{v} \cdot \mathbf{w})^2 \geq 0.$$

Factoring out $(\mathbf{w} \cdot \mathbf{w})$, we see that

$$(\mathbf{w} \cdot \mathbf{w})[(\mathbf{w} \cdot \mathbf{w})(\mathbf{v} \cdot \mathbf{v}) - (\mathbf{v} \cdot \mathbf{w})^2] \geq 0. \tag{13}$$

If $\mathbf{w} \cdot \mathbf{w} = 0$, then $\mathbf{w} = \mathbf{0}$ by the positivity property in Theorem 1.3, and the Schwarz inequality is then true because it reduces to $0 \leq 0$. If $\|\mathbf{w}\|^2 = \mathbf{w} \cdot \mathbf{w} \neq 0$, then the expression in square brackets in relation (13) must also be nonnegative—that is,

$$(\mathbf{w} \cdot \mathbf{w})(\mathbf{v} \cdot \mathbf{v}) - (\mathbf{v} \cdot \mathbf{w})^2 \geq 0,$$

and so

$$(\mathbf{v} \cdot \mathbf{w})^2 \leq (\mathbf{v} \cdot \mathbf{v})(\mathbf{w} \cdot \mathbf{w}) = \|\mathbf{v}\|^2\|\mathbf{w}\|^2.$$

Taking square roots, we obtain the Schwarz inequality. ▲

The Schwarz inequality can be used to prove the triangle inequality that was illustrated in Figure 1.22.

THEOREM 1.5   The Triangle Inequality

Let $\mathbf{v}$ and $\mathbf{w}$ be vectors in $\mathbb{R}^n$. Then $\|\mathbf{v} + \mathbf{w}\| \leq \|\mathbf{v}\| + \|\mathbf{w}\|$.

PROOF   Using properties of the dot product, as well as the Schwarz inequality, we have

$$\begin{aligned}
\|\mathbf{v} + \mathbf{w}\|^2 &= (\mathbf{v} + \mathbf{w}) \cdot (\mathbf{v} + \mathbf{w}) \\
&= (\mathbf{v} \cdot \mathbf{v}) + 2(\mathbf{v} \cdot \mathbf{w}) + (\mathbf{w} \cdot \mathbf{w}) \\
&\leq (\mathbf{v} \cdot \mathbf{v}) + 2\|\mathbf{v}\|\,\|\mathbf{w}\| + (\mathbf{w} \cdot \mathbf{w}) \\
&= \|\mathbf{v}\|^2 + 2\|\mathbf{v}\|\,\|\mathbf{w}\| + \|\mathbf{w}\|^2 \\
&= (\|\mathbf{v}\| + \|\mathbf{w}\|)^2.
\end{aligned}$$

The desired relation follows at once, by taking square roots. ▲

SUMMARY

Let $\mathbf{v} = [v_1, v_2, \ldots, v_n]$ and $\mathbf{w} = [w_1, w_2, \ldots, w_n]$ be vectors in $\mathbb{R}^n$.

1. The *norm* or *magnitude* of $\mathbf{v}$ is $\|\mathbf{v}\| = \sqrt{v_1^2 + v_2^2 + \cdots + v_n^2}$.
2. The norm satisfies the properties given in Theorem 1.2.
3. A *unit vector* is a vector of magnitude 1.
4. The *dot product* of $\mathbf{v}$ and $\mathbf{w}$ is $\mathbf{v} \cdot \mathbf{w} = v_1 w_1 + v_2 w_2 + \cdots + v_n w_n$.
5. The dot product satisfies the properties given in Theorem 1.3.
6. Moreover, we have $\mathbf{v} \cdot \mathbf{v} = \|\mathbf{v}\|^2$ and $|\mathbf{v} \cdot \mathbf{w}| \leq \|\mathbf{v}\|\,\|\mathbf{w}\|$ (*Schwarz inequality*), and also $\|\mathbf{v} + \mathbf{w}\| \leq \|\mathbf{v}\| + \|\mathbf{w}\|$ (*triangle inequality*).
7. The *angle* θ between the vectors $\mathbf{v}$ and $\mathbf{w}$ can be found by using the relation $\mathbf{v} \cdot \mathbf{w} = \|\mathbf{v}\|\,\|\mathbf{w}\| (\cos \theta)$.
8. The vectors $\mathbf{v}$ and $\mathbf{w}$ are *orthogonal* (*perpendicular*) if $\mathbf{v} \cdot \mathbf{w} = 0$.

EXERCISES

In Exercises 1–17, let $\mathbf{u} = [-1, 3, 4]$, $\mathbf{v} = [2, 1, -1]$, *and* $\mathbf{w} = [-2, -1, 3]$. *Find the indicated quantity.*

1. $\|-\mathbf{u}\|$
2. $\|\mathbf{v}\|$
3. $\|\mathbf{u} + \mathbf{v}\|$
4. $\|\mathbf{v} - 2\mathbf{u}\|$
5. $\|3\mathbf{u} - \mathbf{v} + 2\mathbf{w}\|$
6. $\|\frac{4}{5}\mathbf{w}\|$
7. The unit vector parallel to $\mathbf{u}$, having the same direction
8. The unit vector parallel to $\mathbf{w}$, having the opposite direction
9. $\mathbf{u} \cdot \mathbf{v}$
10. $\mathbf{u} \cdot (\mathbf{v} + \mathbf{w})$
11. $(\mathbf{u} + \mathbf{v}) \cdot \mathbf{w}$
12. The angle between $\mathbf{u}$ and $\mathbf{v}$
13. The angle between $\mathbf{u}$ and $\mathbf{w}$
14. The value of x such that $[x, -3, 5]$ is perpendicular to $\mathbf{u}$
15. The value of y such that $[-3, y, 10]$ is perpendicular to $\mathbf{u}$
16. A nonzero vector perpendicular to both $\mathbf{u}$ and $\mathbf{v}$
17. A nonzero vector perpendicular to both $\mathbf{u}$ and $\mathbf{w}$

In Exercises 18–21, use properties of the dot product and norm to compute the indicated quantities mentally, without pencil or paper (or calculator).

18. $\|[42, 14]\|$
19. $\|[10, 20, 25, -15]\|$
20. $[14, 21, 28] \cdot [4, 8, 20]$
21. $[12, -36, 24] \cdot [25, 30, 10]$
22. Find the angle between $[1, -1, 2, 3, 0, 4]$ and $[7, 0, 1, 3, 2, 4]$ in $\mathbb{R}^6$.
23. Prove that $(2, 0, 4)$, $(4, 1, -1)$, and $(6, 7, 7)$ are vertices of a right triangle in $\mathbb{R}^3$.

24. Prove that the angle between two unit vectors $\mathbf{u}_1$ and $\mathbf{u}_2$ in $\mathbb{R}^n$ is $\arccos(\mathbf{u}_1 \cdot \mathbf{u}_2)$.

In Exercises 25–30, classify the vectors as parallel, perpendicular, or neither. If they are parallel, state whether they have the same direction or opposite directions.

25. $[-1, 4]$ and $[8, 2]$
26. $[-2, -1]$ and $[5, 2]$
27. $[3, 2, 1]$ and $[-9, -6, -3]$
28. $[2, 1, 4, -1]$ and $[0, 1, 2, 4]$
29. $[10, 4, -1, 8]$ and $[-5, -2, 3, -4]$
30. $[4, 1, 2, 1, 6]$ and $[8, 2, 4, 2, 3]$
31. The **distance** between points $(v_1, v_2, \ldots, v_n)$ and $(w_1, w_2, \ldots, w_n)$ in $\mathbb{R}^n$ is the norm $\|\mathbf{v} - \mathbf{w}\|$, where $\mathbf{v} = [v_1, v_2, \ldots, v_n]$ and $\mathbf{w} = [w_1, w_2, \ldots, w_n]$. Why is this a reasonable definition of distance?

In Exercises 32–35, use the definition given in Exercise 31 to find the indicated distance.

32. The distance from $(-1, 4, 2)$ to $(0, 8, 1)$ in $\mathbb{R}^3$
33. The distance from $(2, -1, 3)$ to $(4, 1, -2)$ in $\mathbb{R}^3$
34. The distance from $(3, 1, 2, 4)$ to $(-1, 2, 1, 2)$ in $\mathbb{R}^4$
35. The distance from $(-1, 2, 1, 4, 7, -3)$ to $(2, 1, -3, 5, 4, 5)$ in $\mathbb{R}^6$
36. The captain of a barge wishes to get to a point directly across a straight river that runs from north to south. If the current flows directly downstream at 5 knots and the barge steams at 13 knots, in what direction should the captain steer the barge?
37. A 100-lb weight is suspended by a rope passed through an eyelet on top of the weight and making angles of 30° with the vertical, as shown in Figure 1.28. Find the tension (magnitude of the force vector) along the rope. [HINT: The sum of the force vectors

FIGURE 1.28
Both halves of the rope make an angle of 30° with the vertical.

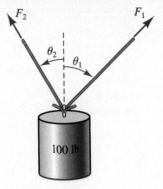

FIGURE 1.29
Two ropes tied at the eyelet and making angles θ_1 and θ_2 with the vertical.

along the two halves of the rope at the eyelet must be an upward vertical vector of magnitude 100.]

38. a. Answer Exercise 37 if each half of the rope makes an angle of θ with the vertical at the eyelet.
   b. Find the tension in the rope if both halves are vertical ($\theta = 0$).
   c. What happens if an attempt is made to stretch the rope out straight (horizontal) while the 100-lb weight hangs on it?

39. Suppose that a weight of 100 lb is suspended by two different ropes tied at an eyelet on top of the weight, as shown in Figure 1.29. Let the angles the ropes make with the vertical be θ_1 and θ_2, as shown in the figure. Let the tensions in the ropes be T_1 for the right-hand rope and T_2 for the left-hand rope.
   a. Show that the force vector $\mathbf{F}_1$ shown in Figure 1.29 is $T_1(\sin \theta_1)\mathbf{i} + T_1(\cos \theta_1)\mathbf{j}$.
   b. Find the corresponding expression for $\mathbf{F}_2$ in terms of T_2 and θ_2.
   c. If the system is in equilibrium, $\mathbf{F}_1 + \mathbf{F}_2 = 100\mathbf{j}$, so $\mathbf{F}_1 + \mathbf{F}_2$ must have $\mathbf{i}$-component 0 and $\mathbf{j}$-component 100. Write two equations reflecting this fact, using the answers to parts (a) and (b).
   d. Find T_1 and T_2 if $\theta_1 = 45°$ and $\theta_2 = 30°$.

40. Mark each of the following True or False.
   ____ **a.** Every nonzero vector in $\mathbb{R}^n$ has nonzero magnitude.
   ____ **b.** Every vector of nonzero magnitude in $\mathbb{R}^n$ is nonzero.
   ____ **c.** The magnitude of $\mathbf{v} + \mathbf{w}$ must be at least as large as the magnitude of either $\mathbf{v}$ or $\mathbf{w}$ in $\mathbb{R}^n$.
   ____ **d.** Every nonzero vector $\mathbf{v}$ in $\mathbb{R}^n$ has exactly one unit vector parallel to it.
   ____ **e.** There are exactly two unit vectors parallel to any given nonzero vector in $\mathbb{R}^n$.
   ____ **f.** There are exactly two unit vectors perpendicular to any given nonzero vector in $\mathbb{R}^n$.
   ____ **g.** The angle between two nonzero vectors in $\mathbb{R}^n$ is less than 90° if and only if the dot product of the vectors is positive.
   ____ **h.** The dot product of a vector with itself yields the magnitude of the vector.
   ____ **i.** For a vector $\mathbf{v}$ in $\mathbb{R}^n$, the magnitude of r times $\mathbf{v}$ is r times the magnitude of $\mathbf{v}$.
   ____ **j.** If $\mathbf{v}$ and $\mathbf{w}$ are vectors in $\mathbb{R}^n$ of the same magnitude, then the magnitude of $\mathbf{v} - \mathbf{w}$ is 0.

41. Prove the indicated property of the norm stated in Theorem 1.2.
   a. The positivity property
   b. The homogeneity property

42. Prove the indicated property of the dot product stated in Theorem 1.3.
   a. The commutative law
   b. The distributive law
   c. The homogeneity property

43. For vectors **v** and **w** in $\mathbb{R}^n$, prove that $\mathbf{v} - \mathbf{w}$ and $\mathbf{v} + \mathbf{w}$ are perpendicular if and only if $\|\mathbf{v}\| = \|\mathbf{w}\|$.

44. For vectors **u**, **v**, and **w** in $\mathbb{R}^n$ and for scalars r and s, prove that, if **w** is perpendicular to both **u** and **v**, then **w** is perpendicular to $r\mathbf{u} + s\mathbf{v}$.

45. Use vector methods to prove that the diagonals of a rhombus (parallelogram with equal sides) are perpendicular. [HINT: Use a figure similar to Figure 1.25 and one of the preceding exercises.]

46. Use vector methods to prove that the midpoint of the hypotenuse of a right triangle is equidistant from the three vertices. [HINT: See Figure 1.30. Show that

$$\left\|\frac{1}{2}(\mathbf{v} + \mathbf{w})\right\| = \left\|\frac{1}{2}(\mathbf{v} - \mathbf{w})\right\|.]$$

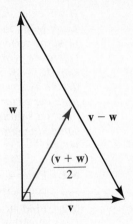

FIGURE 1.30
The vector $\frac{1}{2}(\mathbf{v} + \mathbf{w})$ to the midpoint of the hypotenuse.

MATLAB

MATLAB has a built-in function **norm(x)** for computing the norm of a vector **x**. It has no built-in command for finding a dot product or the angle between two vectors. Because one purpose of these exercises is to give practice at working with MATLAB, we will show how the norm of a vector can be computed without using the built-in function, as well as how to compute dot products and angles between vectors.

It is important to know how to enter data into MATLAB. In Section 1.1, we showed how to enter a vector. We have created M-files on the LINTEK disk that can be used to enter data automatically for our exercises, once practice in manual data entry has been provided. If these M-files have been copied into your MATLAB, you can simply enter **fbc1s2** to create the data vectors **a**, **b**, **c**, **u**, **v**, and **w** for the exercises below. The name of the file containing them is FBC1S2.M, where the FBC1S2 stands for "Fraleigh/Beauregard Chapter 1 Section 2." To view this data file so that you can create data files of your own, if you wish, simply enter **type fbc1s2** when in MATLAB. In addition to saving time, the data files help prevent wrong answers resulting from typos in data entry.

Access MATLAB, and either enter **fbc1s2** *or manually enter the following vectors.*

$$\mathbf{a} = [-2\ 1\ 3\ 5\ 1] \qquad \mathbf{u} = [2/3\ -4/7\ 8/5]$$
$$\mathbf{b} = [4\ -1\ 2\ 3\ 5] \qquad \mathbf{v} = [-1/2\ 13/3\ 17/11]$$
$$\mathbf{c} = [-1\ 0\ 3\ 0\ 4] \qquad \mathbf{w} = [22/7\ 15/2\ -8/3]$$

If you enter these vectors manually, be sure to proofread your data entry for accuracy. Enter **help .** *to see what can be done with the period. Enter* **a .∗ b** *and compare the resulting vector with the vectors* **a** *and* **b**. *(Be sure to get in the habit of putting a space before the period so that MATLAB will never interpret it as a decimal point in this context.) Now enter* **c .^3** *and compare with the vector* **c**. *The symbol* ^ *is used to denote exponentiation. Then enter* **sum(a)** *and note how this number was obtained from the vector* **a**. *Using the . notation, the sum function* **sum(x)**, *the square root function* **sqrt(x)**, *and the arccosine function* **acos(x)**, *we can easily write formulas for computing norms of vectors, dot products of vectors, and the angle between two vectors.*

M1. Enter **x = a** and then enter **normx = sqrt(sum(x .∗ x))** to find $\|a\|$. Compare your answer with the result obtained by entering **norm(a)**.

M2. Find $\|b\|$ by entering **x = b** and then pressing the upward arrow until equation **normx = sqrt(sum(x .∗ x))** is at the cursor, and then pressing the Enter key.

M3. Using the technique outlined in Exercise M2, find $\|u\|$.

M4. Using the appropriate MATLAB commands, compute the dot product **v · w** in (a) short format and (b) rational format.

M5. Repeat Exercise M4 for $(2u - 3v) \cdot (4u - 7v)$.

NOTE: If you are working with your own personal MATLAB, you can add a function **angl(x, y)** for finding the angle between vectors **x** and **y** having the same number of components to MATLAB's supply of available functions. First, enter **help angl** to be sure that MATLAB does not already have a command with the name you will use; otherwise you might delete an existing MATLAB function. Then, assuming that MATLAB has created a subdirectory C:\MATLAB\MATLAB on your disk, get out of MATLAB and either use a word processor that will create ASCII files or skip to the next paragraph. Using a word processor, create an ASCII file designated as C:\MATLAB\MATLAB\ANGL.M by entering each of the following lines.

> **function z = angl(x, y)**
>
> **% ANGL   ANGL (x, y) is the radian angle between vectors x and y.**
>
> **z = acos(sum(x .∗ y)/(norm(x)∗norm(y)))**

Then save the file. This creates a function **angl(x, y)** of two variables **x** and **y** in place of the name anglxy we use in M6. You will now be able to compute the angle between vectors **x** and **y** in MATLAB simply by entering **angl(x, y)**. Note that the file name ANGL.M concludes with the .M suffix. MATLAB comes with a number of these M-files, which are probably in the subdirectory MATLAB\MATLAB of your disk. Remember to enter **help angl** from MATLAB first, to be sure there is not already a file with the name ANGL.M.

   If you do not have a word processor that writes ASCII files, you can still create the file from DOS if your hard disk is the default c-drive. First enter CD C:\MATLAB\MATLAB. Then enter COPY CON ANGL.M and proceed to enter the three lines displayed above. When you have pressed Enter after the final line, press the F6 key and then press Enter again.

   After creating the file, access MATLAB and test your function **angl(x, y)** by finding the angle between the vectors [1, 0] and [−1, 0]. The angle should be $\pi \approx 3.1416$. Then enter **help angl** and you should see the explanatory note on the line of the file that starts with **%** displayed on the screen. Using these directions as a model, you can easily create functions of your own to add to MATLAB.

M6. Enter **x** = **a** and enter **y** = **b**. Then enter

$$\textbf{anglxy} = \textbf{acos(sum(x .* y)/(norm(x)*norm(y)))}$$

to compute the angle (in radians) between **a** and **b**. You should study this formula until you understand why it provides the angle between **a** and **b**.

M7. Compute the angle between **b** and **c** using the technique suggested in Exercise M2. Namely, enter **x** = **b**, enter **y** = **c**, and then use the upward arrow until the cursor is at the formula for **anglxy** and press Enter.

M8. Move the cursor to the formula for **anglxy** and edit the formula so that the angle will be given in degrees rather than in radians. Recall that we multiply by $180/\pi$ to convert radians to degrees. The number π is available as **pi** in MATLAB. Check your editing by computing the angle between the vectors [1, 0] and [0, 1]. Then find the angle between **u** and **w** in degrees.

M9. Find the angle between $3\textbf{u} - 2\textbf{w}$ and $4\textbf{v} + 2\textbf{w}$ in degrees.

1.3

MATRICES AND THEIR ALGEBRA

The Notation $A\textbf{x} = \textbf{b}$

We saw in Section 1.1 that we can write a linear system such as

$$x_1 - 2x_2 = -1$$
$$3x_1 + 5x_2 = 19 \tag{1}$$

in the unknowns x_1 and x_2 as a single column vector equation—namely,

$$x_1 \begin{bmatrix} 1 \\ 3 \end{bmatrix} + x_2 \begin{bmatrix} -2 \\ 5 \end{bmatrix} = \begin{bmatrix} -1 \\ 19 \end{bmatrix}. \tag{2}$$

Another useful way to abbreviate this linear system is

$$\underset{A}{\begin{bmatrix} 1 & -2 \\ 3 & 5 \end{bmatrix}} \underset{\textbf{x}}{\begin{bmatrix} x_1 \\ x_2 \end{bmatrix}} = \underset{\textbf{b}}{\begin{bmatrix} -1 \\ 19 \end{bmatrix}}. \tag{3}$$

Let us denote by A the bracketed array on the left containing the coefficients of the linear system. This array A is followed by the column vector **x** of unknowns, and let the column vector of constants after the equal sign be denoted by **b**. We can then symbolize the linear system as

$$A\textbf{x} = \textbf{b}. \tag{4}$$

There are several reasons why notation (4) is a convenient way to write a linear system. It is much easier to denote a general linear system by $A\textbf{x} = \textbf{b}$ than to write out several linear equations with unknowns $x_1, x_2, \ldots, x_n$, subscripted letters for the coefficients of the unknowns, and constants $b_1, b_2, \ldots, b_m$ to the right of the equal signs. [Just look ahead at Eq. (1) on page 54.] Also, a single linear equation in just one unknown can be written in the form $ax = b$ ($2x = 6$, for example), and the notation $A\mathbf{x} = \mathbf{b}$ is suggestively similar. Furthermore, we will see in Section 2.3 that we can regard such an array A as defining a *function* whose value at $\mathbf{x}$ we will write as $A\mathbf{x}$, much as we write sin x. Solving a linear system $A\mathbf{x} = \mathbf{b}$ can thus be regarded as finding the vector $\mathbf{x}$ such that this function applied to $\mathbf{x}$ yields the vector $\mathbf{b}$. For all of these reasons, the notation $A\mathbf{x} = \mathbf{b}$ for a linear system is one of the most useful notations in mathematics.

It is very important to remember that

$A\mathbf{x}$ is equal to a linear combination of the *column vectors* of A,

---

as illustrated by Eqs. (2) and (3)—namely,

$$\begin{bmatrix} 1 & -2 \\ 3 & 5 \end{bmatrix} \begin{bmatrix} x_1 \\ x_2 \end{bmatrix} = x_1 \begin{bmatrix} 1 \\ 3 \end{bmatrix} + x_2 \begin{bmatrix} -2 \\ 5 \end{bmatrix}. \tag{5}$$

The Notion of a Matrix

We now introduce the usual terminology and notation for an array of numbers such as the coefficient array A in Eq. (3).

A *matrix* is an ordered rectangular array of numbers, usually enclosed in parentheses or square brackets. For example,

$$A = \begin{bmatrix} 1 & -2 \\ 3 & 5 \end{bmatrix} \quad \text{and} \quad B = \begin{bmatrix} -1 & 0 & 3 \\ 2 & 1 & 4 \\ 4 & 5 & -6 \\ -3 & -1 & -1 \end{bmatrix}$$

are matrices. We will generally use upper-case letters to denote matrices.

The size of a matrix is specified by the number of (horizontal) rows and the number of (vertical) columns that it contains. The matrix A above contains two rows and two columns and is called a 2×2 (read "2 by 2") matrix. Similarly, B is a 4×3 matrix. In writing the notation $m \times n$ to describe the shape of a matrix, we always write the number of rows first. An $n \times n$ matrix has the same number of rows as columns and is said to be a *square matrix*. We recognize that a $1 \times n$ matrix is a row vector with n components, and an $m \times 1$ matrix is a column vector with m components. The rows of a matrix are its *row vectors* and the columns are its *column vectors*.

Double subscripts are commonly used to indicate the location of an entry in a matrix that is not a row or column vector. The first subscript gives the number of the row in which the entry appears (counting from the top), and the second subscript gives the number of the column (counting from the left). Thus an $m \times n$ matrix A may be written as

$$A = [a_{ij}] = \begin{bmatrix} a_{11} & a_{12} & a_{13} & \cdots & a_{1n} \\ a_{21} & a_{22} & a_{23} & \cdots & a_{2n} \\ a_{31} & a_{32} & a_{33} & \cdots & a_{3n} \\ & & \vdots & & \\ a_{m1} & a_{m2} & a_{m3} & \cdots & a_{mn} \end{bmatrix}$$

If we want to express the matrix B on page 36 as $[b_{ij}]$, we would have $b_{11} = -1$, $b_{21} = 2$, $b_{32} = 5$, and so on.

Matrix Multiplication

We are going to consider the expression $A\mathbf{x}$ shown in Eq. (3) to be the *product* of the matrix A and the column vector $\mathbf{x}$. Looking back at Eq. (5), we see that such a product of a matrix A with a column vector $\mathbf{x}$ should be the linear combination of the column vectors of A having as coefficients the components in the vector $\mathbf{x}$. Here is a *nonsquare* example in which we replace the vector $\mathbf{x}$ of unknowns by a specific vector of numbers.

EXAMPLE 1   Write as a linear combination and then compute the product

$$\begin{bmatrix} 2 & -3 & 5 \\ -1 & 4 & -7 \end{bmatrix} \begin{bmatrix} -2 \\ 5 \\ 8 \end{bmatrix}.$$

HISTORICAL NOTE   THE TERM *MATRIX* is first mentioned in mathematical literature in an 1850 paper of James Joseph Sylvester (1814–1897). The standard nontechnical meaning of this term is "a place in which something is bred, produced, or developed." For Sylvester, then, a matrix, which was an "oblong arrangement of terms," was an entity out of which one could form various square pieces to produce determinants. These latter quantities, formed from square matrices, were quite well known by this time.

James Sylvester (his original name was James Joseph) was born into a Jewish family in London, and was to become one of the supreme algebraists of the nineteenth century. Despite having studied for several years at Cambridge University, he was not permitted to take his degree there because he "professed the faith in which the founder of Christianity was educated." Therefore, he received his degrees from Trinity College, Dublin. In 1841 he accepted a professorship at the University of Virginia; he remained there only a short time, however, his horror of slavery preventing him from fitting into the academic community. In 1871 he returned to the United States to accept the chair of mathematics at the newly opened Johns Hopkins University. In between these sojourns, he spent about 10 years as an attorney, during which time he met Arthur Cayley (see the note on p. 3), and 15 years as Professor of Mathematics at the Royal Military Academy, Woolwich. Sylvester was an avid poet, prefacing many of his mathematical papers with examples of his work. His most renowned example was the "Rosalind" poem, a 400-line epic, each line of which rhymed with "Rosalind."

SOLUTION   Using Eq. (5) as a guide, we find that

$$\begin{bmatrix} 2 & -3 & 5 \\ -1 & 4 & -7 \end{bmatrix} \begin{bmatrix} -2 \\ 5 \\ 8 \end{bmatrix} = -2 \begin{bmatrix} 2 \\ -1 \end{bmatrix} + 5 \begin{bmatrix} -3 \\ 4 \end{bmatrix} + 8 \begin{bmatrix} 5 \\ -7 \end{bmatrix} = \begin{bmatrix} 21 \\ -34 \end{bmatrix}.$$ ∎

Note that in Example 1, the first entry 21 of the final column vector is computed as $(-2)(2) + (5)(-3) + (8)(5)$, which is precisely the dot product of the first row vector $[2 \ -3 \ 5]$ of the matrix with the column vector $\begin{bmatrix} -2 \\ 5 \\ 8 \end{bmatrix}$.

Similarly, the second component -34 of our answer is the dot product of the second row vector $[-1 \ 4 \ -7]$ with this column vector.

In a similar fashion, we see that the ith component of a column vector $A\mathbf{b}$ will be equal to the dot product of the ith row of A with the column vector $\mathbf{b}$. We should also note from Example 1 that the number of components in a row of A will have to be equal to the number of components in the column vector $\mathbf{b}$ if we are to compute the product $A\mathbf{b}$.

We have illustrated how to compute a product $A\mathbf{b}$ of an $m \times n$ matrix with an $n \times 1$ column vector. We can extend this notion to a product AB of an $m \times n$ matrix A with an $n \times s$ matrix B.

> The product AB is the matrix whose jth column is the product of A with the jth column vector of B.

Letting $\mathbf{b}_j$ be the jth column vector of B, we write $AB = C$ symbolically as

$$A \underbrace{\begin{bmatrix} | & | & & | \\ \mathbf{b}_1 & \mathbf{b}_2 & \cdots & \mathbf{b}_s \\ | & | & & | \end{bmatrix}}_{B} = \underbrace{\begin{bmatrix} | & | & & | \\ A\mathbf{b}_1 & A\mathbf{b}_2 & \cdots & A\mathbf{b}_s \\ | & | & & | \end{bmatrix}}_{C}.$$

Because B has s columns, C has s columns. The comments after Example 1 indicate that the ith entry in the jth column of AB is the dot product of the ith row of A with the jth column of B. We give a formal definition.

DEFINITION 1.8   Matrix Multiplication

Let $A = [a_{ik}]$ be an $m \times n$ matrix, and let $B = [b_{kj}]$ be an $n \times s$ matrix. The **matrix product** AB is the $m \times s$ matrix $C = [c_{ij}]$, where c_{ij} is the dot product of the ith row vector of A and the jth column vector of B.

We illustrate the choice of row i from A and column j from B to find the element c_{ij} in AB, according to Definition 1.8, by the equation

$$AB = [c_{ij}] = \begin{bmatrix} a_{11} & \!\!\!-\!\!\! & a_{1n} \\ \vdots & & \vdots \\ a_{i1} & \!\!\!-\!\!\! & a_{in} \\ \vdots & & \vdots \\ a_{m1} & \!\!\!-\!\!\! & a_{mn} \end{bmatrix} \begin{bmatrix} b_{11} & \cdots & b_{1j} & \cdots & b_{1s} \\ | & & | & & | \\ b_{n1} & \cdots & b_{nj} & \cdots & b_{ns} \end{bmatrix},$$

where

$$c_{ij} = (i\text{th row vector of } A) \cdot (j\text{th column vector of } B).$$

In summation notation, we have

$$c_{ij} = a_{i1}b_{1j} + a_{i2}b_{2j} + \cdots + a_{in}b_{nj}$$

$$= \sum_{k=1}^{n} a_{ik}b_{kj}. \tag{6}$$

Notice again that AB is defined only when the second size-number (the number of columns) of A is the same as the first size-number (the number of rows) of B. The product matrix has the shape

(First size-number of A) $\times$ (Second size-number of B).

EXAMPLE 2 Let A be a 2×3 matrix, and let B be a 3×5 matrix. Find the sizes of AB and BA, if they are defined.

SOLUTION Because the second size-number, 3, of A equals the first size-number, 3, of B, we see that AB is defined; it is a 2×5 matrix. However, BA is not defined, because the second size-number, 5, of B is not the same as the first size-number, 2, of A. ∎

EXAMPLE 3 Compute the product

$$\begin{bmatrix} -2 & 3 & 2 \\ 4 & 6 & -2 \end{bmatrix} \begin{bmatrix} 4 & -1 & 2 & 5 \\ 3 & 0 & 1 & 1 \\ -2 & 3 & 5 & -3 \end{bmatrix}.$$

SOLUTION The product is defined, because the left-hand matrix is 2×3 and the right-hand matrix is 3×4; the product will have size 2×4. The entry in the first row and first column position of the product is obtained by taking the dot product of the first row vector of the left-hand matrix and the first column vector of the right-hand matrix, as follows:

$$(-2)(4) + (3)(3) + (2)(-2) = -8 + 9 - 4 = -3.$$

The entry in the second row and third column of the product is the dot product of the second row vector of the left-hand matrix and the third column vector of the right-hand one:

$$(4)(2) + (6)(1) + (-2)(5) = 8 + 6 - 10 = 4,$$

and so on, through the remaining row and column positions of the product. Eight such computations show that

$$\begin{bmatrix} -2 & 3 & 2 \\ 4 & 6 & -2 \end{bmatrix} \begin{bmatrix} 4 & -1 & 2 & 5 \\ 3 & 0 & 1 & 1 \\ -2 & 3 & 5 & -3 \end{bmatrix} = \begin{bmatrix} -3 & 8 & 9 & -13 \\ 38 & -10 & 4 & 32 \end{bmatrix}. \quad \blacksquare$$

Examples 2 and 3 show that sometimes AB is defined when BA is not. Even if both AB and BA are defined, however, it need not be true that $AB = BA$:

> Matrix multiplication is not commutative.

EXAMPLE 4   Let

$$A = \begin{bmatrix} 0 & 2 \\ 3 & 5 \end{bmatrix} \quad \text{and} \quad B = \begin{bmatrix} 0 & 1 \\ 2 & 5 \end{bmatrix}.$$

Compute AB and BA.

HISTORICAL NOTE   MATRIX MULTIPLICATION originated in the composition of linear substitutions, fully explored by Carl Friedrich Gauss (1777–1855) in his *Disquisitiones Arithmeticae* of 1801 in connection with his study of quadratic forms. Namely, if $F = Ax^2 + 2Bxy + Cy^2$ is such a form, then the linear substitution

$$x = ax' + by' \qquad y = cx' + dy' \qquad \text{(i)}$$

transforms F into a new form F' in the variables x' and y'. If a second substitution

$$x' = ex'' + fy'' \qquad y' = gx'' + hy'' \qquad \text{(ii)}$$

transforms F' into a form F'' in x'', y'', then the composition of the substitutions, found by replacing x', y' in (i) by their values in (ii), gives a substitution transforming F into F'':

$$x = (ae + bg)x'' + (af + bh)y'' \qquad y = (ce + dg)x'' + (cf + dh)y''. \qquad \text{(iii)}$$

The coefficient matrix of substitution (iii) is the product of the coefficient matrices of substitutions (i) and (ii). Gauss performed an analogous computation in his study of substitutions in forms in three variables, which produced the rule for multiplication of 3×3 matrices.

Gauss, however, did not explicitly refer to this idea of composition as a "multiplication." That was done by his student Ferdinand Gotthold Eisenstein (1823–1852), who introduced the notation $S \times T$ to denote the substitution composed of S and T. About this notation Eisenstein wrote, "An algorithm for calculation can be based on this; it consists of applying the usual rules for the operations of multiplication, division, and exponentiation to symbolical equations between linear systems; correct symbolical equations are always obtained, the sole consideration being that the order of the factors may not be altered."

SOLUTION   We compute that

$$AB = \begin{bmatrix} 4 & 10 \\ 10 & 28 \end{bmatrix}, \quad \text{and} \quad BA = \begin{bmatrix} 3 & 5 \\ 15 & 29 \end{bmatrix}. \quad\blacksquare$$

Of course, for a square matrix A, we denote AA by A^2, AAA by A^3, and so on. It can be shown that matrix multiplication is associative; that is,

$$A(BC) = (AB)C$$

whenever the product is defined. This is not difficult to prove from the definition, although keeping track of subscripts can be a bit challenging. We leave the proof as Exercise 33, whose solution is given in the back of this text.

The $n \times n$ Identity Matrix

Let I be the $n \times n$ matrix $[a_{ij}]$ such that $a_{ii} = 1$ for $i = 1, \ldots, n$ and $a_{ij} = 0$ for $i \neq j$. That is,

$$I = \begin{bmatrix} 1 & 0 & 0 & \cdots & 0 \\ 0 & 1 & 0 & \cdots & 0 \\ 0 & 0 & 1 & \cdots & 0 \\ \cdot & \cdot & \cdot & & \cdot \\ \cdot & \cdot & \cdot & & \cdot \\ \cdot & \cdot & \cdot & & \cdot \\ 0 & 0 & 0 & \cdots & 1 \end{bmatrix} = \begin{bmatrix} 1 & & & & \\ & 1 & & \mathbf{0} & \\ & & 1 & & \\ & \mathbf{0} & & \ddots & \\ & & & & 1 \end{bmatrix},$$

where the large zeros above and below the diagonal in the second matrix indicate that each entry of the matrix in those positions is 0. If A is any $m \times n$ matrix and B is any $n \times s$ matrix, we can show that

$$AI = A \quad \text{and} \quad IB = B.$$

We can understand why this is so if we think about why it is that

$$\begin{bmatrix} 2 & 3 \\ -1 & 7 \end{bmatrix} \begin{bmatrix} 1 & 0 \\ 0 & 1 \end{bmatrix} = \begin{bmatrix} 2 & 3 \\ -1 & 7 \end{bmatrix} = \begin{bmatrix} 1 & 0 \\ 0 & 1 \end{bmatrix} \begin{bmatrix} 2 & 3 \\ -1 & 7 \end{bmatrix}.$$

Because of the relations $AI = A$ and $IB = B$, the matrix I is called the $n \times n$ **identity matrix.** It behaves for multiplication of $n \times n$ matrices exactly as the scalar 1 behaves for multiplication of scalars. We have one such square identity matrix for each integer 1, 2, 3, ... To keep notation simple, we denote them all by I, rather than by $I_1, I_2, I_3, \ldots$. The size of I will be clear from the context.

The identity matrix is an example of a **diagonal matrix**—namely, a square matrix with zero entries except possibly on the **main diagonal**, which extends from the upper left corner to lower right corner.

Other Matrix Operations

Although multiplication is a very important matrix operation for our work, we will have occasion to add and subtract matrices, and to multiply a matrix by a scalar, in later chapters. Matrix addition, subtraction, and scalar multiplication are natural extensions of these same operations for vectors as defined in Section 1.1; they are again performed on entries in corresponding positions.

DEFINITION 1.9   Matrix Addition

Let $A = [a_{ij}]$ and $B = [b_{ij}]$ be two matrices of the same size $m \times n$. The **sum** $A + B$ of these two matrices is the $m \times n$ matrix $C = [c_{ij}]$, where

$$c_{ij} = a_{ij} + b_{ij}.$$

That is, the sum of two matrices of the same size is the matrix of that size obtained by adding corresponding entries.

EXAMPLE 5   Find

$$\begin{bmatrix} 1 & 2 & -4 \\ 0 & 3 & -1 \end{bmatrix} + \begin{bmatrix} -1 & 0 & 2 \\ 1 & -5 & 3 \end{bmatrix}.$$

SOLUTION   The sum is the matrix

$$\begin{bmatrix} 0 & 2 & -2 \\ 1 & -2 & 2 \end{bmatrix}.$$ ∎

EXAMPLE 6   Find

$$\begin{bmatrix} 1 & -3 \\ 2 & 4 \end{bmatrix} + \begin{bmatrix} -5 & 4 & 6 \\ 3 & 7 & -1 \end{bmatrix}.$$

SOLUTION   The sum is undefined, because the matrices are not the same size.   ∎

Let A be an $m \times n$ matrix, and let O be the $m \times n$ matrix all of whose entries are zero. Then,

$$A + O = O + A = A.$$

The matrix O is called the $m \times n$ **zero matrix**; the size of such a zero matrix is made clear by the context.

DEFINITION 1.10   Scalar Multiplication

Let $A = [a_{ij}]$, and let r be a scalar. The **product** rA of the scalar r and the matrix A is the matrix $B = [b_{ij}]$ having the same size as A, where

$$b_{ij} = ra_{ij}.$$

EXAMPLE 7   Find

$$2\begin{bmatrix} -2 & 1 \\ 3 & -5 \end{bmatrix}.$$

SOLUTION   Multiplying each entry of the matrix by 2, we obtain the matrix

$$\begin{bmatrix} -4 & 2 \\ 6 & -10 \end{bmatrix}. \qquad \blacksquare$$

For matrices A and B of the same size, we define the **difference** $A - B$ to be

$$A - B = A + (-1)B.$$

The entries in $A - B$ are obtained by subtracting the entries of B from entries in the corresponding positions in A.

EXAMPLE 8   If

$$A = \begin{bmatrix} 3 & -1 & 4 \\ 0 & 2 & -5 \end{bmatrix} \quad \text{and} \quad B = \begin{bmatrix} -1 & 0 & 5 \\ 4 & -2 & 1 \end{bmatrix},$$

find $2A - 3B$.

SOLUTION   We find that

$$2A - 3B = \begin{bmatrix} 9 & -2 & -7 \\ -12 & 10 & -13 \end{bmatrix}. \qquad \blacksquare$$

We introduced the *transpose operation* to change a row vector to a column vector, or vice versa, in Section 1.1. We generalize this operation for application to matrices, changing all the row vectors to column vectors, which results in all the column vectors becoming row vectors.

DEFINITION 1.11   Transpose of a Matrix; Symmetric Matrix

The matrix B is the **transpose** of the matrix A, written $B = A^T$, if each entry b_{ij} in B is the same as the entry a_{ji} in A, and conversely. If A is a matrix and if $A = A^T$, then the matrix A is **symmetric**.

EXAMPLE 9   Find A^T if

$$A = \begin{bmatrix} 1 & 4 & 5 \\ -3 & 2 & 7 \end{bmatrix}.$$

SOLUTION   We have

$$A^T = \begin{bmatrix} 1 & -3 \\ 4 & 2 \\ 5 & 7 \end{bmatrix}.$$

Notice that the rows of A become the columns of A^T. $\blacksquare$

A symmetric matrix must be square. Symmetric matrices arise in some applications, as we shall see in Chapter 8.

EXAMPLE 10   Fill in the missing entries in the 4×4 matrix

$$\begin{bmatrix} 5 & -6 & & 8 \\ & 3 & & \\ -2 & 1 & 0 & 4 \\ & 11 & & -1 \end{bmatrix}$$

to make it symmetric.

SOLUTION   Because rows must match corresponding columns, we obtain

$$\begin{bmatrix} 5 & -6 & -2 & 8 \\ -6 & 3 & 1 & 11 \\ -2 & 1 & 0 & 4 \\ 8 & 11 & 4 & -1 \end{bmatrix}.$$ ∎

In Example 10, note the symmetry in the main diagonal.

We have explained that we will often regard vectors in $\mathbb{R}^n$ as column vectors. If **a** and **b** are two column vectors in $\mathbb{R}^n$, the dot product $\mathbf{a} \cdot \mathbf{b}$ can be written in terms of the transpose operation and matrix multiplication— namely,

$$\mathbf{a} \cdot \mathbf{b} = \mathbf{a}^T \mathbf{b} = [a_1 \; a_2 \; \cdots \; a_n] \begin{bmatrix} b_1 \\ b_2 \\ \vdots \\ b_n \end{bmatrix}. \tag{7}$$

Strictly speaking, $\mathbf{a}^T\mathbf{b}$ is a 1×1 matrix, and its sole entry is $\mathbf{a} \cdot \mathbf{b}$. Identifying a 1×1 matrix with its sole entry should cause no difficulty. The use of Eq. (7) makes some formulas given later in the text much easier to handle.

Properties of Matrix Operations

For handy reference, we box the properties of matrix algebra and of the transpose operation. These properties are valid for all vectors, scalars, and matrices for which the indicated quantities are defined. The exercises ask for proofs of most of them. The proofs of the properties of matrix algebra not involving matrix multiplication are essentially the same as the proofs of the same properties presented for vector algebra in Section 1.1. We would expect this because those operations are performed just on corresponding entries, and every vector can be regarded as either a $1 \times n$ or an $n \times 1$ matrix.

Properties of Matrix Algebra

$A + B = B + A$	**Commutative law of addition**
$(A + B) + C = A + (B + C)$	**Associative law of addition**
$A + O = O + A = A$	**Identity for addition**
$r(A + B) = rA + rB$	**A left distributive law**
$(r + s)A = rA + sA$	**A right distributive law**
$(rs)A = r(sA)$	**Associative law of scalar multiplication**
$(rA)B = A(rB) = r(AB)$	**Scalars pull through**
$A(BC) = (AB)C$	**Associative law of matrix multiplication**
$IA = A$ and $BI = B$	**Identity for matrix multiplication**
$A(B + C) = AB + AC$	**A left distributive law**
$(A + B)C = AC + BC$	**A right distributive law**

Properties of the Transpose Operation

$(A^T)^T = A$	**Transpose of the transpose**
$(A + B)^T = A^T + B^T$	**Transpose of a sum**
$(AB)^T = B^T A^T$	**Transpose of a product**

EXAMPLE 11   Prove that $A(B + C) = AB + AC$ for any $m \times n$ matrix A and any $n \times s$ matrices B and C.

SOLUTION   Let $A = [a_{ij}]$, $B = [b_{jk}]$ and $C = [c_{jk}]$. Note the use of j, which runs from 1 to n, as both the second index for entries in A and the first index for the entries in B and C. The entry in the ith row and kth column of $A(B + C)$ is

$$\sum_{j=1}^{n} a_{ij}(b_{jk} + c_{jk}).$$

By familiar properties of real numbers, this sum is also equal to

$$\sum_{j=1}^{n} (a_{ij}b_{jk} + a_{ij}c_{jk}) = \sum_{j=1}^{n} a_{ij}b_{jk} + \sum_{j=1}^{n} a_{ij}c_{jk}$$

which we recognize as the sum of the entries in the ith row and kth columns of the matrices AB and AC. This completes the proof.   ■

SUMMARY

1. An $m \times n$ matrix is an ordered rectangular array of numbers containing m rows and n columns.

2. An $m \times 1$ matrix is a column vector with m components, and a $1 \times n$ matrix is a row vector with n components.

3. The product $A\mathbf{b}$ of an $m \times n$ matrix A and a column vector $\mathbf{b}$ with components $b_1, b_2, \ldots, b_n$ is the column vector equal to the linear combination of the column vectors of A where the scalar coefficient of the jth column vector of A is b_j.

4. The product AB of an $m \times n$ matrix A and an $n \times s$ matrix B is the $m \times s$ matrix C whose jth column is A times the jth column of B. The entry c_{ij} in the ith row and jth column of C is the dot product of the ith row vector of A and the jth column vector of B. In general, $AB \neq BA$.

5. If $A = [a_{ij}]$ and $B = [b_{ij}]$ are matrices of the same size, then $A + B$ is the matrix of that size with entry $a_{ij} + b_{ij}$ in the ith row and jth column.

6. For any matrix A and scalar r, the matrix rA is found by multiplying each entry in A by r.

7. The transpose of an $m \times n$ matrix A is the $n \times m$ matrix A^T, which has as its kth row vector the kth column vector of A.

8. Properties of the matrix operations are given in boxed displays on page 45.

EXERCISES

In Exercises 1–16, let

$$A = \begin{bmatrix} -2 & 1 & 3 \\ 4 & 0 & -1 \end{bmatrix}, \quad B = \begin{bmatrix} 4 & 1 & -2 \\ 5 & -1 & 3 \end{bmatrix}, \quad C = \begin{bmatrix} 2 & -1 \\ 0 & 6 \\ -3 & 2 \end{bmatrix}, \quad \text{and} \quad D = \begin{bmatrix} -4 & 2 \\ 3 & 5 \\ -1 & -3 \end{bmatrix}.$$

Compute the indicated quantity, if it is defined.

1. $3A$

2. $0B$

3. $A + B$

4. $B + C$

5. $C - D$

6. $4A - 2B$

7. AB

8. $(CD)^T$

9. $(2A)(5C)$

10. $(5D)(4B)$

11. A^2

12. $(AC)^2$

13. $(2A - B)D$

14. ADB

15. $(A^T)A$

16. BC and CB

17. Let

$$A = \begin{bmatrix} 2 & 0 & 0 \\ 0 & -1 & 0 \\ 0 & 0 & 1 \end{bmatrix}.$$

a. Find A^2.

b. Find A^7.

18. Let

$$A = \begin{bmatrix} 0 & 0 & -1 \\ 0 & 2 & 0 \\ 2 & 0 & 0 \end{bmatrix}.$$

a. Find A^2.

b. Find A^7.

19. Consider the row and column vectors

$$\mathbf{x} = [-2, 3, -1] \text{ and } \mathbf{y} = \begin{bmatrix} 4 \\ -1 \\ 3 \end{bmatrix}.$$

Compute the matrix products $\mathbf{xy}$ and $\mathbf{yx}$.

20. Fill in the missing entries in the 4×4 matrix

$$\begin{bmatrix} 1 & -1 & & 5 \\ & 4 & & 8 \\ 2 & -7 & -1 & \\ & & 6 & 3 \end{bmatrix}$$

so that the matrix is symmetric.

21. Mark each of the following True or False. The statements involve matrices A, B, and C that are assumed to have appropriate size.
___ a. If $A = B$, then $AC = BC$.
___ b. If $AC = BC$, then $A = B$.
___ c. If $AB = O$, then $A = O$ or $B = O$.
___ d. If $A + C = B + C$, then $A = B$.
___ e. If $A^2 = I$, then $A = \pm I$.
___ f. If $B = A^2$ and if A is $n \times n$ and symmetric, then $b_{ii} \geq 0$ for $i = 1, 2, \ldots, n$.
___ g. If $AB = C$ and if two of the matrices are square, then so is the third.
___ h. If $AB = C$ and if C is a column vector, then so is B.
___ i. If $A^2 = I$, then $A^n = I$ for all integers $n \geq 2$.
___ j. If $A^2 = I$, then $A^n = I$ for all even integers $n \geq 2$.

22. a. Prove that, if A is a matrix and $\mathbf{x}$ is a row vector, then $\mathbf{x}A$ (if defined) is again a row vector.
    b. Prove that, if A is a matrix and $\mathbf{y}$ is a column vector, then $A\mathbf{y}$ (if defined) is again a column vector.

23. Let A be an $m \times n$ matrix and let $\mathbf{b}$ and $\mathbf{c}$ be column vectors with n components. Express the dot product $(A\mathbf{b}) \cdot (A\mathbf{c})$ as a product of matrices.

24. The product $A\mathbf{b}$ of a matrix and a column vector is equal to a linear combination of columns of A where the scalar coefficient of the jth column of A is b_j. In a similar fashion, describe the product $\mathbf{c}A$ of a row

vector $\mathbf{c}$ and a matrix A as a linear combination of vectors. [HINT: Consider $((\mathbf{c}A)^T)^T$.]

In Exercises 25–34, prove that the given relation holds for all vectors, matrices, and scalars for which the expressions are defined.

25. $A + B = B + A$
26. $(A + B) + C = A + (B + C)$
27. $(r + s)A = rA + sA$
28. $(rs)A = r(sA)$
29. $A(B + C) = AB + AC$
30. $(A^T)^T = A$
31. $(A + B)^T = A^T + B^T$
32. $(AB)^T = B^T A^T$
33. $(AB)C = A(BC)$
34. $(rA)B = A(rB) = r(AB)$

35. If B is an $m \times n$ matrix and if $B = A^T$, find the size of
    a. A,
    b. AA^T,
    c. $A^T A$.

36. Let $\mathbf{v}$ and $\mathbf{w}$ be column vectors in $\mathbb{R}^n$. What is the size of $\mathbf{v}\mathbf{w}^T$? What relationships hold between $\mathbf{v}\mathbf{w}^T$ and $\mathbf{w}\mathbf{v}^T$?

37. The Hilbert matrix H_n is the $n \times n$ matrix $[h_{ij}]$, where $h_{ij} = 1/(i + j - 1)$. Prove that the matrix H_n is symmetric.

38. Prove that, if A is a square matrix, then the matrix $A + A^T$ is symmetric.

39. Prove that, if A is a matrix, then the matrix AA^T is symmetric.

40. a. Prove that, if A is a square matrix, then $(A^2)^T = (A^T)^2$ and $(A^3)^T = (A^T)^3$. [HINT: Don't try to show that the matrices have equal entries; instead use Exercise 32.]
    b. State the generalization of part (a), and give a proof using mathematical induction (see Appendix A).

41. a. Let A be an $m \times n$ matrix, and let $\mathbf{e}_j$ be the $n \times 1$ column vector whose jth component is 1 and whose other components are 0. Show that $A\mathbf{e}_j$ is the jth column vector of A.

b. Let A and B be matrices of the same size.
   i. Prove that, if $A\mathbf{x} = \mathbf{0}$ (the zero vector) for all $\mathbf{x}$, then $A = O$, the zero matrix. [HINT: Use part (a).]
   ii. Prove that, if $A\mathbf{x} = B\mathbf{x}$ for all $\mathbf{x}$, then $A = B$. [HINT: Consider $A - B$.]

42. Let A and B be square matrices. Is

$$(A + B)^2 = A^2 + 2AB + B^2?$$

If so, prove it; if not, give a counterexample and state under what conditions the equation is true.

43. Let A and B be square matrices. Is

$$(A + B)(A - B) = A^2 - B^2?$$

If so, prove it; if not, give a counterexample and state under what conditions the equation is true.

44. An $n \times n$ matrix C is **skew symmetric** if $C^T = -C$. Prove that every square matrix A can be written *uniquely* as $A = B + C$ where B is symmetric and C is skew symmetric.

Matrix A **commutes** *with matrix B if $AB = BA$.*

45. Find all values of r for which

$$\begin{bmatrix} 2 & 0 & 0 \\ 0 & 1 & 0 \\ 0 & 0 & r \end{bmatrix} \quad \text{commutes with} \quad \begin{bmatrix} 1 & 0 & 1 \\ 0 & 1 & 0 \\ 1 & 0 & 1 \end{bmatrix}.$$

46. Find all values of r for which

$$\begin{bmatrix} 2 & 0 & 0 \\ 0 & r & 0 \\ 0 & 0 & 2 \end{bmatrix} \quad \text{commutes with} \quad \begin{bmatrix} 1 & 0 & 1 \\ 0 & 1 & 0 \\ 1 & 0 & 1 \end{bmatrix}.$$

The software LINTEK includes a routine, MATCOMP, that performs the matrix operations described in this section. Let

$$A = \begin{bmatrix} 4 & 6 & 0 & 1 & -9 \\ 2 & 11 & 5 & 2 & -5 \\ -1 & 2 & -4 & 5 & 7 \\ 0 & 12 & -8 & 4 & 3 \\ 10 & 4 & 6 & 2 & -5 \end{bmatrix}$$

and

$$B = \begin{bmatrix} -8 & 15 & 4 & -11 \\ 3 & 5 & 6 & -2 \\ 0 & -1 & 12 & 5 \\ 1 & 13 & -15 & 7 \\ 6 & -8 & 0 & -5 \end{bmatrix}.$$

Use MATCOMP in LINTEK to enter and store these matrices, and then compute the matrices in Exercises 47–54, if they are defined. Write down to hand in, if requested, the entry in the 3rd row, 4th column of the matrix.

47. $A^4 + A$ **50.** BA^2 **53.** $(2A)^3 - A^5$
48. A^2B **51.** $B^T(2A)$ **54.** $(A^T)^5$
49. $A^3(A^T)^2$ **52.** $AB(AB)^T$

MATLAB

To enter a matrix in MATLAB, start with a left bracket [and then type the entries across the rows, separating the entries by spaces and separating the rows by semicolons. Conclude with a right bracket]. To illustrate, we would enter the matrix

$$A = \begin{bmatrix} -1 & 5 \\ 13 & -4 \\ 7 & 0 \end{bmatrix} \quad \text{as} \quad A = [-1 \; 5; \; 13 \; -4; \; 7 \; 0]$$

and MATLAB would then print it for us to proofread. Recall that to avoid having data printed again on the screen, we type a semicolon at the end of the data before pressing the Enter key. Thus if we enter

$$A = [-1 \; 5; \; 13 \; -4; \; 7 \; 0];$$

the matrix A will not be printed for proofreading. In MATLAB, we can enter

> **A + B** to find the sum
>
> **A − B** to find the difference
>
> **A * B** to find the product AB
>
> **A ^ n** to find the power A^n
>
> **r*A** to find the scalar multiple rA
>
> **A'** to take the transpose of A
>
> **eye(n)** for the $n \times n$ identity matrix
>
> **zeros(m,n)** for an $m \times n$ matrix of 0's, or **zeros(n)** if square
>
> **ones(m,n)** for an $m \times n$ matrix of 1's, or **ones(n)** if square
>
> **rand(m,n)** for a matrix of random numbers from 0 to 1, or **rand(n)** if square.

In MATLAB, **A(i, j)** is the entry in the ith row and jth column of A, while **A(k)** is the kth entry in A where entries are numbered consecutively starting at the upper left and proceeding down the first column, then down the second column, etc.

Access your MATLAB, and enter the matrices A and B given before Exercises 47–54. (We ask you to enter them manually this time to be sure you know how to enter matrices.) Proofread the data for A and B. If you find, for example, that you entered 6 rather than 5 for the entry in the 2nd row, 3rd column of A, you can correct your error by entering **A(2,3) = 5.**

M1. Exercises 47–54 are much easier to do with MATLAB than with LINTEK, because operations in LINTEK must be specified one at a time. Find the element in the 3rd row, 4th column of the given matrix.
   a. $B^T(2A)$
   b. $AB(AB)^T$
   c. $(2A)^3 - A^5$

M2. Enter **B(8).** What answer did you get? Why did MATLAB give that answer?

M3. Enter **help :** to review the uses of the colon with matrices. Mastery of use of the colon is a real timesaver in MATLAB. Use the colon to set C equal to the 5×3 matrix consisting of the 3rd through the 5th columns of A. Then compute $C^T C$ and write down your answer.

M4. Form a 5×9 matrix D whose first five columns are those of A and whose last four columns are those of B by
   a. entering **D = [A B]**, which works when A and B have the same number of rows,
   b. first entering **D = A** and then using the colon to specify that columns 6 through 9 of **D** are equal to **B.** Use the fact that **A(:, j)** gives the jth column of A.

   Write down the entry in the 2nd row, 5th column of $D^T D$.

M5. Form a matrix E consisting of B with two rows of zeros put at the bottom and write down the entry in the 2nd row, 3rd column of $E^T E$ by
   a. entering **Z = zeros (2,4)** and then **E = [B; Z]**, which works when B and Z have the same number of columns,
   b. first entering **E = B** and then using the colon to specify that rows 6 through 7 of E are equal to Z. Use the fact that **A(i, :)** gives the ith row of A.

M6. In mathematics, "mean" stands for "average," so the mean of the numbers 2, 4, and 9 is their average $(2 + 4 + 9)/3 = 5$. In MATLAB, enter **help mean** to see what that function gives, and then enter **mean(A)**. Figure out a way to have MATLAB find the mean (average) of all 25 numbers in the matrix A. Find that mean and write down your answer.

M7. If F is a 10×10 matrix with random entries from 0 to 1, approximately what would you expect the mean value of those entries to be? Enter **help rand,** read what it says, and then generate such a matrix F. Using the idea in Exercise M6, compute the mean of the entries in F, and write down your answer. Repeat this several times.

M8. Enter **help ones** and read what it says. Write down a statement that you could enter in MATLAB to form from the matrix F of Exercise M7 a 10×10 matrix G that has random entries from -4 to 4. Using the ideas in Exercise M6, find and write down the mean of the entries in the matrix G. Repeat this several times.

M9. In MATLAB, entering **mesh(X)** will draw a three-dimensional picture indicating the relative values of the entries in a matrix X, much as entering **plot(a)** draws a two-dimensional picture for the entries in a vector **a**. Enter **I = eye(16); mesh(I)** to see a graphic for the 16×16 identity matrix. Then enter **mesh(rot90(I))**. Enter **help rot90** and **help triu**. Enter **X = triu(ones(14));** **mesh(X)**. Then enter **mesh(rot90(X))** and finally **mesh(rot90(X,−1))**.

MATLAB has the capability to draw surface graphs of a function $z = f(x, y)$ using the **mesh** function. This provides an excellent illustration of the use of a matrix to store data. As you experiment on your own, you may run out of computer memory, or try to form matrices larger than your MATLAB will accept. Entering **clear A** will erase a matrix A from memory to free up some space, and entering **clear** will erase all data previously entered. We suggest that you enter **clear** now before proceeding.

MATLAB can draw a surface of $z = f(x, y)$ over a rectangular region $a \leq x \leq b$, $c \leq y \leq d$ of the x,y-plane by computing values z of the function at points on a grid in the region. We can describe the region and the grid of points where we want values computed using the function **meshdom**. Review the use of the colon, using **help :** if you need to, and notice that entering $-3{:}1{:}3$ will form a vector with first entry -3 and successive entries incremented by 1 until 3 is reached. If we enter

$$[X, Y] = \text{meshdom}(-3{:}1{:}3, -2{:}.5{:}2)$$

then MATLAB will create two matrices X and Y containing, respectively, the x-coordinates and y-coordinates of a grid of points in the region $-3 \leq x \leq 3$ and $-2 \leq y \leq 2$. Because the x-increment is 1 and the y-increment is 0.5, we see that both X and Y will be 9×7 matrices.

Enter now

$$[X, Y] = \text{meshdom}(-3{:}1{:}3, -2{:}.5{:}2); \tag{8}$$

and then enter **X** *to see that matrix X and similarly view the matrix Y. We have specified where we want the function values computed.*

Enter **help .** *to recall that entering* **A .* A** *will produce the matrix whose entries are the squares of the entries in A. Thus entering* **Z = X .* X + Y .* Y** *in MATLAB will produce a matrix Z whose entry at a position corresponding to a grid point (x, y)*

will be $x^2 + y^2$. *Entering* **mesh(Z)** *will then create the mesh graph over the region* $-3 \le x \le 3$, $-2 \le y \le 2$.

Enter now

$$Z = X ._* X + Y ._* Y; \tag{9}$$

$$\textbf{mesh(Z)} \tag{10}$$

to see this graph.

M10. Using the up arrow, modify Eq. (8) to make both the x-increment and the y-increment 0.2. After pressing the Enter key, use the up arrow to get Eq. (9) and press the Enter key to form the larger matrix Z for these new grid points. Then create the mesh graph using Eq. (10).

M11. Modify Eq. (9) and create the mesh graph for $z = x^2 - y^2$.

M12. Change Eq. (8) so that the region will be $-3 \le x \le 3$ and $-3 \le y \le 3$ still with 0.2 for increments. Form the mesh graph for $z = 9 - x^2 - y^2$.

M13. Mesh graphs for cylinders are especially nice. Draw the mesh graphs for
 a. $z = x^2$
 b. $z = y^2$.

M14. Change the mesh domain to $-4\pi \le x \le 4\pi$, $-3 \le y \le 3$ with x-increment 0.2 and y-increment 6. Recall that π can be entered as **pi**. Draw the mesh graphs for
 a. the cylinder $z = \sin(x)$,
 b. the cylinder $z = x \sin(x)$, remembering to use **._***, and
 c. the function $z = y \sin(x)$, which is not a cylinder but is pretty.

| 1.4 |

SOLVING SYSTEMS OF LINEAR EQUATIONS

As we have indicated, solving a system of linear equations is a fundamental problem of linear algebra. Many of the computational exercises in this text involve solving such linear systems. This section presents an algorithm for finding all solutions of any linear system.

The solution set of any system of equations is the intersection of the solution sets of the individual equations. That is, any solution of a system must be a solution of each equation in the system; and conversely, any solution of every equation in the system is considered to be a solution of the system. Bearing this in mind, we start with an intuitive discussion of the geometry of linear systems; a more detailed study of this geometry appears in Section 2.5.

The Geometry of Linear Systems

Frequently, students are under the impression that a linear system containing the same number of equations as unknowns always has a unique solution, whereas a system having more equations than unknowns never has a solution. The geometric interpretation of the problem shows that these statements are not true.

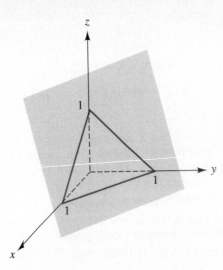

FIGURE 1.31
The plane $x + y + z = 1$.

We know that a single linear equation in two unknowns has a line in the plane as its solution set. Similarly, a single linear equation in three unknowns has a plane in space as its solution set. The solution set of $x + y + z = 1$ is the plane sketched in Figure 1.31. This geometric analysis can be extended to an equation that has more than three variables, but it is difficult for us to represent the solution set of such an equation graphically.

Two lines in the plane usually intersect at a single point; here the word *usually* means that, if the lines are selected in some random way, the chance

HISTORICAL NOTE SYSTEMS OF LINEAR EQUATIONS are found in ancient Babylonian and Chinese texts dating back well over 2000 years. The problems are generally stated in real-life terms, but it is clear that they are artificial and designed simply to train students in mathematical procedures. As an example of a Babylonian problem, consider the following, which has been slightly modified from the original found on a clay tablet from about 300 B.C.: There are two fields whose total area is 1800 square yards. One produces grain at a rate of $\frac{2}{3}$ bushel per square yard, the other at a rate of $\frac{1}{2}$ bushel per square yard. The total yield of the two fields is 1100 bushels. What is the size of each field? This problem leads to the system

$$x + \quad y = 1800$$

$$\frac{2}{3}x + \frac{1}{2}y = 1100.$$

A typical Chinese problem, taken from the Han dynasty text *Nine Chapters of the Mathematical Art* (about 200 B.C.), reads as follows: There are three classes of corn, of which three bundles of the first class, two of the second, and one of the third make 39 measures. Two of the first, three of the second, and one of the third make 34 measures. And one of the first, two of the second, and three of the third make 26 measures. How many measures of grain are contained in one bundle of each class? The system of equations here is

$$3x + 2y + \quad z = 39$$

$$2x + 3y + \quad z = 34$$

$$x + 2y + 3z = 26.$$

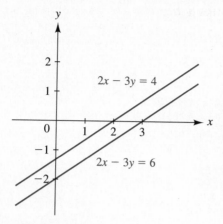

FIGURE 1.32
$2x - 3y = 4$ is parallel to
$2x - 3y = 6$.

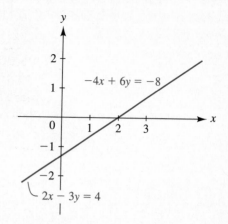

FIGURE 1.33
$2x - 3y = 4$ and $-4x + 6y = -8$
are the same line.

that they either are parallel (have empty intersection) or coincide (have an infinite number of points in their intersection) is very small. Thus we see that a system of two randomly selected equations in two unknowns can be expected to have a unique solution. However, it is *possible* for the system to have no solutions or an infinite number of solutions. For example, the equations

$$2x - 3y = 4$$
$$2x - 3y = 6$$

correspond to distinct parallel lines, as shown in Figure 1.32, and the system consisting of these equations has no solutions. Moreover, the equations

$$2x - 3y = 4$$
$$-4x + 6y = -8$$

correspond to the same line, as shown in Figure 1.33. All points on this line are solutions of this system of two equations. And because it is possible to have any number of lines in the plane—say, fifty lines—pass through a single point, it is possible for a system of fifty equations in only two unknowns to have a unique solution.

Similar illustrations can be made in space, where a linear equation has as its solution set a plane. Three randomly chosen planes can be expected to have a unique point in common. Two of them can be expected to intersect in a line (see Figure 1.34), which in turn can be expected to meet the third plane at a single point. However, it is possible for three planes to have no point in common, giving rise to a linear system with no solutions. It is also possible for all three planes to contain a common line, in which case the corresponding linear system will have an infinite number of solutions.

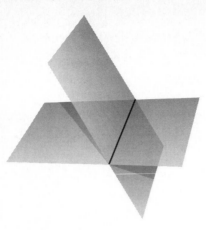

FIGURE 1.34
Two planes intersecting in a line.

Elementary Row Operations

We now describe operations that can be used to modify the equations of a linear system to obtain a system having the same solutions, but whose solutions are obvious. The most general type of linear system can have m equations in n unknowns. Such a system can be written as

$$a_{11}x_1 + a_{12}x_2 + \cdots + a_{1n}x_n = b_1$$
$$a_{21}x_1 + a_{22}x_2 + \cdots + a_{2n}x_n = b_2$$
$$\vdots \tag{1}$$
$$a_{m1}x_1 + a_{m2}x_2 + \cdots + a_{mn}x_n = b_m.$$

System (1) is completely determined by its $m \times n$ **coefficient matrix** $A = [a_{ij}]$ and by the column vector $\mathbf{b}$ with ith component b_i. The system can be written as the single matrix equation

$$A\mathbf{x} = \mathbf{b}, \tag{2}$$

where $\mathbf{x}$ is the column vector with ith component x_i. Any column vector $\mathbf{s}$ such that $A\mathbf{s} = \mathbf{b}$ is a **solution** of system (1).

The **augmented matrix** or **partitioned matrix**

$$\begin{bmatrix} a_{11} & a_{21} & \cdots & a_{1n} & b_1 \\ a_{21} & a_{22} & \cdots & a_{2n} & b_2 \\ & & \vdots & & \vdots \\ a_{m1} & a_{m2} & \cdots & a_{mn} & b_m \end{bmatrix} \tag{3}$$

is a shorthand summary of system (1). The coefficient matrix has been *augmented* by the column vector of constants. We denote matrix (3) by $[A \mid \mathbf{b}]$.

We shall see how to determine all solutions of system (1) by manipulating augmented matrix (3) using *elementary row operations*. The elementary row operations correspond to the following familiar operations with equations of system (1):

R1 Interchange two equations in system (1).

R2 Multiply an equation in system (1) by a nonzero constant.

R3 Replace an equation in system (1) with the sum of itself and a multiple of a different equation of the system.

It is clear that operations R1 and R2 do not change the solution sets of the equations they affect. Therefore, they do not change the intersection of the solution sets of the equations; that is, the solution set of the system is unchanged. The fact that R2 does not change the solution set and the familiar algebraic principle, "Equals added to equals yield equals," show that any solution of both the ith and jth equations is also a solution of a new jth equation obtained by adding s times the ith equation to the jth equation. Thus operation R3 yields a system having all the solutions of the original one. Because the original system can be recovered from the new one by multiplying

---

HISTORICAL NOTE   A MATRIX-REDUCTION METHOD of solving a system of linear equations occurs in the ancient Chinese work, *Nine Chapters of the Mathematical Art*. The author presents the following solution to the system

$$3x + 2y + z = 39$$
$$2x + 3y + z = 34$$
$$x + 2y + 3z = 26.$$

The diagram of the coefficients is to be set up on a "counting board":

$$\begin{array}{ccc} 1 & 2 & 3 \\ 2 & 3 & 2 \\ 3 & 1 & 1 \\ 26 & 34 & 39 \end{array}$$

The author then instructs the reader to multiply the middle column by 3 and subsequently to subtract the right column "as many times as possible"; the same is to be done to the left column. The new diagrams are then

$$\begin{array}{ccc} 1 & 0 & 3 \\ 2 & 5 & 2 \\ 3 & 1 & 1 \\ 26 & 24 & 39 \end{array} \quad \text{and} \quad \begin{array}{ccc} 0 & 0 & 3 \\ 4 & 5 & 2 \\ 8 & 1 & 1 \\ 39 & 24 & 39 \end{array}$$

The next instruction is to multiply the left column by 5 and then to subtract the middle column as many times as possible. This gives

$$\begin{array}{ccc} 0 & 0 & 3 \\ 0 & 5 & 2 \\ 36 & 1 & 1 \\ 99 & 24 & 39 \end{array}$$

The system has thus been reduced to the system $3x + 2y + z = 39$, $5y + z = 24$, $36z = 99$, from which the complete solution is easily found.

the ith equation by $-s$ and adding it to the new jth equation (an R3 operation), we see that the original system has all the solutions of the new one. Hence R3, too, does not alter the solution set of system (1).

These procedures applied to system (1) correspond to **elementary row operations** applied to augmented matrix (3). We list these in a box together with a suggestive notation for each.

Elementary Row Operations	Notations
(*Row interchange*) Interchange the ith and jth row vectors in a matrix.	$R_i \leftrightarrow R_j$
(*Row scaling*) Multiply the ith row vector in a matrix by a nonzero scalar s.	$R_i \rightarrow sR_i$
(*Row addition*) Add to the ith row vector of a matrix s times the jth row vector.	$R_i \rightarrow R_i + sR_j$

If a matrix B can be obtained from a matrix A by means of a sequence of elementary row operations, then A is **row equivalent** to B. Each elementary row operation can be undone by another of the same type. A row-addition operation $R_i \rightarrow R_i + sR_j$ can be undone by $R_i \rightarrow R_i - sR_j$. Row scaling, $R_i \rightarrow sR_i$ for $s \neq 0$, can be undone using $R_i \rightarrow (1/s)R_i$, while a row-interchange operation undoes itself. Thus, if B is row equivalent to A, then A is row equivalent to B; we can simply speak of *row-equivalent matrices* A and B, which we denote by $A \sim B$. (See Exercise 55 in this regard.) We have just seen that the operations on a linear system $A\mathbf{x} = \mathbf{b}$ corresponding to these elementary row operations on the augmented matrix $[A \mid \mathbf{b}]$ do not change the solution set of the system. This gives us at once the following theorem, which is the foundation for the algorithm we will present for solving linear systems.

THEOREM 1.6 Invariance of Solution Sets Under Row Equivalence

If $[A \mid \mathbf{b}]$ and $[H \mid \mathbf{c}]$ are row-equivalent augmented matrices, then the linear systems $A\mathbf{x} = \mathbf{b}$ and $H\mathbf{x} = \mathbf{c}$ have the same solution sets.

Row-Echelon Form

We will solve a linear system $A\mathbf{x} = \mathbf{b}$ by row-reducing the augmented matrix $[A \mid \mathbf{b}]$ to an augmented matrix $[H \mid \mathbf{c}]$, where H is a matrix in *row-echelon form* (which we now define).

DEFINITION 1.12   Row-Echelon Form, Pivot

A matrix is in **row-echelon form** if it satisfies two conditions:

1. All rows containing only zeros appear below rows with nonzero entries.
2. The first nonzero entry in any row appears in a column to the right of the first nonzero entry in any preceding row.

For such a matrix, the first nonzero entry in a row is the **pivot** for that row.

EXAMPLE 1   Determine which of the matrices

$$A = \begin{bmatrix} 1 & 3 & 2 \\ 0 & 0 & 0 \\ 0 & 0 & 1 \end{bmatrix}, \qquad B = \begin{bmatrix} 2 & 4 & 0 \\ 1 & 3 & 2 \\ 0 & 0 & 0 \end{bmatrix}, \qquad C = \begin{bmatrix} 0 & -1 & 2 \\ 0 & 0 & 3 \\ 0 & 0 & 0 \\ 0 & 0 & 0 \end{bmatrix},$$

$$D = \begin{bmatrix} 1 & 3 & 2 & 5 \\ 0 & 0 & 1 & 3 \\ 0 & 0 & 0 & 1 \\ 0 & 0 & 0 & 0 \end{bmatrix}$$

are in row-echelon form.

SOLUTION   Matrix A is not in row-echelon form, because the second row (consisting of all zero entries) is not below the third row (which has a nonzero entry).

Matrix B is not in row-echelon form, because the first nonzero entry in the second row does not appear in a column to the right of the first nonzero entry in the first row.

Matrix C is in row-echelon form, because both conditions of Definition 1.12 are satisfied. The pivots are -1 and 3.

Matrix D satisfies both conditions as well, and is in row-echelon form. The pivots are the entries 1. ∎

Solutions of $H\mathbf{x} = \mathbf{c}$

We illustrate by examples that, if a linear system $H\mathbf{x} = \mathbf{c}$ has coefficient matrix H in row-echelon form, it is easy to determine all solutions of the system. We color the pivots in H in these examples.

EXAMPLE 2   Find all solutions of $H\mathbf{x} = \mathbf{c}$, where

$$[H \mid \mathbf{c}] = \begin{bmatrix} -5 & -1 & 3 & \mid & 3 \\ 0 & 3 & 5 & \mid & 8 \\ 0 & 0 & 2 & \mid & -4 \end{bmatrix}.$$

SOLUTION   The equations corresponding to this augmented matrix are

$$\begin{aligned} -5x_1 - x_2 + 3x_3 &= 3 \\ 3x_2 + 5x_3 &= 8 \\ 2x_3 &= -4. \end{aligned}$$

From the last equation, we obtain $x_3 = -2$. Substituting into the second equation, we have

$$3x_2 + 5(-2) = 8, \qquad 3x_2 = 18, \qquad x_2 = 6.$$

Finally, we substitute these values for x_2 and x_3 into the top equation, obtaining

$$-5x_1 - 6 + 3(-2) = 3, \qquad -5x_1 = 15, \qquad x_1 = -3.$$

Thus the only solution is

$$\mathbf{x} = \begin{bmatrix} -3 \\ 6 \\ -2 \end{bmatrix},$$

or equivalently, $x_1 = -3$, $x_2 = 6$, $x_3 = -2$. ∎

The procedure for finding the solution of $H\mathbf{x} = \mathbf{c}$ illustrated in Example 2 is called **back substitution**, because the values of the variables are found in backward order, starting with the variable with the largest subscript.

By multiplying each nonzero row in $[H \mid \mathbf{b}]$ by the reciprocal of its pivot, we can assume that each pivot in H is 1. We assume that this is the case in the next two examples.

EXAMPLE 3   Use back substitution to find all solutions of $H\mathbf{x} = \mathbf{c}$, where

$$[H \mid \mathbf{c}] = \begin{bmatrix} 1 & -3 & 5 & \bigm| & 3 \\ 0 & 1 & 2 & \bigm| & 2 \\ 0 & 0 & 0 & \bigm| & -1 \end{bmatrix}.$$

SOLUTION   The equation corresponding to the last row of this augmented matrix is

$$0x_1 + 0x_2 + 0x_3 = -1.$$

This equation has no solutions, because the left side is 0 for any values of the variables and the right side is -1. ∎

DEFINITION 1.13   Consistent Linear System

A linear system having no solutions is **inconsistent**. If it has one or more solutions, the linear system is said to be **consistent**.

Now we illustrate a many-solutions case.

EXAMPLE 4   Use back substitution to find all solutions of $H\mathbf{x} = \mathbf{c}$, where

$$[H \mid \mathbf{c}] = \begin{bmatrix} 1 & -3 & 0 & 5 & 0 & 4 \\ 0 & 0 & 1 & 2 & 0 & -7 \\ 0 & 0 & 0 & 0 & 1 & 1 \\ 0 & 0 & 0 & 0 & 0 & 0 \end{bmatrix}.$$

SOLUTION   The linear system corresponding to this augmented matrix is

$$\begin{aligned} x_1 - 3x_2 \quad + 5x_4 \quad &= \quad 4 \\ x_3 + 2x_4 \quad &= -7 \\ x_5 &= \quad 1. \end{aligned}$$

We solve each equation for the variable corresponding to the colored pivot in the matrix. Thus we obtain

$$\begin{aligned} x_1 &= 3x_2 - 5x_4 + 4 \\ x_3 &= -2x_4 - 7 \\ x_5 &= 1. \end{aligned} \tag{4}$$

Notice that x_2 and x_4 correspond to columns of H containing no pivot. We can assign any value r we please to x_2 and any value s to x_4, and we can then use system (4) to determine corresponding values for x_1, x_3, and x_5. Thus the system has an infinite number of solutions. We describe all solutions by the vector equation

$$\mathbf{x} = \begin{bmatrix} x_1 \\ x_2 \\ x_3 \\ x_4 \\ x_5 \end{bmatrix} = \begin{bmatrix} 3r - 5s + 4 \\ r \\ -2s - 7 \\ s \\ 1 \end{bmatrix} \quad \text{for any scalars } r \text{ and } s. \tag{5}$$

We call x_2 and x_4 **free variables**, and we refer to Eq. (5) as the **general solution** of the system. We obtain **particular solutions** by setting r and s equal to specific values. For example, we obtain

$$\begin{bmatrix} 2 \\ 1 \\ -9 \\ 1 \\ 1 \end{bmatrix} \text{ for } r = s = 1 \quad \text{and} \quad \begin{bmatrix} 25 \\ 2 \\ -1 \\ -3 \\ 1 \end{bmatrix} \text{ for } r = 2, s = -3. \qquad \blacksquare$$

Gauss Reduction of $A\mathbf{x} = \mathbf{b}$ to $H\mathbf{x} = \mathbf{c}$

We now show how to reduce an augmented matrix $[A \mid \mathbf{b}]$ to $[H \mid \mathbf{c}]$, where H is in row-echelon form, using a sequence of elementary row operations. Examples 2 through 4 illustrated how to use back substitution afterward to find solutions of the system $H\mathbf{x} = \mathbf{c}$, which are the same as the solutions of $A\mathbf{x} = \mathbf{b}$, by Theorem 1.6. This procedure for solving $A\mathbf{x} = \mathbf{b}$ is known as **Gauss**

reduction with back substitution. In the box below, we give an outline for reducing a matrix A to row-echelon form.

Reducing a Matrix A to Row-Echelon Form H

1. If the first column of A contains only zero entries, cross it off mentally. Continue in this fashion until the left column of the remaining matrix has a nonzero entry or until the columns are exhausted.

2. Use row interchange, if necessary, to obtain a nonzero entry (pivot) p in the top row of the first column of the remaining matrix. For each row below that has a nonzero entry r in the first column, add $-r/p$ times the top row to that row, to create a zero in the first column. In this fashion, create zeros below p in the entire first column of the remaining matrix.

3. Mentally cross off this first column and the first row of the matrix, to obtain a smaller matrix. (See the shaded portion of the third matrix in the solution of Example 5.) Go back to step 1, and repeat the process with this smaller matrix until either no rows or no columns remain.

EXAMPLE 5   Reduce the matrix

$$\begin{bmatrix} 2 & -4 & 2 & -2 \\ 2 & -4 & 3 & -4 \\ 4 & -8 & 3 & -2 \\ 0 & 0 & -1 & 2 \end{bmatrix}$$

to row-echelon form, making all pivots 1.

SOLUTION   We follow the boxed outline and color the pivots of 1. Remember that the symbol $\sim$ denotes row-equivalent matrices.

$$\begin{bmatrix} 2 & -4 & 2 & -2 \\ 2 & -4 & 3 & -4 \\ 4 & -8 & 3 & -2 \\ 0 & 0 & -1 & 2 \end{bmatrix}$$

Multiply the first row by $\frac{1}{2}$, to produce a pivot of 1 in the next matrix.

$$R_1 \to \frac{1}{2}R_1$$

$$\sim \begin{bmatrix} 1 & -2 & 1 & -1 \\ 2 & -4 & 3 & -4 \\ 4 & -8 & 3 & -2 \\ 0 & 0 & -1 & 2 \end{bmatrix}$$

Add -2 times row 1 to row 2, and then add -4 times row 1 to row 3, to obtain the next matrix.

$$R_2 \to R_2 - 2R_1; \quad R_3 \to R_3 - 4R_1$$

$$\sim \begin{bmatrix} 1 & -2 & 1 & -1 \\ 0 & 0 & 1 & -2 \\ 0 & 0 & -1 & 2 \\ 0 & 0 & -1 & 2 \end{bmatrix}$$

Cross off the first shaded column of zeros (mentally), to obtain the next shaded matrix.

$$\sim \begin{bmatrix} 1 & -2 & 1 & -1 \\ 0 & 0 & 1 & -2 \\ 0 & 0 & -1 & 2 \\ 0 & 0 & -1 & 2 \end{bmatrix}$$

Add row 2 to rows 3 and 4, to obtain the final matrix.

$$R_3 \rightarrow R_3 + 1R_2; \quad R_4 \rightarrow R_4 + 1R_2$$

$$\sim \begin{bmatrix} 1 & -2 & 1 & -1 \\ 0 & 0 & 1 & -2 \\ 0 & 0 & 0 & 0 \\ 0 & 0 & 0 & 0 \end{bmatrix}.$$

This last matrix is row-echelon form, with both pivots equal to 1. ■

To solve a linear system $A\mathbf{x} = \mathbf{b}$, we form the augmented matrix $[A \mid \mathbf{b}]$ and row-reduce it to $[H \mid \mathbf{c}]$, where H is in row-echelon form. We can follow the steps outlined in the box preceding Example 5 for row-reducing A to H. Of course, we always perform the elementary row operations on the full augmented matrix, including the entries in the column to the right of the partition.

EXAMPLE 6   Solve the linear system

$$\begin{aligned} x_2 - 3x_3 &= -5 \\ 2x_1 + 3x_2 - x_3 &= 7 \\ 4x_1 + 5x_2 - 2x_3 &= 10. \end{aligned}$$

SOLUTION   We reduce the corresponding augmented matrix, using elementary row operations. Pivots are colored.

$$\begin{bmatrix} 0 & 1 & -3 & -5 \\ 2 & 3 & -1 & 7 \\ 4 & 5 & -2 & 10 \end{bmatrix} \sim \begin{bmatrix} 2 & 3 & -1 & 7 \\ 0 & 1 & -3 & -5 \\ 4 & 5 & -2 & 10 \end{bmatrix} \qquad R_1 \leftrightarrow R_2$$

$$\begin{bmatrix} 2 & 3 & -1 & 7 \\ 0 & 1 & -3 & -5 \\ 4 & 5 & -2 & 10 \end{bmatrix} \sim \begin{bmatrix} 2 & 3 & -1 & 7 \\ 0 & 1 & -3 & -5 \\ 0 & -1 & 0 & -4 \end{bmatrix} \qquad R_3 \rightarrow R_3 - 2R_1$$

$$\begin{bmatrix} 2 & 3 & -1 & 7 \\ 0 & 1 & -3 & -5 \\ 0 & -1 & 0 & -4 \end{bmatrix} \sim \begin{bmatrix} 2 & 3 & -1 & 7 \\ 0 & 1 & -3 & -5 \\ 0 & 0 & -3 & -9 \end{bmatrix}. \qquad R_3 \rightarrow R_3 + 1R_2$$

HISTORICAL NOTE   THE GAUSS SOLUTION METHOD is so named because Gauss described it in a paper detailing the computations he made to determine the orbit of the asteroid Pallas. The parameters of the orbit had to be determined by observations of the asteroid over a 6-year period from 1803 to 1809. These led to six linear equations in six unknowns with quite complicated coefficients. Gauss showed how to solve these equations by systematically replacing them with a new system in which only the first equation had all six unknowns, the second equation included five unknowns, the third equation only four, and so on, until the sixth equation had but one. This last equation could, of course, be easily solved; the remaining unknowns were then found by back substitution.

From the last augmented matrix, we could proceed to write the corresponding equations (as in Example 2) and to solve in succession for x_3, x_2, and x_1 by back substitution. However, it makes sense to keep using our shorthand, without writing out variables, and to do our back substitution in terms of augmented matrices. Starting with the final augmented matrix in the preceding set, we obtain

$$\begin{bmatrix} 2 & 3 & -1 & | & 7 \\ 0 & 1 & -3 & | & -5 \\ 0 & 0 & -3 & | & -9 \end{bmatrix} \sim \begin{bmatrix} 2 & 3 & -1 & | & 7 \\ 0 & 1 & -3 & | & -5 \\ 0 & 0 & 1 & | & 3 \end{bmatrix}$$

$R_3 \rightarrow -\frac{1}{3}R_3$
(This shows that $x_3 = 3$.)

$$\begin{bmatrix} 2 & 3 & -1 & | & 7 \\ 0 & 1 & -3 & | & -5 \\ 0 & 0 & 1 & | & 3 \end{bmatrix} \sim \begin{bmatrix} 2 & 3 & 0 & | & 10 \\ 0 & 1 & 0 & | & 4 \\ 0 & 0 & 1 & | & 3 \end{bmatrix}$$

$R_1 \rightarrow R_1 + 1R_3; \quad R_2 \rightarrow R_2 + 3R_3$
(This shows that $x_2 = 4$.)

$$\begin{bmatrix} 2 & 3 & 0 & | & 10 \\ 0 & 1 & 0 & | & 4 \\ 0 & 0 & 1 & | & 3 \end{bmatrix} \sim \begin{bmatrix} 2 & 0 & 0 & | & -2 \\ 0 & 1 & 0 & | & 4 \\ 0 & 0 & 1 & | & 3 \end{bmatrix}.$$

$R_1 \rightarrow R_1 - 3R_2$
(This shows that $x_1 = -1$.)

We have found the solution: $\mathbf{x} = [x_1, x_2, x_3] = [-1, 4, 3]$. ∎

In Example 6, we had to write down as many matrices to execute the back substitution as we wrote to reduce the original augmented matrix to row-echelon form. We can avoid this by creating zeros *above* as well as below each pivot as we reduce the matrix to row-echelon form. This is known as the **Gauss–Jordan method**. We show in Chapter 10 that, for a large system, it takes about 50% more time for a computer to use the Gauss–Jordan method than to use the Gauss method with back substitution illustrated in Example 6; there are actually about 50% more arithmetic operations involved. Creating the zeros above the pivots requires less computation if we do it *after* the matrix is reduced to row-echelon form than if we do it as we go along. However, when one is working with pencil and paper, fixing up a whole column in a single step avoids writing so many matrices. Our next example illustrates the Gauss–Jordan procedure.

EXAMPLE 7   Determine whether the vector $\mathbf{b} = [1, -7, -4]$ is in the span of the vectors $\mathbf{v} = [2, 1, 1]$ and $\mathbf{w} = [1, 3, 2]$.

SOLUTION   We know that $\mathbf{b}$ is in sp($\mathbf{v}, \mathbf{w}$) if and only if $\mathbf{b} = x_1\mathbf{v} + x_2\mathbf{w}$ for some scalars x_1 and x_2. This vector equation is equivalent to the linear system

$$\begin{bmatrix} 2 & 1 \\ 1 & 3 \\ 1 & 2 \end{bmatrix} \begin{bmatrix} x_1 \\ x_2 \end{bmatrix} = \begin{bmatrix} 1 \\ -7 \\ -4 \end{bmatrix}.$$

Reducing the appropriate augmented matrix, we obtain

$$
\begin{bmatrix} 2 & 1 & | & 1 \\ 1 & 3 & | & -7 \\ 1 & 2 & | & -4 \end{bmatrix} \sim \begin{bmatrix} 1 & 3 & | & -7 \\ 2 & 1 & | & 1 \\ 1 & 2 & | & -4 \end{bmatrix} \sim \begin{bmatrix} 1 & 3 & | & -7 \\ 0 & -5 & | & 15 \\ 0 & -1 & | & 3 \end{bmatrix} \sim \begin{bmatrix} 1 & 0 & | & 2 \\ 0 & 1 & | & -3 \\ 0 & 0 & | & 0 \end{bmatrix}.
$$

$$
\begin{array}{cccc}
R_1 \leftrightarrow R_2 & R_2 \rightarrow R_2 - 2R_1 & R_2 \rightarrow \tfrac{1}{5} R_2 \\
\text{(to avoid} & R_3 \rightarrow R_3 - 1R_1 & R_1 \rightarrow R_1 - 3R_2 \\
\text{fractions)} & & R_3 \rightarrow R_3 + 1R_2
\end{array}
$$

The left side of the final augmented matrix is in *reduced row-echelon form*. From the solution $x_1 = 2$ and $x_2 = -3$, we see that $\mathbf{b} = 2\mathbf{v} - 3\mathbf{w}$, which is indeed in sp $(\mathbf{v}, \mathbf{w})$. ∎

The linear system $A\mathbf{x} = \mathbf{b}$ displayed in Eq. (1) can be written in the form

$$
x_1 \begin{bmatrix} a_{11} \\ a_{21} \\ \cdot \\ \cdot \\ \cdot \\ a_{m1} \end{bmatrix} + x_2 \begin{bmatrix} a_{12} \\ a_{22} \\ \cdot \\ \cdot \\ \cdot \\ a_{m2} \end{bmatrix} + \cdots + x_n \begin{bmatrix} a_{1n} \\ a_{2n} \\ \cdot \\ \cdot \\ \cdot \\ a_{mn} \end{bmatrix} = \begin{bmatrix} b_1 \\ b_2 \\ \cdot \\ \cdot \\ \cdot \\ b_m \end{bmatrix}.
$$

This equation expresses a typical column vector $\mathbf{b}$ in $\mathbb{R}^m$ as a linear combination of the column vectors of the matrix A if and only if scalars $x_1, x_2, \ldots, x_n$ can be found to satisfy that equation. Example 7 illustrates this. We phrase this result as follows:

Let A be an $m \times n$ matrix. The linear system $A\mathbf{x} = \mathbf{b}$ is consistent if and only if the vector $\mathbf{b}$ in $\mathbb{R}^m$ is in the span of the column vectors of A.

A matrix in row-echelon form with all pivots equal to 1 and with zeros above as well as below each pivot is said to be in **reduced row-echelon form**. Thus the Gauss–Jordan method consists of using elementary row operations on an augmented matrix $[A \mid \mathbf{b}]$ to bring the coefficient matrix A into reduced

---

HISTORICAL NOTE   THE JORDAN HALF OF THE GAUSS–JORDAN METHOD is essentially a systematic technique of back substitution. In this form, it was first described by Wilhelm Jordan (1842–1899), a German professor of geodesy, in the third (1888) edition of his *Handbook of Geodesy*. Although Jordan's arrangement of his calculations is different from the one presented here, partly because he was always applying the method to the symmetric system of equations arising out of a least-squares application in geodesy (see Section 6.5), Jordan's method uses the same arithmetic and arrives at the same answers for the unknowns.

Wilhelm Jordan was prominent in his field in the late nineteenth century, being involved in several geodetic surveys in Germany and in the first major survey of the Libyan desert. He was the founding editor of the German geodesy journal and was widely praised as a teacher of his subject. His interest in finding a systematic method of solving large systems of linear equations stems from their frequent appearance in problems of triangulation.

row-echelon form. It can be shown that the reduced row-echelon form of a matrix A is unique. (See Section 2.3, Exercise 33.)

The examples we have given illustrate the three possibilities for solutions of a linear system—namely, no solutions (inconsistent system), a unique solution, or an infinite number of solutions. We state this formally in a theorem and prove it.

THEOREM 1.7   Solutions of $A\mathbf{x} = \mathbf{b}$

Let $A\mathbf{x} = \mathbf{b}$ be a linear system, and let $[A \mid \mathbf{b}] \sim [H \mid \mathbf{c}]$, where H is in row-echelon form.

1. The system $A\mathbf{x} = \mathbf{b}$ is inconsistent if and only if the augmented matrix $[H \mid \mathbf{c}]$ has a row with all entries 0 to the left of the partition and a nonzero entry to the right of the partition.

2. If $A\mathbf{x} = \mathbf{b}$ is consistent and every column of H contains a pivot, the system has a unique solution.

3. If $A\mathbf{x} = \mathbf{b}$ is consistent and some column of H has no pivot, the system has infinitely many solutions, with as many free variables as there are pivot-free columns in H.

PROOF   If $[H \mid \mathbf{c}]$ has an ith row with all entries 0 to the left of the partition and a nonzero entry c_i to the right of the partition, the corresponding ith equation in the system $H\mathbf{x} = \mathbf{c}$ is $0x_1 + 0x_2 + \cdots + 0x_n = c_i$, which has no solutions; therefore, the system $A\mathbf{x} = \mathbf{b}$ has no solutions, by Theorem 1.6. The next paragraph shows that, if H contains no such row, we can find a solution to the system. Thus the system is inconsistent if and only if H contains such a row.

Assume now that $[H \mid \mathbf{c}]$ has no row with all entries 0 to the left of the partition and a nonzero entry to the right. If the ith row of $[H \mid \mathbf{c}]$ is a zero row vector, the corresponding equation $0x_1 + 0x_2 + \cdots + 0x_n = 0$ is satisfied for all values of the variables x_j, and thus it can be deleted from the system $H\mathbf{x} = \mathbf{c}$. Assume that this has been done wherever possible, so that $[H \mid \mathbf{c}]$ has no zero row vectors. For each j such that the jth column has no pivot, we can set x_j equal to any value we please (as in Example 4) and then, starting from the last remaining equation of the system and working back to the first, solve in succession for the variables corresponding to the columns containing the pivots. If some column j has no pivot, there are an infinite number of solutions, because x_j can be set equal to any value. On the other hand, if every column has a pivot (as in Examples 2, 6, and 7), the value of each x_j is uniquely determined.   ▲

With reference to item (3) of Theorem 1.7, the number of free variables in the solution set of a system $A\mathbf{x} = \mathbf{b}$ depends only on the system, and not on the way in which the matrix A is reduced to row-echelon form. This follows from the uniqueness of the reduced row-echelon form. (See Exercise 33 in Section 2.3.)

Elementary Matrices

The elementary row operations we have performed can actually be carried out by means of matrix multiplication. Although it is not efficient to row-reduce a matrix by multiplying it by other matrices, representing row reduction as a product of matrices is a useful theoretical tool. For example, we use elementary matrices in Section 1.5 to show that, for square matrices A and C, if $AC = I$, then $CA = I$. We use them again in Section 4.2 to demonstrate the multiplicative property of determinants, and again in Section 10.2 to exhibit a factorization of some square matrices A into a product LU of a lower-triangular matrix L and an upper-triangular matrix U.

If we interchange its second and third rows, the 3×3 identity matrix

$$I = \begin{bmatrix} 1 & 0 & 0 \\ 0 & 1 & 0 \\ 0 & 0 & 1 \end{bmatrix} \quad \text{becomes} \quad E = \begin{bmatrix} 1 & 0 & 0 \\ 0 & 0 & 1 \\ 0 & 1 & 0 \end{bmatrix}.$$

If $A = [a_{ij}]$ is a 3×3 matrix, we can compute EA, and we find that

$$EA = \begin{bmatrix} 1 & 0 & 0 \\ 0 & 0 & 1 \\ 0 & 1 & 0 \end{bmatrix} \begin{bmatrix} a_{11} & a_{12} & a_{13} \\ a_{21} & a_{22} & a_{23} \\ a_{31} & a_{32} & a_{33} \end{bmatrix} = \begin{bmatrix} a_{11} & a_{12} & a_{13} \\ a_{31} & a_{32} & a_{33} \\ a_{21} & a_{22} & a_{23} \end{bmatrix}.$$

We have interchanged the second and third rows of A by multiplying A on the left by E.

DEFINITION 1.14   Elementary Matrix

Any matrix that can be obtained from an identity matrix by means of one elementary row operation is an **elementary matrix**.

We leave the proof of the following theorem as Exercises 52 through 54.

THEOREM 1.8   Use of Elementary Matrices

Let A be an $m \times n$ matrix, and let E be an $m \times m$ elementary matrix. Multiplication of A on the left by E effects the same elementary row operation on A that was performed on the identity matrix to obtain E.

Thus row reduction of a matrix to row-echelon form can be accomplished by successive multiplication on the left by elementary matrices. In other words, if A can be reduced to H through elementary row operations, there exist elementary matrices $E_1, E_2, \ldots, E_t$ such that

$$H = (E_t \cdots E_2 E_1)A.$$

Again, this is by no means an efficient way to execute row reduction, but such an algebraic representation of H in terms of A is sometimes handy in proving theorems.

EXAMPLE 8   Let

$$A = \begin{bmatrix} 0 & 1 & -3 \\ 2 & 3 & -1 \\ 4 & 5 & -2 \end{bmatrix}.$$

Find a matrix C such that CA is a matrix in row-echelon form that is row equivalent to A.

SOLUTION   We row reduce A to row-echelon form H and write down, for each row operation, the elementary matrix obtained by performing the same operation on the 3×3 identity matrix.

Reduction of A	Row Operation	Elementary Matrix
$A = \begin{bmatrix} 0 & 1 & -3 \\ 2 & 3 & -1 \\ 4 & 5 & -2 \end{bmatrix}$	$R_1 \leftrightarrow R_2$	$E_1 = \begin{bmatrix} 0 & 1 & 0 \\ 1 & 0 & 0 \\ 0 & 0 & 1 \end{bmatrix}$
$\sim \begin{bmatrix} 2 & 3 & -1 \\ 0 & 1 & -3 \\ 4 & 5 & -2 \end{bmatrix}$	$R_3 \rightarrow R_3 - 2R_1$	$E_2 = \begin{bmatrix} 1 & 0 & 0 \\ 0 & 1 & 0 \\ -2 & 0 & 1 \end{bmatrix}$
$\sim \begin{bmatrix} 2 & 3 & -1 \\ 0 & 1 & -3 \\ 0 & -1 & 0 \end{bmatrix}$	$R_3 \rightarrow R_3 + 1R_2$	$E_3 = \begin{bmatrix} 1 & 0 & 0 \\ 0 & 1 & 0 \\ 0 & 1 & 1 \end{bmatrix}$
$\sim \begin{bmatrix} 2 & 3 & -1 \\ 0 & 1 & -3 \\ 0 & 0 & -3 \end{bmatrix} = H.$		

Thus, we must have $E_3(E_2(E_1A)) = H$; so the desired matrix C is

$$C = E_3 E_2 E_1 = \begin{bmatrix} 0 & 1 & 0 \\ 1 & 0 & 0 \\ 1 & -2 & 1 \end{bmatrix}.$$

To compute C, we do not actually have to multiply out $E_3 E_2 E_1$. We know that multiplication of E_1 on the left by E_2 simply adds -2 times row 1 of E_1 to its row 3, and subsequent multiplication on the left by E_3 adds row 2 to row 3 of the matrix E_2E_1. Thus we can find C by executing the same row-reduction steps on I that we executed to change A to H—namely,

$$
\underset{I}{\begin{bmatrix} 1 & 0 & 0 \\ 0 & 1 & 0 \\ 0 & 0 & 1 \end{bmatrix}} \sim \underset{E_1}{\begin{bmatrix} 0 & 1 & 0 \\ 1 & 0 & 0 \\ 0 & 0 & 1 \end{bmatrix}} \sim \underset{E_2E_1}{\begin{bmatrix} 0 & 1 & 0 \\ 1 & 0 & 0 \\ 0 & -2 & 1 \end{bmatrix}} \sim \underset{C = E_3E_2E_1}{\begin{bmatrix} 0 & 1 & 0 \\ 1 & 0 & 0 \\ 1 & -2 & 1 \end{bmatrix}}.
$$

∎

We can execute analogous *elementary column* operations on a matrix by multiplying the matrix on the *right* by an elementary matrix. Column reduction of a matrix A is not important for us in this chapter, because it does not preserve the solution set of $Ax = b$ when applied to the augmented matrix $[A \mid b]$. However, we will have occasion to refer to column reduction when computing determinants. The effect of multiplication of a matrix A on the right by elementary matrices is explored in Exercises 36–38 in Section 1.5.

SUMMARY

1. A linear system has an associated augmented (or partitioned) matrix, having the coefficient matrix of the system on the left of the partition and the column vector of constants on the right of the partition.

2. The elementary row operations on a matrix are as follows:

    (*Row interchange*) Interchange of two rows;

    (*Row scaling*) Multiplication of a row by a nonzero scalar;

    (*Row addition*) Addition of a multiple of a row to a different row.

3. Matrices A and B are row equivalent (written $A \sim B$) if A can be transformed into B by a sequence of elementary row operations.

4. If $Ax = b$ and $Hx = c$ are systems such that the augmented matrices $[A \mid b]$ and $[H \mid c]$ are row equivalent, the systems $Ax = b$ and $Hx = c$ have the same solution set.

5. A matrix is in row-echelon form if:
    a. All rows containing only zero entries are grouped together at the bottom of the matrix.

    b. The first nonzero element (the pivot) in any row appears in a column to the right of the first nonzero element in any preceding row.

6. A matrix is in reduced row-echelon form if it is in row-echelon form and, in addition, each pivot is 1 and is the only nonzero element in its column. Every matrix is row equivalent to a unique matrix in reduced row-echelon form.

7. In the Gauss method with back substitution, we solve a linear system by reducing the augmented matrix so that the portion to the left of the

partition is in row-echelon form. The solution is then found by back substitution.

8. The Gauss–Jordan method is similar to the Gauss method, except that pivots are adjusted to be 1 and zeros are created above as well as below the pivots.

9. A linear system $A\mathbf{x} = \mathbf{b}$ has no solutions if and only if, after $[A \mid \mathbf{b}]$ is row-reduced so that A is transformed into row-echelon form, there exists a row with only zero entries to the left of the partition but with a nonzero entry to the right of the partition. The linear system is then *inconsistent*.

10. If $A\mathbf{x} = \mathbf{b}$ is a consistent linear system and if a row-echelon form H of A has at least one column containing no (nonzero) pivot, the system has an infinite number of solutions. The free variables corresponding to the columns containing no pivots can be assigned any values, and the reduced linear system can then be solved for the remaining variables.

11. An elementary matrix E is one obtained by applying a single elementary row operation to an identity matrix I. Multiplication of a matrix A on the left by E effects the same elementary row operation on A.

EXERCISES

In Exercises 1–6, reduce the matrix to (a) row-echelon form, and (b) reduced row-echelon form. Answers to (a) are not unique, so your answer may differ from the one at the back of the text.

1. $\begin{bmatrix} 2 & 1 & 4 \\ 1 & 3 & 2 \\ 3 & -1 & 6 \end{bmatrix}$

2. $\begin{bmatrix} 2 & 4 & -2 \\ 4 & 8 & 3 \\ -1 & -3 & 0 \end{bmatrix}$

3. $\begin{bmatrix} 0 & 2 & -1 & 3 \\ -1 & 1 & 2 & 0 \\ 1 & 1 & -3 & 3 \\ 1 & 5 & 5 & 9 \end{bmatrix}$

4. $\begin{bmatrix} 0 & 0 & 3 & -2 \\ 0 & 0 & 1 & 2 \\ 1 & 3 & 2 & -4 \end{bmatrix}$

5. $\begin{bmatrix} -1 & 3 & 0 & 1 & 4 \\ 1 & -3 & 0 & 0 & -1 \\ 2 & -6 & 2 & 4 & 0 \\ 0 & 0 & 1 & 3 & -4 \end{bmatrix}$

6. $\begin{bmatrix} 0 & 0 & 1 & 2 & -1 & 4 \\ 0 & 0 & 0 & 1 & -1 & 3 \\ 2 & 4 & -1 & 3 & 2 & -1 \end{bmatrix}$

In Exercises 7–12, describe all solutions of a linear system whose corresponding augmented matrix can be row-reduced to the given matrix. If requested, also give the indicated particular solution, if it exists.

7. $\left[\begin{array}{cc|c} 1 & -1 & 2 \\ 0 & 1 & 4 \end{array}\middle| \begin{array}{c} 3 \\ 2 \end{array}\right]$, solution with $x_3 = 2$

8. $\left[\begin{array}{ccc|c} 1 & 2 & 3 & 3 \\ 0 & 1 & 2 & -1 \\ 0 & 0 & 2 & 4 \end{array}\right]$

9. $\left[\begin{array}{cccc|c} 1 & 0 & 2 & 0 & 1 \\ 0 & 1 & 1 & 3 & -2 \\ 0 & 0 & 0 & 0 & 0 \end{array}\right]$,
   solution with $x_3 = 3$, $x_4 = -2$

10. $\left[\begin{array}{ccccc|c} 1 & 1 & 0 & 3 & 0 & -4 \\ 0 & 0 & 1 & -1 & 0 & 0 \\ 0 & 0 & 0 & 0 & 1 & -2 \\ 0 & 0 & 0 & 0 & 0 & 0 \end{array}\right]$,
    solution with $x_2 = 2$, $x_3 = 1$

11. $\left[\begin{array}{cccc|c} 1 & 0 & 0 & 0 & 0 \\ 0 & 1 & 0 & 0 & 2 \\ 0 & 0 & 1 & 0 & -5 \\ 0 & 0 & 0 & 1 & 2 \\ 0 & 0 & 0 & 0 & 0 \end{array}\right]$

12. $\begin{bmatrix} 1 & -1 & 2 & 0 & 3 & | & 1 \\ 0 & 0 & 0 & 1 & 4 & | & 2 \\ 0 & 0 & 0 & 0 & 0 & | & -1 \\ 0 & 0 & 0 & 0 & 0 & | & 0 \end{bmatrix}$

In Exercises 13–20, find all solutions of the given linear system, using the Gauss method with back substitution.

13. $2x - y = 8$
$6x - 5y = 32$

14. $4x_1 - 3x_2 = 10$
$8x_1 - x_2 = 10$

15. $y + z = 6$
$3x - y + z = -7$
$x + y - 3z = -13$

16. $2x + y - 3z = 0$
$6x + 3y - 8z = 0$
$2x - y + 5z = -4$

17. $x_1 - 2x_2 = 3$
$3x_1 - x_2 = 14$
$x_1 - 7x_2 = -2$

18. $x_1 - 3x_2 + x_3 = 2$
$3x_1 - 8x_2 + 2x_3 = 5$

19. $x_1 + 4x_2 - 2x_3 = 4$
$2x_1 + 7x_2 - x_3 = -2$
$2x_1 + 9x_2 - 7x_3 = 1$

20. $x_1 - 3x_2 + 2x_3 - x_4 = 8$
$3x_1 - 7x_2 \qquad + x_4 = 0$

In Exercises 21–24, find all solutions of the linear system, using the Gauss–Jordan method.

21. $3x_1 - 2x_2 = -8$
$4x_1 + 5x_2 = -3$

22. $2x_1 + 8x_2 = 16$
$5x_1 - 4x_2 = -8$

23. $x_1 \qquad - 2x_3 + x_4 = 6$
$2x_1 - x_2 + x_3 - 3x_4 = 0$
$9x_1 - 3x_2 - x_3 - 7x_4 = 4$

24. $x_1 + 2x_2 - 3x_3 + x_4 = 2$
$3x_1 + 6x_2 - 8x_3 - 2x_4 = 1$

In Exercises 25–28, determine whether the vector **b** *is in the span of the vectors* $\mathbf{v}_i$.

25. $\mathbf{b} = \begin{bmatrix} 3 \\ 5 \\ 3 \end{bmatrix}$, $\mathbf{v}_1 = \begin{bmatrix} 0 \\ 2 \\ 4 \end{bmatrix}$, $\mathbf{v}_2 = \begin{bmatrix} 1 \\ 4 \\ -2 \end{bmatrix}$, $\mathbf{v}_3 = \begin{bmatrix} -3 \\ -1 \\ 5 \end{bmatrix}$

26. $\mathbf{b} = \begin{bmatrix} 8 \\ 26 \\ 14 \end{bmatrix}$, $\mathbf{v}_1 = \begin{bmatrix} 1 \\ 3 \\ 2 \end{bmatrix}$, $\mathbf{v}_2 = \begin{bmatrix} -4 \\ -12 \\ -9 \end{bmatrix}$, $\mathbf{v}_3 = \begin{bmatrix} 1 \\ 5 \\ -1 \end{bmatrix}$

27. $\mathbf{b} = \begin{bmatrix} 8 \\ 17 \\ -8 \\ 3 \end{bmatrix}$, $\mathbf{v}_1 = \begin{bmatrix} 1 \\ 2 \\ -1 \\ 0 \end{bmatrix}$, $\mathbf{v}_2 = \begin{bmatrix} 2 \\ 5 \\ -2 \\ 5 \end{bmatrix}$, $\mathbf{v}_3 = \begin{bmatrix} -3 \\ -6 \\ 1 \\ -8 \end{bmatrix}$,

$\mathbf{v}_4 = \begin{bmatrix} 0 \\ 0 \\ -1 \\ -4 \end{bmatrix}$

28. $\mathbf{b} = \begin{bmatrix} 2 \\ -1 \\ 3 \\ 7 \end{bmatrix}$, $\mathbf{v}_1 = \begin{bmatrix} 1 \\ 1 \\ 2 \\ 3 \end{bmatrix}$, $\mathbf{v}_2 = \begin{bmatrix} -3 \\ -2 \\ -8 \\ -9 \end{bmatrix}$, $\mathbf{v}_3 = \begin{bmatrix} 1 \\ 2 \\ -1 \\ 4 \end{bmatrix}$,

$\mathbf{v}_4 = \begin{bmatrix} 2 \\ 4 \\ 0 \\ 0 \end{bmatrix}$

29. Mark each of the following True or False.

___ **a.** Every linear system with the same number of equations as unknowns has a unique solution.

___ **b.** Every linear system with the same number of equations as unknowns has at least one solution.

___ **c.** A linear system with more equations than unknowns may have an infinite number of solutions.

___ **d.** A linear system with fewer equations than unknowns may have no solution.

___ **e.** Every matrix is row equivalent to a unique matrix in row-echelon form.

___ **f.** Every matrix is row equivalent to a unique matrix in reduced row-echelon form.

___ **g.** If $[A \mid \mathbf{b}]$ and $[B \mid \mathbf{c}]$ are row-equivalent partitioned matrices, the linear systems

$A\mathbf{x} = \mathbf{b}$ and $B\mathbf{x} = \mathbf{c}$ have the same solution set.

___ **h.** A linear system with a square coefficient matrix A has a unique solution if and only if A is row equivalent to the identity matrix.

___ **i.** A linear system with coefficient matrix A has an infinite number of solutions if and only if A can be row-reduced to an echelon matrix that includes some column containing no pivot.

___ **j.** A consistent linear system with coefficient matrix A has an infinite number of solutions if and only if A can be row-reduced to an echelon matrix that includes some column containing no pivot.

In Exercises 30–37, describe all possible values for the unknowns x_i so that the matrix equation is valid.

30. $2[x_1 \ x_2] - [4 \ 7] = [-2 \ 11]$

31. $4[x_1 \ x_2] + 2[x_1 \ 3] = [-6 \ 18]$

32. $[x_1 \ x_2]\begin{bmatrix} 1 \\ -3 \end{bmatrix} = [2]$

33. $[x_1 \ x_2]\begin{bmatrix} 3 & -1 \\ 2 & 4 \end{bmatrix} = [0 \ -14]$

34. $\begin{bmatrix} 1 & -5 \\ 3 & 2 \end{bmatrix}\begin{bmatrix} x_1 \\ x_2 \end{bmatrix} = \begin{bmatrix} 13 \\ 5 \end{bmatrix}$

35. $\begin{bmatrix} x_1 & x_2 \\ x_3 & x_4 \end{bmatrix}\begin{bmatrix} 1 & 1 \\ 1 & 0 \end{bmatrix} = \begin{bmatrix} 0 & 1 \\ 3 & 1 \end{bmatrix}$

36. $[x_1 \ x_2]\begin{bmatrix} 3 & 0 & 4 \\ 2 & 1 & -1 \end{bmatrix} = [3 \ 3 \ -7]$

37. $\begin{bmatrix} x_1 & x_2 \\ x_3 & x_4 \end{bmatrix}\begin{bmatrix} 3 & 5 \\ 2 & 3 \end{bmatrix} = \begin{bmatrix} 1 & 0 \\ 0 & 1 \end{bmatrix}$

38. Determine all values of the b_i that make the linear system

$$x_1 + 2x_2 = b_1$$
$$3x_1 + 6x_2 = b_2$$

consistent.

39. Determine all values b_1 and b_2 such that $\mathbf{b} = [b_1, b_2]$ is a linear combination of $\mathbf{v}_1 = [1, 3]$ and $\mathbf{v}_2 = [5, -1]$.

40. Determine all values of the b_i that make the linear system

$$x_1 + x_2 - x_3 = b_1$$
$$2x_2 + x_3 = b_2$$
$$x_2 - x_3 = b_3$$

consistent.

41. Determine all values b_1, b_2, and b_3 such that $\mathbf{b} = [b_1, b_2, b_3]$ lies in the span of $\mathbf{v}_1 = [1, 1, 0]$, $\mathbf{v}_2 = [3, -1, 4]$, and $\mathbf{v}_3 = [-1, 2, -3]$.

42. Find an elementary matrix E such that

$$E\begin{bmatrix} 1 & 3 & 1 & 4 \\ 0 & 1 & 2 & 1 \\ 3 & 4 & 5 & 1 \end{bmatrix} = \begin{bmatrix} 1 & 3 & 1 & 4 \\ 0 & 1 & 2 & 1 \\ 0 & -5 & 2 & -11 \end{bmatrix}.$$

43. Find an elementary matrix E such that

$$E\begin{bmatrix} 1 & 3 & 1 & 4 \\ 0 & 1 & 2 & 1 \\ 3 & 4 & 5 & 1 \end{bmatrix} = \begin{bmatrix} 1 & 3 & 1 & 4 \\ 2 & 7 & 4 & 9 \\ 3 & 4 & 5 & 1 \end{bmatrix}.$$

44. Find a matrix C such that

$$C\begin{bmatrix} 1 & 2 \\ 3 & 4 \\ 4 & 2 \end{bmatrix} = \begin{bmatrix} 1 & 2 \\ 0 & -2 \\ 0 & -6 \end{bmatrix}.$$

45. Find a matrix C such that

$$C\begin{bmatrix} 1 & 2 \\ 3 & 4 \\ 4 & 2 \end{bmatrix} = \begin{bmatrix} 3 & 4 \\ 4 & 2 \\ 1 & 2 \end{bmatrix}.$$

In Exercises 46–51, let A be a 4×4 matrix. Find a matrix C such that the result of applying the given sequence of elementary row operations to A can also be found by computing the product CA.

46. Interchange row 1 and row 2.

47. Interchange row 1 and row 3; multiply row 3 by 4.

48. Multiply row 1 by 5; interchange rows 2 and 3; add 2 times row 3 to row 4.

49. Add 4 times row 2 to row 4; multiply row 4 by -3; add 5 times row 4 to row 1.

50. Interchange rows 1 and 4; add 6 times row 2 to row 1; add -3 times row 1 to row 3; add -2 times row 4 to row 2.

51. Add 3 times row 2 to row 4; add -2 times row 4 to row 3; add 5 times row 3 to row 1; add -4 times row 1 to row 2.

Exercise 24 in Section 1.3 is useful for the next three exercises.

52. Prove Theorem 1.8 for the row-interchange operation.

53. Prove Theorem 1.8 for the row-scaling operation.

54. Prove Theorem 1.8 for the row-addition operation.

55. Prove that row equivalence $\sim$ is an equivalence relation by verifying the following for $m \times n$ matrices A, B, and C.
   a. $A \sim A$.   (Reflexive Property)
   b. If $A \sim B$, then $B \sim A$.
   (Symmetric Property)
   c. If $A \sim B$ and $B \sim C$, then $A \sim C$.
   (Transitive Property)

56. Find a, b, and c such that the parabola $y = ax^2 + bx + c$ passes through the points $(1, -4)$, $(-1, 0)$, and $(2, 3)$.

57. Find a, b, c, and d such that the quartic curve $y = ax^4 + bx^3 + cx^2 + d$ passes through $(1, 2)$, $(-1, 6)$, $(-2, 38)$, and $(2, 6)$.

58. Let A be an $m \times n$ matrix, and let $\mathbf{c}$ be a column vector such that $A\mathbf{x} = \mathbf{c}$ has a unique solution.
   a. Prove that $m \geq n$.
   b. If $m = n$, must the system $A\mathbf{x} = \mathbf{b}$ be consistent for every choice of $\mathbf{b}$?
   c. Answer part (b) for the case where $m > n$.

A problem we meet when reducing a matrix with the aid of a computer involves determining when a computed entry should be 0. The computer might give an entry as 0.00000001, because of roundoff error, when it should be 0. If the computer uses this entry as a pivot in a future step, the result is chaotic! For this reason, it is common practice to program the computer to replace all sufficiently small computed entries with

0, where the meaning of "sufficiently small" must be specified in terms of the size of the nonzero entries in the original matrix. The routine YUREDUCE in LINTEK provides drill on the steps involved in reducing a matrix without requiring burdensome computation. The program computes the smallest nonzero coefficient magnitude m and asks the user to enter a number r (for ratio); all computed entries of magnitude less than rm produced during reduction of the coefficient matrix will be set equal to zero. In Exercises 59–64, use the routine YUREDUCE, specifying $r = 0.0001$, to solve the linear system.

59. $\begin{aligned} 3x_1 - x_2 &= -10 \\ 7x_1 + 2x_2 &= 7 \\ 2x_1 - 5x_2 &= -37 \end{aligned}$

60. $\begin{aligned} 5x_1 - 2x_2 &= 11 \\ 8x_1 + x_2 &= 3 \\ 6x_1 - 5x_2 &= -4 \end{aligned}$

61. $\begin{aligned} 7x_1 - 2x_2 + x_3 &= -14 \\ -4x_1 + 5x_2 - 3x_3 &= 17 \\ 5x_1 - x_2 + 2x_3 &= -7 \end{aligned}$

62. $\begin{aligned} -3x_1 + 5x_2 + 2x_3 &= 12 \\ 5x_1 - 7x_2 + 6x_3 &= -16 \\ 11x_1 - 17x_2 + 2x_3 &= -40 \end{aligned}$

63. $\begin{aligned} x_1 - 2x_2 + x_3 - x_4 + 2x_5 &= 1 \\ 2x_1 + x_2 - 4x_3 - x_4 + 5x_5 &= 16 \\ 8x_1 - x_2 + 3x_3 - x_4 - x_5 &= 1 \\ 4x_1 - 2x_2 + 3x_3 - 8x_4 + 2x_5 &= -5 \\ 5x_1 + 3x_2 - 4x_3 + 7x_4 - 6x_5 &= 7 \end{aligned}$

64. $\begin{aligned} x_1 - 2x_2 + x_3 - x_4 + 2x_5 &= 1 \\ 2x_1 + x_2 - 4x_3 - x_4 + 5x_5 &= 10 \\ 8x_1 - x_2 + 3x_3 - x_4 - x_5 &= -5 \\ 4x_1 - 2x_2 + 3x_3 - 8x_4 + 2x_5 &= -3 \\ 5x_1 + 3x_2 - 4x_3 + 7x_4 - 6x_5 &= 1 \end{aligned}$

The routine MATCOMP in LINTEK can also be used to find the solutions of a linear system. MATCOMP will bring the left portion of the augmented matrix to reduced row-echelon form and display the result on the screen. The user can *then find the solutions. Use MATCOMP in the remaining exercises.*

65. Find the reduced row-echelon form of the matrix in Exercise 6, by taking it as a coefficient matrix for zero systems.

66. Solve the linear system in Exercise 61.

67. Solve the linear system in Exercise 62.

68. Solve the linear system in Exercise 63.

MATLAB

When reducing a matrix X to reduced row-echelon form, we may need to swap row i with row k. This can be done in MATLAB using the command

$$X([i \ k],:) = X([k \ i],:).$$

If we wish to multiply the ith row by the reciprocal of x_{ij} to create a pivot 1 in the ith row and jth column, we can give the command

$$X(i,:) = X(i,:)/X(i,j).$$

When we have made pivots 1 and wish to make the entry in row k, column j equal to zero using the pivot in row i, column j, we always multiply row i by the negative of the entry that we wish to make zero, and add the result to row k. In MATLAB, this has the form

$$X(k,:) = X(k,:) - X(k, j)*X(i,:).$$

Access MATLAB and enter the lines

$$X = \text{ones}(4); \ i = 1; \ j = 2; \ k = 3;$$
$$X([i \ k],:) = X([k \ i],:)$$
$$X(i,:) = X(i,:)/X(i,j)$$
$$X(k,:) = X(k,:) - X(k, j)*X(i,:),$$

which you can then access using the up-arrow key and edit repeatedly to row-reduce a matrix X. MATLAB will not show a partition in X—you have to supply the partition mentally. If your installation contains the data files for our text, enter **fbc1s4** *now. We will be asking you to work with some of the augmented matrices used in the exercises for this section. In our data file, the augmented matrix for Exercise 63 is called E63, etc. Solve the indicated system by setting X equal to the appropriate matrix and reducing it using the up-arrow key and editing repeatedly the three basic commands above. In MATLAB, only the commands executed most recently can be accessed by using the up-arrow key. To avoid losing the command to interchange rows, which is seldom necessary, execute it at least once in each exercise even if it is not needed. (Interchanging the same rows twice leaves a matrix unchanged.) Solve the indicated exercises listed below.*

M1. Exercise 21

M2. Exercise 23

M3. Exercise 60

M4. Exercise 61

M5. Exercise 62

The command **rref(A)** *in MATLAB will reduce the matrix A to reduced row-echelon form. Use this command to solve the following exercises.*

M6.   Exercise 6   **M8.**   Exercise 63

M7.   Exercise 24   **M9.**   Exercise 64

(MATLAB contains a demo command **rrefmovie(A)** designed to show the step-by-step reduction of *A*, but with our copy and a moderately fast computer, the demo goes so fast that it is hard to catch it with the Pause key in order to view it. If you are handy with a word processor, you might copy the file *rrefmovi.m* as *rrefmovp.m*, and then edit *rrefmovp.m* to supply *,pause* at the end of each of the four lines that start with   *A(* . Entering **rrefmovp(A)** from MATLAB will then run the altered demo, which will pause after each change of a matrix. Strike any key to continue after a pause. You may notice that there seems to be unnecessary row swapping to create pivots. Look at the paragraph on partial pivoting in Section 10.3 to understand the reason for this.)

1.5   INVERSES OF SQUARE MATRICES

Matrix Equations and Inverses

A system of n equations in n unknowns $x_1, x_2, \ldots, x_n$ can be expressed in matrix form as

$$A\mathbf{x} = \mathbf{b}, \tag{1}$$

where A is the $n \times n$ coefficient matrix, $\mathbf{x}$ is the $n \times 1$ column vector with ith entry x_i, and $\mathbf{b}$ is an $n \times 1$ column vector with constant entries. The analogous equation using scalars is

$$ax = b \tag{2}$$

for scalars a and b. If $a \neq 0$, we usually think of solving Eq. (2) for x by dividing by a, but we can just as well think of solving it by multiplying by $1/a$. Breaking the solution down into small steps, we have

$$\left(\frac{1}{a}\right)ax = \left(\frac{1}{a}\right)b \quad \textbf{Multiplication by 1/a}$$

$$\left[\left(\frac{1}{a}\right)a\right]x = \left(\frac{1}{a}\right)b \quad \textbf{Associativity of multiplication}$$

$$1x = \left(\frac{1}{a}\right)b \quad \textbf{Property of 1/a}$$

$$x = \left(\frac{1}{a}\right)b. \quad \textbf{Property of 1}$$

Let us see whether we can solve Eq. (1) similarly if A is a nonzero matrix. Matrix multiplication is associative, and the $n \times n$ identity matrix I plays the same role for multiplication of $n \times n$ matrices that the number 1 plays for multiplication of numbers. The crucial step is to find an $n \times n$ matrix C such that $CA = I$, so that C plays for matrices the role that $1/a$ does for numbers. If such a matrix C exists, we can obtain from Eq. (1)

$$C(A\mathbf{x}) = C\mathbf{b} \qquad \text{Multiplication by } C$$
$$(CA)\mathbf{x} = C\mathbf{b} \qquad \text{Associativity of multiplication}$$
$$I\mathbf{x} = C\mathbf{b} \qquad \text{Property of } C$$
$$\mathbf{x} = C\mathbf{b}, \qquad \text{Property of } I$$

which shows that our column vector $\mathbf{x}$ of unknowns must be the column vector $C\mathbf{b}$. Now an interesting problem arises. When we substitute $\mathbf{x} = C\mathbf{b}$ back into our equation $A\mathbf{x} = \mathbf{b}$ to verify that we do indeed have a solution, we obtain $A\mathbf{x} = A(C\mathbf{b}) = (AC)\mathbf{b}$. But how do we know that $AC = I$ from our assumption that $CA = I$? Matrix multiplication is not a commutative operation. This problem does not arise with our scalar equation $ax = b$, because multiplication of real numbers is commutative. It is indeed true that for square matrices, if $CA = I$ then $AC = I$, and we will work toward a proof of this. For example, the equation

$$\begin{bmatrix} -4 & 9 \\ 1 & -2 \end{bmatrix} \begin{bmatrix} 2 & 9 \\ 1 & 4 \end{bmatrix} = \begin{bmatrix} 1 & 0 \\ 0 & 1 \end{bmatrix} = \begin{bmatrix} 2 & 9 \\ 1 & 4 \end{bmatrix} \begin{bmatrix} -4 & 9 \\ 1 & -2 \end{bmatrix}$$

illustrates that

$$A = \begin{bmatrix} 2 & 9 \\ 1 & 4 \end{bmatrix} \quad \text{and} \quad C = \begin{bmatrix} -4 & 9 \\ 1 & -2 \end{bmatrix}$$

satisfy $CA = I = AC$.

Unfortunately, it is not true that, for each nonzero $n \times n$ matrix A, we can find an $n \times n$ matrix C such that $CA = AC = I$. For example, if the first column of A has only zero entries, then the first column of CA also has only zero entries for any matrix C, so $CA \neq I$ for any matrix C. However, for many important $n \times n$ matrices A, there does exist an $n \times n$ matrix C such that $CA = AC = I$. Let us show that when such a matrix exists, it is unique.

THEOREM 1.9 Uniqueness of an Inverse

Let A be an $n \times n$ matrix. If C and D are matrices such that $AC = DA = I$, then $C = D$. In particular, if $AC = CA = I$, then C is the *unique* matrix with this property.

PROOF Let C and D be matrices such that $AC = DA = I$. Because matrix multiplication is associative, we have

$$D(AC) = (DA)C.$$

But, because $AC = I$ and $DA = I$, we find that

$$D(AC) = DI = D \quad \text{and} \quad (DA)C = IC = C.$$

Therefore, $C = D$.

Now suppose that $AC = CA = I$, and let us show that C is the unique matrix with this property. To this end, suppose also that $AD = DA = I$. Then we have $AC = I = DA$, so $D = C$, as we just showed.   ▲

From the title of the preceding theorem, we anticipate the following definition.

DEFINITION 1.15   Invertible Matrix

> An $n \times n$ matrix A is **invertible** if there exists an $n \times n$ matrix C such that $CA = AC = I$, the $n \times n$ identity matrix. The matrix C is the inverse of A and is denoted by A^{-1}. If A is not invertible, it is **singular**.

Although A^{-1} plays the same role arithmetically as $a^{-1} = 1/a$ (as we showed at the start of this section), we will never write A^{-1} as $1/A$. The powers of an invertible $n \times n$ matrix A are now defined for all integers. That is, for $m > 0$, A^m is the product of m factors A, and A^{-m} is the product of m factors A^{-1}. We consider A^0 to be the $n \times n$ identity matrix I.

Inverses of Elementary Matrices

In Section 1.4, we saw that each elementary row operation can be undone by another (possibly the same) elementary row operation. Let us see how this fact

---

HISTORICAL NOTE   THE NOTION OF THE INVERSE OF A MATRIX first appears in an 1855 note of Arthur Cayley (1821–1895) and is made more explicit in an 1858 paper entitled "A Memoir on the Theory of Matrices." In that work, Cayley outlines the basic properties of matrices, noting that most of these derive from work with sets of linear equations. In particular, the inverse comes from the idea of solving a system

$$X = a\,x + b\,y + c\,z$$
$$Y = a'x + b'y + c'z$$
$$Z = a''x + b''y + c''z$$

for x, y, z in terms of X, Y, Z. Cayley gives an explicit construction for the inverse in terms of the determinants of the original matrix and of the minors.

In 1842, Arthur Cayley graduated from Trinity College, Cambridge, but could not find a suitable teaching post. So, like Sylvester, he studied law and was called to the bar in 1849. During his 14 years as a lawyer, he wrote about 300 mathematical papers; finally, in 1863 he became a professor at Cambridge, where he remained until his death. It was during his stint as a lawyer that he met Sylvester; their discussions over the next 40 years were extremely fruitful for the progress of algebra. Over his lifetime, Cayley produced about 1000 papers in pure mathematics, theoretical dynamics, and mathematical astronomy.

translates to the invertibility of the elementary matrices, and a description of their inverses.

Let E_1 be an elementary row-interchange matrix, obtained from the identity matrix I by the interchanging of rows i and k. Recall that E_1A effects the interchange of rows i and k of A for any matrix A such that E_1A is defined. In particular, taking $A = E_1$, we see that E_1E_1 interchanges rows i and k of E_1, and hence changes E_1 back to I. Thus,

$$E_1E_1 = I.$$

Consequently, E_1 is an invertible matrix and is its own inverse.

Now let E_2 be an elementary row-scaling matrix, obtained from the identity matrix by the multiplication of row i by a nonzero scalar r. Let E_2' be the matrix obtained from the identity matrix by the multiplication of row i by $1/r$. It is clear that

$$E_2'E_2 = E_2E_2' = I,$$

so E_2 is invertible, with inverse E_2'.

Finally, let E_3 be an elementary row-addition matrix, obtained from I by the addition of r times row i to row k. If E_3' is obtained from I by the addition of $-r$ times row i to row k, then

$$E_3'E_3 = E_3E_3' = I.$$

We have established the following fact:

Every elementary matrix is invertible.

EXAMPLE 1   Find the inverses of the elementary matrices

$$E_1 = \begin{bmatrix} 0 & 1 & 0 \\ 1 & 0 & 0 \\ 0 & 0 & 1 \end{bmatrix}, \quad E_2 = \begin{bmatrix} 3 & 0 & 0 \\ 0 & 1 & 0 \\ 0 & 0 & 1 \end{bmatrix}, \quad \text{and} \quad E_3 = \begin{bmatrix} 1 & 0 & 4 \\ 0 & 1 & 0 \\ 0 & 0 & 1 \end{bmatrix}.$$

SOLUTION   Because E_1 is obtained from

$$I = \begin{bmatrix} 1 & 0 & 0 \\ 0 & 1 & 0 \\ 0 & 0 & 1 \end{bmatrix}$$

by the interchanging of the first and second rows, we see that $E_1^{-1} = E_1$.

The matrix E_2 is obtained from I by the multiplication of the first row by 3, so we must multiply the first row of I by $\frac{1}{3}$ to form

$$E_2^{-1} = \begin{bmatrix} \frac{1}{3} & 0 & 0 \\ 0 & 1 & 0 \\ 0 & 0 & 1 \end{bmatrix}.$$

Finally, E_3 is obtained from I by the addition of 4 times row 3 to row 1. To form E_3^{-1} from I, we add -4 times row 3 to row 1, so

$$E_3^{-1} = \begin{bmatrix} 1 & 0 & -4 \\ 0 & 1 & 0 \\ 0 & 0 & 1 \end{bmatrix}.$$

■

Inverses of Products

The next theorem is fundamental in work with inverses.

THEOREM 1.10   Inverses of Products

Let A and B be invertible $n \times n$ matrices. Then AB is invertible, and $(AB)^{-1} = B^{-1}A^{-1}$.

PROOF   By assumption, there exist matrices A^{-1} and B^{-1} such that $AA^{-1} = A^{-1}A = I$ and $BB^{-1} = B^{-1}B = I$. Making use of the associative law for matrix multiplication, we find that

$$(AB)(B^{-1}A^{-1}) = [A(BB^{-1})]A^{-1} = (AI)A^{-1} = AA^{-1} = I.$$

A similar computation shows that $(B^{-1}A^{-1})(AB) = I$. Therefore, the inverse of AB is $B^{-1}A^{-1}$; that is, $(AB)^{-1} = B^{-1}A^{-1}$.   ▲

It is instructive to apply Theorem 1.10 to a product $E_t \cdots E_3E_2E_1$ of elementary matrices. In the expression

$$(E_t \ldots E_3E_2E_1)A,$$

the product $E_t \cdots E_3E_2E_1$ performs a sequence of elementary row operations on A. First, E_1 acts on A; then, E_2 acts on E_1A; and so on. To undo this sequence, we must first undo the last elementary row operation, performed by E_t. This is accomplished by using E_t^{-1}. Continuing, we should perform the sequence of operations given by

$$E_1^{-1}E_2^{-1}E_3^{-1} \cdots E_t^{-1}$$

in order to effect $(E_t \cdots E_3E_2E_1)^{-1}$.

A Commutativity Property

We are now in position to show that if $CA = I$, then $AC = I$. First we prove a lemma (a result preliminary to the main result).

LEMMA 1.1   Condition for $A\mathbf{x} = \mathbf{b}$ to Be Solvable for All b

> Let A be an $n \times n$ matrix. The linear system $A\mathbf{x} = \mathbf{b}$ has a solution for every choice of column vector $\mathbf{b} \in \mathbb{R}^n$ if and only if A is row equivalent to the $n \times n$ identity matrix I.

PROOF   Let $\mathbf{b}$ be any column vector in $\mathbb{R}^n$ and let the augmented matrix $[A \mid \mathbf{b}]$ be row-reduced to $[H \mid \mathbf{c}]$ where H is in reduced row-echelon form. If H is the identity matrix I, then the linear system $A\mathbf{x} = \mathbf{b}$ has the solution $\mathbf{x} = \mathbf{c}$.

For the converse, suppose that reduction of A to reduced row-echelon form yields a matrix H that is not the identity matrix. Then the bottom row of H must have every entry equal to 0. Now there exist elementary matrices $E_1, E_2, \ldots, E_t$ such that $(E_t \cdots E_2 E_1)A = H$. Recall that every elementary matrix is invertible, and that a product of elementary matrices is invertible. Let $\mathbf{b} = (E_t \cdots E_2 E_1)^{-1}\mathbf{e}_n$, where $\mathbf{e}_n$ is the column vector with 1 in its nth component and zeros elsewhere. Reduction of the augmented matrix $[A \mid \mathbf{b}]$ can be accomplished by multiplying both A and $\mathbf{b}$ on the left by $E_t \cdots E_2 E_1$, so the reduction will yield $[H \mid \mathbf{e}_n]$, which represents a system of equations with no solution because the bottom row has entries 0 to the left of the partition and 1 to the right of the partition. This shows that if H is not the identity matrix, then $A\mathbf{x} = \mathbf{b}$ does not have a solution for some $\mathbf{b} \in \mathbb{R}^n$.   ▲

THEOREM 1.11   A Commutativity Property

> Let A and C be $n \times n$ matrices. Then $CA = I$ if and only if $AC = I$.

PROOF   To prove that $CA = I$ if and only if $AC = I$, it suffices to prove that if $AC = I$, then $CA = I$, because the converse statement is obtained by reversing the roles of A and C.

Suppose now that we do have $AC = I$. Then the equation $A\mathbf{x} = \mathbf{b}$ has a solution for every column vector $\mathbf{b}$ in $\mathbb{R}^n$; we need only notice that $\mathbf{x} = C\mathbf{b}$ is a solution because $A(C\mathbf{b}) = (AC)\mathbf{b} = I\mathbf{b} = \mathbf{b}$. By Lemma 1.1, we know that A is row equivalent to the $n \times n$ identity matrix I, so there exists a sequence of elementary matrices $E_1, E_2, \ldots, E_t$ such that $(E_t \cdots E_2 E_1)A = I$. By Theorem 1.9, the two equations

$$(E_t \cdots E_2 E_1)A = I \quad \text{and} \quad AC = I$$

imply that $E_t \cdots E_2 E_1 = C$, so we have $CA = I$ also.   ▲

Computation of Inverses

Let $A = [a_{ij}]$ be an $n \times n$ matrix. To find A^{-1}, if it exists, we must find an $n \times n$ matrix $X = [x_{ij}]$ such that $AX = I$—that is, such that

$$
\begin{bmatrix}
a_{11} & a_{12} & \cdots & a_{1n} \\
a_{21} & a_{22} & \cdots & a_{2n} \\
& \vdots & & \\
a_{n1} & a_{n2} & \cdots & a_{nn}
\end{bmatrix}
\begin{bmatrix}
x_{11} & x_{12} & \cdots & x_{1n} \\
x_{21} & x_{22} & \cdots & x_{2n} \\
& \vdots & & \\
x_{n1} & x_{n2} & \cdots & x_{nn}
\end{bmatrix}
=
\begin{bmatrix}
1 & 0 & \cdots & 0 \\
0 & 1 & \cdots & 0 \\
& \vdots & & \\
0 & 0 & \cdots & 1
\end{bmatrix}. \tag{3}
$$

Matrix equation (3) corresponds to n^2 linear equations in the n^2 unknowns x_{ij}; there is one linear equation for each of the n^2 positions in an $n \times n$ matrix. For example, equating the entries in the row 2, column 1 position on each side of Eq. (3), we obtain the linear equation

$$ a_{21}x_{11} + a_{22}x_{21} + \cdots + a_{2n}x_{n1} = 0. $$

Of these n^2 linear equations, n of them involve the n unknowns x_{i1} for $i = 1, 2, \ldots, n$; and these equations are given by the column-vector equation

$$
A
\begin{bmatrix}
x_{11} \\
x_{21} \\
\vdots \\
x_{n1}
\end{bmatrix}
=
\begin{bmatrix}
1 \\
0 \\
\vdots \\
0
\end{bmatrix}, \tag{4}
$$

which is a square system of equations. There are also n equations involving the n unknowns x_{i2}, for $i = 1, 2, \ldots, n$; and so on. In addition to solving system (4), we must solve the systems

$$
A
\begin{bmatrix}
x_{12} \\
x_{22} \\
\vdots \\
x_{n2}
\end{bmatrix}
=
\begin{bmatrix}
0 \\
1 \\
\vdots \\
0
\end{bmatrix}, \ldots, A
\begin{bmatrix}
x_{1n} \\
x_{2n} \\
\vdots \\
x_{nn}
\end{bmatrix}
=
\begin{bmatrix}
0 \\
0 \\
\vdots \\
1
\end{bmatrix}, \tag{5}
$$

where each system has the same coefficient matrix A. Whenever we want to solve several systems $A\mathbf{x} = \mathbf{b}_i$ with the same coefficient matrix but different vectors $\mathbf{b}_i$, we solve them all at once, rather than one at a time. The main job in solving a linear system is reducing the coefficient matrix to row-echelon or reduced row-echelon form, and we don't want to repeat that work over and over. We simply reduce one augmented matrix, where we line up all the vectors $\mathbf{b}_i$ to the right of the partition. Thus, to solve all the linear systems in Eqs. (4) and (5), we form the augmented matrix

$$
\left[
\begin{array}{cccc|cccc}
a_{11} & a_{12} & \cdots & a_{1n} & 1 & 0 & \cdots & 0 \\
a_{21} & a_{22} & \cdots & a_{2n} & 0 & 1 & \cdots & 0 \\
& \vdots & & & & \vdots & & \\
a_{n1} & a_{n2} & \cdots & a_{nn} & 0 & 0 & \cdots & 1
\end{array}
\right], \tag{6}
$$

which we abbreviate by $[A \mid I]$. The matrix A is to the left of the partition, and the identity matrix I is to the right. We then perform a Gauss–Jordan reduction on this augmented matrix. By Theorem 1.9, we know that if A^{-1} exists, it is *unique*, so that every column in the reduced row-echelon form of A has a pivot. Thus, A^{-1} exists if and only if the augmented matrix (6) can be reduced to

$$\left[\begin{array}{cccc|cccc} 1 & 0 & \cdots & 0 & c_{11} & c_{12} & \cdots & c_{1n} \\ 0 & 1 & \cdots & 0 & c_{21} & c_{22} & \cdots & c_{2n} \\ \vdots & & & & & \vdots & & \\ 0 & 0 & \cdots & 1 & c_{n1} & c_{n2} & \cdots & c_{nn} \end{array}\right],$$

where the $n \times n$ identity matrix I is to the left of the partition. The $n \times n$ solution matrix $C = [c_{ij}]$ to the right of the partition then satisfies $AC = I$, so $A^{-1} = C$. This is an efficient way to compute A^{-1}. We summarize the computation in the following box, and we state the theory in Theorem 1.12.

Computation of A^{-1}

To find A^{-1}, if it exists, proceed as follows:

Step 1 Form the augmented matrix $[A \mid I]$.

Step 2 Apply the Gauss–Jordan method to attempt to reduce $[A \mid I]$ to $[I \mid C]$. If the reduction can be carried out, then $A^{-1} = C$. Otherwise, A^{-1} does not exist.

EXAMPLE 2   For the matrix $A = \begin{bmatrix} 2 & 9 \\ 1 & 4 \end{bmatrix}$, compute the inverse we exhibited at the start of this section, and use this inverse to solve the linear system

$$2x + 9y = -5$$
$$x + 4y = 7.$$

SOLUTION   Reducing the augmented matrix, we have

$$\left[\begin{array}{cc|cc} 2 & 9 & 1 & 0 \\ 1 & 4 & 0 & 1 \end{array}\right] \sim \left[\begin{array}{cc|cc} 1 & 4 & 0 & 1 \\ 2 & 9 & 1 & 0 \end{array}\right] \sim \left[\begin{array}{cc|cc} 1 & 4 & 0 & 1 \\ 0 & 1 & 1 & -2 \end{array}\right]$$

$$\sim \left[\begin{array}{cc|cc} 1 & 0 & -4 & 9 \\ 0 & 1 & 1 & -2 \end{array}\right].$$

Therefore,

$$A^{-1} = \begin{bmatrix} -4 & 9 \\ 1 & -2 \end{bmatrix}.$$

If A^{-1} exists, the solution of $A\mathbf{x} = \mathbf{b}$ is $\mathbf{x} = A^{-1}\mathbf{b}$. Consequently, the solution of our system

$$\begin{bmatrix} 2 & 9 \\ 1 & 4 \end{bmatrix}\begin{bmatrix} x \\ y \end{bmatrix} = \begin{bmatrix} -5 \\ 7 \end{bmatrix} \quad \text{is} \quad \begin{bmatrix} x \\ y \end{bmatrix} = \begin{bmatrix} -4 & 9 \\ 1 & -2 \end{bmatrix}\begin{bmatrix} -5 \\ 7 \end{bmatrix} = \begin{bmatrix} 83 \\ -19 \end{bmatrix}. \qquad \blacksquare$$

We emphasize that the computation of the solution of the linear system in Example 2, using the inverse of the coefficient matrix, was for illustration only. When faced with the problem of solving a square system, $A\mathbf{x} = \mathbf{b}$, one should never start by finding the inverse of the coefficient matrix. To do so would involve row reduction of $[A \mid I]$ and subsequent computation of $A^{-1}\mathbf{b}$, whereas the shorter reduction of $[A \mid \mathbf{b}]$ provides the desired solution at once. The inverse of a matrix is often useful in symbolic computations. For example, if A is an invertible matrix and we know that $AB = AC$, then we can deduce that $B = C$ by multiplying both sides of $AB = AC$ on the left by A^{-1}. If we have r systems of equations

$$A\mathbf{x}_1 = \mathbf{b}_1, \qquad A\mathbf{x}_2 = \mathbf{b}_2, \qquad \ldots, \qquad A\mathbf{x}_r = \mathbf{b}_r,$$

all with the same invertible $n \times n$ coefficient matrix A, it might seem to be more efficient (for large r) to solve all the systems by finding A^{-1} and computing the column vectors

$$\mathbf{x}_1 = A^{-1}\mathbf{b}_1, \qquad \mathbf{x}_2 = A^{-1}\mathbf{b}_2, \qquad \ldots, \qquad \mathbf{x}_r = A^{-1}\mathbf{b}_r.$$

Section 10.1 will show that using the Gauss method with back substitution on the augmented matrix $[A \mid \mathbf{b}_1 \ \mathbf{b}_2 \cdots \mathbf{b}_r]$ remains more efficient. Thus, inversion of a coefficient matrix is not a good numerical way to solve a linear system. However, we will find inverses very useful for solving other kinds of problems.

THEOREM 1.12 Conditions for A^{-1} to Exist

The following conditions for an $n \times n$ matrix A are equivalent:

(i) A is invertible.

(ii) A is row equivalent to the identity matrix I.

(iii) The system $A\mathbf{x} = \mathbf{b}$ has a solution for each n-component column vector $\mathbf{b}$.

(iv) A can be expressed as a product of elementary matrices.

(v) The span of the column vectors of A is $\mathbb{R}^n$.

PROOF Step 2 in the box preceding Example 2 shows that parts (i) and (ii) of Theorem 1.12 are equivalent. For the equivalence of (ii) with (iii), Lemma 1.1 shows that $A\mathbf{x} = \mathbf{b}$ has a solution for each $\mathbf{b} \in \mathbb{R}^n$ if and only if (ii) is true. Thus, (ii) and (iii) are equivalent. The equivalence of (iii) and (v) follows from the box on page 63.

Turning to the equivalence of parts (ii) and (iv), we know that the matrix A is row equivalent to I if and only if there is a sequence of elementary matrices $E_1, E_2, \ldots, E_t$ such that $E_t \cdots E_2 E_1 A = I$; and this is the case if and only if A is expressible as a product $A = E_1^{-1} E_2^{-1} \cdots E_t^{-1}$ of elementary matrices.   ▲

EXAMPLE 3   Using Example 2, express $\begin{bmatrix} 2 & 9 \\ 1 & 4 \end{bmatrix}$ as a product of elementary matrices.

SOLUTION   The steps we performed in Example 2 can be applied in sequence to 2×2 identity matrices to generate elementary matrices:

$$E_1 = \begin{bmatrix} 0 & 1 \\ 1 & 0 \end{bmatrix} \qquad \textbf{Interchange rows 1 and 2.}$$

$$E_2 = \begin{bmatrix} 1 & 0 \\ -2 & 1 \end{bmatrix} \qquad \textbf{Add } -2 \textbf{ times row 1 to row 2.}$$

$$E_3 = \begin{bmatrix} 1 & -4 \\ 0 & 1 \end{bmatrix}. \qquad \textbf{Add } -4 \textbf{ times row 2 to row 1.}$$

Thus we see that $E_3 E_2 E_1 A = I$, so

$$A = E_1^{-1} E_2^{-1} E_3^{-1} = \begin{bmatrix} 0 & 1 \\ 1 & 0 \end{bmatrix} \begin{bmatrix} 1 & 0 \\ 2 & 1 \end{bmatrix} \begin{bmatrix} 1 & 4 \\ 0 & 1 \end{bmatrix}. \qquad ■$$

Example 3 illustrates the following boxed rule for expressing an invertible matrix A as a product of elementary matrices.

Expressing an Invertible Matrix A as a Product of Elementary Matrices

Write in left-to-right order the inverses of the elementary matrices corresponding to successive row operations that reduce A to I.

EXAMPLE 4   Determine whether the matrix

$$A = \begin{bmatrix} 1 & 3 & -2 \\ 2 & 5 & -3 \\ -3 & 2 & -4 \end{bmatrix}$$

is invertible, and find its inverse if it is.

SOLUTION   We have

$$\left[\begin{array}{ccc|ccc} 1 & 3 & -2 & 1 & 0 & 0 \\ 2 & 5 & -3 & 0 & 1 & 0 \\ -3 & 2 & -4 & 0 & 0 & 1 \end{array}\right] \sim \left[\begin{array}{ccc|ccc} 1 & 3 & -2 & 1 & 0 & 0 \\ 0 & -1 & 1 & -2 & 1 & 0 \\ 0 & 11 & -10 & 3 & 0 & 1 \end{array}\right]$$

$$\sim \left[\begin{array}{ccc|ccc} 1 & 0 & 1 & -5 & 3 & 0 \\ 0 & 1 & -1 & 2 & -1 & 0 \\ 0 & 0 & 1 & -19 & 11 & 1 \end{array}\right] \sim \left[\begin{array}{ccc|ccc} 1 & 0 & 0 & 14 & -8 & -1 \\ 0 & 1 & 0 & -17 & 10 & 1 \\ 0 & 0 & 1 & -19 & 11 & 1 \end{array}\right].$$

Therefore, A is an invertible matrix, and

$$A^{-1} = \begin{bmatrix} 14 & -8 & -1 \\ -17 & 10 & 1 \\ -19 & 11 & 1 \end{bmatrix}.$$ ∎

EXAMPLE 5   Express the matrix A of Example 4 as a product of elementary matrices.

SOLUTION   In accordance with the box that follows Example 3, we write in left-to-right order the successive inverses of the elementary matrices corresponding to the row reduction of A in Example 4. We obtain

$$A = \begin{bmatrix} 1 & 0 & 0 \\ 2 & 1 & 0 \\ 0 & 0 & 1 \end{bmatrix} \begin{bmatrix} 1 & 0 & 0 \\ 0 & 1 & 0 \\ -3 & 0 & 1 \end{bmatrix} \begin{bmatrix} 1 & 0 & 0 \\ 0 & -1 & 0 \\ 0 & 0 & 1 \end{bmatrix} \begin{bmatrix} 1 & 3 & 0 \\ 0 & 1 & 0 \\ 0 & 0 & 1 \end{bmatrix}$$

$$\times \begin{bmatrix} 1 & 0 & 0 \\ 0 & 1 & 0 \\ 0 & 11 & 1 \end{bmatrix} \begin{bmatrix} 1 & 0 & 1 \\ 0 & 1 & 0 \\ 0 & 0 & 1 \end{bmatrix} \begin{bmatrix} 1 & 0 & 0 \\ 0 & 1 & -1 \\ 0 & 0 & 1 \end{bmatrix}.$$ ∎

EXAMPLE 6   Determine whether the span of the vectors $[1, -2, 1]$, $[3, -5, 4]$, and $[4, -3, 9]$ is all of $\mathbb{R}^3$.

SOLUTION   Let

$$A = \begin{bmatrix} 1 & 3 & 4 \\ -2 & -5 & -3 \\ 1 & 4 & 9 \end{bmatrix}.$$

We have

$$\begin{bmatrix} 1 & 3 & 4 \\ -2 & -5 & -3 \\ 1 & 4 & 9 \end{bmatrix} \sim \begin{bmatrix} 1 & 3 & 4 \\ 0 & 1 & 5 \\ 0 & 1 & 5 \end{bmatrix} \sim \begin{bmatrix} 1 & 3 & 4 \\ 0 & 1 & 5 \\ 0 & 0 & 0 \end{bmatrix}.$$

We do not have a pivot in the row 3, column 3 position, so we are not able to reduce A to the identity matrix. By Theorem 1.12, the span of the given vectors is not all of $\mathbb{R}^3$. ∎

SUMMARY

1. Let A be a square matrix. A square matrix C such that $CA = AC = I$ is the inverse of A and is denoted by $C = A^{-1}$. If such an inverse of A exists, then A is said to be *invertible*. The inverse of an invertible matrix A is unique. A square matrix that has no inverse is called *singular*.

2. The inverse of a square matrix A exists if and only if A can be reduced to the identity matrix I by means of elementary row operations or (equivalently) if and only if A is a product of elementary matrices. In this case, A is equal to the product, in left-to-right order, of the inverses of the successive

elementary matrices corresponding to the sequence of row operations used to reduce A to I.

3. To find A^{-1}, if it exists, form the augmented matrix $[A \mid I]$ and apply the Gauss–Jordan method to reduce this matrix to $[I \mid C]$. If this can be done, then $A^{-1} = C$. Otherwise, A is not invertible.

4. The inverse of a product of invertible matrices is the product of the inverses in the reverse order.

EXERCISES

In Exercises 1–8, (a) find the inverse of the square matrix, if it exists, and (b) express each invertible matrix as a product of elementary matrices.

1. $\begin{bmatrix} 1 & 1 \\ 0 & 1 \end{bmatrix}$

2. $\begin{bmatrix} 3 & 6 \\ 3 & 8 \end{bmatrix}$

3. $\begin{bmatrix} 3 & 6 \\ 4 & 8 \end{bmatrix}$

4. $\begin{bmatrix} 6 & 7 \\ 8 & 9 \end{bmatrix}$

5. $\begin{bmatrix} 1 & 0 & 1 \\ 0 & 1 & 1 \\ 0 & 0 & -1 \end{bmatrix}$

6. $\begin{bmatrix} 1 & 1 & -1 \\ 2 & 0 & 3 \\ -3 & 1 & -7 \end{bmatrix}$

7. $\begin{bmatrix} 2 & 1 & 4 \\ 3 & 2 & 5 \\ 0 & -1 & 1 \end{bmatrix}$

8. $\begin{bmatrix} -1 & 2 & 1 \\ 2 & -3 & 5 \\ 1 & 0 & 12 \end{bmatrix}$

In Exercises 9 and 10, find the inverse of the matrix, if it exists.

9. $\begin{bmatrix} 1 & 0 & 0 & 0 & 0 & 0 \\ 0 & -1 & 0 & 0 & 0 & 0 \\ 0 & 0 & 2 & 0 & 0 & 0 \\ 0 & 0 & 0 & 3 & 0 & 0 \\ 0 & 0 & 0 & 0 & 4 & 0 \\ 0 & 0 & 0 & 0 & 0 & 5 \end{bmatrix}$

10. $\begin{bmatrix} 0 & 0 & 0 & 0 & 0 & 6 \\ 0 & 0 & 0 & 0 & 5 & 0 \\ 0 & 0 & 0 & 4 & 0 & 0 \\ 0 & 0 & 3 & 0 & 0 & 0 \\ 0 & 2 & 0 & 0 & 0 & 0 \\ 1 & 0 & 0 & 0 & 0 & 0 \end{bmatrix}$

In Exercises 11 and 12, determine whether the span of the column vectors of the given matrix is $\mathbb{R}^4$.

11. $\begin{bmatrix} 1 & 0 & 1 & -1 \\ 0 & -1 & -3 & 4 \\ 1 & 0 & -1 & 2 \\ -3 & 0 & 0 & -1 \end{bmatrix}$

12. $\begin{bmatrix} 1 & -2 & 1 & 0 \\ -3 & 5 & 0 & 2 \\ 0 & 1 & 2 & -4 \\ -1 & 2 & 4 & -2 \end{bmatrix}$

13. **a.** Show that the matrix

$$A = \begin{bmatrix} 2 & -3 \\ 5 & -7 \end{bmatrix}$$

is invertible, and find its inverse.

b. Use the result in (a) to find the solution of the system of equations

$$2x_1 - 3x_2 = 4, \qquad 5x_1 - 7x_2 = -3.$$

14. Using the inverse of the matrix in Exercise 7, find the solution of the system of equations

$$2x_1 + x_2 + 4x_3 = 5$$
$$3x_1 + 2x_2 + 5x_3 = 3$$
$$-x_2 + x_3 = 8.$$

15. Find three linear equations that express x, y, z in terms of r, s, t, if

$$2x + y + 4z = r$$
$$3x + 2y + 5z = s$$
$$-y + z = t.$$

[HINT: See Exercise 14.]

16. Let

$$A^{-1} = \begin{bmatrix} 1 & 2 & 1 \\ 0 & 3 & 1 \\ 4 & 1 & 2 \end{bmatrix}.$$

If possible, find a matrix C such that

$$AC = \begin{bmatrix} 1 & 2 \\ 0 & 1 \\ 4 & 1 \end{bmatrix}.$$

17. Let

$$A^{-1} = \begin{bmatrix} 1 & 2 & 1 \\ 0 & 3 & 1 \\ 4 & 1 & 2 \end{bmatrix}.$$

If possible, find a matrix C such that

$$ACA = \begin{bmatrix} 2 & 1 & 3 \\ -1 & 2 & 2 \\ 2 & 1 & 4 \end{bmatrix}.$$

18. Let

$$A = \begin{bmatrix} 4 & 2 & 2 \\ 0 & 3 & 1 \\ 2 & 0 & 1 \end{bmatrix}.$$

If possible, find a matrix B such that $AB = 2A$.

19. Let

$$A = \begin{bmatrix} 1 & 2 & 1 \\ 0 & 1 & 2 \\ 1 & 3 & 2 \end{bmatrix}.$$

If possible, find a matrix B such that $AB = A^2 + 2A$.

20. Find all numbers r such that

$$\begin{bmatrix} 2 & 4 & 2 \\ 1 & r & 3 \\ 1 & 2 & 1 \end{bmatrix}$$

is invertible.

21. Find all numbers r such that

$$\begin{bmatrix} 2 & 4 & 2 \\ 1 & r & 3 \\ 1 & 1 & 2 \end{bmatrix}$$

is invertible.

22. Let A and B be two $m \times n$ matrices. Show that A and B are row equivalent if and only if there exists an invertible $m \times m$ matrix C such that $CA = B$.

23. Mark each of the following True or False. The statements involve matrices A, B, and C, which are assumed to be of appropriate size.

____ **a.** If $AC = BC$ and C is invertible, then $A = B$.

____ **b.** If $AB = O$ and B is invertible, then $A = O$.

____ **c.** If $AB = C$ and two of the matrices are invertible, then so is the third.

____ **d.** If $AB = C$ and two of the matrices are singular, then so is the third.

____ **e.** If A^2 is invertible, then A^3 is invertible.

____ **f.** If A^3 is invertible, then A^2 is invertible.

____ **g.** Every elementary matrix is invertible.

____ **h.** Every invertible matrix is an elementary matrix.

____ **i.** If A and B are invertible matrices, then so is $A + B$, and $(A + B)^{-1} = A^{-1} + B^{-1}$.

____ **j.** If A and B are invertible, then so is AB, and $(AB)^{-1} = A^{-1}B^{-1}$.

24. Show that, if A is an invertible $n \times n$ matrix, then A^T is invertible. Describe $(A^T)^{-1}$ in terms of A^{-1}.

25. **a.** If A is invertible, is $A + A^T$ always invertible?

b. If A is invertible, is $A + A$ always invertible?

26. Let A be a matrix such that A^2 is invertible. Prove that A is invertible.

27. Let A and B be $n \times n$ matrices with A invertible.

a. Show that $AX = B$ has the unique solution $X = A^{-1}B$.

b. Show that $X = A^{-1}B$ can be found by the following row reduction:

$$[A \mid B] \sim [I \mid X].$$

That is, if the matrix A is reduced to the identity matrix I, then the matrix B will be reduced to $A^{-1}B$.

28. Note that

$$\frac{1}{a} + \frac{1}{b} = \frac{(a + b)}{(ab)}$$

for nonzero scalars $a, b \in \mathbb{R}$. Find an analogous equality for invertible $n \times n$ matrices A and B.

29. An $n \times n$ matrix A is **nilpotent** if $A^r = O$ (the $n \times n$ zero matrix) for some positive integer r.

   a. Give an example of a nonzero nilpotent 2×2 matrix.

   b. Show that, if A is an invertible $n \times n$ matrix, then A is not nilpotent.

30. A square matrix A is said to be **idempotent** if $A^2 = A$.

   a. Give an example of an idempotent matrix other than O and I.

   b. Show that, if a matrix A is both idempotent and invertible, then $A = I$.

31. Show that

$$\begin{bmatrix} 0 & a_1 & a_2 & a_3 \\ 0 & 0 & b_1 & b_2 \\ 0 & 0 & 0 & c_1 \\ 0 & 0 & 0 & 0 \end{bmatrix}$$

is nilpotent. (See Exercise 29.)

32. A square matrix is **upper triangular** if all entries below the main diagonal are zero. **Lower triangular** is defined symmetrically. Give an example of a nilpotent 4×4 matrix that is not upper or lower triangular. (See Exercises 29 and 31.)

33. Give an example of two invertible 4×4 matrices whose sum is singular.

34. Give an example of two singular 3×3 matrices whose sum is invertible.

35. Consider the 2×2 matrix

$$A = \begin{bmatrix} a & b \\ c & d \end{bmatrix},$$

and let $h = ad - bc$.

   a. Show that, if $h \neq 0$, then

$$\begin{bmatrix} d/h & -b/h \\ -c/h & a/h \end{bmatrix}$$

is the inverse of A.

   b. Show that A is invertible if and only if $h \neq 0$.

Exercises 36–38 develop elementary column operations.

36. For each type of elementary matrix E, explain how E can be obtained from the identity matrix by means of operations on columns.

37. Let A be a square matrix, and let E be an elementary matrix of the same size. Find the effect on A of multiplying A on the right by E. [HINT: Use Exercise 36.]

38. Let A be an invertible square matrix. Recall that $(BA)^{-1} = A^{-1}B^{-1}$, and use Exercise 37 to answer the following questions:

   a. If two rows of A are interchanged, how does the inverse of the resulting matrix compare with A^{-1}?

   b. Answer the question in part (a) if, instead, a row of A is multiplied by a nonzero scalar r.

   c. Answer the question in part (a) if, instead, r times the ith row of A is added to the jth row.

In Exercises 39–42, use the routine YUREDUCE in LINTEK to find the inverse of the matrix, if it exists. If a printer is available, make a copy of the results. Otherwise, copy down the answers to three significant figures.

39. $\begin{bmatrix} 3 & -1 & 2 \\ 1 & 2 & 1 \\ 0 & 3 & -4 \end{bmatrix}$

40. $\begin{bmatrix} -2 & 1 & 4 \\ 3 & 6 & 7 \\ 13 & 15 & -2 \end{bmatrix}$

41. $\begin{bmatrix} 2 & -1 & 3 & 4 \\ -5 & 2 & 0 & 11 \\ 12 & 13 & -6 & 8 \\ 18 & -10 & 3 & 0 \end{bmatrix}$

42. $\begin{bmatrix} 4 & -10 & 3 & 17 \\ 2 & 0 & -3 & 11 \\ 14 & 2 & 12 & -15 \\ 0 & -10 & 9 & -5 \end{bmatrix}$

In Exercises 43–48, follow the instructions for Exercises 39–42, but use the routine MATCOMP in LINTEK. Check to ensure that $AA^{-1} = I$ for each matrix A whose inverse is found.

43. The matrix in Exercise 9

44. The matrix in Exercise 10

45. The matrix in Exercise 41

46. The matrix in Exercise 40

47. $\begin{bmatrix} 4 & 1 & -3 & 2 & 6 \\ 0 & 1 & 5 & 2 & 1 \\ 3 & 8 & -11 & 4 & 6 \\ 2 & 1 & -8 & 7 & 2 \\ 1 & 3 & -1 & 4 & 8 \end{bmatrix}$

48. $\begin{bmatrix} 2 & -1 & 0 & 1 & 6 \\ 3 & -1 & 2 & 4 & 6 \\ 0 & 1 & 3 & 4 & 8 \\ -1 & 1 & 1 & 1 & 8 \\ 3 & 1 & 4 & -11 & 10 \end{bmatrix}$

MATLAB

Access MATLAB and, if the data files for our text are accessible, enter **fbc1s5**. Otherwise, enter these four matrices by hand. [In MATLAB, ln(x) is denoted by **log(x)**.]

$$A = \begin{bmatrix} -2 & 3 & 2/7 \\ \pi/2 & 1 & 3.2 \\ 5 & -6 & 1.3 \end{bmatrix}, \quad B = \begin{bmatrix} 3\pi & \cos 2 & 21/8 \\ \sqrt{7} & \ln 4 & 2/3 \\ \sqrt{2} & \sin 4 & 8.3 \end{bmatrix},$$

$$C = \begin{bmatrix} -3.2 & 1.4 & 5.3 \\ 1.7 & -3.6 & 4.1 \\ 10.3 & 8.5 & -7.6 \end{bmatrix}$$

As you work the problems, write down the entry in the 2nd row, 3rd column position of the answer, with four-significant-figure accuracy, to hand in.

Enter **help inv**, read the information, and then use the function **inv** to work problems M1 through M4.

M1. Compute C^{-3}.

M2. Compute $A^3B^{-2}C$.

M3. Find the matrix X such that $XB = C$.

M4. Find the matrix X such that $B^2XC = A$.

Enter **help /** and then **help **, read the information, and then use **/** and **** rather than the function **inv** to work problems M5 through M8.

M5. Compute $A^{-1}B^2C^{-1}B$.

M6. Compute $B^{-2}CA^{-3}B^3$.

M7. Find the matrix X such that $CX = B^{-2}$.

M8. Find the matrix X such that $AXC^3 = B^4$.

1.6

HOMOGENEOUS SYSTEMS, SUBSPACES, AND BASES

The Solution Set of a Homogeneous System

A linear system $A\mathbf{x} = \mathbf{b}$ is **homogeneous** if $\mathbf{b} = \mathbf{0}$. A homogeneous linear system $A\mathbf{x} = \mathbf{0}$ is always consistent, because $\mathbf{x} = \mathbf{0}$, the zero vector, is certainly a solution. The zero vector is called the **trivial solution**. Other solutions are **nontrivial solutions**. A homogeneous system is special in that its solution set has a self-contained algebraic structure of its own, as we now show.

THEOREM 1.13 Structure of the Solution Set of $A\mathbf{x} = \mathbf{0}$

Let $A\mathbf{x} = \mathbf{0}$ be a homogeneous linear system. If $\mathbf{h}_1$ and $\mathbf{h}_2$ are solutions of $A\mathbf{x} = \mathbf{0}$, then so is the linear combination $r\mathbf{h}_1 + s\mathbf{h}_2$ for any scalars r and s.

PROOF Let $\mathbf{h}_1$ and $\mathbf{h}_2$ be solutions of $A\mathbf{x} = \mathbf{0}$, so that $A\mathbf{h}_1 = \mathbf{0}$ and $A\mathbf{h}_2 = \mathbf{0}$. By the distributive and scalars-pull-through properties of matrix algebra, we have

$$A(r\mathbf{h}_1 + s\mathbf{h}_2) = A(r\mathbf{h}_1) + A(s\mathbf{h}_2)$$
$$= r(A\mathbf{h}_1) + s(A\mathbf{h}_2)$$
$$= r\mathbf{0} + s\mathbf{0} = \mathbf{0}$$

for all scalars r and s. Thus the vector $r\mathbf{h}_1 + s\mathbf{h}_2$ is a solution of the system $A\mathbf{x} = \mathbf{0}$. ▲

Notice how easy it was to write down the proof of Theorem 1.13 in matrix notation. What a chore it would have been to write out the proof using equations with their subscripted variables and coefficients to denote a general $m \times n$ homogeneous system!

Although we stated and proved Theorem 1.13 for just two solutions of $A\mathbf{x} = \mathbf{0}$, either induction or the same proof using k solutions shows that:

Every linear combination of solutions of a homogeneous system $A\mathbf{x} = \mathbf{0}$ is again a solution of the system.

Subspaces

The solution set of a homogeneous system $A\mathbf{x} = \mathbf{0}$ in n unknowns is an example of a subset W of $\mathbb{R}^n$ with the property that every linear combination of vectors in W is again in W. Note that W contains all linear combinations of its vectors if and only if it contains every sum of two of its vectors and every scalar multiple of each of its vectors. We now give a formal definition of a subset of $\mathbb{R}^n$ having such a self-contained algebraic structure. Rather than phrase the definition in terms of linear combinations, we state it in terms of the two basic vector operations, vector addition and scalar multiplication.

DEFINITION 1.16   Closure and Subspace

A subset W of $\mathbb{R}^n$ is **closed under vector addition** if for all $\mathbf{u}, \mathbf{v} \in W$ the sum $\mathbf{u} + \mathbf{v}$ is in W. If $r\mathbf{v} \in W$ for all $\mathbf{v} \in W$ and all scalars r, then W is **closed under scalar multiplication**. A nonempty subset W of $\mathbb{R}^n$ that is closed under both vector addition and scalar multiplication is a **subspace** of $\mathbb{R}^n$.

Theorem 1.13 shows that the solution set of every homogeneous system with n unknowns is a subspace of $\mathbb{R}^n$. We give an example of a subset of $\mathbb{R}^2$ that is a subspace and an example of a subset that is not a subspace.

EXAMPLE 1   Show that $W = \{[x, 2x] \mid x \in \mathbb{R}\}$ is a subspace of $\mathbb{R}^2$.

SOLUTION   Of course, W is a nonempty subset of $\mathbb{R}^2$. Let $\mathbf{u}, \mathbf{v} \in W$ so that $\mathbf{u} = [a, 2a]$ and $\mathbf{v} = [b, 2b]$. Then $\mathbf{u} + \mathbf{v} = [a, 2a] + [b, 2b] = [a + b, 2(a + b)]$ is again of the form $[x, 2x]$, and consequently is in W. This shows that W is closed under vector addition. Because $c\mathbf{u} = c[a, 2a] = [ca, 2(ca)]$ is in W, we see that W is also closed under scalar multiplication, so W is a subspace of $\mathbb{R}^2$.   ∎

You might recognize the subspace W of Example 1 as a line passing through the origin. However, not all lines in $\mathbb{R}^2$ are subspaces. (See Exercises 11 and 14.)

EXAMPLE 2   Determine whether $W = \{[x, y] \in \mathbb{R}^2 \mid xy \geq 0\}$ is a subspace of $\mathbb{R}^2$.

SOLUTION   Here W consists of the vectors in the first or third quadrants (including the coordinate axes), as shown in Figure 1.35. As the figure illustrates, the sum of a vector in the first quadrant and a vector in the third quadrant may be a vector in the second quadrant, so W is not closed under vector addition, and is not a subspace. For a numerical example, $[1, 2] + [-2, -1] = [-1, 1]$, which is not in W.   ∎

Note that the set $\{\mathbf{0}\}$ consisting of just the zero vector in $\mathbb{R}^n$ is a subspace of $\mathbb{R}^n$, because $\mathbf{0} + \mathbf{0} = \mathbf{0}$ and $r\mathbf{0} = \mathbf{0}$ for all scalars r. We refer to $\{\mathbf{0}\}$ as the **zero subspace**. Of course, $\mathbb{R}^n$ itself is a subspace of $\mathbb{R}^n$, because it is closed under vector addition and scalar multiplication. The two subspaces $\{\mathbf{0}\}$ and $\mathbb{R}^n$ represent extremes in size for subspaces of $\mathbb{R}^n$. The next theorem shows one way to form subspaces of various sizes.

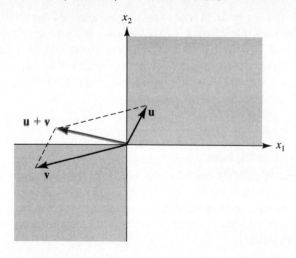

FIGURE 1.35
The shaded subset is not closed under addition.

THEOREM 1.14   Subspace Property of a Span

Let $W = \text{sp}(\mathbf{w}_1, \mathbf{w}_2, \ldots, \mathbf{w}_k)$ be the span of $k > 0$ vectors in $\mathbb{R}^n$. Then W is a subspace of $\mathbb{R}^n$.

PROOF   Let

$$\mathbf{u} = r_1\mathbf{w}_1 + r_2\mathbf{w}_2 + \cdots + r_k\mathbf{w}_k \quad \text{and} \quad \mathbf{v} = s_1\mathbf{w}_1 + s_2\mathbf{w}_2 + \cdots + s_k\mathbf{w}_k$$

be two elements of W. Their sum is

$$\mathbf{u} + \mathbf{v} = (r_1 + s_1)\mathbf{w}_1 + (r_2 + s_2)\mathbf{w}_2 + \cdots + (r_k + s_k)\mathbf{w}_k,$$

which is again a linear combination of $\mathbf{w}_1, \mathbf{w}_2, \ldots, \mathbf{w}_k$, so $\mathbf{u} + \mathbf{v}$ is in W. Thus W is closed under vector addition. Similarly, for any scalar c,

$$c\mathbf{u} = (cr_1)\mathbf{w}_1 + (cr_2)\mathbf{w}_2 + \cdots + (cr_k)\mathbf{w}_k$$

is again in W—that is, W is closed under scalar multiplication. Because $k > 0$, W is also nonempty, so W is a subspace of $\mathbb{R}^n$.   ▲

We say that the vectors $\mathbf{w}_1, \mathbf{w}_2, \ldots, \mathbf{w}_k$ **span** or **generate** the subspace $\text{sp}(\mathbf{w}_1, \mathbf{w}_2, \ldots, \mathbf{w}_k)$ of $\mathbb{R}^n$.

We will see in Section 2.1 that every subspace in $\mathbb{R}^n$ can be described as the span of at most n vectors in $\mathbb{R}^n$. In particular, the solution set of a homogeneous system $A\mathbf{x} = \mathbf{0}$ can always be described as a span of some of the solution vectors. We illustrate how to describe the solution set this way in an example.

EXAMPLE 3   Express the solution set of the homogeneous system

$$\begin{aligned} x_1 - 2x_2 + x_3 - x_4 &= 0 \\ 2x_1 - 3x_2 + 4x_3 - 3x_4 &= 0 \\ 3x_1 - 5x_2 + 5x_3 - 4x_4 &= 0 \\ -x_1 + x_2 - 3x_3 + 2x_4 &= 0 \end{aligned}$$

as a span of solution vectors.

SOLUTION   We reduce the augmented matrix $[A \mid \mathbf{0}]$ to transform the coefficient matrix A of the given system into reduced row-echelon form. We have

$$\left[\begin{array}{cccc|c} 1 & -2 & 1 & -1 & 0 \\ 2 & -3 & 4 & -3 & 0 \\ 3 & -5 & 5 & -4 & 0 \\ -1 & 1 & -3 & 2 & 0 \end{array}\right] \sim \left[\begin{array}{cccc|c} 1 & -2 & 1 & -1 & 0 \\ 0 & 1 & 2 & -1 & 0 \\ 0 & 1 & 2 & -1 & 0 \\ 0 & -1 & -2 & 1 & 0 \end{array}\right] \sim \left[\begin{array}{cccc|c} 1 & 0 & 5 & -3 & 0 \\ 0 & 1 & 2 & -1 & 0 \\ 0 & 0 & 0 & 0 & 0 \\ 0 & 0 & 0 & 0 & 0 \end{array}\right].$$

The reduction is complete. Notice that we didn't really need to insert the column vector $\mathbf{0}$ in the augmented matrix, because it never changes.

From the reduced matrix, we find that the homogeneous system has two free variables and has a solution set described by the *general solution vector*

$$\mathbf{x} = \begin{bmatrix} x_1 \\ x_2 \\ x_3 \\ x_4 \end{bmatrix} = \begin{bmatrix} -5r + 3s \\ -2r + s \\ r \\ s \end{bmatrix} = r\begin{bmatrix} -5 \\ -2 \\ 1 \\ 0 \end{bmatrix} + s\begin{bmatrix} 3 \\ 1 \\ 0 \\ 1 \end{bmatrix}. \tag{1}$$

Thus the solution set is

$$\text{sp}\left(\begin{bmatrix} -5 \\ -2 \\ 1 \\ 0 \end{bmatrix}, \begin{bmatrix} 3 \\ 1 \\ 0 \\ 1 \end{bmatrix}\right).$$

We chose these two generating vectors from Eq. (1) by taking $r = 1$, $s = 0$ for the first and $r = 0$, $s = 1$ for the second.   ∎

The preceding example indicates how we can always express the entire solution set of a homogeneous system with k free variables as the span of k solution vectors.

Given an $m \times n$ matrix A, there are three natural subspaces of $\mathbb{R}^m$ or $\mathbb{R}^n$ associated with it. Recall (Theorem 1.14) that a span of vectors in $\mathbb{R}^n$ is always a subspace of $\mathbb{R}^n$. The span of the row vectors of A is the **row space** of A, and is of course a subspace of $\mathbb{R}^n$. The span of the column vectors of A is the **column space** of A and is a subspace of $\mathbb{R}^m$. The solution set of $A\mathbf{x} = \mathbf{0}$, which we have been discussing, is the **nullspace** of A and is a subspace of $\mathbb{R}^n$. For example, if

$$A = \begin{bmatrix} 1 & 0 & 3 \\ 0 & 1 & -1 \end{bmatrix},$$

we see that

the row space of A is sp($[1, 0, 3]$, $[0, 1, -1]$) in $\mathbb{R}^3$,

the column space of A is sp$\left(\begin{bmatrix} 1 \\ 0 \end{bmatrix}, \begin{bmatrix} 0 \\ 1 \end{bmatrix}, \begin{bmatrix} 3 \\ -1 \end{bmatrix}\right)$ in $\mathbb{R}^2$, and

the nullspace of A is sp$\left(\begin{bmatrix} -3 \\ 1 \\ 1 \end{bmatrix}\right)$ in $\mathbb{R}^3$.

The nullspace of A was readily found because A is in reduced row-echelon form.

In Section 1.3, we emphasized that for an $m \times n$ matrix A and $\mathbf{x} \in \mathbb{R}^n$, the vector $A\mathbf{x}$ is a linear combination of the column vectors of A. In Section 1.4 we saw that the system $A\mathbf{x} = \mathbf{b}$ has a solution if and only if $\mathbf{b}$ is equal to some linear combination of the column vectors of A. We rephrase this criterion for existence of a solution of $A\mathbf{x} = \mathbf{b}$ in terms of the column space of A.

Column Space Criterion

A linear system $A\mathbf{x} = \mathbf{b}$ has a solution if and only if $\mathbf{b}$ is in the column space of A.

We have discussed the significance of the nullspace and the column space of a matrix. The row space of A is significant because the row vectors of A are orthogonal to the vectors in the nullspace of A, as the ith equation in the system $A\mathbf{x} = \mathbf{0}$ shows. This observation will be useful when we compute projections in Section 6.1.

Bases

We have seen how the solution set of a homogeneous linear system can be expressed as the span of certain selected solution vectors. Look again at Eq. (1), which shows the solution set of the linear system in Example 3 to be sp($\mathbf{w}_1, \mathbf{w}_2$) for

$$\mathbf{w}_1 = \begin{bmatrix} -5 \\ -2 \\ 1 \\ 0 \end{bmatrix} \quad \text{and} \quad \mathbf{w}_2 = \begin{bmatrix} 3 \\ 1 \\ 0 \\ 1 \end{bmatrix}.$$

The last two components of these vectors are 1, 0 for $\mathbf{w}_1$ and 0, 1 for $\mathbf{w}_2$. These components mirror the vectors $\mathbf{i} = [1, 0]$ and $\mathbf{j} = [0, 1]$ in the plane. Now the vector $[r, s]$ in the plane can be expressed *uniquely* as a linear combination of $\mathbf{i}$ and $\mathbf{j}$—namely, as $r\mathbf{i} + s\mathbf{j}$. Thus we see that every solution vector in Eq. (1) of the linear system in Example 3 is a *unique* linear combination of $\mathbf{w}_1$ and $\mathbf{w}_2$—namely, $r\mathbf{w}_1 + s\mathbf{w}_2$. We can think of (r, s) as being *coordinates* of the solution relative to $\mathbf{w}_1$ and $\mathbf{w}_2$. Because we regard all ordered pairs of numbers as filling a plane, this indicates how we might regard the solution set of this system as a plane in $\mathbb{R}^4$. We can think of $\{\mathbf{w}_1, \mathbf{w}_2\}$ as a set of reference vectors for the plane. We have depicted this plane in Figure 1.36

More generally, if every vector $\mathbf{w}$ in the subspace $W = \mathrm{sp}(\mathbf{w}_1, \mathbf{w}_2, \ldots, \mathbf{w}_k)$ of $\mathbb{R}^n$ can be expressed as a linear combination $\mathbf{w} = c_1\mathbf{w}_1 + c_2\mathbf{w}_2 + \cdots + c_k\mathbf{w}_k$ in a *unique* way, then we can consider the ordered k-tuple $(c_1, c_2, \ldots, c_k)$ in $\mathbb{R}^k$ to be the *coordinates* of $\mathbf{w}$. The set $\{\mathbf{w}_1, \mathbf{w}_2, \ldots, \mathbf{w}_k\}$ is considered to be a set of reference vectors for the subspace W. Such a set is known as a *basis* for W, as the next definition indicates.

DEFINITION 1.17   Basis for a Subspace

Let W be a subspace of $\mathbb{R}^n$. A subset $\{\mathbf{w}_1, \mathbf{w}_2, \ldots, \mathbf{w}_k\}$ of W is a **basis for** W if every vector in W can be expressed *uniquely* as a linear combination of $\mathbf{w}_1, \mathbf{w}_2, \ldots, \mathbf{w}_k$.

Our discussion following Example 3 shows that the two vectors

$$\mathbf{w}_1 = \begin{bmatrix} -5 \\ -2 \\ 1 \\ 0 \end{bmatrix} \quad \text{and} \quad \mathbf{w}_2 = \begin{bmatrix} 3 \\ 1 \\ 0 \\ 1 \end{bmatrix}$$

form a basis for the solution space of the homogeneous system there.

If $\{\mathbf{w}_1, \mathbf{w}_2, \ldots, \mathbf{w}_k\}$ is a basis for W, then we have $W = \mathrm{sp}(\mathbf{w}_1, \mathbf{w}_2, \ldots, \mathbf{w}_k)$ as well as the uniqueness requirement. Remember that we called $\mathbf{e}_1, \mathbf{e}_2, \ldots, \mathbf{e}_n$ *standard basis vectors* for $\mathbb{R}^n$. The reason for this is that every element of $\mathbb{R}^n$ can be expressed *uniquely* as a linear combination of these vectors $\mathbf{e}_i$. We call $\{\mathbf{e}_1, \mathbf{e}_2, \ldots, \mathbf{e}_n\}$ the **standard basis** for $\mathbb{R}^n$.

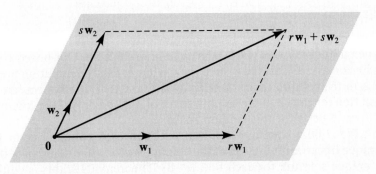

FIGURE 1.36
The plane $\mathrm{sp}(\mathbf{w}_1, \mathbf{w}_2)$

We would like to be able to determine whether $\{\mathbf{w}_1, \mathbf{w}_2, \ldots, \mathbf{w}_k\}$ is a basis for the subspace $W = \mathrm{sp}(\mathbf{w}_1, \mathbf{w}_2, \ldots, \mathbf{w}_k)$ of $\mathbb{R}^n$—that is, whether the *uniqueness* criterion holds. The next theorem will be helpful. It shows that we need only examine uniqueness at the zero vector.

THEOREM 1.15 Unique Linear Combinations

The set $\{\mathbf{w}_1, \mathbf{w}_2, \ldots, \mathbf{w}_k\}$ is a basis for $W = \mathrm{sp}(\mathbf{w}_1, \mathbf{w}_2, \ldots, \mathbf{w}_k)$ in $\mathbb{R}^n$ if and only if the zero vector is a *unique* linear combination of the $\mathbf{w}_i$—that is, if and only if $r_1\mathbf{w}_1 + r_2\mathbf{w}_2 + \cdots + r_k\mathbf{w}_k = \mathbf{0}$ implies that $r_1 = r_2 = \cdots = r_k = 0$.

PROOF   If $\{\mathbf{w}_1, \mathbf{w}_2, \ldots, \mathbf{w}_k\}$ is a basis for W, then the expression for every vector in W as a linear combination of the $\mathbf{w}_i$ is unique, so, in particular, the linear combination that gives the zero vector must be unique. Because $0\mathbf{w}_1 + 0\mathbf{w}_2 + \cdots + 0\mathbf{w}_k = \mathbf{0}$, it follows that $r_1\mathbf{w}_1 + r_2\mathbf{w}_2 + \cdots + r_k\mathbf{w}_k = \mathbf{0}$ implies that each r_i must be 0.

Conversely, suppose that $0\mathbf{w}_1 + 0\mathbf{w}_2 + \cdots + 0\mathbf{w}_k$ is the only linear combination giving the zero vector. If we have two linear combinations

$$\mathbf{w} = c_1\mathbf{w}_1 + c_2\mathbf{w}_2 + \cdots + c_k\mathbf{w}_k$$
$$\mathbf{w} = d_1\mathbf{w}_1 + d_2\mathbf{w}_2 + \cdots + d_k\mathbf{w}_k$$

for a vector $\mathbf{w} \in W$, then, subtracting these two equations, we obtain

$$\mathbf{0} = (c_1 - d_1)\mathbf{w}_1 + (c_2 - d_2)\mathbf{w}_2 + \cdots + (c_k - d_k)\mathbf{w}_k.$$

From the unique linear combination giving the zero vector, we see that

$$c_1 - d_1 = c_2 - d_2 = \cdots = c_k - d_k = 0,$$

and so $c_i = d_i$ for $i = 1, 2, \ldots, k$, showing that the linear combination giving $\mathbf{w}$ is unique.   ▲

The Unique Solution Case for $A\mathbf{x} = \mathbf{b}$

The preceding theorem immediately focuses our attention on determining when a linear system $A\mathbf{x} = \mathbf{b}$ has a unique solution. Our boxed column space criterion asserts that the system $A\mathbf{x} = \mathbf{b}$ has *at least one* solution precisely when $\mathbf{b}$ is in the column space of A. By Definition 1.17, the system has *exactly one* solution for each $\mathbf{b}$ in the column space of A if and only if the column vectors of A form a basis for this column space.

Let A be a square $n \times n$ matrix. Then each column vector $\mathbf{b}$ in $\mathbb{R}^n$ is a unique linear combination of the column vectors of A if and only if $A\mathbf{x} = \mathbf{b}$ has a unique solution for each $\mathbf{b} \in \mathbb{R}^n$. By Theorem 1.12, this is equivalent to A being row-reducible to the identity matrix, so that A is invertible. We have now established another equivalent condition to add to those in Theorem 1.12. We summarize the main ones in a new theorem for easy reference.

THEOREM 1.16 The Square Case, $m = n$

Let A be an $n \times n$ matrix. The following are equivalent.

1. The linear system $A\mathbf{x} = \mathbf{b}$ has a unique solution for each $\mathbf{b} \in \mathbb{R}^n$.
2. The matrix A is row equivalent to the identity matrix I.
3. The matrix A is invertible.
4. The column vectors of A form a basis for $\mathbb{R}^n$.

In describing examples and stating exercises, we often write vectors in $\mathbb{R}^n$ as row vectors to save space. However, the vector $\mathbf{b} = A\mathbf{x}$ in Theorem 1.16 is necessarily a column vector. Thus, solutions to examples and exercises that use results such as Theorem 1.16 are done using column-vector notation, as our next example illustrates.

EXAMPLE 4　Determine whether the vectors $\mathbf{v}_1 = [1, 1, 3]$, $\mathbf{v}_2 = [3, 0, 4]$, and $\mathbf{v}_3 = [1, 4, -1]$ form a basis for $\mathbb{R}^3$.

SOLUTION　We must see whether the matrix A having $\mathbf{v}_1$, $\mathbf{v}_2$, and $\mathbf{v}_3$ as column vectors is row equivalent to the identity matrix. We need only create zeros below pivots to determine if this is the case. We obtain

$$A = \begin{bmatrix} 1 & 3 & 1 \\ 1 & 0 & 4 \\ 3 & 4 & -1 \end{bmatrix} \sim \begin{bmatrix} 1 & 3 & 1 \\ 0 & -3 & 3 \\ 0 & -5 & -4 \end{bmatrix} \sim \begin{bmatrix} 1 & 3 & 1 \\ 0 & 1 & -1 \\ 0 & 0 & -9 \end{bmatrix}.$$

There is no point in going further. We see that we will be able to get the identity matrix, so $\{\mathbf{v}_1, \mathbf{v}_2, \mathbf{v}_3\}$ is a basis for $\mathbb{R}^3$. ■

A linear system having the same number n of equations as unknowns is called a **square system**, because the coefficient matrix is a square $n \times n$ matrix. When a square matrix is reduced to echelon form, the result is a square matrix having only zero entries below the *main diagonal*, which runs from the upper left-hand corner to the lower right-hand corner. This follows at once from the fact that the pivot in a nonzero row—say, the ith row—is always in a column j, where $j \geq i$. Such a square matrix U with zero entries below the main diagonal is called **upper triangular**. The final matrix displayed in Example 4 is upper triangular.

For a general linear system $A\mathbf{x} = \mathbf{b}$ of m equations in n unknowns, we consult Theorem 1.7. It tells us that a consistent system $A\mathbf{x} = \mathbf{b}$ has a *unique* solution if and only if a row-echelon form H of A has a pivot in each of its n columns. Because no two pivots appear in the same row of H, we see that H has at least as many rows as columns; that is, $m \geq n$. Consequently, the reduced row-echelon form for A must consist of the identity matrix, followed by $m - n$

zero rows. For example, if $m = 5$ and $n = 3$, the reduced row-echelon form for A in this unique solution case must be

$$\begin{bmatrix} 1 & 0 & 0 \\ 0 & 1 & 0 \\ 0 & 0 & 1 \\ 0 & 0 & 0 \\ 0 & 0 & 0 \end{bmatrix}.$$

We summarize these observations as a theorem.

THEOREM 1.17 The General Unique Solution Case

Let A be an $m \times n$ matrix. The following are equivalent.

1. Each consistent system $A\mathbf{x} = \mathbf{b}$ has a unique solution.
2. The reduced row-echelon form of A consists of the $n \times n$ identity matrix followed by $m - n$ rows of zeros.
3. The column vectors of A form a basis for the column space of A.

EXAMPLE 5   Determine whether the vectors $\mathbf{w}_1 = [1, 2, 3, -1]$, $\mathbf{w}_2 = [-2, -3, -5, 1]$, and $\mathbf{w}_3 = [-1, -3, -4, 2]$ form a basis for the subspace $\text{sp}(\mathbf{w}_1, \mathbf{w}_2, \mathbf{w}_3)$ in $\mathbb{R}^4$.

SOLUTION   By Theorem 1.17, we need to determine whether the reduced row-echelon form of the matrix A with $\mathbf{w}_1$, $\mathbf{w}_2$, and $\mathbf{w}_3$ as column vectors consists of the 3×3 identity matrix followed by a row of zeros. Again, we can determine this using just the row-echelon form, without creating zeros above the pivots. We obtain

$$A = \begin{bmatrix} 1 & -2 & -1 \\ 2 & -3 & -3 \\ 3 & -5 & -4 \\ -1 & 1 & 2 \end{bmatrix} \sim \begin{bmatrix} 1 & -2 & -1 \\ 0 & 1 & -1 \\ 0 & 1 & -1 \\ 0 & -1 & 1 \end{bmatrix} \sim \begin{bmatrix} 1 & -2 & -1 \\ 0 & 1 & -1 \\ 0 & 0 & 0 \\ 0 & 0 & 0 \end{bmatrix}.$$

We cannot obtain the 3×3 identity matrix. Thus the vectors do not form a basis for the subspace which is the column space of A.   ∎

A linear system having an infinite number of solutions is called **underdetermined**. We now prove a corollary of the preceding theorem: that a consistent system is underdetermined if it has fewer equations than unknowns.

COROLLARY 1   Fewer Equations than Unknowns, $m < n$

If a linear system $A\mathbf{x} = \mathbf{b}$ is consistent and has fewer equations than unknowns, then it has an infinite number of solutions.

PROOF   If $m < n$ in Theorem 1.17, the reduced row-echelon form of A cannot contain the $n \times n$ identity matrix, so we cannot be in the unique solution case. Because we are assuming that the system is consistent, there are an infinite number of solutions.   ▲

The next corollary follows at once from Corollary 1 and Theorem 1.17.

COROLLARY 2   The Homogeneous Case

1. A homogeneous linear system $A\mathbf{x} = \mathbf{0}$ having fewer equations than unknowns has a nontrivial solution—that is, a solution other than the zero vector.
2. A square homogeneous system $A\mathbf{x} = \mathbf{0}$ has a nontrivial solution if and only if A is not row equivalent to the identity matrix of the same size.

EXAMPLE 6   Show that a basis for $\mathbb{R}^n$ cannot contain more than n vectors.

SOLUTION   If $\{\mathbf{v}_1, \mathbf{v}_2, \ldots, \mathbf{v}_k\}$ is a basis for $\mathbb{R}^n$, then, by Theorem 1.15, the only linear combination of the $\mathbf{v}_j$ equal to the zero vector is the one for which the coefficient of each $\mathbf{v}_j$ is the scalar 0. In terms of a matrix equation, the homogeneous linear system $A\mathbf{x} = \mathbf{0}$, where $\mathbf{v}_j$ is the jth column vector of the $n \times k$ matrix A, must have only the trivial solution. If $k > n$, then this linear system has fewer equations than unknowns, and therefore a nontrivial solution by Corollary 2. Consequently, we must have $k \leq n$.   ■

The Solution Set of $A\mathbf{x} = \mathbf{b}$

Theorem 1.13 tells us the structure of the solution set of a homogeneous linear system. To conclude the section, we now describe the solution set of $A\mathbf{x} = \mathbf{b}$ in terms of the solution set of the *corresponding homogeneous system* $A\mathbf{x} = \mathbf{0}$. It is customary to refer to an equation that describes the whole solution set as the **general solution**, and to refer to each element of the solution set as a **particular solution**.

THEOREM 1.18   Structure of the Solution Set of $A\mathbf{x} = \mathbf{b}$

Let $A\mathbf{x} = \mathbf{b}$ be a linear system. If $\mathbf{p}$ is any particular solution of $A\mathbf{x} = \mathbf{b}$ and $\mathbf{h}$ is a solution of the corresponding homogeneous system $A\mathbf{x} = \mathbf{0}$, then $\mathbf{p} + \mathbf{h}$ is a solution of $A\mathbf{x} = \mathbf{b}$. Moreover, every solution of $A\mathbf{x} = \mathbf{b}$ has this form $\mathbf{p} + \mathbf{h}$, so that the general solution is $\mathbf{x} = \mathbf{p} + \mathbf{h}$ where $A\mathbf{h} = \mathbf{0}$.

PROOF   Let $\mathbf{p}$ be a solution of $A\mathbf{x} = \mathbf{b}$, so that $A\mathbf{p} = \mathbf{b}$, and let $\mathbf{h}$ be a solution of $A\mathbf{x} = \mathbf{0}$, so that $A\mathbf{h} = \mathbf{0}$. Then

$$A(\mathbf{p} + \mathbf{h}) = A\mathbf{p} + A\mathbf{h} = \mathbf{b} + \mathbf{0} = \mathbf{b},$$

and so $\mathbf{p} + \mathbf{h}$ is indeed a solution. Moreover, if $\mathbf{q}$ is any solution of $A\mathbf{x} = \mathbf{b}$, then

$$A(\mathbf{q} - \mathbf{p}) = A\mathbf{q} - A\mathbf{p} = \mathbf{b} - \mathbf{b} = \mathbf{0},$$

and so $\mathbf{q} - \mathbf{p}$ is a solution $\mathbf{h}$ of $A\mathbf{x} = \mathbf{0}$. From $\mathbf{q} - \mathbf{p} = \mathbf{h}$, it follows that $\mathbf{q} = \mathbf{p} + \mathbf{h}$. This completes the proof.   ▲

Students who have studied differential equations may be familiar with a similar theorem describing the general solution of a linear differential equation.

EXAMPLE 7   Illustrate Theorem 1.18 for the linear system $A\mathbf{x} = \mathbf{b}$ given by

$$
\begin{aligned}
x_1 - 2x_2 + x_3 - x_4 &= 4 \\
2x_1 - 3x_2 + 4x_3 - 3x_4 &= -1 \\
3x_1 - 5x_2 + 5x_3 - 4x_4 &= 3 \\
-x_1 + x_2 - 3x_3 + 2x_4 &= 5.
\end{aligned}
$$

SOLUTION   We reduce the augmented matrix $[A \mid \mathbf{b}]$ to transform the coefficient matrix A of the given system into reduced row-echelon form. (The coefficient matrix A is the same as in Example 3.) We have

$$
\begin{bmatrix}
1 & -2 & 1 & -1 & 4 \\
2 & -3 & 4 & -3 & -1 \\
3 & -5 & 5 & -4 & 3 \\
-1 & 1 & -3 & 2 & 5
\end{bmatrix}
\sim
\begin{bmatrix}
1 & -2 & 1 & -1 & 4 \\
0 & 1 & 2 & -1 & -9 \\
0 & 1 & 2 & -1 & -9 \\
0 & -1 & -2 & 1 & 9
\end{bmatrix}
\sim
\begin{bmatrix}
1 & 0 & 5 & -3 & -14 \\
0 & 1 & 2 & -1 & -9 \\
0 & 0 & 0 & 0 & 0 \\
0 & 0 & 0 & 0 & 0
\end{bmatrix}.
$$

Writing the general solution in the usual form and then as described in Theorem 1.18, we have

$$
\mathbf{x} = \begin{bmatrix} x_1 \\ x_2 \\ x_3 \\ x_4 \end{bmatrix} = \begin{bmatrix} -14 - 5r + 3s \\ -9 - 2r + s \\ r \\ s \end{bmatrix} = \underbrace{\begin{bmatrix} -14 \\ -9 \\ 0 \\ 0 \end{bmatrix}}_{\mathbf{p}} + \underbrace{\begin{bmatrix} -5r + 3s \\ -2r + s \\ r \\ s \end{bmatrix}}_{\mathbf{h}}.
$$

General solution   ■

SUMMARY

1. A linear system $A\mathbf{x} = \mathbf{b}$ is *homogeneous* if $\mathbf{b} = \mathbf{0}$.

2. Every linear combination of solutions of a homogeneous system $A\mathbf{x} = \mathbf{0}$ is again a solution of the system.

3. A subset W of $\mathbb{R}^n$ is *closed under vector addition* if the sum of two vectors in W is again in W. The subset W is *closed under scalar multiplication* if every scalar multiple of every vector in W is in W. If W is nonempty and closed under both operations, then W is a *subspace* of $\mathbb{R}^n$.

4. The span of any k vectors in $\mathbb{R}^n$ is a subspace of $\mathbb{R}^n$. If A is an $m \times n$ matrix, the *row space* of A is the span in $\mathbb{R}^n$ of the row vectors of A, the *column space* of A is the span in $\mathbb{R}^m$ of the column vectors, and the *nullspace* of A is the solution set of $A\mathbf{x} = \mathbf{0}$ in $\mathbb{R}^n$.

5. A subset $\{\mathbf{w}_1, \mathbf{w}_2, \ldots, \mathbf{w}_k\}$ of a subspace W of $\mathbb{R}^n$ is a *basis for W* if every vector in W can be expressed uniquely as a linear combination of $\mathbf{w}_1, \mathbf{w}_2, \ldots, \mathbf{w}_k$.

6. The set $\{\mathbf{w}_1, \mathbf{w}_2, \ldots, \mathbf{w}_k\}$ is a basis for $\mathrm{sp}(\mathbf{w}_1, \mathbf{w}_2, \ldots, \mathbf{w}_k)$ if and only if $0\mathbf{w}_1 + 0\mathbf{w}_2 + \cdots + 0\mathbf{w}_k$ is the unique linear combination of the $\mathbf{w}_i$ that is equal to the zero vector.

7. A consistent linear system $A\mathbf{x} = \mathbf{b}$ of m equations in n unknowns has a unique solution if and only if the reduced row-echelon form of A appears as the $n \times n$ identity matrix followed by $m - n$ rows of zeros.

8. A consistent linear system having fewer equations than unknowns is underdetermined—that is, it has an infinite number of solutions.

9. A square linear system has a unique solution if and only if its coefficient matrix is row equivalent to the identity matrix.

10. The solutions of any consistent linear system $A\mathbf{x} = \mathbf{b}$ are precisely the vectors $\mathbf{p} + \mathbf{h}$, where $\mathbf{p}$ is any one particular solution of $A\mathbf{x} = \mathbf{b}$ and $\mathbf{h}$ varies through the solution set of the homogeneous system $A\mathbf{x} = \mathbf{0}$.

EXERCISES

In Exercises 1–10, determine whether the indicated subset is a subspace of the given Euclidean space $\mathbb{R}^n$.

1. $\{[r, -r] \mid r \in \mathbb{R}\}$ in $\mathbb{R}^2$
2. $\{[x, x + 1] \mid x \in \mathbb{R}\}$ in $\mathbb{R}^2$
3. $\{[n, m] \mid n \text{ and } m \text{ are integers}\}$ in $\mathbb{R}^2$
4. $\{[x, y] \mid x,y \in \mathbb{R} \text{ and } x,y \geq 0\}$ (the first quadrant of $\mathbb{R}^2$)
5. $\{[x, y, z] \mid x,y,z \in \mathbb{R} \text{ and } z = 3x + 2\}$ in $\mathbb{R}^3$
6. $\{[x, y, z] \mid x,y,z \in \mathbb{R} \text{ and } x = 2y + z\}$ in $\mathbb{R}^3$
7. $\{[x, y, z] \mid x,y,z \in \mathbb{R} \text{ and } z = 1, y = 2x\}$ in $\mathbb{R}^3$
8. $\{[2x, x + y, y] \mid x,y \in \mathbb{R}\}$ in $\mathbb{R}^3$
9. $\{[2x_1, 3x_2, 4x_3, 5x_4] \mid x_i \in \mathbb{R}\}$ in $\mathbb{R}^4$
10. $\{[x_1, x_2, \ldots, x_n] \mid x_i \in \mathbb{R}, x_2 = 0\}$ in $\mathbb{R}^n$
11. Prove that the line $y = mx$ is a subspace of $\mathbb{R}^2$. [HINT: Write the line as $W = \{[x, mx] \mid x \in \mathbb{R}\}$.]

12. Let a, b, and c be scalars such that $abc \neq 0$. Prove that the plane $ax + by + cz = 0$ is a subspace of $\mathbb{R}^3$.

13. a. Give a geometric description of all subspaces of $\mathbb{R}^2$.
    b. Repeat part (a) for $\mathbb{R}^3$.

14. Prove that every subspace of $\mathbb{R}^n$ contains the zero vector.

15. Is the zero vector a basis for the subspace $\{\mathbf{0}\}$ of $\mathbb{R}^n$? Why or why not?

In Exercises 16–21, find a basis for the solution set of the given homogeneous linear system.

16. $\begin{aligned} x - y &= 0 \\ 2x - 2y &= 0 \end{aligned}$

17. $\begin{aligned} 3x_1 + x_2 + x_3 &= 0 \\ 6x_1 + 2x_2 + 2x_3 &= 0 \\ -9x_1 - 3x_2 - 3x_3 &= 0 \end{aligned}$

18. $\begin{aligned} x_1 - x_2 + x_3 - x_4 &= 0 \\ x_2 + x_3 &= 0 \\ x_1 + 2x_2 - x_3 + 3x_4 &= 0 \end{aligned}$

19. $\begin{aligned} 2x_1 + x_2 + x_3 + x_4 &= 0 \\ x_1 - 6x_2 + x_3 &= 0 \\ 3x_1 - 5x_2 + 2x_3 + x_4 &= 0 \\ 5x_1 - 4x_2 + 3x_3 + 2x_4 &= 0 \end{aligned}$

20. $\begin{aligned} 2x_1 + x_2 + x_3 + x_4 &= 0 \\ 3x_1 + x_2 - x_3 + 2x_4 &= 0 \\ x_1 + x_2 + 3x_3 &= 0 \\ x_1 - x_2 - 7x_3 + 2x_4 &= 0 \end{aligned}$

21. $\begin{aligned} x_1 - x_2 + 6x_3 + x_4 - x_5 &= 0 \\ 3x_1 + 2x_2 - 3x_3 + 2x_4 + 5x_5 &= 0 \\ 4x_1 + 2x_2 - x_3 + 3x_4 - x_5 &= 0 \\ 3x_1 - 2x_2 + 14x_3 + x_4 - 8x_5 &= 0 \\ 2x_1 - x_2 + 8x_3 + 2x_4 - 7x_5 &= 0 \end{aligned}$

In Exercises 22–30, determine whether the set of vectors is a basis for the subspace of $\mathbb{R}^n$ that the vectors span.

22. $\{[-1, 1], [1, 2]\}$ in $\mathbb{R}^2$

23. $\{[-1, 3, 1], [2, 1, 4]\}$ in $\mathbb{R}^3$

24. $\{[-1, 3, 4], [1, 5, -1], [1, 13, 2]\}$ in $\mathbb{R}^3$

25. $\{[2, 1, -3], [4, 0, 2], [2, -1, 3]\}$ in $\mathbb{R}^3$

26. $\{[2, 1, 0, 2], [2, -3, 1, 0], [3, 2, 0, 0]\}$ in $\mathbb{R}^4$

27. The set of row vectors of the matrix

$$\begin{bmatrix} 2 & -6 & 1 \\ 1 & -3 & 4 \end{bmatrix}.$$

28. The set of column vectors of the matrix in Exercise 27.

29. The set of row vectors of the matrix

$$\begin{bmatrix} 1 & -1 & 0 \\ 0 & 1 & 2 \\ 2 & 1 & -3 \\ 1 & -3 & 4 \end{bmatrix}.$$

30. The set of column vectors of the matrix in Exercise 29.

31. Find a basis for the nullspace of the matrix

$$\begin{bmatrix} 2 & 3 & 1 \\ 5 & 2 & 1 \\ 1 & 7 & 2 \\ 6 & -2 & 0 \end{bmatrix}.$$

32. Find a basis for the nullspace of the matrix

$$\begin{bmatrix} 1 & 3 & 5 & 7 \\ 2 & 0 & 4 & 2 \\ 3 & 2 & 8 & 7 \end{bmatrix}.$$

33. Let $v_1, v_2, \ldots, v_k$ and $w_1, w_2, \ldots, w_m$ be vectors in a vector space V. Give a necessary and sufficient condition, involving linear combinations, for

$$\mathrm{sp}(v_1, v_2, \ldots, v_k) = \mathrm{sp}(w_1, w_2, \ldots, w_m).$$

In Exercises 34–37, solve the given linear system and express the solution set in a form that illustrates Theorem 1.18.

34. $x_1 - 2x_2 + x_3 + 5x_4 = 7$

35. $\begin{aligned} 2x_1 - x_2 + 3x_3 &= -3 \\ 4x_1 + 2x_2 - x_4 &= 1 \end{aligned}$

36. $\begin{aligned} x_1 - 2x_2 + x_3 + x_4 &= 4 \\ 2x_1 + x_2 - 3x_3 - x_4 &= 6 \\ x_1 - 7x_2 - 6x_3 + 2x_4 &= 6 \end{aligned}$

37. $\begin{aligned} 2x_1 + x_2 + 3x_3 &= 5 \\ x_1 - x_2 + 2x_3 + x_4 &= 0 \\ 4x_1 - x_2 + 7x_3 + 2x_4 &= 5 \\ -x_1 - 2x_2 - x_3 + x_4 &= -5 \end{aligned}$

38. Mark each of the following True or False.

___ **a.** A linear system with fewer equations than unknowns has an infinite number of solutions.

___ **b.** A consistent linear system with fewer equations than unknowns has an infinite number of solutions.

___ **c.** If a square linear system $Ax = b$ has a solution for *every* choice of column vector b, then the solution is unique for each b.

___ **d.** If a square system $Ax = 0$ has only the trivial soltuion, then $Ax = b$ has a unique solution for every column vector b with the appropriate number of components.

___ **e.** If a linear system $Ax = 0$ has only the trivial solution, then $Ax = b$ has a unique solution for every column vector b with the appropriate number of components.

___ **f.** The sum of two solution vectors of any linear system is also a solution vector of the system.

____ **g.** The sum of two solution vectors of any homogeneous linear system is also a solution vector of the system.

____ **h.** A scalar multiple of a solution vector of any homogeneous linear system is also a solution vector of the system.

____ **i.** Every line in $\mathbb{R}^2$ is a subspace of $\mathbb{R}^2$ generated by a single vector.

____ **j.** Every line through the origin in $\mathbb{R}^2$ is a subspace of $\mathbb{R}^2$ generated by a single vector.

39. We have defined a linear system to be *underdetermined* if it has an infinite number of solutions. Explain why this is a reasonable term to use for such a system.

40. A linear system is **overdetermined** if it has more equations than unknowns. Explain why this is a reasonable term to use for such a system.

41. Referring to Exercises 39 and 40, give an example of an overdetermined underdetermined linear system!

42. Use Theorem 1.13 to explain why a homogeneous system of linear equations has either a unique solution or an infinite number of solutions.

43. Use Theorem 1.18 to explain why no system of linear equations can have exactly two solutions.

44. Let A be an $m \times n$ matrix such that the homogeneous system $A\mathbf{x} = \mathbf{0}$ has only the trivial solution.

    a. Does it follow that every system $A\mathbf{x} = \mathbf{b}$ is consistent?

    b. Does it follow that every consistent system $A\mathbf{x} = \mathbf{b}$ has a unique solution?

45. Let $\mathbf{v}_1$ and $\mathbf{v}_2$ be vectors in $\mathbb{R}^n$. Prove the following set equalities by showing that each of the spans is contained in the other.

    a. $\mathrm{sp}(\mathbf{v}_1, \mathbf{v}_2) = \mathrm{sp}(\mathbf{v}_1, 2\mathbf{v}_1 + \mathbf{v}_2)$

    b. $\mathrm{sp}(\mathbf{v}_1, \mathbf{v}_2) = \mathrm{sp}(\mathbf{v}_1 + \mathbf{v}_2, \mathbf{v}_1 - \mathbf{v}_2)$

46. Referring to Exercise 45, prove that if $\{v_1, v_2\}$ is a basis for $\mathrm{sp}(\mathbf{v}_1, \mathbf{v}_2)$, then

    a. $\{\mathbf{v}_1, 2\mathbf{v}_1 + \mathbf{v}_2\}$ is also a basis.

    b. $\{\mathbf{v}_1 + \mathbf{v}_2, \mathbf{v}_1 - \mathbf{v}_2\}$ is also a basis.

    c. $\{\mathbf{v}_1 + \mathbf{v}_2, \mathbf{v}_1 - \mathbf{v}_2, 2\mathbf{v}_1 - 3v_2\}$ is not a basis.

47. Let W_1 and W_2 be two subspaces of $\mathbb{R}^n$. Prove that their intersection $W_1 \cap W_2$ is also a subspace.

In Exercises 48–51, use LINTEK or MATLAB to determine whether the given vectors form a basis for the subspace of $\mathbb{R}^n$ that they span.

48. $\mathbf{a}_1 = [1, 1, -1, 0]$

    $\mathbf{a}_2 = [5, 1, 1, 2]$

    $\mathbf{a}_3 = [5, -3, 2, -1]$

    $\mathbf{a}_4 = [9, 3, 0, 3]$

49. $\mathbf{b}_1 = [3, -4, 0, 0, 1]$

    $\mathbf{b}_2 = [4, 0, 2, -6, 2]$

    $\mathbf{b}_3 = [0, 1, 1, -3, 0]$

    $\mathbf{b}_4 = [1, 4, -1, 3, 0]$

50. $\mathbf{v}_1 = [4, -1, 2, 1]$

    $\mathbf{v}_2 = [10, -2, 5, 1]$

    $\mathbf{v}_3 = [-9, 1, -6, -3]$

    $\mathbf{v}_4 = [1, -1, 0, 0]$

51. $\mathbf{w}_1 = [1, 4, -8, 16]$

    $\mathbf{w}_2 = [1, 1, -1, 1]$

    $\mathbf{w}_3 = [1, 4, 8, 16]$

    $\mathbf{w}_4 = [1, 1, 1, 1]$

MATLAB

Access MATLAB and enter **fbc1s6** *if our text data files are available; otherwise, enter the vectors in Exercises 48–51 by hand. Use MATLAB matrix commands to form the necessary matrix and reduce it in problems M1–M4.*

M1. Solve Exercise 48.

M2. Solve Exercise 49.

M3. Solve Exercise 50.

M4. Solve Exercise 51.

M5. What do you think the probability would be that if n vectors in $\mathbb{R}^n$ were selected at random, they would form a basis for $\mathbb{R}^n$. (The *probability* of an event is a number from 0 to 1. An impossible event has probability 0, a certain event has probability 1, an event as likely to occur as not to occur has probability 0.5, etc.)

M6. As a way of testing your answer to the preceding exercise, you might experiment by asking MATLAB to generate "random" $n \times n$ matrices for some value of n, and reducing them to see if their column vectors form a basis for $\mathbb{R}^n$. Enter **rand(8)** to view an 8×8 matrix with "random" entries between 0 and 1. The column vectors cannot be considered to be random vectors in $\mathbb{R}^8$, because all their components lie between 0 and 1. Do you think the probability that such column vectors form a basis for $\mathbb{R}^8$ is the same as in the preceding exercise? As an experimental check, execute the command **rref(rand(8))** ten times to row-reduce ten such 8×8 matrices, examining each reduced matrix to see if the column vectors of the matrix generated by **rand(8)** did form a basis for $\mathbb{R}^8$.

M7. Note that **4*rand(8)−2*ones(8)** will produce an 8×8 matrix with "random" entries between -2 and 2. Again, its column vectors cannot be regarded as random vectors in $\mathbb{R}^8$, but at least the components of the vectors need not all be positive, as they were in the preceding exercise. Do you think the probability that such column vectors form a basis for $\mathbb{R}^8$ is the same as in Exercise M5? As an experimental check, row-reduce ten such matrices.

1.7   APPLICATION TO POPULATION DISTRIBUTION (OPTIONAL)

Linear algebra has proved to be a valuable tool for many practical and mathematical problems. In this section, we present an application to population distribution (Markov chains).

Consider situations in which people are split into two or more categories. For example, we might split the citizens of the United States according to income into categories of

poor,   middle income,   rich.

We might split the inhabitants of North America into categories according to the climate in which they live:

hot,   temperate,   cold.

In this book, we will speak of a *population* split into *states*. In the two illustrations above, the populations and states are given by the following:

Population	States
Citizens of the United States	poor, middle income, rich
People in North America	hot, temperate, cold

Our populations will often consist of people, but this is not essential. For example, at any moment we can classify the population of cars as operational or not operational.

We are interested in how the distribution of a population between (or among) states may change over a period of time. Matrices and their multiplication can play an important role in such considerations.

Transition Matrices

The tendency of a population to move among n states can sometimes be described using an $n \times n$ matrix. Consider a population distributed among $n = 3$ states, which we call state 1, state 2, and state 3. Suppose that we know the *proportion* t_{ij} of the population of state j that moves to state i over a given fixed time period. Notice that the direction of movement from state j to state i is the right-to-left order of the subscripts in t_{ij}. The matrix $T = [t_{ij}]$ is called a **transition matrix**. (Do not confuse our use of T as a transition matrix in this one section with our use of T as a linear transformation elsewhere in the text.)

EXAMPLE 1   Let the population of a country be classified according to income as

State 1: poor,

State 2: middle income,

State 3: rich.

Suppose that, over each 20-year period (about one generation), we have the following data for people and their offspring:

Of the poor people, 19% become middle income and 1% rich.

Of the middle income people, 15% become poor and 10% rich.

Of the rich people, 5% become poor and 30% middle income.

Give the transition matrix describing these data.

SOLUTION   The entry t_{ij} in the transition matrix T represents the proportion of the population moving from state j to state i, not the percentage. Thus, because 19% of the poor (state 1) will become middle income (state 2), we should take $t_{21} = .19$. Similarly, because 1% of the people in state 1 move to state 3 (rich), we should take $t_{31} = .01$. Now t_{11} represents the proportion of the poor people who remain poor at the end of 20 years. Because this is 80%, we should take $t_{11} = .80$. Continuing in this fashion, starting in state 2 and then in state 3, we obtain the matrix

$$T = \begin{matrix} & \begin{matrix} \text{poor} & \text{mid} & \text{rich} \end{matrix} & \\ \begin{bmatrix} .80 & .15 & .05 \\ .19 & .75 & .30 \\ .01 & .10 & .65 \end{bmatrix} & \begin{matrix} \text{poor} \\ \text{mid} \\ \text{rich} \end{matrix} \end{matrix}.$$

We have labeled the columns and rows with the names of the states. Notice that an entry of the matrix gives the proportion of the population in the state above the entry that moves to the state at the right of the entry during one 20-year period.   ■

In Example 1, the sum of the entries in each column of T is 1, because the sum reflects the movement of the entire population for the state listed at the top of the column. Now suppose that the proportions of the entire population in Example 1 that fall into the various states at the start of a time period are given in the column vector

$$\mathbf{p} = \begin{bmatrix} p_1 \\ p_2 \\ p_3 \end{bmatrix}.$$

For example, we would have

$$\mathbf{p} = \begin{bmatrix} \frac{1}{3} \\ \frac{1}{3} \\ \frac{1}{3} \end{bmatrix}$$

if the whole population were initially equally divided among the states. The entries in such a **population distribution vector p** must be nonnegative and must have a sum equal to 1.

Let us find the proportion of the entire population that is in state 1 after one time period of 20 years, knowing that initially the proportion in state 1 is p_1. The proportion of the state-1 population that remains in state 1 is t_{11}. This gives a contribution of $t_{11}p_1$ to the proportion of the entire population that will be found in state 1 at the end of 20 years. Of course, we also get contributions to state 1 at the end of 20 years from states 2 and 3. These two states contribute proportions $t_{12}p_2$ and $t_{13}p_3$ of the entire population to state 1. Thus, after 20 years, the proportion in state 1 is

$$t_{11}p_1 + t_{12}p_2 + t_{13}p_3.$$

This is precisely the first entry in the column vector given by the product

$$T\mathbf{p} = \begin{bmatrix} t_{11} & t_{12} & t_{13} \\ t_{21} & t_{22} & t_{23} \\ t_{31} & t_{32} & t_{33} \end{bmatrix} \begin{bmatrix} p_1 \\ p_2 \\ p_3 \end{bmatrix}.$$

In a similar fashion, we find that the second and third components of $T\mathbf{p}$ give the proportions of population in state 2 and in state 3 after one time period.

For an initial population distribution vector **p** and transition matrix T, the product vector $T\mathbf{p}$ is the population distribution vector after one time period.

Markov Chains

In Example 1, we found the transition matrix governing the flow of a population among three states over a period of 20 years. Suppose that the same transition matrix is valid over the next 20-year period, and for the next 20 years after that, and so on. That is, suppose that there is a sequence or *chain* of 20-year periods over which the transition matrix is valid. Such a situation is called a **Markov chain**. Let us give a formal definition of a transition matrix for a Markov chain.

DEFINITION 1.18   Transition Matrix

An $n \times n$ matrix T is the **transition matrix** for an n-state Markov chain if all entries in T are nonnegative and the sum of the entries in each column of T is 1.

Markov chains arise naturally in biology, psychology, economics, and many other sciences. Thus they are an important application of linear algebra and of probability. The entry t_{ij} in a transition matrix T is known as the **probability** of moving from state j to state i over one time period.

EXAMPLE 2   Show that the matrix

$$T = \begin{bmatrix} 0 & 0 & 1 \\ 1 & 0 & 0 \\ 0 & 1 & 0 \end{bmatrix}$$

is a transition matrix for a three-state Markov chain, and explain the significance of the zeros and the ones.

SOLUTION   The entries are all nonnegative, and the sum of the entries in each column is 1. Thus the matrix is a transition matrix for a Markov chain.

At least for finite populations, a transition probability $t_{ij} = 0$ means that there is no movement from state j to state i over the time period. That is,

---

HISTORICAL NOTE   MARKOV CHAINS are named for the Russian mathematician Andrei Andreevich Markov (1856–1922), who first defined them in a paper of 1906 dealing with the Law of Large Numbers and subsequently proved many of the standard results about them. His interest in these sequences stemmed from the needs of probability theory; Markov never dealt with their applications to the sciences. The only real examples he used were from literary texts, where the two possible states were vowels and consonants. To illustrate his results, he did a statistical study of the alternation of vowels and consonants in Pushkin's *Eugene Onegin*.

Andrei Markov taught at St. Petersburg University from 1880 to 1905, when he retired to make room for younger mathematicians. Besides his work in probability, he contributed to such fields as number theory, continued fractions, and approximation theory. He was an active participant in the liberal movement in the pre–World War I era in Russia; on many occasions he made public criticisms of the actions of state authorities. In 1913, when as a member of the Academy of Sciences he was asked to participate in the pompous ceremonies celebrating the 300th anniversary of the Romanov dynasty, he instead organized a celebration of the 200th anniversary of Jacob Bernoulli's publication of the Law of Large Numbers.

transition from state j to state i over the time period is impossible. On the other hand, if $t_{ij} = 1$, the entire population of state j moves to state i over the time period. That is, transition from state j to state i in the time period is certain.

For the given matrix, we see that, over one time period, the entire population of state 1 moves to state 2, the entire population of state 2 moves to state 3, and the entire population of state 3 moves to state 1. ∎

If T is an $n \times n$ transition matrix and $\mathbf{p}$ is a population distribution column vector with n components, then we can readily see that $T\mathbf{p}$ is again a population distribution vector. We illustrate the general argument with the case $n = 2$, avoiding summation notation and saving space. We have

$$T\mathbf{p} = \begin{bmatrix} t_{11} & t_{12} \\ t_{21} & t_{22} \end{bmatrix} \begin{bmatrix} p_1 \\ p_2 \end{bmatrix} = \begin{bmatrix} t_{11}p_1 + t_{12}p_2 \\ t_{21}p_1 + t_{22}p_2 \end{bmatrix}.$$

To show that the sum of the components of $T\mathbf{p}$ is 1, we simply rearrange the sum of the four products involved so that the terms involving p_1 appear first, followed by the terms involving p_2. We obtain

$$t_{11}p_1 + t_{21}p_1 + t_{12}p_2 + t_{22}p_2 = p_1 (t_{11} + t_{21}) + p_2(t_{21} + t_{22})$$
$$= p_1(1) + p_2(1) = p_1 + p_2 = 1.$$

The proof for the $n \times n$ case is identical; we would have n^2 products rather than four. Note that it follows that if T is a transition matrix, then so is T^2; we need only observe that the jth column $\mathbf{c}$ of T is itself a population distribution vector, so the jth column of T^2, which is $T\mathbf{c}$, has a component sum equal to 1.

Let T be the transition matrix over a time period—say, 20 years—in a Markov chain. We can form a new Markov chain by looking at the flow of the population over a time period twice as long—that is, over 40 years. Let us see the relationship of the transition matrix for the 40-year time period to the one for the 20-year time period. We might guess that the transition matrix for 40 years is T^2. This is indeed the case. First, note that the jth column vector of an $n \times n$ matrix A is $A\mathbf{e}_j$, where $\mathbf{e}_j$ is the jth standard basis vector of $\mathbb{R}^n$, regarded as a column vector. Now $\mathbf{e}_j$ is a population distribution vector, so the jth column of the two-period transition matrix is $T(T\mathbf{e}_j) = T^2\mathbf{e}_j$, showing that T^2 is indeed the two-period matrix.

If we extend the argument above, we find that the three-period transition matrix for the Markov chain is T^3, and so on. This exhibits another situation in which matrix multiplication is useful. Although raising even a small matrix to a power using pencil and paper is tedious, a computer can do it easily. LINTEK and MATLAB can be used to compute a power of a matrix.

---

m-Period Transition Matrix

A Markov chain with transition matrix T has T^m as its m-period transition matrix.

EXAMPLE 3   For the transition matrix

$$T = \begin{bmatrix} 0 & 0 & 1 \\ 1 & 0 & 0 \\ 0 & 1 & 0 \end{bmatrix}$$

in Example 2, show that, after three time periods, the distribution of population among the three states is the same as the initial population distribution.

SOLUTION   After three time periods, the population distribution vector is $T^3\mathbf{p}$, and we easily compute that $T^3 = I$, the 3×3 identity matrix. Thus $T^3\mathbf{p} = \mathbf{p}$, as asserted. Alternatively, we could note that the entire population of state 1 moves to state 2 in the first time period, then to state 3 in the next time period, and finally back to state 1 in the third time period. Similarly, the populations of the other two states move around and then back to the beginning state over the three periods.   ∎

Regular Markov Chains

We now turn to Markov chains where there exists some fixed number m of time periods in which it is possible to get from any state to any other state. This means that the mth power T^m of the transition matrix has no zero entries.

DEFINITION 1.19   Regular Transition Matrix, Regular Chain

A transition matrix T is **regular** if T^m has no zero entries for some integer m. A Markov chain having a regular transition matrix is called a **regular chain**.

EXAMPLE 4   Show that the transition matrix

$$T = \begin{bmatrix} 0 & 0 & 1 \\ 1 & 0 & 0 \\ 0 & 1 & 0 \end{bmatrix}$$

of Example 2 is not regular.

SOLUTION   A computation shows that T^2 still has zero entries. We saw in Example 3 that $T^3 = I$, the 3×3 identity matrix, so we must have $T^4 = T$, $T^5 = T^2$, $T^6 = T^3 = I$, and the powers of T repeat in this fashion. We never eliminate all the zeros. Thus T is not a regular transition matrix.   ∎

If T^m has no zero entries, then $T^{m+1} = (T^m)T$ has no zero entries, because the entries in any column vector of T are nonnegative with at least one nonzero entry. In determining whether a transition matrix is regular, it is not necessary to compute the entries in powers of the matrix. We need only determine whether or not they are zero.

EXAMPLE 5   If $\times$ denotes a nonzero entry, determine whether a transition matrix T with the zero and nonzero configuration given by

$$T = \begin{bmatrix} \times & 0 & \times & 0 \\ 0 & 0 & \times & 0 \\ \times & 0 & 0 & \times \\ 0 & \times & 0 & 0 \end{bmatrix}$$

is regular.

SOLUTION   We compute configurations of high powers of T as rapidly as we can, because once a power has no zero entries, all higher powers must have nonzero entries. We find that

$$T^2 = \begin{bmatrix} \times & 0 & \times & \times \\ \times & 0 & 0 & \times \\ \times & \times & \times & 0 \\ 0 & 0 & \times & 0 \end{bmatrix}, \quad T^4 = \begin{bmatrix} \times & \times & \times & \times \\ \times & 0 & \times & \times \\ \times & \times & \times & \times \\ \times & \times & \times & 0 \end{bmatrix}, \quad T^8 = \begin{bmatrix} \times & \times & \times & \times \\ \times & \times & \times & \times \\ \times & \times & \times & \times \\ \times & \times & \times & \times \end{bmatrix},$$

so the matrix T is indeed regular. ∎

It can be shown that, if a Markov chain is regular, the distribution of population among the states over many time periods approaches a fixed *steady-state distribution vector* **s**. That is, the distribution of population among the states no longer changes significantly as time progresses. This is not to say that there is no longer movement of population between states; the transition matrix T continues to effect changes. But the movement of population out of any state over one time period is balanced by the population moving into that state, so the proportion of the total population in that state remains constant. This is a consequence of the following theorem, whose proof is beyond the scope of this book.

THEOREM 1.19   Achievement of Steady State

Let T be a regular transition matrix. There exists a unique column vector **s** with strictly positive entries whose sum is 1 such that the following hold:

1. As m becomes larger and larger, all columns of T^m approach the column vector **s**.
2. $T\mathbf{s} = \mathbf{s}$, and **s** is the unique column vector with this property and whose components add up to 1.

From Theorem 1.19 we can show that, if **p** is any initial population distribution vector for a regular Markov chain with transition matrix T, the population distribution vector after many time periods approaches the vector s described in the theorem. Such a vector is called a **steady-state distribution vector**. We indicate the argument using a 3×3 matrix T. We know that the population distribution vector after m time periods is $T^m\mathbf{p}$. If we let

$$\mathbf{p} = \begin{bmatrix} p_1 \\ p_2 \\ p_3 \end{bmatrix} \quad \text{and} \quad \mathbf{s} = \begin{bmatrix} s_1 \\ s_2 \\ s_3 \end{bmatrix},$$

then Theorem 1.19 tells us that $T^m\mathbf{p}$ is approximately

$$\begin{bmatrix} s_1 & s_1 & s_1 \\ s_2 & s_2 & s_2 \\ s_3 & s_3 & s_3 \end{bmatrix} \begin{bmatrix} p_1 \\ p_2 \\ p_3 \end{bmatrix} = \begin{bmatrix} s_1p_1 + s_1p_2 + s_1p_3 \\ s_2p_1 + s_2p_2 + s_2p_3 \\ s_3p_1 + s_3p_2 + s_3p_3 \end{bmatrix}.$$

Because $p_1 + p_2 + p_3 = 1$, this vector becomes

$$\begin{bmatrix} s_1 \\ s_2 \\ s_3 \end{bmatrix}.$$

Thus, after many time periods, the population distribution vector is approximately equal to the steady-state vector $\mathbf{s}$ for any choice of initial population distribution vector $\mathbf{p}$.

There are two ways we can attempt to compute the steady-state vector $\mathbf{s}$ of a regular transition matrix T. If we have a computer handy, we can simply raise T to a sufficiently high power so that all column vectors are the same, as far as the computer can print them. The software LINTEK or MATLAB can be used to do this. Alternatively, we can use part (2) of Theorem 1.19 and solve for $\mathbf{s}$ in the equation

$$T\mathbf{s} = \mathbf{s}. \tag{1}$$

In solving Eq. (1), we will be finding our first *eigenvector* in this text. We will have a lot more to say about eigenvectors in Chapter 5.

Using the identity matrix I, we can rewrite Eq. (1) as

$$T\mathbf{s} = I\mathbf{s}$$

$$T\mathbf{s} - I\mathbf{s} = \mathbf{0}$$

$$(T - I)\mathbf{s} = \mathbf{0}.$$

The last equation represents a homogeneous system of linear equations with coefficient matrix $(T - I)$ and column vector $\mathbf{s}$ of unknowns. From all the solutions of this homogeneous system, choose the solution vector with positive entries that add up to 1. Theorem 1.19 assures us that this solution exists and is unique. Of course, the homogeneous system can be solved easily using a computer. Either LINTEK or MATLAB will reduce the augmented matrix to a form from which the solutions can be determined easily. We illustrate both methods with examples.

EXAMPLE 6   Use the routine MATCOMP in LINTEK, and raise the transition matrix to powers to find the steady-state distribution vector for the Markov chain in Example 1, having states labeled

poor,   middle income,   rich.

SOLUTION   Using MATCOMP and experimenting a bit with powers of the matrix T in Example 1, we find that

$$T^{60} \approx \begin{matrix} \textbf{poor} & \textbf{mid} & \textbf{rich} \\ \begin{bmatrix} .3872054 & .3872054 & .3872054 \\ .4680135 & .4680135 & .4680135 \\ .1447811 & .1447811 & .1447811 \end{bmatrix} & & \end{matrix} \begin{matrix} \textbf{poor} \\ \textbf{mid} \\ \textbf{rich} \end{matrix}$$

Thus eventually about 38.7% of the population is poor, about 46.8% is middle income, and about 14.5% is rich, and these percentages no longer change as time progresses further over 20-year periods.   ■

EXAMPLE 7   The inhabitants of a vegetarian-prone community agree on the following rules:

1. No one will eat meat two days in a row.
2. A person who eats no meat one day will flip a fair coin and eat meat on the next day if and only if a head appears.

Determine whether this Markov-chain situation is regular; and if so, find the steady-state distribution vector for the proportions of the population eating no meat and eating meat.

SOLUTION   The transition matrix T is

$$T = \begin{matrix} \textbf{no meat} & \textbf{meat} \\ \begin{bmatrix} \frac{1}{2} & 1 \\ \frac{1}{2} & 0 \end{bmatrix} & \end{matrix} \begin{matrix} \textbf{no meat} \\ \textbf{meat} \end{matrix}$$

Because T^2 has no zero entries, the Markov chain is regular. We solve

$$(T - I)\mathbf{s} = \mathbf{0}, \quad \text{or} \quad \begin{bmatrix} -\frac{1}{2} & 1 \\ \frac{1}{2} & -1 \end{bmatrix} \begin{bmatrix} s_1 \\ s_2 \end{bmatrix} = \begin{bmatrix} 0 \\ 0 \end{bmatrix}.$$

We reduce the augmented matrix:

$$\begin{bmatrix} -\frac{1}{2} & 1 & \Big| & 0 \\ \frac{1}{2} & -1 & \Big| & 0 \end{bmatrix} \sim \begin{bmatrix} 1 & -2 & \Big| & 0 \\ \frac{1}{2} & -1 & \Big| & 0 \end{bmatrix} \sim \begin{bmatrix} 1 & -2 & \Big| & 0 \\ 0 & 0 & \Big| & 0 \end{bmatrix}.$$

Thus we have

$$\begin{bmatrix} s_1 \\ s_2 \end{bmatrix} = \begin{bmatrix} 2r \\ r \end{bmatrix} \quad \text{for some scalar } r.$$

But we must have $s_1 + s_2 = 1$, so $2r + r = 1$ and $r = \frac{1}{3}$. Consequently, the steady-state population distribution is given by the vector $\begin{bmatrix} \frac{2}{3} \\ \frac{1}{3} \end{bmatrix}$. We see that, eventually, on each day about $\frac{2}{3}$ of the people eat no meat and the other $\frac{1}{3}$ eat meat. This is independent of the initial distribution of population between the states. All might eat meat the first day, or all might eat no meat; the steady-state vector remains $\begin{bmatrix} \frac{2}{3} \\ \frac{1}{3} \end{bmatrix}$ in either case. ∎

If we were to solve Example 7 by reducing an augmented matrix with a computer, we should add a new row corresponding to the condition $s_1 + s_2 = 1$ for the desired steady-state vector. Then the unique solution could be seen at once from the reduction of the augmented matrix. This can be done using pencil-and-paper computations just as well. If we insert this as the first condition on **s** and rework Example 7, our work appears as follows:

$$\begin{bmatrix} 1 & 1 & | & 1 \\ -\frac{1}{2} & 1 & | & 0 \\ \frac{1}{2} & -1 & | & 0 \end{bmatrix} \sim \begin{bmatrix} 1 & 1 & | & 1 \\ 0 & \frac{3}{2} & | & \frac{1}{2} \\ 0 & -\frac{3}{2} & | & -\frac{1}{2} \end{bmatrix} \sim \begin{bmatrix} 1 & 0 & | & \frac{2}{3} \\ 0 & 1 & | & \frac{1}{3} \\ 0 & 0 & | & 0 \end{bmatrix}.$$

Again, we obtain the steady-state vector $\begin{bmatrix} \frac{2}{3} \\ \frac{1}{3} \end{bmatrix}$.

SUMMARY

1. A transition matrix for a Markov chain is a square matrix with nonnegative entries such that the sum of the entries in each column is 1.

2. The entry in the ith row and jth column of a transition matrix is the proportion of the population in state j that moves to state i during one time period of the chain.

3. If the column vector **p** is the initial population distribution vector between states in a Markov chain with transition matrix T, the population distribution vector after one time period of the chain is $T\mathbf{p}$.

4. If T is the transition matrix for one time period of a Markov chain, then T^m is the transition matrix for m time periods.

5. A Markov chain and its associated transition matrix T are called *regular* if there exists an integer m such that T^m has no zero entries.

6. If T is a regular transition matrix for a Markov chain,

   a. The columns of T^m all approach the same probability distribution vector **s** as m becomes large;

   b. **s** is the unique probability distribution vector satisfying $T\mathbf{s} = \mathbf{s}$; and

   c. As the number of time periods increases, the population distribution vectors approach **s** regardless of the initial population distribution vector **p**. Thus **s** is the steady-state population distribution vector.

EXERCISES

In Exercises 1–8, determine whether the given matrix is a transition matrix. If it is, determine whether it is regular.

1. $\begin{bmatrix} \frac{1}{2} & \frac{1}{2} \\ \frac{1}{3} & \frac{2}{3} \end{bmatrix}$

2. $\begin{bmatrix} 0 & \frac{1}{4} \\ 1 & \frac{3}{4} \end{bmatrix}$

3. $\begin{bmatrix} .2 & .1 & .3 \\ .4 & .5 & -.1 \\ .4 & .4 & .8 \end{bmatrix}$

4. $\begin{bmatrix} .1 & .2 \\ .3 & .4 \\ .6 & .4 \end{bmatrix}$

5. $\begin{bmatrix} .5 & 0 & 0 & .5 \\ .5 & .5 & 0 & 0 \\ 0 & .5 & .5 & 0 \\ 0 & 0 & .5 & .5 \end{bmatrix}$

6. $\begin{bmatrix} .3 & .2 & 0 & .5 \\ .4 & .2 & 0 & .1 \\ .1 & .2 & 1 & .2 \\ .2 & .4 & 0 & .2 \end{bmatrix}$

7. $\begin{bmatrix} 0 & .5 & .2 & .2 & .1 \\ .3 & 0 & .1 & .8 & .5 \\ 0 & 0 & .4 & 0 & .1 \\ .7 & .5 & .1 & 0 & .1 \\ 0 & 0 & .2 & 0 & .2 \end{bmatrix}$

8. $\begin{bmatrix} 0 & .1 & 0 & 0 & .9 \\ 0 & .2 & 0 & 0 & .1 \\ 0 & .3 & 0 & 1 & 0 \\ 1 & .1 & 0 & 0 & 0 \\ 0 & .3 & 1 & 0 & 0 \end{bmatrix}$

In Exercises 9–12, let $T = \begin{bmatrix} .3 & .7 & .4 \\ .4 & .2 & .1 \\ .3 & .1 & .5 \end{bmatrix}$ be the transition matrix for a Markov chain, and let $\mathbf{p} = \begin{bmatrix} .3 \\ .2 \\ .5 \end{bmatrix}$ be the initial population distribution vector.

9. Find the proportion of the state 2 population that is in state 3 after two time periods.

10. Find the proportion of the state 3 population that is in state 2 after two time periods.

11. Find the proportion of the total population that is in state 3 after two time periods.

12. Find the population distribution vector after two time periods.

In Exercises 13–18, determine whether the given transition matrix with the indicated distribution of zero entries and nonzero × entries is regular.

13. $\begin{bmatrix} \times & \times & \times \\ 0 & \times & \times \\ 0 & \times & \times \end{bmatrix}$

14. $\begin{bmatrix} \times & 0 & \times \\ 0 & 0 & \times \\ \times & \times & 0 \end{bmatrix}$

15. $\begin{bmatrix} 0 & \times & 0 \\ \times & \times & \times \\ 0 & \times & 0 \end{bmatrix}$

16. $\begin{bmatrix} \times & 0 & \times & \times \\ \times & \times & \times & \times \\ \times & 0 & \times & \times \\ \times & 0 & \times & \times \end{bmatrix}$

17. $\begin{bmatrix} 0 & \times & \times & \times \\ \times & 0 & \times & \times \\ 0 & 0 & \times & \times \\ 0 & 0 & \times & \times \end{bmatrix}$

18. $\begin{bmatrix} 0 & 0 & \times & \times \\ \times & 0 & 0 & \times \\ 0 & \times & 0 & \times \\ 0 & 0 & 0 & \times \end{bmatrix}$

In Exercises 19–24, find the steady-state distribution vector for the given transition matrix of a Markov chain.

19. $\begin{bmatrix} \frac{2}{3} & 1 \\ \frac{1}{3} & 0 \end{bmatrix}$

20. $\begin{bmatrix} \frac{3}{4} & \frac{1}{2} \\ \frac{1}{4} & \frac{1}{2} \end{bmatrix}$

21. $\begin{bmatrix} 0 & \frac{1}{3} & \frac{1}{3} \\ 0 & \frac{2}{3} & \frac{1}{3} \\ 1 & 0 & \frac{1}{3} \end{bmatrix}$

22. $\begin{bmatrix} 1 & \frac{1}{4} & \frac{1}{2} \\ 0 & \frac{1}{4} & 0 \\ 0 & \frac{1}{2} & \frac{1}{2} \end{bmatrix}$

23. $\begin{bmatrix} \frac{1}{4} & \frac{1}{2} & \frac{1}{3} \\ \frac{1}{4} & 0 & \frac{1}{3} \\ \frac{1}{2} & \frac{1}{2} & \frac{1}{3} \end{bmatrix}$

24. $\begin{bmatrix} \frac{1}{5} & \frac{1}{5} & \frac{1}{3} \\ \frac{2}{3} & \frac{1}{5} & \frac{1}{3} \\ \frac{2}{5} & \frac{3}{5} & \frac{1}{3} \end{bmatrix}$

25. Mark each of the following True or False.

____ a. All entries in a transition matrix are nonnegative.

____ b. Every matrix whose entries are all nonnegative is a transition matrix.

____ **c.** The sum of all the entries in an $n \times n$ transition matrix is 1.

____ **d.** The sum of all the entries in an $n \times n$ transition matrix is n.

____ **e.** If a transition matrix contains no zero entries, it is regular.

____ **f.** If a transition matrix is regular, it contains no nonzero entries.

____ **g.** Every power of a transition matrix is again a transition matrix.

____ **h.** If a transition matrix is regular, its square has equal column vectors.

____ **i.** If a transition matrix T is regular, there exists a unique vector **s** such that $T\mathbf{s} = \mathbf{s}$.

____ **j.** If a transition matrix T is regular, there exists a unique population distribution vector **s** such that $T\mathbf{s} = \mathbf{s}$.

26. Estimate A^{100}, if A is the matrix in Exercise 20.

27. Estimate A^{100}, if A is the matrix in Exercise 23.

Exercises 28–33 deal with the following Markov chain. We classify the women in a country according as to whether they live in an urban (U), suburban (S), or rural (R) area. Suppose that each woman has just one daughter, who in turn has just one daughter, and so on. Suppose further that the following are true:

For urban women, 10% of the daughters settle in rural areas, and 50% in suburban areas.

For suburban women, 20% of the daughters settle in rural areas, and 30% in urban areas.

For rural women, 20% of the daughters settle in the suburbs, and 70% in rural areas.

Let this Markov chain have as its period the time required to produce the next generation.

28. Give the transition matrix for this Markov chain, taking states in the order U, S, R.

29. Find the proportion of urban women whose granddaughters are suburban women.

30. Find the proportion of rural women whose granddaughters are also rural women.

31. If the initial population distribution vector for all the women is $\begin{bmatrix} .4 \\ .5 \\ .1 \end{bmatrix}$, find the population distribution vector for the next generation.

32. Repeat Exercise 31, but find the population distribution vector for the following (third) generation.

33. Show that this Markov chain is regular, and find the steady-state probability distribution vector.

Exercises 34–39 deal with a simple genetic model involving just two types of genes, G and g. Suppose that a physical trait, such as eye color, is controlled by a pair of these genes, one inherited from each parent. A person may be classified as being in one of three states:

> *Dominant (type GG),   Hybrid (type Gg), Recessive (type gg).*

We assume that the gene inherited from a parent is a random choice from the parent's two genes—that is, the gene inherited is just as likely to be one of the parent's two genes as to be the other. We form a Markov chain by starting with a population and always crossing with hybrids to produce offspring. We take the time required to produce a subsequent generation as the time period for the chain.

34. Give an intuitive argument in support of the idea that the transition matrix for this Markov chain is

$$T = \begin{array}{c} \\ \\ \end{array} \begin{array}{ccc} \mathbf{D} & \mathbf{H} & \mathbf{R} \end{array} \\ \begin{bmatrix} \frac{1}{2} & \frac{1}{4} & 0 \\ \frac{1}{2} & \frac{1}{2} & \frac{1}{2} \\ 0 & \frac{1}{4} & \frac{1}{2} \end{bmatrix} \begin{array}{c} \mathbf{D} \\ \mathbf{H} \\ \mathbf{R} \end{array}$$

35. What proportion of the third-generation offspring (after two time periods) of the recessive (gg) population is dominant (GG)?

36. What proportion of the third-generation offspring (after two time periods) of the hybrid (Gg) population is not hybrid?

37. If initially the entire population is hybrid, find the population distribution vector in the next generation.

38. If initially the population is evenly divided among the three states, find the population distribution vector in the third generation (after two time periods).

39. Show that this Markov chain is regular, and find the steady-state population distribution vector.

40. A state in a Markov chain is called **absorbing** if it is impossible to leave that state over the next time period. What characterizes the transition matrix of a Markov chain with an absorbing state? Can a Markov chain with an absorbing state be regular?

41. Consider the genetic model for Exercises 34–39. Suppose that, instead of always crossing with hybrids to produce offspring, we always cross with recessives. Give the transition matrix for this Markov chain, and show that there is an absorbing state. (See Exercise 40.)

42. A Markov chain is termed **absorbing** if it contains at least one absorbing state (see Exercise 40) and if it is possible to get from any state to an absorbing state in some number of time periods.

    a. Give an example of a transition matrix for a three-state absorbing Markov chain.

    b. Give an example of a transition matrix for a three-state Markov chain that is not absorbing but has an absorbing state.

43. With reference to Exercise 42, consider an absorbing Markov chain with transition matrix T and a single absorbing state. Argue that, for any initial distribution vector $\mathbf{p}$, the vectors $T^n\mathbf{p}$ for large n approach the vector containing 1 in the component that corresponds to the absorbing state and zeros elsewhere. [SUGGESTION: Let m be such that it is possible to reach the absorbing state from any state in m time periods, and let q be the smallest entry in the row of T^m corresponding to the absorbing state. Form a new chain with just two states, Absorbing (A) and Free (F), which has as time period m time periods of the original chain, and with probability q of moving from state F to state

A in that time period. Argue that, by starting in any state in the original chain, you are more likely to reach an absorbing state in mr time periods than you are by starting from state F in the new chain and going for r time periods. Using the fact that large powers of a positive number less than 1 are almost 0, show that for the two-state chain, the population distribution vector approaches $\begin{bmatrix} 1 \\ 0 \end{bmatrix}$ as the number of time periods increases, regardless of the initial population distribution vector.]

44. Let T be an $n \times n$ transition matrix. Show that, if every row and every column have fewer than $n/2$ zero entries, the matrix is regular.

In Exercises 45–49, find the steady-state population distribution vector for the given transition matrix. See the comment following Example 7.

45. $\begin{bmatrix} 0 & \frac{1}{4} \\ 1 & \frac{3}{4} \end{bmatrix}$ 46. $\begin{bmatrix} \frac{1}{2} & \frac{1}{4} \\ \frac{1}{2} & \frac{3}{4} \end{bmatrix}$ 47. $\begin{bmatrix} \frac{1}{5} & 1 \\ \frac{4}{5} & 0 \end{bmatrix}$

48. $\begin{bmatrix} 0 & \frac{1}{4} & \frac{1}{2} \\ \frac{1}{2} & 0 & \frac{1}{2} \\ \frac{1}{2} & \frac{3}{4} & 0 \end{bmatrix}$ 49. $\begin{bmatrix} \frac{1}{5} & \frac{3}{4} & \frac{1}{8} \\ \frac{4}{5} & 0 & \frac{3}{8} \\ 0 & \frac{1}{4} & \frac{1}{2} \end{bmatrix}$

In Exercises 50–54, find the steady-state population distribution vector by (a) raising the matrix to a power and (b) solving a linear system. Use LINTEK or MATLAB.

50. $\begin{bmatrix} .1 & .3 & .4 \\ .2 & 0 & .2 \\ .7 & .7 & .4 \end{bmatrix}$ 51. $\begin{bmatrix} .3 & .3 & .1 \\ .1 & .3 & .5 \\ .6 & .4 & .4 \end{bmatrix}$

52. $\begin{bmatrix} .1 & 0 & .2 & .5 \\ .4 & 0 & .2 & .5 \\ .2 & .5 & .4 & 0 \\ .3 & .5 & .2 & 0 \end{bmatrix}$ 53. $\begin{bmatrix} .5 & 0 & 0 & 0 & .9 \\ .5 & .4 & 0 & 0 & 0 \\ 0 & .6 & .3 & 0 & 0 \\ 0 & 0 & .7 & .2 & 0 \\ 0 & 0 & 0 & .8 & .1 \end{bmatrix}$

54. The matrix in Exercise 8

APPLICATION TO BINARY LINEAR CODES (OPTIONAL)

We are not concerned here with secret codes. Rather, we discuss the problem of encoding information for transmission so that errors occurring during transmission or reception have a good chance of being detected, and perhaps even being corrected, by an appropriate decoding procedure. The diagram

$$\text{message} \to \boxed{\text{encode}} \to \boxed{\text{transmit}} \to \boxed{\text{receive}} \to \boxed{\text{decode}} \to \text{message}$$

shows the steps with which we are concerned. Errors could be caused at any stage of the process by equipment malfunction, human error, lightning, sunspots, cross talk interference, etc.

Numerical Representation of Information

All information can be reduced to sequences of numbers. For example, we could number the letters of our alphabet and represent every word in our language as a finite sequence of numbers. We concentrate on how to encode numbers to detect and possibly correct errors.

We are accustomed to expressing numbers in decimal (base 10) notation, using as *alphabet* the set $\{0, 1, 2, 3, 4, 5, 6, 7, 8, 9\}$. However, they also can be expressed using any integer base greater than or equal to 2. A computer works in binary (base 2) notation, which uses the smaller alphabet $\{0, 1\}$; the number 1 can be represented by the presence of an electric charge or by current flowing, whereas the absence of a charge or current can represent 0. The American National Standard Code for Information Interchange (ASCII) has 256 characters and is widely used in personal computers. It includes all the characters that we customarily find on typewriter keyboards, such as

$$A\,a\,B\,b\,Z\,z\,1\,2\,3\,4\,5\,6\,7\,8\,9\,0\,, ; ? * \& \# ! + - / ' ".$$

The 256 characters are assigned numbers from 0 to 255. For example, S is assigned the number 83 (decimal), which is 1010011 (binary) because, reading 1010011 from left to right, we see that

$$1(2^6) + 0(2^5) + 1(2^4) + 0(2^3) + 0(2^2) + 1(2^1) + 1(2^0) = 83.$$

The ASCII code number for the character 7 is 55 (decimal), which is 110111 (binary). Because $2^8 = 256$, each character in the ASCII code can be represented by a sequence of eight 0's or 1's; the S is represented by 01010011 and 7 by 00110111. This discussion makes it clear that all information can be encoded using just the *binary alphabet* $\mathbb{B} = \{0, 1\}$.

Message Words and Code Words

An algebraist refers to a sequence of characters from some alphabet, such as *01010011* using the alphabet $\mathbb{B}$ or the sequence of letters *glypt* using our usual letter alphabet, as a **word**; the computer scientist refers to this as a *string*. As we discuss encoding words so that transmission errors can be detected, it is convenient to use as small an alphabet as possible, so we restrict ourselves to the **binary words** using the alphabet $\mathbb{B} = \{0, 1\}$. Rather than give illustrations using words of eight characters as in the ASCII code, let us use words of just four characters; the 16 possible words are

$$0000 \quad 0001 \quad 0010 \quad 0011 \quad 0100 \quad 0101 \quad 0110 \quad 0111$$

$$1000 \quad 1001 \quad 1010 \quad 1011 \quad 1100 \quad 1101 \quad 1110 \quad 1111.$$

An error in transmitting a word occurs when a 1 is changed to a 0 or vice versa during transmission. The first two illustrations exhibit an inefficient way and an efficient way of detecting a transmission error in which *only one* erroneous interchange of the characters 0 and 1 occurs. In each illustration, a binary *message word* is encoded to form the *code word* to be transmitted.

ILLUSTRATION 1   Suppose we wish to send 1011 as the binary message word. To detect any single-error transmission, we could send each character twice—that is, we could *encode* 1011 as 11001111 when we send it. If a single error is made in transmission of the code word 11001111 and the recipient knows the encoding scheme, then the error will be detected. For illustration, if the received code word is 11001011, the recipient knows there is an error because the fifth and sixth characters are different. Of course, the recipient does not know whether 0 or 1 is the correct character. But note that not all double-error transmissions can be detected. For example, if the received code word is 11000011, the recipient perceives no error, and obtains 1001 upon decoding, which was not the message sent.   ■

One problem with encoding a word by repeating every character as in Illustration 1 is that the code word is twice as long as the original message word. There is a lot of *redundancy*. The next illustration shows how we can more efficiently achieve the goal of warning the receiver whenever a single error has been committed.

ILLUSTRATION 2   Suppose again that we wish to transmit a four-character word on the alphabet $\mathbb{B}$. Let us denote the word symbolically by $x_1x_2x_3x_4$, where each x_i is either 0 or 1. We make use of *modulo 2 arithmetic* on $\mathbb{B}$, where

$$0 + 0 = 0, \quad 1 + 0 = 0 + 1 = 1, \quad \text{and} \quad 1 + 1 = 0 \quad \text{(modulo 2 sums)}$$

and where subtraction is the same as addition, so that $0 - 1 = 0 + 1 = 1$. Multiplication is as usual: $1 \cdot 0 = 0 \cdot 1 = 0$ and $1 \cdot 1 = 1$. We append to the word $x_1x_2x_3x_4$ the modulo 2 sum

$$x_5 = x_1 + x_2 + x_3 + x_4. \tag{1}$$

This amounts to appending the character 0 if the message word contains an even number of characters 1, and appending a 1 if the number of 1's in the word is odd. Note that the result is a five-character code word $x_1x_2x_3x_4x_5$ definitely containing an even number of 1's. Thus we have

$$x_1 + x_2 + x_3 + x_4 + x_5 = 0 \quad \text{(modulo 2)}.$$

If the message word is 1011, as in Illustration 1, then the encoded word is 10111. If a single error is made in transmitting this code word, changing a single 0 to 1 or a single 1 to 0, then the modulo 2 sum of the five characters will be 1 rather than 0, and the recipient will recognize that there has been an error in transmitting the code word.   ■

Illustration 2 attained the goal of recognizing single-error transmissions with less redundancy than in Illustration 1. In Illustration 2, we used just one extra character, whereas in Illustration 1 we used four extra characters. However, using the scheme in Illustration 1, we were able to identify which character of the message was affected, whereas the technique in Illustration 2 showed only that at least one error had been made.

Computers use the scheme in Illustration 2 when storing the ASCII code number of one of the 256 ASCII characters. An extra 0 or 1 is appended to the binary form of the code number, so that the number of 1's in the augmented word is even. When the encoded ASCII character is retrieved, the computer checks that the number of 1's is indeed even. If it is not, it can try to read that binary word again. The user may be warned that there is a PARITY CHECK problem if the computer is not successful.

Terminology

Equation 1 in Illustration 2 is known as a **parity-check equation**. In general, starting with a **message word** $x_1x_2 \cdots x_k$ of k characters, we encode it as a **code word** $x_1x_2 \cdots x_k \cdots x_n$ of n characters. The first k characters are the **information portion** of the encoded word, and the final $n - k$ characters are the **redundancy portion** or **parity-check portion**.

We introduce more notation and terminology to make our discussion easier. Let $\mathbb{B}^n$ be the set of all binary words of n consecutive 0's or 1's. A **binary code** C is any subset of $\mathbb{B}^n$. We can identify a vector of n components with each word in $\mathbb{B}^n$—namely, the vector whose ith component is the ith character in the word. For example, we can identify the word 1101 with the row vector $[1, 1, 0, 1]$. It is convenient to denote the set of all of these row vectors with n components by $\mathbb{B}^n$ also. This notation is similar to the notation $\mathbb{R}^n$ for all n-component vectors of real numbers. On occasion, we may find it convenient to use column vectors rather than row vectors.

The **length** of a word u in $\mathbb{B}^n$ is n, the number of its components. The **Hamming weight** $\text{wt}(u)$ of u is the number of components that are 1. Given two binary words u and v in $\mathbb{B}^n$, the **distance between them**, denoted by $d(u, v)$, is the number of components in which the entries in u and v are different, so that one of the words has a 0 where the other has a 1.

ILLUSTRATION 3   Consider the binary words $u = 11010011$ and $v = 01110111$. Both words have length 8. Also, wt$(u) = 5$, whereas wt$(v) = 6$. The associated vectors differ in the first, third, and sixth components, so d$(u, v) = 3$. ∎

We can define addition on the set $\mathbb{B}^n$ by adding modulo 2 the characters in the corresponding positions. Remembering that $1 + 1 = 0$, we add 0011101010 and 1010110001 as follows.

$$\begin{array}{r} 0011101010 \\ + \ 1010110001 \\ \hline 1001011011 \end{array}$$

We refer to this operation as **word addition**. Word subtraction is the same as word addition. Exercise 17 shows that $\mathbb{B}^n$ is closed under word addition. A **binary group code** is any nonempty subset C of $\mathbb{B}^n$ that is closed under word addition. It can be shown that C has precisely 2^k elements for some integer k where $0 \leq k \leq n$. We refer to such a code as an (n, k) binary group code.

Encoding to Enable Correcting a Single-Error Transmission

We now show how, using more than one parity-check equation, we can not only detect but actually correct a single-error transmission of a code word. This method of encoding was developed by Richard Hamming in 1948.

Suppose we wish to encode the 16 message words

0000 0001 0010 0011 0100 0101 0110 0111

1000 1001 1010 1011 1100 1101 1110 1111

in $\mathbb{B}^4$ so that any single-error transmission of a code word not only can be detected but also corrected. The basic idea is simple. We append to the message word $x_1x_2x_3x_4$ some additional binary characters given by parity-check equations such as the equation $x_5 = x_1 + x_2 + x_3 + x_4$ in Illustration 2, and try to design the equations so that the minimum distance between the 16 code words created will be at least 3. Now, with a single-error transmission of a code word, the distance from the received word to that code word will be 1. If we can make our code words all at least three units apart, the pretransmission code word will be the unique code word at distance 1 from the received word. (If there were two such code words, the distance between them would be at most 2.)

In order to detect the error in a single-error transmission of a code word, including not only message word characters but also the redundant parity-check characters, we need to have each component x_i in the code word appear at least once in *some* parity-check equation. Note that in Illustration 1, each component $x_1, x_2, x_3, x_4,$ and x_5 appears in the parity-check equation $x_5 = x_1 + x_2 + x_3 + x_4$. The parity-check equations

$$x_5 = x_1 + x_2 + x_3, \quad x_6 = x_1 + x_3 + x_4, \quad \text{and} \quad x_7 = x_2 + x_3 + x_4, \quad \text{(2)}$$

which we will show accomplish our goal, also satisfy this condition.

Let us see how to get a distance of at least 3 between each pair of the 16 code words. Of course, the distance between any two of the original 16 four-character message words is at least 1 because they are all different. Suppose now that two message words u and v differ in just one component, say x_2. A single parity-check equation containing x_2 then yields a different character for u than for v. This shows that if each x_i in our original message word appears in at least two parity-check equations, then any message words at a distance of 1 are encoded into code words of distance at least 3. Note that the three parity-check equations [Eqs. (2)] satisfy this condition. It remains to ensure that two message words at a distance of 2 are encoded to increase this distance by at least 1. Suppose two message words u and v differ in only the ith and jth components. Now a parity-check equation containing both x_i and x_j will create the same parity-check character for u as for v. Thus, for each such combination i,j of positions in our message word, we need some parity-check equation to contain either x_i but not x_j or x_j but not x_i. We see that this condition is satisfied by the three parity-check equations [Eqs. (2)] for all possible combinations i,j—namely,

$$1,2 \quad 1,3 \quad 1,4 \quad 2,3 \quad 2,4 \quad \text{and} \quad 3,4.$$

Thus, these equations accomplish our goal. The 16 binary words of length 7, obtained by encoding the 16 binary words

$$0000 \quad 0001 \quad 0010 \quad 0011 \quad 0100 \quad 0101 \quad 0110 \quad 0111$$

$$1000 \quad 1001 \quad 1010 \quad 1011 \quad 1100 \quad 1101 \quad 1110 \quad 1111$$

of length 4 using Eqs. (2) form a subset of $\mathbb{B}^7$ called the Hamming (7, 4) code. In Exercise 16, we ask you to show that the Hamming (7, 4) code is a binary group code.

We can encode each of the 16 binary words of length 4 to form the Hamming (7, 4) code by multiplying the vector form $[x_1, x_2, x_3, x_4]$ of the word on the right by a 4×7 matrix G, called the **standard generator matrix**— namely, we compute

$$[x_1, x_2, x_3, x_4] \begin{bmatrix} 1 & 0 & 0 & 0 & 1 & 1 & 0 \\ 0 & 1 & 0 & 0 & 1 & 0 & 1 \\ 0 & 0 & 1 & 0 & 1 & 1 & 1 \\ 0 & 0 & 0 & 1 & 0 & 1 & 1 \end{bmatrix}.$$
$$G$$

To see this, note that the first four columns of G give the 4×4 identity matrix I, so the first four entries in the encoded word will yield precisely the message word $x_1 x_2 x_3 x_4$. In columns 5, 6, and 7, we put the coefficients of x_1, x_2, x_3, and x_4 as they appear in the parity-check equations defining x_5, x_6, and x_7, respectively. Table 1.1 shows the 16 message words and the code words obtained using this generator matrix G. Note that the message words 0011 and 0111, which are at distance 1, have been encoded as 0011100 and 0111001, which are at distance 3. Also, the message words 0101 and 0011, which are at distance 2, have been encoded as 0101110 and 0011100, which are at distance 3.

Decoding a received word w by selecting a code word at minimum distance from w (in some codes, more than one code word might be at the minimum distance) is known as **nearest-neighbor decoding**. If transmission errors are independent from each other, it can be shown that this is equivalent to what is known as *maximum-likelihood decoding*.

ILLUSTRATION 4   Suppose the Hamming (7, 4) code shown in Table 1.1 is used. If the received word is 1011010, then the decoded message word consists of the first four characters 1011, because 1011010 is a code word. However, suppose the word 0110101 is received. This is not a code word. The closest code word is 0100101, which is at distance 1 from 0110101. Thus we decode 0110101 as

TABLE 1.1
The Hamming (7, 4) code

Message	Code Word
0000	0000000
0001	0001011
0010	0010111
0011	0011100
0100	0100101
0101	0101110
0110	0110010
0111	0111001
1000	1000110
1001	1001101
1010	1010001
1011	1011010
1100	1100011
1101	1101000
1110	1110100
1111	1111111

HISTORICAL NOTE   RICHARD HAMMING (*b.* 1915) had his interest in the question of coding stimulated in 1947 when he was using an early Bell System relay computer on weekends only (because he did not have priority use of the machine). During the week, the machine sounded an alarm when it discovered an error so that an operator could attempt to correct it. On weekends, however, the machine was unattended and would dump any problem in which it discovered an error and proceed to the next one. Hamming's frustration with this behavior of the machine grew when errors cost him two consecutive weekends of work. He decided that if the machine could discover errors—it used a fairly simple error-detecting code—there must be a way for it to correct them and proceed with the solution. He therefore worked on this idea for the next year and discovered several different methods of creating error-correcting codes. Because of patent considerations, Hamming did not publish his solutions until 1950. A brief description of his (7, 4) code, however, appeared in a paper of Claude Shannon (*b.* 1916) in 1948.

Hamming, in fact, developed some of the parity-check ideas discussed in the text as well as the geometric model in which the distance between code words is the number of coordinates in which they differ. He also, in essence, realized that the set of actual code words embedded in $\mathbb{B}^7$ was a four-dimensional subspace of that space.

0100, which differs from the first four characters of the received word. On the other hand, if we receive the noncode word 1100111, we decode it as 1100, because the closest code word to 1100111 is 1100011. ∎

Note in Illustration 4 that if the code word 0001011 is transmitted and is received as 0011001, with two errors, then the recipient knows that an error has been made, because 0011001 is not a code word. However, nearest-neighbor decoding yields the code word 0111001, which corresponds to a message word 0111 rather than the intended 0001. When retransmission is practical, it may be better to request it when an error is detected rather than blindly using nearest-neighbor decoding.

Of course, if errors are generated independently and transmission is of high quality, it should be much less likely for a word to be transmitted with two errors than with one error. If the probability of an error in transmission of a single character is p and errors are generated independently, then probability theory shows that in transmitting a word of length n,

the probability of no error is $(1 - p)^n$,

the probability of exactly one error is $np(1 - p)^{n-1}$, and

the probability of exactly two errors is $\frac{n(n-1)}{2}p^2(1 - p)^{n-2}$.

For example, if $p = 0.0001$ so that we can expect about one character out of every 10,000 to be changed and if the length of the word is $n = 10$, then the probabilities of no error, one error, and two errors, respectively, are approximately 0.999, 0.000999, and 0.0000004.

Parity-Check Matrix Decoding

You can imagine that if we encoded all of the 256 ASCII characters in an $(n, 8)$ linear code and used nearest-neighbor decoding, it would be a job to pore over the list of 256 encoded characters to determine the nearest neighbor to a received code word. There is an easier way, which we illustrate using the Hamming (7, 4) code developed before Illustration 4. Recall that the parity-check equations for that code are

$$x_5 = x_1 + x_2 + x_3, \quad x_6 = x_1 + x_3 + x_4, \quad \text{and} \quad x_7 = x_2 + x_3 + x_4.$$

Let us again concern ourselves with detecting and correcting just single-error transmissions. If these parity-check equations hold for the received word, then no such single error has occurred. Suppose, on the other hand, that the first two equations fail and the last one holds. The only character appearing in both of the first two equations but not in the last is x_1, so we could simply change the character x_1 from 0 to 1, or vice versa, to decode. Note that each of x_1, x_2, and x_4 is omitted just once but in different equations, x_3 is the only character that appears in all three equations, and each of x_5, x_6, and x_7 appears just once but in different equations. This allows us to identify the character in a single-error transmission easily.

We can be even more systematic. Let us rewrite the equations as

$$x_1 + x_2 + x_3 + x_5 = 0, \quad x_1 + x_3 + x_4 + x_6 = 0, \quad \text{and}$$
$$x_2 + x_3 + x_4 + x_7 = 0.$$

We form the **parity-check matrix** H whose ith row contains the seven coefficients of $x_1, x_2, x_3, x_4, x_5, x_6, x_7$ in the ith equation—namely,

$$H = \begin{bmatrix} 1 & 1 & 1 & 0 & 1 & 0 & 0 \\ 1 & 0 & 1 & 1 & 0 & 1 & 0 \\ 0 & 1 & 1 & 1 & 0 & 0 & 1 \end{bmatrix}.$$

Let w be a received word, written as a column vector. Exercise 26 shows that w is a code word if and only if Hw is the zero column vector, where we are always using modulo 2 arithmetic. If w resulted from a single-error transmission in which the character in the jth position was changed, then Hw would have 1 in its ith component if and only if x_j appeared in the ith parity-check equation, so that the column vector Hw would be the jth column of H. Thus we can decode a received word w as follows in the Hamming (7, 4) code of Illustration 4, and be confident of detecting and correcting any single-position error, as follows.

Parity-Check Matrix Decoding

1. Compute Hw.
2. If Hw is the zero vector, decode as the first four characters of w.
3. If Hw is the jth column of H, then:

    a. if $j > 4$, decode as the first four characters of w;

    b. otherwise, decode as the first four characters with the jth character changed.

4. If Hw is neither the zero vector nor the jth column of H, then more than one error has been made: ask for retransmission.

ILLUSTRATION 5   Suppose the Hamming (7, 4) code shown in Table 1.1 on page 120 is used and the word $w = 0110101$ is received. We compute that

$$Hw = \begin{bmatrix} 1 & 1 & 1 & 0 & 1 & 0 & 0 \\ 1 & 0 & 1 & 1 & 0 & 1 & 0 \\ 0 & 1 & 1 & 1 & 0 & 0 & 1 \end{bmatrix} \begin{bmatrix} 0 \\ 1 \\ 1 \\ 0 \\ 1 \\ 0 \\ 1 \end{bmatrix} = \begin{bmatrix} 1 \\ 1 \\ 1 \end{bmatrix}.$$

Because this is the third column of H, we change the third character in the message portion 0110 and decode as 0100. Note that this is what we obtained in Illustration 4 when we decoded this word using Table 1.1. ∎

Just as we did after Illustration 4, we point out that if two errors are made in transmission, the preceding outline may lead to incorrect decoding. If the code word $v = 0001011$ is transmitted and received as $w = 0011001$ with *two errors, then*

$$H\mathbf{w} = \begin{bmatrix} 1 \\ 0 \\ 1 \end{bmatrix},$$

which is column 2 of the matrix H. Thus, decoding by the steps above leads to the incorrect message word 0111. Note that item 4 in the list above does not say that if more than one error has been made, then $H\mathbf{w}$ is neither the zero vector nor the jth column of H.

EXERCISES

1. Let the binary numbers 1 through 15 stand for the letters $A\ B\ C\ D\ E\ F\ G\ H\ I\ J\ K\ L\ M\ N\ O$, in that order. Using Table 1.1 for the Hamming (7, 4) code and letting 0000 stand for a space between words, encode the message *A GOOD DOG*.

2. With the same understanding as in the preceding exercise, use nearest-neighbor decoding to decode this received message.

    0111001   1111111   1010100   0101110
    0000000   1100110   1111111   1101000
    1101110

In Exercises 3 through 11, consider the (6, 3) linear code C with standard generator matrix

$$G = \begin{bmatrix} 1 & 0 & 0 & 1 & 1 & 0 \\ 0 & 1 & 0 & 1 & 0 & 1 \\ 0 & 0 & 1 & 0 & 1 & 1 \end{bmatrix}.$$

3. Give the parity-check equations for this code.

4. List the code words in C.

5. How many errors can always be detected using this code?

6. How many errors can always be corrected using this code?

7. Assuming that the given word has been received, decode it using nearest-neighbor decoding, using your list of code words in Exercise 4. (Recall that in the case where more than one code word is at minimum distance from the received word, a code word of minimum distance is selected arbitrarily.)

    a. 110111
    b. 001011
    c. 111011
    d. 101010
    e. 100101

8. Give the parity-check matrix for this code.

9. Use the parity-check matrix to decode the received words in Exercise 7.

10. Let $u = 1101010111$ and $v = 0111001110$. Find

    a. wt(u)
    b. wt(v)
    c. $u + v$
    d. d(u, v)

11. Show that for word addition of binary words u and v of the same length, we have $u + v = u - v$.

12. If a binary code word u is transmitted and the received word is w, then the sum $u + w$ given by word addition modulo 2 is called the *error pattern*. Explain why this is a descriptive name for this sum.

13. Show that for two binary words of the same length, we have d(u, v) = wt($u - v$).

14. Prove the following properties of the distance function for binary words u, v, and w of the same length.
a. $d(u, v) = 0$ if and only if $u = v$
b. $d(u, v) = d(v, u)$ (Symmetry)
c. $d(u, w) \leq d(u, v) + d(v, w)$ (Triangle inequality)
d. $d(u, v) = d(u + w, v + w)$ (Invariance under translation)

15. Show that $\mathbb{B}^n$ is closed under word addition.

16. Recall that we call a nonempty subset C of $\mathbb{B}^n$ a *binary group code* if C is closed under addition. Show that the Hamming (7, 4) code is a group code. [HINT: To show closure under word addition, use the fact that the words in the Hamming (7, 4) code can be formed from those in $\mathbb{B}^4$ by multiplying by a generator matrix.]

17. Show that in a binary group code C, the minimum distance between code words is equal to the minimum weight of the nonzero code words.

18. Suppose that you want to be able to recognize that a received word is incorrect when m or fewer of its characters have been changed during transmission. What must be the minimum distance between code words to accomplish this?

19. Suppose that you want to be able to find a unique nearest neighbor for a received word that has been transmitted with m or fewer of its characters changed. What must be the minimum distance between code words to accomplish this?

20. Show that if the minimum nonzero weight of code words in a group code C is at least $2t + 1$, then the code can detect any $2t$ errors and correct any t errors. (Compare the result stated in this exercise with your answers to the two preceding ones.)

21. Show that if the minimum distance between the words in an (n, k) binary group code C is at least 3, we must have

$$2^{n-k} \geq 1 + n.$$

[HINT: Let e_i be the word in $\mathbb{B}^n$ with 1 in the ith position and 0's elsewhere. Show that e_i is not in C and that, for any two distinct words v and w in C, we have $v + e_i \neq w + e_j$

if $i \neq j$. Then use the fact that n must be large enough so that $\mathbb{B}^n$ contains C and all words whose distance from some word in C is 1.]

22. Show that if the minimum distance between the words in an (n, k) binary group code C is at least 5, then we must have

$$2^{n-k} \geq 1 + n + \frac{n(n - 1)}{2}.$$

[HINT: Proceed as suggested by the hint in Exercise 21 to count the words at distance 1 and at distance 2 from some word in C.]

23. Using the formulas in Exercises 21 and 22, find a lower bound for the number of parity-check equations necessary to encode the 2^k words in $\mathbb{B}^k$ so that the minimum distance between different code words is at least m for the given values of m and k. (Note that $k = 8$ would allow us to encode all the ASCII characters, and that $m = 5$ would allow us to detect and correct all single-error and double-error transmissions using nearest-neighbor decoding.)
a. $k = 2$, $m = 3$ **d.** $k = 2$, $m = 5$
b. $k = 4$, $m = 3$ **e.** $k = 4$, $m = 5$
c. $k = 8$, $m = 3$ **f.** $k = 8$, $m = 5$

24. Find parity-check equations for encoding the 32 words in $\mathbb{B}^5$ into an $(n, 5)$ linear code that can be used to detect and correct any single-error transmission of a code word. (Recall that each character x_j must appear in two parity-check equations, and that for each pair x_i, x_j some equation must contain one of them but not the other.) Try to make the number of parity-check equations as small as possible; see Exercise 21. Give the standard generator matrix for your code.

25. The 256 ASCII characters are numbered from 0 to 255, and thus can be represented by the 256 binary words in $\mathbb{B}^8$. Find $n - 8$ parity-check equations that can be used to form an $(n, 8)$ linear code that can be used to detect and correct any single-error transmission of a code word. Try to make n the value found in part (c) of Exercise 23.

26. Let C be an (n, k) linear code with parity-check matrix H. We know that $Hc = 0$ for all $c \in C$. Show conversely that if $w \in \mathbb{B}^n$ and $Hw = 0$, then $w \in C$.

DIMENSION, RANK, AND LINEAR TRANSFORMATIONS

Given a finite set S of vectors that generate a subspace W of $\mathbb{R}^n$, we would like to delete from S any superfluous vectors, obtaining as small a subset B of S as we can that still generates W. We tackle this problem in Section 2.1. In doing so, we encounter the notion of an *independent set* of vectors. We discover that such a minimal subset B of S that generates W is a basis for W, so that every vector in W can be expressed uniquely as a linear combination of vectors in B. We will see that any two bases for W contain the same number of vectors—the *dimension* of W will be defined to be this number. Section 2.2 discusses the relationships among the dimensions of the column space, the row space, and the nullspace of a matrix.

In Section 2.3, we discuss functions mapping $\mathbb{R}^n$ into $\mathbb{R}^m$ that *preserve*, in a sense that we will describe, both vector addition and scalar multiplication. Such functions are known as *linear transformations*. We will see that for a linear transformation, the image of a vector $\mathbf{x}$ in $\mathbb{R}^n$ can be computed by multiplying the column vector $\mathbf{x}$ on the left by a suitable $m \times n$ matrix. Optional Section 2.4 then applies matrix techniques in describing geometrically all linear transformations of the plane $\mathbb{R}^2$ into itself.

As another application to geometry, optional Section 2.5 uses vector techniques to generalize the notions of line and plane to k-dimensional flats in $\mathbb{R}^n$.

2.1   INDEPENDENCE AND DIMENSION

Finding a Basis for a Span of Vectors

Let $\mathbf{w}_1, \mathbf{w}_2, \ldots, \mathbf{w}_k$ be vectors in $\mathbb{R}^n$ and let $W = \text{sp}(\mathbf{w}_1, \mathbf{w}_2, \ldots, \mathbf{w}_k)$. Now W can be characterized as the *smallest* subspace of $\mathbb{R}^n$ containing all of the vectors $\mathbf{w}_1, \mathbf{w}_2, \ldots, \mathbf{w}_k$, because every subspace containing these vectors must contain all linear combinations of them, and consequently must include every vector in W. We set ourselves the problem of finding a basis for W.

Let us assume that $\{\mathbf{w}_1, \mathbf{w}_2, \dots, \mathbf{w}_k\}$ is not a basis for W. Theorem 1.15 then tells us that we can express the zero vector as a linear combination of the $\mathbf{w}_j$ in some nontrivial way. As an illustration, suppose that

$$2\mathbf{w}_2 - 5\mathbf{w}_6 + \frac{1}{3}\mathbf{w}_7 = \mathbf{0}. \tag{1}$$

Using Eq. (1), we can express each of $\mathbf{w}_2$, $\mathbf{w}_6$, and $\mathbf{w}_7$ as a linear combination of the other two. For example, we have

$$\frac{1}{3}\mathbf{w}_7 = -2\mathbf{w}_2 + 5\mathbf{w}_6 \quad \text{so} \quad \mathbf{w}_7 = -6\mathbf{w}_2 + 15\mathbf{w}_6.$$

We claim that we can delete $\mathbf{w}_7$ from our list $\mathbf{w}_1, \mathbf{w}_2, \dots, \mathbf{w}_k$, and the remaining $\mathbf{w}_j$ will still span W. The space spanned by the remaining $\mathbf{w}_j$, which is contained in W, will still contain $\mathbf{w}_7$ because $\mathbf{w}_7 = -6\mathbf{w}_2 + 15\mathbf{w}_6$, and we have seen that W is the *smallest space* containing all the $\mathbf{w}_j$. Thus the vector $\mathbf{w}_7$ in the original list is not needed to span W.

The preceding illustration indicates that we can find a basis for $W = \text{sp}(\mathbf{w}_1, \mathbf{w}_2, \dots, \mathbf{w}_k)$ by repeatedly deleting from the list of $\mathbf{w}_1, \mathbf{w}_2, \dots, \mathbf{w}_k$ one vector that appears with a nonzero coefficient in a linear combination giving the zero vector, such as Eq. (1), until no such nontrivial linear combination for $\mathbf{0}$ exists. The final list of remaining $\mathbf{w}_j$ will still span W and be a basis for W by Theorem 1.15.

EXAMPLE 1   Find a basis for $W = \text{sp}([2, 3], [0, 1], [4, -6])$ in $\mathbb{R}^2$.

SOLUTION   The presence of the vector $[0, 1]$ allows us to spot that $[4, -6] = 2[2, 3] - 12[0, 1]$, so we have a relation like Eq. (1)—namely,

$$2[2, 3] - 12[0, 1] - [4, -6] = [0, 0].$$

Thus we can delete any one of the three vectors, and the remaining two will still span W. For example, we can delete the vector $[4, 6]$ and we will have $W = \text{sp}([2, 3], [0, 1])$. Because neither of these two remaining vectors is a multiple of the other, we see that the zero vector cannot be expressed as a nontrivial linear combination of them. (See Exercise 29.) Thus $\{[2, 3], [0, 1]\}$ is a basis for W, which we realize is actually all of $\mathbb{R}^2$, because any two nonzero and nonparallel vectors span $\mathbb{R}^2$ (Theorem 1.16 in Section 1.6). ∎

Our attention is focused on the existence of a nontrivial linear combination yielding the zero vector, such as Eq. (1). Such an equation is known as a *dependence relation*. We formally define this, and the notions of *dependence* and *independence* for vectors. This is a very important definition in our study of linear algebra.

DEFINITION 2.1   Linear Dependence and Independence

Let $\{\mathbf{w}_1, \mathbf{w}_2, \ldots, \mathbf{w}_k\}$ be a set of vectors in $\mathbb{R}^n$. A **dependence relation** in this set is an equation of the form

$$r_1\mathbf{w}_1 + r_2\mathbf{w}_2 + \cdots + r_k\mathbf{w}_k = \mathbf{0}, \qquad \text{with at least one } r_j \neq 0. \quad (2)$$

If such a dependence relation exists, then $\{\mathbf{w}_1, \mathbf{w}_2, \ldots, \mathbf{w}_k\}$ is a **linearly dependent** set of vectors. Otherwise, the set of vectors is **linearly independent**.

For convenience, we will often drop the word *linearly* from the terms *linearly dependent* and *linearly independent*, and just speak of a *dependent* or *independent set of vectors*. We will sometimes drop the words *set of* and refer to *dependent* or *independent vectors* $\mathbf{w}_1, \mathbf{w}_2, \ldots, \mathbf{w}_k$.

Two nonzero vectors in $\mathbb{R}^n$ are independent if and only if one is not a scalar multiple of the other (see Exercise 29). Figure 2.1(a) shows two independent vectors $\mathbf{w}_1$ and $\mathbf{w}_2$ in $\mathbb{R}^n$. A little thought shows why $r_1\mathbf{w}_1 + r_2\mathbf{w}_2$ in this figure can be the zero vector if and only if $r_1 = r_2 = 0$. Figure 2.1(b) shows three independent vectors $\mathbf{w}_1$, $\mathbf{w}_2$, and $\mathbf{w}_3$ in $\mathbb{R}^n$. Note how $\mathbf{w}_1 \notin \text{sp}(\mathbf{w}_2, \mathbf{w}_3)$. Similarly, $\mathbf{w}_2 \notin \text{sp}(\mathbf{w}_1, \mathbf{w}_3)$ and $\mathbf{w}_3 \notin \text{sp}(\mathbf{w}_1, \mathbf{w}_2)$.

Using our new terminology, Theorem 1.15 shows that $\{\mathbf{w}_1, \mathbf{w}_2, \ldots, \mathbf{w}_k\}$ is a basis for a subspace W of $\mathbb{R}^n$ if and only if the vectors $\mathbf{w}_1, \mathbf{w}_2, \ldots, \mathbf{w}_k$ span W and are independent. This is taken as a definition of a basis in many texts. We chose the "unique linear combination" characterization in Definition 1.17 because it is the most important property of bases and was the natural choice arising from our discussion of the solution set of $A\mathbf{x} = \mathbf{0}$. We state this alternative characterization as a theorem.

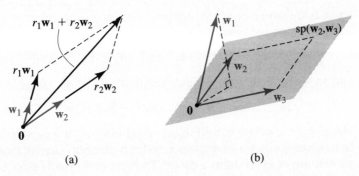

(a)   (b)

FIGURE 2.1
(a) Independent vectors $\mathbf{w}_1$ and $\mathbf{w}_2$; (b) independent vectors $\mathbf{w}_1$, $\mathbf{w}_2$, and $\mathbf{w}_3$.

THEOREM 2.1   Alternative Characterization of a Basis

> Let W be a subspace of $\mathbb{R}^n$. A subset $\{\mathbf{w}_1, \mathbf{w}_2, \ldots, \mathbf{w}_k\}$ of W is a basis for W if and only if the following two conditions are met:
>
> 1. The vectors $\mathbf{w}_1, \mathbf{w}_2, \ldots, \mathbf{w}_k$ span W.
> 2. The vectors are linearly independent.

We turn to a technique for computing a basis for $W = \mathrm{sp}(\mathbf{w}_1, \mathbf{w}_2, \ldots, \mathbf{w}_k)$ in $\mathbb{R}^n$. Determining whether there is a nontrivial dependence relation

$$x_1\mathbf{w}_1 + x_2\mathbf{w}_2 + \cdots + x_k\mathbf{w}_k = \mathbf{0}, \qquad \text{some } x_j \neq 0,$$

amounts to determining whether the linear system $A\mathbf{x} = \mathbf{0}$ has a nontrivial solution, where A is the matrix whose jth column vector is $\mathbf{w}_j$. The ubiquitous row reduction appears again! This time, we will get much more information than just the existence of a dependence relation. Recall that the solutions of $A\mathbf{x} = \mathbf{0}$ are identical with those of the system $H\mathbf{x} = \mathbf{0}$ where $[H \mid \mathbf{0}]$ is obtained from $[A \mid \mathbf{0}]$ by row reduction. Suppose, for illustration, that H is in reduced row-echelon form, and that

$$[H \mid \mathbf{0}] = \begin{bmatrix} 1 & 0 & 2 & 0 & 5 & \mid & 0 \\ 0 & 1 & -3 & 0 & 9 & \mid & 0 \\ 0 & 0 & 0 & 1 & -7 & \mid & 0 \\ 0 & 0 & 0 & 0 & 0 & \mid & 0 \\ 0 & 0 & 0 & 0 & 0 & \mid & 0 \\ 0 & 0 & 0 & 0 & 0 & \mid & 0 \end{bmatrix}.$$

(Normally, we would not bother to write the column of zeros to the right of the partition, but we want to be sure that you realize that, in this context, we imagine the zero column to be there.) The zeros above as well as below the pivots allow us to spot some dependence relations (arising from solutions of $H\mathbf{x} = \mathbf{0}$) immediately, because the columns with pivots are in standard basis vector form. Every nontrivial solution of $H\mathbf{x} = \mathbf{0}$ is a nontrivial solution of $A\mathbf{x} = \mathbf{0}$ and so gives a dependence relation on the column vectors $\mathbf{w}_j$ of A. In particular, we see that

$$\mathbf{w}_3 = 2\mathbf{w}_1 - 3\mathbf{w}_2 \quad \text{and} \quad \mathbf{w}_5 = 5\mathbf{w}_1 + 9\mathbf{w}_2 - 7\mathbf{w}_4,$$

and so we have the dependence relations

$$-2\mathbf{w}_1 + 3\mathbf{w}_2 + \mathbf{w}_3 = \mathbf{0} \quad \text{and} \quad -5\mathbf{w}_1 - 9\mathbf{w}_2 + 7\mathbf{w}_4 + \mathbf{w}_5 = \mathbf{0}.$$

Hence we can delete $\mathbf{w}_3$ and $\mathbf{w}_5$ and retain $\{\mathbf{w}_1, \mathbf{w}_2, \mathbf{w}_4\}$ as a basis for W. In order to be systematic, we have chosen to keep precisely the vectors $\mathbf{w}_j$ such that the jth column of H contains a pivot. Thus we don't really have to obtain *reduced* row-echelon form with zeros above pivots to do this; row-echelon form is enough. We have hit upon the following elegant technique.

Finding a Basis for $W = sp(w_1, w_2, \ldots, w_k)$

1. Form the matrix A whose jth column vector is w_j.
2. Row-reduce A to row-echelon form H.
3. The set of all w_j such that the jth column of H contains a pivot is a basis for W.

EXAMPLE 2   Find a basis for the subspace W of $\mathbb{R}^5$ spanned by

$$w_1 = [1, -1, 0, 2, 1], \qquad w_2 = [2, 1, -2, 0, 0],$$

$$w_3 = [0, -3, 2, 4, 2], \qquad w_4 = [3, 3, -4, -2, -1],$$

$$w_5 = [2, 4, 1, 0, 1], \qquad w_6 = [5, 7, -3, -2, 0].$$

SOLUTION   We reduce the matrix that has w_j as jth column vector, obtaining

$$\begin{bmatrix} 1 & 2 & 0 & 3 & 2 & 5 \\ -1 & 1 & -3 & 3 & 4 & 7 \\ 0 & -2 & 2 & -4 & 1 & -3 \\ 2 & 0 & 4 & -2 & 0 & -2 \\ 1 & 0 & 2 & -1 & 1 & 0 \end{bmatrix} \sim \begin{bmatrix} 1 & 2 & 0 & 3 & 2 & 5 \\ 0 & 3 & -3 & 6 & 6 & 12 \\ 0 & -2 & 2 & -4 & 1 & -3 \\ 0 & -4 & 4 & -8 & -4 & -12 \\ 0 & -2 & 2 & -4 & -1 & -5 \end{bmatrix}$$

$$\sim \begin{bmatrix} 1 & 2 & 0 & 3 & 2 & 5 \\ 0 & 1 & -1 & 2 & 2 & 4 \\ 0 & 0 & 0 & 0 & 5 & 5 \\ 0 & 0 & 0 & 0 & 4 & 4 \\ 0 & 0 & 0 & 0 & 3 & 3 \end{bmatrix} \sim \begin{bmatrix} 1 & 2 & 0 & 3 & 2 & 5 \\ 0 & 1 & -1 & 2 & 2 & 4 \\ 0 & 0 & 0 & 0 & 1 & 1 \\ 0 & 0 & 0 & 0 & 0 & 0 \\ 0 & 0 & 0 & 0 & 0 & 0 \end{bmatrix}.$$

Because there are pivots in columns 1, 2, and 5 of the row-echelon form, the vectors w_1, w_2, and w_5 are retained and are independent. We obtain $\{w_1, w_2, w_5\}$ as a basis for W. ∎

We emphasize that the vectors retained are in columns of the matrix A formed at the start of the reduction process. A common error is to take instead the actual column vectors containing pivots in the row-echelon form H. There is no reason even to expect that the column vectors of H lie in the column space of A. For example,

$$A = \begin{bmatrix} 1 & 1 \\ 1 & 1 \end{bmatrix} \sim \begin{bmatrix} 1 & 1 \\ 0 & 0 \end{bmatrix} = H,$$

and certainly $\begin{bmatrix} 1 \\ 0 \end{bmatrix}$ is not in the column space of A.

Note that we can test whether vectors in $\mathbb{R}^n$ are independent by reducing a matrix having them as column vectors. The vectors are independent if and only if row reduction of the matrix yields a matrix with a pivot in every column. In particular, n vectors in $\mathbb{R}^n$ are independent if and only if row reduction of the matrix having them as column vectors yields the $n \times n$ identity matrix I. On the other hand, more than n vectors in $\mathbb{R}^n$ must be dependent, because an $m \times n$ matrix with $m < n$ cannot have a pivot in every column.

EXAMPLE 3   Determine whether the vectors $\mathbf{v}_1 = [1, 2, 3, 1]$, $\mathbf{v}_2 = [2, 2, 1, 3]$, and $\mathbf{v}_3 = [-1, 2, 7, -3]$ in $\mathbb{R}^4$ are independent.

SOLUTION   Reducing the matrix with jth column vector $\mathbf{v}_j$, we obtain

$$\begin{bmatrix} 1 & 2 & -1 \\ 2 & 2 & 2 \\ 3 & 1 & 7 \\ 1 & 3 & -3 \end{bmatrix} \sim \begin{bmatrix} 1 & 2 & -1 \\ 0 & -2 & 4 \\ 0 & -5 & 10 \\ 0 & 1 & -2 \end{bmatrix} \sim \begin{bmatrix} 1 & 0 & 3 \\ 0 & 1 & -2 \\ 0 & 0 & 0 \\ 0 & 0 & 0 \end{bmatrix}.$$

We see that the vectors are not independent. In fact, $\mathbf{v}_3 = 3\mathbf{v}_1 - 2\mathbf{v}_2$.   ∎

The Dimension of a Subspace

We realize that a basis for a subspace W of $\mathbb{R}^n$ is far from unique. In Example 1, for $W = \text{sp}([2, 3], [0, 1], [4, -6])$ in $\mathbb{R}^2$, we discovered that any two of the three vectors can be used to form a basis for W. We also know that any two nonzero, nonparallel vectors in $\mathbb{R}^2$ form a basis for $\mathbb{R}^2$. Likewise, in Example 2, the vectors we found for a basis for $\text{sp}(\mathbf{w}_1, \mathbf{w}_2, \mathbf{w}_3, \mathbf{w}_4, \mathbf{w}_5, \mathbf{w}_6)$ depended on the order in which we put them as columns in the matrix A that we row-reduced. If we reverse the order of the columns in the matrix A in Example 2, we will wind up with a different basis. However, it is true that given a subspace W of $\mathbb{R}^n$, all bases for W contain the same number of vectors. This is an important result, which will be a quick corollary of the following theorem.

THEOREM 2.2   Relative Sizes of Spanning and Independent Sets

> Let W be a subspace of $\mathbb{R}^n$. Let $\mathbf{w}_1, \mathbf{w}_2, \ldots, \mathbf{w}_k$ be vectors in W that span W, and let $\mathbf{v}_1, \mathbf{v}_2, \ldots, \mathbf{v}_m$ be vectors in W that are independent. Then $k \geq m$.

PROOF   Let us suppose that $k < m$. We will show that the vectors $\mathbf{v}_1, \mathbf{v}_2, \ldots, \mathbf{v}_m$ are dependent, contrary to hypothesis. Because the vectors $\mathbf{w}_1, \mathbf{w}_2, \ldots, \mathbf{w}_k$ span W, there exist scalars a_{ij} such that

$$\mathbf{v}_1 = a_{11}\mathbf{w}_1 + a_{21}\mathbf{w}_2 + \cdots + a_{k1}\mathbf{w}_k,$$
$$\mathbf{v}_2 = a_{12}\mathbf{w}_1 + a_{22}\mathbf{w}_2 + \cdots + a_{k2}\mathbf{w}_k,$$
$$\vdots \qquad\qquad\qquad\qquad (3)$$
$$\mathbf{v}_m = a_{1m}\mathbf{w}_1 + a_{2m}\mathbf{w}_2 + \cdots + a_{km}\mathbf{w}_k.$$

We compute $x_1\mathbf{v}_1 + x_2\mathbf{v}_2 + \cdots + x_m\mathbf{v}_m$ in an attempt to find a dependence relation by multiplying the first equation in Eqs. (3) by x_1, the second by x_2, etc., and then adding the equations. Now the resulting sum is sure to be the zero vector if the total coefficient of each $\mathbf{w}_j$ on the right-hand side in the sum after adding is zero—that is, if we can make

$$x_1 a_{11} + x_2 a_{12} + \cdots + x_m a_{1m} = 0,$$

$$x_1 a_{21} + x_2 a_{22} + \cdots + x_m a_{2m} = 0,$$

$$\vdots$$

$$x_1 a_{k1} + x_2 a_{k2} + \cdots + x_m a_{km} = 0.$$

This gives us a homogeneous linear system of k equations in m unknowns x_i. Corollary 2 of Theorem 1.17 tells us that such a homogeneous system has a nontrivial solution if there are fewer equations than unknowns, and this is the case because we are supposing that $k < m$. Thus we can find scalars $x_1, x_2, \ldots,$ x_m, not all zero, such that $x_1\mathbf{v}_1 + x_2\mathbf{v}_2 + \cdots + x_m\mathbf{v}_m = \mathbf{0}$. That is, the vectors $\mathbf{v}_1, \mathbf{v}_2, \ldots, \mathbf{v}_m$ are dependent if $k < m$, as we wanted to show.   ▲

COROLLARY   Invariance of Dimension

Any two bases of a subspace W of $\mathbb{R}^n$ contain the same number of vectors.

PROOF   Suppose that both a set B with k vectors and a set B' with m vectors are bases for W. Then both B and B' are independent sets of vectors, and the vectors in either set span W. Regarding B as a set of k vectors spanning W and regarding B' as a set of m independent vectors in W, Theorem 2.2 tells us that $k \geq m$. Switching around and regarding B' as a set of m vectors spanning W and regarding B as a set of k independent vectors in W, the theorem tells us that $m \geq k$. Therefore, $k = m$.   ▲

As the title of the corollary indicates, we will consider the number of vectors in a basis for a subspace W of $\mathbb{R}^n$ to be the *dimension* of W. If different bases for W were to have different numbers of vectors, then this notion of dimension would not be well defined (that is, unambiguous). Because different people may come up with different bases, the preceding corollary is necessary in order to define *dimension*.

DEFINITION 2.2   Dimension of a Subspace

Let W be a subspace of $\mathbb{R}^n$. The number of elements in a basis for W is the **dimension** of W, and is denoted by $\dim(W)$.

Thus the dimension of $\mathbb{R}^n$ is n, because we have the standard basis $\{\mathbf{e}_1, \mathbf{e}_2, \ldots, \mathbf{e}_n\}$. Now $\mathbb{R}^n$ cannot be spanned by fewer than n vectors, because a spanning set can always be cut down (if necessary) to form a basis using the technique boxed before Example 2. Theorem 2.2 also tells us that we cannot find a set containing more than n independent vectors in $\mathbb{R}^n$. The same observations hold for any subspace W of $\mathbb{R}^n$ using the same arguments. If $\dim(W) = k$, then W cannot be spanned by fewer than k vectors, and an independent set of vectors in W can contain at most k elements. Perhaps you just assumed that this would be the case; it is gratifying now to have justification for it.

EXAMPLE 4   Find the dimension of the subspace $W = \mathrm{sp}(\mathbf{w}_1, \mathbf{w}_2, \mathbf{w}_3, \mathbf{w}_4)$ of $\mathbb{R}^3$ where $\mathbf{w}_1 = [1, -3, 1]$, $\mathbf{w}_2 = [-2, 6, -2]$, $\mathbf{w}_3 = [2, 1, -4]$, and $\mathbf{w}_4 = [-1, 10, -7]$.

SOLUTION   Clearly, $\dim(W)$ is no larger than 3. To determine its value, we form the matrix

$$A = \begin{bmatrix} 1 & -2 & 2 & -1 \\ -3 & 6 & 1 & 10 \\ 1 & -2 & -4 & -7 \end{bmatrix}.$$

We reduce the matrix A to row-echelon form, obtaining

$$\begin{bmatrix} 1 & -2 & 2 & -1 \\ -3 & 6 & 1 & 10 \\ 1 & -2 & -4 & -7 \end{bmatrix} \sim \begin{bmatrix} 1 & -2 & 2 & -1 \\ 0 & 0 & 7 & 7 \\ 0 & 0 & -6 & -6 \end{bmatrix} \sim \begin{bmatrix} 1 & -2 & 2 & -1 \\ 0 & 0 & 1 & 1 \\ 0 & 0 & 0 & 0 \end{bmatrix}.$$

Thus the column vectors

$$\begin{bmatrix} 1 \\ -3 \\ 1 \end{bmatrix} \quad \text{and} \quad \begin{bmatrix} 2 \\ 1 \\ -4 \end{bmatrix}$$

form a basis for W, the column space of A, and so $\dim(W) = 2$.   ∎

In Section 1.6, we stated that we would show that every subspace W of $\mathbb{R}^n$ is of the form $\mathrm{sp}(\mathbf{w}_1, \mathbf{w}_2, \ldots, \mathbf{w}_k)$. We do this now by showing that every subspace $W \neq \{\mathbf{0}\}$ has a basis. Of course, $\{\mathbf{0}\} = \mathrm{sp}(\mathbf{0})$. To construct a basis for $W \neq \{\mathbf{0}\}$, choose any nonzero vector $\mathbf{w}_1$ in W. If $W = \mathrm{sp}(\mathbf{w}_1)$, we are done. If not, choose a vector $\mathbf{w}_2$ in W that is not in $\mathrm{sp}(\mathbf{w}_1)$. Now the vectors $\mathbf{w}_1, \mathbf{w}_2$ must be independent, for a dependence relation would allow us to express $\mathbf{w}_2$ as a multiple of $\mathbf{w}_1$, contrary to our choice of $\mathbf{w}_2$ not in $\mathrm{sp}(\mathbf{w}_1)$. If $\mathrm{sp}(\mathbf{w}_1, \mathbf{w}_2) = W$, we are done. If not, choose $\mathbf{w}_3 \in W$ that is not in $\mathrm{sp}(\mathbf{w}_1, \mathbf{w}_2)$. Again, no dependence relation can exist for $\mathbf{w}_1, \mathbf{w}_2, \mathbf{w}_3$ because none exists for $\mathbf{w}_1, \mathbf{w}_2$ and because $\mathbf{w}_3$ cannot be a linear combination of $\mathbf{w}_1$ and $\mathbf{w}_2$. Continue in this fashion. Now W cannot contain an independent set with more than n vectors because no independent subset of $\mathbb{R}^n$ can have more than n vectors (Theorem 2.2). The process must stop with $W = \mathrm{sp}(\mathbf{w}_1, \mathbf{w}_2, \ldots, \mathbf{w}_k)$ for some $k \leq n$, which demonstrates our goal. In order to be able to say that every subspace of $\mathbb{R}^n$ has a basis, we *define* the basis of the zero subspace $\{\mathbf{0}\}$ to be the empty set. Note that although $\mathrm{sp}(\mathbf{0}) = \{\mathbf{0}\}$, the zero vector is not a *unique* linear combination of itself, because $r\mathbf{0} = \mathbf{0}$ for all scalars r. In view of our definition, we have $\dim(\{\mathbf{0}\}) = 0$.

The construction technique in the preceding paragraph also shows that every independent subset S of vectors in a subspace W of $\mathbb{R}^n$ can be enlarged, if necessary, to become a basis for W. Namely, if S is not already a basis, we choose a vector in W that is not in the span of the vectors in S, enlarge S by this vector, and continue this process until S becomes a basis.

If we know already that $\dim(W) = k$ and want to check that a subset S containing k vectors of W is a basis, it is not necessary to check both (1) that the vectors in S span W and (2) that they are independent. It suffices to check just one of these conditions. Because if the vectors span S, we know that the set S can be cut down—if necessary, by the technique of Example 2—to become a basis. Because S already has the required number of vectors for a basis, no such cutting down can occur. On the other hand, if we know that S is an independent set, then the preceding paragraph shows that S can be enlarged, if necessary, to become a basis. But because S has the right number of vectors for a basis, no such enlargement is possible.

We collect the observations in the preceding three paragraphs in a theorem for easy reference.

THEOREM 2.3   Existence and Determination of Bases

1. Every subspace W of $\mathbb{R}^n$ has a basis and $\dim(W) \leq n$.
2. Every independent set of vectors in $\mathbb{R}^n$ can be enlarged, if necessary, to become a basis for $\mathbb{R}^n$.
3. If W is a subspace of $\mathbb{R}^n$ and $\dim(W) = k$, then

   a. every independent set of k vectors in W is a basis for W, and

   b. every set of k vectors in W that spans W is a basis for W.

The example that follows illustrates a technique for enlarging an independent set of vectors in $\mathbb{R}^n$ to a basis for $\mathbb{R}^n$.

EXAMPLE 5   Enlarge the independent set $\{[1, 1, -1], [1, 2, -2]\}$ to a basis for $\mathbb{R}^3$.

SOLUTION   Let $\mathbf{v}_1 = [1, 1, -1]$ and $\mathbf{v}_2 = [1, 2, -2]$. We know a spanning set for $\mathbb{R}^3$—namely, $\{\mathbf{e}_1, \mathbf{e}_2, \mathbf{e}_3\}$. We write $\mathbb{R}^3 = \mathrm{sp}(\mathbf{v}_1, \mathbf{v}_2, \mathbf{e}_1, \mathbf{e}_2, \mathbf{e}_3)$ and apply the technique of Example 2 to find a basis. As long as we put $\mathbf{v}_1$ and $\mathbf{v}_2$ first as columns of the matrix to be reduced, pivots will occur in those columns, so $\mathbf{v}_1$ and $\mathbf{v}_2$ will be retained in the basis. We obtain

$$\begin{bmatrix} 1 & 1 & 1 & 0 & 0 \\ 1 & 2 & 0 & 1 & 0 \\ -1 & -2 & 0 & 0 & 1 \end{bmatrix} \sim \begin{bmatrix} 1 & 1 & 1 & 0 & 0 \\ 0 & 1 & -1 & 1 & 0 \\ 0 & -1 & 1 & 0 & 1 \end{bmatrix} \sim \begin{bmatrix} 1 & 1 & 1 & 0 & 0 \\ 0 & 1 & -1 & 1 & 0 \\ 0 & 0 & 0 & 1 & 1 \end{bmatrix}.$$

We see that the pivots occur in columns 1, 2, and 4. Thus a basis containing $\mathbf{v}_1$ and $\mathbf{v}_2$ is $\{[1, 1, -1], [1, 2, -2], [0, 1, 0]\}$.   ∎

SUMMARY

1. A set of vectors $\{\mathbf{w}_1, \mathbf{w}_2, \ldots, \mathbf{w}_k\}$ *in* $\mathbb{R}^n$ is *linearly dependent* if there exists a *dependence relation*

$$r_1\mathbf{w}_1 + r_2\mathbf{w}_2 + \cdots + r_k\mathbf{w}_k = \mathbf{0}, \qquad \text{with at least one } r_j \neq 0.$$

The set is *linearly independent* if no such dependence relation exists, so that a linear combination of the $\mathbf{w}_i$ is the zero vector only if all of the scalar coefficients are zero.

2. A set B of vectors in a subspace W of $\mathbb{R}^n$ is a *basis* for W if and only if the set is independent and the vectors span W. Equivalently, each vector in W can be written *uniquely* as a linear combination of the vectors in B.

3. If $W = \mathrm{sp}(\mathbf{w}_1, \mathbf{w}_2, \ldots, \mathbf{w}_k)$, then the set $\{\mathbf{w}_1, \mathbf{w}_2, \ldots, \mathbf{w}_k\}$ can be cut down, if necessary, to a basis for W by reducing the matrix A having $\mathbf{w}_j$ as the jth column vector to row-echelon form H, and retaining $\mathbf{w}_j$ if and only if the jth column of H contains a pivot.

4. Every subspace W of $\mathbb{R}^n$ has a basis, and every independent set of vectors in W can be enlarged (if necessary) to a basis for W.

5. Let W be a subspace of $\mathbb{R}^n$. All bases of W contain the same number of vectors. The *dimension* of W, denoted by $\dim(W)$, is the number of vectors in any basis for W.

6. Let W be a subspace of $\mathbb{R}^n$ and let $\dim(W) = k$. A subset S of W containing exactly k vectors is a basis for W if either
   a. S is an independent set, or
   b. S spans W.

   That is, it is not necessary to check both conditions in Theorem 2.1 for a basis if S has the right number of elements for a basis.

EXERCISES

1. Give a geometric criterion for a set of two distinct nonzero vectors in $\mathbb{R}^2$ to be dependent.

2. Argue geometrically that any set of three distinct vectors in $\mathbb{R}^2$ is dependent.

3. Give a geometric criterion for a set of two distinct nonzero vectors in $\mathbb{R}^3$ to be dependent.

4. Give a geometric description of the subspace of $\mathbb{R}^3$ generated by an independent set of two vectors.

5. Give a geometric criterion for a set of three distinct nonzero vectors in $\mathbb{R}^3$ to be dependent.

6. Argue geometrically that every set of four distinct vectors in $\mathbb{R}^3$ is dependent.

In Exercises 7–11, use the technique of Example 2, described in the box on page 129, to find a basis for the subspace spanned by the given vectors.

7. $\mathrm{sp}([-3, 1], [6, 4])$ in $\mathbb{R}^2$

8. $\mathrm{sp}([-3, 1], [9, -3])$ in $\mathbb{R}^2$

9. $\mathrm{sp}([2, 1], [-6, -3], [1, 4])$ in $\mathbb{R}^2$

10. $\mathrm{sp}([-2, 3, 1], [3, -1, 2], [1, 2, 3], [-1, 5, 4])$ in $\mathbb{R}^3$

11. $sp([1, 2, 1, 2], [2, 1, 0, -1], [-1, 4, 3, 8],$
$[0, 3, 2, 5])$ in $\mathbb{R}^4$

12. Find a basis for the column space of the matrix

$$A = \begin{bmatrix} 2 & 3 & 1 \\ 5 & 2 & 1 \\ 1 & 7 & 2 \\ 6 & -2 & 0 \end{bmatrix}.$$

13. Find a basis for the row space of the matrix

$$A = \begin{bmatrix} 1 & 3 & 5 & 7 \\ 2 & 0 & 4 & 2 \\ 3 & 2 & 8 & 7 \end{bmatrix}.$$

14. Find a basis for the column space of the matrix A in Exercise 13.

15. Find a basis for the row space of the matrix A in Exercise 12.

In Exercises 16–25, use the technique illustrated in Example 3 to determine whether the given set of vectors is dependent or independent.

16. $\{[1, 3], [-2, -6]\}$ in $\mathbb{R}^2$

17. $\{[1, 3], [2, -4]\}$ in $\mathbb{R}^2$

18. $\{[-3, 1], [6, 4]\}$ in $\mathbb{R}^2$

19. $\{[-3, 1], [9, -3]\}$ in $\mathbb{R}^2$

20. $\{[2, 1], [-6, -3], [1, 4]\}$ in $\mathbb{R}^2$

21. $\{[-1, 2, 1], [2, -4, 3]\}$ in $\mathbb{R}^3$

22. $\{[1, -3, 2], [2, -5, 3], [4, 0, 1]\}$ in $\mathbb{R}^3$

23. $\{[1, -4, 3], [3, -11, 2], [1, -3, -4]\}$ in $\mathbb{R}^3$

24. $\{[1, 4, -1, 3], [-1, 5, 6, 2], [1, 13, 4, 7]\}$ in $\mathbb{R}^4$

25. $\{[-2, 3, 1], [3, -1, 2], [1, 2, 3], [-1, 5, 4]\}$ in $\mathbb{R}^3$

In Exercises 26 and 27, enlarge the given independent set to a basis for the entire space $\mathbb{R}^n$.

26. $\{[1, 2, 1]\}$ in $\mathbb{R}^3$

27. $\{[2, 1, 1, 1], [1, 0, 1, 1]\}$ in $\mathbb{R}^4$

28. Let $S = \{v_1, v_2, \ldots, v_k\}$ be a set of vectors in $\mathbb{R}^n$. Mark each of the following True or False.
___ **a.** A subset of $\mathbb{R}^n$ containing two nonzero distinct parallel vectors is dependent.

___ **b.** If a set of nonzero vectors in $\mathbb{R}^n$ is dependent, then any two vectors in the set are parallel.

___ **c.** Every subset of three vectors in $\mathbb{R}^2$ is dependent.

___ **d.** Every subset of two vectors in $\mathbb{R}^2$ is independent.

___ **e.** If a subset of two vectors in $\mathbb{R}^2$ spans $\mathbb{R}^2$, then the subset is independent.

___ **f.** Every subset of $\mathbb{R}^n$ containing the zero vector is dependent.

___ **g.** If S is independent, then each vector in $\mathbb{R}^n$ can be expressed uniquely as a linear combination of vectors in S.

___ **h.** If S is independent and spans $\mathbb{R}^n$, then each vector in $\mathbb{R}^n$ can be expressed uniquely as a linear combination of vectors in S.

___ **i.** If each vector in $\mathbb{R}^n$ can be expressed uniquely as a linear combination of vectors in S, then S is an independent set.

___ **j.** The subset S is independent if and only if each vector in $sp(v_1, v_2, \ldots, v_k)$ has a unique expression as a linear combination of vectors in S.

___ **k.** The zero subspace of $\mathbb{R}^n$ has dimension 0.

___ **l.** Any two bases of a subspace W of $\mathbb{R}^n$ contain the same number of vectors.

___ **m.** Every independent subset of $\mathbb{R}^n$ is a subset of every basis for $\mathbb{R}^n$.

___ **n.** Every independent subset of $\mathbb{R}^n$ is a subset of some basis for $\mathbb{R}^n$.

29. Let u and v be two different vectors in $\mathbb{R}^n$. Prove that $\{u, v\}$ is linearly dependent if and only if one of the vectors is a multiple of the other.

30. Let v_1, v_2, v_3 be independent vectors in $\mathbb{R}^n$. Prove that $w_1 = 3v_1$, $w_2 = 2v_1 - v_2$, and $w_3 = v_1 + v_3$ are also independent.

31. Let v_1, v_2, v_3 be any vectors in $\mathbb{R}^n$. Prove that $w_1 = 2v_1 + 3v_2$, $w_2 = v_2 - 2v_3$, and $w_3 = -v_1 - 3v_3$ are dependent.

32. Find all scalars s, if any exist, such that $[1, 0, 1], [2, s, 3], [2, 3, 1]$ are independent.

33. Find all scalars s, if any exist, such that $[1, 0, 1], [2, s, 3], [1, -s, 0]$ are independent.

34. Let **v** and **w** be independent column vectors in $\mathbb{R}^3$, and let A be an invertible 3×3 matrix. Prove that the vectors $A\mathbf{v}$ and $A\mathbf{w}$ are independent.

35. Give an example showing that the conclusion of the preceding exercise need not hold if A is nonzero but singular. Can you also find specific independent vectors **v** and **w** and a singular matrix A such that $A\mathbf{v}$ and $A\mathbf{w}$ are still independent?

36. Let **v** and **w** be column vectors in $\mathbb{R}^n$, and let A be an $n \times n$ matrix. Prove that, if $A\mathbf{v}$ and $A\mathbf{w}$ are independent, **v** and **w** are independent.

37. Generalizing Exercise 34, let $\mathbf{v}_1, \mathbf{v}_2, \ldots, \mathbf{v}_k$ be independent column vectors in $\mathbb{R}^n$, and let C be an invertible $n \times n$ matrix. Prove that the vectors $C\mathbf{v}_1, C\mathbf{v}_2, \ldots, C\mathbf{v}_k$ are independent.

38. Prove that if W is a subspace of $\mathbb{R}^n$ and $\dim(W) = n$, then $W = \mathbb{R}^n$.

In Exercises 39–42, use LINTEK to find a basis for the space spanned by the given vectors in $\mathbb{R}^n$.

39. $\mathbf{v}_1 = [5, 4, 3],$ $\mathbf{v}_4 = [6, 1, 4],$
$\mathbf{v}_2 = [2, 1, 6],$ $\mathbf{v}_5 = [1, 1, 1]$
$\mathbf{v}_3 = [4, 5, -12],$

40. $\mathbf{a}_1 = [0, 1, 1, 2],$ $\mathbf{a}_4 = [-1, 4, 6, 11],$
$\mathbf{a}_2 = [-3, -2, 4, 5],$ $\mathbf{a}_5 = [1, 1, 1, 3],$
$\mathbf{a}_3 = [1, 2, 0, 1],$ $\mathbf{a}_6 = [3, 7, 3, 9]$

41. $\mathbf{u}_1 = [3, 1, 2, 4, 1],$
$\mathbf{u}_2 = [3, -2, 6, 7, -3],$
$\mathbf{u}_3 = [3, 4, -2, 1, 5],$
$\mathbf{u}_4 = [1, 2, 3, 2, 1],$
$\mathbf{u}_5 = [7, 1, 11, 13, -1],$
$\mathbf{u}_6 = [2, -1, 2, 3, 1]$

42. $\mathbf{w}_1 = [2, -1, 3, 4, 1, 2],$
$\mathbf{w}_2 = [-2, 5, 3, -2, 1, -4],$
$\mathbf{w}_3 = [2, 4, 6, 5, 2, 1],$
$\mathbf{w}_4 = [1, -1, 1, -1, 2, 2],$
$\mathbf{w}_5 = [1, 8, 10, 2, 5, -1],$
$\mathbf{w}_6 = [3, 0, 0, 2, 1, 5]$

MATLAB

Access MATLAB and work the indicated exercise. If the data files for the text are available, enter **fbc2s1** *for the vector data. Otherwise, enter the vector data by hand.*

M1. Exercise 39
M2. Exercise 40
M3. Exercise 41
M4. Exercise 42

2.2   THE RANK OF A MATRIX

In Section 1.6 we discussed three subspaces associated with an $m \times n$ matrix A: its *column space* in $\mathbb{R}^m$, its *row space* in $\mathbb{R}^n$, and its *nullspace* (solution space of $A\mathbf{x} = \mathbf{0}$) in $\mathbb{R}^n$. In this section we consider how the dimensions of these subspaces are related.

We can find the dimension of the column space of A by row-reducing A to row-echelon form H. This dimension is the number of columns of H having pivots.

Turning to the row space, note that interchange of rows does not change the row space, and neither does multiplication of a row by a nonzero scalar. If we multiply the ith row vector $\mathbf{v}_i$ by a scalar r and add it to the kth row vector $\mathbf{v}_k$, then the new kth row vector is $r\mathbf{v}_i + \mathbf{v}_k$, which is still in the row space of A

because it is a linear combination of rows of A. But the original row vector $\mathbf{v}_k$ is also in the row space of the new matrix, because it is equal to $(r\mathbf{v}_i + \mathbf{v}_k) + (-r)\mathbf{v}_i$. Thus row addition also does not change the row space of a matrix.

Suppose that the reduced row-echelon form of a matrix A is

$$H = \begin{bmatrix} 1 & 0 & 2 & 0 & 5 \\ 0 & 1 & -3 & 0 & 9 \\ 0 & 0 & 0 & 1 & -7 \\ 0 & 0 & 0 & 0 & 0 \\ 0 & 0 & 0 & 0 & 0 \\ 0 & 0 & 0 & 0 & 0 \end{bmatrix}.$$

The configuration of the three nonzero *row* vectors in their 1st, 2nd, and 4th components is the configuration of the *row* vectors $\mathbf{e}_1, \mathbf{e}_2, \mathbf{e}_3$ in $\mathbb{R}^3$, and ensures that the first three row vectors of H are independent. In this way we see that the dimension of the row space of any matrix is the number of nonzero rows in its reduced row-echelon form, or just in its row-echelon form. But this is also the number of pivots in the matrix. Thus the dimension of the column space of A must be equal to the dimension of its row space. This common dimension of the row space and the column space is the **rank** of A, denoted by rank(A). These arguments generalize to prove the following theorem.

THEOREM 2.4   Row Rank Equals Column Rank

> Let A be an $m \times n$ matrix. The dimension of the row space of A is equal to the dimension of the column space of A. The common dimension, the **rank** of A, is the number of pivots in a row-echelon form of A.

We know that a basis for the column space of A consists of the columns of A giving rise to pivots in a row-echelon form of A. We saw how to find a basis for the nullspace of A in Section 1.6. We would like to be able to find a basis for the row space. We could work with the transpose of A, but this would require

---

HISTORICAL NOTE   THE RANK OF A MATRIX was defined in 1879 by Georg Frobenius (1849–1917) as follows: If all determinants of the $(r + 1)$st degree vanish, but not all of the rth degree, then r is the rank of the matrix. Frobenius used this concept to deal with the questions of canonical forms for certain matrices of integers and with the solutions of certain systems of linear congruences.

The nullity was defined by James Sylvester in 1884 for square matrices as follows: The nullity of an $n \times n$ matrix is i if every minor (determinant) of order $n - i + 1$ (and therefore of every higher order) equals 0 and i is the largest such number for which this is true. Sylvester was interested here, as in much of his mathematical career, in discovering invariants—properties of particular mathematical objects that do not change under specified types of transformations. He proceeded to prove what he called one of the cardinal laws in the theory of matrices, that the nullity of the product of two matrices is not less than the nullity of any factor or greater than the sum of the nullities of the factors.

another reduction to row-echelon form. In fact, because the elementary row operations do not change the row space of A, we simply can take as a basis for the row space of A the nonzero rows in a row-echelon form. We summarize in a box.

Finding Bases for Spaces Associated with a Matrix

Let A be an $m \times n$ matrix with row-echelon form H.

1. For a basis of the row space of A, use the nonzero rows of H.

2. For a basis of the column space of A, use the columns of A corresponding to the columns of H containing pivots.

3. For a basis of the nullspace of A, use H and back substitution to solve $H\mathbf{x} = \mathbf{0}$ in the usual way (see Example 3, Section 1.6).

EXAMPLE 1   Find the rank, a basis for the row space, a basis for the column space, and a basis for the nullspace of the matrix

$$A = \begin{bmatrix} 1 & 3 & 0 & -1 & 2 \\ 0 & -2 & 4 & -2 & 0 \\ 3 & 11 & -4 & -1 & 6 \\ 2 & 5 & 3 & -4 & 0 \end{bmatrix}.$$

SOLUTION   We reduce A all the way to reduced row-echelon form, because we also want to find a basis for the nullspace of A. We obtain

$$A = \begin{bmatrix} 1 & 3 & 0 & -1 & 2 \\ 0 & -2 & 4 & -2 & 0 \\ 3 & 11 & -4 & -1 & 6 \\ 2 & 5 & 3 & -4 & 0 \end{bmatrix} \sim \begin{bmatrix} 1 & 3 & 0 & -1 & 2 \\ 0 & -2 & 4 & -2 & 0 \\ 0 & 2 & -4 & 2 & 0 \\ 0 & -1 & 3 & -2 & -4 \end{bmatrix} \sim \begin{bmatrix} 1 & 0 & 6 & -4 & 2 \\ 0 & 1 & -2 & 1 & 0 \\ 0 & 0 & 0 & 0 & 0 \\ 0 & 0 & 1 & -1 & -4 \end{bmatrix} \sim$$

$$\begin{bmatrix} 1 & 0 & 0 & 2 & 26 \\ 0 & 1 & 0 & -1 & -8 \\ 0 & 0 & 1 & -1 & -4 \\ 0 & 0 & 0 & 0 & 0 \end{bmatrix} = H.$$

Because the reduced form H contains three pivots, we see that rank(A) = 3.

As a basis for the row space of A, we take the nonzero row vectors of H, obtaining

$$\{[1, 0, 0, 2, 26], [0, 1, 0, -1, -8], [0, 0, 1, -1, -4]\}.$$

Notice that the next to the last matrix in the reduction shows that the first three row vectors of A are dependent, so we must not take them as a basis for the row space.

Now the columns of A in which pivots appear in H form a basis for the column space, and from H we see that the solution of $A\mathbf{x} = \mathbf{0}$ is

$$\mathbf{x} = \begin{bmatrix} -2r - 26s \\ r + 8s \\ r + 4s \\ r \\ s \end{bmatrix} = r\begin{bmatrix} -2 \\ 1 \\ 1 \\ 1 \\ 0 \end{bmatrix} + s\begin{bmatrix} -26 \\ 8 \\ 4 \\ 0 \\ 1 \end{bmatrix}.$$ Thus we have the following bases:

$$\textit{Column space: } \left\{ \begin{bmatrix} 1 \\ 0 \\ 3 \\ 2 \end{bmatrix}, \begin{bmatrix} 3 \\ -2 \\ 11 \\ 5 \end{bmatrix}, \begin{bmatrix} 0 \\ 4 \\ -4 \\ 3 \end{bmatrix} \right\} \quad \textit{Null space: } \left\{ \begin{bmatrix} -2 \\ 1 \\ 1 \\ 1 \\ 0 \end{bmatrix}, \begin{bmatrix} -26 \\ 8 \\ 4 \\ 0 \\ 1 \end{bmatrix} \right\}. \quad \blacksquare$$

The Rank Equation

Let A be an $m \times n$ matrix. Recall that the nullspace of A—that is, the solution set of $A\mathbf{x} = \mathbf{0}$, has a basis with as many vectors as the number of free scalar variables, like r and s, appearing in the solution vector above. Because we have one free scalar variable for each column without a pivot in a row-echelon form of A, we see that the dimension of the nullspace of A is the number of columns of A that do not contain a pivot. This dimension is called the **nullity** of A, and is denoted by nullity(A). Because the number of columns that *do* have a pivot is the dimension, rank(A), of the column space of A, we see that

$$\text{rank}(A) + \text{nullity}(A) = n, \quad \textbf{Rank equation}$$

where n is the number of columns of A. This equation turns out to be very useful. We summarize this equation, and the method for computing the numbers it involves, in a theorem.

THEOREM 2.5   Rank Equation

Let A be an $m \times n$ matrix with row-echelon form H. Then:

1. nullity(A) = (Number of free variables in the solution space of $A\mathbf{x} = \mathbf{0}$) = (Number of pivot-free columns in H);

2. rank(A) = (Number of pivots in H);

3. (*Rank equation*): rank(A) + nullity(A) = (Number of columns of A).

Because nullity(A) is defined as the number of vectors in a basis of the nullspace of A, the invariance of dimension shows that the number of free variables obtained in the solution of a linear system $A\mathbf{x} = \mathbf{b}$ is independent of the steps in the row reduction to echelon form, as we asserted in Section 1.4.

EXAMPLE 2   Illustrate the rank equation for the matrix A in Example 1.

SOLUTION   The matrix A in Example 1 has $n = 5$ columns, and we saw that rank(A) = 3 and nullity(A) = 2. Thus the rank equation is $3 + 2 = 5$. ∎

Our work has given us still another criterion for the invertibility of a square matrix.

THEOREM 2.6   An Invertibility Criterion

An $n \times n$ matrix A is invertible if and only if rank(A) = n.

SUMMARY

1. Let A be an $m \times n$ matrix. The dimension of the row space of A is equal to the dimension of the column space of A, and is called the *rank* of A, denoted by rank(A). The rank of A is equal to the number of pivots in a row-echelon form H of A. The *nullity* of A, denoted by nullity(A), is the dimension of the nullspace of A—that is, of the solution set of $A\mathbf{x} = \mathbf{0}$.

2. Bases for the row space, the column space, and the nullspace of a matrix A can be found as described in a box in the text.

3. (*Rank Equation*) For an $m \times n$ matrix A, we have

$$\text{rank}(A) + \text{nullity}(A) = n.$$

EXERCISES

For the matrices in Exercises 1–6, find (a) the rank of the matrix, (b) a basis for the row space, (c) a basis for the column space, and (d) a basis for the nullspace.

1. $\begin{bmatrix} 2 & 0 & -3 & 1 \\ 3 & 4 & 2 & 2 \end{bmatrix}$

2. $\begin{bmatrix} 5 & -1 & 0 & 2 \\ 1 & 2 & 1 & 0 \\ 3 & 1 & -2 & 4 \\ 0 & 4 & -1 & 2 \end{bmatrix}$

3. $\begin{bmatrix} 0 & 6 & 6 & 3 \\ 1 & 2 & 1 & 1 \\ 4 & 1 & -3 & 4 \\ 1 & 3 & 2 & 0 \end{bmatrix}$

4. $\begin{bmatrix} 3 & 1 & 4 & 2 \\ -1 & 0 & -1 & 0 \\ 2 & 1 & 0 & 1 \\ 1 & 0 & -1 & 1 \end{bmatrix}$

5. $\begin{bmatrix} 0 & 1 & 2 & 1 \\ 2 & 1 & 0 & 2 \\ 0 & 2 & 1 & 1 \end{bmatrix}$

6. $\begin{bmatrix} 0 & 2 & 3 & 1 \\ -4 & 4 & 1 & 4 \\ 3 & 3 & 2 & 0 \\ -4 & 0 & 1 & 2 \end{bmatrix}$

In Exercises 7–10, determine whether the given matrix is invertible, by finding its rank.

7. $\begin{bmatrix} 0 & -9 & -9 & 2 \\ 1 & 2 & 1 & 1 \\ 4 & 1 & -3 & 4 \\ 1 & 3 & 2 & 0 \end{bmatrix}$

8. $\begin{bmatrix} 2 & 3 & 1 \\ 4 & -1 & 2 \\ 1 & 0 & 1 \end{bmatrix}$

9. $\begin{bmatrix} 2 & 0 & 1 \\ 0 & 0 & 4 \\ 2 & 4 & 0 \end{bmatrix}$

10. $\begin{bmatrix} 3 & 0 & -1 & 2 \\ 4 & 2 & 1 & 8 \\ 1 & 4 & 0 & 1 \\ 2 & 6 & -3 & 1 \end{bmatrix}$

11. Mark each of the following True or False.
___ a. The number of independent row vectors in a matrix is the same as the number of independent column vectors.
___ b. If H is a row-echelon form of a matrix A, then the nonzero column vectors in H form a basis for the column space of A.
___ c. If H is a row-echelon form of a matrix A, then the nonzero row vectors in H are a basis for the row space of A.
___ d. If an $n \times n$ matrix A is invertible, then rank(A) = n.
___ e. For every matrix A, we have rank(A) > 0.
___ f. For all positive integers m and n, the rank of an $m \times n$ matrix might be any number from 0 to the maximum of m and n.
___ g. For all positive integers m and n, the rank of an $m \times n$ matrix might be any number from 0 to the minimum of m and n.
___ h. For all positive integers m and n, the nullity of an $m \times n$ matrix might be any number from 0 to n.
___ i. For all positive integers m and n, the nullity of an $m \times n$ matrix might be any number from 0 to m.
___ j. For all positive integers m and n, with $m \geq n$, the nullity of an $m \times n$ matrix might be any number from 0 to n.

12. Prove that, if A is a square matrix, the nullity of A is the same as the nullity of A^T.

13. Let A be an $m \times n$ matrix, and let $\mathbf{b}$ be an $n \times 1$ vector. Prove that the system of equations $A\mathbf{x} = \mathbf{b}$ has a solution for $\mathbf{x}$ if and only if rank(A) = rank($A \mid \mathbf{b}$), where rank ($A \mid \mathbf{b}$) represents the rank of the associated augmented matrix [$A \mid \mathbf{b}$] of the system.

In Exercises 14–16, let A and C be matrices such that the product AC is defined.

14. Prove that the column space of AC is contained in the column space of A.

15. Is it true that the column space of AC is contained in the column space of C? Explain.

16. State the analogue of Exercise 14 concerning the row spaces of A and C.

17. Give an example of a 3×3 matrix A such that rank(A) = 2 and rank(A^3) = 0.

In Exercises 18–20, let A and C be matrices such that the product AC is defined.

18. Prove that rank(AC) $\leq$ rank(A).

19. Give an example where rank(AC) < rank(A).

20. Is it true that rank(AC) $\leq$ rank(C)? Explain.

It can be shown that rank(A^TA) = rank(A) (see Theorem 6.10). Use this result in Exercises 21–23.

21. Let A be an $m \times n$ matrix. Prove that rank($A(A^T)$) = rank(A).

22. If $\mathbf{a}$ is an $n \times 1$ vector and $\mathbf{b}$ is a $1 \times m$ vector, prove that $\mathbf{ab}$ is an $n \times m$ matrix of rank at most one.

23. Let A be an $m \times n$ matrix. Prove that the column space and row space of $(A^T)A$ are the same.

24. Suppose that you are using computer software, such as LINTEK or MATLAB, that will compute and print the reduced row-echelon form of a matrix but does not indicate any row interchanges it may have made. How can you determine what rows of the *original* matrix form a basis for the row space?

In Exercises 25 and 26, use LINTEK or MATLAB to request a row reduction of the matrix, without seeing intermediate steps. Load data files as usual if they are available. (a) Give the rank of the matrix, and (b) use the software as suggested in Exercise 24 to find the lowest numbered rows, in consecutive order, of the given matrix that form a basis for its row space.

25. $A = \begin{bmatrix} 2 & -3 & 0 & 1 & 4 \\ 1 & 4 & -6 & 3 & -2 \\ 0 & 11 & -12 & 5 & -8 \\ 4 & -1 & 5 & 3 & 7 \end{bmatrix}$

26. $B = \begin{bmatrix} -1 & 1 & 3 & -6 & 8 & -2 \\ -3 & 5 & 3 & 1 & 4 & 8 \\ 1 & -3 & 3 & -13 & 12 & -12 \\ 0 & 2 & -6 & 19 & -20 & 14 \\ 5 & 13 & -21 & 3 & 11 & 6 \end{bmatrix}$

2.3   LINEAR TRANSFORMATIONS OF EUCLIDEAN SPACES

When we introduced the notation $A\mathbf{x} = \mathbf{b}$ and indicated why it is one of the most useful notations in mathematics, we mentioned that we would see that we could regard A as a function and view $A\mathbf{x} = \mathbf{b}$ as meaning that the function maps the vector $\mathbf{x}$ to the vector $\mathbf{b}$. If A is an $m \times n$ matrix and the product $A\mathbf{x}$ is defined, then $\mathbf{x} \in \mathbb{R}^n$ can be viewed as the input variable and $\mathbf{b} \in \mathbb{R}^m$ can be viewed as the output variable.

Functions are used throughout mathematics to study the structures of sets and relationships between sets. You are familiar with the notation $y = f(x)$, where f is a function that acts on numbers, signified by the input variable x, and produces numbers signified by the output variable y. In linear algebra, we are interested in functions $\mathbf{y} = f(\mathbf{x})$, where f acts on vectors, signified by the input variable $\mathbf{x}$, and produces vectors signified by the output variable $\mathbf{y}$.

In general, a function $f: X \to Y$ is a rule that associates with each x in the set X an element $y = f(x)$ in Y. We say that f **maps the set** X into the set Y and **maps the element** x to the element y. The set X is the **domain** of f and the set Y is called the **codomain**. To describe a function, we must give its domain and codomain, and then we must specify the action of the function on each element of its domain. For any subset H of X, we let $f[H] = \{f(h) \mid h \in H\}$; the set $f[H]$ is called the **image** of H under f. The image of the domain of f is the **range** of f. Likewise, for a subset K of Y, the set $f^{-1}[K] = \{x \in X \mid f(x) \in K\}$ is the **inverse image** of K under f. This is illustrated in Figure 2.2. For example, if $f: \mathbb{R} \to \mathbb{R}$ is defined by $f(x) = x^2$, then $f[\{1, 2, 3\}] = \{1, 4, 9\}$ and $f^{-1}[\{1, 4, 9\}] = \{-1, 1, -2, 2, -3, 3\}$. In this section, we study functions known as *linear transformations* T that have as domain $\mathbb{R}^n$ and as codomain $\mathbb{R}^m$, as depicted in Figure 2.3.

The Notion of a Linear Transformation

Notice that for an $m \times n$ matrix A, the function mapping $\mathbf{x} \in \mathbb{R}^n$ into $A\mathbf{x}$ in $\mathbb{R}^m$ satisfies the two conditions in the following definition (see Example 3).

DEFINITION 2.3   Linear Transformation

A function $T: \mathbb{R}^n \to \mathbb{R}^m$ is a **linear transformation** if it satisfies two conditions:

1. $T(\mathbf{u} + \mathbf{v}) = T(\mathbf{u}) + T(\mathbf{v})$ **Preservation of addition**
2. $T(r\mathbf{u}) = rT(\mathbf{u})$ **Preservation of scalar multiplication**

for all vectors $\mathbf{u}$ and $\mathbf{v}$ in $\mathbb{R}^n$ and for all scalars r.

From properties 1 and 2, it follows that $T(r\mathbf{u} + s\mathbf{v}) = rT(\mathbf{u}) + sT(\mathbf{v})$ for all $\mathbf{u}, \mathbf{v} \in \mathbb{R}^n$ and all scalars r and s. (See Exercise 32.) In fact, this equation can be

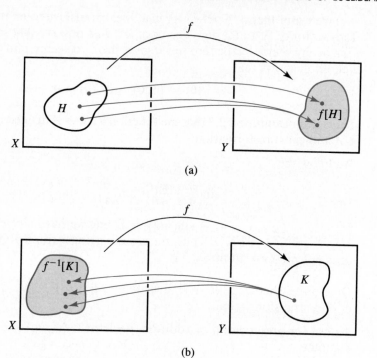

(a)

(b)

FIGURE 2.2
(a) The image of H under f; (b) the inverse image of K under f.

extended to any number of summands by induction—that is, for $\mathbf{v}_1, \mathbf{v}_2, \ldots, \mathbf{v}_k$ in $\mathbb{R}^n$ and scalars $r_1, r_2, \ldots, r_k$, we have

$$T(r_1\mathbf{v}_1 + r_2\mathbf{v}_2 + \cdots + r_k\mathbf{v}_k) = r_1T(\mathbf{v}_1) + r_2T(\mathbf{v}_2) + \cdots + r_kT(\mathbf{v}_k). \quad (1)$$

Equation (1) is often expressed verbally as follows:

Linear transformations preserve linear combinations.

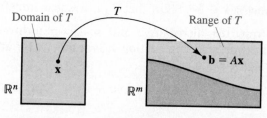

FIGURE 2.3
The linear transformation T(x) = Ax.

We claim that if $T: \mathbb{R}^n \to \mathbb{R}^m$ is a linear transformation, then T maps the zero vector of $\mathbb{R}^n$ to the zero vector of $\mathbb{R}^m$. Just observe that scalar multiplication of any vector by the zero scalar gives the zero vector, and property 2 of the definition shows that

$$T(\mathbf{0}) = T(0\mathbf{0}) = 0T(\mathbf{0}) = \mathbf{0}.$$

EXAMPLE 1   Show from Definition 2.3 that the function $T: \mathbb{R} \to \mathbb{R}$ defined by $T(x) = \sin x$ is not a linear transformation.

SOLUTION   We know that

$$\sin\left(\frac{\pi}{4} + \frac{\pi}{4}\right) \neq \sin\left(\frac{\pi}{4}\right) + \sin\left(\frac{\pi}{4}\right),$$

because $\sin(\pi/4 + \pi/4) = \sin(\pi/2) = 1$, but we have $\sin(\pi/4) + \sin(\pi/4) = 1/\sqrt{2} + 1/\sqrt{2} = 2/\sqrt{2}$. Thus, $\sin x$ is not a linear transformation, because it does not preserve addition.   ∎

EXAMPLE 2   Determine whether $T: \mathbb{R}^2 \to \mathbb{R}^3$ defined by $T([x_1, x_2]) = [x_2, x_1 - x_2, 2x_1 + x_2]$ is a linear transformation.

SOLUTION   To test for preservation of addition, we let $\mathbf{u} = [u_1, u_2]$ and $\mathbf{v} = [v_1, v_2]$, and compute

$$\begin{aligned}
T(\mathbf{u} + \mathbf{v}) &= T([u_1 + v_1, u_2 + v_2]) \\
&= [u_2 + v_2, u_1 + v_1 - u_2 - v_2, 2u_1 + 2v_1 + u_2 + v_2] \\
&= [u_2, u_1 - u_2, 2u_1 + u_2] + [v_2, v_1 - v_2, 2v_1 + v_2] \\
&= T(\mathbf{u}) + T(\mathbf{v}),
\end{aligned}$$

and so vector addition is preserved. To test for preservation of scalar multiplication, we compute

$$\begin{aligned}
T(r\mathbf{u}) &= T([ru_1, ru_2]) = [ru_2, ru_1 - ru_2, 2ru_1 + ru_2] \\
&= r[u_2, u_1 - u_2, 2u_1 + u_2] \\
&= rT(\mathbf{u}).
\end{aligned}$$

Thus, scalar multiplication is also preserved, and so T is a linear transformation.   ∎

EXAMPLE 3   Let A be an $m \times n$ matrix, and let $T_A: \mathbb{R}^n \to \mathbb{R}^m$ be defined by $T_A(\mathbf{x}) = A\mathbf{x}$ for each column vector $\mathbf{x} \in \mathbb{R}^n$. Show that T_A is a linear transformation.

SOLUTION   This follows from the distributive and scalars-pull-through properties of matrix multiplication stated in Section 1.3—namely, for any vectors $\mathbf{u}$ and $\mathbf{v}$ and for any scalar r, we have

$$T_A(\mathbf{u} + \mathbf{v}) = A(\mathbf{u} + \mathbf{v}) = A\mathbf{u} + A\mathbf{v} = T_A(\mathbf{u}) + T_A(\mathbf{v})$$

and

$$T_A(r\mathbf{u}) = A(r\mathbf{u}) = r(A\mathbf{u}) = rT_A(\mathbf{u}).$$

These are precisely the conditions for a linear transformation given in Definition 2.3. ∎

Looking back at Example 2 we see that, in column-vector notation, the transformation there appears as

$$T\left(\begin{bmatrix} x_1 \\ x_2 \end{bmatrix}\right) = \begin{bmatrix} x_2 \\ x_1 - x_2 \\ 2x_1 + x_2 \end{bmatrix} = \begin{bmatrix} 0 & 1 \\ 1 & -1 \\ 2 & 1 \end{bmatrix} \begin{bmatrix} x_1 \\ x_2 \end{bmatrix}.$$

Based on Example 3, we can conclude that T is a linear transformation, obviating the need to check linearity directly as we did in Example 2. In a moment, we will show that *every* linear transformation of $\mathbb{R}^n$ into $\mathbb{R}^m$ has the form $T(\mathbf{x}) = A\mathbf{x}$ for some $m \times n$ matrix A. This is especially easy to see for linear transformations of $\mathbb{R}$ into $\mathbb{R}$.

EXAMPLE 4   Determine all linear transformations of $\mathbb{R}$ into $\mathbb{R}$.

SOLUTION   Let $T: \mathbb{R} \to \mathbb{R}$ be a linear transformation. Each element of $\mathbb{R}$ can be viewed either as a vector or as a scalar. Let $a = T(1)$. Applying Property 2 of Definition 2.3 with $\mathbf{u} = 1$ and $r = x$, we obtain

$$T(x) = T(x(1)) = xT(1) = xa = ax.$$

Identifying a with the 1×1 matrix A having a as its sole entry, we see that we have $T(\mathbf{x}) = A\mathbf{x}$, and we know this transformation satisfies properties 1 and 2 in Definition 2.3. From a geometric viewpoint, we see that the *linear transformations of $\mathbb{R}$ into $\mathbb{R}$* can be described as precisely those functions whose graphs are *lines through the origin*. ∎

Example 4 shows that a linear transformation of $\mathbb{R}$ into $\mathbb{R}$ is completely determined as soon as $T(1)$ is known. More generally, a linear transformation $T: \mathbb{R}^n \to \mathbb{R}^m$ is uniquely determined by its values on any basis for $\mathbb{R}^n$, as we now show.

THEOREM 2.7   Bases and Linear Transformations

Let $T: \mathbb{R}^n \to \mathbb{R}^m$ be a linear transformation, and let $B = \{\mathbf{b}_1, \mathbf{b}_2, \ldots, \mathbf{b}_n\}$ be a basis for $\mathbb{R}^n$. For any vector $\mathbf{v}$ in $\mathbb{R}^n$, the vector $T(\mathbf{v})$ is uniquely determined by the vectors $T(\mathbf{b}_1), T(\mathbf{b}_2), \ldots, T(\mathbf{b}_n)$.

PROOF   Let $\mathbf{v}$ be any vector in $\mathbb{R}^n$. We know that because B is a basis, there exist *unique* scalars $r_1, r_2, \ldots, r_n$ such that

$$\mathbf{v} = r_1\mathbf{b}_1 + r_2\mathbf{b}_2 + \cdots + r_n\mathbf{b}_n.$$

Using Eq. (1), we see that

$$T(\mathbf{v}) = T(r_1\mathbf{b}_1 + r_2\mathbf{b}_2 + \cdots + r_n\mathbf{b}_n) = r_1T(\mathbf{b}_1) + r_2T(\mathbf{b}_2) + \cdots + r_nT(\mathbf{b}_n).$$

Because the coefficients r_i are *uniquely* determined by $\mathbf{v}$, it follows that $T(\mathbf{v})$ is completely determined by the vectors $T(\mathbf{b}_i)$ for $i = 1, 2, \ldots, n$. ▲

Theorem 2.7 shows that if two linear transformations have the same value at each basis vector $\mathbf{b}_i$, then the two transformations have the same value at each vector in $\mathbb{R}^n$, and thus they are the same transformation.

COROLLARY Standard Matrix Representation of a Linear Transformation

---

Let $T\colon \mathbb{R}^n \to \mathbb{R}^m$ be a linear transformation, and let A be the $m \times n$ matrix whose jth column vector is $T(\mathbf{e}_j)$, which we denote symbolically as

$$
A = \begin{bmatrix} | & | & & | \\ T(\mathbf{e}_1) & T(\mathbf{e}_2) & \cdots & T(\mathbf{e}_n) \\ | & | & & | \end{bmatrix}. \tag{2}
$$

Then $T(\mathbf{x}) = A\mathbf{x}$ for each column vector $\mathbf{x} \in \mathbb{R}^n$.

PROOF Recall that for any matrix A, $A\mathbf{e}_j$ is the jth column of A. This shows at once that if A is the matrix described in Eq. (2), then $A\mathbf{e}_j = T(\mathbf{e}_j)$, and so T and the linear transformation T_A given by $T_A(\mathbf{x}) = A\mathbf{x}$ agree on the standard basis $\{\mathbf{e}_1, \mathbf{e}_2, \ldots, \mathbf{e}_n\}$ of $\mathbb{R}^n$. By Theorem 2.7, and the comment following this theorem, we know that then $T(\mathbf{x}) = T_A(\mathbf{x})$ for every $\mathbf{x} \in \mathbb{R}^n$—that is, $T(\mathbf{x}) = A\mathbf{x}$ for every $\mathbf{x} \in \mathbb{R}^n$. ▲

The matrix A in Eq. (2) is the **standard matrix representation** of the linear transformation T.

EXAMPLE 5   Let $T\colon \mathbb{R}^2 \to \mathbb{R}^3$ be the linear transformation such that

$$
T(\mathbf{e}_1) = [2, 1, 4] \quad \text{and} \quad T(\mathbf{e}_2) = [3, 0, -2].
$$

Find the standard matrix representation A of T and find a formula for $T([x_1, x_2])$.

SOLUTION   Equation (2) for the standard matrix representation shows that

$$
A = \begin{bmatrix} 2 & 3 \\ 1 & 0 \\ 4 & -2 \end{bmatrix}, \text{ so } T\left(\begin{bmatrix} x_1 \\ x_2 \end{bmatrix}\right) = \begin{bmatrix} 2 & 3 \\ 1 & 0 \\ 4 & -2 \end{bmatrix} \begin{bmatrix} x_1 \\ x_2 \end{bmatrix} = \begin{bmatrix} 2x_1 + 3x_2 \\ x_1 \\ 4x_1 - 2x_2 \end{bmatrix}.
$$

In row-vector notation, we have the formula

$$T([x_1, x_2]) = [2x_1 + 3x_2, x_1, 4x_1 - 2x_2].$$ ∎

EXAMPLE 6 Find the standard matrix representation of the linear transformation $T: \mathbb{R}^4 \to \mathbb{R}^3$ where

$$T([x_1, x_2, x_3, x_4]) = [x_2 - 3x_3, 2x_1 - x_2 + 3x_4, 8x_1 - 4x_2 + 3x_3 - x_4]. \quad \textbf{(3)}$$

SOLUTION We compute

$$T(\mathbf{e}_1) = T([1, 0, 0, 0]) = [0, 2, 8], \; T(\mathbf{e}_2) = T([0, 1, 0, 0]) = [1, -1, -4]$$

$$T(\mathbf{e}_3) = T([0, 0, 1, 0]) = [-3, 0, 3], \; T(\mathbf{e}_4) = T([0, 0, 0, 1]) = [0, 3, -1].$$

Using Eq. (2), we find that

$$A = \begin{bmatrix} 0 & 1 & -3 & 0 \\ 2 & -1 & 0 & 3 \\ 8 & -4 & 3 & -1 \end{bmatrix}.$$ ∎

Perhaps you noticed in Example 6 that the first row of the matrix A consists of the coefficients of x_1, x_2, x_3, and x_4 in the first component $x_2 - 3x_3$ of $T([x_1, x_2, x_3, x_4])$. The second and third rows can be found similarly. If Eq. (3) is written in column-vector form, the matrix A jumps out at you immediately. Try it! This is often a fast way to write down the standard matrix representation when the transformation is described by a formula as in Example 6. Be sure to remember, however, the Eq. (2) formulation for the standard matrix representation.

We give another example indicating how a linear transformation is determined, as in the proof of Theorem 2.7, if we know its values on a basis for its domain. Note that the vectors $\mathbf{u} = [-1, 2]$ and $\mathbf{v} = [3, -5]$ are two nonparallel vectors in the plane, and form a basis for $\mathbb{R}^2$.

EXAMPLE 7 Let $\mathbf{u} = [-1, 2]$ and $\mathbf{v} = [3, -5]$ be in $\mathbb{R}^2$, and let $T: \mathbb{R}^2 \to \mathbb{R}^3$ be a linear transformation such that $T(\mathbf{u}) = [-2, 1, 0]$ and $T(\mathbf{v}) = [5, -7, 1]$. Find the standard matrix representation A of T and compute $T([-4, 3])$.

SOLUTION To find the standard matrix representation of T, we need to find $T(\mathbf{e}_1)$ and $T(\mathbf{e}_2)$ for $\mathbf{e}_1, \mathbf{e}_2 \in \mathbb{R}^2$. Following the argument in the proof of Theorem 2.7, we express $\mathbf{e}_1$ and $\mathbf{e}_2$ as linear combinations of the basis vectors $\mathbf{u}$ and $\mathbf{v}$ for $\mathbb{R}^2$, where we know the action of T. To express $\mathbf{e}_1$ and $\mathbf{e}_2$ as linear combinations of $\mathbf{u}$ and $\mathbf{v}$, we solve the two linear systems $A\mathbf{x} = \mathbf{e}_1$ and $A\mathbf{x} = \mathbf{e}_2$, where the coefficient matrix A has $\mathbf{u}$ and $\mathbf{v}$ as its column vectors. Because both systems have the same coefficient matrix, we can solve them both at once as follows:

$$\begin{bmatrix} -1 & 3 & | & 1 & 0 \\ 2 & -5 & | & 0 & 1 \end{bmatrix} \sim \begin{bmatrix} 1 & -3 & | & -1 & 0 \\ 0 & 1 & | & 2 & 1 \end{bmatrix} \sim \begin{bmatrix} 1 & 0 & | & 5 & 3 \\ 0 & 1 & | & 2 & 1 \end{bmatrix}.$$

We see that $\mathbf{e}_1 = 5\mathbf{u} + 2\mathbf{v}$ and $\mathbf{e}_2 = 3\mathbf{u} + \mathbf{v}$. Using the linearity properties,

$$T(\mathbf{e}_1) = T(5\mathbf{u} + 2\mathbf{v}) = 5T(\mathbf{u}) + 2T(\mathbf{v}) = 5[-2, 1, 0] + 2[5, -7, 1]$$
$$= [0, -9, 2]$$

and

$$T(\mathbf{e}_2) = T(3\mathbf{u} + \mathbf{v}) = 3T(\mathbf{u}) + T(\mathbf{v}) = 3[-2, 1, 0] + [5, -7, 1]$$
$$= [-1, -4, 1].$$

The standard matrix representation A of T and $T([-4, 3])$ are

$$A = \begin{bmatrix} 0 & -1 \\ -9 & -4 \\ 2 & 1 \end{bmatrix} \quad \text{and} \quad T\left(\begin{bmatrix} -4 \\ 3 \end{bmatrix}\right) = \begin{bmatrix} 0 & -1 \\ -9 & -4 \\ 2 & 1 \end{bmatrix} \begin{bmatrix} -4 \\ 3 \end{bmatrix} = \begin{bmatrix} -3 \\ 24 \\ -5 \end{bmatrix}. \quad \blacksquare$$

Some Terminology of Linear Transformations

The matrix representation A of a linear transformation $T: \mathbb{R}^n \to \mathbb{R}^m$ is a great help in working with T. Let us use column-vector notation. Suppose, for example, we want to find the range of T—that is, the set of all elements of $\mathbb{R}^m$ that are equal to $T(\mathbf{x})$ for some $\mathbf{x} \in \mathbb{R}^n$. Recall that we use the notation $T[\mathbb{R}^n]$ to denote the set of all these elements, so that $T[\mathbb{R}^n] = \{T(\mathbf{x}) \mid \mathbf{x} \in \mathbb{R}^n\}$. Because $T(\mathbf{x}) = A\mathbf{x}$, we have $T[\mathbb{R}^n] = \{A\mathbf{x} \mid \mathbf{x} \in \mathbb{R}^n\}$. Now $A\mathbf{x}$ is a linear combination of the column vectors of A where the coefficient of the jth column of A in the linear combination is x_j, the jth component of the vector $\mathbf{x}$. Thus the range of T is precisely the column space of A.

For another illustration, finding all $\mathbf{x}$ such that $T(\mathbf{x}) = \mathbf{0}$ amounts to solving the linear system $A\mathbf{x} = \mathbf{0}$. We know that the solution of this homogeneous linear system is a subspace of $\mathbb{R}^n$, called the *nullspace* of the matrix A. This nullspace is often called the **kernel** of T as well as the **nullspace** of T, and is denoted $\ker(T)$.

Let W be a subspace of $\mathbb{R}^n$. Then $W = \text{sp}(\mathbf{b}_1, \mathbf{b}_2, \ldots, \mathbf{b}_k)$ where $B = \{\mathbf{b}_1, \mathbf{b}_2, \ldots, \mathbf{b}_k\}$ is a basis for W. Because T preserves linear combinations, we have

$$T(r_1\mathbf{b}_1 + r_2\mathbf{b}_2 + \cdots + r_k\mathbf{b}_k) = r_1 T(\mathbf{b}_1) + r_2 T(\mathbf{b}_2) + \cdots + r_k T(\mathbf{b}_k).$$

This shows that $T[W] = \text{sp}(T(\mathbf{b}_1), T(\mathbf{b}_2), \ldots, T(\mathbf{b}_k))$, which we know is a subspace of $\mathbb{R}^m$.

We summarize the three preceding paragraphs in a box.

Let $T: \mathbb{R}^n \to \mathbb{R}^m$ be a linear transformation with standard matrix representation A.

1. The *range* $T[\mathbb{R}^n]$ of T is the column space of A in $\mathbb{R}^m$.
2. The *kernel* of T is the nullspace of A and is denoted $\ker(T)$.
3. If W is a subspace of $\mathbb{R}^n$, then $T[W]$ is a subspace of $\mathbb{R}^m$; that is, T *preserves subspaces*.

EXAMPLE 8   Find the kernel of the linear transformation $T: \mathbb{R}^3 \to \mathbb{R}^2$ where $T([x_1, x_2, x_3])$ $= [x_1 - 2x_2, x_1 + 4x_3]$.

SOLUTION   We simply find the nullspace of the standard matrix representation A of T. Writing down and then reducing the matrix A, we have

$$A = \begin{bmatrix} 1 & -2 & 0 \\ 1 & 0 & 4 \end{bmatrix} \sim \begin{bmatrix} 1 & -2 & 0 \\ 0 & 2 & 4 \end{bmatrix} \sim \begin{bmatrix} 1 & 0 & 4 \\ 0 & 1 & 2 \end{bmatrix}.$$

Thus we find that $\ker(T) = \text{sp}([-4, -2, 1])$.   ∎

Matrix Operations and Linear Transformations

It is very fruitful to be able to hop back and forth at will between matrices and their associated linear transformations. Every property of matrices has an interpretation for linear transformations, and vice versa. For example, the rank equation, $\text{rank}(A) + \text{nullity}(A) = n$, for an $m \times n$ matrix A becomes

$$\dim(\text{range } T) + \dim(\ker(T)) = \dim(\text{domain } T).$$

The dimension of range T is called the **rank** of T, and the dimension of $\ker(T)$ is called the **nullity** of T.

Also, matrix multiplication and matrix inversion have very significant analogues in terms of transformations. Let $T: \mathbb{R}^n \to \mathbb{R}^m$ and $T': \mathbb{R}^m \to \mathbb{R}^k$ be two linear transformations. We can consider the **composite** function $(T' \circ T): \mathbb{R}^n \to \mathbb{R}^k$ where $(T' \circ T)(\mathbf{x}) = T'(T(\mathbf{x}))$ for $\mathbf{x} \in \mathbb{R}^n$. Figure 2.4 gives a graphic illustration of this composite map.

Now suppose that A is the $m \times n$ matrix associated with T and that B is the $k \times m$ matrix associated with T'. Then we can compute $T'(T(\mathbf{x}))$ as

$$T'(T(\mathbf{x})) = T'(A\mathbf{x}) = B(A\mathbf{x}).$$

But

$$B(A\mathbf{x}) = (BA)\mathbf{x}, \quad \textbf{Associativity of matrix multiplication}$$

so $(T' \circ T)(\mathbf{x}) = (BA)\mathbf{x}$. From Example 3, we see that $T' \circ T$ is again a linear transformation, and that the matrix associated with it is the product of the matrix associated with T' and the matrix associated with T, in that order. Notice how easily this follows from the associativity of matrix multiplication.

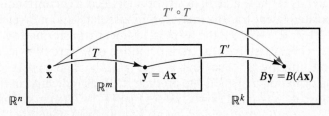

FIGURE 2.4
The composite map $T' \circ T$.

It really makes us appreciate the power of associativity! We can also show directly from Definition 2.3 that the composite of two linear transformations is again a linear transformation. (See Exercise 31.)

Matrix Multiplication and Composite Transformations

A composition of two linear transformations T and T' yields a linear transformation $T' \circ T$ having as its associated matrix the product of the matrices associated with T' and T, in that order.

This result has some surprising uses.

ILLUSTRATION 1　(*The Double Angle Formulas*) It is shown in the next section that rotation of the plane $\mathbb{R}^2$ counterclockwise about the origin through an angle θ is a linear transformation $T: \mathbb{R}^2 \to \mathbb{R}^2$ with standard matrix representation

$$\begin{bmatrix} \cos\theta & -\sin\theta \\ \sin\theta & \cos\theta \end{bmatrix}. \quad \textbf{Counterclockwise rotation through } \theta \qquad (3)$$

Thus, T applied twice—that is, $T \circ T$—rotates the plane through 2θ. Replacing θ by 2θ in matrix (3), we find that the standard matrix representation for $T \circ T$ must be

$$\begin{bmatrix} \cos 2\theta & -\sin 2\theta \\ \sin 2\theta & \cos 2\theta \end{bmatrix}. \qquad (4)$$

On the other hand, we know that the standard matrix representation for the composition $T \circ T$ must be the square of the standard matrix representation for T, and so matrix (4) must be equal to

$$\begin{bmatrix} \cos\theta & -\sin\theta \\ \sin\theta & \cos\theta \end{bmatrix} \begin{bmatrix} \cos\theta & -\sin\theta \\ \sin\theta & \cos\theta \end{bmatrix} = \begin{bmatrix} \cos^2\theta - \sin^2\theta & -2\sin\theta\cos\theta \\ 2\sin\theta\cos\theta & -\sin^2\theta + \cos^2\theta \end{bmatrix}. \qquad (5)$$

Comparing the entries in matrix (4) with the final result in Eq. (5), we obtain the double angle trigonometric identities

$$\sin 2\theta = 2\sin\theta\cos\theta \quad \text{and} \quad \cos 2\theta = \cos^2\theta - \sin^2\theta. \qquad \blacksquare$$

Let us see how matrix invertibility reflects a corresponding property of the associated linear transformation. Suppose that A is an invertible $n \times n$ matrix, and let $T: \mathbb{R}^n \to \mathbb{R}^n$ be the associated linear transformation, so that $\mathbf{y} = T(\mathbf{x}) = A\mathbf{x}$. There exists a linear transformation of $\mathbb{R}^n$ into $\mathbb{R}^n$ associated with A^{-1}; we denote this by T^{-1}, so that $T^{-1}(\mathbf{y}) = A^{-1}\mathbf{y}$. The matrix of the composite transformation $T^{-1} \circ T$ is the product $A^{-1}A$, as indicated in the preceding box. Because $A^{-1}A = I$ and $I\mathbf{x} = \mathbf{x}$, we see that $(T^{-1} \circ T)(\mathbf{x}) = \mathbf{x}$. That is, $T^{-1} \circ T$ is the *identity transformation*, leaving all vectors fixed. (See Fig. 2.5.) Because $AA^{-1} = I$ too, we see that $T \circ T^{-1}$ is also the identity transformation on $\mathbb{R}^n$. If

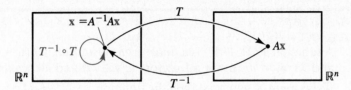

FIGURE 2.5
$T^{-1} \circ T$ is the identity transformation.

$\mathbf{y} = T(\mathbf{x})$, then $\mathbf{x} = T^{-1}(\mathbf{y})$. This transformation T^{-1} is the **inverse transformation of** T, and T is an **invertible** linear transformation.

Invertible Matrices and Inverse Transformations

Let A be an invertible $n \times n$ matrix with associated linear transformation T. The transformation T^{-1} associated with A^{-1} is the inverse transformation of T, and $T \circ T^{-1}$ and $T^{-1} \circ T$ are both the identity transformation on $\mathbb{R}^n$. A linear transformation $T: \mathbb{R}^n \to \mathbb{R}^n$ is invertible if and only if its associated matrix is invertible.

EXAMPLE 9   Show that the linear transformation $T: \mathbb{R}^3 \to \mathbb{R}^3$ defined by $T([x_1, x_2, x_3]) = [x_1 - 2x_2 + x_3, x_2 - x_3, 2x_2 - 3x_3]$ is invertible, and find a formula for its inverse.

SOLUTION   Using column-vector notation, we see that $T(\mathbf{x}) = A\mathbf{x}$, where

$$A = \begin{bmatrix} 1 & -2 & 1 \\ 0 & 1 & -1 \\ 0 & 2 & -3 \end{bmatrix}.$$

Next, we find the inverse of A:

$$\left[\begin{array}{ccc|ccc} 1 & -2 & 1 & 1 & 0 & 0 \\ 0 & 1 & -1 & 0 & 1 & 0 \\ 0 & 2 & -3 & 0 & 0 & 1 \end{array}\right] \sim \left[\begin{array}{ccc|ccc} 1 & 0 & -1 & 1 & 2 & 0 \\ 0 & 1 & -1 & 0 & 1 & 0 \\ 0 & 0 & -1 & 0 & -2 & 1 \end{array}\right] \sim \left[\begin{array}{ccc|ccc} 1 & 0 & 0 & 1 & 4 & -1 \\ 0 & 1 & 0 & 0 & 3 & -1 \\ 0 & 0 & 1 & 0 & 2 & -1 \end{array}\right].$$

Therefore,

$$T^{-1}(\mathbf{x}) = A^{-1}\mathbf{x} = \begin{bmatrix} 1 & 4 & -1 \\ 0 & 3 & -1 \\ 0 & 2 & -1 \end{bmatrix}\begin{bmatrix} x_1 \\ x_2 \\ x_3 \end{bmatrix} = \begin{bmatrix} x_1 + 4x_2 - x_3 \\ 3x_2 - x_3 \\ 2x_2 - x_3 \end{bmatrix},$$

which we express in row notation as

$$T^{-1}([x_1, x_2, x_3]) = [x_1 + 4x_2 - x_3, 3x_2 - x_3, 2x_2 - x_3].$$

In Exercise 30, we ask you to verify that $T^{-1}(T(\mathbf{x})) = \mathbf{x}$, as in Figure 2.5.   ∎

SUMMARY

1. A function $T: \mathbb{R}^n \to \mathbb{R}^m$ is a *linear transformation* if $T(\mathbf{u} + \mathbf{v}) = T(\mathbf{u}) + T(\mathbf{v})$ and $T(r\mathbf{u}) = rT(\mathbf{u})$ for all vectors $\mathbf{u}, \mathbf{v} \in \mathbb{R}^n$ and all scalars r.

2. If A is an $m \times n$ matrix, then the function $T_A: \mathbb{R}^n \to \mathbb{R}^m$ given by $T_A(\mathbf{x}) = A\mathbf{x}$ for all $\mathbf{x} \in \mathbb{R}^n$ is a linear transformation.

3. A linear transformation $T: \mathbb{R}^n \to \mathbb{R}^m$ is uniquely determined by $T(\mathbf{b}_1)$, $T(\mathbf{b}_2), \ldots, T(\mathbf{b}_n)$ for any basis $\{\mathbf{b}_1, \mathbf{b}_2, \ldots, \mathbf{b}_n\}$ of $\mathbb{R}^n$.

4. Let $T: \mathbb{R}^n \to \mathbb{R}^m$ be a linear transformation and let A be the $m \times n$ matrix whose jth column vector is $T(\mathbf{e}_j)$. Then $T(\mathbf{x}) = A\mathbf{x}$ for all $\mathbf{x} \in \mathbb{R}^n$; the matrix A is the standard matrix representation of T. The kernel of T is the nullspace of A, and the range of T is the column space of A.

5. Let $T: \mathbb{R}^n \to \mathbb{R}^m$ and $T': \mathbb{R}^m \to \mathbb{R}^k$ be linear transformations with standard matrix representations A and B, respectively. The composition $T' \circ T$ of the two transformations is a linear transformation, and its standard matrix representation is BA.

6. If $\mathbf{y} = T(\mathbf{x}) = A(\mathbf{x})$ where A is an invertible $n \times n$ matrix, then T is invertible and the transformation T^{-1} defined by $T^{-1}(\mathbf{y}) = A^{-1}\mathbf{y}$ is the inverse of T. Both $T^{-1} \circ T$ and $T \circ T^{-1}$ are the identity transformation of $\mathbb{R}^n$.

EXERCISES

1. Is $T([x_1, x_2, x_3]) = [x_1 + x_2, x_1 - 3x_2]$ a linear transformation of $\mathbb{R}^3$ into $\mathbb{R}^2$? Why or why not?

2. Is $T([x_1, x_2, x_3]) = [0, 0, 0, 0]$ a linear transformation of $\mathbb{R}^3$ into $\mathbb{R}^4$? Why or why not?

3. Is $T([x_1, x_2, x_3]) = [1, 1, 1, 1]$ a linear transformation of $\mathbb{R}^3$ into $\mathbb{R}^4$? Why or why not?

4. Is $T([x_1, x_2]) = [x_1 - x_2, x_2 + 1, 3x_1 - 2x_2]$ a linear transformation of $\mathbb{R}^2$ into $\mathbb{R}^3$? Why or why not?

In Exercises 5–12, assume that T is a linear transformation. Refer to Example 7 for Exercises 9–12, if necessary.

5. If $T([1, 0]) = [3, -1]$ and $T([0, 1]) = [-2, 5]$, find $T([4, -6])$.

6. If $T([-1, 0]) = [2, 3]$ and $T([0, 1]) = [5, 1]$, find $T([-3, -5])$.

7. If $T([1, 0, 0]) = [3, 1, 2]$, $T([0, 1, 0]) = [2, -1, 4]$, and $T([0, 0, 1]) = [6, 0, 1]$, find $T([2, -5, 1])$.

8. If $T([1, 0, 0]) = [-3, 1]$, $T([0, 1, 0]) = [4, -1]$, and $T([0, -1, 1]) = [3, -5]$, find $T([-1, 4, 2])$.

9. If $T([-1, 2]) = [1, 0, 0]$ and $T([2, 1]) = [0, 1, 2]$, find $T([0, 10])$.

10. If $T([-1, 1]) = [2, 1, 4]$ and $T([1, 1]) = [-6, 3, 2]$, find $T([x, y])$.

11. If $T([1, 2, -3]) = [1, 0, 4, 2]$, $T([3, 5, 2]) = [-8, 3, 0, 1]$, and $T([-2, -3, -4]) = [0, 2, -1, 0]$, find $T([5, -1, 4])$.
    [*Computational aid: See* Example 4 in Section 1.5.]

12. If $T([2, 3, 0]) = 8$, $T([1, 2, -1]) = -5$, and $T([4, 5, 1]) = 17$, find $T([-3, 11, -4])$.
    [*Computational aid: See the answer to* Exercise 7 in Section 1.5.]

In Exercises 13–18, the given formula defines a linear transformation. Give its standard matrix representation.

13. $T([x_1, x_2]) = [x_1 + x_2, x_1 - 3x_2]$
14. $T([x_1, x_2]) = [2x_1 - x_2, x_1 + x_2, x_1 + 3x_2]$
15. $T([x_1, x_2, x_3]) = [x_1 + x_2 + x_3, x_1 + x_2, x_1]$
16. $T([x_1, x_2, x_3]) = [2x_1 + x_2 + x_3, x_1 + x_2 + 3x_3]$
17. $T([x_1, x_2, x_3]) = [x_1 - x_2 + 3x_3, x_1 + x_2 + x_3, x_1]$
18. $T([x_1, x_2, x_3]) = x_1 + x_2 + x_3$

19. If $T: \mathbb{R}^2 \to \mathbb{R}^3$ is defined by $T([x_1, x_2]) = [2x_1 + x_2, x_1, x_1 - x_2]$ and $T': \mathbb{R}^3 \to \mathbb{R}^2$ is defined by $T'([x_1, x_2, x_3]) = [x_1 - x_2 + x_3, x_1 + x_2]$, find the standard matrix representation for the linear transformation $T' \circ T$ that carries $\mathbb{R}^2$ into $\mathbb{R}^2$. Find a formula for $(T' \circ T)([x_1, x_2])$.

20. Referring to Exercise 19, find the standard matrix representation for the linear transformation $T \circ T'$ that carries $\mathbb{R}^3$ into $\mathbb{R}^3$. Find a formula for $(T \circ T')([x_1, x_2, x_3])$.

In Exercises 21–28, determine whether the indicated linear transformation T is invertible. If it is, find a formula for $T^{-1}(\mathbf{x})$ in row notation. If it is not, explain why it is not.

21. The transformation in Exercise 13.
22. The transformation in Exercise 14.
23. The transformation in Exercise 15.
24. The transformation in Exercise 16.
25. The transformation in Exercise 17.
26. The transformation in Exercise 18.
27. The transformation in Exercise 19.
28. The transformation in Exercise 20.
29. Mark each of the following True or False.
____ a. Every linear transformation is a function.
____ b. Every function mapping $\mathbb{R}^n$ into $\mathbb{R}^m$ is a linear transformation.
____ c. Composition of linear transformations corresponds to multiplication of their standard matrix representations.
____ d. Function composition is associative.

____ e. An invertible linear transformation mapping $\mathbb{R}^n$ into itself has a unique inverse.
____ f. The same matrix may be the standard matrix representation for several different linear transformations.
____ g. A linear transformation having an $m \times n$ matrix as standard matrix representation maps $\mathbb{R}^n$ into $\mathbb{R}^m$.
____ h. If T and T' are different linear transformations mapping $\mathbb{R}^n$ into $\mathbb{R}^m$, then we may have $T(\mathbf{e}_i) = T'(\mathbf{e}_i)$ for some standard basis vector $\mathbf{e}_i$ of $\mathbb{R}^n$.
____ i. If T and T' are different linear transformations mapping $\mathbb{R}^n$ into $\mathbb{R}^m$, then we may have $T(\mathbf{e}_i) = T'(\mathbf{e}_i)$ for all standard basis vectors $\mathbf{e}_i$ of $\mathbb{R}^n$.
____ j. If $B = \{\mathbf{b}_1, \mathbf{b}_2, \ldots, \mathbf{b}_n\}$ is a basis for $\mathbb{R}^n$ and T and T' are linear transformations mapping $\mathbb{R}^n$ into $\mathbb{R}^m$, then $T(\mathbf{x}) = T'(\mathbf{x})$ for all $\mathbf{x} \in \mathbb{R}^n$ if and only if $T(\mathbf{b}_i) = T'(\mathbf{b}_i)$ for $i = 1, 2, \ldots, n$.

30. Verify that $T^{-1}(T(\mathbf{x})) = \mathbf{x}$ for the linear transformation T in Example 9 of the text.

31. Let $T: \mathbb{R}^n \to \mathbb{R}^m$ and $T': \mathbb{R}^m \to \mathbb{R}^k$ be linear transformations. Prove directly from Definition 2.3 that $(T' \circ T) \mathbb{R}^n \to \mathbb{R}^k$ is also a linear transformation.

32. Let $T: \mathbb{R}^n \to \mathbb{R}^m$ be a linear transformation. Prove from Definition 2.3 that $T(r\mathbf{u} + s\mathbf{v}) = rT(\mathbf{u}) + sT(\mathbf{v})$ for all $\mathbf{u}, \mathbf{v} \in \mathbb{R}^n$ and all scalars r and s.

Exercise 33 shows that the reduced row-echelon form of a matrix is unique.

33. Let A be an $m \times n$ matrix with row-echelon form H, and let V be the row space of A (and thus of H). Let $W_k = \text{sp}(\mathbf{e}_1, \mathbf{e}_2, \ldots, \mathbf{e}_k)$ be the subspace of $\mathbb{R}^n$ generated by the first k rows of the $n \times n$ identity matrix. Consider $T_k: V \to W_k$ defined by

$$T_k([x_1, x_2, \ldots, x_n]) = [x_1, x_2, \ldots, x_k, 0, \ldots, 0].$$

a. Show that T_k is a linear transformation of V into W_k and that $T_k[V] = \{T_k(\mathbf{v}) \mid \mathbf{v} \text{ in } V\}$ is a subspace of W_k.

b. If $T_k[V]$ has dimension d_k, show that, for each $j < n$, we have either $d_{j+1} = d_j$ or $d_{j+1} = d_j + 1$.

c. Assume that A has four columns. Referring to part (b), suppose that $d_1 = d_2 = 1$ and $d_3 = d_4 = 2$. Find the number of pivots in H, and give the location of each.

d. Repeat part (c) for the case where A has six columns and $d_1 = 1$, $d_2 = d_3 = d_4 = 2$, and $d_5 = d_6 = 3$.

e. Argue that, for any matrix A, the number of pivots and the location of each pivot in any row-echelon form of A is always the same.

f. Show that the reduced row-echelon form of a matrix A is unique. [HINT: Consider the nature of the basis for the row space of A given by the nonzero rows of H.]

34. Let $T: \mathbb{R}^n \to \mathbb{R}^m$ be a linear transformation and let U be a subspace of $\mathbb{R}^m$. Prove that the inverse image $T^{-1}[U]$ is a subspace of $\mathbb{R}^n$.

In Exercises 35–38, let T_1, T_2, T_3, and T_4 be linear transformations whose standard matrix representations are

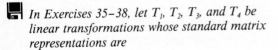

$$A = \begin{bmatrix} -4 & 5 & 7 \\ 2 & 4 & 5 \\ 1 & 8 & -5 \end{bmatrix}, \quad B = \begin{bmatrix} 8 & -3 \\ 1 & 4 \\ 2 & 5 \end{bmatrix},$$

$$C = \begin{bmatrix} -5 & 3 & -6 \\ 11 & 7 & -1 \end{bmatrix}, \quad \text{and}$$

$$D = \begin{bmatrix} 3 & -4 & 1 \\ -2 & 5 & 0 \end{bmatrix},$$

respectively. Use LINTEK or MATLAB to compute the indicated quantity, if it is defined. Load data files for the matrices if the data files are available.

35. $(T_1 \circ T_2 \circ T_4)([1, 2, 1])$

36. $(T_3 \circ T_1^{-1} \circ T_2)([0, -1])$

37. $(T_4 \circ (T_2 \circ T_3)^{-1} \circ T_2)([-1, 0])$

38. $(T_2 \circ (T_3 \circ T_2)^{-1} \circ T_4)([-1, 0, 1])$

39. Work with Topic 4 of the LINTEK routine VECTGRPH until you can consistently achieve a score of at least 80%.

40. Work with Topic 5 of the LINTEK routine VECTGRPH until you can regularly attain a score of at least 82%.

2.4   LINEAR TRANSFORMATIONS OF THE PLANE (OPTIONAL)

From the preceding section, we know that every linear transformation $T: \mathbb{R}^2 \to \mathbb{R}^2$ is given by $T(\mathbf{x}) = T_A(\mathbf{x}) = A\mathbf{x}$, where A is some 2×2 matrix. Different 2×2 matrices A give different transformations because $T(\mathbf{e}_1) = A\mathbf{e}_1$ is the first column vector of A and $T(\mathbf{e}_2) = A\mathbf{e}_2$ is the second column vector. The entire plane is mapped onto the column space of the matrix A. In this section we discuss these linear transformations of the plane $\mathbb{R}^2$ into itself, where we can draw reasonable pictures. We will use the familiar x,y-notation for coordinates in the plane.

The Collapsing (Noninvertible) Transformations

For a 2×2 matrix A to be noninvertible, it must have rank 0 or 1. If rank$(A) = 0$, then A is the zero matrix

$$\begin{bmatrix} 0 & 0 \\ 0 & 0 \end{bmatrix}$$

and we have $A(\mathbf{v}) = \mathbf{0}$ for all $\mathbf{v} \in \mathbb{R}^2$. Geometrically, the entire plane is collapsed to a single point—the origin.

If rank$(A) = 1$, then the column space of A, which is the range of T_A, is a one-dimensional subspace of $\mathbb{R}^2$, which is a line through the origin. The matrix A contains at least one nonzero column vector; if both column vectors are nonzero, then the second one is a scalar multiple of the first one. Examples of such matrices are

$$\begin{bmatrix} 1 & 0 \\ 0 & 0 \end{bmatrix}, \quad \begin{bmatrix} 0 & 0 \\ 0 & 1 \end{bmatrix}, \quad \begin{bmatrix} -1 & 0 \\ 1 & 0 \end{bmatrix}, \quad \begin{bmatrix} 1 & -3 \\ 2 & -6 \end{bmatrix}. \tag{1}$$

Projection   Projection   Collapse   Collapse
on x-axis   on y-axis   onto $y = -x$ onto $y = 2x$

The first two of these matrices produce *projections* on the coordinate axes, as labeled. *Projection* of the plane on a line L through the origin maps each vector $\mathbf{v}$ onto a vector $\mathbf{p}$ represented geometrically by the arrow starting at the origin and having its tip at the point on L that is closest to the tip of $\mathbf{v}$. The line through the tips of $\mathbf{v}$ and $\mathbf{p}$ must be perpendicular to the line L; phrased entirely in terms of vectors, the vector $\mathbf{v} - \mathbf{p}$ must be orthogonal to $\mathbf{p}$. This is illustrated in Figure 2.6. Projection on the x-axis is illustrated in Figure 2.7; we see that we have

$$T\left(\begin{bmatrix} x \\ y \end{bmatrix}\right) = \begin{bmatrix} x \\ 0 \end{bmatrix} = \begin{bmatrix} 1 & 0 \\ 0 & 0 \end{bmatrix}\begin{bmatrix} x \\ y \end{bmatrix}$$

in accord with our labeling of the first matrix in (1). Similarly, the second matrix in (1) gives projection on the y-axis. We refer to such matrices as *projection matrices*. The third and fourth matrices map the plane onto the indicated lines, as we readily see by examination of their column vectors. The transformations represented by these matrices are not projections onto those lines, however. Note that when projecting onto a line, every vector along the

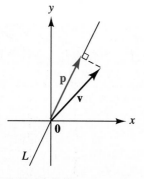

FIGURE 2.6

Projection onto the line *L*.

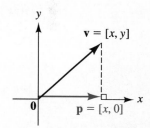

FIGURE 2.7

Projection onto the *x*-axis.

line is left fixed—that is, it is carried into itself. Now $[3, -3]$ is a vector along the line $y = -x$, but

$$\begin{bmatrix} -1 & 0 \\ 1 & 0 \end{bmatrix}\begin{bmatrix} 3 \\ -3 \end{bmatrix} = \begin{bmatrix} -3 \\ 3 \end{bmatrix} \neq \begin{bmatrix} 3 \\ -3 \end{bmatrix},$$

which shows that the third matrix in (1) is not a projection matrix. A similar computation shows that the final matrix in (1) is not a projection matrix. (See Exercise 1.) Chapter 6 discusses larger projection matrices.

Invertible Linear Transformations of the Plane

We know that if $T_A: \mathbb{R}^2 \to \mathbb{R}^2$ given by $T_A(\mathbf{x}) = A\mathbf{x}$ is an invertible linear transformation of the plane into itself, then A is an invertible 2×2 matrix so that $A\mathbf{x} = \mathbf{b}$ has a solution for every $\mathbf{b} \in \mathbb{R}^2$. Thus the range of T_A is all of $\mathbb{R}^2$.

Among the invertible linear transformations of the plane are the *rigid motions* of the plane that carry the origin into itself. *Rotation of the plane about the origin* counterclockwise through an angle θ, as illustrated in Figure 2.8, is an example of such a rigid motion.

EXAMPLE 1   Explain geometrically why $T: \mathbb{R}^2 \to \mathbb{R}^2$, which rotates the plane counterclockwise through an angle θ, is a linear transformation, and find its standard matrix representation. An algebraic proof is outlined in Exercise 23.

SOLUTION   We must show that for all $\mathbf{u}, \mathbf{v}, \mathbf{w} \in \mathbb{R}^2$ and all scalars r, we have $T(\mathbf{u} + \mathbf{v}) = T(\mathbf{u}) + T(\mathbf{v})$ and $T(r\mathbf{w}) = rT(\mathbf{w})$. Figure 2.8(a) indicates that the parallelogram that defines $\mathbf{u} + \mathbf{v}$ is carried into the parallelogram defining $T(\mathbf{u}) + T(\mathbf{v})$ by T, and Figure 2.8(b) similarly shows the lines illustrating $r\mathbf{w}$ and $T(r\mathbf{w})$. Thus T preserves addition and scalar multiplication. Figure 2.9 indicates that

$$T(\mathbf{e}_1) = [\cos\theta, \sin\theta] \quad \text{and} \quad T(\mathbf{e}_2) = [-\sin\theta, \cos\theta].$$

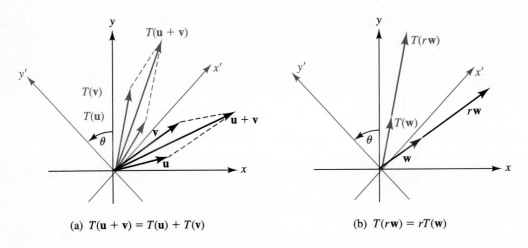

(a) $T(\mathbf{u} + \mathbf{v}) = T(\mathbf{u}) + T(\mathbf{v})$ (b) $T(r\mathbf{w}) = rT(\mathbf{w})$

FIGURE 2.8

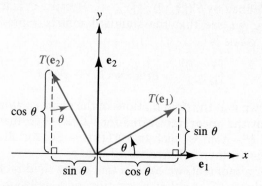

FIGURE 2.9
Counterclockwise rotation of e_1 and e_2
through the angle θ.

Thus

$$\begin{bmatrix} \cos\theta & -\sin\theta \\ \sin\theta & \cos\theta \end{bmatrix}. \quad \text{Counterclockwise rotation through } \theta \qquad (2)$$

is the standard matrix representation of this transformation.   ■

Another type of rigid motion T of the plane consists of "turning the plane over" around a line L through the origin. Turn the plane by holding the ends of the line L and rotating 180°, as you might hold a pencil by the ends with the "No. 2" designation on top and rotate it 180° so that the "No. 2" is on the underside. In analogy with the rotation in Figure 2.8, the parallelogram defining $\mathbf{u} + \mathbf{v}$ is carried into one defining $T(\mathbf{u}) + T(\mathbf{v})$, and similarly for the arrows defining the scalar product $r\mathbf{w}$. This type of rigid motion of the plane is called a *reflection in the line L*, because if we think of holding a mirror perpendicular to the plane with its bottom edge falling on L, then $T(\mathbf{v})$ is the reflection of $\mathbf{v}$ in this mirror, as indicated in Figure 2.10. Every vector $\mathbf{w}$ along L is carried into itself. As indicated in Figure 2.11, the reflection T of the plane

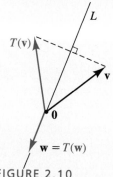

FIGURE 2.10
Reflection in the line L.

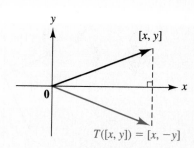

FIGURE 2.11
Reflection in the x-axis.

in the x-axis is defined by $T([x, y]) = [x, -y]$. Because $\mathbf{e}_1$ is left fixed and $\mathbf{e}_2$ is carried into $-\mathbf{e}_2$, we see that the standard matrix representation of this reflection in the x-axis is

$$\begin{bmatrix} 1 & 0 \\ 0 & -1 \end{bmatrix}. \quad \text{Reflection in the } x\text{-axis} \tag{3}$$

It can be shown that the rigid motions of the plane carrying the origin into itself are precisely the linear transformations $T: \mathbb{R}^2 \to \mathbb{R}^2$ that *preserve lengths* of all vectors in $\mathbb{R}^2$—that is, such that $\|T(\mathbf{x})\| = \|\mathbf{x}\|$ for all $\mathbf{x} \in \mathbb{R}^2$. We will discuss such ideas further in Exercises 17–22.

Thinking for a moment, we can see that every rigid motion of the plane leaving the origin fixed is either a rotation or a reflection followed by a rotation. Namely, if the plane is not turned over, all we can do is rotate it about the origin. If the plane has been turned over, we can achieve its final position by reflection in the x-axis, turning it over horizontally, followed by a rotation about the origin to obtain the desired position. We will use this last fact in the second solution of the next example. (Actually, every rigid motion leaving the origin fixed and turning the plane over is a reflection in some line through the origin, although this is not quite as easy to see.) The first solution of the next example illustrates that bases for $\mathbb{R}^2$ other than the standard basis can be useful.

EXAMPLE 2   Find the standard matrix representation A for the reflection of the plane in the line $y = 2x$.

SOLUTION 1   Let $\mathbf{b}_1 = [1, 2]$, which lies along the line $y = 2x$, and let $\mathbf{b}_2 = [-2, 1]$, which is orthogonal to $\mathbf{b}_1$ because $\mathbf{b}_1 \cdot \mathbf{b}_2 = 0$. These vectors are shown in Figure 2.12. If $T: \mathbb{R}^2 \to \mathbb{R}^2$ is reflection in the line $y = 2x$, then we have

$$T(\mathbf{b}_1) = \mathbf{b}_1 \quad \text{and} \quad T(\mathbf{b}_2) = -\mathbf{b}_2.$$

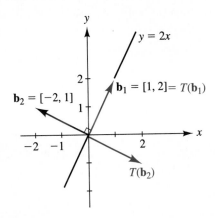

FIGURE 2.12
Reflection in the line $y = 2x$.

Now $\{\mathbf{b}_1, \mathbf{b}_2\}$ is a basis for $\mathbb{R}^2$, and Theorem 2.7 tells us that T is completely determined by its action on this basis. To find $T(\mathbf{e}_1)$ and $T(\mathbf{e}_2)$ for the column vectors in the standard matrix representation A of T, we first express $\mathbf{e}_1$ and $\mathbf{e}_2$ as linear combinations of $\mathbf{b}_1$ and $\mathbf{b}_2$. To do this, we solve the two linear systems with $\mathbf{e}_1$ and $\mathbf{e}_2$ as column vectors of constants and $\mathbf{b}_1$ and $\mathbf{b}_2$ as columns of the coefficient matrix, as follows:

$$\left[\begin{array}{rr|rr} 1 & -2 & 1 & 0 \\ 2 & 1 & 0 & 1 \end{array}\right] \sim \left[\begin{array}{rr|rr} 1 & -2 & 1 & 0 \\ 0 & 5 & -2 & 1 \end{array}\right] \sim \left[\begin{array}{rr|rr} 1 & 0 & \frac{1}{5} & \frac{2}{5} \\ 0 & 1 & -\frac{2}{5} & \frac{1}{5} \end{array}\right].$$

Thus we have

$$\mathbf{e}_1 = \frac{1}{5}\mathbf{b}_1 - \frac{2}{5}\mathbf{b}_2 \quad \text{and} \quad \mathbf{e}_2 = \frac{2}{5}\mathbf{b}_1 + \frac{1}{5}\mathbf{b}_2.$$

Applying the transformation T to both sides of these equations, we obtain

$$T(\mathbf{e}_1) = \frac{1}{5}T(\mathbf{b}_1) - \frac{2}{5}T(\mathbf{b}_2) = \frac{1}{5}\mathbf{b}_1 + \frac{2}{5}\mathbf{b}_2 = \frac{1}{5}[1, 2] + \frac{2}{5}[-2, 1] = \left[-\frac{3}{5}, \frac{4}{5}\right]$$

and

$$T(\mathbf{e}_2) = \frac{2}{5}T(\mathbf{b}_1) + \frac{1}{5}T(\mathbf{b}_2) = \frac{2}{5}\mathbf{b}_1 - \frac{1}{5}\mathbf{b}_2 = \frac{2}{5}[1, 2] - \frac{1}{5}[-2, 1] = \left[\frac{4}{5}, \frac{3}{5}\right].$$

Thus the standard matrix representation is

$$A = \left[\begin{array}{rr} -\frac{3}{5} & \frac{4}{5} \\ \frac{4}{5} & \frac{3}{5} \end{array}\right].$$

SOLUTION 2   The three parts of Figure 2.13 show that we can attain the reflection of the plane in the line $y = 2x$ as follows: First reflect in the x-axis, taking us from part (a) to part (b) of the figure, and then rotate counterclockwise through the angle 2θ, where θ is the angle from the x-axis to the line $y = 2x$, measured counterclockwise. Using the double angle formulas derived in Section 2.3, we see from the right triangle in Figure 2.13(a) that

$$\sin 2\theta = 2 \sin\theta \cos\theta = 2\left(\frac{2}{\sqrt{5}}\right)\left(\frac{1}{\sqrt{5}}\right) = \frac{4}{5}$$

and

$$\cos 2\theta = \cos^2\theta - \sin^2\theta = \frac{1}{5} - \frac{4}{5} = -\frac{3}{5}.$$

Replacing θ by 2θ in the matrix in Example 1, we see that the standard matrix representation for rotation through the angle 2θ is

$$\left[\begin{array}{rr} -\frac{3}{5} & -\frac{4}{5} \\ \frac{4}{5} & -\frac{3}{5} \end{array}\right]. \tag{4}$$

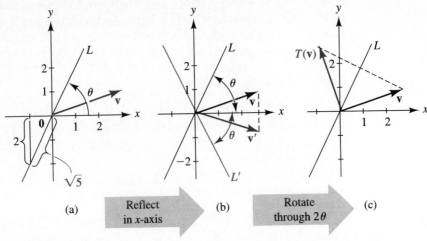

FIGURE 2.13
(a) The vector v   (b) Reflected   (c) Rotated

Multiplying matrices (4) and (3), we obtain

$$A = \begin{bmatrix} -\frac{3}{5} & -\frac{4}{5} \\ \frac{4}{5} & -\frac{3}{5} \end{bmatrix} \begin{bmatrix} 1 & 0 \\ 0 & -1 \end{bmatrix} = \begin{bmatrix} -\frac{3}{5} & \frac{4}{5} \\ \frac{4}{5} & \frac{3}{5} \end{bmatrix}.$$

Rotate   Reflect   ∎

In Example 2, note that because we first reflect in the x-axis and then rotate through 2θ, the matrix for the reflection is the one on the *right*, which acts on a vector $\mathbf{v} \in \mathbb{R}^2$ *first* when computing $A\mathbf{v}$.*

A Geometric Description of All Invertible Transformations of the Plane

We exploit matrix techniques to describe geometrically all invertible linear transformations of the plane into itself. Recall that every invertible matrix is a product of elementary matrices. If we can interpret geometrically the effect on the plane of the linear transformations having elementary 2×2 matrices as their standard representations, we will gain insight into all invertible linear transformations of $\mathbb{R}^2$ into $\mathbb{R}^2$.

---

*This right-to-left order for composite transformations occurs because we write functions on the left of the elements of the domain on which they act, writing $f(x)$ rather than $(x)f$. From a pedagogical standpoint, writing functions on the left must be regarded as a peculiarity in the development of mathematical notations in a society where text is read from left to right. If we wrote functions on the right side, then we would take the transpose of $A\mathbf{x}$ and write

$$\mathbf{x}^T A^T = \mathbf{x}^T[\text{reflection matrix}]^T[\text{rotation matrix}]^T,$$

except that we would of course have developed things in terms of row vectors so that the transpose notation would not appear as it does here.

EXAMPLE 3   Describe geometrically the effect on the plane of the linear transformation T_E where E is an elementary matrix obtained by multiplying a row of the 2×2 identity matrix I by -1.

SOLUTION   The matrix obtained by multiplying the second row of I by -1 is

$$E = \begin{bmatrix} 1 & 0 \\ 0 & -1 \end{bmatrix},$$

which is the matrix given as matrix (3). We saw there that T_E is the reflection in the x-axis and $T_E([x, y]) = [x, -y]$. Similarly, we see that the elementary matrix obtained by multiplying the first row of I by -1 represents the transformation that changes the sign of the first component of a vector, carrying $[x, y]$ into $[-x, y]$. This is the reflection in the y-axis.   ∎

EXAMPLE 4   Describe geometrically the effect on the plane of the linear transformation T_E where E is an elementary matrix obtained by interchanging the rows of the 2×2 identity matrix I.

SOLUTION   Here we have

$$E = \begin{bmatrix} 0 & 1 \\ 1 & 0 \end{bmatrix} \quad \text{and} \quad \begin{bmatrix} 0 & 1 \\ 1 & 0 \end{bmatrix} \begin{bmatrix} x \\ y \end{bmatrix} = \begin{bmatrix} y \\ x \end{bmatrix}.$$

In row-vector notation, we have $T_E([x, y]) = [y, x]$. Figure 2.14 indicates that this transformation, which interchanges the components of a vector in the plane, is the reflection in the line $y = x$.   ∎

EXAMPLE 5   Describe geometrically the linear transformation $T\left(\begin{bmatrix} x \\ y \end{bmatrix}\right) = E\begin{bmatrix} x \\ y \end{bmatrix}$, where E is a 2×2 elementary matrix corresponding to row scaling.

SOLUTION   The matrix E has the form

$$\begin{bmatrix} r & 0 \\ 0 & 1 \end{bmatrix} \quad \text{or} \quad \begin{bmatrix} 1 & 0 \\ 0 & r \end{bmatrix}$$

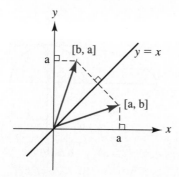

FIGURE 2.14
Reflection in the line $y = x$.

for some nonzero scalar r. We discuss the first case and leave the second as Exercise 8. The transformation is given by

$$T\left(\begin{bmatrix} x \\ y \end{bmatrix}\right) = \begin{bmatrix} r & 0 \\ 0 & 1 \end{bmatrix}\begin{bmatrix} x \\ y \end{bmatrix} = \begin{bmatrix} rx \\ y \end{bmatrix},$$

or, in row notation, $T([x, y]) = [rx, y]$. The second component of $[x, y]$ is unchanged. However, the first component is multiplied by the scalar r, resulting in a horizontal expansion if $r > 1$ or in a horizontal contraction if $0 < r < 1$. In Figure 2.15, we illustrate the effect of such a horizontal expansion or contraction on the points of the unit circle. If $r < 0$, we have an expansion or contraction followed by a reflection in the y-axis. For example,

$$\begin{bmatrix} -3 & 0 \\ 0 & 1 \end{bmatrix} = \underbrace{\begin{bmatrix} -1 & 0 \\ 0 & 1 \end{bmatrix}}_{\text{Reflection}}\underbrace{\begin{bmatrix} 3 & 0 \\ 0 & 1 \end{bmatrix}}_{\text{Horizontal expansion}},$$

indicating a horizontal expansion by a factor of 3, followed by a reflection in the y-axis.   ■

EXAMPLE 6   Describe geometrically the linear transformation $T\left(\begin{bmatrix} x \\ y \end{bmatrix}\right) = E\begin{bmatrix} x \\ y \end{bmatrix}$, where E is a 2×2 elementary matrix corresponding to row addition.

SOLUTION   The matrix E has the form

$$\begin{bmatrix} 1 & 0 \\ r & 1 \end{bmatrix} \quad \text{or} \quad \begin{bmatrix} 1 & r \\ 0 & 1 \end{bmatrix}$$

for some nonzero scalar r. We discuss the first case, and leave the second as Exercise 10. The transformation is given by

$$T\left(\begin{bmatrix} x \\ y \end{bmatrix}\right) = \begin{bmatrix} 1 & 0 \\ r & 1 \end{bmatrix}\begin{bmatrix} x \\ y \end{bmatrix} = \begin{bmatrix} x \\ rx + y \end{bmatrix},$$

or, in row-vector notation, $T([x, y])$ and $[x, rx + y]$. The first component of the vector $[x, y]$ is unchanged. However, the second component is changed by the

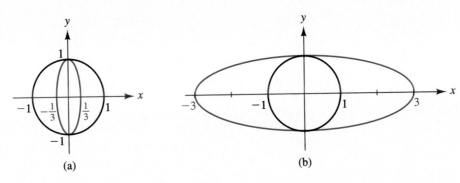

FIGURE 2.15

(a) $T([x, y]) = \left[\frac{1}{3}x, y\right]$ contracts horizontally; (b) $T([x, y]) = [3x, y]$ expands horizontally.

addition of *rx*. For example, [1, 0] is carried onto [1, *r*], and [1, 1] is carried onto [1, 1 + *r*], while [0, 0] and [0, 1] are carried onto themselves. Notice that every vector along the *y*-axis remains fixed. Figure 2.16 illustrates the effect of this transformation. The squares shaded in black are carried onto the parallelograms shaded in color. This transformation is called a *vertical shear*. Exercise 10 deals with the case of a horizontal shear.   ∎

We have noted that a square matrix *A* is invertible if and only if it is the product of elementary matrices. We also know that a product of matrices corresponds to the composition of the associated linear transformations, and we have seen the effect of transformations associated with elementary matrices on the plane. Putting all of these ideas together, we obtain the following.

Geometric Description of Invertible Transformations of $\mathbb{R}^2$

A linear transformation *T* of the plane $\mathbb{R}^2$ into itself is invertible if and only if *T* consists of a finite sequence of:

Reflections in the *x*-axis, the *y*-axis, or the line *y* = *x*;

Vertical or horizontal expansions or contractions; and

Vertical or horizontal shears.

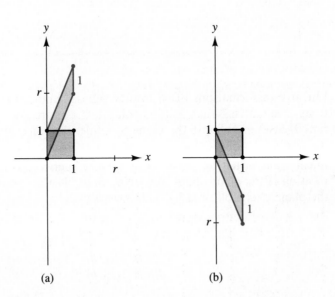

FIGURE 2.16
(a) The vertical shear *T*([*x*, *y*]) = [*x*, *rx* + *y*], *r* > 0
(b) the vertical shear *T*([*x*, *y*]) = [*x*, *rx* + *y*], *r* < 0

EXAMPLE 7   Illustrate the result relating to the boxed description above for the invertible linear transformation $T([x, y]) = [x + 2y, 3x + 4y]$.

SOLUTION   We reduce the standard matrix representation A of T, obtaining

$$A = \begin{bmatrix} 1 & 2 \\ 3 & 4 \end{bmatrix} \overset{E_1}{\sim} \begin{bmatrix} 1 & 2 \\ 0 & -2 \end{bmatrix} \overset{E_2}{\sim} \begin{bmatrix} 1 & 2 \\ 0 & 1 \end{bmatrix} \overset{E_3}{\sim} \begin{bmatrix} 1 & 0 \\ 0 & 1 \end{bmatrix}.$$

$$R_2 \to R_2 - 3R_1 \qquad R_2 \to -\frac{1}{2}R_2 \qquad R_1 \to R_1 - 2R_2$$

In terms of elementary matrices, this reduction becomes

$$\underset{E_3}{\begin{bmatrix} 1 & -2 \\ 0 & 1 \end{bmatrix}} \underset{E_2}{\begin{bmatrix} 1 & 0 \\ 0 & -\frac{1}{2} \end{bmatrix}} \underset{E_1}{\begin{bmatrix} 1 & 0 \\ -3 & 1 \end{bmatrix}} A = I,$$

and so

$$A = E_1^{-1} E_2^{-1} E_3^{-1} = \begin{bmatrix} 1 & 0 \\ 3 & 1 \end{bmatrix} \begin{bmatrix} 1 & 0 \\ 0 & -2 \end{bmatrix} \begin{bmatrix} 1 & 2 \\ 0 & 1 \end{bmatrix}.$$

This shows that T consists of a horizontal shear (matrix E_3^{-1}) followed by an expansion and reflection (matrix E_2^{-1}), followed in turn by a vertical shear (matrix E_1^{-1}). ▪

SUMMARY

1. Linear transformations of $\mathbb{R}^2$ into $\mathbb{R}^2$ whose standard matrix representations have rank less than 2 either collapse the entire plane to the origin (the rank 0 case) or collapse the plane to a line (the rank 1 case).

2. A rigid motion of the plane into itself that leaves the origin fixed gives a linear transformation $T: \mathbb{R}^2 \to \mathbb{R}^2$. Every such rigid motion is either a rotation of the plane about the origin, or a reflection in the x-axis (to turn the plane over) followed by such a rotation.

3. The standard matrix representation for the rotation of the plane counter-clockwise about the origin through an angle θ is

$$\begin{bmatrix} \cos \theta & -\sin \theta \\ \sin \theta & \cos \theta \end{bmatrix}.$$

4. An invertible linear transformation of $\mathbb{R}^2$ into itself can be described geometrically, using elementary matrices, as indicated in the box preceding Example 7.

EXERCISES

1. Explain why the linear transformation
   $T_A: \mathbb{R}^2 \to \mathbb{R}^2$, where $A = \begin{bmatrix} 1 & -3 \\ 2 & -6 \end{bmatrix}$, has the line
   $y = 2x$ as range, but is not the projection
   of $\mathbb{R}^2$ onto that line.

2. Give the standard matrix representation of
   the rotation of the plane *counterclockwise*
   about the origin through an angle of
   a. 30°,
   b. 90°,
   c. 135°.

3. Give the standard matrix representation of
   the rotation of the plane *clockwise* about the
   origin through an angle of
   a. 45°,
   b. 60°,
   c. 150°.

4. Use the rotation matrix in item 3 of the
   Summary to derive trigonometric identities
   for $\sin 3\theta$ and $\cos 3\theta$ in terms of $\sin \theta$ and
   $\cos \theta$. (See Illustration 1, Section 2.3.)

5. Use the rotation matrix in item 3 of the
   Summary to derive trigonometric identities
   for $\sin(\theta + \phi)$ and $\cos(\theta + \phi)$ in terms of
   $\sin \theta$, $\sin \phi$, $\cos \theta$, and $\cos \phi$. (See
   Illustration 1, Section 2.3.)

6. Find the general matrix representation for
   the reflection of the plane in the line $y = mx$, using the method for the case $m = 2$ in
   Solution 1 of Example 2 in the text.

7. Repeat Exercise 6, but use the method for
   the case $m = 2$ in Solution 2 of Example 2
   in the text.

8. Show that the linear transformation

   $$T\left(\begin{bmatrix} x \\ y \end{bmatrix}\right) = \begin{bmatrix} 1 & 0 \\ 0 & r \end{bmatrix} \begin{bmatrix} x \\ y \end{bmatrix}$$

   affects the plane $\mathbb{R}^2$ as follows:
   (i) A vertical expansion, if $r > 1$;
   (ii) A vertical contraction, if $0 < r < 1$;
   (iii) A vertical expansion followed by a
        reflection in the x-axis, if $r < -1$;
   (iv) A vertical contraction followed by a
        reflection in the x-axis, if $-1 < r < 0$.

9. Referring to Exercise 8, explain algebraically
   why cases (iii) and (iv) can be described by
   the reflection followed by the expansion or
   contraction, in that order.

10. Show that the linear transformation

    $$T\left(\begin{bmatrix} x \\ y \end{bmatrix}\right) = \begin{bmatrix} 1 & r \\ 0 & 1 \end{bmatrix} \begin{bmatrix} x \\ y \end{bmatrix}$$

    corresponds to a horizontal shear of the
    plane.

*In Exercises 11–15, express the standard matrix
representation of the given invertible
transformation of $\mathbb{R}^2$ into itself as a product of
elementary matrices. Use this expression to
describe the transformation as a product of one or
more reflections, horizontal or vertical expansions
or contractions, and shears.*

11. $T([x, y]) = [-y, x]$ (Rotation
    counterclockwise through 90°)

12. $T([x, y]) = [2x, 2y]$ (Expansion away from
    the origin by a factor of 2)

13. $T([x, y]) = [-x, -y]$ (Rotation through
    180°)

14. $T([x, y]) = [x + y, 2x - y]$

15. $T([x, y]) = [x + y, 3x + 5y]$

16. Mark each of the following True or False.
    ___ **a.** Every rotation of the plane is a linear
       transformation.
    ___ **b.** Every rotation of the plane about the
       origin is a linear transformation.
    ___ **c.** Every reflection of the plane in a line L is
       a rigid motion of the plane.
    ___ **d.** Every reflection of the plane in a line L is
       a linear transformation of the plane.
    ___ **e.** Every rigid motion of the plane that
       carries the origin into itself is a linear
       transformation.
    ___ **f.** Every invertible linear transformation of
       the plane is a rigid motion.
    ___ **g.** If a linear transformation $T: \mathbb{R}^2 \to \mathbb{R}^2$ is a
       rigid motion of the plane, then
       $\|T(\mathbf{x})\| = \|\mathbf{x}\|$ for all $\mathbf{x} \in \mathbb{R}^2$.

_____ **h.** The geometric effect of all invertible linear transformations of $\mathbb{R}^3$ into itself can be described in terms of the geometric effect of the linear transformations of $\mathbb{R}^3$ having elementary matrices as standard matrix representations.

_____ **i.** Every linear transformation of the plane into itself can be achieved through a succession of reflections, expansions, contractions, and shears.

_____ **j.** Every invertible linear transformation of the plane into itself can be achieved through a succession of reflections, expansions, contractions, and shears.

A linear transformation $T: \mathbb{R}^2 \to \mathbb{R}^2$ **preserves length** if $\|T(\mathbf{x})\| = \|\mathbf{x}\|$ for all $\mathbf{x} \in \mathbb{R}^2$. It **preserves angle** if the angle between $\mathbf{u}$ and $\mathbf{v}$ is the same as the angle between $T(\mathbf{u})$ and $T(\mathbf{v})$ for all $\mathbf{u}, \mathbf{v} \in \mathbb{R}^2$. It **preserves the dot product** if $T(\mathbf{u}) \cdot T(\mathbf{v}) = \mathbf{u} \cdot \mathbf{v}$ for all $\mathbf{u}, \mathbf{v} \in \mathbb{R}^2$.

We recommend that Exercises 17–22 be worked sequentially, or at least be read sequentially.

17. Use the familiar equation that describes the dot product $\mathbf{u} \cdot \mathbf{v}$ geometrically to prove that if a linear transformation $T: \mathbb{R}^2 \to \mathbb{R}^2$ preserves both length and angle, then it also preserves the dot product.

18. Use algebraic properties of the dot product to compute $\|\mathbf{u} - \mathbf{v}\|^2 = (\mathbf{u} - \mathbf{v}) \cdot (\mathbf{u} - \mathbf{v})$, and prove from the resulting equation that a linear transformation $T: \mathbb{R}^2 \to \mathbb{R}^2$ that preserves length also preserves the dot product.

19. Express both the length of a vector $\mathbf{v} \in \mathbb{R}^2$ and the angle between two nonzero vectors $\mathbf{u}, \mathbf{v} \in \mathbb{R}^2$ in terms of the dot product only. (From this we may conclude that if a linear transformation $T: \mathbb{R}^2 \to \mathbb{R}^2$ preserves the dot product, then it preserves length and angle.)

20. Suppose that $T_A: \mathbb{R}^2 \to \mathbb{R}^2$ preserves both length and angle. Prove that the two column vectors of the matrix A are orthogonal unit vectors.

21. Prove that the two column vectors of a 2×2 matrix A are orthogonal unit vectors if and only if $(A^T)A = I$. Demonstrate that the matrix representations for the rigid motions given in Examples 1 and 2 satisfy this condition.

22. Let A be a 2×2 matrix such that $(A^T)A = I$. Prove that the linear transformation T_A preserves the dot product, and hence also preserves length and angle. [HINT: Note that the dot product of two column vectors $\mathbf{u}, \mathbf{v} \in \mathbb{R}^n$ is the entry in the 1×1 matrix $(\mathbf{u}^T)\mathbf{v}$. Compute the dot product $T_A(\mathbf{u}) \cdot T_A(\mathbf{v})$ by computing $(A\mathbf{u})^T(A\mathbf{v})$.]

23. This exercise outlines an algebraic proof that rotation of the plane about the origin is a linear transformation. Let $T: \mathbb{R}^2 \to \mathbb{R}^2$ be the function that rotates the plane counterclockwise through an angle θ as in Example 1.

   a. Prove algebraically that each vector $\mathbf{v} \in \mathbb{R}^2$ can be written in the *polar form* $\mathbf{v} = r[\cos \alpha, \sin \alpha]$. [HINT: Each unit vector has this form with $r = 1$.]

   b. For $\mathbf{v} = r[\cos \alpha, \sin \alpha]$, express $T(\mathbf{v})$ in this polar form.

   c. Using column-vector notation and appropriate trigonometric identities, find a matrix A such that $T(\mathbf{v}) = A\mathbf{v}$. The existence of such a matrix A proves that T is a linear transformation.

LINES, PLANES, AND OTHER FLATS (Optional)

We turn to geometry in this section, and generalize the notions of a *line* in the plane or in space and of a *plane* in space. Our work in the preceding sections will enable us to describe geometrically the solution set of any consistent linear system.

The Notion of a Line in $\mathbb{R}^n$

For a nonzero vector **d** in $\mathbb{R}^2$, we visualize the one-dimensional subspace sp(**d**) as a line through the origin, as shown in Figure 2.17. Similarly, Figure 2.18 indicates that for a nonzero vector **d** in $\mathbb{R}^n$, we can view sp(**d**) = $\{t\mathbf{d} \mid t \in \mathbb{R}\}$ as a line through the origin. Every subspace of $\mathbb{R}^n$ contains the origin (zero vector), but we surely want to consider lines in $\mathbb{R}^n$ that do not pass through the origin, such as line L in Figure 2.19. As indicated in Figure 2.19, if **a** is a vector to a point on the line L, then every point on L is at the tip of a vector $\mathbf{x} = t\mathbf{d} + \mathbf{a}$, where t is a scalar and **d** is any fixed nonzero vector that we regard intuitively as *parallel* to the line L. This line is thus obtained from the line sp(**d**) by *translation*. Geometers consider a translation of a subset S of $\mathbb{R}^n$ to be a sliding of every point in S in the same direction and for the same distance. The direction and distance for a translation can be specified by a vector (such as vector **a** above) pointing in the direction of the translation and having magnitude equal to the distance the points are moved. The image of S under such a translation is a *translate* of S. We will give a formal definition of a translate of a subset, present an example, and then define a line in $\mathbb{R}^n$.

DEFINITION 2.4   Translate of a Subset of $\mathbb{R}^n$

> Let S be a subset of $\mathbb{R}^n$ and let **a** be a vector in $\mathbb{R}^n$. The set $\{\mathbf{x} + \mathbf{a} \mid \mathbf{x} \in S\}$ is the **translate** of S by **a**, and is denoted by $S + \mathbf{a}$. The vector **a** is the **translation vector**.

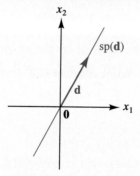

FIGURE 2.17
A line through the origin in $\mathbb{R}^2$.

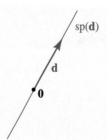

FIGURE 2.18
A line through the origin in $\mathbb{R}^n$.

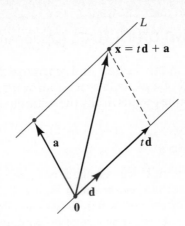

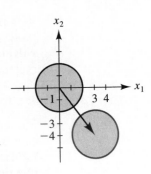

FIGURE 2.19
A general line *L* in $\mathbb{R}^n$.

FIGURE 2.20
Translate of $\{x \in \mathbb{R}^2 \mid \|x\| \le 2\}$ in $\mathbb{R}^2$
by $[3, -4]$.

EXAMPLE 1   Sketch the translate of the subset $S = \{x \in \mathbb{R}^2 \mid \|x\| \le 2\}$ of $\mathbb{R}^2$ by the vector $[3, -4]$.

SOLUTION   The subset S of $\mathbb{R}^2$ is a disk with center at the origin and a radius of 2. As shown in Figure 2.20, its translate by $[3, -4]$ is the disk with center at the point $(3, -4)$ and a radius of 2.   ∎

DEFINITION 2.5   Line in $\mathbb{R}^n$

A **line** in $\mathbb{R}^n$ is a translate of a one-dimensional subspace of $\mathbb{R}^n$.

Although our definition defines a line to be a set of vectors, it is customary in geometry to consider a line as the set of *points* in $\mathbb{R}^n$ whose coordinates correspond to the components of the vectors.

We can specify a line L in $\mathbb{R}^n$ by giving a point $(a_1, a_2, \ldots, a_n)$ on the line and a vector **d** parallel to the line. We consider **d** to be a *direction vector* for the line, whereas **a** is a *translation vector*. In the terminology of Definition 2.5, L is the translate of the subspace sp(**d**) by the vector $\mathbf{a} = [a_1, a_2, \ldots, a_n]$—that is, $L = \{t\mathbf{d} + \mathbf{a} \mid t \in \mathbb{R}\}$. We can describe the line by the single equation

$$\mathbf{x} = t\mathbf{d} + \mathbf{a} \quad \text{Vector equation of } L$$

or by the equations

$$x_1 = td_1 + a_1$$
$$x_2 = td_2 + a_2$$
$$\quad \cdot \qquad\qquad \textbf{Component equations of } L$$
$$\quad \cdot$$
$$\quad \cdot$$
$$x_n = td_n + a_n.$$

In classical geometry, component equations for L are also called **parametric equations** for L, and the variable t is a **parameter**. Of course, this same parameter t appears in the vector equation also.

EXAMPLE 2   Find a vector equation and component equations for the line in $\mathbb{R}^2$ through $(2, 1)$ having direction vector $[3, 4]$. Then find the point on the line having -4 as its x_1-coordinate.

SOLUTION   The line can be characterized as the translate of $\text{sp}([3, 4])$ by the vector $[2, 1]$, and so a vector equation of the line is

$$[x_1, x_2] = t[3, 4] + [2, 1].$$

The component equations are

$$x_1 = 3t + 2, \qquad x_2 = 4t + 1.$$

Because t runs through all real numbers, we obtain all points (x_1, x_2) on the line from these component equations. In order to find the point on the line with -4 as x_1-coordinate, we set $-4 = 3t + 2$ and obtain $t = -2$. Thus the x_2-coordinate is $4(-2) + 1 = -7$, and so the desired point is $(-4, -7)$.   ∎

EXAMPLE 3   Find parametric equations of the line in $\mathbb{R}^3$ that passes through the points $(2, -1, 3)$ and $(1, 3, 5)$.

SOLUTION   We arbitrarily choose $\mathbf{a} = [2, -1, 3]$ as the translation vector corresponding to the point $(2, -1, 3)$ on the line. A direction vector is given by

$$\mathbf{d} = [1, 3, 5] - [2, -1, 3] = [-1, 4, 2],$$

as indicated in Figure 2.21. We obtain

$$[x_1, x_2, x_3] = t[-1, 4, 2] + [2, -1, 3]$$

as a vector equation for the line. The corresponding parametric (component) equations are

$$x_1 = -t + 2, \qquad x_2 = 4t - 1, \qquad x_3 = 2t + 3.$$ ∎

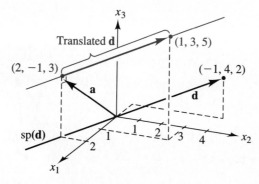

FIGURE 2.21
The line passing through $(2, -1, 3)$ and $(1, 3, 5)$.

Line Segments

Consider the line in $\mathbb{R}^n$ that passes through the two points $(a_1, a_2, \ldots, a_n)$ and $(b_1, b_2, \ldots, b_n)$. Letting **a** and **b** be the vectors corresponding to these points, we see as in Example 3 that $\mathbf{d} = \mathbf{b} - \mathbf{a}$ is a direction vector for the line. The vector equation

$$\mathbf{x} = t\mathbf{d} + \mathbf{a} = t(\mathbf{b} - \mathbf{a}) + \mathbf{a} \tag{1}$$

for the line in effect presents the line as a t-axis whose origin is at the point $(a_1, a_2, \ldots, a_n)$ and on which a one-unit change in t corresponds to $\|\mathbf{d}\|$ units distance in $\mathbb{R}^n$. This is illustrated in Figure 2.22.

As illustrated in Figure 2.23, each point in $\mathbb{R}^n$ on the line segment that joins the tip of **a** to the tip of **b** lies at the tip of a vector **x** obtained in Eq. (1) for some value of t for which $0 \leq t \leq 1$. Note that $t = 0$ yields the point at the tip of **a** and $t = 1$ yields the point at the tip of **b**. By choosing t between 0 and 1 appropriately, we can find the coordinates of any point on this line segment. In particular, the coordinates of the **midpoint** of the line segment are the components of the vector

$$\mathbf{a} + \frac{1}{2}(\mathbf{b} - \mathbf{a}) = \frac{1}{2}(\mathbf{a} + \mathbf{b}).$$

EXAMPLE 4   Find the points that divide into five equal parts the line segment that joins $(1, 2, 1, 3)$ to $(2, 1, 4, 2)$ in $\mathbb{R}^4$.

SOLUTION   We obtain $\mathbf{d} = [2, 1, 4, 2] - [1, 2, 1, 3] = [1, -1, 3, -1]$ as a direction vector for the line through the two given points. The corresponding vector equation of the line is

$$[x_1, x_2, x_3, x_4] = t[1, -1, 3, -1] + [1, 2, 1, 3].$$

By choosing $t = 0, \frac{1}{5}, \frac{2}{5}, \frac{3}{5}, \frac{4}{5}$, and 1, we obtain coordinates of the points that divide the segment as required, as shown in Table 2.1.   ∎

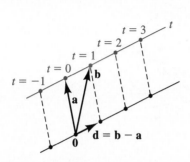

FIGURE 2.22
Equation (1) sets up a t-axis.

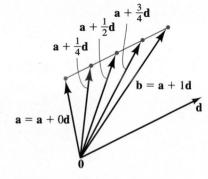

FIGURE 2.23
Points on a line segment.

TABLE 2.1

t	Equally Spaced Points
0	(1, 2, 1, 3)
$\frac{1}{5}$	(1.2, 1.8, 1.6, 2.8)
$\frac{2}{5}$	(1.4, 1.6, 2.2, 2.6)
$\frac{3}{5}$	(1.6, 1.4, 2.8, 2.4)
$\frac{4}{5}$	(1.8, 1.2, 3.4, 2.2)
1	(2, 1, 4, 2)

Flats in $\mathbb{R}^n$

Just as a line is a translate of a one-dimensional subspace in $\mathbb{R}^n$, a *plane* in $\mathbb{R}^n$ is a translate of a two-dimensional subspace $\mathrm{sp}(\mathbf{d}_1, \mathbf{d}_2)$, where $\mathbf{d}_1$ and $\mathbf{d}_2$ are nonzero, nonparallel vectors in $\mathbb{R}^n$. A plane appears as a *flat* piece of $\mathbb{R}^n$, as illustrated in Figure 2.24. We have no word analogous to "straight" or "flat" in our language to denote that $\mathbb{R}^3$ is not "curved." We borrow the term "flat" when generalizing to higher dimensions, and describe a translate of a k-dimensional subspace of $\mathbb{R}^n$ for $k < n$ as being "flat." Let us give a formal definition.

DEFINITION 2.6   A k-Flat in $\mathbb{R}^n$

A **k-flat** in $\mathbb{R}^n$ is a translate of a k-dimensional subspace of $\mathbb{R}^n$. In particular, a 1-flat is a **line**, a 2-flat is a **plane**, and an $(n - 1)$-flat is a **hyperplane**. We consider each point of $\mathbb{R}^n$ to be a **zero-flat**.

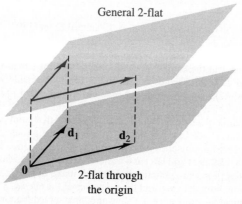

General 2-flat

$\mathbf{d}_1$ $\mathbf{d}_2$

0

2-flat through
the origin

FIGURE 2.24
Planes or 2-flats in $\mathbb{R}^n$

Just as for a line, it is conventional in geometry to speak of the translate $W + \mathbf{a}$ of a k-dimensional subspace W of $\mathbb{R}^n$ as the *k-flat through the point* $(a_1, a_2, \ldots, a_n)$ *parallel to* W. If $\{\mathbf{d}_1, \mathbf{d}_2, \ldots, \mathbf{d}_k\}$ is a basis for W, then

$$\mathbf{x} = t_1\mathbf{d}_1 + t_2\mathbf{d}_2 + \cdots + t_k\mathbf{d}_k + \mathbf{a} \tag{2}$$

is the **vector equation** of the k-flat. (We use the letter $\mathbf{d}$ because W determines the direction of the k-flat as being parallel to W.) The corresponding component equations are again called **parametric equations** for the k-flat.

EXAMPLE 5   Find parametric equations of the plane in $\mathbb{R}^4$ passing through the points $(1, 1, 1, 1)$, $(2, 1, 1, 0)$, and $(3, 2, 1, 0)$.

SOLUTION   We arbitrarily choose $\mathbf{a} = [1, 1, 1, 1]$ as the translation vector corresponding to the point $(1, 1, 1, 1)$ on the desired plane. Two vectors that (when translated) start at this point and reach to the other two points are

$$\mathbf{d}_1 = [2, 1, 1, 0] - [1, 1, 1, 1] = [1, 0, 0, -1]$$

and

$$\mathbf{d}_2 = [3, 2, 1, 0] - [1, 1, 1, 1] = [2, 1, 0, -1].$$

Because these vectors are nonparallel, they form a basis for the 2-flat through the origin and parallel to the desired plane. See Figure 2.25. The vector equation of the plane is $\mathbf{x} = s\mathbf{d}_1 + t\mathbf{d}_2 + \mathbf{a}$, or, written out,

$$[x_1, x_2, x_3, x_4] = s[1, 0, 0, -1] + t[2, 1, 0, -1] + [1, 1, 1, 1].$$

HISTORICAL NOTE   THE EQUATION OF A PLANE IN $\mathbb{R}^3$ appears as early as 1732 in a paper of Jacob Hermann (1678–1733). He was able to determine the plane's position by using intercepts, and he also noted that the sine of the angle between the plane and the one coordinate plane he dealt with (what we call the x_1, x_2-plane) was

$$\frac{\sqrt{d_1^2 + d_2^2}}{\sqrt{d_1^2 + d_2^2 + d_3^2}}.$$

In his 1748 *Introduction to Infinitesimal Analysis*, Leonhard Euler (1707–1783) used, instead, the cosine of this angle, $d_3/\sqrt{d_1^2 + d_2^2 + d_3^2}$.

At the end of the eighteenth century, Gaspard Monge (1746–1818), in his notes for a course on solid analytic geometry at the Ecole Polytechnique, related the equation of a plane to all three coordinate planes and gave the cosines of the angles the plane made with each of these (the so-called direction cosines). He also presented many of the standard problems of solid analytic geometry, examples of which appear in the exercises. For instance, he showed how to find the plane passing through three given points, the line passing through a point perpendicular to a plane, the distance between two parallel planes, and the angle between a line and a plane.

Known as "the greatest geometer of the eighteenth century," Monge developed new graphical geometric techniques as a student and later as a professor at a military school. The first problem he solved had to do with a procedure enabling soldiers to make quickly a fortification capable of shielding a position from both the view and the firepower of the enemy. Monge served the French revolutionary government as minister of the navy and later served Napoleon in various scientific offices. Ultimately, he was appointed senator for life by the emperor.

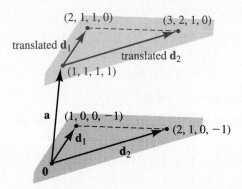

FIGURE 2.25
A 2-flat in $\mathbb{R}^4$.

Parametric equations obtained by equating components are

$$
\begin{aligned}
x_1 &= & s + 2t + 1\\
x_2 &= & t + 1\\
x_3 &= & 1\\
x_4 &= -s - & t + 1.
\end{aligned}
$$ ∎

The Geometry of Linear Systems

Let $A\mathbf{x} = \mathbf{b}$ be any system of m equations in n unknowns that has at least one solution $\mathbf{x} = \mathbf{p}$. Theorem 1.18 on p. 97 shows that the solution set of the system consists of all vectors of the form $\mathbf{x} = \mathbf{p} + \mathbf{h}$, where $\mathbf{h}$ is a solution of the homogeneous system $A\mathbf{x} = \mathbf{0}$. Because the solution set of $A\mathbf{x} = \mathbf{0}$ is a subspace of $\mathbb{R}^n$, we see that the solution set of $A\mathbf{x} = \mathbf{b}$ is the translate of this subspace by the vector $\mathbf{p}$. That is, the solution set of $A\mathbf{x} = \mathbf{b}$ is a k-flat, where k is the nullity of A. If the system of equations has a unique solution, its solution set is a zero-flat.

EXAMPLE 6   Show that the linear equation $c_1x_1 + c_2x_2 + c_3x_3 = b$, where at least one of c_1, c_2, c_3 is nonzero, represents a plane in $\mathbb{R}^3$.

SOLUTION   Let us assume that $c_1 \neq 0$. A particular solution of the given equation is $\mathbf{a} = [b/c_1, 0, 0]$. The corresponding homogeneous equation has a solution space generated by $\mathbf{d}_1 = [c_3, 0, -c_1]$ and $\mathbf{d}_2 = [c_2, -c_1, 0]$. Thus the solution set of the linear equation is a 2-flat in $\mathbb{R}^3$ with equation $\mathbf{x} = s\mathbf{d}_1 + t\mathbf{d}_2 + \mathbf{a}$—that is, a plane in $\mathbb{R}^3$.   ∎

Reasoning as in Example 6, we see that every linear equation

$$c_1x_1 + c_2x_2 + \cdots + c_nx_n = b$$

represents a hyperplane—that is, an $(n - 1)$-flat in $\mathbb{R}^n$.

EXAMPLE 7   Solve the system of equations

$$x_1 + 2x_2 - 2x_3 + x_4 + 3x_5 = 1$$
$$2x_1 + 5x_2 - 3x_3 - x_4 + 2x_5 = 2$$
$$-3x_1 - 8x_2 + 6x_3 - x_4 - 5x_5 = 1,$$

and write the solution set as a k-flat.

SOLUTION   Reducing the corresponding partitioned matrix, we have

$$\begin{bmatrix} 1 & 2 & -2 & 1 & 3 & | & 1 \\ 2 & 5 & -3 & -1 & 2 & | & 2 \\ -3 & -8 & 6 & -1 & -5 & | & 1 \end{bmatrix} \sim \begin{bmatrix} 1 & 2 & -2 & 1 & 3 & | & 1 \\ 0 & 1 & 1 & -3 & -4 & | & 0 \\ 0 & -2 & 0 & 2 & 4 & | & 4 \end{bmatrix}$$

$$\sim \begin{bmatrix} 1 & 0 & -4 & 7 & 11 & | & 1 \\ 0 & 1 & 1 & -3 & -4 & | & 0 \\ 0 & 0 & 2 & -4 & -4 & | & 4 \end{bmatrix} \sim \begin{bmatrix} 1 & 0 & 0 & -1 & 3 & | & 9 \\ 0 & 1 & 0 & -1 & -2 & | & -2 \\ 0 & 0 & 1 & -2 & -2 & | & 2 \end{bmatrix}.$$

Thus, $\mathbf{a} = [9, -2, 2, 0, 0]$ is a particular solution to the given system, and $\mathbf{d}_1 = [1, 1, 2, 1, 0]$ and $\mathbf{d}_2 = [-3, 2, 2, 0, 1]$ form a basis for the solution space of the corresponding homogeneous system. The solution set of the given system is the 2-flat in $\mathbb{R}^5$ with vector equation $\mathbf{x} = \mathbf{a} + t_1\mathbf{d}_1 + t_2\mathbf{d}_2$, which can be written in the form

$$\begin{bmatrix} x_1 \\ x_2 \\ x_3 \\ x_4 \\ x_5 \end{bmatrix} = \begin{bmatrix} 9 \\ -2 \\ 2 \\ 0 \\ 0 \end{bmatrix} + t_1 \begin{bmatrix} 1 \\ 1 \\ 2 \\ 1 \\ 0 \end{bmatrix} + t_2 \begin{bmatrix} -3 \\ 2 \\ 2 \\ 0 \\ 1 \end{bmatrix}. \qquad \blacksquare$$

In the preceding example, we described the solution set of a system of equations as a 2-flat in $\mathbb{R}^5$. Notice that the original form of the system represents an intersection of three hyperplanes in $\mathbb{R}^5$—one for each equation. We generalize this example to the solution set of any consistent linear system.

Consider a system of m equations in n unknowns. Let the rank of the coefficient matrix be r so that, when the matrix is reduced to row-echelon form, there are r nonzero rows. According to the rank equation, the number of free variables is then $n - r$; the corresponding homogeneous system has as its solution set an $(n - r)$-dimensional subspace—that is, an $(n - r)$-flat through the origin. The solution set of the original nonhomogeneous system is a translate of this subspace and is an $(n - r)$-flat in $\mathbb{R}^n$. In particular, a single consistent linear equation has as its solution set an $(n - 1)$-flat in $\mathbb{R}^n$. In general, if we adjoin an additional linear equation to a given linear system, we expect the dimension of the solution flat to be reduced by 1. This is the case precisely when the new system is still consistent and when the new equation is independent of the others (in the sense that it yields a new nonzero row when the augmented matrix is row-reduced to echelon form).

We have shown that a system $A\mathbf{x} = \mathbf{b}$ of m equations in n unknowns has as its solution set an $(n - r)$-flat, where r is the rank of A. Conversely, it can be shown that a k-flat in $\mathbb{R}^n$ is the solution set of some system of $n - k$ linear equations in n unknowns. That is, a k-flat in $\mathbb{R}^n$ is the intersection of $n - k$ hyperplanes. Thus there are two ways to view a k-flat in $\mathbb{R}^n$:

1. As a translate of a k-dimensional subspace of $\mathbb{R}^n$, described using parametric equations
2. As an intersection of $n - k$ hyperplanes, described with a system of linear equations.

EXAMPLE 8   Describe the line (1-flat) in $\mathbb{R}^3$ that passes through $(2, -1, 3)$ and $(1, 3, 5)$ in terms of

(1) parametric equations, and

(2) a system of linear equations.

SOLUTION   (1) In Example 3, we found the parametric equations for the line:

$$x_1 = -t + 2, \qquad x_2 = 4t - 1, \qquad x_3 = 2t + 3. \qquad (3)$$

(2) In order to describe the line with a system of linear equations, we eliminate the parameter t from Eqs. (3):

$$\begin{aligned} 4x_1 + x_2 \quad &= \quad 7 \qquad \text{Add four times the first to the second.} \\ x_2 - 2x_3 &= -7 \qquad \text{Subtract twice the third from the second.} \end{aligned} \qquad (4)$$

This system describes the line as an intersection of two planes. The line can be represented as the intersection of any two distinct planes, each containing the line. This is illustrated by the equivalent systems we have at the various stages in the Gauss reduction of system (4) to obtain solution (3). ∎

SUMMARY

1. The translate of a subset S of $\mathbb{R}^n$ by a vector $\mathbf{a} \in \mathbb{R}^n$ is the set of all vectors in $\mathbb{R}^n$ of the form $\mathbf{x} + \mathbf{a}$ for $\mathbf{x} \in S$, and is denoted by $S + \mathbf{a}$.

2. A k-flat in $\mathbb{R}^n$ is a translate of a k-dimensional subspace and has the form $\mathbf{a} + \mathrm{sp}(\mathbf{d}_1, \mathbf{d}_2, \ldots, \mathbf{d}_k)$, where $\mathbf{a}$ is a vector in $\mathbb{R}^n$ and $\mathbf{d}_1, \mathbf{d}_2, \ldots, \mathbf{d}_k$ are independent vectors in $\mathbb{R}^n$. The vector equation of the k-flat is $\mathbf{x} = \mathbf{a} + t_1\mathbf{d}_1 + t_2\mathbf{d}_2 + \cdots + t_k\mathbf{d}_k$ for scalars t_i in $\mathbb{R}$.

3. A line in $\mathbb{R}^n$ is a 1-flat. The line passing through the point $\mathbf{a}$ with parallel vector $\mathbf{d}$ is given by $\mathbf{x} = \mathbf{a} + t\mathbf{d}$, where t runs through all scalars. Parametric equations of the line are the component equations $x_i = a_i + d_i t$ for $i = 1, 2, \ldots, n$.

4. Let $\mathbf{a}$ and $\mathbf{b}$ be vectors in $\mathbb{R}^n$. Vectors to points on the line segment from the tip of $\mathbf{a}$ to the tip of $\mathbf{b}$ are vectors of the form $\mathbf{x} = t(\mathbf{b} - \mathbf{a}) + \mathbf{a}$ for $0 \leq t \leq 1$.

5. A plane in $\mathbb{R}^n$ is a 2-flat; a hyperplane in $\mathbb{R}^n$ is an $(n - 1)$-flat.

6. The solution set of a consistent linear system in n variables with coefficient matrix of rank r is an $(n - r)$-flat in $\mathbb{R}^n$.

7. Every k-flat in $\mathbb{R}^n$ can be viewed both as a translate of a k-dimensional subspace and as the intersection of $n - k$ hyperplanes.

EXERCISES

In keeping with classical geometry, many of the exercises that follow are phrased in terms of points rather than in terms of vectors.

In Exercises 1–6, sketch the indicated translate of the subset of $\mathbb{R}^n$ in an appropriate figure.

1. The translate of the line $x_2 = 2x_1 + 3$ in $\mathbb{R}^2$ by the vector $[-3, 0]$

2. The translate of $\{(t, t^2) \mid t \in \mathbb{R}\}$ in $\mathbb{R}^2$ by the vector $[-1, -2]$

3. The translate of $\{\mathbf{x} \in \mathbb{R}^2 \mid |x_i| \leq 1 \text{ for } i = 1, 2\}$ by the vector $[1, 2]$

4. The translate of $\{\mathbf{x} \in \mathbb{R}^2 \mid \|\mathbf{x}\| \geq 3\}$ by the vector $[2, 3]$

5. The translate of $\{\mathbf{x} \in \mathbb{R}^3 \mid \|\mathbf{x}\| \leq 1\}$ by the vector $[2, 4, 3]$

6. The translate of the plane $x_1 + x_2 = 2$ in $\mathbb{R}^3$ by the vector $[-1, 2, 3]$

7. Give parametric equations for the line in $\mathbb{R}^2$ through $(3, -3)$ with direction vector $\mathbf{d} = [-8, 4]$. Sketch the line in an appropriate figure.

8. Give parametric equations for the line in $\mathbb{R}^3$ through $(-1, 3, 0)$ with direction vector $\mathbf{d} = [-2, -1, 4]$. Sketch the line in an appropriate figure.

9. Consider the line in $\mathbb{R}^2$ that is given by the equation $d_1x_1 + d_2x_2 = c$ for numbers d_1, d_2, and c in $\mathbb{R}$, where d_1 and d_2 are not both zero. Find parametric equations of the line.

10. Find parametric equations for the line in $\mathbb{R}^2$ through $(5, -1)$ and orthogonal to the line with parametric equations $x_1 = 4 - 2t$, $x_2 = 7 + t$.

11. For each pair of points, find parametric equations of the line containing them.
    a. $(-2, 4)$ and $(3, -1)$ in $\mathbb{R}^2$
    b. $(3, -1, 6)$ and $(0, -3, -1)$ in $\mathbb{R}^3$
    c. $(2, 0, 4)$ and $(-1, 5, -8)$ in $\mathbb{R}^3$

12. For each of the given pairs of lines in $\mathbb{R}^3$, determine whether the lines intersect. If they do intersect, find the point of intersection, and determine whether the lines are orthogonal.
    a. $x_1 = 4 + t, \quad x_2 = 2 - 3t,$
       $x_3 = -3 + 5t$
       and
       $x_1 = 11 + 3s, \quad x_2 = -9 - 4s,$
       $x_3 = -4 - 3s$
    b. $x_1 = 11 + 3t, \quad x_2 = -3 - t,$
       $x_3 = 4 + 3t$
       and
       $x_1 = 6 - 2s, \quad x_2 = -2 + s,$
       $x_3 = -15 + 7s$

13. Find all points in common to the lines in $\mathbb{R}^2$ given by $x_1 = 5 - 3t, x_2 = -1 + t$ and $x_1 = -7 + 6s, x_2 = 3 - 2s$.

14. Find parametric equations for the line in $\mathbb{R}^3$ through $(-1, 2, 3)$ that is orthogonal to each of the two lines having parametric equations $x_1 = -2 + 3t, x_2 = 4, x_3 = 1 - t$ and $x_1 = 7 - t, x_2 = 2 + 3t, x_3 = 4 + t$.

15. Find the midpoint of the line segment joining each pair of points.
    a. $(-2, 4)$ and $(3, -1)$ in $\mathbb{R}^2$
    b. $(3, -1, 6)$ and $(0, -3, -1)$ in $\mathbb{R}^3$
    c. $(0, 4, 8)$ and $(-4, 5, 9)$ in $\mathbb{R}^3$

16. Find the point in $\mathbb{R}^2$ on the line segment joining $(-1, 3)$ and $(2, 5)$ that is twice as close to $(-1, 3)$ as to $(2, 5)$.

17. Find the point in $\mathbb{R}^3$ on the line segment joining $(-2, 1, 3)$ and $(0, -5, 6)$ that is one-fourth of the way from $(-2, 1, 3)$ to $(0, -5, 6)$.

18. Find the points that divide the line segment between $(2, 1, 3, 4)$ and $(-1, 2, 1, 3)$ in $\mathbb{R}^4$ into three equal parts.

19. Find the midpoint of the line segment between $(2, 1, 3, 4, 0)$ and $(1, 2, -1, 3, -1)$ in $\mathbb{R}^5$.

20. Find the intersection in $\mathbb{R}^3$ of the line given by

$$x_1 = 5 + t, \qquad x_2 = -3t, \qquad x_3 = -2 + 4t$$

and the plane with equation $x_1 - 3x_2 + 2x_3 = -25$.

21. Find the intersection in $\mathbb{R}^3$ of the line given by

$$x_1 = 2, \qquad x_2 = 5 - t, \qquad x_3 = 2t$$

and the plane with equation $x_1 + 2x_3 = 10$.

22. Find parametric equations of the plane that passes through the unit coordinate points $(1, 0, 0)$, $(0, 1, 0)$, and $(0, 0, 1)$ in $\mathbb{R}^3$.

23. Find a single linear equation in three variables whose solution set is the plane in Exercise 22. [HINT: We suggest two general methods of attack: (1) eliminate the parameters from your answer to Exercise 22, or (2) solve an appropriate linear system. Actually, this particular answer can be found by inspection.]

24. Find parametric equations of the plane in $\mathbb{R}^3$ that passes through $(1, 0, 0)$, $(0, 1, -1)$, and $(1, 1, 1)$.

25. Find a single linear equation in three variables whose solution set is the plane in Exercise 24. [See the hint for Exercise 23.]

26. Find a vector equation of the plane that passes through the points $(1, 2, 1)$, $(-1, 2, 3)$, and $(2, 1, 4)$ in $\mathbb{R}^3$.

27. Find a single linear equation in three variables whose solution set is the plane in Exercise 26. [See the hint for Exercise 23.]

28. Find a vector equation for the plane in $\mathbb{R}^4$ that passes through the points $(1, 2, 1, 3)$, $(4, 1, 2, 1)$, and $(3, 1, 2, 0)$.

29. Find a linear system with two equations in four variables whose solution set is the plane in Exercise 28. [See the hint for Exercise 23.]

30. Find a vector equation of the hyperplane that passes through the points $(1, 2, 1, 2, 3)$, $(0, 1, 2, 1, 3)$, $(0, 0, 3, 1, 2)$, $(0, 0, 0, 1, 4)$, and $(0, 0, 0, 0, 2)$ in $\mathbb{R}^5$.

31. Find a single linear equation in five variables whose solution set is the hyperplane in Exercise 30. [See the hint for Exercise 23.]

32. Find a vector equation of the hyperplane in $\mathbb{R}^6$ through the endpoints of $e_1, e_2, \ldots, e_6$.

33. Find a single linear equation in six variables whose solution set is the hyperplane in Exercise 32. [See the hint for Exercise 23.]

In Exercises 34–42, solve the given system of linear equations and write the solution set as a k-flat.

34. $x_1 - 2x_2 = 3$
    $3x_1 - x_2 = 14$

35. $x_1 + 2x_2 - x_3 = -3$
    $3x_1 + 7x_2 + 2x_3 = 1$
    $4x_1 - 2x_2 + x_3 = -2$

36. $x_1 + 4x_2 - 2x_3 = 4$
    $2x_1 + 7x_2 - x_3 = -2$
    $x_1 + 3x_2 + x_3 = -6$

37. $x_1 - 3x_2 + x_3 = 2$
    $3x_1 - 8x_2 + 2x_3 = 5$
    $3x_1 - 7x_2 + x_3 = 4$

38. $x_1 - 3x_2 + 2x_3 - x_4 = 8$
    $3x_1 - 7x_2 + x_4 = 0$

39. $x_1 - 2x_3 + x_4 = 6$
    $2x_1 - x_2 + x_3 - 3x_4 = 0$
    $9x_1 - 3x_2 - x_3 - 7x_4 = 4$

40. $x_1 + 2x_2 - 3x_3 + x_4 = 2$
    $3x_1 + 6x_2 - 8x_3 - 2x_4 = 1$

41.
$$x_1 - 3x_2 + x_3 + 2x_4 = 2$$
$$x_1 - 2x_2 + 2x_3 + 4x_4 = -1$$
$$2x_1 - 8x_2 - x_3 \qquad = 3$$
$$3x_1 - 9x_2 + 4x_3 \qquad = 7$$

42. $2x_1 - 5x_2 + x_3 - 10x_4 + 15x_5 = 60$

43. Mark each of the following True or False.

___ **a.** The solution set of a linear equation in x_1 and x_2 can be regarded as a hyperplane in $\mathbb{R}^2$.

___ **b.** Every line and hyperplane in $\mathbb{R}^n$ intersect in a single point.

___ **c.** The intersection of two distinct hyperplanes in $\mathbb{R}^5$ is a line, if the intersection is nonempty.

___ **d.** The Euclidean space $\mathbb{R}^5$ has no physical existence, but exists only in our minds.

___ **e.** The Euclidean spaces $\mathbb{R}$, $\mathbb{R}^2$, and $\mathbb{R}^3$ have no physical existence, but exist only in our minds.

___ **f.** The mathematical existence of Euclidean 5-space is as substantial as the mathematical existence of Euclidean 3-space.

___ **g.** Every plane in $\mathbb{R}^n$ is a two-dimensional subspace of $\mathbb{R}^n$.

___ **h.** Every plane through the origin in $\mathbb{R}^n$ is a two-dimensional subspace of $\mathbb{R}^n$.

___ **i.** Every k-flat in $\mathbb{R}^n$ contains the origin.

___ **j.** Every k-flat in $\mathbb{R}^n$ is a translate of a k-dimensional subspace.

VECTOR SPACES

For the sake of efficiency, mathematicians often study objects just in terms of their mathematical structure, deemphasizing such things as particular symbols used, names of things, and applications. Any properties derived exclusively from mathematical structure will hold for all objects having that structure. Organizing mathematics in this way avoids repeating the same arguments in different contexts. Viewed from this perspective, linear algebra is the study of all objects that have a *vector-space* structure. The Euclidean spaces $\mathbb{R}^n$ that we treated in Chapters 1 and 2 serve as our guide.

Section 3.1 defines the general notion of a *vector space*, motivated by the familiar algebraic structure of the spaces $\mathbb{R}^n$. Our examples focus mainly on spaces other than $\mathbb{R}^n$, such as function spaces. Unlike the first two chapters in our text, this chapter draws on calculus for many of its illustrations.

Section 3.2 explains how the linear-algebra terminology introduced in Chapter 1 for $\mathbb{R}^n$ carries over to a general vector space V. The definitions given in Chapters 1 and 2 for linear combinations, spans, subspaces, bases, dependent vectors, independent vectors, and dimension can be left mostly unchanged, except for replacing "$\mathbb{R}^n$" by "a vector space V." Indeed, with this replacement, many of the theorems and proofs in Chapter 1 have word-for-word validity for general vector spaces.

Section 3.3 shows that every finite-dimensional (real) vector space can be *coordinatized* to become algebraically indistinguishable from one of the spaces $\mathbb{R}^n$. This coordinatization allows us to apply the matrix techniques developed in Chapters 1 and 2 to any finite-dimensional vector space for such things as determining whether vectors are independent or form a basis.

Linear transformations of one vector space into another are the topic of Section 3.4. We will see that some of the basic operations of calculus, such as differentiation, can be viewed as linear transformations.

To conclude the chapter, optional Section 3.5 describes how we try to access such geometric notions as length and angle even in infinite-dimensional vector spaces.

3.1   VECTOR SPACES

The Vector-Space Operations

In each Euclidean space $\mathbb{R}^n$, we know how to add two vectors and how to perform scalar multiplication of a vector by a real number (scalar). These are the two *vector-space operations*. The first requirement that a set V, whose elements we will call "vectors," must satisfy in order to be a vector space is that it have two well-defined algebraic operations, each of which yields an element of V—namely:

Addition of two elements of V	**Vector addition**
Multiplication of an element of V by a scalar.	**Scalar multiplication**

For example, we know how to add the two functions x^2 and $\sin x$, and we know how to multiply them by a real number. We require that, whenever addition or scalar multiplication is performed with elements in V, *the answers obtained lie again in V*. That is, we require that V be **closed under vector addition** and **closed under scalar multiplication**. This notion of *closure* under an operation is familiar to us from Chapter 1.

Definition of a Vector Space

The definition of a *vector space* which follows incorporates the ideas we have just discussed. It also requires that the vector addition and scalar multiplication satisfy the algebraic properties that hold in $\mathbb{R}^n$—namely, those listed in Theorem 1.1.

DEFINITION 3.1   Vector Space

A **(real) vector space** is a set V of objects called **vectors**, together with a rule for adding any two vectors $\mathbf{v}$ and $\mathbf{w}$ to produce a vector $\mathbf{v} + \mathbf{w}$ in V and a rule for multiplying any vector $\mathbf{v}$ in V by any scalar r in $\mathbb{R}$ to produce a vector $r\mathbf{v}$ in V. Moreover, there must exist a vector $\mathbf{0}$ in V, and for each $\mathbf{v}$ in V there must exist a vector $-\mathbf{v}$ in V such that properties A1 through A4 and S1 through S4 below are satisfied for all choices of vectors $\mathbf{u}, \mathbf{v}, \mathbf{w} \in V$ and scalars $r, s \in \mathbb{R}$.

Properties of Vector Addition

A1	$(\mathbf{u} + \mathbf{v}) + \mathbf{w} = \mathbf{u} + (\mathbf{v} + \mathbf{w})$	**An associative law**
A2	$\mathbf{v} + \mathbf{w} = \mathbf{w} + \mathbf{v}$	**A commutative law**
A3	$\mathbf{0} + \mathbf{v} = \mathbf{v}$	**0 as additive identity**
A4	$\mathbf{v} + (-\mathbf{v}) = \mathbf{0}$	**$-\mathbf{v}$ as additive inverse of v**

Properties Involving Scalar Multiplication

S1 $r(\mathbf{v} + \mathbf{w}) = r\mathbf{v} + r\mathbf{w}$ A distributive law

S2 $(r + s)\mathbf{v} = r\mathbf{v} + s\mathbf{v}$ A distributive law

S3 $r(s\mathbf{v}) = (rs)\mathbf{v}$ An associative law

S4 $1\mathbf{v} = \mathbf{v}$ Preservation of scale

In a moment we will show that there is only one vector $\mathbf{0}$ in V satisfying condition A3; this vector is called the **zero vector**. Similarly, we will see that the vector $-\mathbf{v}$ in condition A4 is uniquely determined by $\mathbf{v}$; it is called the **additive inverse** of $\mathbf{v}$, and is usually read "minus $\mathbf{v}$." We write $\mathbf{v} - \mathbf{w}$ for $\mathbf{v} + (-\mathbf{w})$.

The adjective "real" in parentheses in the first line of Definition 3.1 signifies that it is sometimes necessary to allow the scalars to be complex numbers, $a + bi$ where $a, b \in \mathbb{R}$ and $i^2 = -1$. Linear algebra using complex scalars is the topic of Chapter 9. There are several branches of mathematics in which one does not gain full insight without using complex numbers, and linear algebra is one of them. Unfortunately, pencil and paper computation with complex numbers is cumbersome, and so it is customary in a first course to work mostly with real scalars. Fortunately, many of the concepts of linear algebra can be adequately explained in terms of real vector spaces.

Because our definition of a vector space was modeled on the algebraic structure of the Euclidean spaces $\mathbb{R}^n$ discussed in Chapter 1, we see that $\mathbb{R}^n$ is a vector space for each positive integer n. We now proceed to illustrate this concept with other examples.

HISTORICAL NOTE   ALTHOUGH THE OBJECTS WE CALL VECTOR SPACES were well known in the late nineteenth century, the first mathematician to give an abstract definition of a vector space was Giuseppe Peano (1858–1932) in his *Calcolo Geometrico* of 1888. Peano's aim in the book, as the title indicates, was to develop a geometric calculus. Such a calculus "consists of a system of operations analogous to those of algebraic calculus but in which the objects with which the calculations are performed are, instead of numbers, geometrical objects." Much of the book consists of calculations dealing with points, lines, planes, and volumes. But in the ninth chapter, Peano defines what he called a *linear system*. This was a set of objects that was provided with operations of addition and scalar multiplication. These operations were to satisfy axioms A1–A4 and S1–S4 presented in this section. Peano also defined the *dimension* of a linear system to be the maximum number of linearly independent objects in the system and noted that the set of polynomial functions in one variable forms a linear system of infinite dimension.

Curiously, Peano's work had no immediate effect on the mathematical community. The definition was even forgotten. It only entered the mathematical mainstream through the book *Space–Time–Matter* (1918) by Hermann Weyl (1885–1955). Weyl wrote this book as an introduction to Einstein's general theory of relativity. In Chapter 1 he discusses the nature of a Euclidean space and, as part of that discussion, formulates the same standard axioms as Peano did earlier. He also gives a philosophic reason for adopting such a definition:

> Not only in geometry, but to a still more astonishing degree in physics, has it become more and more evident that as soon as we have succeeded in unraveling fully the natural laws which govern reality, we find them to be expressible by mathematical relations of surpassing simplicity and architectonic perfection. . . . Analytical geometry [the axiom system which he presented] . . . conveys an idea, even if inadequate, of this perfection of form.

EXAMPLE 1   Show that the set $M_{m,n}$ of all $m \times n$ matrices is a vector space, using as vector addition and scalar multiplication the usual addition of matrices and multiplication of a matrix by a scalar.

SOLUTION   We have seen that addition of $m \times n$ matrices and multiplication of an $m \times n$ matrix by a scalar again yield an $m \times n$ matrix. Thus, $M_{m,n}$ is closed under vector addition and scalar multiplication. We take as zero vector in $M_{m,n}$ the usual zero matrix, all of whose entries are zero. For any matrix A in $M_{m,n}$ we consider $-A$ to be the matrix $(-1)A$. The properties of matrix arithmetic on page 45 show that all eight properties A1–A4 and S1–S4 required of a vector space are satisfied.   ■

The preceding example introduced the notation $M_{m,n}$ for the vector space of all $m \times n$ matrices. We use M_n for the vector space of all square $n \times n$ matrices.

EXAMPLE 2   Show that the set P of all polynomials in the variable x with coefficients in $\mathbb{R}$ is a vector space, using for vector addition and scalar multiplication the usual addition of polynomials and multiplication of a polynomial by a scalar.

SOLUTION   Let p and q be polynomials

$$p = a_0 + a_1 x + a_2 x^2 + \cdots + a_n x^n$$

and

$$q = b_0 + b_1 x + b_2 x^2 + \cdots + b_m x^m.$$

If $m \geq n$, the **sum** of p and q is given by

$$p + q = (a_0 + b_0) + (a_1 + b_1)x + \cdots + (a_n + b_n)x^n \\ + b_{n+1}x^{n+1} + \cdots + b_m x^m.$$

For example, if $p = 1 + 2x + 3x^2$ and $q = x + x^3$, then $p + q = 1 + 3x + 3x^2 + x^3$. A similar definition is made if $m < n$. The **product** of p and a scalar r is given by

$$rp = ra_0 + ra_1 x + ra_2 x^2 + \cdots + ra_n x^n.$$

Taking the usual notions of the zero polynomial and of $-p$, we recognize that the eight properties A1–A4 and S1–S4 required of a vector space are familiar properties for these polynomial operations. Thus, P is a vector space.   ■

EXAMPLE 3   Let F be the set of all real-valued functions with domain $\mathbb{R}$; that is, let F be the set of all functions mapping $\mathbb{R}$ into $\mathbb{R}$. The vector **sum** $f + g$ of two functions f and g in F is defined in the usual way to be the function whose value at any x in $\mathbb{R}$ is $f(x) + g(x)$; that is,

$$(f + g)(x) = f(x) + g(x).$$

For any scalar r in $\mathbb{R}$ and function f in F, the **product** rf is the function whose value at x is $rf(x)$, so that

$$(rf)(x) = rf(x).$$

Show that F with these operations is a vector space.

SOLUTION   We observe that, for f and g in F, both $f + g$ and rf are functions mapping $\mathbb{R}$ into $\mathbb{R}$, so $f + g$ and rf are in F. Thus, F is closed under vector addition and under scalar multiplication. We take as zero vector in F the constant function whose value at each x in $\mathbb{R}$ is 0. For each function f in F, we take as $-f$ the function $(-1)f$ in F.

There are four vector-addition properties to verify, and they are all easy. We illustrate by verifying condition A4. For f in F, the function $f + (-f) = f + (-1)f$ has as its value at x in $\mathbb{R}$ the number $f(x) + (-1)f(x)$, which is 0. Consequently, $f + (-f)$ is the zero function, and A4 is verified.

The scalar multiplicative properties are just as easy to verify. For example, to verify S4, we must compute $1f$ at any x in $\mathbb{R}$ and compare the result with $f(x)$. We obtain $(1f)(x) = 1f(x) = f(x)$, and so $1f = f$.   ∎

EXAMPLE 4   Show that the set P_∞ of *formal power series* in x of the form

$$\sum_{n=0}^{\infty} a_n x^n = a_0 + a_1 x + a_2 x^2 + \cdots + a_n x^n + \cdots,$$

with addition and scalar multiplication defined by

$$\left(\sum_{n=0}^{\infty} a_n x^n\right) + \left(\sum_{n=0}^{\infty} b_n x^n\right) = \sum_{n=0}^{\infty} (a_n + b_n)x^n \quad \text{and} \quad r\left(\sum_{n=0}^{\infty} a_n x^n\right) = \sum_{n=0}^{\infty} ra_n x^n,$$

is a vector space.*

SOLUTION   The reasoning here is precisely the same as for the space P of polynomials in Example 2. The zero series is $\sum_{n=0}^{\infty} 0x^n$, and the additive inverse of $\sum_{n=0}^{\infty} a_n x^n$ is $\sum_{n=0}^{\infty} (-a_n)x^n$. All the other axioms follow from the associative, commutative, and distributive laws for the real-number coefficients a_n in the series.   ∎

The examples of vector spaces that we have presented so far are all based on algebraic structures familiar to us—namely, the algebra of matrices, polynomials, and functions. You may be thinking, "How can anything with an

---

*We can add only a *finite* number of vectors in a vector space. Thus we do not regard these formal power series as infinite sums in P_∞ of monomials. Also, we are not concerned with questions of convergence or divergence, as studied in calculus. This is the significance of the word *formal*.

addition and scalar multiplication defined on it fail to be a vector space?" To answer this, we give two more-esoteric examples, where checking the axioms is not as natural a process.

EXAMPLE 5   Let $\mathbb{R}^2$ have the usual operation of addition, but define scalar multiplication $r[x, y]$ by $r[x, y] = [0, 0]$. Determine whether $\mathbb{R}^2$ with these operations is a vector space.

SOLUTION   Because conditions A1–A4 of Definition 3.1 do not involve scalar multiplication, and because addition is the usual operation, we need only check conditions S1–S4. We see that all of these hold, except for condition S4: because $1[x, y] = [0, 0]$, the scale is not preserved. Thus, $\mathbb{R}^2$ is not a vector space with these particular two operations.   ∎

EXAMPLE 6   Let $\mathbb{R}^2$ have the usual scalar multiplication, but let addition $\dotplus$ be defined on $\mathbb{R}^2$ by the formula

$$[x, y] \dotplus [r, s] = [x + r, 2y + s].$$

Determine whether $\mathbb{R}^2$ with these operations is a vector space. (We use the symbol $\dotplus$ for the warped addition of vectors in $\mathbb{R}^2$, to distinguish it from the usual addition.)

SOLUTION   We check the associative law for $\dotplus$:

$$([x, y] \dotplus [r, s]) \dotplus [a, b] = [x + r, 2y + s] \dotplus [a, b]$$
$$= [(x + r) + a, 2(2y + s) + b]$$
$$= [x + r + a, 4y + 2s + b],$$

whereas

$$[x, y] \dotplus ([r, s] \dotplus [a, b]) = [x, y] \dotplus (r + a, 2s + b]$$
$$= [x + (r + a), 2y + (2s + b)]$$
$$= [x + r + a, 2y + 2s + b].$$

Because the two colored scalars are not equal, we expect that $\dotplus$ is not associative. We can find a specific violation of the associative law by choosing $y \neq 0$; for example,

$$([0, 1] \dotplus [0, 0]) \dotplus [0, 0] = [0, 4],$$

whereas

$$[0, 1] \dotplus ([0, 0] \dotplus [0, 0]) = [0, 2].$$

Therefore, $\mathbb{R}^2$ is not a vector space with these two operations.   ∎

We now indicate that vector addition and scalar multiplication possess still more of the properties we are accustomed to expect. It is important to realize that everything has to be proved using just the axioms A1 through A4 and S1 through S4. Of course, once we have proved something from the axioms, we can then use it in the proofs of other things. The properties that appear in Theorem 3.1 below are listed in a convenient order for proof; for example, we will see that it is convenient to know property 3 in order to prove property 4. Property 4 states that in a vector space V, we have $0\mathbf{v} = \mathbf{0}$ for all $\mathbf{v} \in V$. Students often attempt to prove this by saying,

> "Let $\mathbf{v} = (a_1, a_2, \ldots, a_n)$. Then $0\mathbf{v} = 0(a_1, a_2, \ldots, a_n)$
> $= (0, 0, \ldots, 0) = \mathbf{0}$."

This is a fine argument if V is $\mathbb{R}^n$, but we have now expanded our concept of *vector space*, and we can no longer assume that $\mathbf{v} \in V$ is some n-tuple of real numbers.

THEOREM 3.1   Elementary Properties of Vector Spaces

Every vector space V has the following properties:

1. The vector $\mathbf{0}$ is the *unique* vector $\mathbf{x}$ satisfying the equation $\mathbf{x} + \mathbf{v} = \mathbf{v}$ for all vectors $\mathbf{v}$ in V.
2. For each vector $\mathbf{v}$ in V, the vector $-\mathbf{v}$ is the *unique* vector $\mathbf{y}$ satisfying $\mathbf{v} + \mathbf{y} = \mathbf{0}$.
3. If $\mathbf{u} + \mathbf{v} = \mathbf{u} + \mathbf{w}$ for vectors $\mathbf{u}$, $\mathbf{v}$, and $\mathbf{w}$ in V, then $\mathbf{v} = \mathbf{w}$.
4. $0\mathbf{v} = \mathbf{0}$ for all vectors $\mathbf{v}$ in V.
5. $r\mathbf{0} = \mathbf{0}$ for all scalars r in $\mathbb{R}$.
6. $(-r)\mathbf{v} = r(-\mathbf{v}) = -(r\mathbf{v})$ for all scalars r in $\mathbb{R}$ and vectors $\mathbf{v}$ in V.

PROOF   We prove only properties 1 and 4, leaving proofs of the remaining properties as Exercises 19 through 22. In proving property 4, we assume that properties 2 and 3 have been proved.

Turning to property 1, the standard way to prove that something is unique is to suppose that there are two of them, and then show that they must be equal. Suppose, therefore, that there exist vectors $\mathbf{0}$ and $\mathbf{0}'$ satisfying

$$\mathbf{0} + \mathbf{v} = \mathbf{v} \quad \text{and} \quad \mathbf{0}' + \mathbf{v} = \mathbf{v} \quad \text{for all } \mathbf{v} \in V.$$

Taking $\mathbf{v} = \mathbf{0}'$ in the first equation and $\mathbf{v} = \mathbf{0}$ in the second equation, we obtain

$$\mathbf{0} + \mathbf{0}' = \mathbf{0}' \quad \text{and} \quad \mathbf{0}' + \mathbf{0} = \mathbf{0}.$$

By the commutative law A2, we know that $\mathbf{0} + \mathbf{0}' = \mathbf{0}' + \mathbf{0}$, and we conclude that $\mathbf{0} = \mathbf{0}'$.

Turning to property 4, notice that the equation $0\mathbf{v} = \mathbf{0}$ which we want to prove involves both scalar multiplication (namely, $0\mathbf{v}$) and vector addition ($\mathbf{0}$ is an *additive* concept, given by axiom A3). To prove a relationship between these two algebraic operations, we must use an axiom that involves *both* of them—namely, one of the distributive laws S1 or S2. Using distributive law S2, we have for $\mathbf{v} \in V$,

$$0\mathbf{v} = (0 + 0)\mathbf{v} = 0\mathbf{v} + 0\mathbf{v}.$$

By the additive identity axiom A3 and the commutative law A2, we know that

$$0\mathbf{v} = \mathbf{0} + 0\mathbf{v} = 0\mathbf{v} + \mathbf{0}.$$

Therefore,

$$0\mathbf{v} + 0\mathbf{v} = 0\mathbf{v} + \mathbf{0},$$

and, by property (3), we conclude that $0\mathbf{v} = \mathbf{0}$. ▲

The Universality of Function Spaces (Optional)

In Example 3, we showed that the set F of all functions mapping $\mathbb{R}$ into $\mathbb{R}$ is a vector space, where we define for $f, g \in F$ and for $r \in \mathbb{R}$

$$(f + g)(x) = f(x) + g(x) \quad \text{and} \quad (rf)(x) = rf(x). \tag{1}$$

Note that these definitions of addition and scalar multiplication for functions having $\mathbb{R}$ as both domain and codomain use only the algebraic structure of the codomain $\mathbb{R}$ and not the algebraic structure of the domain $\mathbb{R}$. That is, the defining addition and scalar multiplication appearing on the right-hand sides of Eqs. (1) take place in the codomain. We do not see anything like $f(a + b)$, which involves addition in the domain $\mathbb{R}$, or like $f(rx)$, which involves scalar multiplication in the domain. This suggests that if S is any set and we let F be the set of all functions mapping S into $\mathbb{R}$, then Example 3 might still go through, and show that F is a vector space. We show that this is the case in our next example.

EXAMPLE 7 Let F be the set of all real-valued functions on a (nonempty) set S; that is, let F be the set of all functions mapping S into $\mathbb{R}$. For $f, g \in F$, let the sum $f + g$ of two functions f and g in F be defined by

$$(f + g)(x) = f(x) + g(x) \text{ for all } x \in S,$$

and, for any scalar r, let scalar multiplication be defined by

$$(rf)(x) = rf(x) \text{ for all } x \in S.$$

Show that F with these operations is a vector space.

SOLUTION The solution of Example 3 is valid word for word if we replace each reference $\mathbb{R}$ to the domain by the new domain S. That is, the functions map S into $\mathbb{R}$, and we let $x \in S$ rather than $x \in \mathbb{R}$ in this solution. ■

We now indicate why we headed this discussion "The Universality of Function Spaces." Let $S = \{1, 2\}$, so that F is the set of all functions mapping $\{1, 2\}$ into $\mathbb{R}$. Let us abbreviate the description of a function f as $[f(1), f(2)]$. For example, we consider $[-3, 8]$ to denote the function in F that maps 1 into -3 and maps 2 into 8. In this way, we identify each vector $[a, b]$ in $\mathbb{R}^2$ with a function f mapping $\{1, 2\}$ into $\mathbb{R}$—namely, $f(1) = a$ and $f(2) = b$. If we view $[a, b]$ as the function f and $[c, d]$ as the function g, then

$$(f + g)(1) = f(1) + g(1) = a + c \quad \text{and} \quad (f + g)(2) = f(2) + g(2) = b + d.$$

Thus, the vector $[a + c, b + d] = [a, b] + [c, d]$ in $\mathbb{R}^2$ is the function $f + g$. Similarly, we see that the vector $[ra, rb] = r[a, b]$ in $\mathbb{R}^2$ is the function rf. In this way, we can regard $\mathbb{R}^2$ as the vector space of functions mapping $\{1, 2\}$ into $\mathbb{R}$. We realize that we can equally well consider each vector of $\mathbb{R}^n$ as a function mapping $\{1, 2, 3, \ldots, n\}$ into $\mathbb{R}$. For example, the vector $[-3, 5, 2, 7]$ can be considered as the function $f: \{1, 2, 3, 4\} \to \mathbb{R}$, where $f(1) = -3, f(2) = 5, f(3) = 2$, and $f(4) = 7$.

In a similar fashion, we can view the vector space $M_{m,n}$ of matrices in Example 1 as the vector space of functions mapping the positive integers from 1 to mn into $\mathbb{R}$. For example, taking $m = 2$ and $n = 3$, we can view the matrix

$$\begin{bmatrix} a_1 & a_3 & a_5 \\ a_2 & a_4 & a_6 \end{bmatrix}$$

as the function $f: \{1, 2, 3, 4, 5, 6\} \to \mathbb{R}$, where $f(i) = a_i$. Addition of functions as defined in Example 7 again corresponds to addition of matrices, and the same is true for scalar multiplication.

The vector space P of all polynomials in Example 2 is not quite as easy to present as a function space, because not all the polynomials have the same number of terms. However, we can view the vector space P_∞ of formal power series in x as the space of all functions mapping $\{0, 1, 2, 3, \ldots\}$ into $\mathbb{R}$. Namely, if f is such a function and if $f(n) = a_n$ for $n \in \{0, 1, 2, 3, \ldots\}$, then we can denote this function symbolically by $\sum_{n=0}^{\infty} a_n x^n$. We see that function addition and multiplication by a scalar will produce precisely the addition and multiplication of power series defined in Example 4. We will show in the next section that we can view the vector space P of polynomials as a *subspace* of P_∞.

We have now freed the domain of our function spaces from having to be the set $\mathbb{R}$. Let's see how much we can free the codomain. The definitions

$$(f + g)(x) = f(x) + g(x) \quad \text{and} \quad (rf)(x) = rf(x)$$

for addition and scalar multiplication of functions show that we need a notion of addition and scalar multiplication for the codomain. For function addition to be associative and commutative, we need the addition in the codomain to be associative and commutative. For there to be a zero vector, we need an additive identity—let's call it **0**—in the codomain, so that we will have a "zero constant function" to serve as additive identity, etc. It looks as though what we need is to have the codomain itself have a vector space structure! This is indeed the case. If S is a set and V is a vector space, then the set F of all functions $f: S \to V$ with this notion of function addition and multiplication by a scalar is again a vector space. Note that $\mathbb{R}$ itself is a vector space, so the set of functions $f: S \to \mathbb{R}$ in Example 7 is a special case of this construction. For another example, the set of all functions mapping $\mathbb{R}^3$ into $\mathbb{R}^3$ has a vector space structure. In third-semester calculus, we sometimes speak of a "vector-valued function," meaning a function whose codomain is a vector space, although we usually don't talk about vector spaces there.

Thus, starting with a set S, we could consider the vector space V_1 of all functions mapping S into $\mathbb{R}$, and then the vector space V_2 of all functions mapping S into V_1, and then the vector space V_3 of all functions mapping S into V_2, and then—oops! We had better stop now. People who do too much of this stuff are apt to start climbing the walls. (However, mathematicians do sometimes build cumulative structures in this fashion.)

SUMMARY

1. A vector space is a nonempty set V of objects called *vectors*, together with rules for adding any two vectors $\mathbf{v}$ and $\mathbf{w}$ in V and for multiplying any vector $\mathbf{v}$ in V by any scalar r in $\mathbb{R}$. Furthermore, V must be closed under this vector addition and scalar multiplication so that $\mathbf{v} + \mathbf{w}$ and $r\mathbf{v}$ are both in V. Moreover, the following axioms must be satisfied for all vectors $\mathbf{u}$, $\mathbf{v}$, and $\mathbf{w}$ in V and all scalars r and s in $\mathbb{R}$:

   A1 $(\mathbf{u} + \mathbf{v}) + \mathbf{w} = \mathbf{u} + (\mathbf{v} + \mathbf{w})$
   A2 $\mathbf{v} + \mathbf{w} = \mathbf{w} + \mathbf{v}$
   A3 There exists a zero vector $\mathbf{0}$ in V such that $\mathbf{0} + \mathbf{v} = \mathbf{v}$ for all $\mathbf{v} \in V$.
   A4 Each $\mathbf{v} \in V$ has an additive inverse $-\mathbf{v}$ in V such that $\mathbf{v} + (-\mathbf{v}) = \mathbf{0}$.
   S1 $r(\mathbf{v} + \mathbf{w}) = r\mathbf{v} + r\mathbf{w}$
   S2 $(r + s)\mathbf{v} = r\mathbf{v} + s\mathbf{v}$
   S3 $r(s\mathbf{v}) = (rs)\mathbf{v}$
   S4 $1\mathbf{v} = \mathbf{v}$

2. Elementary properties of vector spaces are listed in Theorem 3.1.

3. Examples of vector spaces include $\mathbb{R}^n$, the space $M_{m,n}$ of all $m \times n$ matrices, the space of all polynomials in the variable x, and the space of all functions $f: \mathbb{R} \to \mathbb{R}$. For all of these examples, vector addition and scalar multiplication are the addition and multiplication by a real number with which we are already familiar.

4. (*Optional*) For any set S, the set of all functions mapping S into $\mathbb{R}$ forms a vector space with the usual definitions of addition and scalar multiplication of functions. The Euclidean space $\mathbb{R}^n$ can be viewed as the vector space of functions mapping $\{1, 2, 3, \ldots, n\}$ into $\mathbb{R}$, where a vector $\mathbf{a} = [a_1, a_2, a_3, \ldots, a_n]$ is viewed as the function f such that $f(i) = a_i$.

5. (*Optional*) Still more generally, for any set S and any vector space V, the set of all functions mapping S into V forms a vector space with the usual definitions of addition and scalar multiplication of functions.

EXERCISES

In Exercises 1–8, decide whether or not the given set, together with the indicated operations of addition and scalar multiplication, is a (real) vector space.

1. The set $\mathbb{R}^2$, with the usual addition but with scalar multiplication defined by $r[x, y] = [ry, rx]$.

2. The set $\mathbb{R}^2$, with the usual scalar multiplication but with addition defined by $[x, y] \dotplus [r, s] = [y + s, x + r]$.

3. The set $\mathbb{R}^2$, with addition defined by $[x, y] \dotplus [a, b] = [x + a + 1, y + b]$ and with scalar multiplication defined by $r[x, y] = [rx + r - 1, ry]$.

4. The set of all 2×2 matrices, with the usual scalar multiplication but with addition defined by $A \dotplus B = O$, the 2×2 zero matrix.

5. The set of all 2×2 matrices, with the usual addition but with scalar multiplication defined by $rA = O$, the 2×2 zero matrix.

6. The set F of all functions mapping $\mathbb{R}$ into $\mathbb{R}$, with scalar multiplication defined as in Example 7 but with addition defined by $(f \dotplus g)(x) = \max\{f(x), g(x)\}$.

7. The set F of all functions mapping $\mathbb{R}$ into $\mathbb{R}$, with scalar multiplication defined as in Example 7 but with addition defined by $(f \dotplus g)(x) = f(x) + 2g(x)$.

8. The set F of all functions mapping $\mathbb{R}$ into $\mathbb{R}$, with scalar multiplication defined as in Example 7 but with addition defined by $(f \dotplus g)(x) = 2f(x) + 2g(x)$.

In Exercises 9–16, determine whether the given set is closed under the usual operations of addition and scalar multiplication, and is a (real) vector space.

9. The set of all upper-triangular $n \times n$ matrices.

10. The set of all 2×2 matrices of the form
$$\begin{bmatrix} \times & 1 \\ 1 & \times \end{bmatrix},$$
where each $\times$ may be any scalar.

11. The set of all diagonal $n \times n$ matrices.

12. The set of all 3×3 matrices of the form
$$\begin{bmatrix} \times & 0 & \times \\ 0 & \times & 0 \\ \times & 0 & \times \end{bmatrix},$$
where each $\times$ may be any scalar.

13. The set $\{0\}$ consisting only of the number 0.

14. The set $\mathbb{Q}$ of all rational numbers.

15. The set $\mathbb{C}$ of complex numbers; that is,
$$\mathbb{C} = \{a + b\sqrt{-1} \mid a, b \text{ in } \mathbb{R}\},$$
with the usual addition of complex numbers and with scalar multiplication defined in the usual way by $r(a + b\sqrt{-1}) = ra + rb\sqrt{-1}$ for any numbers a, b, and r in $\mathbb{R}$.

16. The set P_n of all polynomials in x, with real coefficients and of degree less than or equal to n, together with the zero polynomial.

17. Your answer to Exercise 3 should be that $\mathbb{R}^2$ with the given operations is a vector space.
    a. Describe the "zero vector" in this vector space.
    b. Explain why the relations $r[0, 0] = [r - 1, 0] \neq [0, 0]$ do not violate property (5) of Theorem 3.1.

18. Mark each of the following True or False.
    ____ a. Matrix multiplication is a vector-space operation on the set $M_{m \times n}$ of all $m \times n$ matrices.
    ____ b. Matrix multiplication is a vector-space operation on the set M_n of all square $n \times n$ matrices.
    ____ c. Multiplication of any vector by the zero scalar always yields the zero vector.
    ____ d. Multiplication of a nonzero vector by a nonzero scalar never yields the zero vector.
    ____ e. No vector is its own additive inverse.

____ f. The zero vector is the only vector that is its own additive inverse.

____ g. Multiplication of two scalars is of no concern in the definition of a vector space.

____ h. One of the axioms for a vector space relates addition of scalars, multiplication of a vector by scalars, and addition of vectors.

____ i. Every vector space has at least two vectors.

____ j. Every vector space has at least one vector.

19. Prove property 2 of Theorem 3.1.

20. Prove property 3 of Theorem 3.1.

21. Prove property 5 of Theorem 3.1.

22. Prove property 6 of Theorem 3.1.

23. Let V be a vector space. Prove that, if $\mathbf{v}$ is in V and if r is a scalar and if $r\mathbf{v} = \mathbf{0}$, then either $r = 0$ or $\mathbf{v} = \mathbf{0}$.

24. Let V be a vector space and let $\mathbf{v}$ and $\mathbf{w}$ be nonzero vectors in V. Prove that if $\mathbf{v}$ is not a scalar multiple of $\mathbf{w}$, then $\mathbf{v}$ is not a scalar multiple of $\mathbf{v} + \mathbf{w}$.

25. Let V be a vector space, and let $\mathbf{v}$ and $\mathbf{w}$ be vectors in V. Prove that there is a *unique* vector $\mathbf{x}$ in V such that $\mathbf{x} + \mathbf{v} = \mathbf{w}$.

Exercises 26–29 are based on the optional subsection on the universality of function spaces.

26. Using the discussion of function spaces at the end of this section, explain how we can view the Euclidean vector space $\mathbb{R}^{mn}$ and the vector space $M_{m,n}$ of all $m \times n$ matrices as essentially the same vector space with just a different notation for the vectors.

27. Repeat Exercise 26 for the vector space $M_{2,6}$ of 2×6 matrices and the vector space $M_{3,4}$ of 3×4 matrices.

28. If you worked Exercise 16 correctly, you found that the polynomials in x of degree at most n do form a vector space P_n. Explain how P_n and $\mathbb{R}^{n+1}$ can be viewed as essentially the same vector space, with just a different notation for the vectors.

29. Referring to the three preceding exercises, list the vector spaces $\mathbb{R}^{24}$, $\mathbb{R}^{25}$, $\mathbb{R}^{26}$, P_{24}, P_{25}, P_{26}, $M_{4,7}$, $M_{3,8}$, $M_{3,9}$, $M_{2,12}$, $M_{2,13}$, $M_{4,6}$, and $M_{5,5}$ in two or more columns in such a way that any two vector spaces listed in the same column can be viewed as the same vector space with just different notation for vectors, but two vector spaces that appear in different columns cannot be so viewed.

| 3.2 | ## BASIC CONCEPTS OF VECTOR SPACES |

We now extend the terminology we developed in Chapters 1 and 2 for the Euclidean spaces $\mathbb{R}^n$ to general vector spaces V. That is, we discuss linear combinations of vectors, the span of vectors, subspaces, dependent and independent vectors, bases, and dimension. Definitions of most of these concepts, and the theorems concerning them, can be lifted with minor changes from Chapters 1 and 2, replacing "in $\mathbb{R}^n$" wherever it occurs by "in a vector space V." Where we do make changes, the reasons for them are explained.

Linear Combinations, Spans, and Subspaces

The major change from Chapter 1 is that our vector spaces may now be so large that they cannot be spanned by a *finite* number of vectors. Each Euclidean space $\mathbb{R}^n$ and each subspace of $\mathbb{R}^n$ can be spanned by a finite set of vectors, but this is not the case for a general vector space. For example, no *finite* set of polynomials can span the space P of all polynomials in x, because a finite set of polynomials cannot contain polynomials of arbitrarily high degree. Because we can add only a finite number of vectors, our definition of a linear combination in Chapter 1 will be unchanged. However, surely we want to consider the space P of all polynomials to be spanned by the monomials in the *infinite* set $\{1, x, x^2, x^3, \ldots \}$, because every polynomial is a linear combination of these monomials. Thus we must modify the definition of the span of vectors to include the case where the number of vectors may be infinite.

DEFINITION 3.2   Linear Combinations

Given vectors $\mathbf{v}_1, \mathbf{v}_2, \ldots, \mathbf{v}_k$ in a vector space V and scalars $r_1, r_2, \ldots, r_k$ in $\mathbb{R}$, the vector

$$r_1\mathbf{v}_1 + r_2\mathbf{v}_2 + \cdots + r_k\mathbf{v}_k$$

is a **linear combination** of the vectors $\mathbf{v}_1, \mathbf{v}_2, \ldots, \mathbf{v}_k$ with **scalar coefficients** $r_1, r_2, \ldots, r_k$.

DEFINITION 3.3   Span of a Subset X of V

Let X be a subset of a vector space V. The **span** of X is the set of all linear combinations of vectors in X, and is denoted by $\mathrm{sp}(X)$. If X is a finite set, so that $X = \{\mathbf{v}_1, \mathbf{v}_2, \ldots, \mathbf{v}_k\}$, then we also write $\mathrm{sp}(X)$ as $\mathrm{sp}(\mathbf{v}_1, \mathbf{v}_2, \ldots, \mathbf{v}_k)$. If $W = \mathrm{sp}(X)$, the vectors in X **span** or **generate** W. If $V = \mathrm{sp}(X)$ for some *finite* subset X of V, then V is **finitely generated**.

HISTORICAL NOTE   A COORDINATE-FREE TREATMENT of vector-space concepts appeared in 1862 in the second version of *Ausdehnungslehre* (*The Calculus of Extension*) by Hermann Grassman (1809–1877). In this version he was able to suppress somewhat the philosophical bias that had made his earlier work so unreadable and to concentrate on his new mathematical ideas. These included the basic ideas of the theory of n-dimensional vector spaces, including linear combinations, linear independence, and the notions of a subspace and a basis. He developed the idea of the dimension of a subspace as the maximal number of linearly independent vectors and proved the fundamental relation for two subspaces V and W that $\dim(V + W) = \dim(V) + \dim(W) - \dim(V \cap W)$.

Grassmann's notions derived from the attempt to translate geometric ideas about n-dimensional space into the language of algebra without dealing with coordinates, as is done in ordinary analytic geometry. He was the first to produce a complete system in which such concepts as points, line segments, planes, and their analogues in higher dimensions are represented as single elements. Although his ideas were initially difficult to understand, ultimately they entered the mathematical mainstream in such fields as vector analysis and the exterior algebra. Grassmann himself, unfortunately, never attained his goal of becoming a German university professor, spending most of his professional life as a mathematics teacher at a gymnasium (high school) in Stettin. In the final decades of his life, he turned away from mathematics and established himself as an expert in linguistics.

In general, we suppress Euclidean space illustrations here, because they are already familiar from Chapter 1.

ILLUSTRATION 1 Let P be the vector space of all polynomials, and let $M = \{1, x, x^2, x^3, \ldots\}$ be this subset of monomials. Then $P = \mathrm{sp}(M)$. Our remarks above Definition 3.2 indicate that P is not finitely generated. ∎

ILLUSTRATION 2 Let $M_{m,n}$ be the vector space of all $m \times n$ matrices, and let E be the set consisting of the matrices $E_{i,j}$, where $E_{i,j}$ is the $m \times n$ matrix having entry 1 in the ith row and jth column and entries 0 elsewhere. There are mn of these matrices in the set E. Then $M_{m,n} = \mathrm{sp}(E)$ and is finitely generated. ∎

The notion of closure of a subset of a vector space V under vector addition or scalar multiplication is the same as for a subset of $\mathbb{R}^n$. Namely, a subset W of a vector space V is **closed under vector addition** if for all $\mathbf{u}, \mathbf{v} \in W$ the sum $\mathbf{u} + \mathbf{v}$ is in W. If for all $\mathbf{v} \in W$ and all scalars r we have $r\mathbf{v} \in W$, then W is **closed under scalar multiplication**.

We will call a subset W of V a subspace precisely when it is nonempty and closed under vector addition and scalar multiplication, just as in Definition 1.16. However, we really should define it here in a different fashion, reflecting the fact that we have given an *axiomatic definition* of a vector space. A vector space is just one of many mathematical structures that are defined axiomatically. (Some other such axiomatic structures are groups, rings, fields, topological spaces, fiber bundles, sheaves, and manifolds.) A substructure is always understood to be a subset of the original structure set that satisfies, all by itself, the axioms for that type of structure, using *inherited* features, such as operations of addition and scalar multiplication, from the original structure. We give a definition of a subspace of a vector space V that reflects this point of view, and then prove as a theorem that a subset of V is a subspace if and only if it is nonempty and closed under vector addition and scalar multiplication.

DEFINITION 3.4 Subspace

A subset W of a vector space V is a **subspace** of V if W itself fulfills the requirements of a vector space, where addition and scalar multiplication of vectors in W produce the same vectors as these operations did in V.

In order for a nonempty subset W of a vector space V to be a subspace, the subset (together with the operations of vector addition and scalar multiplication) must form a self-contained system. That is, any addition or scalar multiplication using vectors in the subset W must always yield a vector that lies again in W. Then taking any $\mathbf{v}$ in W, we see that $0\mathbf{v} = \mathbf{0}$ and $(-1)\mathbf{v} = -\mathbf{v}$ are also in W. The eight properties A1–A4 and S1–S4 required of a vector space in Definition 3.1 are sure to be true for the subset, because they hold in all of V.

That is, if W is nonempty and is closed under addition and scalar multiplication, it is sure to be a vector space in its own right. We have arrived at an efficient test for determining whether a subset is a subspace of a vector space.

THEOREM 3.2 Test for a Subspace

A subset W of a vector space V is a subspace of V if and only if W is nonempty and satisfies the following two conditions:

1. If $\mathbf{v}$ and $\mathbf{w}$ are in W, then $\mathbf{v} + \mathbf{w}$ is in W.   Closure under vector addition

2. If r is any scalar in $\mathbb{R}$ and $\mathbf{v}$ is in W, then $r\mathbf{v}$ is in W.   Closure under scalar multiplication

Condition (2) of Theorem 3.2 with $r = 0$ shows that the zero vector lies in every subspace. Recall that a subspace of $\mathbb{R}^n$ always contains the origin.

The entire vector space V satisfies the conditions of Theorem 3.2. That is, V is a subspace of itself. Other subspaces of V are called **proper subspaces**. One such subspace is the subset $\{\mathbf{0}\}$, consisting of only the zero vector. We call $\{\mathbf{0}\}$ the **zero subspace** of V.

Note that if V is a vector space and X is any nonempty subset of V, then $\mathrm{sp}(X)$ is a subspace of V, because the sum of two linear combinations of vectors in X is again a linear combination of vectors in X, as is any scalar multiple of such a linear combination. Thus the closure conditions of Theorem 3.2 are satisfied. A moment of thought shows that $\mathrm{sp}(X)$ is the smallest subspace of V containing all the vectors in X.

ILLUSTRATION 3   The space P of all polynomials in x is a subspace of the vector space P_∞ of power series in x, described in Example 4 of Section 3.1. Exercise 16 in Section 3.1 shows that the set consisting of all polynomials in x of degree at most n, together with the zero polynomial, is a vector space P_n. This space P_n is a subspace both of P and of P_∞. ∎

ILLUSTRATION 4   The set of invertible $n \times n$ matrices is not a subspace of the vector space M_n of all $n \times n$ matrices, because the sum of two invertible matrices may not be invertible; also, the zero matrix is not invertible. ∎

ILLUSTRATION 5   The set of all upper-triangular $n \times n$ matrices is a subspace of the space M_n of all $n \times n$ matrices, because sums and scalar multiples of upper-triangular matrices are again upper triangular. ∎

ILLUSTRATION 6   Let F be the vector space of all functions mapping $\mathbb{R}$ into $\mathbb{R}$. Because sums and scalar multiples of continuous functions are continuous, the subset C of F consisting of all continuous functions mapping $\mathbb{R}$ into $\mathbb{R}$ is a subspace of F. Because sums and scalar multiples of differentiable functions are differentiable, the subset D of F consisting of all differentiable functions mapping $\mathbb{R}$ into

$\mathbb{R}$ is also a subspace of F. Because every differentiable function is continuous, we see that D is also a subspace of C. Let D_∞ be the set of all functions mapping $\mathbb{R}$ into $\mathbb{R}$ that have derivatives of all orders. Note that D_∞ is closed under addition and scalar multiplication and is a subspace of D, C, and F.   ■

EXAMPLE 1   Let F be the vector space of all functions mapping $\mathbb{R}$ into $\mathbb{R}$. Show that the set S of all solutions in F of the differential equation

$$f'' + f = 0$$

is a subspace of F.

SOLUTION   We note that the zero function in F is a solution, and so the set S is nonempty. If f and g are in S, then $f'' + f = 0$ and $g'' + g = 0$, and so $(f + g)'' + (f + g) = f'' + g'' + f + g = (f'' + f) + (g'' + g) = 0 + 0$, which shows that S is closed under addition. Similarly, $(rf)'' + rf = rf'' + rf = r(f'' + f) = r0 = 0$, so S is closed under scalar multiplication. Thus S is a subspace of F.   ■

The preceding example is a special case of a general theorem stating that all solutions in F of a homogeneous linear differential equation form a subspace of F. We ask you to write out the proof of the general theorem in Exercise 40. Recall that all solutions of the homogeneous linear system $A\mathbf{x} = \mathbf{0}$, where A is an $m \times n$ matrix, form a subspace of $\mathbb{R}^n$.

Independence

We wish to extend the notions of dependence and independence that were given in Chapter 2. We restricted our consideration to finite sets of vectors in Chapter 2 because we can't have more than n vectors in an independent subset of $\mathbb{R}^n$. In this chapter, we have to worry about larger sets, because it may take an infinite set to span a vector space V. Recall that the vector space P of all polynomials cannot be spanned by a finite set of vectors. We make the following slight modification to Definition 2.1.

DEFINITION 3.5   Linear Dependence and Independence

Let X be a set of vectors in a vector space V. A **dependence relation** in this set X is an equation of the form

$$r_1\mathbf{v}_1 + r_2\mathbf{v}_2 + \cdots + r_k\mathbf{v}_k = \mathbf{0}, \qquad \text{some } r_j \neq 0$$

where $\mathbf{v}_i \in X$ for $i = 1, 2, \ldots, k$. If such a dependence relation exists, then X is a **linearly dependent** set of vectors. Otherwise, the set X of vectors is **linearly independent**.

ILLUSTRATION 7   The subset $\{1, x, x^2, \ldots, x^n, \ldots\}$ of monomials in the vector space P of all polynomials is an independent set.   ■

ILLUSTRATION 8   The subset $\{\sin^2 x, \cos^2 x, 1\}$ of the vector space F of all functions mapping $\mathbb{R}$ into $\mathbb{R}$ is dependent. A dependence relation is

$$1(\sin^2 x) + 1(\cos^2 x) + (-1)1 = 0.$$

Similarly, the subset $\{\sin^2 x, \cos^2 x, \cos 2x\}$ is dependent, because we have the trigonometric identity $\cos 2x = \cos^2 x - \sin^2 x$. Thus $1(\cos 2x) + (-1)(\cos^2 x) + 1(\sin^2 x) = 0$ is a dependence relation.   ∎

We know a mechanical procedure for determining whether a finite set of vectors in $\mathbb{R}^n$ is independent. We simply put the vectors as column vectors in a matrix and reduce the matrix to row-echelon form. The set is independent if and only if every column in the matrix contains a pivot. There is no such mechanical procedure for determining whether a finite set of vectors in a general vector space V is independent. We illustrate two methods that are used in function spaces in the next two examples.

EXAMPLE 2   Show that $\{\sin x, \cos x\}$ is an independent set of functions in the space F of all functions mapping $\mathbb{R}$ into $\mathbb{R}$.

SOLUTION   We show that there is no dependence relation of the form

$$r(\sin x) + s(\cos x) = 0, \tag{1}$$

where the 0 on the right of the equation is the function that has the value 0 for all x. If Eq. (1) holds for all x, then setting $x = 0$ and $x = \pi/2$, we obtain the linear system

$$r(0) + s(1) = 0 \quad \text{Setting } x = 0$$

$$r(1) + s(0) = 0, \quad \text{Setting } x = \frac{\pi}{2}$$

whose only solution is $r = s = 0$. Thus Eq. (1) holds only if $r = s = 0$, and so the functions $\sin x$ and $\cos x$ are independent.   ∎

From Example 2, we see that one way to try to show independence of k functions $f_1(x), f_2(x), \ldots, f_k(x)$ is to substitute k different values of x in a *dependence relation format*

$$r_1 f_1(x) + r_2 f_2(x) + \cdots + r_k f_k(x) = 0.$$

This will lead to a homogeneous system of k equations in the k unknowns $r_1, r_2, \ldots, r_k$. If that system has only the zero solution, then the functions are independent. If there is a nontrivial solution, we can't draw any conclusion. For example, substituting $x = 0$ and $x = \pi$ in Eq. (1) in Example 2 yields the system

$$r(0) + s = 0 \quad \text{Setting } x = 0$$

$$r(0) - s = 0, \quad \text{Setting } x = \pi$$

which has a nontrivial solution—for example, $r = 10$ and $s = 0$. This occurs because we just chose the wrong values for x. The values 0 and $\pi/2$ for x do demonstrate independence, as we saw in Example 2.

EXAMPLE 3   Show that the functions e^x and e^{2x} are independent in the vector space F of all functions mapping $\mathbb{R}$ into $\mathbb{R}$.

SOLUTION   We set up the dependence relation format

$$re^x + se^{2x} = 0,$$

and try to determine if we must have $r = s = 0$. Illustrating a different technique than in Example 2, we write this equation and its derivative:

$$re^x + se^{2x} = 0$$
$$re^x + 2se^{2x} = 0. \quad \textbf{Differentiating}$$

Setting $x = 0$ in both equations, we obtain the homogeneous linear system

$$r + s = 0$$
$$r + 2s = 0,$$

which has only the trivial solution $r = s = 0$. Thus the functions are independent.   ∎

In summary, we can try to show independence of functions by starting with an equation in dependence relation format, and then substituting different values of the variable, or differentiating (possibly several times) and substituting values, or a combination of both, to obtain a square homogeneous system with the coefficients in the dependence relation format as unknowns. If the system has only the zero solution, the functions are independent. If there are nontrivial solutions, we can't come to a conclusion without more work.

Bases and Dimension

Recall that we defined a subset $\{\mathbf{w}_1, \mathbf{w}_2, \ldots, \mathbf{w}_k\}$ to be a *basis* for the subspace $W = \mathrm{sp}(\mathbf{w}_1, \mathbf{w}_2, \ldots, \mathbf{w}_k)$ of $\mathbb{R}^n$ if every vector in W can be expressed *uniquely* as a linear combination of $\mathbf{w}_1, \mathbf{w}_2, \ldots, \mathbf{w}_k$. Theorem 1.15 shows that to demonstrate this uniqueness, we need only show that the only linear combination that yields the zero vector is the one with all coefficients 0—that is, the uniqueness condition can be replaced by the condition that the set $\{\mathbf{w}_1, \mathbf{w}_2, \ldots, \mathbf{w}_k\}$ be independent. This led us to an alternate characterization of a basis for W (Theorem 2.1) as a subset of W that is *independent* and that *spans W*. It is this alternate description that is traditional for general vector spaces. The uniqueness condition then becomes a theorem; it remains the most important aspect of a basis and forms the foundation for the next section of our text. For a general vector space, we may need an *infinite* set of vectors to form a basis; for example, a basis for the space P of all polynomials is the set

$\{1, x, x^2, \ldots, x^n, \ldots\}$ of monomials. The following definition takes this possibility into account. Also, because a subspace of a vector space is again a vector space, it is unnecessary now to explicitly include the word "subspace" in the definition.

DEFINITION 3.6   Basis for a Vector Space

Let V be a vector space. A set of vectors in V is a **basis** for V if the following conditions are met:

1. The set of vectors spans V.
2. The set of vectors is linearly independent.

ILLUSTRATION 9   The set $X = \{1, x, x^2, \ldots, x^n, \ldots\}$ of monomials is a basis for the vector space P of all polynomials. It is not a basis for the vector space P_∞ of formal power series in x, discussed in Example 4 of Section 3.1, because a series $\displaystyle\sum_{n=0}^{\infty} a_n x^n$ cannot be expressed as a *finite* sum of scalar multiples of the monomials unless all but a finite number of the coefficients a_n are 0. For example, $1 + x + x^2 + x^3 + \ldots$ is not a finite sum of monomials. Remember that all linear combinations are *finite* sums.

The vector space P_n of polynomials of degree at most n, together with the zero polynomial, has as a basis $\{1, x, x^2, \ldots, x^n\}$. ∎

We now prove as a theorem the uniqueness which was the defining criterion in Definition 1.17 in Section 1.6. Namely, we show that a subset B of nonzero vectors in a vector space V is a basis for V if and only if each vector $\mathbf{v}$ in V can be expressed *uniquely* in the form

$$\mathbf{v} = r_1\mathbf{b}_1 + r_2\mathbf{b}_2 + \cdots + r_k\mathbf{b}_k \qquad (2)$$

for scalars r_i and vectors $\mathbf{b}_i$ in B. Because B can be an infinite set, we need to elaborate on the meaning of uniqueness. Suppose that there are two expressions in the form of Eq. (2) for $\mathbf{v}$. The two expressions might involve some of the same vectors from B and might involve some different vectors from B. The important thing is that each involves only a finite number of vectors. Thus if we take all vectors in B appearing in one expression or the other, or in both, we have just a finite list of vectors to be concerned with. We may assume that both expressions contain each vector in this list by inserting any missing vector with a zero coefficient. Assuming now that Eq. (2) is the result of this adjustment of the first expression for $\mathbf{v}$, the second expression for $\mathbf{v}$ may be written as

$$\mathbf{v} = s_1\mathbf{b}_1 + s_2\mathbf{b}_2 + \cdots + s_k\mathbf{b}_k. \qquad (3)$$

Uniqueness asserts that $s_i = r_i$ for each i.

THEOREM 3.3   Unique Combination Criterion for a Basis

> Let B be a set of nonzero vectors in a vector space V. Then B is a basis for V if and only if each vector $\mathbf{v}$ in V can be *uniquely* expressed in the form of Eq. (2) for scalars r_i and vectors $\mathbf{b}_i \in B$.

PROOF   Assume that B is a basis for V. Condition 1 of Definition 3.6 tells us that a vector $\mathbf{v}$ in V can be expressed in the form of Eq. (2). Suppose now that $\mathbf{v}$ can also be written in the form of Eq. (3). Subtracting Eq. (3) from Eq. (2), we obtain

$$(r_1 - s_1)\mathbf{b}_1 + (r_2 - s_2)\mathbf{b}_2 + \cdots + (r_k - s_k)\mathbf{b}_k = \mathbf{0}.$$

Because B is independent, we see that $r_1 - s_1 = 0, r_2 - s_2 = 0, \ldots, r_k - s_k = 0$, and so $r_i = s_i$ for each i. Thus we have established uniqueness.

Now assume that each vector in V can be expressed uniquely in the form of Eq. (2). In particular, this is true of the zero vector. This means that no dependence relation in B is possible. That is, if

$$r_1\mathbf{b}_1 + r_2\mathbf{b}_2 + \cdots + r_k\mathbf{b}_k = \mathbf{0}$$

for vectors $\mathbf{b}_i$ in B and scalars r_i, then, because we always have

$$0\mathbf{b}_1 + 0\mathbf{b}_2 + \cdots + 0\mathbf{b}_k = \mathbf{0},$$

each $r_i = 0$ by uniqueness. This shows that B is independent. Because B also generates V (by hypothesis), it is a basis.   ▲

Dimension

We will have to restrict our treatment here to **finitely generated vector spaces**—that is, those that can be spanned by a finite number of vectors. We can argue just as we did for $\mathbb{R}^n$ that every finitely generated vector space V contains a basis.* Namely, if $V = \text{sp}(\mathbf{v}_1, \mathbf{v}_2, \ldots, \mathbf{v}_k)$, then we can examine the vectors $\mathbf{v}_i$ in turn, and delete at each step any that can be expressed as linear combinations of those that remain. We would also like to know that any two bases have the same number of elements, so that we can have a well-defined concept of dimension for a finitely generated vector space. The main tool is Theorem 2.2 on page 130, which we can restate for general vector spaces, rather than for subspaces of $\mathbb{R}^n$.

---

*If we are willing to assume the **Axiom of Choice**:

> Given a collection of nonempty sets, no two of which have an element in common, there exists a "choice set" C that contains exactly one element from each set in the collection.

then we can prove that *every* vector space has a basis, and that given a vector space V, every basis has the same number of elements, although we may be totally unable to actually specify a basis. This kind of work is regarded as magnificent mathematics by some and as abstract nonsense by others. For example, using the Axiom of Choice, we can prove that the space P_∞ of all formal power series in x has a basis. But we are still unable to exhibit one! It has been shown that the Axiom of Choice is independent of the other axioms of set theory.

THEOREM 3.4 Relative Size of Spanning and Independent Sets

> Let V be a vector space. Let $\mathbf{w}_1, \mathbf{w}_2, \ldots, \mathbf{w}_k$ be vectors in V that span V, and let $\mathbf{v}_1, \mathbf{v}_2, \ldots, \mathbf{v}_m$ be vectors in V that are independent. Then $k \geq m$.

PROOF The proof is the same, word for word, as the proof of Theorem 2.2. ▲

It is not surprising that the proof of the preceding theorem is the same as that of Theorem 2.2. The next section will show that we can expect Chapter 2 arguments to be valid whenever we deal just with finitely generated vector spaces.

The same arguments as those in the corollary to Theorem 2.2 give us the following corollary to Theorem 3.4.

COROLLARY Invariance of Dimension for Finitely Generated Spaces

> Let V be a finitely generated vector space. Then any two bases of V have the same number of elements.

We can now rewrite the definition of dimension for finitely generated vector spaces.

DEFINITION 3.7 Dimension of a Finitely Generated Vector Space

> Let V be a finitely generated vector space. The number of elements in a basis for V is the **dimension** of V, and is denoted by $\dim(V)$.

ILLUSTRATION 10 Let P_n be the vector space of polynomials in x of degree at most n. Because $\{1, x, x^2, \ldots, x^n\}$ is a basis for P_n, we see that $\dim(P_n) = n + 1$. ∎

ILLUSTRATION 11 The set E of matrices $E_{i,j}$ in Illustration 2 is a basis for the vector space $M_{m,n}$ of all $m \times n$ matrices, so $\dim(M_{m,n}) = mn$. ∎

By the same arguments that we used for $\mathbb{R}^n$ (page 132), we see that for a finitely generated vector space V, every independent set of vectors in V can be enlarged, if necessary, to a basis. Also, if $\dim(V) = k$, then every independent set of k vectors in V is a basis for V, and every set of k vectors that span V is a basis for V. (See Theorem 2.3 on page 133.)

EXAMPLE 4 Determine whether $S = \{1 - x, 2 - 3x^2, x + 2x^2\}$ is a basis for the vector space P_2 of polynomials of degree at most 2, together with the zero polynomial.

SOLUTION We know that $\dim(P_2) = 3$ because $\{1, x, x^2\}$ is a basis for P_2. Thus S will be a basis if and only if S is an independent set. We can rewrite the dependence relation format

$$r(1 - x) + s(2 - 3x^2) + t(x + 2x^2) = 0$$

as

$$(r + 2s)1 + (-r + t)x + (-3s + 2t)x^2 = 0.$$

Because $\{1, x, x^2\}$ is independent, this relation can hold if and only if

$$
\begin{aligned}
r + \quad 2s \quad\quad &= 0 \\
-r \quad\quad + \quad t &= 0 \\
-3s + 2t &= 0.
\end{aligned}
$$

Reducing the coefficient matrix of this homogeneous system, we obtain

$$
\begin{bmatrix} 1 & 2 & 0 \\ -1 & 0 & 1 \\ 0 & -3 & 2 \end{bmatrix}
\sim
\begin{bmatrix} 1 & 2 & 0 \\ 0 & 2 & 1 \\ 0 & -3 & 2 \end{bmatrix}
\sim
\begin{bmatrix} 1 & 2 & 0 \\ 0 & 1 & \frac{1}{2} \\ 0 & 0 & \frac{7}{2} \end{bmatrix}.
$$

We see at once that the homogeneous system with this coefficient matrix has only the trivial solution. Thus no dependence relation exists, and so S is an independent set with the necessary number of vectors, and is thus a basis. ■

EXAMPLE 5 Find a basis for the vector space P_3 (polynomials of degree at most 3, and 0) containing the polynomials $x^2 + 1$ and $x^2 - 1$.

SOLUTION First we observe that the two given polynomials are independent because neither is a scalar multiple of the other. The vectors

$$x^2 + 1, \ x^2 - 1, \ 1, \ x, \ x^2, \ x^3$$

generate P_3 because the last four of these form a basis for P_3. We can reduce this list of vectors to a basis by deleting any vector that is a linear combination of others in the list, being sure to retain the first two. For this example, it is actually easier to notice that surely x is not in $\text{sp}(x^2 + 1, x^2 - 1)$ and x^3 is not in $\text{sp}(x^2 - 1, x^2 + 1, x)$. Thus the set $\{x^2 + 1, x^2 - 1, x, x^3\}$ is independent. Because $\dim(P_3) = 4$ and this independent set contains four vectors, it must be a basis for P_3. Alternatively, we could have deleted 1 and x by noticing that

$$1 = \frac{1}{2}(x^2 + 1) - \frac{1}{2}(x^2 - 1) \quad \text{and} \quad x^2 = \frac{1}{2}(x^2 + 1) + \frac{1}{2}(x^2 - 1). \quad ■$$

1. A subset W of a vector space V is a *subspace* of V if and only if it is nonempty and satisfies the two closure properties:

   $\mathbf{v} + \mathbf{w}$ is contained in W for all vectors $\mathbf{v}$ and $\mathbf{w}$ in W, and
   $r\mathbf{v}$ is contained in W for all vectors $\mathbf{v}$ in W and all scalars r.

2. Let X be a subset of a vector space V. The set $sp(X)$ of all linear combinations of vectors in X is a subspace of V called the *span* of X, or the *subspace of V generated by the vectors in X*. It is the smallest subspace of V containing all the vectors in X.

3. A vector space V is *finitely generated* if $V = sp(X)$ for some set $X = \{\mathbf{v}_1, \mathbf{v}_2, \ldots, \mathbf{v}_k\}$ containing only a finite number of vectors in V.

4. A set X of vectors in a vector space V is *linearly dependent* if there exists a dependence relation

$$r_1\mathbf{v}_1 + r_2\mathbf{v}_2 + \cdots + r_k\mathbf{v}_k = \mathbf{0}, \qquad \text{at least one } r_j \neq 0,$$

   where each $\mathbf{v}_i \in X$ and each $r_i \in \mathbb{R}$. The set X is *linearly independent* if no such dependence relation exists, and so a linear combination of vectors in X is the zero vector only if all scalar coefficients are zero.

5. A set B of vectors in a vector space V is a *basis* for V if B spans V and is independent.

6. A subset B of nonzero vectors in a vector space V is a basis for V if and only if every nonzero vector in V can be expressed as a linear combination of vectors in B in a *unique* way.

7. If X is a finite set of vectors spanning a vector space V, then X can be reduced, if necessary, to a basis for V by deleting in turn any vector that can be expressed as a linear combination of those remaining.

8. If a vector space V has a finite basis, then all bases for V have the same number of vectors. The number of vectors in a basis for V is the *dimension* of V, denoted by $\dim(V)$.

9. The following are equivalent for n vectors in a vector space V where $\dim(V) = n$.
   a. The vectors are linearly independent.
   b. The vectors generate V.

EXERCISES

In Exercises 1–6, determine whether the indicated subset is a subspace of the given vector space.

1. The set of all polynomials of degree greater than 3 together with the zero polynomial in the vector space P of all polynomials with coefficients in $\mathbb{R}$

2. The set of all polynomials of degree 4 together with the zero polynomial in the vector space P of all polynomials in x

3. The set of all functions f such that $f(0) = 1$ in the vector space F of all functions mapping $\mathbb{R}$ into $\mathbb{R}$

4. The set of all functions f such that $f(1) = 0$ in the vector space F of all functions mapping $\mathbb{R}$ into $\mathbb{R}$

5. The set of all functions f in the vector space W of differentiable functions mapping $\mathbb{R}$ into $\mathbb{R}$ (see Illustration 6) such that $f'(2) = 0$

6. The set of all functions f in the vector space W of differentiable functions mapping $\mathbb{R}$ into $\mathbb{R}$ (see Illustration 6) such that f has derivatives of all orders

7. Let F be the vector space of functions mapping $\mathbb{R}$ into $\mathbb{R}$. Show that
   a. $\mathrm{sp}(\sin^2 x, \cos^2 x)$ contains all constant functions,
   b. $\mathrm{sp}(\sin^2 x, \cos^2 x)$ contains the function $\cos 2x$,
   c. $\mathrm{sp}(7, \sin^2 2x)$ contains the function $8 \cos 4x$.

8. Let P be the vector space of polynomials. Prove that $\mathrm{sp}(1, x) = \mathrm{sp}(1 + 2x, x)$. [HINT: Show that each of these subspaces is a subset of the other.]

9. Let V be a vector space, and let $\mathbf{v}_1$ and $\mathbf{v}_2$ be vectors in V. Follow the hint of Exercise 8 to prove that
   a. $\mathrm{sp}(\mathbf{v}_1, \mathbf{v}_2) = \mathrm{sp}(\mathbf{v}_1, 2\mathbf{v}_1 + \mathbf{v}_2)$,
   b. $\mathrm{sp}(\mathbf{v}_1, \mathbf{v}_2) = \mathrm{sp}(\mathbf{v}_1 + \mathbf{v}_2, \mathbf{v}_1 - \mathbf{v}_2)$.

10. Let $\mathbf{v}_1, \mathbf{v}_2, \ldots, \mathbf{v}_k$ and $\mathbf{w}_1, \mathbf{w}_2, \ldots, \mathbf{w}_m$ be vectors in a vector space V. Give a necessary and sufficient condition, involving linear combinations, for
$$\mathrm{sp}(\mathbf{v}_1, \mathbf{v}_2, \ldots, \mathbf{v}_k) = \mathrm{sp}(\mathbf{w}_1, \mathbf{w}_2, \ldots, \mathbf{w}_m).$$

In Exercises 11–13, determine whether the given set of vectors is dependent or independent.

11. $\{x^2 - 1, x^2 + 1, 4x, 2x - 3\}$ in P
12. $\{1, 4x + 3, 3x - 4, x^2 + 2, x - x^2\}$ in P
13. $\{1, \sin^2 x, \cos 2x, \cos^2 x\}$ in F

In Exercises 14–19, use the technique discussed following Example 3 to determine whether the given set of functions in the vector space F is independent or dependent.

14. $\{\sin x, \cos x\}$
15. $\{1, x, x^2\}$
16. $\{\sin x, \sin 2x, \sin 3x\}$
17. $\{\sin x, \sin(-x)\}$
18. $\{e^{2x}, e^{3x}, e^{4x}\}$
19. $\{1, e^x + e^{-x}, e^x - e^{-x}\}$

In Exercises 20 and 21, determine whether or not the given set of vectors is a basis for the indicated vector space.

20. $\{x, x^2 + 1, (x - 1)^2\}$ for P_2
21. $\{x, (x + 1)^2, (x - 1)^2\}$ for P_2

In Exercises 22–24, find a basis for the given subspace of the vector space.

22. $\mathrm{sp}(x^2 - 1, x^2 + 1, 4, 2x - 3)$ in P
23. $\mathrm{sp}(1, 4x + 3, 3x - 4, x^2 + 2, x - x^2)$ in P
24. $\mathrm{sp}(1, \sin^2 x, \cos 2x, \cos^2 x)$ in F

25. Mark each of the following True or False.
___ a. The set consisting of the zero vector is a subspace for every vector space.
___ b. Every vector space has at least two distinct subspaces.
___ c. Every vector space with a nonzero vector has at least two distinct subspaces.
___ d. If $\{\mathbf{v}_1, \mathbf{v}_2, \ldots, \mathbf{v}_n\}$ is a subset of a vector space V, then $\mathbf{v}_i$ is in $\mathrm{sp}(\mathbf{v}_1, \mathbf{v}_2, \ldots, \mathbf{v}_n)$ for $i = 1, 2, \ldots, n$.
___ e. If $\{\mathbf{v}_1, \mathbf{v}_2, \ldots, \mathbf{v}_n\}$ is a subset of a vector space V, then the sum $\mathbf{v}_i + \mathbf{v}_j$ is in $\mathrm{sp}(\mathbf{v}_1, \mathbf{v}_2, \ldots, \mathbf{v}_n)$ for all choices of i and j from 1 to n.
___ f. If $\mathbf{u} + \mathbf{v}$ lies in a subspace W of a vector space V, then both $\mathbf{u}$ and $\mathbf{v}$ lie in W.
___ g. Two subspaces of a vector space V may have empty intersection.
___ h. If S is independent, each vector in V can be expressed uniquely as a linear combination of vectors in S.
___ i. If S is independent and generates V, each vector in V can be expressed uniquely as a linear combination of vectors in S.
___ j. If each vector in V can be expressed uniquely as a linear combination of vectors in S, then S is an independent set.

26. Let V be a vector space. Mark each of the following True or False.

_____ **a.** Every independent set of vectors in V is a basis for the subspace the vectors span.

_____ **b.** If $\{\mathbf{v}_1, \mathbf{v}_2, \ldots, \mathbf{v}_n\}$ generates V, then each $\mathbf{v} \in V$ is a linear combination of the vectors in this set.

_____ **c.** If $\{\mathbf{v}_1, \mathbf{v}_2, \ldots, \mathbf{v}_n\}$ generates V, then each $\mathbf{v} \in V$ is a unique linear combination of the vectors in this set.

_____ **d.** If $\{\mathbf{v}_1, \mathbf{v}_2, \ldots, \mathbf{v}_n\}$ generates V and is independent, then each $\mathbf{v} \in V$ is a unique linear combination of the vectors in this set.

_____ **e.** If $\{\mathbf{v}_1, \mathbf{v}_2, \ldots, \mathbf{v}_n\}$ generates V, then this set of vectors is independent.

_____ **f.** If each vector in V is a unique linear combination of the vectors in the set $\{\mathbf{v}_1, \mathbf{v}_2, \ldots, \mathbf{v}_n\}$, then this set is independent.

_____ **g.** If each vector in V is a unique linear combination of the vectors in the set $\{\mathbf{v}_1, \mathbf{v}_2, \ldots, \mathbf{v}_n\}$, then this set is a basis for V.

_____ **h.** All vector spaces having a basis are finitely generated.

_____ **i.** Every independent subset of a finitely generated vector space V is a part of some basis for V.

_____ **j.** Any two bases in a finite-dimensional vector space V have the same number of elements.

27. Let W_1 and W_2 be subspaces of a vector space V. Prove that the intersection $W_1 \cap W_2$ is also a subspace of V.

28. Let $W_1 = \mathrm{sp}((1, 2, 3), (2, 1, 1))$ and $W_2 = \mathrm{sp}((1, 0, 1), (3, 0, -1))$ in $\mathbb{R}^3$. Find a set of generating vectors for $W_1 \cap W_2$.

29. Let V be a vector space with basis $\{\mathbf{v}_1, \mathbf{v}_2, \mathbf{v}_3\}$. Prove that $\{\mathbf{v}_1, \mathbf{v}_1 + \mathbf{v}_2, \mathbf{v}_1 + \mathbf{v}_2 + \mathbf{v}_3\}$ is also a basis for V.

30. Let V be a vector space with basis $\{\mathbf{v}_1, \mathbf{v}_2, \ldots, \mathbf{v}_n\}$, and let $W = \mathrm{sp}(\mathbf{v}_3, \mathbf{v}_4, \ldots, \mathbf{v}_n)$. If $\mathbf{w} = r_1\mathbf{v}_1 + r_2\mathbf{v}_2$ is in W, show that $\mathbf{w} = \mathbf{0}$.

31. Let $\{\mathbf{v}_1, \mathbf{v}_2, \mathbf{v}_3\}$ be a basis for a vector space V. Prove that the vectors $\mathbf{w}_1 = \mathbf{v}_1 + \mathbf{v}_2$, $\mathbf{w}_2 = \mathbf{v}_2 + \mathbf{v}_3$, $\mathbf{w}_3 = \mathbf{v}_1 - \mathbf{v}_3$ do not generate V.

32. Let $\{\mathbf{v}_1, \mathbf{v}_2, \mathbf{v}_3\}$ be a basis for a vector space V. Prove that, if $\mathbf{w}$ is not in $\mathrm{sp}(\mathbf{v}_1, \mathbf{v}_2)$, then $\{\mathbf{v}_1, \mathbf{v}_2, \mathbf{w}\}$ is also a basis for V.

33. Let $\{\mathbf{v}_1, \mathbf{v}_2, \ldots, \mathbf{v}_n\}$ be a basis for a vector space V, and let $\mathbf{w} = t_1\mathbf{v}_1 + t_2\mathbf{v}_2 + \cdots + t_k\mathbf{v}_k$, with $t_k \neq 0$. Prove that

$$\{\mathbf{v}_1, \mathbf{v}_2, \ldots, \mathbf{v}_{k-1}, \mathbf{w}, \mathbf{v}_{k+1}, \ldots, \mathbf{v}_n\}$$

is a basis for V.

34. Let W and U be subspaces of a vector space V, and let $W \cap U = \{\mathbf{0}\}$. Let $\{\mathbf{w}_1, \mathbf{w}_2, \ldots, \mathbf{w}_k\}$ be a basis for W, and let $\{\mathbf{u}_1, \mathbf{u}_2, \ldots, \mathbf{u}_m\}$ be a basis for U. Prove that, if each vector $\mathbf{v}$ in V is expressible in the form $\mathbf{v} = \mathbf{w} + \mathbf{u}$ for $\mathbf{w} \in W$ and $\mathbf{u} \in U$, then

$$\{\mathbf{w}_1, \mathbf{w}_2, \ldots, \mathbf{w}_k, \mathbf{u}_1, \mathbf{u}_2, \ldots, \mathbf{u}_m\}$$

is a basis for V.

35. Illustrate Exercise 34 with nontrivial subspaces W and U of $\mathbb{R}^5$.

36. Prove that, if W is a subspace of an n-dimensional vector space V and $\dim(W) = n$, then $W = V$.

37. Let $\mathbf{v}_1, \mathbf{v}_2, \ldots, \mathbf{v}_n$ be a list of nonzero vectors in a vector space V such that no vector in this list is a linear combination of its predecessors. Show that the vectors in the list form an independent set.

38. Exercise 37 indicates that a finite generating set for a vector space can be reduced to a basis by deleting, from left to right in a list of the vectors, each vector that is a linear combination of its predecessors. Use this technique to find a basis for the subspace

$$\mathrm{sp}(x^2 + 1, x^2 + x - 1, 3x - 6,$$
$$x^3 + x^2 + 1, x^3)$$

of the polynomial space P.

39. We once watched a speaker in a lecture derive the equation $f(x) \sin x + g(x) \cos x = 0$, and then say, "Now everyone knows that $\sin x$ and $\cos x$ are independent functions, so $f(x) = 0$ and $g(x) = 0$." Was the statement correct or incorrect? Give a proof or a counterexample.

40. A homogeneous linear nth-order differential equation has the form

$$f_n(x)y^{(n)} + f_{n-1}(x)y^{(n-1)} + \cdots + f_2(x)y'' + f_1(x)y' + f_0(x)y = 0.$$

Show that the set of all solutions of this equation that lie in the space F of all functions mapping $\mathbb{R}$ into $\mathbb{R}$ is a subspace of F.

41. Referring to Exercise 40, suppose that the differential equation

$$f_n(x)y^{(n)} + f_{n-1}(x)y^{(n-1)} + \cdots + f_2(x)y'' + f_1(x)y' + f_0(x)y = g(x)$$

does have a solution $y = p(x)$ in the space F of all functions mapping $\mathbb{R}$ into $\mathbb{R}$. By analogy with Theorem 1.18 on p. 97, describe the structure of the set of solutions of this equation that lie in F.

42. Solve the differential equation $y' = 2x$ and describe your solution in terms of your answer to Exercise 41, or in terms of the answer in the back of the text.

It is a theorem of differential equations that if the functions $f_i(x)$ of the differential equation in Exercise 40 are all constant, then all the solutions of the equation lie in the vector space F of all functions mapping $\mathbb{R}$ into $\mathbb{R}$ and form a subspace of F of dimension n. Thus every solution can be written as a linear combination of n independent functions in F that form a basis for the solution space.

In Exercises 43–45, use your knowledge of calculus and the solution of Exercise 41 to describe the solution set of the given differential equation. You should be able to work these problems without having had a course in differential equations, using the hints.

43. a. $y'' + y = 0$ [HINT: You need to find two independent functions such that when you differentiate twice, you get the negative of the function you started with.]
  b. $y'' + y = x$ [HINT: Find one solution by experimentation.]

44. a. $y'' - 4y = 0$ [HINT: What two independent functions, when differentiated twice, give 4 times the original function?]
  b. $y'' - 4y = x$ [HINT: Find one solution by experimentation.]

45. a. $y^{(3)} - 9y' = 0$ [HINT: Try to find values of m such that $y = e^{mx}$ is a solution.]
  b. $y^{(3)} - 9y' = x^2 + 2x$ [HINT: Find one solution by experimentation.]

46. Let S be any set and let F be the set of all functions mapping S into $\mathbb{R}$. Let W be the subset of F consisting of all functions $f \in F$ such that $f(s) = 0$ for all but a finite number of elements s in S.
  a. Show that W is a subspace of F.
  b. What condition must be satisfied to have $W = F$?

47. Referring to Exercise 46, describe a basis B for the subspace W of F. Explain why B is not a basis for F unless $F = W$.

3.3   COORDINATIZATION OF VECTORS

Much of the work in this text is phrased in terms of the Euclidean vector spaces $\mathbb{R}^n$ for $n = 1, 2, 3, \ldots$. In this section we show that, for finite-dimensional vector spaces, no loss of generality results from restricting ourselves to the spaces $\mathbb{R}^n$. Specifically, we will see that if a vector space V has dimension n, then V can be coordinatized so that it will look just like $\mathbb{R}^n$. We can then work with these coordinates by utilizing the matrix techniques we have developed for the space $\mathbb{R}^n$. Throughout this section, we consider V to be a finite-dimensional vector space.

Ordered Bases

The vector [2, 5] in $\mathbb{R}^2$ can be expressed in terms of the standard basis vectors as $2\mathbf{e}_1 + 5\mathbf{e}_2$. The components of [2, 5] are precisely the coefficients of these basis vectors. The vector [2, 5] is different from the vector [5, 2], just as the point (2, 5) is different from the point (5, 2). We regard the standard basis vectors as having a natural order, $\mathbf{e}_1 = [1, 0]$ and $\mathbf{e}_2 = [0, 1]$. In a nonzero vector space V with a basis $B = \{\mathbf{b}_1, \mathbf{b}_2, \ldots, \mathbf{b}_n\}$, there is usually no natural order for the basis vectors. For example, the vectors $\mathbf{b}_1 = [-1, 5]$ and $\mathbf{b}_2 = [3, 2]$ form a basis for $\mathbb{R}^2$, but there is no natural order for these vectors. If we want the vectors to have an order, we must specify their order. By convention, set notation does not denote order; for example, $\{\mathbf{b}_1, \mathbf{b}_2\} = \{\mathbf{b}_2, \mathbf{b}_1\}$. To describe order, we use parentheses, ( ), in place of set braces, { }; we are used to paying attention to order in the notation $(\mathbf{b}_1, \mathbf{b}_2)$. We denote an *ordered basis* of n vectors in V by $B = (\mathbf{b}_1, \mathbf{b}_2, \ldots, \mathbf{b}_n)$. For example, the standard basis $\{\mathbf{e}_1, \mathbf{e}_2, \mathbf{e}_3\}$ of $\mathbb{R}^3$ gives rise to six different ordered bases—namely,

$$(\mathbf{e}_1, \mathbf{e}_2, \mathbf{e}_3) \quad (\mathbf{e}_2, \mathbf{e}_1, \mathbf{e}_3) \quad (\mathbf{e}_3, \mathbf{e}_1, \mathbf{e}_2) \quad (\mathbf{e}_1, \mathbf{e}_3, \mathbf{e}_2) \quad (\mathbf{e}_2, \mathbf{e}_3, \mathbf{e}_1) \quad (\mathbf{e}_3, \mathbf{e}_2, \mathbf{e}_1).$$

These correspond to the six possible orders for the unit coordinate vectors. The ordered basis $(\mathbf{e}_1, \mathbf{e}_2, \mathbf{e}_3)$ is the *standard ordered basis* for $\mathbb{R}^3$, and in general, the basis $E = (\mathbf{e}_1, \mathbf{e}_2, \ldots, \mathbf{e}_n)$ is the *standard ordered basis* for $\mathbb{R}^n$.

Coordinatization of Vectors

Let V be a finite-dimensional vector space, and let $B = (\mathbf{b}_1, \mathbf{b}_2, \ldots, \mathbf{b}_n)$ be a basis for V. By Theorem 3.3, every vector $\mathbf{v}$ in V can be expressed in the form

$$\mathbf{v} = r_1\mathbf{b}_1 + r_2\mathbf{b}_2 + \cdots + r_n\mathbf{b}_n$$

for unique scalars $r_1, r_2, \ldots, r_n$. We associate the vector $[r_1, r_2, \ldots, r_n]$ in $\mathbb{R}^n$ with $\mathbf{v}$. This gives us a way of coordinatizing V.

DEFINITION 3.8 Coordinate Vector Relative to an Ordered Basis

Let $B = (\mathbf{b}_1, \mathbf{b}_2, \ldots, \mathbf{b}_n)$ be an ordered basis for a finite-dimensional vector space V, and let

$$\mathbf{v} = r_1\mathbf{b}_1 + r_2\mathbf{b}_2 + \cdots + r_n\mathbf{b}_n.$$

The vector $[r_1, r_2, \ldots, r_n]$ is the **coordinate vector of v relative to the ordered basis** B, and is denoted by $\mathbf{v}_B$.

ILLUSTRATION 1   Let P_n be the vector space of polynomials of degree at most n. There are two natural choices for an ordered basis for P_n—namely,

$$B = (1, x, x^2, \ldots, x^n) \quad \text{and} \quad B' = (x^n, x^{n-1}, \ldots, x^2, x, 1).$$

Taking $n = 4$, we see that for the polynomial $p(x) = -x + x^3 + 2x^4$ we have

$$p(x)_B = [0, -1, 0, 1, 2] \quad \text{and} \quad p(x)_{B'} = [2, 1, 0, -1, 0]. \qquad \blacksquare$$

EXAMPLE 1   Find the coordinate vectors of $[1, -1]$ and of $[-1, -8]$ relative to the ordered basis $B = ([1, -1], [1, 2])$ of $\mathbb{R}^2$.

SOLUTION   We see that $[1, -1]_B = [1, 0]$, because

$$[1, -1] = 1[1, -1] + 0[1, 2].$$

To find $[-1, -8]_B$, we must find r_1 and r_2 such that $[-1, -8] = r_1[1, -1] + r_2[1, 2]$. Equating components of this vector equation, we obtain the linear system

$$r_1 + r_2 = -1$$
$$-r_1 + 2r_2 = -8.$$

The solution of this system is $r_1 = 2$, $r_2 = -3$, so we have $[-1, 8]_B = [2, -3]$. Figure 3.1 indicates the geometric meaning of these coordinates.   ∎

EXAMPLE 2   Find the coordinate vector of $[1, 2, -2]$ relative to the ordered basis $B = ([1, 1, 1], [1, 2, 0], [1, 0, 1])$ in $\mathbb{R}^3$.

SOLUTION   We must express $[1, 2, -2]$ as a linear combination of the basis vectors in B. Working with column vectors, we must solve the equation

$$r_1\begin{bmatrix} 1 \\ 1 \\ 1 \end{bmatrix} + r_2\begin{bmatrix} 1 \\ 2 \\ 0 \end{bmatrix} + r_3\begin{bmatrix} 1 \\ 0 \\ 1 \end{bmatrix} = \begin{bmatrix} 1 \\ 2 \\ -2 \end{bmatrix}$$

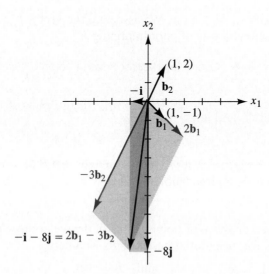

FIGURE 3.1
$[-1, -8]_B = [2, -3]$.

for r_1, r_2, and r_3. We find the unique solution by a Gauss–Jordan reduction:

$$\begin{bmatrix} 1 & 1 & 1 & | & 1 \\ 1 & 2 & 0 & | & 2 \\ 1 & 0 & 1 & | & -2 \end{bmatrix} \sim \begin{bmatrix} 1 & 1 & 1 & | & 1 \\ 0 & 1 & -1 & | & 1 \\ 0 & -1 & 0 & | & -3 \end{bmatrix}$$

$$\sim \begin{bmatrix} 1 & 0 & 2 & | & 0 \\ 0 & 1 & -1 & | & 1 \\ 0 & 0 & -1 & | & -2 \end{bmatrix} \sim \begin{bmatrix} 1 & 0 & 0 & | & -4 \\ 0 & 1 & 0 & | & 3 \\ 0 & 0 & 1 & | & 2 \end{bmatrix}.$$

Therefore, $[1, 2, -2]_B = [-4, 3, 2]$. ∎

We now box the procedure illustrated by Example 2.

Finding the Coordinate Vector of v in $\mathbb{R}^n$ Relative to an Ordered Basis $B = (\mathbf{b}_1, \mathbf{b}_2, \ldots, \mathbf{b}_n)$

Step 1 Writing vectors as column vectors, form the augmented matrix $[\mathbf{b}_1 \ \mathbf{b}_2 \ldots \mathbf{b}_n \ | \ \mathbf{v}]$.

Step 2 Use a Gauss–Jordan reduction to obtain the augmented matrix $[I \ | \ \mathbf{v}_B]$, where I is the $n \times n$ identity matrix and $\mathbf{v}_B$ is the desired coordinate vector.

Coordinatization of a Finite-Dimensional Vector Space

We can coordinatize a finite-dimensional vector space V by selecting an ordered basis $B = (\mathbf{b}_1, \mathbf{b}_2, \ldots, \mathbf{b}_n)$ and associating with each vector in V its unique coordinate vector relative to B. This gives a one-to-one correspondence between all the vectors in V and all the vectors in $\mathbb{R}^n$. To show that we may now work in $\mathbb{R}^n$ rather than in V, we have to show that the vector-space operations of vector addition and scalar multiplication in V are mirrored by those operations on coordinate vectors in $\mathbb{R}^n$. That is, we must show that

$$(\mathbf{v} + \mathbf{w})_B = \mathbf{v}_B + \mathbf{w}_B \quad \text{and} \quad (t\mathbf{v})_B = t\mathbf{v}_B \tag{1}$$

for all vectors $\mathbf{v}$ and $\mathbf{w}$ in V and for all scalars t in $\mathbb{R}$. To do this, suppose that

$$\mathbf{v} = r_1\mathbf{b}_1 + r_2\mathbf{b}_2 + \cdots + r_n\mathbf{b}_n$$

and

$$\mathbf{w} = s_1\mathbf{b}_1 + s_2\mathbf{b}_2 + \cdots + s_n\mathbf{b}_n.$$

Because

$$\mathbf{v} + \mathbf{w} = (r_1 + s_1)\mathbf{b}_1 + (r_2 + s_2)\mathbf{b}_2 + \cdots + (r_n + s_n)\mathbf{b}_n,$$

we see that the coordinate vector of $\mathbf{v} + \mathbf{w}$ is

$$(\mathbf{v} + \mathbf{w})_B = [r_1 + s_1, r_2 + s_2, \ldots, r_n + s_n]$$
$$= [r_1, r_2, \ldots, r_n] + [s_1, s_2, \ldots, s_n]$$
$$= \mathbf{v}_B + \mathbf{w}_B,$$

which is the sum of the coordinate vectors of $\mathbf{v}$ and of $\mathbf{w}$. Similarly, for any scalar t, we have

$$t\mathbf{v} = t(r_1\mathbf{b}_1 + r_2\mathbf{b}_2 + \cdots + r_n\mathbf{b}_n)$$
$$= (tr_1)\mathbf{b}_1 + (tr_2)\mathbf{b}_2 + \cdots + (tr_n)\mathbf{b}_n,$$

so the coordinate vector of $t\mathbf{v}$ is

$$(t\mathbf{v})_B = [tr_1, tr_2, \ldots, tr_n]$$
$$= t[r_1, r_2, \ldots, r_n] = t\mathbf{v}_B.$$

This completes the demonstration of relations (1). These relations tell us that, when we rename the vectors in V by coordinates relative to B, the resulting vector space of coordinates—namely, $\mathbb{R}^n$—has the same vector-space structure as V. Whenever the vectors in a vector space V can be renamed to make V appear structurally identical to a vector space W, we say that V and W are *isomorphic vector spaces*. Our discussion shows that every real vector space having a basis of n vectors is isomorphic to $\mathbb{R}^n$. For example, the space P_n of all polynomials of degree at most n is isomorphic to $\mathbb{R}^{n+1}$, because P_n has an ordered basis $B = (x^n, x^{n-1}, \ldots, x^2, x, 1)$ of $n + 1$ vectors. Each polynomial

$$a_n x^n + a_{n-1} x^{n-1} + \cdots + a_1 x + a_0$$

can be renamed by its coordinate vector

$$[a_n, a_{n-1}, \ldots, a_1, a_0]$$

relative to B. The adjective *isomorphic* is used throughout algebra to signify that two algebraic structures are identical except in the names of their elements.

For a vector space V isomorphic to a vector space W, all of the algebraic properties of vectors in V that can be derived solely from the axioms of a vector space correspond to identical properties in W. However, we cannot expect other features—such as whether the vectors are functions, matrices, or n-tuples—to be carried over from one space to the other. But a generating set of vectors or an independent set of vectors in one space corresponds to a set with the same property in the other space. Here is an example showing how we can simplify computations in a finite-dimensional vector space V, by working instead in $\mathbb{R}^n$.

EXAMPLE 3   Determine whether $x^2 - 3x + 2$, $3x^2 + 5x - 4$, and $7x^2 + 21x - 16$ are independent in the vector space P_2 of polynomials of degree at most 2.

SOLUTION We take $B = (x^2, x, 1)$ as an ordered basis for P_2. The coordinate vectors relative to B of the given polynomials are

$$(x^2 - 3x + 2)_B = [1, -3, 2],$$
$$(3x^2 + 5x - 4)_B = [3, 5, -4],$$
$$(7x^2 + 21x - 16)_B = [7, 21, -16].$$

We can determine whether the polynomials are independent by determining whether the corresponding coordinate vectors in $\mathbb{R}^3$ are independent. We set up the usual matrix, with these vectors as column vectors, and then we row-reduce it, obtaining

$$\begin{bmatrix} 1 & 3 & 7 \\ -3 & 5 & 21 \\ 2 & -4 & -16 \end{bmatrix} \sim \begin{bmatrix} 1 & 3 & 7 \\ 0 & 14 & 42 \\ 0 & -10 & -30 \end{bmatrix} \sim \begin{bmatrix} 1 & 3 & 7 \\ 0 & 1 & 3 \\ 0 & 0 & 0 \end{bmatrix}.$$

Because the third column in the echelon form has no pivot, these three coordinate vectors in $\mathbb{R}^3$ are dependent, and so the three polynomials are dependent. ∎

Continuing Example 3, to further illustrate working with coordinates in $\mathbb{R}^n$, we can reduce the final matrix further to

$$\begin{bmatrix} 1 & 0 & -2 \\ 0 & 1 & 3 \\ 0 & 0 & 0 \end{bmatrix}.$$

If we imagine a partition between the second and third columns, we see that

$$[7, 21, -16] = -2[1, -3, 2] + 3[3, 5, -4].$$

Thus

$$7x^2 + 21x - 16 = -2(x^2 - 3x + 2) + 3(3x^2 + 5x - 4).$$

EXAMPLE 4 It can be shown that the set $\{1, \sin x, \sin 2x, \ldots, \sin nx\}$ is an independent subset of the vector space F of all functions mapping $\mathbb{R}$ into $\mathbb{R}$. Find a basis for the subspace of F spanned by the functions

$$f_1(x) = 3 - \sin x + 3 \sin 2x - \sin 3x + 5 \sin 4x,$$
$$f_2(x) = 1 + 2 \sin x + 4 \sin 2x - \sin 4x$$
$$f_3(x) = -1 + 5 \sin x + 5 \sin 2x + \sin 3x - 7 \sin 4x$$
$$f_4(x) = 3 \sin 2x - \sin 4x.$$

SOLUTION We see that all these functions lie in the subspace W of F given by $W = \text{sp}(1, \sin x, \sin 2x, \sin 3x, \sin 4x)$, and we take

$$B = (1, \sin x, \sin 2x, \sin 3x, \sin 4x)$$

as an ordered basis for this subspace. Working with coordinates relative to B, the problem reduces to finding a basis for the subspace of $\mathbb{R}^5$ spanned by $[3, -1, 3, -1, 5]$, $[1, 2, 4, 0, -1]$, $[-1, 5, 5, 1, -7]$, and $[0, 0, 3, 0, -1]$. We reduce the matrix having these as column vectors, and begin this by switching minus the fourth row with the first row to create the pivot 1 in the first column. We obtain

$$
\begin{bmatrix}
3 & 1 & -1 & 0 \\
-1 & 2 & 5 & 0 \\
3 & 4 & 5 & 3 \\
-1 & 0 & 1 & 0 \\
5 & -1 & -7 & -1
\end{bmatrix}
\sim
\begin{bmatrix}
1 & 0 & -1 & 0 \\
0 & 2 & 4 & 0 \\
0 & 4 & 8 & 3 \\
0 & 1 & 2 & 0 \\
0 & -1 & -2 & -1
\end{bmatrix}
\sim
\begin{bmatrix}
1 & 0 & -1 & 0 \\
0 & 1 & 2 & 0 \\
0 & 0 & 0 & 3 \\
0 & 0 & 0 & 0 \\
0 & 0 & 0 & -1
\end{bmatrix}
\sim
\begin{bmatrix}
1 & 0 & -1 & 0 \\
0 & 1 & 2 & 0 \\
0 & 0 & 0 & 1 \\
0 & 0 & 0 & 0 \\
0 & 0 & 0 & 0
\end{bmatrix}.
$$

Because the first, second, and fourth columns have pivots, we should keep the first, second, and fourth of the original column vectors, so that the set $\{f_1(x), f_2(x), f_4(x)\}$ is a basis for the subspace W. ∎

We give an example showing the utility of different bases in the study of polynomial functions. We know that a polynomial function

$$y = p(x) = a_n x^n + \cdots + a_2 x^2 + a_1 x + a_0$$

has a graph that passes through the origin if $a_0 = 0$. If both a_0 and a_1 are zero, then the graph not only goes through the origin, but is quite flat there; indeed, if $a_2 \neq 0$, then the function behaves approximately like $a_2 x^2$ very near $x = 0$, because for very small values of x, such as 0.00001, the value of x^2 is much greater than the values of x^3, x^4, and other higher powers of x. If $a_2 = 0$ also, but $a_3 \neq 0$, then the function behaves very much like $a_3 x^3$ for such values of x very close to 0, etc. If instead of studying a polynomial function near $x = 0$, we want to study it near some other x-value, say near $x = a$, then we would like to express the polynomial as a linear combination of powers $(x - a)^i$—that is,

$$p(x) = b_n(x - a)^n + \cdots + b_2(x - a)^2 + b_1(x - a) + b_0.$$

Both $B = (x^n, \ldots, x^2, x, 1)$ and $B' = ((x - a)^n, \ldots, (x - a)^2, x - a, 1)$ are ordered bases for the space P_n of polynomials of degree at most n. (We leave the demonstration that B' is a basis as Exercise 20.) We give an example illustrating a method for expressing the polynomial $x^3 + x^2 - x - 1$ as a linear combination of $(x + 1)^3$, $(x + 1)^2$, $x + 1$, and 1.

EXAMPLE 5   Find the coordinate vector of $p(x) = x^3 + x^2 - x - 1$ relative to the ordered basis $B' = ((x + 1)^3, (x + 1)^2, x + 1, 1)$.

SOLUTION   Multiplying out the powers of $x + 1$, to express the vectors in B' in terms of our usual ordered basis $B = (x^3, x^2, x, 1)$ for P_3, we see that

$$B' = (x^3 + 3x^2 + 3x + 1, x^2 + 2x + 1, x + 1, 1).$$

Using coordinates relative to the ordered basis B, our problem reduces to expressing the vector $[1, 1, -1, -1]$ as a linear combination of the vectors

[1, 3, 3, 1], [0, 1, 2, 1], [0, 0, 1, 1], and [0, 0, 0, 1]. Reducing the matrix corresponding to the associated linear system, we obtain

$$\left[\begin{array}{cccc|c} 1 & 0 & 0 & 0 & 1 \\ 3 & 1 & 0 & 0 & 1 \\ 3 & 2 & 1 & 0 & -1 \\ 1 & 1 & 1 & 1 & -1 \end{array}\right] \sim \left[\begin{array}{cccc|c} 1 & 0 & 0 & 0 & 1 \\ 0 & 1 & 0 & 0 & -2 \\ 0 & 2 & 1 & 0 & -4 \\ 0 & 1 & 1 & 1 & -2 \end{array}\right] \sim \left[\begin{array}{cccc|c} 1 & 0 & 0 & 0 & 1 \\ 0 & 1 & 0 & 0 & -2 \\ 0 & 0 & 1 & 0 & 0 \\ 0 & 0 & 1 & 1 & 0 \end{array}\right] \sim \left[\begin{array}{cccc|c} 1 & 0 & 0 & 0 & 1 \\ 0 & 1 & 0 & 0 & -2 \\ 0 & 0 & 1 & 0 & 0 \\ 0 & 0 & 0 & 1 & 0 \end{array}\right].$$

Thus the required coordinate vector is $p(x)_{B'} = [1, -2, 0, 0]$, and so

$$x^3 + x^2 - x - 1 = (x + 1)^3 - 2(x + 1)^2.$$ ∎

Linear algebra is not the only tool that can be used to solve the problem in Example 5. Exercise 13 suggests a polynomial algebra solution, and Exercise 16 describes a calculus solution.

SUMMARY

Let V be a vector space with basis $\{\mathbf{b}_1, \mathbf{b}_2, \ldots, \mathbf{b}_n\}$.

1. $B = (\mathbf{b}_1, \mathbf{b}_2, \ldots, \mathbf{b}_n)$ is an ordered basis; the vectors are regarded as being in a specified order in this n-tuple notation.

2. Each vector $\mathbf{v}$ in V has a unique expression as a linear combination:
$$\mathbf{v} = r_1\mathbf{b}_1 + r_2\mathbf{b}_2 + \cdots + r_n\mathbf{b}_n.$$

3. The vector $\mathbf{v}_B = [r_1, r_2, \ldots, r_n]$ for the uniquely determined scalars r_i in the preceding equation (summary item 2) is the coordinate vector of $\mathbf{v}$ relative to B.

4. The vector space V can be coordinatized, using summary item 3, so that V is *isomorphic* to $\mathbb{R}^n$.

EXERCISES

In Exercises 1–10, find the coordinate vector of the given vector relative to the indicated ordered basis.

1. $[-1, 1]$ in $\mathbb{R}^2$ relative to $([0, 1], [1, 0])$

2. $[-2, 4]$ in $\mathbb{R}^2$ relative to $([0, -2], [-\frac{1}{2}, 0])$

3. $[4, 6, 2]$ in $\mathbb{R}^3$ relative to $([2, 0, 0], [0, 1, 1], [0, 0, 1])$

4. $[4, -2, 1]$ in $\mathbb{R}^3$ relative to $([0, 1, 1], [2, 0, 0], [0, 3, 0])$

5. $[3, 13, -1]$ in $\mathbb{R}^3$ relative to $([1, 3, -2], [4, 1, 3], [-1, 2, 0])$

6. $[9, 6, 11, 0]$ in $\mathbb{R}^4$ relative to $([1, 0, 1, 0], [2, 1, 1, -1], [0, 1, 1, -1], [2, 1, 3, 1])$

7. $x^3 + x^2 - 2x + 4$ in P_3 relative to $(1, x^2, x, x^3)$

8. $x^3 + 3x^2 - 4x + 2$ in P_3 relative to $(x, x^2 - 1, x^3, 2x^2)$

9. $x + x^4$ in P_4 relative to $(1, 2x - 1, x^3 + x^4, 2x^3, x^2 + 2)$

10. $\begin{bmatrix} 1 & -2 \\ 3 & 4 \end{bmatrix}$ in M_2 relative to
$$\left(\begin{bmatrix} 0 & 1 \\ 1 & 0 \end{bmatrix}, \begin{bmatrix} 0 & -1 \\ 0 & 0 \end{bmatrix}, \begin{bmatrix} 1 & -1 \\ 0 & 3 \end{bmatrix}, \begin{bmatrix} 0 & 1 \\ 0 & 1 \end{bmatrix} \right)$$

11. Find the coordinate vector of the polynomial $x^3 - 4x^2 + 3x + 7$ relative to the ordered basis $B' = ((x - 2)^3, (x - 2)^2, (x - 2), 1)$ of

the vector space P_3 of polynomials of degree at most 3. Use the method illustrated in Example 5.

12. Find the coordinate vector of the polynomial $4x^3 - 9x^2 + x$ relative to the ordered basis $B' = ((x - 1)^3, (x - 1)^2, (x - 1), 1)$ of the vector space P_3 of polynomials of degree at most 3. Use the method illustrated in Example 5.

13. Example 5 showed how to use *linear algebra* to rewrite the polynomial $p(x) = x^3 + x^2 - x - 1$ in powers of $x + 1$ rather than in powers of x. This exercise indicates a *polynomial algebra* solution to this problem. Replace x in $p(x)$ by $[(x + 1) - 1]$, and expand using the binomial theorem, keeping the $(x + 1)$ intact. Check your answer with that in Example 5.

14. Repeat Exercise 11 using the polynomial algebra method indicated in Exercise 13.

15. Repeat Exercise 12 using the polynomial algebra method indicated in Exercise 13.

16. Example 5 showed how to use *linear algebra* to rewrite the polynomial $p(x) = x^3 + x^2 - x - 1$ in powers of $x + 1$ rather than in powers of x. This exercise indicates a *calculus* solution to this problem. Form the equation

$$x^3 + x^2 - x - 1 = b_3(x + 1)^3 + b_2(x + 1)^2 + b_1(x + 1) + b_0.$$

Find b_0 by substituting $x = -1$ in this equation. Then equate the derivatives of both sides, and substitute $x = -1$ to find b_1. Continue differentiating both sides and substituting $x = -1$ to find b_2 and b_3. Check your answer with that in Example 5.

17. Repeat Exercise 11 using the calculus method indicated in Exercise 16.

18. Repeat Exercise 12 using the calculus method indicated in Exercise 16.

19. **a.** Prove that $\{1, \sin x, \cos x, \sin 2x, \cos 2x\}$ is an independent set of functions in the vector space F of all functions mapping $\mathbb{R}$ into $\mathbb{R}$.

    b. Find a basis for the subspace of F generated by the functions

$f_1(x) = 1 - 2 \sin x + 4 \cos x - \sin 2x - 3 \cos 2x,$

$f_2(x) = 2 - 3 \sin x - \cos x + 4 \sin 2x + 5 \cos 2x$

$f_3(x) = 5 - 8 \sin x + 2 \cos x + 7 \sin 2x + 7 \cos 2x$

$f_4(x) = -1 + 14 \cos x - 11 \sin 2x - 19 \cos 2x$

20. Prove that for every positive integer n and every $a \in \mathbb{R}$, the set

$$\{(x - a)^n, (x - a)^{n-1}, \ldots, (x - a)^2, x - a, 1\}$$

is a basis for the vector space P_n of polynomials of degree at most n.

21. Find the polynomial in P_2 whose coordinate vector relative to the ordered basis $B = (x + x^2, x - x^2, 1 + x)$ is [3, 1, 2].

22. Let V be a nonzero finite-dimensional vector space. Mark each of the following True or False.

___ **a.** The vector space V is isomorphic to $\mathbb{R}^n$ for some positive integer n.

___ **b.** There is a unique coordinate vector associated with each vector $\mathbf{v} \in V$.

___ **c.** There is a unique coordinate vector associated with each vector $\mathbf{v} \in V$ relative to a basis for V.

___ **d.** There is a unique coordinate vector associated with each vector $\mathbf{v} \in V$ relative to an ordered basis for V.

___ **e.** Distinct vectors in V have distinct coordinate vectors relative to the same ordered basis B for V.

___ **f.** The same vector in V cannot have the same coordinate vector relative to different ordered bases for V.

___ **g.** There are six possible ordered bases for $\mathbb{R}^3$.

___ **h.** There are six possible ordered bases for $\mathbb{R}^3$, consisting of the standard unit coordinate vectors in $\mathbb{R}^3$.

___ **i.** A reordering of elements in an ordered basis for V corresponds to a similar reordering of components in coordinate vectors with respect to the basis.

___ **j.** Addition and multiplication by scalars in V can be computed in terms of coordinate vectors with respect to any fixed ordered basis for V.

LINEAR TRANSFORMATIONS

Linear transformations mapping $\mathbb{R}^n$ into $\mathbb{R}^m$ were defined in Section 2.3. Now that we have considered vectors in more general spaces than $\mathbb{R}^n$, it is natural to extend the notion to linear transformations of other vector spaces, not necessarily finite-dimensional. In this section, we introduce linear transformations in a general setting. Recall that a linear transformation $T: \mathbb{R}^n \to \mathbb{R}^m$ is a function that satisfies

$$T(\mathbf{u} + \mathbf{v}) = T(\mathbf{u}) + T(\mathbf{v}) \tag{1}$$

and

$$T(r\mathbf{u}) = rT(\mathbf{u}) \tag{2}$$

for all vectors $\mathbf{u}$ and $\mathbf{v}$ in $\mathbb{R}^n$ and for all scalars r.

Linear Transformations $T: V \to V'$

The definition of a linear transformation of a vector space V into a vector space V' is practically identical to the definition for the Euclidean vector spaces in Section 2.3. We need only replace $\mathbb{R}^n$ by V and $\mathbb{R}^m$ by V'.

DEFINITION 3.9 Linear Transformation

A function T that maps a vector space V into a vector space V' is a **linear transformation** if it satisfies two criteria:

1. $T(\mathbf{u} + \mathbf{v}) = T(\mathbf{u}) + T(\mathbf{v})$,   **Preservation of addition**
2. $T(r\mathbf{u}) = rT(\mathbf{u})$,   **Preservation of scalar multiplication**

for all vectors $\mathbf{u}$ and $\mathbf{v}$ in V and for all scalars r in $\mathbb{R}$.

Exercise 35 shows that the two conditions of Definition 3.9 may be combined into the single condition

$$T(r\mathbf{u} + s\mathbf{v}) = rT(\mathbf{u}) + sT(\mathbf{v}) \tag{3}$$

for all vectors $\mathbf{u}$ and $\mathbf{v}$ in V and for all scalars r and s in $\mathbb{R}$. Mathematical induction can be used to verify the analogous relation for n summands:

$$T(r_1\mathbf{v}_1 + r_2\mathbf{v}_2 + \cdots + r_n\mathbf{v}_n) = r_1T(\mathbf{v}_1) + r_2T(\mathbf{v}_2) + \cdots + r_nT(\mathbf{v}_n). \tag{4}$$

We remind you of some terminology and notation defined in Section 2.3 for functions in general and linear transformations in particular. We state things here in the language of linear transformations. For a linear transformation $T: V \to V'$, the set V is the **domain** of T and the set V' is the **codomain** of T. If W is a subset of V, then $T[W] = \{T(\mathbf{w}) \mid \mathbf{w} \in W\}$ is the **image** of W under T. In particular, $T[V]$ is the **range** of T. Similarly, if W' is a subset of V', then

$T^{-1}[W'] = \{\mathbf{v} \in V \mid T(\mathbf{v}) \in W'\}$ is the **inverse image** of W' under T. In particular, $T^{-1}[\{\mathbf{0}'\}]$ is the **kernel** of T, denoted by $\ker(T)$. It consists of all of the vectors in V that T maps into $\mathbf{0}'$.

Let V, V', and V'' be vector spaces, and let $T: V \to V'$ and $T': V' \to V''$ be linear transformations. The **composite transformation** $T' \circ T: V \to V''$ is defined by $(T' \circ T)(\mathbf{v}) = T'(T(\mathbf{v}))$ for $\mathbf{v}$ in V. Exercise 36 shows that $T' \circ T$ is again a linear transformation.

EXAMPLE 1 Let F be the vector space of all functions $f: \mathbb{R} \to \mathbb{R}$, and let D be its subspace of all differentiable functions. Show that differentiation is a linear transformation of D into F.

SOLUTION Let $T: D \to F$ be defined by $T(f) = f'$, the derivative of f. Using the familiar rules

$$(f + g)' = f' + g' \quad \text{and} \quad (rf)' = r(f')$$

for differentiation from calculus, we see that

$$T(f + g) = (f + g)' = f' + g' = T(f) + T(g)$$

and

$$T(rf) = (rf)' = r(f') = rT(f)$$

for all functions f and g in D and scarlars r. In other words, these two rules for differentiating a sum of functions and for differentiating a scalar times a function constitute precisely the assertion that differentiation is a linear transformation. ∎

EXAMPLE 2 Let F be the vector space of all functions $f: \mathbb{R} \to \mathbb{R}$, and let c be in $\mathbb{R}$. Show that the evaluation function $T: F \to \mathbb{R}$ defined by $T(f) = f(c)$, which maps each function f in F into its value at c, is a linear transformation.

HISTORICAL NOTE The concept of a linear substitution dates back to the eighteenth century. But it was only after physicists became used to dealing with vectors that the idea of a function of vectors became explicit. One of the founders of vector analysis, Oliver Heaviside (1850–1925), introduced the idea of a linear vector operator in one of his works on electromagnetism in 1885. He defined it using coordinates: $\vec{B}$ comes from $\vec{H}$ by a linear vector operator if, when $\vec{B}$ has components B_1, B_2, B_3 and $\vec{H}$ has components H_1, H_2, H_3, there are numbers μ_{ij} for $i, j = 1, 2, 3$, where

$$B_1 = \mu_{11}H_1 + \mu_{12}H_2 + \mu_{13}H_3$$
$$B_2 = \mu_{21}H_1 + \mu_{22}H_2 + \mu_{23}H_3$$
$$B_3 = \mu_{31}H_1 + \mu_{32}H_2 + \mu_{33}H_3.$$

In his lectures at Yale, which were published in 1901, J. Willard Gibbs called this same transformation a *linear vector function*. But he also defined this more abstractly as a continuous function f such that $f(\vec{v} + \vec{w}) = f(\vec{v}) + f(\vec{w})$. A fully abstract definition, exactly like Definition 3.9, was given by Hermann Weyl in *Space–Time–Matter* (1918).

Oliver Heaviside was a self-taught expert on mathematical physics who played an important role in the development of electromagnetic theory and especially its practical applications. In 1901 he predicted the existence of a reflecting ionized region surrounding the earth; the existence of this layer, now called the *ionosphere*, was soon confirmed.

SOLUTION   We show that T preserves addition and scalar multiplication. If f and g are functions in the vector space F, then, evaluating $f + g$ at c, we obtain

$$(f + g)(c) = f(c) + g(c).$$

Therefore,

$$
\begin{aligned}
T(f + g) &= (f + g)(c) &&\text{Definition of } T \\
&= f(c) + g(c) &&\text{Definition of } f + g \text{ in } F \\
&= T(f) + T(g). &&\text{Definition of } T
\end{aligned}
$$

This shows that T preserves addition. In a similar manner, the computation

$$
\begin{aligned}
T(rf) &= (rf)(c) &&\text{Definition of } T \\
&= r(f(c)) &&\text{Definition of } rf \text{ in } F \\
&= r(T(f)) &&\text{Definition of } T
\end{aligned}
$$

shows that T preserves scalar multiplication. ∎

EXAMPLE 3   Let $C_{a,b}$ be the vector space of all continuous functions mapping the closed interval $a \leq x \leq b$ of $\mathbb{R}$ into $\mathbb{R}$. Show that $T: C_{a,b} \to \mathbb{R}$ defined by $T(f) = \int_a^b f(x)\, dx$ is a linear transformation.

SOLUTION   From properties of the definite integral, we know that for $f, g \in C_{a,b}$ and for any scalar r, we have

$$T(f + g) = \int_a^b [f(x) + g(x)]\, dx = \int_a^b f(x)\, dx + \int_a^b g(x)\, dx = T(f) + T(g)$$

and

$$T(rf) = \int_a^b rf(x)\, dx = r \int_a^b f(x)\, dx = rT(f).$$

This shows that T is indeed a linear transformation. ∎

EXAMPLE 4   Let C be the vector space of all continuous functions mapping $\mathbb{R}$ into $\mathbb{R}$. Let $a \in \mathbb{R}$ and let $T_a: C \to C$ be defined by $T_a(f) = \int_a^x f(t)\, dt$. Show that T_a is a linear transformation.

SOLUTION   This follows from the same properties of the integral that we used in the solution of the preceding exercise. From calculus, we know that the range of T_a is actually a subset of the vector space of differentiable functions mapping $\mathbb{R}$ into $\mathbb{R}$. (Theorem 3.7, which follows shortly, will show that the range is actually a subspace.) ∎

EXAMPLE 5   Let D_∞ be the space of functions mapping $\mathbb{R}$ into $\mathbb{R}$ that have derivatives of all orders, and let $a_0, a_1, a_2, \ldots, a_n \in \mathbb{R}$. Show that $T: D_\infty \to D_\infty$ defined by $T(f) = a_n f^{(n)}(x) + \cdots + a_2 f''(x) + a_1 f'(x) + a_0 f(x)$ is a linear transformation.

SOLUTION   This follows from the fact that the ith derivative of a sum of functions is the sum of their ith derivatives—that is, $(f + g)^{(i)}(x) = f^{(i)}(x) + g^{(i)}(x)$—and that the ith derivative of $rf(x)$ is $rf^{(i)}(x)$, together with the fact that $T(f)$ is defined to be a *linear combination* of these derivatives. (We consider $f(x)$ to be the 0th derivative.) ∎

Note that the computation of $T(f)$ in Example 5 amounts to the computation of the left-hand side of the general linear differential equation with constant coefficients

$$a_n y^{(n)} + \cdots + a_2 y'' + a_1 y' + a_0 y = g(x) \tag{5}$$

for $y = f(x)$. Thus, solving this differential equation amounts to finding all $f \in D_\infty$ such that $T(f) = g(x)$.

Properties of Linear Transformations

The two properties of linear transformations in our next theorem are useful and easy to prove.

THEOREM 3.5   Preservation of Zero and Subtraction

Let V and V' be vector spaces, and let $T: V \to V'$ be a linear transformation. Then

1. $T(\mathbf{0}) = \mathbf{0}'$, and   **Preservation of zero**
2. $T(\mathbf{v}_1 - \mathbf{v}_2) = T(\mathbf{v}_1) - T(\mathbf{v}_2)$ **Preservation of subtraction**

for any vectors $\mathbf{v}_1$ and $\mathbf{v}_2$ in V.

PROOF   We establish preservation of zero by taking $r = 0$ and $\mathbf{v} = \mathbf{0}$ in condition 2 of Definition 3.9 for a linear transformation. Condition 2 and the property $0\mathbf{v} = \mathbf{0}$ (see Theorem 3.1) yield

$$T(\mathbf{0}) = T(0\mathbf{0}) = 0T(\mathbf{0}) = \mathbf{0}'.$$

Preservation of subtraction follows from Eq. (3), as follows:

$$\begin{aligned}
T(\mathbf{v}_1 - \mathbf{v}_2) &= T(\mathbf{v}_1 + (-1)\mathbf{v}_2) \\
&= T(\mathbf{v}_1) + (-1)T(\mathbf{v}_2) \\
&= T(\mathbf{v}_1) - T(\mathbf{v}_2).
\end{aligned}$$
▲

EXAMPLE 6   Let D be the vector space of all differentiable functions and F the vector space of all functions mapping $\mathbb{R}$ into $\mathbb{R}$. Determine whether $T: D \to F$ defined by $T(f) = 2f''(x) + 3f'(x) + x^2$ is a linear transformation.

SOLUTION   Because for the zero constant function we have $T(0) = 2(0'') + 3(0') + x^2 = 2(0) + 3(0) + x^2 = x^2$, and x^2 is not the zero constant function, we see that T does not preserve zero, and so it is not a linear transformation.   ■

Just as for a linear transformation $T: \mathbb{R}^n \to \mathbb{R}^m$ discussed in Section 2.3, a linear transformation of a vector space is determined by its action on any basis for the domain. (See Exercises 40 and 41 as well.) Because our bases here may be infinite, we restate the result and give the short proof.

THEOREM 3.6   Bases and Linear Transformations

Let $T: V \to V'$ be a linear transformation, and let B be a basis for V. For any vector $\mathbf{v}$ in V, the vector $T(\mathbf{v})$ is uniquely determined by the vectors $T(\mathbf{b})$ for all $\mathbf{b} \in B$. In other words, if two linear transformations have the same value at each basis vector $\mathbf{b} \in B$, the two transformations have the same value at each vector in V; that is, they are the same transformation.

PROOF   Let T and $\overline{T}$ be two linear transformations such that $T(\mathbf{b}_i) = \overline{T}(\mathbf{b}_i)$ for each vector $\mathbf{b}_i$ in B. Let $\mathbf{v} \in V$. Then there exist vectors $\mathbf{b}_1, \mathbf{b}_2, \ldots, \mathbf{b}_k$ in B and scalars $r_1, r_2, \ldots, r_k$ such that

$$\mathbf{v} = r_1\mathbf{b}_1 + r_2\mathbf{b}_2 + \cdots + r_k\mathbf{b}_k.$$

We then have

$$
\begin{aligned}
T(\mathbf{v}) &= T(r_1\mathbf{b}_1 + r_2\mathbf{b}_2 + \cdots + r_k\mathbf{b}_k) \\
&= r_1T(\mathbf{b}_1) + r_2T(\mathbf{b}_2) + \cdots + r_kT(\mathbf{b}_k). \\
&= r_1\overline{T}(\mathbf{b}_1) + r_2\overline{T}(\mathbf{b}_2) + \cdots + r_k\overline{T}(\mathbf{b}_k). \\
&= \overline{T}(r_1\mathbf{b}_1 + r_2\mathbf{b}_2 + \cdots + r_k\mathbf{b}_k) \\
&= \overline{T}(\mathbf{v}).
\end{aligned}
$$

Thus T and $\overline{T}$ are the same transformation.   ▲

The next theorem also generalizes results that appear in the text and exercises of Section 2.3 for linear transformations mapping $\mathbb{R}^n$ into $\mathbb{R}^m$.

THEOREM 3.7   Preservation of Subspaces

Let V and V' be vector spaces, and let $T: V \to V'$ be a linear transformation.

1. If W is a subspace of V, then $T[W]$ is a subspace of V'.
2. If W' is a subspace of V', then $T^{-1}[W']$ is a subspace of V.

PROOF

1. Because $T(0) = 0'$, we need only show that $T[W]$ is closed under vector addition and under scalar multiplication. Let $T(\mathbf{w}_1)$ and $T(\mathbf{w}_2)$ be any vectors in $T(W)$, where $\mathbf{w}_1$ and $\mathbf{w}_2$ are vectors in W. Then

$$T(\mathbf{w}_1) + T(\mathbf{w}_2) = T(\mathbf{w}_1 + \mathbf{w}_2),$$

by preservation of addition. Now $\mathbf{w}_1 + \mathbf{w}_2$ is in W because W is itself closed under addition, and so $T(\mathbf{w}_1 + \mathbf{w}_2)$ is in $T[W]$. This shows that $T[W]$ is closed under vector addition. If r is any scalar, then $r\mathbf{w}_1$ is in W, and

$$rT(\mathbf{w}_1) = T(r\mathbf{w}_1).$$

This shows that $rT(\mathbf{w}_1)$ is in $T[W]$, and so $T[W]$ is closed under scalar multiplication. Thus, $T[W]$ is a subspace of V'.

2. Notice that $0 \in T^{-1}[W']$. Let $\mathbf{v}_1$ and $\mathbf{v}_2$ be any vectors in $T^{-1}[W']$, so that $T(\mathbf{v}_1)$ and $T(\mathbf{v}_2)$ are in W'. Then

$$T(\mathbf{v}_1 + \mathbf{v}_2) = T(\mathbf{v}_1) + T(\mathbf{v}_2)$$

is also in the subspace W', and so $\mathbf{v}_1 + \mathbf{v}_2$ is in $T^{-1}[W']$. For any scalar r, we know that

$$rT(\mathbf{v}_1) = T(r\mathbf{v}_1),$$

and $rT(\mathbf{v}_1)$ is in W'. Thus, $r\mathbf{v}_1$ is also in $T^{-1}[W']$. This shows that $T^{-1}[W']$ is closed under addition and under scalar multiplication, and so $T^{-1}[W']$ is a subspace of V.   ▲

The Equation $T(\mathbf{x}) = \mathbf{b}$

Let $T: V \to V'$ be a linear transformation. From Theorem 3.7, we know that $\ker(T) = T^{-1}[\{0'\}]$ is a subspace of V. This subspace is the solution set of the *homogeneous transformation equation* $T(\mathbf{x}) = 0'$. The structure of the solution set of $T(\mathbf{x}) = \mathbf{b}$ exactly parallels the structure of the solution set of $A\mathbf{x} = \mathbf{b}$ described on p. 97. Namely,

---

Solution Set of $T(\mathbf{x}) = \mathbf{b}$

Let $T: V \to V'$ be a linear transformation and let $T(\mathbf{p}) = \mathbf{b}$ for a particular vector $\mathbf{p}$ in V. The solution set of $T(\mathbf{x}) = \mathbf{b}$ is the set $\{\mathbf{p} + \mathbf{h} \mid \mathbf{h} \in \ker(T)\}$.

The proof is essentially the same as that of Theorem 1.18; we ask you to write it out for this case in Exercise 46. This boxed result shows that if $\ker(T) = \{0\}$, then $T(\mathbf{x}) = \mathbf{b}$ has at most one solution, and so T is **one-to-one**, meaning that $T(\mathbf{v}_1) = T(\mathbf{v}_2)$ implies that $\mathbf{v}_1 = \mathbf{v}_2$. Conversely, if T is one-to-one, then $T(\mathbf{x}) = 0'$ has only one solution—namely, $\mathbf{0}$—so $\ker(T) = \{0\}$. We box this fact.

> **Condition for T to Be One-to-One**
>
> A linear transformation T is one-to-one if and only if $\ker(T) = \{\mathbf{0}\}$.

ILLUSTRATION 1   The differential equation $y'' - 4y = x^2$ is linear with constant coefficients and is of the form of Eq. (5). In Example 5, we showed that the left-hand side of such an equation defines a linear transformation of D_∞ into itself. In differential equations, it is shown that the kernel of the transformation $T(f) = f'' - 4f$ is two-dimensional. We can check that the independent functions $h_1(x) = e^{2x}$ and $h_2(x) = e^{-2x}$ both satisfy the differential equation $y'' - 4y = 0$. Thus $\{e^{2x}, e^{-2x}\}$ is a basis for the kernel of T. All solutions of the homogeneous system are of the form $c_1 e^{2x} + c_2 e^{-2x}$. We find a particular solution $p(x)$ of $y'' - 4y = x^2$ by inspection: if $y = -x^2/4$, then the term $-4y$ yields x^2 but the second derivative of $-x^2/4$ is $-\frac{1}{2}$; we can kill off this unwanted $-\frac{1}{2}$ by taking

$$p(x) = -\frac{1}{4}x^2 - \frac{1}{8}$$

(note that the second derivative of $-\frac{1}{8}$ is 0). Thus the general solution of this differential equation is

$$y = c_1 e^{2x} + c_2 e^{-2x} - \frac{1}{4}x^2 - \frac{1}{8}. \qquad \blacksquare$$

Invertible Transformations

In Section 2.3, we saw that, if A is an invertible $n \times n$ matrix, the linear transformation $T: \mathbb{R}^n \to \mathbb{R}^n$ defined by $T(\mathbf{x}) = A\mathbf{x}$ has an inverse $T^{-1}: \mathbb{R}^n \to \mathbb{R}^n$, where $T^{-1}(\mathbf{y}) = A^{-1}\mathbf{y}$, and that both composite transformations $T^{-1} \circ T$ and $T \circ T^{-1}$ are the identity transformation of $\mathbb{R}^n$. We now extend this idea to linear transformations of vector spaces in general.

DEFINITION 3.10   Invertible Transformation

> Let V and V' be vector spaces. A linear transformation $T: V \to V'$ is **invertible** if there exists a linear transformation $T^{-1}: V' \to V$ such that $T^{-1} \circ T$ is the identity transformation on V and $T \circ T^{-1}$ is the identity transformation on V'.

EXAMPLE 7   Determine whether the evaluation transformation $T: F \to \mathbb{R}$ of Example 2, where $T(f) = f(c)$ for some fixed c in $\mathbb{R}$, is invertible.

SOLUTION   Consider the polynomial functions f and g, where $f(x) = 2x + c$ and $g(x) = 4x - c$. Then $T(f) = T(g) = 3c$. If T were invertible, there would have to be a linear transformation $T^{-1}: \mathbb{R} \to F$ such that $T^{-1}(3c)$ is both f and g. But this is impossible. $\blacksquare$

Example 7 illustrates that, if $T: V \to V'$ is an *invertible* linear transformation, T must satisfy the following property:

$$\text{if } \mathbf{v}_1 \neq \mathbf{v}_2, \text{ then } T(\mathbf{v}_1) \neq T(\mathbf{v}_2). \quad \textbf{One-to-one} \tag{6}$$

As in the argument of Example 7, if $T(\mathbf{v}_1) = T(\mathbf{v}_2) = \mathbf{v}'$ and T is invertible, there would have to be a linear transformation $T^{-1}: V' \to V$ such that $T^{-1}(\mathbf{v}')$ is simultaneously $\mathbf{v}_1$ and $\mathbf{v}_2$, which is impossible when $\mathbf{v}_1 \neq \mathbf{v}_2$.

An invertible linear transformation $T: V \to V'$ must also satisfy another property:

$$\text{if } \mathbf{v}' \text{ is in } V', \text{ then } T(\mathbf{v}) = \mathbf{v}' \text{ for some } \mathbf{v} \text{ in } V. \quad \textbf{Onto} \tag{7}$$

This follows at once from the fact that, for T^{-1} with the properties in Definition 3.10 and for $\mathbf{v}'$ in V', we have $T^{-1}(\mathbf{v}') = \mathbf{v}$ for some $\mathbf{v}$ in V. But then $\mathbf{v}' = (T \circ T^{-1})(\mathbf{v}') = T(T^{-1}(\mathbf{v}')) = T(\mathbf{v})$. A transformation $T: V \to V'$ satisfying property (7) is **onto** V'; in this case the range of T is all of V'. We have thus proved half of the following theorem.*

THEOREM 3.8 Invertibility of T

> A linear transformation $T: V \to V'$ is invertible if and only if it is one-to-one and onto V'.

PROOF We have just shown that, if T is invertible, it must be one-to-one and onto V'.

Suppose now that T is one-to-one and onto V'. Because T is onto V', for each $\mathbf{v}'$ in V' we can find $\mathbf{v}$ in V such that $T(\mathbf{v}) = \mathbf{v}'$. Because T is one-to-one, this vector $\mathbf{v}$ in V is *unique*. Let $T^{-1}: V' \to V$ be defined by $T^{-1}(\mathbf{v}') = \mathbf{v}$, where $\mathbf{v}$ is the unique vector in V such that $T(\mathbf{v}) = \mathbf{v}'$. Then

$$(T \circ T^{-1})(\mathbf{v}') = T(T^{-1}(\mathbf{v}')) = T(\mathbf{v}) = \mathbf{v}'$$

and

$$(T^{-1} \circ T)(\mathbf{v}) = T^{-1}(T(\mathbf{v})) = T^{-1}(\mathbf{v}') = \mathbf{v},$$

which shows that $T \circ T^{-1}$ is the identity map of V' and that $T^{-1} \circ T$ is the identity map of V. It only remains to show that T^{-1} is indeed a linear transformation—that is, that

$$T^{-1}(\mathbf{v}_1' + \mathbf{v}_2') = T^{-1}(\mathbf{v}_1') + T^{-1}(\mathbf{v}_2') \quad \text{and} \quad T^{-1}(r\mathbf{v}_1') = rT^{-1}(\mathbf{v}_1')$$

---

*This theorem is valid for functions in general.

for all $\mathbf{v}'_1$ and $\mathbf{v}'_2$ in V' and for all scalars r. Let $\mathbf{v}_1$ and $\mathbf{v}_2$ be the unique vectors in V such that $T(\mathbf{v}_1) = \mathbf{v}'_1$ and $T(\mathbf{v}_2) = \mathbf{v}'_2$. Remembering that $T^{-1} \circ T$ is the identity map, we have

$$T^{-1}(\mathbf{v}'_1 + \mathbf{v}'_2) = T^{-1}(T(\mathbf{v}_1) + T(\mathbf{v}_2)) = T^{-1}(T(\mathbf{v}_1 + \mathbf{v}_2))$$
$$= (T^{-1} \circ T)(\mathbf{v}_1 + \mathbf{v}_2) = \mathbf{v}_1 + \mathbf{v}_2 = T^{-1}(\mathbf{v}'_1) + T^{-1}(\mathbf{v}'_2).$$

Similarly,

$$T^{-1}(r\mathbf{v}'_1) = T^{-1}(rT(\mathbf{v}_1)) = T^{-1}(T(r\mathbf{v}_1)) = r\mathbf{v}_1 = rT^{-1}(\mathbf{v}'_1).\qquad\blacktriangle$$

The proof of Theorem 3.8 shows that, if $T: V \to V'$ is invertible, the linear transformation $T^{-1}: V' \to V$ described in Definition 3.10 is unique. This transformation T^{-1} is the **inverse transformation** of T.

ILLUSTRATION 2   Let D be the vector space of differentiable functions and F the vector space of all functions mapping $\mathbb{R}$ into $\mathbb{R}$. Then $T: D \to F$, where $T(f) = f'$, the derivative of f, is not an invertible transformation. Specifically, we have $T(x) = T(x + 17) = 1$, showing that T is not one-to-one. Notice also that the kernel of T is not $\{\mathbf{0}\}$, but consists of all constant functions.   ∎

ILLUSTRATION 3   Let P be the vector space of all polynomials in x and let W be the subspace of all polynomials in x with constant term 0, so that $q(0) = 0$ for $q(x) \in W$. Let $T: P \to W$ be defined by $T(p(x)) = xp(x)$. Then T is a linear transformation. Because $xp_1(x) = xp_2(x)$ if and only if $p_1(x) = p_2(x)$, we see that T is one-to-one. Every polynomial $q(x)$ in W contains no constant term, and so it can be factored as $q(x) = xp(x)$; because $T(p(x)) = q(x)$, we see that T maps P onto W. Thus T is an invertible linear transformation.   ∎

Isomorphism

An **isomorphism** is a linear transformation $T: V \to V'$ that is one-to-one and onto V'. Theorem 3.8 shows that isomorphisms are precisely the invertible linear transformations $T: V \to V'$. If an isomorphism T exists, then it is invertible and its inverse is also an isomorphism. The vector spaces V and V' are said to be **isomorphic** in this case. We view isomorphic vector spaces V and V' as being structurally identical in the following sense. Let $T: V \to V'$ be an isomorphism. Rename each $\mathbf{v}$ in V by the $\mathbf{v}' = T(\mathbf{v})$ in V'. Because T is one-to-one, no two different elements of V get the same name from V', and because T is onto V', all names in V' are used. The renamed V and V' then appear identical as sets. But they also have the same algebraic structure as vector spaces, as Figure 3.2 illustrates. We discussed a special case of this concept before Example 3 in Section 3.3, indicating informally that every finite-dimensional vector space is structurally the same as $\mathbb{R}^n$ for some n. We are now in a position to state this formally.

THEOREM 3.9 Coordinatization of Finite-Dimensional Spaces

Let V be a finite-dimensional vector space with ordered basis $B = (\mathbf{b}_1, \mathbf{b}_2, \ldots, \mathbf{b}_n)$. The map $T: V \to \mathbb{R}^n$ defined by $T(\mathbf{v}) = \mathbf{v}_B$, the coordinate vector of $\mathbf{v}$ relative to B, is an isomorphism.

PROOF   Equation 1 in Section 3.3 shows that T preserves addition and scalar multiplication. Moreover, T is one-to-one, because the coordinate vector $\mathbf{v}_B$ of $\mathbf{v}$ uniquely determines $\mathbf{v}$, and the range of T is all of $\mathbb{R}^n$. Therefore, T is an isomorphism.   ▲

The isomorphism of V with $\mathbb{R}^n$, described in Theorem 3.9, is by no means unique. There is one such isomorphism for each choice of an ordered basis B of V.

Let V and V' be vector spaces of dimensions n and m, respectively. By Theorem 3.9, we can choose ordered bases B for V and B' for V', and essentially convert V into $\mathbb{R}^n$ and V' into $\mathbb{R}^m$ by renaming vectors by their coordinate vectors relative to these bases. Then each linear transformation $T: V \to V'$ corresponds to a linear transformation $\overline{T}: \mathbb{R}^n \to \mathbb{R}^m$ in a natural way, and we can answer questions about T by studying $\overline{T}$ instead. But we can study $\overline{T}$ in turn by studying its standard matrix representation A, as we will

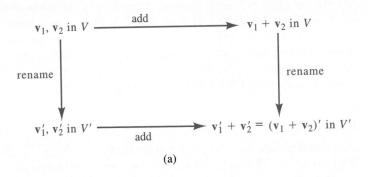

(a)

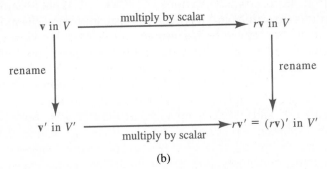

(b)

FIGURE 3.2
(a) The equal sign shows that the renaming preserves vector addition.
(b) The equal sign shows that the renaming preserves scalar multiplication.

illustrate shortly. This matrix changes if we change the ordered bases B or B'. A good deal of the remainder of our text is devoted to studying how to choose B and B' when $m = n$ so that the square matrix A has a simple form that illuminates the structure of the transformation T. This is the thrust of Chapters 5 and 7.

Matrix Representations of Transformations

In Section 2.3, we saw that every linear transformation $T: \mathbb{R}^n \to \mathbb{R}^m$ can be computed using its standard matrix representation A, where A is the $m \times n$ matrix having as jth column vector $T(\mathbf{e}_j)$. Namely, for this matrix A and any column vector $\mathbf{x} \in \mathbb{R}^n$, we have $T(\mathbf{x}) = A\mathbf{x}$.

We just showed that finite-dimensional vector spaces V and V', where $\dim(V) = n$ and $\dim(V') = m$, are isomorphic to $\mathbb{R}^n$ and $\mathbb{R}^m$, respectively. We can rename vectors in V and V' by their coordinate vectors relative to ordered bases $B = (\mathbf{b}_1, \mathbf{b}_2, \ldots, \mathbf{b}_n)$ of V and $B' = (\mathbf{b}'_1, \mathbf{b}'_2, \ldots, \mathbf{b}'_m)$ of V', and work with the coordinates, doing computations in $\mathbb{R}^n$ and $\mathbb{R}^m$. In particular, a linear transformation $T: V \to V'$ and a choice of ordered bases B and B' then gives a linear transformation $\overline{T}: \mathbb{R}^n \to \mathbb{R}^m$ such that for $\mathbf{v} \in V$ we have $\overline{T}(\mathbf{v}_B) = T(\mathbf{v})_{B'}$. This transformation $\overline{T}$ has an $m \times n$ standard matrix representation whose jth column vector is $\overline{T}(\mathbf{e}_j)$. Now $\mathbf{e}_j$ is the coordinate vector, relative to B, of the jth vector $\mathbf{b}_j$ in the ordered basis B, so $\overline{T}(\mathbf{e}_j) = \overline{T}((\mathbf{b}_j)_B) = T(\mathbf{b}_j)_{B'}$. Thus we see that the standard matrix representation of $\overline{T}$ is

$$A = \begin{bmatrix} | & | & & | \\ T(\mathbf{b}_1)_{B'} & T(\mathbf{b}_2)_{B'} & \cdots & T(\mathbf{b}_n)_{B'} \\ | & | & & | \end{bmatrix}. \tag{8}$$

We summarize in a theorem.

THEOREM 3.10   Matrix Representations of Linear Transformations

Let V and V' be finite-dimensional vector spaces and let $B = (\mathbf{b}_1, \mathbf{b}_2, \ldots, \mathbf{b}_n)$ and $B' = (\mathbf{b}'_1, \mathbf{b}'_2, \ldots, \mathbf{b}'_m)$ be ordered bases for V and V', respectively. Let $T: V \to V'$ be a linear transformation, and let $\overline{T}: \mathbb{R}^n \to \mathbb{R}^m$ be the linear transformation such that for each $\mathbf{v} \in V$, we have $\overline{T}(\mathbf{v}_B) = T(\mathbf{v})_{B'}$. Then the standard matrix representation of $\overline{T}$ is the matrix A whose jth column vector is $T(\mathbf{b}_j)_{B'}$, and $T(\mathbf{v})_{B'} = A\mathbf{v}_B$ for all vectors $\mathbf{v} \in V$.

DEFINITION 3.11   Matrix Representation of T Relative to B,B'

The matrix A in Eq. (8) and described in Theorem 3.10 is the **matrix representation of T relative to B,B'**.

EXAMPLE 8   Let V be the subspace sp(sin x cos x, sin^2x, cos^2x) of the vector space D of all differentiable functions mapping $\mathbb{R}$ into $\mathbb{R}$. Differentiation gives a linear transformation T of V into itself. Find the matrix representation A of T relative to B,B' where $B = B' = $ (sin x cos x, sin^2x, cos^2x). Use A to compute the derivative of

$$f(x) = 3 \sin x \cos x - 5 \sin^2x + 7 \cos^2x.$$

SOLUTION   We find that

$$T(\sin x \cos x) = -\sin^2x + \cos^2x \quad \text{so} \quad T(\sin x \cos x)_{B'} = [0, -1, 1],$$

$$T(\sin^2x) = 2 \sin x \cos x \quad \text{so} \quad T(\sin^2x)_{B'} = [2, 0, 0], \quad \text{and}$$

$$T(\cos^2x) = -2 \sin x \cos x \quad \text{so} \quad T(\cos^2x)_{B'} = [-2, 0, 0].$$

Using Eq. (8), we have

$$A = \begin{bmatrix} 0 & 2 & -2 \\ -1 & 0 & 0 \\ 1 & 0 & 0 \end{bmatrix} \quad \text{so} \quad A(f(x)_B) = \begin{bmatrix} 0 & 2 & -2 \\ -1 & 0 & 0 \\ 1 & 0 & 0 \end{bmatrix} \begin{bmatrix} 3 \\ -5 \\ 7 \end{bmatrix} = \begin{bmatrix} -24 \\ -3 \\ 3 \end{bmatrix}.$$

Thus $f'(x) = -24 \sin x \cos x - 3 \sin^2x + 3 \cos^2x$.   ∎

EXAMPLE 9   Letting P_n be the vector space of polynomials of degree at most n, we note that $T: P_2 \rightarrow P_3$ defined by $T(p(x)) = (x + 1)p(x - 2)$ is a linear transformation. Find the matrix representation A of T relative to the ordered bases $B = (x^2, x, 1)$ and $B' = (x^3, x^2, x, 1)$ for P_2 and P_3, respectively. Use A to compute $T(p(x))$ for $p(x) = 5x^2 - 7x + 18$.

SOLUTION   We compute

$$T(x^2) = (x + 1)(x - 2)^2 = (x + 1)(x^2 - 4x + 4) = x^3 - 3x^2 + 4,$$

$$T(x) = (x + 1)(x - 2) = x^2 - x - 2, \quad \text{and} \quad T(1) = (x + 1)1 = x + 1.$$

Thus $T(x^2)_{B'} = [1, -3, 0, 4]$, $T(x)_{B'} = [0, 1, -1, -2]$, and $T(1)_{B'} = [0, 0, 1, 1]$. Consequently,

$$A = \begin{bmatrix} 1 & 0 & 0 \\ -3 & 1 & 0 \\ 0 & -1 & 1 \\ 4 & -2 & 1 \end{bmatrix} \quad \text{so} \quad A(p(x)_B) = \begin{bmatrix} 1 & 0 & 0 \\ -3 & 1 & 0 \\ 0 & -1 & 1 \\ 4 & -2 & 1 \end{bmatrix} \begin{bmatrix} 5 \\ -7 \\ 18 \end{bmatrix} = \begin{bmatrix} 5 \\ -22 \\ 25 \\ 52 \end{bmatrix}.$$

Thus $T(p(x)) = 5x^3 - 22x^2 + 25x + 52$.   ∎

Let V and V' be n-dimensional vector spaces with ordered bases B and B', respectively. Theorem 3.8 tells us that a linear transformation $T: V \rightarrow V'$ is invertible if and only if it is one-to-one and onto V'. Under these circumstances, we might expect the following to be true:

Matrix Representation of T^{-1}

The matrix representation of T^{-1} relative to B',B is the inverse of the matrix representation of T relative to B,B'.

This is indeed the case in view of Theorem 3.10 and the fact that linear transformations of $\mathbb{R}^n$ have this property. (See the box on page 151.) Exercise 20 gives an illustration of this.

It is important to note that the matrix representation of the linear transformation $T: V \to V'$ depends on the particular bases B for V and B' for V'. We really should use some notation such as $A_{B,B'}$ or $R_{B,B'}$ to denote this dependency. Such notations appear cumbersome and complicated, so we avoid them for now. We will use such a notation in Chapter 7, where we will consider this dependency in more detail.

SUMMARY

Let V and V' be vector spaces, and let T be a function mapping V into V'.

1. The function T is a linear transformation if it preserves addition and scalar multiplication—that is, if

$$T(\mathbf{v}_1 + \mathbf{v}_2) = T(\mathbf{v}_1) + T(\mathbf{v}_2)$$

and

$$T(r\mathbf{v}_1) = rT(\mathbf{v}_1)$$

for all vectors $\mathbf{v}_1$ and $\mathbf{v}_2$ in V and for scalars r in $\mathbb{R}$.

2. If T is a linear transformation, then $T(\mathbf{0}) = \mathbf{0}'$ and also $T(\mathbf{v}_1 - \mathbf{v}_2) = T(\mathbf{v}_1) - T(\mathbf{v}_2)$.

3. A function $T: \mathbb{R}^n \to \mathbb{R}^m$ is a linear transformation if and only if T has the form $T(\mathbf{x}) = A\mathbf{x}$ for some $m \times n$ matrix A.

4. The matrix A in summary item 3 is the standard matrix representation of the transformation T.

5. A linear transformation $T: V \to V'$ is invertible if and only if it is one-to-one and onto V'. Such transformations are isomorphisms.

6. Every nonzero finite-dimensional real vector space V is isomorphic to $\mathbb{R}^n$, where $n = \dim(V)$.

Now let $T: V \to V'$ be a linear transformation.

7. If W is a subspace of V, then $T[W]$ is a subspace of V'. If W' is a subspace of V', then $T^{-1}[W']$ is a subspace of V.

8. The range of T is the subspace $\{T(\mathbf{v}) \mid \mathbf{v} \text{ in } V\}$ of V', and the kernel $\ker(T)$ is the subspace $\{\mathbf{v} \in V \mid T(\mathbf{v}) = \mathbf{0}'\}$ of V.

9. T is one-to-one if and only if $\ker(T) = \{\mathbf{0}\}$. If T is one-to-one and has as its range all of V', then $T^{-1}: V' \rightarrow V$ is well-defined and is a linear transformation. In this case, both T and T^{-1} are isomorphisms.

10. If $T: V \rightarrow V'$ is a linear transformation and $T(\mathbf{p}) = \mathbf{b}$, then the solution set of $T(\mathbf{x}) = \mathbf{b}$ is $\{\mathbf{p} + \mathbf{h} \mid \mathbf{h} \in \ker(T)\}$.

11. Let $B = (\mathbf{b}_1, \mathbf{b}_2, \ldots, \mathbf{b}_n)$ and $B' = (\mathbf{b}'_1, \mathbf{b}'_2, \ldots, \mathbf{b}'_m)$ be ordered bases for V and V', respectively. The *matrix representation of T relative to B,B'* is the $m \times n$ matrix A having $T(\mathbf{b}_j)_{B'}$ as its jth column vector. We have $T(\mathbf{v})_{B'} = A(\mathbf{v}_B)$ for all $\mathbf{v} \in V$.

EXERCISES

In Exercises 1–5, let F be the vector space of all functions mapping $\mathbb{R}$ into $\mathbb{R}$. Determine whether the given function T is a linear transformation. If it is a linear transformation, describe the kernel of T and determine whether the transformation is invertible.

1. $T: F \rightarrow \mathbb{R}$ defined by $T(f) = f(-4)$

2. $T: F \rightarrow \mathbb{R}$ defined by $T(f) = f(5)^2$

3. $T: F \rightarrow F$ defined by $T(f) = f + f$

4. $T: F \rightarrow F$ defined by $T(f) = f + 3$, where 3 is the constant function with value 3 for all $x \in \mathbb{R}$.

5. $T: F \rightarrow F$ defined by $T(f) = -f$

6. Let $C_{0,2}$ be the space of continuous functions mapping the interval $0 \le x \le 2$ into $\mathbb{R}$. Let $T: C_{0,2} \rightarrow \mathbb{R}$ be defined by $T(f) = \int_0^2 f(x)\, dx$. See Example 3. If possible, give three different functions in $\ker(T)$.

7. Let C be the space of all continuous functions mapping $\mathbb{R}$ into $\mathbb{R}$, and let $T: C \rightarrow C$ be defined by $T(f) = \int_1^x f(t)\, dt$. See Example 4. If possible, give three different functions in $\ker(T)$.

8. Let F be the vector space of all functions mapping $\mathbb{R}$ into $\mathbb{R}$, and let $T: F \rightarrow F$ be a linear transformation such that $T(e^{2x}) = x^2$, $T(e^{3x}) = \sin x$, and $T(1) = \cos 5x$. Find the following, if it is determined by this data.
   a. $T(e^{5x})$
   b. $T(3 + 5e^{3x})$
   c. $T(3e^{4x})$
   d. $T\left(\dfrac{e^{4x} + 2e^{5x}}{e^{2x}}\right)$

9. Note that one solution of the differential equation $y'' - 4y = \sin x$ is $y = -\frac{1}{5} \sin x$. Use summary item 10 and Illustration 1 to describe all solutions of this equation.

Let D_∞ be the vector space of functions mapping $\mathbb{R}$ into $\mathbb{R}$ that have derivatives of all orders. It can be shown that the kernel of a linear transformation $T: D_\infty \rightarrow D_\infty$ of the form $T(f) = a_n f^{(n)} + \cdots + a_1 f' + a_0 f$ where $a_n \neq 0$ is an n-dimensional subspace of D_∞.

In Exercises 10–15, use the preceding information, summary item 10, and your knowledge of calculus to find all solutions in D_∞ of the given differential equation. See Illustration 1 in the text.

10. $y' = \sin 2x$

11. $y'' = -\cos x$

12. $y' - y = x$

13. $y'' + 4y = x^2$

14. $y'' + y' = 3e^x$

15. $y^{(3)} - 2y'' = x$

16. Let V and V' be vector spaces having ordered bases $B = (\mathbf{b}_1, \mathbf{b}_2, \mathbf{b}_3)$ and $B' = (\mathbf{b}'_1, \mathbf{b}'_2, \mathbf{b}'_3, \mathbf{b}'_4)$, respectively. Let $T: V \rightarrow V'$ be a linear transformation such that

$$\begin{aligned} T(\mathbf{b}_1) &= 3\mathbf{b}'_1 + \mathbf{b}'_2 + 4\mathbf{b}'_3 - \mathbf{b}'_4 \\ T(\mathbf{b}_2) &= \mathbf{b}'_1 + 2\mathbf{b}'_2 - \mathbf{b}'_3 + 2\mathbf{b}'_4 \\ T(\mathbf{b}_3) &= -2\mathbf{b}'_1 - \mathbf{b}'_2 + 2\mathbf{b}'_3. \end{aligned}$$

Find the matrix representation A of T relative to B,B'.

In Exercises 17–19, let V and V' be vector spaces with ordered bases $B = (\mathbf{b}_1, \mathbf{b}_2, \mathbf{b}_3)$ and $B' = (\mathbf{b}_1', \mathbf{b}_2', \mathbf{b}_3', \mathbf{b}_4')$, respectively, and let $T: V \to V'$ be the linear transformation having the given matrix A as matrix representation relative to B,B'. Find $T(\mathbf{v})$ for the given vector $\mathbf{v}$.

17. $A = \begin{bmatrix} 4 & 1 & -1 \\ 2 & 2 & 0 \\ 0 & 6 & 1 \\ 2 & 1 & 3 \end{bmatrix}$, $\mathbf{v} = \mathbf{b}_1 + \mathbf{b}_2 + \mathbf{b}_3$

18. A as in Exercise 17, $\mathbf{v} = 3\mathbf{b}_3 - \mathbf{b}_1$

19. $A = \begin{bmatrix} 0 & 4 & -1 \\ 1 & 1 & 2 \\ 2 & 0 & 1 \\ 0 & 1 & 1 \end{bmatrix}$, $\mathbf{v} = 6\mathbf{b}_1 - 4\mathbf{b}_2 + \mathbf{b}_3$

In Exercises 20–33, we consider A to be the matrix representation of the indicated linear transformation T (you may assume it is linear) relative to the indicated ordered bases B and B'. Starting with Exercise 21, we let D denote differentiating once, D^2 differentiating twice, and so on.

20. Let V and V' be vector spaces with ordered bases $B = (\mathbf{b}_1, \mathbf{b}_2, \mathbf{b}_3)$ and $B' = (\mathbf{b}_1', \mathbf{b}_2', \mathbf{b}_3')$, respectively. Let $T: V \to V'$ be the linear transformation such that

$$T(\mathbf{b}_1) = \quad \mathbf{b}_1' + 2\mathbf{b}_2' - 3\mathbf{b}_3'$$
$$T(\mathbf{b}_2) = \quad 3\mathbf{b}_1' + 5\mathbf{b}_2' + 2\mathbf{b}_3'$$
$$T(\mathbf{b}_3) = -2\mathbf{b}_1' - 3\mathbf{b}_2' - 4\mathbf{b}_3'.$$

a. Find the matrix A.
b. Use A to find $T(\mathbf{v})_{B'}$, if $\mathbf{v}_B = [2, -5, 1]$.
c. Show that T is invertible, and find the matrix representation of T^{-1} relative to B',B.
d. Find $T^{-1}(\mathbf{v}')_B$ if $\mathbf{v}_{B'}' = [-1, 1, 3]$.
e. Express $T^{-1}(\mathbf{b}_1')$, $T^{-1}(\mathbf{b}_2')$, and $T^{-1}(\mathbf{b}_3')$ as linear combinations of the vectors in B.

21. Let $T: P_3 \to P_3$ be defined by $T(p(x)) = D(p(x))$, the derivative of $p(x)$. Let the ordered bases for P_3 be $B = B' = (x^3, x^2, x, 1)$.
a. Find the matrix A.
b. Use A to find the derivative of $4x^3 - 5x^2 + 10x - 13$.

c. Noting that $T \circ T = D^2$, find the second derivative of $-5x^3 + 8x^2 - 3x + 4$ by multiplying a column vector by an appropriate matrix.

22. Let $T: P_3 \to P_3$ be defined by $T(p(x)) = x\, D(p(x))$ and let the ordered bases B and B' be as in Exercise 21.
a. Find the matrix representation A relative to B,B'.
b. Working with the matrix A and coordinate vectors, find all solutions $p(x)$ of $T(p(x)) = x^3 - 3x^2 + 4x$.
c. The transformation T can be decomposed into $T = T_2 \circ T_1$, where $T_1: P_3 \to P_2$ is defined by $T_1(p(x)) = D(p(x))$ and $T_2: P_2 \to P_3$ is defined by $T_2(p(x)) = xp(x)$. Find the matrix representations of T_1 and T_2 using the ordered bases B of P_3 and $B'' = (x^2, x, 1)$ of P_2. Now multiply these matrix representations for T_1 and T_2 to obtain a 4×4 matrix, and compare with the matrix A. What do you notice?
d. Multiply the two matrices found in part (c) to obtain a 3×3 matrix. Let $T_3: P_2 \to P_2$ be the linear transformation having this matrix as matrix representation relative to B'',B'' for the ordered basis B'' of part (c). Find $T_3(a_2x^2 + a_1x + a_0)$. How is T_3 related to T_1 and T_2?

23. Let V be the subspace $\text{sp}(x^2e^x, xe^x, e^x)$ of the vector space of all differentiable functions mapping $\mathbb{R}$ into $\mathbb{R}$. Let $T: V \to V$ be the linear transformation of V into itself given by taking second derivatives, so $T = D^2$, and let $B = B' = (x^2e^x, xe^x, e^x)$. Find the matrix A by
a. following the procedure in summary item 11, and
b. finding and then squaring the matrix A_1 that represents the transformation D corresponding to taking first derivatives.

24. Let $T: P_3 \to P_2$ be defined by $T(p(x)) = p'(2x + 1)$, where $p'(x) = D(p(x))$, and let $B = (x^3, x^2, x, 1)$ and $B' = (x^2, x, 1)$.
a. Find the matrix A.
b. Use A to compute $T(4x^3 - 5x^2 + 4x - 7)$.

25. Let $V = \text{sp}(\sin^2 x, \cos^2 x)$ and let $T: V \to V$ be defined by taking second derivatives.

Taking $B = B' = (\sin^2 x, \cos^2 x)$, find A in two ways.

  a. Compute A as described in summary item 11.

  b. Find the space W spanned by the first derivatives of the vectors in B, choose an ordered basis for W, and compute A as a product of the two matrices representing the differentiation map from V into W followed by the differentiation map from W into V.

26. Let $T: P_3 \to P_3$ be the linear transformation defined by $T(p(x)) = D^2(p(x)) - 4D(p(x)) + p(x)$. Find the matrix representation A of T, where $B = (x, 1 + x, x + x^2, x^3)$.

27. Let $W = \mathrm{sp}(e^{2x}, e^{4x}, e^{8x})$ be the subspace of the vector space of all real-valued functions with domain $\mathbb{R}$, and let $B = (e^{2x}, e^{4x}, e^{8x})$. Find the matrix representation A relative to B, B of the linear transformation $T: W \to W$ defined by $T(f) = D^2(f) + 2D(f) + f$.

28. For W and B in Exercise 27, find the matrix representation A of the linear transformation $T: W \to W$ defined by $T(f) = \int_{-\infty}^{x} f(t)\, dt$.

29. For W and B in Exercise 27, find $T(ae^{2x} + be^{4x} + ce^{8x})$ for the linear transformation T whose matrix representation relative to B, B is

$$A = \begin{bmatrix} 1 & 0 & 1 \\ 0 & 1 & 0 \\ 1 & 0 & 1 \end{bmatrix}.$$

30. Repeat Exercise 29, given that

$$A = \begin{bmatrix} 6 & 0 & 0 \\ 0 & 5 & 0 \\ 0 & 0 & -3 \end{bmatrix}.$$

31. Let W be the subspace $\mathrm{sp}(\sin 2x, \cos 2x)$ of the vector space of all real-valued functions with domain $\mathbb{R}$, and let $B = (\sin 2x, \cos 2x)$. Find the matrix representation A relative to B, B for the linear transformation $T: W \to W$ defined by $T(f) = D^2(f) + 2D(f) + f$.

32. For W and B in Exercise 31, find $T(a \sin 2x + b \cos 2x)$ for $T: W \to W$ whose matrix representation is

$$A = \begin{bmatrix} 1 & 1 \\ 1 & -1 \end{bmatrix}.$$

33. For W and B in Exercise 31, find $T(a \sin 2x + b \cos 2x)$ for $T: W \to W$ whose matrix representation is

$$A = \begin{bmatrix} 0 & -2 \\ 2 & 0 \end{bmatrix}.$$

34. Let V and V' be vector spaces. Mark each of the following True or False.

  ___ **a.** A linear transformation of vector spaces preserves the vector-space operations.

  ___ **b.** Every function mapping V into V' relates the algebraic structure of V to that of V'.

  ___ **c.** A linear transformation $T: V \to V'$ carries the zero vector of V into the zero vector of V'.

  ___ **d.** A linear transformation $T: V \to V'$ carries a pair $\mathbf{v}, -\mathbf{v}$ in V into a pair $\mathbf{v}', -\mathbf{v}'$ in V'.

  ___ **e.** For every vector $\mathbf{b}'$ in V', the function $T_{\mathbf{b}'}: V \to V'$ defined by $T_{\mathbf{b}'}(\mathbf{v}) = \mathbf{b}'$ for all $\mathbf{v}$ in V is a linear transformation.

  ___ **f.** The function $T_0: V \to V'$ defined by $T_0(\mathbf{v}) = \mathbf{0}'$, the zero vector of V', for all $\mathbf{v}$ in V is a linear transformation.

  ___ **g.** The vector space P_{10} of polynomials of degree ≤ 10 is isomorphic to $\mathbb{R}^{10}$.

  ___ **h.** There is exactly one isomorphism $T: P_{10} \to \mathbb{R}^{11}$.

  ___ **i.** Let V and V' be vector spaces of dimensions n and m, respectively. A linear transformation $T: V \to V'$ is invertible if and only if $m = n$.

  ___ **j.** If T in part (i) is an invertible transformation, then $m = n$.

35. Prove that the two conditions in Definition 3.9 for a linear transformation are equivalent to the single condition in Eq. (3).

36. Let V, V', and V'' be vector spaces, and let $T: V \to V'$ and $T': V' \to V''$ be linear transformations. Prove that the composite function $(T' \circ T): V \to V''$ defined by $(T' \circ T)(\mathbf{v}) = T'(T(\mathbf{v}))$ for each $\mathbf{v}$ in V is again a linear transformation.

37. Prove that, if T and T' are invertible linear transformations of vector spaces such that $T' \circ T$ is defined, $T' \circ T$ is also invertible.

38. State conditions for an $m \times n$ matrix A that are equivalent to the condition that the linear transformation $T(\mathbf{x}) = A\mathbf{x}$ for $\mathbf{x}$ in $\mathbb{R}^n$ is an isomorphism.

39. Let **v** and **w** be independent vectors in V, and let $T: V \rightarrow V'$ be a one-to-one linear transformation of V into V'. Prove that $T(\mathbf{v})$ and $T(\mathbf{w})$ are independent vectors in V'.

40. Let V and V' be vector spaces, let $B = \{\mathbf{b}_1, \mathbf{b}_2, \ldots, \mathbf{b}_n\}$ be a basis for V, and let $\mathbf{c}'_1, \mathbf{c}'_2, \ldots, \mathbf{c}'_n \in V'$. Prove that there exists a linear transformation $T: V \rightarrow V'$ such that $T(\mathbf{b}_i) = \mathbf{c}'_i$ for $i = 1, 2, \ldots, n$.

41. State and prove a generalization of Exercise 40 for any vector spaces V and V', where V has a basis B.

42. If the matrix representation of $T: \mathbb{R}^n \rightarrow \mathbb{R}^n$ relative to B,B is a diagonal matrix, describe the effect of T on the basis vectors in B.

Exercises 43 and 44 show that the set L(V, V') of all linear transformations mapping a vector space V into a vector space V' is a subspace of the vector space of all functions mapping V into V'. (See summary item 5 in Section 3.1.)

43. Let T_1 and T_2 be in $L(V, V')$, and let $(T_1 + T_2): V \rightarrow V'$ be defined by

$$(T_1 + T_2)(\mathbf{v}) = T_1(\mathbf{v}) + T_2(\mathbf{v})$$

for each vector **v** in V. Prove that $T_1 + T_2$ is again a linear transformation of V into V'.

44. Let T be in $L(V, V')$, let r be any scalar in $\mathbb{R}$, and let $rT: V \rightarrow V'$ be defined by

$$(rT)(\mathbf{v}) = r(T(\mathbf{v}))$$

for each vector **v** in V. Prove that rT is again a linear transformation of V into V'.

45. If V and V' are the finite-dimensional spaces in Exercises 43 and 44 and have ordered bases B and B', respectively, describe the matrix representations of $T_1 + T_2$ in Exercise 43 and rT in Exercise 44 in terms of the matrix representations of $T_1, T_2,$ and T relative to B,B'.

46. Prove that if $T: V \rightarrow V'$ is a linear transformation and $T(\mathbf{p}) = \mathbf{b}$, then the solution set of $T(x) = \mathbf{b}$ is $\{\mathbf{p} + \mathbf{h} \mid \mathbf{h} \in \ker(T)\}$.

47. Prove that, for any five linear transformations T_1, T_2, T_3, T_4, T_5 mapping $\mathbb{R}^2$ into $\mathbb{R}^2$, there exist scalars c_1, c_2, c_3, c_4, c_5 (not all of which are zero) such that $T = c_1 T_1 + c_2 T_2 + c_3 T_3 + c_4 T_4 + c_5 T_5$ has the property that $T(\mathbf{x}) = \mathbf{0}$ for all **x** in $\mathbb{R}^2$.

48. Let $T: \mathbb{R}^n \rightarrow \mathbb{R}^n$ be a linear transformation. Prove that, if $T(T(\mathbf{x})) = T(\mathbf{x}) + T(\mathbf{x}) + 3\mathbf{x}$ for all **x** in $\mathbb{R}^n$, then T is a one-to-one mapping of $\mathbb{R}^n$ into $\mathbb{R}^n$.

49. Let V and V' be vector spaces having the same finite dimension, and let $T: V \rightarrow V'$ be a linear transformation. Prove that T is one-to-one if and only if range$(T) = V'$. [HINT: Use Exercise 36 in Section 3.2.]

50. Give an example of a vector space V and a linear transformation $T: V \rightarrow V$ such that T is one-to-one but range$(T) \neq V$. [HINT: By Exercise 49, what must be true of the dimension of V?]

51. Repeat Exercise 50, but this time make range$(T) = V$ for a transformation T that is not one-to-one.

| 3.5 | **INNER-PRODUCT SPACES (Optional)** |

In Section 1.2, we introduced the concepts of the length of a vector and the angle between vectors in $\mathbb{R}^n$. *Length* and *angle* are defined and computed in $\mathbb{R}^n$ using the dot product of vectors. In this section, we discuss these notions for more general vector spaces. We start by recalling the properties of the dot product in $\mathbb{R}^n$, listed in Theorem 1.3, Section 1.2.

Properties of the Dot Product in $\mathbb{R}^n$

Let $\mathbf{u}$, $\mathbf{v}$, and $\mathbf{w}$ be vectors in $\mathbb{R}^n$ and let r be any scalar in $\mathbb{R}$. The following properties hold:

D1	$v \cdot w = w \cdot v$,	Commutative law
D2	$\mathbf{u} \cdot (\mathbf{v} + \mathbf{w}) = \mathbf{u} \cdot \mathbf{v} + \mathbf{u} \cdot \mathbf{w}$,	Distributive law
D3	$r(\mathbf{v} \cdot \mathbf{w}) = (r\mathbf{v}) \cdot \mathbf{w} = \mathbf{v} \cdot (r\mathbf{w})$,	Homogeneity
D4	$\mathbf{v} \cdot \mathbf{v} \geq 0$, and $\mathbf{v} \cdot \mathbf{v} = 0$ if and only if $\mathbf{v} = \mathbf{0}$.	Positivity

There are many vector spaces for which we can define useful dot products satisfying properties D1–D4. In phrasing a general definition for such vector spaces, it is customary to speak of an *inner product* rather than of a dot product, and to use the notation $\langle \mathbf{v}, \mathbf{w} \rangle$ in place of $\mathbf{v} \cdot \mathbf{w}$. One such example would be $\langle [2, 3], [4, -1] \rangle = 2(4) + 3(-1) = 5$.

DEFINITION 3.12   Inner-Product Space

An **inner product** on a vector space V is a function that associates with each ordered pair of vectors $\mathbf{v}$, $\mathbf{w}$ in V a real number, written $\langle \mathbf{v}, \mathbf{w} \rangle$, satisfying the following properties for all $\mathbf{u}$, $\mathbf{v}$, and $\mathbf{w}$ in V and for all scalars r:

P1	$\langle \mathbf{v}, \mathbf{w} \rangle = \langle \mathbf{w}, \mathbf{v} \rangle$,	Symmetry
P2	$\langle \mathbf{u}, \mathbf{v} + \mathbf{w} \rangle = \langle \mathbf{u}, \mathbf{v} \rangle + \langle \mathbf{u}, \mathbf{w} \rangle$,	Additivity
P3	$r\langle \mathbf{v}, \mathbf{w} \rangle = \langle r\mathbf{v}, \mathbf{w} \rangle = \langle \mathbf{v}, r\mathbf{w} \rangle$,	Homogeneity
P4	$\langle \mathbf{v}, \mathbf{v} \rangle \geq 0$, and $\langle \mathbf{v}, \mathbf{v} \rangle = 0$ if and only if $\mathbf{v} = \mathbf{0}$.	Positivity

An **inner-product space** is a vector space V together with an inner product on V.

EXAMPLE 1   Determine whether $\mathbb{R}^2$ is an inner-product space if, for $\mathbf{v} = [v_1, v_2]$ and $\mathbf{w} = [w_1, w_2]$, we define

$$\langle \mathbf{v}, \mathbf{w} \rangle = 2v_1 w_1 + 5v_2 w_2.$$

SOLUTION   We check each of the four properties in Definition 3.12.

P1: Because $\langle \mathbf{v}, \mathbf{w} \rangle = 2v_1 w_1 + 5v_2 w_2$ and because $\langle \mathbf{w}, \mathbf{v} \rangle = 2w_1 v_1 + 5w_2 v_2$, the first property holds.

P2: We compute

$$\langle \mathbf{u}, \mathbf{v} + \mathbf{w} \rangle = 2u_1(v_1 + w_1) + 5u_2(v_2 + w_2),$$
$$\langle \mathbf{u}, \mathbf{v} \rangle = 2u_1 v_1 + 5u_2 v_2,$$
$$\langle \mathbf{u}, \mathbf{w} \rangle = 2u_1 w_1 + 5u_2 w_2.$$

The sum of the right-hand sides of the last two equations equals the right-hand side of the first equation. This establishes property P2.

P3: We compute

$$r\langle \mathbf{v}, \mathbf{w} \rangle = r(2v_1 w_1 + 5v_2 w_2),$$
$$\langle r\mathbf{v}, \mathbf{w} \rangle = 2(rv_1)w_1 + 5(rv_2)w_2,$$
$$\langle \mathbf{v}, r\mathbf{w} \rangle = 2v_1(r_1 w_1) + 5v_2(rw_2).$$

Because the three right-hand expressions are equal, property P3 holds.

P4: We see that $\langle \mathbf{v}, \mathbf{v} \rangle = 2v_1^2 + 5v_2^2 \geq 0$. Because both terms of the sum are nonnegative, the sum is zero if and only if $v_1 = v_2 = 0$—that is, if and only if $\mathbf{v} = \mathbf{0}$. Therefore, we have an inner-product space.   ∎

EXAMPLE 2   Determine whether $\mathbb{R}^2$ is an inner-product space when, for $\mathbf{v} = [v_1, v_2]$ and $\mathbf{w} = [w_1, w_2]$, we define

$$\langle \mathbf{v}, \mathbf{w} \rangle = 2v_1 w_1 - 5v_2 w_2.$$

SOLUTION   A solution similar to the one for Example 1 goes along smoothly until we check property P4, which fails:

$$\langle [1, 1], [1, 1] \rangle = 2 - 5 = -3 < 0.$$

Therefore, $\mathbb{R}^2$ with this definition of $\langle \, , \, \rangle$ is not an inner-product space.   ∎

EXAMPLE 3   Determine whether the space $P_{0,1}$ of all polynomial functions with real coefficients and domain $0 \leq x \leq 1$ is an inner-product space if for p and q in $P_{0,1}$ we define

$$\langle p, q \rangle = \int_0^1 p(x)q(x)\, dx.$$

SOLUTION   We check the properties for an inner product.

P1: Clearly $\langle p, q \rangle = \langle q, p \rangle$ because

$$\int_0^1 p(x)q(x)\, dx = \int_0^1 q(x)p(x)\, dx.$$

P2: For polynomial functions p, q, and h, we have

$$\langle p, q + h \rangle = \int_0^1 p(x)(q(x) + h(x))\, dx$$

$$= \int_0^1 p(x)q(x)\, dx + \int_0^1 p(x)h(x)\, dx$$
$$= \langle p, q \rangle + \langle p, h \rangle.$$

P3: We have

$$r \int_0^1 p(x)q(x)\,dx = \int_0^1 r(p(x))q(x)\,dx$$

$$= \int_0^1 p(x)r(q(x))\,dx,$$

so P3 holds.

P4: Because $\langle p, p \rangle = \int_0^1 p(x)^2\,dx$ and because $p(x)^2 \geq 0$ for all x, we have $\langle p, p \rangle = \int_0^1 p(x)^2\,dx \geq 0$. Now $p(x)^2$ is a continuous nonnegative polynomial function and can be zero only at a finite number of points unless $p(x)$ is the zero polynomial. It follows that

$$\langle p, p \rangle = \int_0^1 p(x)^2\,dx > 0,$$

unless $p(x)$ is the zero polynomial. This establishes P4.

Because all four properties of Definition 3.12 hold, the space $P_{0,1}$ is an inner-product space with the given inner product. ∎

We mention that Example 3 is a very famous inner product, and the same definition

$$\langle f, g \rangle = \int_0^1 f(x)g(x)\,dx$$

gives an inner product on the space $C_{0,1}$ of all continuous real-valued functions with domain of the interval $0 \leq x \leq 1$. The hypothesis of continuity is essential for the demonstration of P4, as is shown in advanced calculus. Of course, there is nothing unique about the interval $0 \leq x \leq 1$. Any interval $a \leq x \leq b$ can be used in its place. The choice of interval depends on the application.

Magnitude

The condition $\langle \mathbf{v}, \mathbf{v} \rangle \geq 0$ in the definition of an inner-product space allows us to define the magnitude of a vector, just as we did in Section 1.2 using the dot product.

DEFINITION 3.13 Magnitude or Norm of a Vector

Let V be an inner-product space. The **magnitude** or **norm** of a vector $\mathbf{v}$ in V is $\|\mathbf{v}\| = \sqrt{\langle \mathbf{v}, \mathbf{v} \rangle}$.

This definition of *norm* reduces to the usual definition of magnitude of vectors in $\mathbb{R}^n$ when the dot product is used. That is, $\|\mathbf{v}\| = \sqrt{\mathbf{v} \cdot \mathbf{v}}$.

Recall that in $\mathbb{R}^n$ we can visualize the vector $\mathbf{v} - \mathbf{w}$ geometrically as an arrow reaching from the tip of the arrow representing $\mathbf{w}$ to the tip of the arrow representing $\mathbf{v}$, so that $\|\mathbf{v} - \mathbf{w}\|$ is the distance between the tip of $\mathbf{v}$ and the tip of $\mathbf{w}$. This leads us to define the **distance** between $\mathbf{v}$ and $\mathbf{w}$ in an inner-product space V to be $d(\mathbf{v}, \mathbf{w}) = \|\mathbf{v} - \mathbf{w}\|$.

EXAMPLE 4   In the inner-product space $P_{0,1}$ of all polynomial functions with real coefficients and domain $0 \leq x \leq 1$, and with inner product defined by

$$\langle p, q \rangle = \int_0^1 p(x)q(x)\, dx,$$

(a) find the magnitude of the polynomial $p(x) = x + 1$, and (b) compute the distance $d(x^2, x)$ from x^2 to x.

SOLUTION   For part (a), we have

$$\|x + 1\|^2 = \langle x + 1, x + 1 \rangle = \int_0^1 (x + 1)^2\, dx$$

$$= \int_0^1 (x^2 + 2x + 1)\, dx = \left(\frac{x^3}{3} + x^2 + x \right) \Big|_0^1 = \frac{7}{3}.$$

Therefore, $\|x + 1\| = \sqrt{7/3}$.

For part (b), we have $d(x^2, x) = \|x^2 - x\|$. We compute

$$\|x^2 - x\|^2 = \langle x^2 - x, x^2 - x \rangle = \int_0^1 (x^2 - x)^2\, dx$$

$$= \int_0^1 (x^4 - 2x^3 + x^2)\, dx = \frac{1}{5} - \frac{1}{2} + \frac{1}{3} = \frac{1}{30}.$$

Therefore, $d(x^2, x) = 1/\sqrt{30}$.   ∎

The inner product used in Example 4 was not contrived just for this illustration. Another measure of the distance between the functions x and x^2 over $0 \leq x \leq 1$ is the maximum vertical distance between their graphs over the interval $0 \leq x \leq 1$, which calculus easily shows to be $\frac{1}{4}$ where $x = \frac{1}{2}$. In contrast, the inner product in this example uses an *integral* to measure the distance between the functions, not only at one point of the interval, but over the interval as a whole. Notice that the distance $1/\sqrt{30}$ we obtained is less than the maximum distance $\frac{1}{4}$ between the graphs, reflecting the fact that the graphs are less than $\frac{1}{4}$ unit apart over much of the interval. The functions are shown in Figure 3.3. The notion (used in Example 4) of distance between functions over an interval is very important in advanced mathematics, where it is used in approximating a complicated function over an interval as closely as possible by a function that is easier to handle.

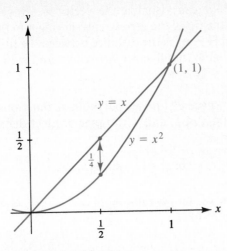

FIGURE 3.3
Graphs of x and x^2 over $0 \leq x \leq 1$.

It is often best, when working with the norm of a vector $\mathbf{v}$, to work with $\|\mathbf{v}\|^2 = \langle \mathbf{v}, \mathbf{v} \rangle$ and to introduce the radical in the final stages of the computation.

EXAMPLE 5   Let V be an inner-product space. Verify that

$$\|r\mathbf{v}\| = |r|\,\|\mathbf{v}\|$$

for any vector $\mathbf{v}$ in V and for any scalar r.

SOLUTION   We have

$$\|r\mathbf{v}\|^2 = \langle r\mathbf{v}, r\mathbf{v} \rangle \qquad \textbf{Applying homogeneity}$$
$$= r^2 \langle \mathbf{v}, \mathbf{v} \rangle \qquad \textbf{(property P3) twice}$$
$$= r^2 \|\mathbf{v}\|^2.$$

On taking square roots, we have $\|r\mathbf{v}\| = |r|\,\|\mathbf{v}\|$. ∎

The property of norms in the preceding example can be useful in computing the magnitude of a vector and in establishing relations between magnitudes, as illustrated in the following two examples.

EXAMPLE 6   Using the standard inner product in $\mathbb{R}^n$, find the magnitude of the vector

$$\mathbf{v} = [-6, -12, 6, 18, -6].$$

SOLUTION   Because $\mathbf{v} = -6[1, 2, -1, -3, 1]$, the property of norms in Example 5 tells us that

$$\|\mathbf{v}\| = |-6|\,\|[1, 2, -1, -3, 1]\| = 6\sqrt{16} = 24.$$ ∎

The Schwarz and Triangle Inequalities

In Section 1.2, we defined the angle θ between two nonzero vectors $\mathbf{v}$ and $\mathbf{w}$ in $\mathbb{R}^n$ to be

$$\theta = \arccos \frac{\mathbf{v} \cdot \mathbf{w}}{\|\mathbf{v}\| \, \|\mathbf{w}\|}.$$

The validity of this definition rested on the fact that for $\mathbf{v}, \mathbf{w} \in \mathbb{R}^n$, we have

$$|\mathbf{v} \cdot \mathbf{w}| \leq \|\mathbf{v}\| \, \|\mathbf{w}\| \qquad \textbf{Schwarz inequality}$$

so that

$$-1 \leq \frac{\mathbf{v} \cdot \mathbf{w}}{\|\mathbf{v}\| \, \|\mathbf{w}\|} \leq 1.$$

The Schwarz inequality with $\mathbf{v} \cdot \mathbf{w}$ replaced by $\langle \mathbf{v}, \mathbf{w} \rangle$ holds in any inner-product space.

THEOREM 3.11   Schwarz Inequality

> Let V be an inner-product space, and let $\mathbf{v}$ and $\mathbf{w}$ be vectors in V. Then $|\langle \mathbf{v}, \mathbf{w} \rangle| \leq \|\mathbf{v}\| \, \|\mathbf{w}\|$.

PROOF   Because the properties required for an inner-product space are modeled on those for the dot product in $\mathbb{R}^n$, we expect the proof here of the Schwarz inequality to be essentially the same as in Theorem 1.4, Section 1.2 for $\mathbb{R}^n$. This is indeed the case. Just replace every occurrence, such as $\mathbf{v} \cdot \mathbf{w}$, of a dot product in the proof of Theorem 1.4 by the corresponding inner product, such as $\langle \mathbf{v}, \mathbf{w} \rangle$.   ▲

We now define the **angle between two vectors** $\mathbf{v}$ and $\mathbf{w}$ in any inner-product space to be

$$\theta = \arccos \frac{\langle \mathbf{v}, \mathbf{w} \rangle}{\|\mathbf{v}\| \, \|\mathbf{w}\|}.$$

In particular, we define $\mathbf{v}$ and $\mathbf{w}$ to be **orthogonal** (or **perpendicular**) if $\langle \mathbf{v}, \mathbf{w} \rangle = 0$.

Recall that another important inequality in $\mathbb{R}^n$ that follows readily from the Schwarz inequality is

$$\|\mathbf{v} + \mathbf{w}\| \leq \|\mathbf{v}\| + \|\mathbf{w}\|. \qquad \textbf{Triangle inequality}$$

See Theorem 1.5 in Section 1.2. The triangle inequality is also valid in any inner-product space; its proof can also be obtained from the proof of Theorem 1.5 by replacing a dot product such as $\mathbf{v} \cdot \mathbf{w}$ by the corresponding inner product, $\langle \mathbf{v}, \mathbf{w} \rangle$.

SUMMARY

1. An inner-product space is a vector space V with an inner product $\langle\,,\,\rangle$ that associates with each pair of vectors $\mathbf{v}$, $\mathbf{w}$ in V a scalar $\langle \mathbf{v}, \mathbf{w} \rangle$ that satisfies the following conditions for all vectors $\mathbf{u}$, $\mathbf{v}$, and $\mathbf{w}$ in V and all scalars r:

   P1 $\langle \mathbf{v}, \mathbf{w} \rangle = \langle \mathbf{w}, \mathbf{v} \rangle$,
   P2 $\langle \mathbf{u}, \mathbf{v} + \mathbf{w} \rangle = \langle \mathbf{u}, \mathbf{v} \rangle + \langle \mathbf{u}, \mathbf{w} \rangle$,
   P3 $r\langle \mathbf{v}, \mathbf{w} \rangle = \langle r\mathbf{v}, \mathbf{w} \rangle = \langle \mathbf{v}, r\mathbf{w} \rangle$,
   P4 $\langle \mathbf{v}, \mathbf{v} \rangle \geq 0$, and $\langle \mathbf{v}, \mathbf{v} \rangle = 0$ if and only if $\mathbf{v} = \mathbf{0}$.

2. $\mathbb{R}^n$ is an inner-product space using the usual dot product as inner product.

3. In an inner-product space V, the norm of a vector is $\|\mathbf{v}\| = \sqrt{\langle \mathbf{v}, \mathbf{v} \rangle}$ and satisfies $\|r\mathbf{v}\| = |r|\,\|\mathbf{v}\|$. The distance between vectors $\mathbf{v}$ and $\mathbf{w}$ is $\|\mathbf{v} - \mathbf{w}\|$.

4. For all vectors $\mathbf{v}$ and $\mathbf{w}$ in an inner-product space, we have the following two inequalities:

   Schwarz inequality: $|\langle \mathbf{v}, \mathbf{w} \rangle| \leq \|\mathbf{v}\|\,\|\mathbf{w}\|$,
   Triangle inequality: $\|\mathbf{v} + \mathbf{w}\| \leq \|\mathbf{v}\| + \|\mathbf{w}\|$.

5. Vectors $\mathbf{v}$ and $\mathbf{w}$ in an inner-product space are orthogonal (perpendicular) if and only if $\langle \mathbf{v}, \mathbf{w} \rangle = 0$.

EXERCISES

In Exercises 1–9, determine whether or not the indicated product satisfies the conditions for an inner product in the given vector space.

1. In $\mathbb{R}^2$, let $\langle [x_1, x_2], [y_1, y_2] \rangle = x_1 y_1 - x_2 y_2$.

2. In $\mathbb{R}^2$, let $\langle [x_1, x_2], [y_1, y_2] \rangle = x_1 x_2 + y_1 y_2$.

3. In $\mathbb{R}^2$, let $\langle [x_1, x_2], [y_1, y_2] \rangle = x_1^2 y_1 + x_2 y_2$.

4. In $\mathbb{R}^2$, let $\langle [x_1, x_2], [y_1, y_2] \rangle = x_1 y_2$.

5. In $\mathbb{R}^3$, let $\langle [x_1, x_2, x_3], [y_1, y_2, y_3] \rangle = x_1 y_1$.

6. In $\mathbb{R}^3$, let $\langle [x_1, x_2, x_3], [y_1, y_2, y_3] \rangle = x_1 + y_1$.

7. In the vector space M_2 of all 2×2 matrices, let
$$\left\langle \begin{bmatrix} a_1 & a_2 \\ a_3 & a_4 \end{bmatrix}, \begin{bmatrix} b_1 & b_2 \\ b_3 & b_4 \end{bmatrix} \right\rangle = a_1 b_1 + a_2 b_2 + a_3 b_3 + a_4 b_4.$$

8. Let $C_{-1,1}$ be the vector space of all continuous functions mapping the interval $-1 \leq x \leq 1$ into $\mathbb{R}$, and let $\langle f, g \rangle = \int_{-1}^{1} f(x)g(x)\,dx$.

9. Let C be as in Exercise 8, and let $\langle f, g \rangle = f(0)g(0)$.

10. Let $C_{a,b}$ be the vector space of all continuous real-valued functions with domain $a \leq x \leq b$. Prove that $\langle\,,\,\rangle$, defined in $C_{a,b}$ by $\langle f, g \rangle = \int_a^b f(x)g(x)\,dx$, is an inner product in $C_{a,b}$.

11. Let $C_{0,1}$ be the vector space of all continuous real-valued functions with domain $0 \leq x \leq 1$. Let $\langle\,,\,\rangle$ be defined in $C_{0,1}$ by $\langle f, g \rangle = \int_0^1 f(x)g(x)\,dx$.
    a. Find $\langle (x + 1), x \rangle$.
    b. Find $\|x\|$.
    c. Find $\|x^2 - x\|$.
    d. Find $\|\sin \pi x\|$.

12. Prove that $\sin x$ and $\cos x$ are orthogonal functions in the vector space $C_{0,\pi}$ of Exercise 10, with the inner product defined there.

13. Let $\langle\,,\,\rangle$ be defined in $C_{0,1}$ as in Exercise 11. Find a set of two independent functions in $C_{0,1}$, each of which is orthogonal to the constant function 1.

14. Let $\mathbf{u}$ and $\mathbf{v}$ be vectors in an inner-product space, and suppose that $\|\mathbf{u}\| = 3$ and $\|\mathbf{v}\| = 5$. Find $\langle \mathbf{u} + 2\mathbf{v}, \mathbf{u} - 2\mathbf{v} \rangle$.

15. Suppose that the vectors **u** and **v** in Exercise 14 are perpendicular. Find $\langle \mathbf{u} + 2\mathbf{v}, 3\mathbf{u} + \mathbf{v} \rangle$.

16. Let V be an inner-product space. Mark each of the following True or False.

_____ **a.** The norm of every vector in V is a positive real number.

_____ **b.** The norm of every nonzero vector in V is a positive real number.

_____ **c.** We have $\|r\mathbf{v}\| = r\|\mathbf{v}\|$ for every scalar r and vector **v** in V.

_____ **d.** We have $\|\mathbf{u} + \mathbf{v}\|^2 = \|\mathbf{u}\|^2 + \|\mathbf{v}\|^2$ for all vectors **u** and **v** in V.

_____ **e.** Two nonzero orthogonal vectors in V are independent.

_____ **f.** If $\|\mathbf{u} + \mathbf{v}\|^2 = \|\mathbf{u}\|^2 + \|\mathbf{v}\|^2$ for two nonzero vectors **u** and **v** in V, then **u** and **v** are orthogonal.

_____ **g.** An inner product can be defined on every finite-dimensional real vector space.

_____ **h.** Let r be any real scalar. Then $\langle \, , \, \rangle'$, defined by $\langle \mathbf{u}, \mathbf{v} \rangle' = r\langle \mathbf{u}, \mathbf{v} \rangle$ for vectors **u** and **v** in V, is also an inner product on V.

_____ **i.** $\langle \, , \, \rangle'$, defined in part (h), is an inner product on V if r is nonzero.

_____ **j.** The distance between two vectors **u** and **v** in V is given by $|\langle \mathbf{u} - \mathbf{v}, \mathbf{u} - \mathbf{v} \rangle|$.

17. For vectors **v** and **w** in an inner-product space, prove that $\mathbf{v} - \mathbf{w}$ and $\mathbf{v} + \mathbf{w}$ are perpendicular if and only if $\|\mathbf{v}\| = \|\mathbf{w}\|$.

18. For vectors **u**, **v**, and **w** in an inner-product space and for scalars r and s, prove that, if **w** is perpendicular to both **u** and **v**, then **w** is perpendicular to $r\mathbf{u} + s\mathbf{v}$.

19. Let S be a subset of nonzero vectors in an inner-product space V, and suppose that any two different vectors in S are orthogonal. Prove that S is an independent set.

20. Let V be an inner-product space, and let S be a subset of V. Prove that

$$S^{\perp} = \{\mathbf{v} \in V \mid \mathbf{v} \text{ is orthogonal to each vector in } S\}$$

is a subspace of V.

21. Referring to Exercise 20, prove that $S \subseteq (S^{\perp})^{\perp}$.

22. Give an example of an inner-product space V for which there exists a subspace W such that $(W^{\perp})^{\perp} \neq W$.

23. (*Pythagorean theorem*) Let **u** and **v** be orthogonal vectors in an inner-product space V. Prove that $\|\mathbf{u} + \mathbf{v}\|^2 = \|\mathbf{u}\|^2 + \|\mathbf{v}\|^2$.

24. Use the triangle inequality to prove that

$$\|\mathbf{v} - \mathbf{w}\| \leq \|\mathbf{v}\| + \|\mathbf{w}\|$$

for any vectors **v** and **w** in an inner-product space V.

25. Prove that, for any vectors **v** and **w** in an inner-product space V, we have

$$\|\mathbf{v} - \mathbf{w}\| \geq \|\mathbf{v}\| - \|\mathbf{w}\|.$$

26. Prove that the vectors $\|\mathbf{v}\|\mathbf{w} + \|\mathbf{w}\|\mathbf{v}$ and $\|\mathbf{v}\|\mathbf{w} - \|\mathbf{w}\|\mathbf{v}$ in an inner-product space V are perpendicular.

27. Consider the space $C_{a,b}$ of continuous functions with domain the closed interval $a \leq x \leq b$, and let $w(x)$ be a positive continuous *weight function*, so that $w(x) > 0$ for $a \leq x \leq b$. Prove that for f and g in $C_{a,b}$, the weighted integral

$$\langle f, g \rangle = \int_a^b w(x)f(x)g(x)\, dx$$

defines an inner product on $C_{a,b}$. (Such a weight function has the effect of making the portions of the interval $a \leq x \leq b$ where $w(x)$ is large more significant than portions where $w(x)$ is smaller in computing inner products and norms.)

4

DETERMINANTS

Each square matrix has associated with it a number called the *determinant* of the matrix. In Section 4.1 we introduce determinants of 2×2 and 3×3 matrices, motivated by computations of area and volume. Section 4.2 discusses determinants of $n \times n$ matrices and their properties.

Section 4.3 opens with an efficient way to compute determinants and then presents Cramer's rule as well as a formula for the inverse of an invertible square matrix in terms of determinants. Cramer's rule expresses, in terms of determinants, the solution of a square linear system having a unique solution. This method is primarily of theoretical interest because the methods presented in Chapter 1 are much more efficient for solving a square system with more than two or three equations. Because references and formulas involving Cramer's rule appear in advanced calculus and other fields, we believe that students should at least read the statement of Cramer's rule and look at an illustration.

The chapter concludes with optional Section 4.4, which discusses the significance of the determinant of the standard matrix representation of a linear transformation mapping $\mathbb{R}^n$ into $\mathbb{R}^n$. The ideas in that section form the foundation for the change-of-variable formulas for definite integrals of functions of one or more variables.

4.1 AREAS, VOLUMES, AND CROSS PRODUCTS

We introduce determinants by discussing one of their most important applications: finding areas and volumes. We will find areas and volumes of very simple boxlike regions. In calculus, one finds areas and volumes of regions having more general shapes, using formulas that involve determinants.

The Area of a Parallelogram

The parallelogram determined by two nonzero and nonparallel vectors $\mathbf{a} = [a_1, a_2]$ and $\mathbf{b} = [b_1, b_2]$ in $\mathbb{R}^2$ is shown in Figure 4.1. This parallelogram has a vertex at the origin, and we regard the arrows representing $\mathbf{a}$ and $\mathbf{b}$ as forming the two sides of the parallelogram having the origin as a common vertex.

We can find the area of this parallelogram by multiplying the length $\|\mathbf{a}\|$ of its base by the altitude h, obtaining

$$\text{Area} = \|\mathbf{a}\|\, h = \|\mathbf{a}\|\, \|\mathbf{b}\|(\sin \theta) = \|\mathbf{a}\|\, \|\mathbf{b}\|\sqrt{1 - \cos^2\theta}.$$

Recall from page 24 of Section 1.2 that $\mathbf{a} \cdot \mathbf{b} = \|\mathbf{a}\|\, \|\mathbf{b}\|(\cos \theta)$. Squaring our area equation, we have

$$\begin{aligned}
(\text{Area})^2 &= \|\mathbf{a}\|^2\|\mathbf{b}\|^2 - \|\mathbf{a}\|^2\|\mathbf{b}\|^2 \cos^2\theta \\
&= \|\mathbf{a}\|^2\|\mathbf{b}\|^2 - (\mathbf{a} \cdot \mathbf{b})^2 \\
&= (a_1^2 + a_2^2)(b_1^2 + b_2^2) - (a_1b_1 + a_2b_2)^2 \\
&= (a_1b_2 - a_2b_1)^2.
\end{aligned} \tag{1}$$

The last equality should be checked using pencil and paper. On taking square roots, we obtain

$$\text{Area} = |a_1b_2 - a_2b_1|.$$

The number within the absolute value bars is known as the **determinant** of the matrix

$$A = \begin{bmatrix} a_1 & a_2 \\ b_1 & b_2 \end{bmatrix}$$

and is denoted by $|A|$ or $\det(A)$, so that

$$\det(A) = \begin{vmatrix} a_1 & a_2 \\ b_1 & b_2 \end{vmatrix}.$$

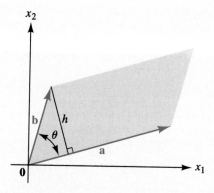

FIGURE 4.1
The parallelogram determined by a and b.

That is, if

$$A = \begin{bmatrix} a_1 & a_2 \\ b_1 & b_2 \end{bmatrix},$$

then

$$\det(A) = \begin{vmatrix} a_1 & a_2 \\ b_1 & b_2 \end{vmatrix} = a_1 b_2 - a_2 b_1. \tag{2}$$

We can remember this formula for the determinant by taking the product of the black entries on the main diagonal of the matrix, minus the product of the colored entries on the other diagonal.

EXAMPLE 1   Find the determinant of the matrix

$$\begin{bmatrix} 2 & 3 \\ 1 & 4 \end{bmatrix}.$$

SOLUTION   We have

$$\begin{vmatrix} 2 & 3 \\ 1 & 4 \end{vmatrix} = (2)(4) - (3)(1) = 5. \qquad \blacksquare$$

EXAMPLE 2   Find the area of the parallelogram in $\mathbb{R}^2$ with vertices $(1, 1), (2, 3), (2, 1), (3, 3)$.

SOLUTION   The parallelogram is sketched in Figure 4.2. The sides having $(1, 1)$ as common vertex can be regarded as the vectors

$$\mathbf{a} = [2, 1] - [1, 1] = [1, 0]$$

and

$$\mathbf{b} = [2, 3] - [1, 1] = [1, 2],$$

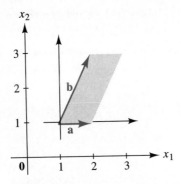

FIGURE 4.2
The parallelogram determined by $\mathbf{a} = [1, 0]$ and $\mathbf{b} = [1, 2]$.

as shown in the figure. Therefore, the area of the parallelogram is given by the determinant

$$\begin{vmatrix} 1 & 0 \\ 1 & 2 \end{vmatrix} = (1)(2) - (0)(1) = 2.$$

■

The Cross Product

Equation (2) defines a **second-order determinant**, associated with a 2×2 matrix. Another application of these second-order determinants appears when we find a vector in $\mathbb{R}^3$ that is perpendicular to each of two given independent vectors $\mathbf{b} = [b_1, b_2, b_3]$ and $\mathbf{c} = [c_1, c_2, c_3]$. Recall that the unit coordinate vectors in $\mathbb{R}^3$ are $\mathbf{i} = [1, 0, 0]$, $\mathbf{j} = [0, 1, 0]$, and $\mathbf{k} = [0, 0, 1]$. We leave as an exercise the verification that

$$\mathbf{p} = \begin{vmatrix} b_2 & b_3 \\ c_2 & c_3 \end{vmatrix} \mathbf{i} - \begin{vmatrix} b_1 & b_3 \\ c_1 & c_3 \end{vmatrix} \mathbf{j} + \begin{vmatrix} b_1 & b_2 \\ c_1 & c_2 \end{vmatrix} \mathbf{k} \qquad (3)$$

is a vector perpendicular to both $\mathbf{b}$ and $\mathbf{c}$. (See Exercise 5.) This can be seen by computing $\mathbf{p} \cdot \mathbf{b} = \mathbf{p} \cdot \mathbf{c} = 0$. The vector $\mathbf{p}$ in formula (3) is known as the **cross product** of $\mathbf{b}$ and $\mathbf{c}$, and is denoted $\mathbf{p} = \mathbf{b} \times \mathbf{c}$.

There is a very easy way to remember formula (3) for the cross product $\mathbf{b} \times \mathbf{c}$. Form the 3×3 *symbolic matrix*

$$\begin{bmatrix} \mathbf{i} & \mathbf{j} & \mathbf{k} \\ b_1 & b_2 & b_3 \\ c_1 & c_2 & c_3 \end{bmatrix}.$$

HISTORICAL NOTE   The notion of a cross product grew out of Sir William Rowan Hamilton's attempt to develop a multiplication for "triplets"—that is, vectors in $\mathbb{R}^3$. He wanted this multiplication to satisfy the associative and commutative properties as well as the distributive law. He wanted division to be always possible, except by $\mathbf{0}$. And he wanted the lengths to multiply—that is, if $(a_1, a_2, a_3)(b_1, b_2, b_3) = (c_1, c_2, c_3)$, then $\|(a_1, a_2, a_3)\| \, \|(b_1, b_2, b_3)\| = \|(c_1, c_2, c_3)\|$. After struggling with this problem for 13 years, Hamilton finally succeeded in solving it on October 16, 1843, although not in the way he had hoped. Namely, he discovered an analogous result—not for triples, but for quadruples. As he walked that day in Dublin, he wrote, he could not "resist the impulse . . . to cut with a knife on a stone of Brougham Bridge . . . the fundamental formula with the symbols i, j, k; namely, $i^2 = j^2 = k^2 = ijk = -1$." This formula symbolized his discovery of *quaternions*, elements of the form $Q = w + xi + yj + zk$ with w, x, y, z real numbers, whose multiplication obeys the laws just given, as well as the other laws Hamilton desired, except for the commutative law.

Hamilton noted the convenience of writing a quaternion Q in two parts: the scalar part w and the vector part $xi + yj + zk$. Then the product of two quaternions $\alpha = xi + yj + zk$ and $\beta = x'i + y'j + z'k$ with scalar parts 0 is given as

$$\alpha\beta = (-xx' - yy' - zz') + (yz' - zy')i + (zx' - xz')j + (xy' - yx')k.$$

The vector part of this product is our modern cross product of the "vectors" α and β, while the scalar part is the negative of the modern dot product (Section 1.2).

Although Hamilton and others pushed for the use of quaternions in physics, physicists realized by the end of the nineteenth century that the only parts of the subject necessary for their work were the two types of products of vectors. It was Josiah Willard Gibbs (1839–1903), professor of mathematical physics at Yale, who introduced our modern notation for both the dot product and the cross product and developed their properties in detail in his classes at Yale and finally in his *Vector Analysis* of 1901.

Formula (3) can be obtained from this matrix in a simple way. Multiply the vector **i** by the determinant of the 2×2 matrix obtained by crossing out the row and column containing **i**, as in

$$\begin{bmatrix} i & j & k \\ b_1 & b_2 & b_3 \\ c_1 & c_2 & c_3 \end{bmatrix}.$$

Similarly, multiply $(-\mathbf{j})$ by the determinant of the matrix obtained by crossing out the row and column in which **j** appears. Finally, multiply **k** by the determinant of the matrix obtained by crossing out the row and column containing **k**, and add these multiples of **i**, **j**, and **k** to obtain formula (3).

EXAMPLE 3   Find a vector perpendicular to both [2, 1, 1] and [1, 2, 3] in $\mathbb{R}^3$.

SOLUTION   We form the symbolic matrix

$$\begin{bmatrix} i & j & k \\ 2 & 1 & 1 \\ 1 & 2 & 3 \end{bmatrix}$$

and find that

$$[2, 1, 1] \times [1, 2, 3] = \begin{vmatrix} 1 & 1 \\ 2 & 3 \end{vmatrix} i - \begin{vmatrix} 2 & 1 \\ 1 & 3 \end{vmatrix} j + \begin{vmatrix} 2 & 1 \\ 1 & 2 \end{vmatrix} k$$

$$= i - 5j + 3k = [1, -5, 3]. \qquad \blacksquare$$

The cross product $\mathbf{p} = \mathbf{b} \times \mathbf{c}$ as defined in Eq. (3) not only is perpendicular to both **b** and **c** but points in the direction determined by the familiar *right-hand rule*: when the fingers of the right hand curve in the direction from **b** to **c**, the thumb points in the direction of $\mathbf{b} \times \mathbf{c}$. (See Figure 4.3.) We do not attempt to prove this.

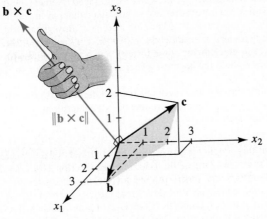

FIGURE 4.3
The area of the parallelogram determined by b and c is $\|b \times c\|$.

The magnitude of the vector $\mathbf{p} = \mathbf{b} \times \mathbf{c}$ in formula (3) is of interest as well: it is the area of the parallelogram with a vertex at the origin in $\mathbb{R}^3$ and edges at that vertex given by the vectors $\mathbf{b}$ and $\mathbf{c}$. To see this, we refer to a diagram such as the one in Figure 4.1, but with $\mathbf{a}$ replaced by $\mathbf{c}$, and we repeat the computation for area. This time, Eq. (1) takes the form

$$(\text{Area})^2 = \|\mathbf{c}\|^2 \|\mathbf{b}\|^2 - (\mathbf{c} \cdot \mathbf{b})^2$$
$$= (c_1^2 + c_2^2 + c_3^2)(b_1^2 + b_2^2 + b_3^2) - (c_1 b_1 + c_2 b_2 + c_3 b_3)^2$$
$$= \begin{vmatrix} b_2 & b_3 \\ c_2 & c_3 \end{vmatrix}^2 + \begin{vmatrix} b_1 & b_3 \\ c_1 & c_3 \end{vmatrix}^2 + \begin{vmatrix} b_1 & b_2 \\ c_1 & c_2 \end{vmatrix}^2.$$

Again, pencil and paper are needed to check this last equality. Taking square roots, we obtain

$$\|\mathbf{b} \times \mathbf{c}\| = \text{Area of the parallelogram in } \mathbb{R}^3 \text{ determined by } \mathbf{b} \text{ and } \mathbf{c}.$$

EXAMPLE 4   Find the area of the parallelogram in $\mathbb{R}^3$ determined by the vectors $\mathbf{b} = [3, 1, 0]$ and $\mathbf{c} = [1, 3, 2]$.

SOLUTION   From the symbolic matrix

$$\begin{bmatrix} \mathbf{i} & \mathbf{j} & \mathbf{k} \\ 3 & 1 & 0 \\ 1 & 3 & 2 \end{bmatrix},$$

we find that

$$\mathbf{b} \times \mathbf{c} = \begin{vmatrix} 1 & 0 \\ 3 & 2 \end{vmatrix} \mathbf{i} - \begin{vmatrix} 3 & 0 \\ 1 & 2 \end{vmatrix} \mathbf{j} + \begin{vmatrix} 3 & 1 \\ 1 & 3 \end{vmatrix} \mathbf{k}$$
$$= [2, -6, 8].$$

Therefore, the area of the parallelogram shown in Figure 4.3 is

$$\|\mathbf{b} \times \mathbf{c}\| = 2\sqrt{1 + 9 + 16} = 2\sqrt{26}. \qquad \blacksquare$$

EXAMPLE 5   Find the area of the triangle in $\mathbb{R}^3$ with vertices $(-1, 2, 0)$, $(2, 1, 3)$, and $(1, 1, -1)$.

SOLUTION   We think of $(-1, 2, 0)$ as a local origin, and we take translated vectors starting there and reaching to $(2, 1, 3)$ and to $(1, 1, -1)$—namely,

$$\mathbf{a} = [2, 1, 3] - [-1, 2, 0] = [3, -1, 3]$$

and

$$\mathbf{b} = [1, 1, -1] - [-1, 2, 0] = [2, -1, -1].$$

Now $\|\mathbf{a} \times \mathbf{b}\|$ is the area of the parallelogram determined by these two vectors, and the area of the triangle is half the area of the parallelogram, as shown in Figure 4.4. We form the symbolic matrix

$$\begin{bmatrix} \mathbf{i} & \mathbf{j} & \mathbf{k} \\ 3 & -1 & 3 \\ 2 & -1 & -1 \end{bmatrix},$$

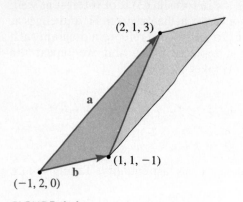

FIGURE 4.4
The triangle constitutes half the parallelogram.

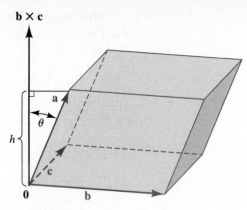

FIGURE 4.5
The box in $\mathbb{R}^3$ determined by a, b, and c.

and we find that $\mathbf{a} \times \mathbf{b} = 4\mathbf{i} + 9\mathbf{j} - \mathbf{k}$. Thus,

$$\|\mathbf{a} \times \mathbf{b}\| = \sqrt{16 + 81 + 1} = \sqrt{98} = 7\sqrt{2},$$

so the area of the triangle is $7\sqrt{2}/2$. ∎

The Volume of a Box

The cross product is useful in finding the volume of the box, or parallelepiped, determined by the three nonzero vectors $\mathbf{a} = [a_1, a_2, a_3]$, $\mathbf{b} = [b_1, b_2, b_3]$, and $\mathbf{c} = [c_1, c_2, c_3]$ in $\mathbb{R}^3$, as shown in Figure 4.5. The volume of the box can be computed by multiplying the area of the base by the altitude h.

HISTORICAL NOTE THE VOLUME INTERPRETATION of a determinant first appeared in a 1773 paper on mechanics by Joseph Louis Lagrange (1736–1813). He noted that if the points M, M', M'' have coordinates (x, y, z), (x', y', z'), (x'', y'', z''), respectively, then the tetrahedron with vertices at the origin and at those three points will have volume

$$\frac{1}{6}[z(x'y'' - y'x'') + z'(yx'' - xy'') + z''(xy' - yx')];$$

that is,

$$\frac{1}{6} \det \begin{bmatrix} x & y & z \\ x' & y' & z' \\ x'' & y'' & z'' \end{bmatrix}.$$

Lagrange was born in Turin, but spent most of his mathematical career in Berlin and in Paris. He contributed important results to such varied fields as the calculus of variation, celestial mechanics, number theory, and the theory of equations. Among his most famous works are the *Treatise on Analytical Mechanics* (1788), in which he presented the various principles of mechanics from a single point of view, and the *Theory of Analytic Functions* (1797), in which he attempted to base the differential calculus on the theory of power series.

We have just seen that the area of the base of the box is equal to $\|\mathbf{b} \times \mathbf{c}\|$, and the altitude can be found by computing

$$h = \|\mathbf{a}\| \, |\cos \theta| = \frac{\|\mathbf{b} \times \mathbf{c}\| \, \|\mathbf{a}\| \, |\cos \theta|}{\|\mathbf{b} \times \mathbf{c}\|} = \frac{|(\mathbf{b} \times \mathbf{c}) \cdot \mathbf{a}|}{\|\mathbf{b} \times \mathbf{c}\|}.$$

The absolute value is used in case $\cos \theta$ is negative. This would be the case if the direction of $\mathbf{b} \times \mathbf{c}$ were opposite to that shown in Figure 4.5. Thus,

$$\text{Volume} = (\text{Area of base})h = \|\mathbf{b} \times \mathbf{c}\| \frac{|(\mathbf{b} \times \mathbf{c}) \cdot \mathbf{a}|}{\|\mathbf{b} \times \mathbf{c}\|} = |(\mathbf{b} \times \mathbf{c}) \cdot \mathbf{a}|.$$

That is, referring to formula (3), which defines $\mathbf{b} \times \mathbf{c}$, we see that

$$\text{Volume} = |a_1(b_2c_3 - b_3c_2) - a_2(b_1c_3 - b_3c_1) + a_3(b_1c_2 - b_2c_1)|. \qquad (4)$$

The number within the absolute value bars is known as a **third-order determinant**. It is the **determinant of the matrix**

$$A = \begin{bmatrix} a_1 & a_2 & a_3 \\ b_1 & b_2 & b_3 \\ c_1 & c_2 & c_3 \end{bmatrix}$$

and is denoted by

$$\det(A) = \begin{vmatrix} a_1 & a_2 & a_3 \\ b_1 & b_2 & b_3 \\ c_1 & c_2 & c_3 \end{vmatrix}.$$

It can be computed as

$$\det(A) = a_1 \begin{vmatrix} b_2 & b_3 \\ c_2 & c_3 \end{vmatrix} - a_2 \begin{vmatrix} b_1 & b_3 \\ c_1 & c_3 \end{vmatrix} + a_3 \begin{vmatrix} b_1 & b_2 \\ c_1 & c_2 \end{vmatrix}. \qquad (5)$$

Notice the similarity of formula (5) to our computation of the cross product $\mathbf{b} \times \mathbf{c}$ in formula (3). We simply replace $\mathbf{i}, \mathbf{j},$ and $\mathbf{k}$ by $a_1, a_2,$ and a_3, respectively.

EXAMPLE 6   Find the determinant of the matrix

$$A = \begin{bmatrix} 2 & 1 & 3 \\ 4 & 1 & 2 \\ 1 & 2 & -3 \end{bmatrix}.$$

SOLUTION   Using formula (5), we have

$$\begin{vmatrix} 2 & 1 & 3 \\ 4 & 1 & 2 \\ 1 & 2 & -3 \end{vmatrix} = 2 \begin{vmatrix} 1 & 2 \\ 2 & -3 \end{vmatrix} - 1 \begin{vmatrix} 4 & 2 \\ 1 & -3 \end{vmatrix} + 3 \begin{vmatrix} 4 & 1 \\ 1 & 2 \end{vmatrix}$$

$$= 2(-7) - (-14) + 3(7) = 21. \qquad \blacksquare$$

EXAMPLE 7   Find the volume of the box with vertex at the origin determined by the vectors $\mathbf{a} = [4, 1, 1]$, $\mathbf{b} = [2, 1, 0]$, and $\mathbf{c} = [0, 2, 3]$, and sketch the box in a figure.

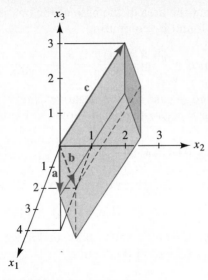

FIGURE 4.6
The box determined by a, b, and c.

SOLUTION   The box is shown in Figure 4.6. Its volume is given by the absolute value of the determinant

$$\begin{vmatrix} 4 & 1 & 1 \\ 2 & 1 & 0 \\ 0 & 2 & 3 \end{vmatrix} = 4\begin{vmatrix} 1 & 0 \\ 2 & 3 \end{vmatrix} - \begin{vmatrix} 2 & 0 \\ 0 & 3 \end{vmatrix} + \begin{vmatrix} 2 & 1 \\ 0 & 2 \end{vmatrix}$$

$$= 10. \qquad \blacksquare$$

The computation of $\det(A)$ in Eq. (5) is referred to as an *expansion of the determinant on the first row*. It is a particular case of a more general procedure for computing $\det(A)$, which is described in the next section.

We list the results of our work with the cross product and some of its algebraic properties in one place as a theorem. The algebraic properties can be checked through computation with the components of the vectors. Example 8 gives an illustration.

THEOREM 4.1   Properties of the Cross Product

Let **a**, **b**, and **c** be vectors in $\mathbb{R}^3$.

1. $\mathbf{b} \times \mathbf{c} = -(\mathbf{c} \times \mathbf{b})$.   Anticommutativity

2. $\mathbf{a} \times (\mathbf{b} \times \mathbf{c})$ is generally different   Nonassociativity of $\times$
   from $(\mathbf{a} \times \mathbf{b}) \times \mathbf{c}$.

3. $\mathbf{a} \times (\mathbf{b} + \mathbf{c}) = (\mathbf{a} \times \mathbf{b}) + (\mathbf{a} \times \mathbf{c})$ Distributive properties
   $(\mathbf{a} + \mathbf{b}) \times \mathbf{c} = (\mathbf{a} \times \mathbf{c}) + (\mathbf{b} \times \mathbf{c})$.

4. $\mathbf{b} \cdot (\mathbf{b} \times \mathbf{c}) = (\mathbf{b} \times \mathbf{c}) \cdot \mathbf{c} = 0$.   Perpendicularity of $\mathbf{b} \times \mathbf{c}$ to both $\mathbf{b}$ and $\mathbf{c}$

5. $\|\mathbf{b} \times \mathbf{c}\| = $ Area of the parallelogram determined by $\mathbf{b}$ and $\mathbf{c}$.   Area property

6. $\mathbf{a} \cdot (\mathbf{b} \times \mathbf{c}) = (\mathbf{a} \times \mathbf{b}) \cdot \mathbf{c} = \pm$ Volume of the box determined by $\mathbf{a}$, $\mathbf{b}$, and $\mathbf{c}$.   Volume property

7. $\mathbf{a} \times (\mathbf{b} \times \mathbf{c}) = (\mathbf{a} \cdot \mathbf{c})\mathbf{b} - (\mathbf{a} \cdot \mathbf{b})\mathbf{c}$.   Formula for computation of $\mathbf{a} \times (\mathbf{b} \times \mathbf{c})$

EXAMPLE 8   Show that $\mathbf{b} \times \mathbf{c} = -(\mathbf{c} \times \mathbf{b})$ for any vectors $\mathbf{b}$ and $\mathbf{c}$ in $\mathbb{R}^3$.

SOLUTION   We compute

$$\mathbf{c} \times \mathbf{b} = \begin{vmatrix} \mathbf{i} & \mathbf{j} & \mathbf{k} \\ c_1 & c_2 & c_3 \\ b_1 & b_2 & b_3 \end{vmatrix}$$

$$= \begin{vmatrix} c_2 & c_3 \\ b_2 & b_3 \end{vmatrix} \mathbf{i} - \begin{vmatrix} c_1 & c_3 \\ b_1 & b_3 \end{vmatrix} \mathbf{j} + \begin{vmatrix} c_1 & c_2 \\ b_1 & b_2 \end{vmatrix} \mathbf{k}.$$

A simple computation shows that interchanging the rows of a 2×2 matrix having a determinant d gives a matrix with determinant $-d$ (see Exercise 11). Comparison of the preceding formula for $\mathbf{c} \times \mathbf{b}$ with the formula for $\mathbf{b} \times \mathbf{c}$ in Eq. (3) then shows that $\mathbf{b} \times \mathbf{c} = -(\mathbf{c} \times \mathbf{b})$.   ∎

SUMMARY

1. A second-order determinant is defined by

$$\begin{vmatrix} a & b \\ c & d \end{vmatrix} = ad - bc.$$

A third-order determinant is defined by Eq. (5).

2. The area of the parallelogram with vertex at the origin determined by nonzero vectors $\mathbf{a}$ and $\mathbf{b}$ in $\mathbb{R}^2$ is the absolute value of the determinant of the matrix having row vectors $\mathbf{a}$ and $\mathbf{b}$.

3. The cross product of vectors $\mathbf{b}$ and $\mathbf{c}$ in $\mathbb{R}^3$ can be computed by using the symbolic determinant

$$\begin{vmatrix} \mathbf{i} & \mathbf{j} & \mathbf{k} \\ b_1 & b_2 & b_3 \\ c_1 & c_2 & c_3 \end{vmatrix}.$$

This vector $\mathbf{b} \times \mathbf{c}$ is perpendicular to both $\mathbf{b}$ and $\mathbf{c}$.

4. The area of the parallelogram determined by nonzero vectors $\mathbf{b}$ and $\mathbf{c}$ in $\mathbb{R}^3$ is $\|\mathbf{b} \times \mathbf{c}\|$.

5. The volume of the box determined by nonzero vectors **a**, **b**, and **c** in $\mathbb{R}^3$ is the absolute value of the determinant of the matrix having row vectors **a**, **b**, and **c**. This determinant is also equal to $\mathbf{a} \cdot (\mathbf{b} \times \mathbf{c})$.

EXERCISES

In Exercises 1–4, find the indicated determinant.

1. $\begin{vmatrix} -1 & 3 \\ 5 & 0 \end{vmatrix}$ **2.** $\begin{vmatrix} -1 & 0 \\ 0 & 7 \end{vmatrix}$

3. $\begin{vmatrix} 0 & -3 \\ 5 & 0 \end{vmatrix}$ **4.** $\begin{vmatrix} 21 & -4 \\ 10 & 7 \end{vmatrix}$

5. Show that the vector $\mathbf{p} = \mathbf{b} \times \mathbf{c}$ given in Eq. (3) is perpendicular to both **b** and **c**.

In Exercises 6–9, find the indicated determinant.

6. $\begin{vmatrix} 1 & 4 & -2 \\ 5 & 13 & 0 \\ 2 & -1 & 3 \end{vmatrix}$ **7.** $\begin{vmatrix} 2 & -5 & 3 \\ 1 & 3 & 4 \\ -2 & 3 & 7 \end{vmatrix}$

8. $\begin{vmatrix} 1 & -2 & 7 \\ 0 & 1 & 4 \\ 1 & 0 & 3 \end{vmatrix}$ **9.** $\begin{vmatrix} 2 & -1 & 1 \\ -1 & 0 & 3 \\ 2 & 1 & -4 \end{vmatrix}$

10. Show by direct computation that:

a. $\begin{vmatrix} a_1 & a_2 & a_3 \\ a_1 & a_2 & a_3 \\ c_1 & c_2 & c_3 \end{vmatrix} = 0;$

b. $\begin{vmatrix} a_1 & a_2 & a_3 \\ b_1 & b_2 & b_3 \\ a_1 & a_2 & a_3 \end{vmatrix} = 0.$

11. Show by direct computation that

$$\begin{vmatrix} a_1 & a_2 \\ b_1 & b_2 \end{vmatrix} = -\begin{vmatrix} b_1 & b_2 \\ a_1 & a_2 \end{vmatrix}.$$

12. Show by direct computation that

$$\begin{vmatrix} a_1 & a_2 & a_3 \\ b_1 & b_2 & b_3 \\ c_1 & c_2 & c_3 \end{vmatrix} = -\begin{vmatrix} a_1 & a_2 & a_3 \\ c_1 & c_2 & c_3 \\ b_1 & b_2 & b_3 \end{vmatrix}.$$

In Exercises 13–18, find $\mathbf{a} \times \mathbf{b}$.

13. $\mathbf{a} = 2\mathbf{i} - \mathbf{j} + 3\mathbf{k}, \mathbf{b} = \mathbf{i} + 2\mathbf{j}$

14. $\mathbf{a} = -5\mathbf{i} + \mathbf{j} + 4\mathbf{k}, \mathbf{b} = 2\mathbf{i} + \mathbf{j} - 3\mathbf{k}$

15. $\mathbf{a} = -\mathbf{i} + 2\mathbf{j} + 4\mathbf{k}, \mathbf{b} = 2\mathbf{i} - 4\mathbf{j} - 8\mathbf{k}$

16. $\mathbf{a} = \mathbf{i} - \mathbf{j} + \mathbf{k}, \mathbf{b} = 3\mathbf{i} - 2\mathbf{j} + 7\mathbf{k}$

17. $\mathbf{a} = 2\mathbf{i} - 3\mathbf{j} + 5\mathbf{k}, \mathbf{b} = 4\mathbf{i} - 5\mathbf{j} + \mathbf{k}$

18. $\mathbf{a} = -2\mathbf{i} + 3\mathbf{j} - \mathbf{k}, \mathbf{b} = 4\mathbf{i} - 6\mathbf{j} + \mathbf{k}$

19. Mark each of the following True or False.

___ **a.** The determinant of a 2×2 matrix is a vector.

___ **b.** If two rows of a 3×3 matrix are interchanged, the sign of the determinant is changed.

___ **c.** The determinant of a 3×3 matrix is zero if two rows of the matrix are parallel vectors in $\mathbb{R}^3$.

___ **d.** In order for the determinant of a 3×3 matrix to be zero, two rows of the matrix must be parallel vectors in $\mathbb{R}^3$.

___ **e.** The determinant of a 3×3 matrix is zero if the points in $\mathbb{R}^3$ given by the rows of the matrix lie in a plane.

___ **f.** The determinant of a 3×3 matrix is zero if the points in $\mathbb{R}^3$ given by the rows of the matrix lie in a plane through the origin.

___ **g.** The parallelogram in $\mathbb{R}^2$ determined by nonzero vectors **a** and **b** is a square if and only if $\mathbf{a} \cdot \mathbf{b} = 0$.

___ **h.** The box in $\mathbb{R}^3$ determined by vectors **a**, **b**, and **c** is a cube if and only if $\mathbf{a} \cdot \mathbf{b} = \mathbf{a} \cdot \mathbf{c} = \mathbf{b} \cdot \mathbf{c} = 0$ and $\mathbf{a} \cdot \mathbf{a} = \mathbf{b} \cdot \mathbf{b} = \mathbf{c} \cdot \mathbf{c}$.

___ **i.** If the angle between vectors **a** and **b** in $\mathbb{R}^3$ is $\pi/4$, then $\|\mathbf{a} \times \mathbf{b}\| = |\mathbf{a} \cdot \mathbf{b}|$.

___ **j.** For any vector **a** in $\mathbb{R}^3$, we have $\|\mathbf{a} \times \mathbf{a}\| = \|\mathbf{a}\|^2$.

In Exercises 20–24, find the area of the parallelogram with vertex at the origin and with the given vectors as edges.

20. $-\mathbf{i} + 4\mathbf{j}$ and $2\mathbf{i} + 3\mathbf{j}$

21. $-5\mathbf{i} + 3\mathbf{j}$ and $\mathbf{i} + 7\mathbf{j}$

22. $\mathbf{i} + 3\mathbf{j} - 5\mathbf{k}$ and $2\mathbf{i} + 4\mathbf{j} - \mathbf{k}$

23. $2\mathbf{i} - \mathbf{j} + \mathbf{k}$ and $\mathbf{i} + 3\mathbf{j} - \mathbf{k}$

24. $\mathbf{i} - 4\mathbf{j} + \mathbf{k}$ and $2\mathbf{i} + 3\mathbf{j} - 2\mathbf{k}$

In Exercises 25–32, find the area of the given geometric configuration.

25. The triangle with vertices $(-1, 2)$, $(3, -1)$, and $(4, 3)$

26. The triangle with vertices $(3, -4)$, $(1, 1)$ and $(5, 7)$

27. The triangle with vertices $(2, 1, -3)$, $(3, 0, 4)$, and $(1, 0, 5)$

28. The triangle with vertices $(3, 1, -2)$, $(1, 4, 5)$, and $(2, 1, -4)$

29. The triangle in the plane $\mathbb{R}^2$ bounded by the lines $y = x$, $y = -3x + 8$, and $3y + 5x = 0$

30. The parallelogram with vertices $(1, 3)$, $(-2, 6)$, $(1, 11)$, and $(4, 8)$

31. The parallelogram with vertices $(1, 0, 1)$, $(3, 1, 4)$, $(0, 2, 9)$, and $(-2, 1, 6)$

32. The parallelogram in the plane $\mathbb{R}^2$ bounded by the lines $x - 2y = 3$, $x - 2y = 10$, $2x + 3y = -1$, and $2x + 3y = -8$

In Exercises 33–36, find $\mathbf{a} \cdot (\mathbf{b} \times \mathbf{c})$ and $\mathbf{a} \times (\mathbf{b} \times \mathbf{c})$.

33. $\mathbf{a} = \mathbf{i} + 2\mathbf{j} - 3\mathbf{k}$, $\mathbf{b} = 4\mathbf{i} - \mathbf{j} + 2\mathbf{k}$, $\mathbf{c} = 3\mathbf{i} + \mathbf{k}$

34. $\mathbf{a} = -\mathbf{i} + \mathbf{j} + 2\mathbf{k}$, $\mathbf{b} = \mathbf{i} + \mathbf{k}$, $\mathbf{c} = 3\mathbf{i} - 2\mathbf{j} + 5\mathbf{k}$

35. $\mathbf{a} = \mathbf{i} - 3\mathbf{k}$, $\mathbf{b} = -\mathbf{i} + 4\mathbf{j}$, $\mathbf{c} = \mathbf{i} + \mathbf{j} + \mathbf{k}$

36. $\mathbf{a} = 4\mathbf{i} - \mathbf{j} + 2\mathbf{k}$, $\mathbf{b} = 3\mathbf{i} + 5\mathbf{j} - 2\mathbf{k}$, $\mathbf{c} = \mathbf{i} - 3\mathbf{j} + \mathbf{k}$

In Exercises 37–40, find the volume of the box having the given vectors as adjacent edges.

37. $-\mathbf{i} + 4\mathbf{j} + 7\mathbf{k}$, $3\mathbf{i} - 2\mathbf{j} - \mathbf{k}$, $4\mathbf{i} + 2\mathbf{k}$

38. $2\mathbf{i} + \mathbf{j} - 4\mathbf{k}$, $3\mathbf{i} - \mathbf{j} + 2\mathbf{k}$, $\mathbf{i} + 3\mathbf{j} - 8\mathbf{k}$

39. $-2\mathbf{i} + \mathbf{j}$, $3\mathbf{i} - 4\mathbf{j} + \mathbf{k}$, $\mathbf{i} - 2\mathbf{k}$

40. $3\mathbf{i} - \mathbf{j} + 4\mathbf{k}$, $\mathbf{i} - 2\mathbf{j} + 7\mathbf{k}$, $5\mathbf{i} - 3\mathbf{j} + 10\mathbf{k}$

In Exercises 41–44, find the volume of the tetrahedron having the given vertices. (Consider how the volume of a tetrahedron having three vectors from one point as edges is related to the volume of the box having the same three vectors as adjacent edges.)

41. $(-3, 0, 1)$, $(4, 2, 1)$, $(0, 1, 7)$, $(1, 1, 1)$

42. $(0, 1, 1)$, $(8, 2, -7)$, $(3, 1, 6)$, $(-4, -2, 0)$

43. $(-1, 1, 2)$, $(3, 1, 4)$, $(-1, 6, 0)$, $(2, -1, 5)$

44. $(-1, 2, 4)$, $(2, -3, 0)$, $(-4, 2, -1)$, $(0, 3, -2)$

In Exercises 45–48, use a determinant to ascertain whether the given points lie on a line in $\mathbb{R}^2$. [HINT: What is the area of a "parallelogram" with collinear vertices?]

45. $(0, 0)$, $(3, 5)$, $(6, 9)$

46. $(0, 0)$, $(4, 2)$, $(-6, -3)$

47. $(1, 5)$, $(3, 7)$, $(-3, 1)$

48. $(2, 3)$, $(1, -4)$, $(6, 2)$

In Exercises 49–52, use a determinant to ascertain whether the given points lie in a plane in $\mathbb{R}^3$. [HINT: What is the "volume" of a box with coplanar vertices?]

49. $(0, 0, 0)$, $(1, 4, 3)$, $(2, 5, 8)$, $(-1, 2, -5)$

50. $(0, 0, 0)$, $(2, 1, 1)$, $(3, -2, 1)$, $(-1, 2, 3)$

51. $(1, -1, 3)$, $(4, 2, 3)$, $(3, 1, -2)$, $(5, 5, -5)$

52. $(1, 2, 1)$, $(3, 3, 4)$, $(2, 2, 2)$, $(4, 3, 5)$

Let $\mathbf{a}$, $\mathbf{b}$, and $\mathbf{c}$ be any vectors in $\mathbb{R}^3$. In Exercises 53–56, simplify the given expression.

53. $\mathbf{a} \cdot (\mathbf{a} \times \mathbf{b})$

54. $(\mathbf{b} \times \mathbf{c}) - (\mathbf{c} \times \mathbf{b})$

55. $\|\mathbf{a} \times \mathbf{b}\|^2 + (\mathbf{a} \cdot \mathbf{b})^2$

56. $\mathbf{a} \times (\mathbf{b} \times \mathbf{c}) + \mathbf{b} \times (\mathbf{c} \times \mathbf{a}) + \mathbf{c} \times (\mathbf{a} \times \mathbf{b})$

57. Prove property (2) of Theorem 4.1.

58. Prove property (3) of Theorem 4.1.

59. Prove property (6) of Theorem 4.1.

60. Option 7 of the routine VECTGRPH in LINTEK provides drill on the determinant of a 2×2 matrix as the area of the parallelogram determined by its row vectors, with an associated plus or minus sign. Run this option until you can regularly achieve a score of 80% or better.

*MATLAB has a function **det(A)** which gives the determinant of a matrix A. In Exercises 61–63, use the routine MATCOMP in LINTEK or MATLAB to find the volume of the box having the given vectors in $\mathbb{R}^3$ as adjacent edges. (We have not supplied matrix files for these problems.)*

61. $-\mathbf{i} + 7\mathbf{j} + 3\mathbf{k}$, $4\mathbf{i} + 23\mathbf{j} - 13\mathbf{k}$, $12\mathbf{i} - 17\mathbf{j} - 31\mathbf{k}$

62. $4.1\mathbf{i} - 2.3\mathbf{k}$, $5.3\mathbf{j} - 2.1\mathbf{k}$, $6.1\mathbf{i} + 5.7\mathbf{j}$

63. $2.13\mathbf{i} + 4.71\mathbf{j} - 3.62\mathbf{k}$, $5\mathbf{i} - 3.2\mathbf{j} + 6.32\mathbf{k}$, $8.3\mathbf{i} - 0.45\mathbf{j} + 1.13\mathbf{k}$

MATLAB

M1. Enter the data vectors $\mathbf{x} = [1\ \ 5\ \ 7]$ and $\mathbf{y} = [-3\ \ 2\ \ 4]$ into MATLAB. Then enter a line **crossxy = [** **]**, which will compute the cross product $\mathbf{x} \times \mathbf{y}$ of vectors $[\mathbf{x}(1)\ \ \mathbf{x}(2)\ \ \mathbf{x}(3)]$ and $[\mathbf{y}(1)\ \ \mathbf{y}(2)\ \ \mathbf{y}(3)]$. [HINT: The first component in [ ] will be $\mathbf{x}(2)*\mathbf{y}(3)-\mathbf{y}(2)*\mathbf{x}(3)$.] Be sure you use no spaces except one between the vector components. Check that the value given for **crossxy** is the correct vector $6\mathbf{i} - 25\mathbf{j} + 17\mathbf{k}$ for the data vectors entered.

M2. Use the **norm** function in MATLAB to find the area of the parallelogram in $\mathbb{R}^3$ having the vectors $\mathbf{x}$ and $\mathbf{y}$ in the preceding exercise as adjacent edges.

M3. Enter the vectors $\mathbf{x} = 4.2\mathbf{i} - 3.7\mathbf{j} + 5.6\mathbf{k}$ and $\mathbf{y} = -7.3\mathbf{i} + 4.5\mathbf{j} + 11.4\mathbf{k}$.
   a. Using the up-arrow key to access your line defining **crossxy**, find $\mathbf{x} \times \mathbf{y}$.
   b. Find the area of the parallelogram in $\mathbb{R}^3$ having $\mathbf{x}$ and $\mathbf{y}$ as adjacent edges.

M4. Find the area of the triangle in $\mathbb{R}^3$ having vertices $(-1.2, 3.4, -6.7)$, $(2.3, -5.2, 9.4)$, and $(3.1, 8.3, -3.6)$. [HINT: Enter vectors **a**, **b**, and **c** from the origin to these points and set $\mathbf{x}$ and $\mathbf{y}$ equal to appropriate differences of them.]

NOTE: If you want to add a function **cross(x, y)** to your own personal MATLAB, do so following a procedure analogous to that described at the very end of Section 1.2 for adding the function **angl(x, y)**.

4.2

THE DETERMINANT OF A SQUARE MATRIX

The Definition

We defined a third-order determinant in terms of second-order determinants in Eq. (5) on page 245. A second-order determinant can be defined in terms of first-order determinants if we interpret the determinant of a 1×1 matrix to be its sole entry. We define an nth-order determinant in terms of determinants of order $n - 1$. In order to facilitate this, we introduce the **minor matrix** A_{ij} of an $n \times n$ matrix $A = [a_{ij}]$; it is the $(n - 1) \times (n - 1)$ matrix obtained by crossing out the ith row and jth column of A. The minor matrix is the portion shown in color shading in the matrix

$$A_{ij} = \begin{bmatrix} a_{11} & \cdots & a_{1j} & \cdots & a_{1n} \\ \vdots & & \vdots & & \vdots \\ a_{i1} & \cdots & a_{ij} & \cdots & a_{in} \\ \vdots & & \vdots & & \vdots \\ a_{n1} & \cdots & a_{nj} & \cdots & a_{nn} \end{bmatrix} \; i\text{th row.} \tag{1}$$

j**th column**

Using $|A_{ij}|$ as notation for the determinant of the minor matrix A_{ij}, we can express the determinant of a 3×3 matrix A as

$$\begin{vmatrix} a_{11} & a_{12} & a_{13} \\ a_{21} & a_{22} & a_{23} \\ a_{31} & a_{32} & a_{33} \end{vmatrix} = a_{11}|A_{11}| - a_{12}|A_{12}| + a_{13}|A_{13}|.$$

The numbers $a'_{11} = |A_{11}|$, $a'_{12} = -|A_{12}|$, and $a'_{13} = |A_{13}|$ are appropriately called the **cofactors** of a_{11}, a_{12}, and a_{13}. We now proceed to define the *determinant of any square matrix*, using mathematical induction. (See Appendix A for a discussion of mathematical induction.)

HISTORICAL NOTE   THE FIRST APPEARANCE OF THE DETERMINANT OF A SQUARE MATRIX in Western Europe occurred in a 1683 letter from Gottfried von Leibniz (1646–1716) to the Marquis de L'Hôpital (1661–1704). Leibniz wrote a system of three equations in two unknowns with abstract "numerical" coefficients,

$$10 + 11x + 12y = 0$$
$$20 + 21x + 22y = 0$$
$$30 + 31x + 32y = 0,$$

in which he noted that each coefficient number has "two characters, the first marking in which equation it occurs, the second marking which letter it belongs to." He then proceeded to eliminate first y and then x to show that the criterion for the system of equations to have a solution is that

$$10 \cdot 21 \cdot 32 + 11 \cdot 22 \cdot 30 + 12 \cdot 20 \cdot 31 = 10 \cdot 22 \cdot 31 + 11 \cdot 20 \cdot 32 + 12 \cdot 21 \cdot 30.$$

This is equivalent to the modern condition that the determinant of the matrix of coefficients must be zero.

Determinants also appeared in the contemporaneous work of the Japanese mathematician Seki Takakazu (1642–1708). Seki's manuscript of 1683 includes his detailed calculations of determinants of 2×2, 3×3, and 4×4 matrices—although his version was the negative of the version used today. Seki applied the determinant to the solving of certain types of equations, but evidently not to the solving of systems of linear equations. Seki spent most of his life as an accountant working for two feudal lords, Tokugawa Tsunashige and Tokugawa Tsunatoyo, in Kofu, a city in the prefecture of Yamanashi, west of Tokyo.

DEFINITION 4.1 Cofactors and Determinants

The **determinant** of a 1×1 matrix is its sole entry; it is a **first-order** determinant. Let $n > 1$, and assume that determinants of order less than n have been defined. Let $A = [a_{ij}]$ be an $n \times n$ matrix. The **cofactor** of a_{ij} in A is

$$a'_{ij} = (-1)^{i+j} \det(A_{ij}), \tag{2}$$

where A_{ij} given in Eq. (1) is the minor matrix of A. The **determinant** of A is

$$\det(A) = \begin{vmatrix} a_{11} & a_{12} & \cdots & a_{1n} \\ a_{21} & a_{22} & \cdots & a_{2n} \\ \cdot & \cdot & & \cdot \\ \cdot & \cdot & & \cdot \\ \cdot & \cdot & & \cdot \\ a_{n1} & a_{n2} & \cdots & a_{nn} \end{vmatrix}$$

$$= a_{11}a'_{11} + a_{12}a'_{12} + \cdots + a_{1n}a'_{1n}, \tag{3}$$

and is an **nth-order** determinant.

In addition to the notation $\det(A)$ for the determinant of A, we will sometimes use $|A|$ when determinants appear in equations, to make the equations easier to read.

EXAMPLE 1   Find the cofactor of the entry 3 in the matrix

$$A = \begin{bmatrix} 2 & 1 & 0 & 1 \\ 3 & 2 & 1 & 2 \\ 4 & 0 & 1 & 4 \\ 1 & 0 & 2 & 1 \end{bmatrix}.$$

SOLUTION   Because 3 is in the row 2, column 1 position of A, we cross out the second row and first column of A and find the cofactor of 3 to be

$$a'_{21} = (-1)^{2+1} \begin{vmatrix} 1 & 0 & 1 \\ 0 & 1 & 4 \\ 0 & 2 & 1 \end{vmatrix} = -\begin{vmatrix} 1 & 4 \\ 2 & 1 \end{vmatrix} - \begin{vmatrix} 0 & 1 \\ 0 & 2 \end{vmatrix}$$

$$= -(-7 - 0) = 7. \qquad \blacksquare$$

EXAMPLE 2   Use Eq. (3) in Definition 4.1 to find the determinant of the matrix

$$A = \begin{bmatrix} 5 & -2 & 4 & -1 \\ 0 & 1 & 5 & 2 \\ 1 & 2 & 0 & 1 \\ -3 & 1 & -1 & 1 \end{bmatrix}.$$

SOLUTION   We have

$$\det(A) = \begin{vmatrix} 5 & -2 & 4 & -1 \\ 0 & 1 & 5 & 2 \\ 1 & 2 & 0 & 1 \\ -3 & 1 & -1 & 1 \end{vmatrix}$$

$$= 5(-1)^2 \begin{vmatrix} 1 & 5 & 2 \\ 2 & 0 & 1 \\ 1 & -1 & 1 \end{vmatrix} + (-2)(-1)^3 \begin{vmatrix} 0 & 5 & 2 \\ 1 & 0 & 1 \\ -3 & -1 & 1 \end{vmatrix}$$

$$+ 4(-1)^4 \begin{vmatrix} 0 & 1 & 2 \\ 1 & 2 & 1 \\ -3 & 1 & 1 \end{vmatrix} + (-1)(-1)^5 \begin{vmatrix} 0 & 1 & 5 \\ 1 & 2 & 0 \\ -3 & 1 & -1 \end{vmatrix}.$$

Computing the third-order determinants, we have

$$\begin{vmatrix} 1 & 5 & 2 \\ 2 & 0 & 1 \\ 1 & -1 & 1 \end{vmatrix} = 1 \begin{vmatrix} 0 & 1 \\ -1 & 1 \end{vmatrix} - 5 \begin{vmatrix} 2 & 1 \\ 1 & 1 \end{vmatrix} + 2 \begin{vmatrix} 2 & 0 \\ 1 & -1 \end{vmatrix}$$

$$= 1(1) - 5(1) + 2(-2) = -8;$$

$$\begin{vmatrix} 0 & 5 & 2 \\ 1 & 0 & 1 \\ -3 & -1 & 1 \end{vmatrix} = 0 \begin{vmatrix} 0 & 1 \\ -1 & 1 \end{vmatrix} - 5 \begin{vmatrix} 1 & 1 \\ -3 & 1 \end{vmatrix} + 2 \begin{vmatrix} 1 & 0 \\ -3 & -1 \end{vmatrix}$$

$$= 0(1) - 5(4) + 2(-1) = -22;$$

$$\begin{vmatrix} 0 & 1 & 2 \\ 1 & 2 & 1 \\ -3 & 1 & 1 \end{vmatrix} = 0 \begin{vmatrix} 2 & 1 \\ 1 & 1 \end{vmatrix} - 1 \begin{vmatrix} 1 & 1 \\ -3 & 1 \end{vmatrix} + 2 \begin{vmatrix} 1 & 2 \\ -3 & 1 \end{vmatrix}$$

$$= 0(1) - 1(4) + 2(7) = 10;$$

$$\begin{vmatrix} 0 & 1 & 5 \\ 1 & 2 & 0 \\ -3 & 1 & -1 \end{vmatrix} = 0 \begin{vmatrix} 2 & 0 \\ 1 & -1 \end{vmatrix} - 1 \begin{vmatrix} 1 & 0 \\ -3 & -1 \end{vmatrix} + 5 \begin{vmatrix} 1 & 2 \\ -3 & 1 \end{vmatrix}$$

$$= 0(-2) - 1(-1) + 5(7) = 36.$$

Therefore, $\det(A) = 5(-8) + 2(-22) + 4(10) + 1(36) = -8.$ ∎

The preceding example makes one thing plain:

Computation of determinants of matrices of even moderate size using only Definition 4.1 is a tremendous chore.

According to modern astronomical theory, our solar system would be dead long before a present-day computer could find the determinant of a 50×50 matrix using just the inductive Definition 4.1. (See Exercise 37.) Section 4.3 gives an alternative, efficient method for computing determinants.

Determinants play an important role in calculus for functions of several variables. In some cases where primary values depend on some secondary values, the single number that best measures the rate at which the primary values change as the secondary values change is given by a determinant. This is closely connected with the geometric interpretation of a determinant as a *volume*. In Section 4.1, we motivated the determinant of a 2×2 matrix by using area, and we motivated the determinant of a 3×3 matrix by using volume. In Section 4.4, we show how an nth-order determinant can be interpreted as a "volume."

It is desirable to have an efficient way to compute a determinant. We will spend the remainder of this section developing properties of determinants that will enable us to find a good method for their computation. The computation of $\det(A)$ using Eq. (3) is called **expansion by minors** on the first row. Appendix B gives a proof by mathematical induction that $\det(A)$ can be obtained by using an expansion by minors *on any row or on any column*. We state this more precisely in a theorem.

THEOREM 4.2 General Expansion by Minors

Let A be an $n \times n$ matrix, and let r and s be any selections from the list of numbers $1, 2, \ldots, n$. Then

$$\det(A) = a_{r1}a'_{r1} + a_{r2}a'_{r2} + \cdots + a_{rn}a'_{rn}, \tag{4}$$

and also

$$\det(A) = a_{1s}a'_{1s} + a_{2s}a'_{2s} + \cdots + a_{ns}a'_{ns}, \tag{5}$$

where a'_{ij} is the cofactor of a_{ij} given in Definition 4.1.

Equation (4) is the **expansion of** $\det(A)$ **by minors on the rth row** of A, and Eq. (5) is the **expansion of** $\det(A)$ **by minors on the sth column** of A. Theorem 4.2 thus says that $\det(A)$ can be found by expanding by minors on any row or on any column of A.

EXAMPLE 3   Find the determinant of the matrix

$$A = \begin{bmatrix} 3 & 2 & 0 & 1 & 3 \\ -2 & 4 & 1 & 2 & 1 \\ 0 & -1 & 0 & 1 & -5 \\ -1 & 2 & 0 & -1 & 2 \\ 0 & 0 & 0 & 0 & 2 \end{bmatrix}.$$

SOLUTION   A recursive computation such as the one in Definition 4.1 is still the only way we have of computing $\det(A)$ at the moment, but we can expedite the computation if we expand by minors at each step on the row or column containing the most zeros. We have

$$\det(A) = 2(-1)^{5+5} \begin{vmatrix} 3 & 2 & 0 & 1 \\ -2 & 4 & 1 & 2 \\ 0 & -1 & 0 & 1 \\ -1 & 2 & 0 & -1 \end{vmatrix}$$ Expanding on row 5

$$= (2)(1)(-1)^{2+3} \begin{vmatrix} 3 & 2 & 1 \\ 0 & -1 & 1 \\ -1 & 2 & -1 \end{vmatrix}$$ Expanding on column 3

$$= -2\left(3(-1)^{1+1} \begin{vmatrix} -1 & 1 \\ 2 & -1 \end{vmatrix} - 1(-1)^{3+1} \begin{vmatrix} 2 & 1 \\ -1 & 1 \end{vmatrix}\right)$$ Expanding on column 1

$$= -2(3(1-2) - 1(2+1)) = -2(-3-3) = 12.$$ ∎

EXAMPLE 4   Show that the determinant of an upper- or lower-triangular square matrix is the product of its diagonal elements.

SOLUTION   We work with an upper-triangular matrix, the other case being analogous. If

$$U = \begin{bmatrix} u_{11} & u_{12} & \cdots & u_{1n} \\ & u_{22} & \cdots & u_{2n} \\ & & \ddots & \vdots \\ \mathbf{0} & & & u_{nn} \end{bmatrix}$$

is an upper-triangular matrix, then by expanding on first columns each time, we have

$$\det(U) = u_{11} \begin{vmatrix} u_{22} & u_{23} & \cdots & u_{2n} \\ 0 & u_{33} & \cdots & u_{3n} \\ \vdots & & & \vdots \\ 0 & 0 & \cdots & u_{nn} \end{vmatrix} = u_{11}u_{22} \begin{vmatrix} u_{33} & \cdots & u_{3n} \\ 0 & \cdots & u_{4n} \\ \vdots & & \vdots \\ 0 & \cdots & u_{nn} \end{vmatrix}$$

$$= \cdots = u_{11}u_{22} \cdots u_{nn}.$$ ∎

Properties of the Determinant

Using Theorem 4.2, we can establish several properties of the determinant that will be of tremendous help in its computation. Because Definition 4.1 was an inductive one, we use mathematical induction as we start to prove properties of the determinant. (Again, mathematical induction is reviewed in Appendix A.) We will always consider A to be an $n \times n$ matrix.

PROPERTY 1   The Transpose Property

For any square matrix A, we have $\det(A) = \det(A^T)$.

PROOF   Verification of this property is trivial for determinants of orders 1 or 2. Let $n > 2$, and assume that the property holds for square matrices of size smaller than $n \times n$. We proceed to prove Property 1 for an $n \times n$ matrix A. We have

$$\det(A) = a_{11}|A_{11}| - a_{12}|A_{12}| + \cdots + (-1)^{n+1}a_{1n}|A_{1n}|. \quad \text{Expanding on row 1 of } A$$

Writing $B = A^T$, we have

$$\det(B) = b_{11}|B_{11}| - b_{21}|B_{21}| + \cdots + (-1)^{n+1}b_{n1}|B_{n1}|. \quad \text{Expanding on column 1 of } B$$

However, $a_{1j} = b_{j1}$ and $B_{j1} = A_{1j}{}^T$, because $B = A^T$. Applying our induction hypothesis to the $(n-1)$st-order determinant $|A_{1j}|$, we have $|A_{1j}| = |B_{j1}|$. We conclude that $\det(A) = \det(B) = \det(A^T)$.   ▲

This transpose property has a very useful consequence. It guarantees that any property of the determinant involving rows of a matrix is equally valid if we replace *rows* by *columns* in the statement of that property. For example, the next property has an analogue for columns.

PROPERTY 2   The Row-Interchange Property

If two different rows of a square matrix A are interchanged, the determinant of the resulting matrix is $-\det(A)$.

PROOF   Again we find that the proof is trivial for the case $n = 2$. Assume that $n > 2$, and that this row-interchange property holds for matrices of size smaller than $n \times n$. Let A be an $n \times n$ matrix, and let B be the matrix obtained from A by interchanging the ith and rth rows, leaving the other rows unchanged. Because $n > 2$, we can choose a kth row for expansion by minors, where k is different from both r and i. Consider the cofactors

$$(-1)^{k+j}|A_{kj}| \quad \text{and} \quad (-1)^{k+j}|B_{kj}|.$$

These numbers must have opposite signs, by our induction hypothesis, because the minor matrices A_{kj} and B_{kj} have size $(n-1) \times (n-1)$, and B_{kj} can be obtained from A_{kj} by interchanging two rows. That is, $|B_{kj}| = -|A_{kj}|$. Expanding by minors on the kth row to find $\det(A)$ and $\det(B)$, we see that $\det(A) = -\det(B)$.   ▲

PROPERTY 3   The Equal-Rows Property

If two rows of a square matrix A are equal, then $\det(A) = 0$.

PROOF   Let B be the matrix obtained from A by interchanging the two equal rows of A. By the row-interchange property, we have $\det(B) = -\det(A)$. On the other hand, $B = A$, so $\det(A) = -\det(A)$. Therefore, $\det(A) = 0$.   ▲

PROPERTY 4   The Scalar-Multiplication Property

If a single row of a square matrix A is multiplied by a scalar r, the determinant of the resulting matrix is $r \cdot \det(A)$.

PROOF   Let r be any scalar, and let B be the matrix obtained from A by replacing the kth row $[a_{k1}, a_{k2}, \ldots, a_{kn}]$ of A by $[ra_{k1}, ra_{k2}, \ldots, ra_{kn}]$. Since the rows of B are equal to those of A except possibly for the kth row, it follows that the minor matrices A_{kj} and B_{kj} are equal for each j. Therefore, $a'_{kj} = b'_{kj}$, and computing $\det(B)$ by expanding by minors on the kth row, we have

$$
\begin{aligned}
\det(B) &= b_{k1}b'_{k1} + b_{k2}b'_{k2} + \cdots + b_{kn}b'_{kn} \\
&= r \cdot a_{k1}a'_{k1} + r \cdot a_{k2}a'_{k2} + \cdots + r \cdot a_{kn}a'_{kn} \\
&= r \cdot \det(A).
\end{aligned}
$$
   ▲

HISTORICAL NOTE   THE THEORY OF DETERMINANTS grew from the efforts of many mathematicians of the late eighteenth and early nineteenth centuries. Besides Gabriel Cramer, whose work we will discuss in the note on page 267, Etienne Bezout (1739–1783) in 1764 and Alexandre-Theophile Vandermonde (1735–1796) in 1771 gave various methods for computing determinants. In a work on integral calculus, Pierre Simon Laplace (1749–1827) had to deal with systems of linear equations. He repeated the work of Cramer, but he also stated and proved the rule that interchanging two adjacent columns of the determinant changes the sign and showed that a determinant with two equal columns will be 0.

The most complete of the early works on determinants is that of Augustin-Louis Cauchy (1789–1857) in 1812. In this work, Cauchy introduced the name *determinant* to replace several older terms, used our current double-subscript notation for a square array of numbers, defined the array of adjoints (or minors) to a given array, and showed that one can calculate the determinant by expanding on any row or column. In addition, Cauchy re-proved many of the standard theorems on determinants that had been more or less known for the past 50 years.

Cauchy was the most prolific mathematician of the nineteenth century, contributing to such areas as complex analysis, calculus, differential equations, and mechanics. In particular, he wrote the first calculus text using our modern ϵ, δ-approach to continuity. Politically he was a conservative; when the July Revolution of 1830 replaced the Bourbon king Charles X with the Orleans king Louis-Philippe, Cauchy refused to take the oath of allegiance, thereby forfeiting his chairs at the Ecole Polytechnique and the Collège de France and going into exile in Turin and Prague.

EXAMPLE 5   Find the determinant of the matrix

$$A = \begin{bmatrix} 2 & 1 & 3 & 4 & 2 \\ 6 & 2 & 1 & 4 & 1 \\ 6 & 3 & 9 & 12 & 6 \\ 2 & 1 & 3 & 4 & 1 \\ 1 & 4 & 2 & 1 & 1 \end{bmatrix}.$$

SOLUTION   We note that the third row of A is three times the first row. Therefore, we have

$$\det(A) = 3 \begin{vmatrix} 2 & 1 & 3 & 4 & 2 \\ 6 & 2 & 1 & 4 & 1 \\ 2 & 1 & 3 & 4 & 2 \\ 2 & 1 & 3 & 4 & 1 \\ 1 & 4 & 2 & 1 & 1 \end{vmatrix} \qquad \text{Property 4}$$

$$= 3(0) = 0. \qquad \text{Property 3} \qquad\blacksquare$$

The row-interchange property and the scalar-multiplication property indicate how the determinant of a matrix changes when two of the three elementary row operations are used. The next property deals with the most complicated of the elementary row operations, and lies at the heart of the efficient computation of determinants given in the next section.

PROPERTY 5   The Row-Addition Property

> If the product of one row of a square matrix A by a scalar is added to a different row of A, the determinant of the resulting matrix is the same as $\det(A)$.

PROOF   Let $\mathbf{a}_i = [a_{i1}, a_{i2}, \dots, a_{in}]$ be the ith row of A. Suppose that $r\mathbf{a}_i$ is added to the kth row $\mathbf{a}_k$ of A, where r is any scalar and $k \neq i$. We obtain a matrix B whose rows are the same as the rows of A except possibly for the kth row, which is

$$\mathbf{b}_k = [ra_{i1} + a_{k1}, ra_{i2} + a_{k2}, \dots, ra_{in} + a_{kn}].$$

Clearly the minor matrices A_{kj} and B_{kj} are equal for each j. Therefore, $a'_{kj} = b'_{kj}$, and computing $\det(B)$ by expanding by minors on the kth row, we have

$$\begin{aligned} \det(B) &= b_{k1}b'_{k1} + b_{k2}b'_{k2} + \cdots + b_{kn}b'_{kn} \\ &= (ra_{i1} + a_{k1})a'_{k1} + (ra_{i2} + a_{k2})a'_{k2} + \cdots + (ra_{in} + a_{kn})a'_{kn} \\ &= (ra_{i1}a'_{k1} + ra_{i2}a'_{k2} + \cdots + ra_{in}a'_{kn}) \\ &\quad + (a_{k1}a'_{k1} + a_{k2}a'_{k2} + \cdots + a_{kn}a'_{kn}) \\ &= r \cdot \det(C) + \det(A), \end{aligned}$$

where C is the matrix obtained from A by replacing the kth row of A with the ith row of A. Because C has two equal rows, its determinant is zero, so $\det(B) = \det(A)$. ▲

We now know how the three types of elementary row operations affect the determinant of a matrix A. In particular, if we reduce A to an echelon form H and avoid the use of row scaling, then $\det(A) = \pm\det(H)$, and $\det(H)$ is the product of its diagonal entries. (See Example 4.) We know that an echelon form of A has only nonzero entries on its main diagonal if and only if A is invertible. Thus, $\det(A) \neq 0$ if and only if A is invertible. We state this new condition for invertibility as a theorem.

THEOREM 4.3 Determinant Criterion for Invertibility

A square matrix A is invertible if and only if $\det(A) \neq 0$. Equivalently, A is singular if and only if $\det(A) = 0$.

We conclude with a multiplicative property of determinants. Section 4.4 indicates that this property has important geometric significance. Rather than labeling it "Property 6," we emphasize its increased level of importance over Properties 1 through 5 by stating it as a theorem.

THEOREM 4.4 The Multiplicative Property

If A and B are $n \times n$ matrices, then $\det(AB) = \det(A) \cdot \det(B)$.

PROOF First we note that, if A is a diagonal matrix, the result follows easily, because the product

$$\begin{bmatrix} a_{11} & & & \\ & a_{22} & & \mathbf{0} \\ & & \ddots & \\ \mathbf{0} & & & a_{nn} \end{bmatrix} \begin{bmatrix} b_{11} & b_{12} & \cdots & b_{1n} \\ b_{21} & b_{22} & \cdots & b_{2n} \\ & & \vdots & \\ b_{n1} & b_{n2} & \cdots & b_{nn} \end{bmatrix}$$

has its ith row equal to a_{ii} times the ith row of B. Using the scalar-multiplication property in each of these rows, we obtain

$$\det(AB) = (a_{11}a_{22} \cdots a_{nn}) \cdot \det(B) = \det(A) \cdot \det(B).$$

To deal with the nondiagonal case, we begin by reducing the problem to the case in which the matrix A is invertible. For if A is singular, then so is AB

(see Exercise 30); so both A and AB have a zero determinant, by Theorem 4.3, and $\det(A) \cdot \det(B) = 0$, too.

If we assume that A is invertible, it can be row-reduced through row-interchange and row-addition operations to an upper-triangular matrix with nonzero entries on the diagonal. We continue such row reduction analogous to the Gauss–Jordan method but without making pivots 1, and finally we reduce A to a diagonal matrix D with nonzero diagonal entries. We can write $D = EA$, where E is the product of elementary matrices corresponding to the row interchanges and row additions used to reduce A to D. By the properties of determinants, we have $\det(A) = (-1)^r \cdot \det(D)$, where r is the number of row interchanges. The same sequence of steps will reduce the matrix AB to the matrix $E(AB) = (EA)B = DB$, so $\det(AB) = (-1)^r \cdot \det(DB)$. Therefore,

$$\det(AB) = (-1)^r \cdot \det(DB) = (-1)^r \cdot \det(D) \cdot \det(B) = \det(A) \cdot \det(B),$$

and the proof is complete. ▲

EXAMPLE 6   Find $\det(A)$ if

$$A = \begin{bmatrix} 2 & 0 & 0 \\ 1 & 3 & 0 \\ 4 & 2 & 1 \end{bmatrix} \begin{bmatrix} 1 & 2 & 3 \\ 0 & 1 & 2 \\ 0 & 0 & 2 \end{bmatrix}.$$

SOLUTION   Because the determinant of an upper- or lower-triangular matrix is the product of the diagonal elements (see Example 4), Theorem 4.4 shows that

$$\det(A) = \begin{vmatrix} 2 & 0 & 0 \\ 1 & 3 & 0 \\ 4 & 2 & 1 \end{vmatrix} \begin{vmatrix} 1 & 2 & 3 \\ 0 & 1 & 2 \\ 0 & 0 & 2 \end{vmatrix} = (6)(2) = 12.$$ ■

EXAMPLE 7   If $\det(A) = 3$, find $\det(A^5)$ and $\det(A^{-1})$.

SOLUTION   Applying Theorem 4.4 several times, we have

$$\det(A^5) = [\det(A)]^5 = 3^5 = 243.$$

From $AA^{-1} = I$, we obtain

$$1 = \det(I) = \det(AA^{-1}) = \det(A)\det(A^{-1}) = 3[\det(A^{-1})],$$

so $\det(A^{-1}) = \frac{1}{3}$. ■

Exercise 31 asks you to prove that if A is invertible, then

$$\det(A^{-1}) = \frac{1}{\det(A)}.$$

SUMMARY

1. The cofactor of an element a_{ij} in a square matrix A is $(-1)^{i+j}|A_{ij}|$, where A_{ij} is the matrix obtained from A by deleting the ith row and the jth column.

2. The *determinant* of an $n \times n$ matrix may be defined inductively by expansion by minors on the first row. The determinant can be computed by expansion by minors on any row or on any column; it is the sum of the products of the entries in that row or column by the cofactors of the entries. For large matrices, such a computation is hopelessly long.

3. The elementary row operations have the following effect on the determinant of a square matrix A.
    a. If two different rows of A are interchanged, the sign of the determinant is changed.
    b. If a single row of A is multiplied by a scalar, the determinant is multiplied by the scalar.
    c. If a multiple of one row is added to a different row, the determinant is not changed.

4. We have $\det(A) = \det(A^T)$. As a consequence, the properties just listed for elementary row operations are also true for elementary column operations.

5. If two rows or two columns of a matrix are the same, the determinant of the matrix is zero.

6. The determinant of an upper-triangular matrix or of a lower-triangular matrix is the product of the diagonal entries.

7. An $n \times n$ matrix A is invertible if and only if $\det(A) \neq 0$.

8. If A and B are $n \times n$ matrices, then $\det(AB) = \det(A) \cdot \det(B)$.

EXERCISES

In Exercises 1–10, find the determinant of the given matrix.

1. $\begin{bmatrix} 5 & 2 & 1 \\ 1 & -1 & 4 \\ 3 & 0 & 2 \end{bmatrix}$

2. $\begin{bmatrix} 1 & 0 & 6 \\ 4 & 1 & -1 \\ 5 & 0 & 1 \end{bmatrix}$

3. $\begin{bmatrix} 3 & 2 & 4 \\ 0 & 1 & 2 \\ 1 & 4 & 1 \end{bmatrix}$

4. $\begin{bmatrix} 4 & -1 & 2 \\ 3 & 1 & 0 \\ -1 & 2 & 1 \end{bmatrix}$

5. $\begin{bmatrix} 0 & 1 & 4 \\ 2 & 3 & 1 \\ 1 & 4 & 1 \end{bmatrix}$

6. $\begin{bmatrix} 6 & 2 & 1 \\ 0 & 4 & 1 \\ 0 & 0 & 5 \end{bmatrix}$

7. $\begin{bmatrix} 2 & 3 & 4 & 6 \\ 2 & 0 & -9 & 6 \\ 4 & 1 & 0 & 2 \\ 0 & 1 & -1 & 0 \end{bmatrix}$

8. $\begin{bmatrix} 2 & 0 & -1 & 7 \\ 6 & 1 & 0 & 4 \\ 8 & -2 & 1 & 0 \\ 4 & 1 & 0 & 2 \end{bmatrix}$

9. $\begin{bmatrix} 1 & 2 & 0 & -1 & 2 & 4 \\ 6 & 2 & 8 & 1 & -1 & 1 \\ 4 & 2 & 1 & 2 & 2 & -5 \\ 4 & 5 & 4 & 5 & 1 & 2 \\ 1 & 2 & 0 & -1 & 2 & 4 \\ 1 & 0 & 1 & 8 & 1 & 5 \end{bmatrix}$

1. Row reduce

$|A| = (-1)^r$ (product of pivots)
$r = $ # of row interchanges

10. $\begin{bmatrix} 1 & 0 & 1 & 2 \\ 3 & 4 & 1 & 2 \\ 6 & 1 & 0 & 0 \\ 0 & 1 & 2 & 1 \end{bmatrix}$

11. Find the cofactor of 5 for the matrix in Exercise 2.

12. Find the cofactor of 3 for the matrix in Exercise 4.

13. Find the cofactor of 7 for the matrix in Exercise 8.

14. Find the cofactor of -5 for the matrix in Exercise 9.

In Exercises 15–20, let A be a 3 × 3 matrix with det(A) = 2.

15. Find $\det(A^2)$.   **16.** Find $\det(A^k)$.

17. Find $\det(3A)$.   **18.** Find $\det(A + A)$.

19. Find $\det(A^{-1})$.   **20.** Find $\det(A^T)$.

21. Mark each of the following True or False.

___ **a.** The determinant $\det(A)$ is defined for any matrix A.

___ **b.** The determinant $\det(A)$ is defined for each square matrix A.

___ **c.** The determinant of a square matrix is a scalar.

___ **d.** If a matrix A is multiplied by a scalar c, the determinant of the resulting matrix is $c \cdot \det(A)$.

___ **e.** If an $n \times n$ matrix A is multiplied by a scalar c, the determinant of the resulting matrix is $c^n \cdot \det(A)$.

___ **f.** For every square matrix A, we have $\det(AA^T) = \det(A^TA) = [\det(A)]^2$.

___ **g.** If two rows and also two columns of a square matrix A are interchanged, the determinant changes sign.

___ **h.** The determinant of an elementary matrix is nonzero.

___ **i.** If $\det(A) = 2$ and $\det(B) = 3$, then $\det(A + B) = 5$.

___ **j.** If $\det(A) = 2$ and $\det(B) = 3$, then $\det(AB) = 6$.

In Exercises 22–25, let A be a 3 × 3 matrix with row vectors a, b, c and with determinant equal to 3. Find the determinant of the matrix having the indicated row vectors.

22. $\mathbf{a} + \mathbf{a}, \mathbf{a} + \mathbf{b}, \mathbf{a} + \mathbf{c}$

23. $\mathbf{a}, \mathbf{b}, 2\mathbf{a} + 3\mathbf{b}$

24. $\mathbf{a}, \mathbf{b}, 2\mathbf{a} + 3\mathbf{b} + 2\mathbf{c}$

25. $\mathbf{a} + \mathbf{b}, \mathbf{b} + \mathbf{c}, \mathbf{c} + \mathbf{a}$

In Exercises 26–29, find the values of λ for which the given matrix is singular.

26. $\begin{bmatrix} 1 - \lambda & 2 \\ 3 & 2 - \lambda \end{bmatrix}$ **27.** $\begin{bmatrix} -\lambda & 5 \\ 2 & 3 - \lambda \end{bmatrix}$

28. $\begin{bmatrix} 2 - \lambda & 0 & 0 \\ 0 & 1 - \lambda & 4 \\ 0 & 1 & 1 - \lambda \end{bmatrix}$

29. $\begin{bmatrix} 1 - \lambda & 0 & 2 \\ 0 & 4 - \lambda & 3 \\ 0 & 4 & -\lambda \end{bmatrix}$

30. If A and B are $n \times n$ matrices and if A is singular, prove (without using Theorem 4.4) that AB is also singular. [HINT: Assume that AB is invertible, and derive a contradiction.]

31. Prove that if A is invertible, then $\det(A^{-1}) = 1/\det(A)$.

32. If A and C are $n \times n$ matrices, with C invertible, prove that $\det(A) = \det(C^{-1}AC)$.

33. Without using the multiplicative property of determinants (Theorem 4.4), prove that $\det(AB) = \det(A) \cdot \det(B)$ for the case where B is a diagonal matrix.

34. Continuing Exercise 33, find two other types of matrices B for which it is easy to show that $\det(AB) = \det(A) \cdot \det(B)$.

35. Prove that, if three $n \times n$ matrices A, B, and C are identical except for the kth rows $\mathbf{a}_k$, $\mathbf{b}_k$, and $\mathbf{c}_k$, respectively, which are related by $\mathbf{a}_k = \mathbf{b}_k + \mathbf{c}_k$, then

$$\det(A) = \det(B) + \det(C).$$

36. Notice that

$$\begin{vmatrix} a_{11} & a_{12} \\ a_{21} & a_{22} \end{vmatrix} = (a_{11}a_{22}) = (-a_{12}a_{21})$$

is a sum of signed products, where each product contains precisely one factor from each row and one factor from each column of the corresponding matrix. Prove by induction that this is true for an $n \times n$ matrix $A = [a_{ij}]$.

37. (*Application to permutation theory*) Consider an arrangement of n objects, lined up in a column. A rearrangement of the order of the objects is called a *permutation* of the objects. Every such permutation can be achieved by successively swapping the positions of pairs of the objects. For example, the first swap might be to interchange the first object with whatever one you want to be first in the new arrangement, and then continuing this procedure with the second, the third, etc. However, there are many possible sequences of swaps that will achieve a given permutation. Use the theory of determinants to prove that it is impossible to achieve the same permutation using both an even number and an odd number of swaps. [HINT: It doesn't matter what the objects actually are—think of them as being the rows of an $n \times n$ matrix.]

38. This exercise is for the reader who is skeptical of our assertion that the solar system would be dead long before a present-day computer could find the determinant of a 50×50 matrix using just Definition 4.1 with expansion by minors.

   a. Recall that $n! = n(n-1) \cdots (3)(2)(1)$. Show by induction that expansion of an $n \times n$ matrix by minors requires at least $n!$ multiplications for $n > 1$.

   b. Run the routine EBYMTIME in LINTEK and find the time required to perform $n!$ multiplications for $n = 8, 12, 16, 20, 25, 30, 40, 50, 70,$ and 100.

39. Use MATLAB or the routine MATCOMP in LINTEK to check Example 2 and Exercises 5–10. Load the appropriate file of matrices if it is accessible. The determinant of a matrix A is found in MATLAB using the command **det(A)**.

4.3   COMPUTATION OF DETERMINANTS AND CRAMER'S RULE

We have seen that computation of determinants of high order is an unreasonable task if it is done directly from Definition 4.1, relying entirely on repeated expansion by minors. In the special case where a square matrix is triangular, Example 4 in Section 4.2 shows that the determinant is simply the product of the diagonal entries. We know that a matrix can be reduced to row-echelon form by means of elementary row operations, and row-echelon form for a square matrix is always triangular. The discussion leading to Theorem 4.3 in the previous section actually shows how the determinant of a matrix can be computed by a row reduction to echelon form. We rephrase part of this discussion in a box as an algorithm that a computer might follow to find a determinant.

Computation of a Determinant

The determinant of an $n \times n$ matrix A can be computed as follows:

1. Reduce A to an echelon form, using only row addition and row interchanges.
2. If any of the matrices appearing in the reduction contains a row of zeros, then det(A) = 0.

3. Otherwise,

$$\det(A) = (-1)^r \cdot (\text{Product of pivots}),$$

where r is the number of row interchanges performed.

When doing a computation with pencil and paper rather than with a computer, we often use row scaling to make pivots 1, in order to ease calculations. As you study the following example, notice how the pivots accumulate as factors when the scalar-multiplication property of determinants is repeatedly used.

EXAMPLE 1   Find the determinant of the following matrix by reducing it to row-echelon form.

$$A = \begin{bmatrix} 2 & 2 & 0 & 4 \\ 3 & 3 & 2 & 2 \\ 0 & 1 & 3 & 2 \\ 2 & 0 & 2 & 1 \end{bmatrix}$$

SOLUTION   We find that

$$\begin{vmatrix} 2 & 2 & 0 & 4 \\ 3 & 3 & 2 & 2 \\ 0 & 1 & 3 & 2 \\ 2 & 0 & 2 & 1 \end{vmatrix} = 2\begin{vmatrix} 1 & 1 & 0 & 2 \\ 3 & 3 & 2 & 2 \\ 0 & 1 & 3 & 2 \\ 2 & 0 & 2 & 1 \end{vmatrix}$$ **Scalar-multiplication property**

$$= 2\begin{vmatrix} 1 & 1 & 0 & 2 \\ 0 & 0 & 2 & -4 \\ 0 & 1 & 3 & 2 \\ 0 & -2 & 2 & -3 \end{vmatrix}$$ **Row-addition property twice**

$$= -2\begin{vmatrix} 1 & 1 & 0 & 2 \\ 0 & 1 & 3 & 2 \\ 0 & 0 & 2 & -4 \\ 0 & -2 & 2 & -3 \end{vmatrix}$$ **Row-interchange property**

$$= -2\begin{vmatrix} 1 & 1 & 0 & 2 \\ 0 & 1 & 3 & 2 \\ 0 & 0 & 2 & -4 \\ 0 & 0 & 8 & 1 \end{vmatrix}$$ **Row-addition property**

$$= (-2)(2)\begin{vmatrix} 1 & 1 & 0 & 2 \\ 0 & 1 & 3 & 2 \\ 0 & 0 & 1 & -2 \\ 0 & 0 & 8 & 1 \end{vmatrix}$$ **Scalar-multiplication property**

$$= (-2)(2) \begin{vmatrix} 1 & 1 & 0 & 2 \\ 0 & 1 & 3 & 2 \\ 0 & 0 & 1 & -2 \\ 0 & 0 & 0 & 17 \end{vmatrix}. \qquad \textbf{Row-addition property}$$

Therefore, $\det(A) = (-2)(2)(17) = -68.$ ∎

In our written work, we usually don't write out the shaded portion of the computation in the preceding example.

Row reduction offers an efficient way to program a computer to compute a determinant. If we are using pencil and paper, a further modification is more practical. We can use elementary row or column operations and the properties of determinants to reduce the computation to the determinant of a matrix having some row or column with a sole nonzero entry. A computer program generally modifies the matrix so that the first column has a single nonzero entry, but we can look at the matrix and choose the row or column where this can be achieved most easily. Expanding by minors on that row or column reduces the computation to a determinant of order one less, and we can continue the process until we are left with the computation of a determinant of a 2×2 matrix. Here is an illustration.

EXAMPLE 2   Find the determinant of the matrix

$$A = \begin{bmatrix} 2 & -1 & 3 & 5 \\ 2 & 0 & 1 & 0 \\ 6 & 1 & 3 & 4 \\ -7 & 3 & -2 & 8 \end{bmatrix}.$$

SOLUTION   It is easiest to create zeros in the second row and then expand by minors on that row. We start by adding -2 times the third column to the first column, and we continue in this fashion:

$$\begin{vmatrix} 2 & -1 & 3 & 5 \\ 2 & 0 & 1 & 0 \\ 6 & 1 & 3 & 4 \\ -7 & 3 & -2 & 8 \end{vmatrix} = \begin{vmatrix} -4 & -1 & 3 & 5 \\ 0 & 0 & 1 & 0 \\ 0 & 1 & 3 & 4 \\ -3 & 3 & -2 & 8 \end{vmatrix} = -\begin{vmatrix} -4 & -1 & 5 \\ 0 & 1 & 4 \\ -3 & 3 & 8 \end{vmatrix}$$

$$= -\begin{vmatrix} -4 & -1 & 9 \\ 0 & 1 & 0 \\ -3 & 3 & -4 \end{vmatrix} = -\begin{vmatrix} -4 & 9 \\ -3 & -4 \end{vmatrix}$$

$$= -(16 + 27) = -43.$$ ∎

Cramer's Rule

We now exhibit formulas in terms of determinants for the components in the solution vector of a square linear system $A\mathbf{x} = \mathbf{b}$, where A is an invertible matrix. The formulas are contained in the following theorem.

THEOREM 4.5 Cramer's Rule

Consider the linear system $A\mathbf{x} = \mathbf{b}$, where $A = [a_{ij}]$ is an $n \times n$ invertible matrix,

$$\mathbf{x} = \begin{bmatrix} x_1 \\ \cdot \\ \cdot \\ \cdot \\ x_n \end{bmatrix}, \text{ and } \mathbf{b} = \begin{bmatrix} b_1 \\ \cdot \\ \cdot \\ \cdot \\ b_n \end{bmatrix}.$$

The system has a unique solution given by

$$x_k = \frac{\det(B_k)}{\det(A)} \quad \text{for} \quad k = 1, \ldots, n, \tag{1}$$

where B_k is the matrix obtained from A by replacing the kth-column vector of A by the column vector $\mathbf{b}$.

PROOF   Because A is invertible, we know that the linear system $A\mathbf{x} = \mathbf{b}$ has a unique solution, and we let $\mathbf{x}$ be this solution. Let X_k be the matrix obtained from the $n \times n$ identity matrix by replacing its kth-column vector by the column vector $\mathbf{x}$, so that

$$X_k = \begin{bmatrix} 1 & 0 & 0 & \cdots & x_1 & 0 & 0 & \cdots & 0 \\ 0 & 1 & 0 & \cdots & x_2 & 0 & 0 & \cdots & 0 \\ & & & & \cdot \\ & & & & \cdot \\ 0 & 0 & 0 & \cdots & x_k & 0 & 0 & \cdots & 0 \\ & & & & \cdot \\ & & & & \cdot \\ 0 & 0 & 0 & \cdots & x_n & 0 & 0 & \cdots & 1 \end{bmatrix}.$$

Let us compute the product AX_k. If $j \neq k$, then the jth column of AX_k is the product of A and the jth column of the identity matrix, which also yields the jth column of A. If $j = k$, then the jth column of AX_k is $A\mathbf{x} = \mathbf{b}$. Thus AX_k is the matrix obtained from A by replacing the kth column of A by the column vector $\mathbf{b}$. That is, AX_k is the matrix B_k described in the statement of the theorem. From the equation $AX_k = B_k$ and the multiplicative property of determinants, we obtain

$$\det(A) \cdot \det(X_k) = \det(B_k).$$

Computing $\det(X_k)$ by expanding by minors across the kth row, we see that $\det(X_k) = x_k$, and thus $\det(A) \cdot x_k = \det(B_k)$. Because A is invertible, we know that $\det(A) \neq 0$ and so $x_k = \det(B_k)/\det(A)$ as asserted in Equation (1). ▲

EXAMPLE 3   Solve the linear system

$$5x_1 - 2x_2 + x_3 = 1$$
$$3x_1 + 2x_2 \quad\quad = 3$$
$$x_1 + x_2 - x_3 = 0,$$

using Cramer's rule.

SOLUTION   Using the notation in Theorem 4.5, we find that

$$\det(A) = \begin{vmatrix} 5 & -2 & 1 \\ 3 & 2 & 0 \\ 1 & 1 & -1 \end{vmatrix} = -15, \quad \det(B_1) = \begin{vmatrix} 1 & -2 & 1 \\ 3 & 2 & 0 \\ 0 & 1 & -1 \end{vmatrix} = -5,$$

$$\det(B_2) = \begin{vmatrix} 5 & 1 & 1 \\ 3 & 3 & 0 \\ 1 & 0 & -1 \end{vmatrix} = -15, \quad \det(B_3) = \begin{vmatrix} 5 & -2 & 1 \\ 3 & 2 & 3 \\ 1 & 1 & 0 \end{vmatrix} = -20.$$

Hence,

$$x_1 = \frac{-5}{-15} = \frac{1}{3},$$

$$x_2 = \frac{-15}{-15} = 1,$$

$$x_3 = \frac{-20}{-15} = \frac{4}{3}.$$ ∎

HISTORICAL NOTE   CRAMER'S RULE appeared for the first time in full generality in the *Introduction to the Analysis of Algebraic Curves* (1750) by Gabriel Cramer (1704–1752). Cramer was interested in the problem of determining the equation of a plane curve of given degree passing through a certain number of given points. For example, the general second-degree curve, whose equation is

$$A + By + Cx + Dy^2 + Exy + x^2 = 0,$$

is determined by five points. To determine A, B, C, D, and E, given the five points, Cramer substituted the coordinates of each of the points into the equation for the second-degree curve and found five linear equations for the five unknown coefficients. Cramer then referred to the appendix of the work, in which he gave his general rule: "One finds the value of each unknown by forming n fractions of which the common denominator has as many terms as there are permutations of n things." He went on to explain exactly how one calculates these terms as products of certain coefficients of the n equations, how one determines the appropriate sign for each term, and how one determines the n numerators of the fractions by replacing certain coefficients in this calculation by the constant terms of the system.

Cramer did not, however, explain why his calculations work. An explanation of the rule for the cases $n = 2$ and $n = 3$ did appear, however, in *A Treatise of Algebra* by Colin Maclaurin (1698–1746). This work was probably written in the 1730s, but was not published until 1748, after his death. In it, Maclaurin derived Cramer's rule for the two-variable case by going through the standard elimination procedure. He then derived the three-variable version by solving two pairs of equations for one unknown and equating the results, thus reducing the problem to the two-variable case. Maclaurin then described the result for the four-variable case, but said nothing about any further generalization. Interestingly, Leonhard Euler, in his *Introduction of Algebra* of 1767, does not mention Cramer's rule at all in his section on solving systems of linear equations.

The most efficient way we have presented for computing a determinant is to row-reduce a matrix to triangular form. This is also the way we solve a square linear system. If A is a 10×10 invertible matrix, solving $Ax = \mathbf{b}$ using Cramer's rule involves row-reducing eleven 10×10 matrices $A, B_1, B_2, \ldots, B_{10}$ to triangular form. Solving the linear system by the method of Section 1.4 requires row-reducing just one 10×11 matrix so that the first ten columns are in upper-triangular form. This illustrates the folly of using Cramer's rule to solve linear systems. However, the structure of the components of the solution vector, as given by the Cramer's rule formula $x_k = \det(B_k)/\det(A)$, is of interest in the study of advanced calculus, for example.

The Adjoint Matrix

We conclude this section by finding a formula in terms of determinants for the inverse of an invertible $n \times n$ matrix $A = [a_{ij}]$. Recall the definition of the cofactor a'_{ij} from Eq. (2) of Section 4.2. Let $A_{i \to j}$ be the matrix obtained from A by replacing the jth row of A by the ith row. That is,

$$
A_{i \to j} = \begin{bmatrix}
a_{11} & a_{12} & \cdots & a_{1n} \\
\cdot & \cdot & & \cdot \\
\cdot & \cdot & & \cdot \\
\cdot & \cdot & & \cdot \\
a_{i1} & a_{i2} & \cdots & a_{in} \\
\cdot & \cdot & & \cdot \\
\cdot & \cdot & & \cdot \\
a_{i1} & a_{i2} & \cdots & a_{in} \\
\cdot & \cdot & & \cdot \\
\cdot & \cdot & & \cdot \\
a_{n1} & a_{n2} & \cdots & a_{nn}
\end{bmatrix}
\begin{matrix}
\\ \\ \\ \\ i\text{th row} \\ \\ \\ j\text{th row} \\ \\ \\
\end{matrix}
$$

Then

$$
\det(A_{i \to j}) = \begin{cases} \det(A) & \text{if } i = j, \\ 0 & \text{if } i \neq j. \end{cases}
$$

If we expand $\det(A_{i \to j})$ by minors on the jth row, we have

$$
\det(A_{i \to j}) = \sum_{s=1}^{n} a_{is} a'_{js},
$$

and we obtain the important relation

$$
\sum_{s=1}^{n} a_{is} a'_{js} = \begin{cases} \det(A) & \text{if } i = j, \\ 0 & \text{if } i \neq j. \end{cases} \tag{2}
$$

The term on the left-hand side in Eq. (2) is the entry in the ith row and jth column in the product $A(A')^T$, where $A' = [a'_{ij}]$ is the matrix whose entries are the cofactors of the entries of A. Thus Eq. (2) can be written in matrix form as

$$
A(A')^T = (\det(A))I,
$$

where I is the $n \times n$ identity matrix. Similarly, replacing the ith column of A by the jth column and by expanding on the ith column, we have

$$\sum_{r=1}^{n} a'_{ri}a_{rj} = \begin{cases} \det(A) & \text{if } i = j, \\ 0 & \text{if } i \neq j. \end{cases} \qquad (3)$$

Relation (3) yields $(A')^T A = (\det(A))I$.

The matrix $(A')^T$ is called the **adjoint** of A and is denoted by adj(A). We have established an important relationship between a matrix and its adjoint.

THEOREM 4.6 Property of the Adjoint

Let A be an $n \times n$ matrix. The adjoint adj(A) = $(A')^T$ of A satisfies

$$(\text{adj}(A))A = A(\text{adj}(A)) = (\det(A))I,$$

where I is the $n \times n$ identity matrix.

Theorem 4.6 provides a formula for the inverse of an invertible matrix, which we present as a corollary.

COROLLARY A Formula for the Inverse of an Invertible Matrix

Let $A = [a_{ij}]$ be an $n \times n$ matrix with $\det(A) \neq 0$. Then A is invertible, and

$$A^{-1} = \frac{1}{\det(A)} \text{adj}(A),$$

where adj(A) = $[a'_{ij}]^T$ is the transposed matrix of cofactors.

EXAMPLE 4 Find the inverse of

$$A = \begin{bmatrix} 4 & 0 & 1 \\ 2 & 2 & 0 \\ 3 & 1 & 1 \end{bmatrix}$$

if the matrix is invertible, using the corollary of Theorem 4.6.

SOLUTION We find that $\det(A) = 4$, so A is invertible. The cofactors a'_{ij} are

$$a'_{11} = (-1)^2 \begin{vmatrix} 2 & 0 \\ 1 & 1 \end{vmatrix} = 2, \qquad a'_{12} = (-1)^3 \begin{vmatrix} 2 & 0 \\ 3 & 1 \end{vmatrix} = -2,$$

$$a'_{13} = (-1)^4 \begin{vmatrix} 2 & 2 \\ 3 & 1 \end{vmatrix} = -4, \qquad a'_{21} = (-1)^3 \begin{vmatrix} 0 & 1 \\ 1 & 1 \end{vmatrix} = 1,$$

$$a'_{22} = (-1)^4 \begin{vmatrix} 4 & 1 \\ 3 & 1 \end{vmatrix} = 1, \qquad a'_{23} = (-1)^5 \begin{vmatrix} 4 & 0 \\ 3 & 1 \end{vmatrix} = -4,$$

$$a'_{31} = (-1)^4 \begin{vmatrix} 0 & 1 \\ 2 & 0 \end{vmatrix} = -2, \qquad a'_{32} = (-1)^5 \begin{vmatrix} 4 & 1 \\ 2 & 0 \end{vmatrix} = 2,$$

$$a'_{33} = (-1)^6 \begin{vmatrix} 4 & 0 \\ 2 & 2 \end{vmatrix} = 8.$$

Hence,

$$A' = [a'_{ij}] = \begin{bmatrix} 2 & -2 & -4 \\ 1 & 1 & -4 \\ -2 & 2 & 8 \end{bmatrix}, \quad \text{so} \quad \text{adj}(A) = \begin{bmatrix} 2 & 1 & -2 \\ -2 & 1 & 2 \\ -4 & -4 & 8 \end{bmatrix}$$

and

$$A^{-1} = \frac{1}{\det(A)} \text{adj}(A) = \frac{1}{4} \begin{bmatrix} 2 & 1 & -2 \\ -2 & 1 & 2 \\ -4 & -4 & 8 \end{bmatrix} = \begin{bmatrix} \frac{1}{2} & \frac{1}{4} & -\frac{1}{2} \\ -\frac{1}{2} & \frac{1}{4} & \frac{1}{2} \\ -1 & -1 & 2 \end{bmatrix}. \qquad \blacksquare$$

The method described in Section 1.5 for finding the inverse of an invertible matrix is more efficient than the method illustrated in the preceding example, especially if the matrix is large. The corollary is often used to find the inverse of a 2 × 2 matrix. We see that if $ad - bc \neq 0$, then

$$\begin{bmatrix} a & b \\ c & d \end{bmatrix}^{-1} = \frac{1}{ad - bc} \begin{bmatrix} d & -b \\ -c & a \end{bmatrix}.$$

SUMMARY

1. A computationally feasible algorithm for finding the determinant of a matrix is to reduce the matrix to echelon form, using just row-addition and row-interchange operations. If a row of zeros is formed during the process, the determinant is zero. Otherwise, the determinant of the original matrix is found by computing $(-1)^r \cdot$ (Product of pivots) in the echelon form, where r is the number of row interchanges performed. This is one way to program a computer to find a determinant.

2. The determinant of a matrix can be found by row or column reduction of the matrix to a matrix having a sole nonzero entry in some column or row. One then expands by minors on that column or row, and continues this process. If a matrix having a zero row or column is encountered, the determinant is zero. Otherwise, one continues until the computation is reduced to the determinant of a 2 × 2 matrix. This is a good way to find a determinant when working with pencil and paper.

3. If A is invertible, the linear system $A\mathbf{x} = \mathbf{b}$ has the unique solution $\mathbf{x}$ whose kth component is given explicitly by the formula

$$x_k = \frac{\det(B_k)}{\det(A)},$$

where the matrix B_k is obtained from matrix A by replacing the kth column of A by $\mathbf{b}$.

4. The methods of Chapter 1 are far more efficient than those described in this section for actual computation of both the inverse of A and the solution of the system $A\mathbf{x} = \mathbf{b}$.

5. Let A be an $n \times n$ matrix, and let A' be its matrix of cofactors. The adjoint adj(A) is the matrix $(A')^T$ and satisfies $(\text{adj}(A))A = A(\text{adj}(A)) = (\det(A))I$, where I is the $n \times n$ identity matrix.

6. The inverse of an invertible matrix A is given by the explicit formula

$$A^{-1} = \frac{1}{\det(A)} \text{adj}(A).$$

EXERCISES

In Exercises 1–10, find the determinant of the given matrix.

1. $\begin{bmatrix} 2 & 3 & -1 \\ 5 & -7 & 1 \\ -3 & 2 & -1 \end{bmatrix}$

2. $\begin{bmatrix} 4 & -3 & 2 \\ -1 & -1 & 1 \\ -5 & 5 & 7 \end{bmatrix}$

3. $\begin{bmatrix} 5 & 2 & 4 & 0 \\ 2 & -3 & -1 & 2 \\ 3 & -4 & 3 & 7 \\ 1 & -1 & 0 & 1 \end{bmatrix}$

4. $\begin{bmatrix} 3 & -5 & -1 & 7 \\ 0 & 3 & 1 & -6 \\ 2 & -5 & -1 & 8 \\ -8 & 8 & 2 & -9 \end{bmatrix}$

5. $\begin{bmatrix} 2 & 1 & 0 & 0 & 0 \\ 3 & -1 & 2 & 0 & 0 \\ 0 & 4 & 1 & -1 & 2 \\ 0 & 0 & -3 & 2 & 4 \\ 0 & 0 & 0 & -1 & 3 \end{bmatrix}$

6. $\begin{bmatrix} 3 & 2 & 0 & 0 & 0 \\ -1 & 4 & 1 & 0 & 0 \\ 0 & -3 & 5 & 2 & 0 \\ 0 & 0 & 0 & 1 & 4 \\ 0 & 0 & 0 & -1 & 2 \end{bmatrix}$

7. $\begin{bmatrix} 0 & 0 & 0 & 3 & 1 \\ 0 & 0 & 2 & 0 & -3 \\ 0 & -2 & 1 & 0 & 0 \\ 5 & -3 & 2 & 0 & 0 \\ -3 & 4 & 0 & 0 & 0 \end{bmatrix}$

8. $\begin{bmatrix} 2 & -1 & 0 & 0 \\ 4 & 5 & 0 & 0 \\ 0 & 0 & 3 & 6 \\ 0 & 0 & -4 & 2 \end{bmatrix}$

9. $\begin{bmatrix} 2 & -1 & 3 & 0 & 0 \\ 0 & 1 & 4 & 0 & 0 \\ -5 & 2 & 6 & 0 & 0 \\ 0 & 0 & 0 & 1 & 4 \\ 0 & 0 & 0 & -2 & 8 \end{bmatrix}$

10. $\begin{bmatrix} 0 & 0 & 0 & 3 & -4 \\ 0 & 0 & 0 & 2 & 1 \\ -1 & 2 & 4 & 0 & 0 \\ 3 & 1 & -2 & 0 & 0 \\ 5 & 1 & 5 & 0 & 0 \end{bmatrix}$

11. The matrices in Exercises 8 and 9 have zero entries except for entries in an $r \times r$ submatrix R and a separate $s \times s$ submatrix S whose main diagonals lie on the main diagonal of the whole $n \times n$ matrix, and where $r + s = n$. Prove that, if A is such a matrix with submatrices R and S, then $\det(A) = \det(R) \cdot \det(S)$.

12. The matrix A in Exercise 10 has a structure similar to that discussed in Exercise 11, except that the square submatrices R and S lie along the other diagonal. State and prove a result similar to that in Exercise 11 for such a matrix.

13. State and prove a generalization of the result in Exercise 11, when the matrix A has zero entries except for entries in k submatrices positioned along the diagonal.

In Exercises 14–19, use the corollary to Theorem 4.6 to find A^{-1} if A is invertible.

14. $A = \begin{bmatrix} 2 & 0 \\ 1 & -1 \end{bmatrix}$

15. $A = \begin{bmatrix} 4 & 1 \\ 2 & 1 \end{bmatrix}$

16. $A = \begin{bmatrix} 2 & 1 & 1 \\ 0 & 1 & 1 \\ -2 & 1 & 1 \end{bmatrix}$ **17.** $A = \begin{bmatrix} 3 & 0 & 4 \\ -2 & 1 & 1 \\ 3 & 1 & 2 \end{bmatrix}$

18. $A = \begin{bmatrix} 3 & 0 & 3 \\ 4 & 1 & -2 \\ -5 & 1 & 4 \end{bmatrix}$ **19.** $A = \begin{bmatrix} 2 & 1 & 3 \\ 0 & 1 & 4 \\ 1 & 2 & 1 \end{bmatrix}$

20. Find the adjoint of the matrix $\begin{bmatrix} 4 & 5 \\ -3 & 6 \end{bmatrix}$.

21. Find the adjoint of the matrix $\begin{bmatrix} 2 & 1 & 0 \\ 3 & 1 & 4 \\ 0 & 2 & 1 \end{bmatrix}$.

22. Given that $A^{-1} = \begin{bmatrix} a & b \\ c & d \end{bmatrix}$ and $\det(A^{-1}) = 3$, find the matrix A.

23. If A is a matrix with integer entries and if $\det(A) = \pm 1$, prove that A^{-1} also has the same properties.

In Exercises 24–31, solve the given system of linear equations by Cramer's rule wherever it is possible.

24. $\begin{aligned} x_1 - 2x_2 &= 1 \\ 3x_1 + 4x_2 &= 3 \end{aligned}$ **25.** $\begin{aligned} 2x_1 - 3x_2 &= 1 \\ -4x_1 + 6x_2 &= -2 \end{aligned}$

26. $\begin{aligned} 3x_1 + x_2 &= 5 \\ 2x_1 + x_2 &= 0 \end{aligned}$ **27.** $\begin{aligned} x_1 + x_2 &= 1 \\ x_1 + 2x_2 &= 2 \end{aligned}$

28. $\begin{aligned} 5x_1 - 2x_2 + x_3 &= 1 \\ x_2 + x_3 &= 0 \\ x_1 + 6x_2 - x_3 &= 4 \end{aligned}$

29. $\begin{aligned} x_1 + 2x_2 - x_3 &= -2 \\ 2x_1 + x_2 + x_3 &= 0 \\ 3x_1 - x_2 + 5x_3 &= 1 \end{aligned}$

30. $\begin{aligned} x_1 - x_2 + x_3 &= 0 \\ x_1 + 2x_2 - x_3 &= 1 \\ x_1 - x_2 + 2x_3 &= 0 \end{aligned}$

31. $\begin{aligned} 3x_1 + 2x_2 - x_3 &= 1 \\ x_1 - 4x_2 + x_3 &= -2 \\ 5x_1 + 2x_2 &= 1 \end{aligned}$

In Exercises 32 and 33, find the component x_2 of the solution vector for the given linear system.

32. $\begin{aligned} x_1 + x_2 - 3x_3 + x_4 &= 1 \\ 2x_1 + x_2 \qquad\quad + 2x_4 &= 0 \\ x_2 - 6x_3 - x_4 &= 5 \\ 3x_1 + x_2 \qquad\quad + x_4 &= 1 \end{aligned}$

33. $\begin{aligned} 6x_1 + x_2 - x_3 \qquad\quad &= 4 \\ x_1 - x_2 \qquad\quad + 5x_4 &= -2 \\ -x_1 + 3x_2 + x_3 \qquad\quad &= 2 \\ x_1 + x_2 - x_3 + 2x_4 &= 0 \end{aligned}$

34. Find the unique solution (assuming that it exists) of the system of equations represented by the partitioned matrix

$$\begin{bmatrix} a_1 & b_1 & c_1 & d_1 & | & 3b_1 \\ a_2 & b_2 & c_2 & d_2 & | & 3b_2 \\ a_3 & b_3 & c_3 & d_3 & | & 3b_3 \\ a_4 & b_4 & c_4 & d_4 & | & 3b_4 \end{bmatrix}.$$

35. Let A be a square matrix. Mark each of the following True or False.

____ **a.** The determinant of a square matrix is the product of the entries on its main diagonal.

____ **b.** The determinant of an upper-triangular square matrix is the product of the entries on its main diagonal.

____ **c.** The determinant of a lower-triangular square matrix is the product of the entries on its main diagonal.

____ **d.** A square matrix is nonsingular if and only if its determinant is positive.

____ **e.** The column vectors of an $n \times n$ matrix are independent if and only if the determinant of the matrix is nonzero.

____ **f.** A homogeneous square linear system has a nontrivial solution if and only if the determinant of its coefficient matrix is zero.

____ **g.** The product of a square matrix and its adjoint is the identity matrix.

____ **h.** The product of a square matrix and its adjoint is equal to some scalar times the identity matrix.

____ **i.** The transpose of the adjoint of A is the matrix of cofactors of A.

____ **j.** The formula $A^{-1} = (1/\det(A))\text{adj}(A)$ is of practical use in computing the inverse of a large nonsingular matrix.

36. Prove that the inverse of a nonsingular upper-triangular matrix is upper triangular.

37. Prove that a square matrix is invertible if and only if its adjoint is an invertible matrix.

38. Let A be an $n \times n$ matrix. Prove that $\det(\text{adj}(A)) = \det(A)^{n-1}$.

39. Let A be an invertible $n \times n$ matrix with $n > 1$. Using Exercises 37 and 38, prove that $\text{adj}(\text{adj}(A)) = (\det(A))^{n-2}A$.

The routine YUREDUCE in LINTEK has a menu option D that will compute and display the product of the diagonal elements of a square matrix. The routine MATCOMP has a menu option D to compute a determinant. Use YUREDUCE or MATLAB to compute the determinant of the matrices in Exercises 40–42. Write down your results. If you used YUREDUCE, use MATCOMP to compute the determinants of the same matrices again and compare the answers.

40. $\begin{bmatrix} 11 & -9 & 28 \\ 32 & -24 & 21 \\ 10 & 13 & -19 \end{bmatrix}$ **41.** $\begin{bmatrix} 13 & -15 & 33 \\ -15 & 25 & 40 \\ 12 & -33 & 27 \end{bmatrix}$

42. $\begin{bmatrix} 7.6 & 2.8 & -3.9 & 19.3 & 25.0 \\ -33.2 & 11.4 & 13.2 & 22.4 & 18.3 \\ 21.4 & -32.1 & 45.7 & -8.9 & 12.5 \\ 17.4 & 11.0 & -6.8 & 20.3 & -35.1 \\ 22.7 & 11.9 & 33.2 & 2.5 & 7.8 \end{bmatrix}$

43. MATCOMP computes determinants in essentially the way described in this section. The matrix

$$A = \begin{bmatrix} 3 & 4 \\ 2 & 3 \end{bmatrix}$$

has determinant 1, so every power of it should have determinant 1. Use MATCOMP with single-precision printing and with the default roundoff control ratio r. Start computing determinants of powers of A. Find the smallest positive integer m such that $\det(A^m) \neq 1$, according to MATCOMP. How bad is the error? What does MATCOMP give for $\det(A^{20})$? At what integer exponent does the break occur between the incorrect value 0 and incorrect values of large magnitude?

Repeat the above, taking zero for roundoff control ratio r. Try to explain why the results are different and what is happening in each case.

44. Using MATLAB, find the smallest positive integer m such that $\det(A^m) \neq 1$, according to MATLAB, for the matrix A in Exercise 43.

In Exercises 45–47, use MATCOMP in LINTEK or MATLAB and the corollary of Theorem 4.6 to find the matrix of cofactors of the given matrix.

45. $\begin{bmatrix} 1 & 2 & -3 \\ 2 & 3 & 0 \\ 3 & 1 & 4 \end{bmatrix}$

46. $\begin{bmatrix} -52 & 31 & 47 \\ 21 & -11 & 28 \\ 43 & -71 & 87 \end{bmatrix}$

47. $\begin{bmatrix} 6 & -3 & 2 & 14 \\ -3 & 7 & 8 & 1 \\ 4 & 9 & -5 & 3 \\ -8 & -40 & 47 & 29 \end{bmatrix}$

[HINT: Entries in the matrix of cofactors are integers. The cofactors of a matrix are continuous functions of its entries; that is, changing an entry by a very slight amount will change a cofactor only slightly. Change some entry just a bit to make the determinant nonzero.]

4.4 LINEAR TRANSFORMATIONS AND DETERMINANTS (OPTIONAL)

We continue our program of exhibiting the relationship between matrices and linear transformations. Associated with an $m \times n$ matrix A is the linear transformation T mapping $\mathbb{R}^n$ into $\mathbb{R}^m$, where $T(\mathbf{x}) = A\mathbf{x}$ for $\mathbf{x}$ in $\mathbb{R}^n$. If $n = m$, so that A is a square matrix, then $\det(A)$ is defined. But what is the meaning of the number $\det(A)$ for the transformation T? We now tackle this question, and the answer is so important that it merits a section all to itself. The notion of the determinant associated with a linear transformation T mapping $\mathbb{R}^n$ into $\mathbb{R}^n$ lies at the heart of variable substitution in integral calculus. This section presents an informal and intuitive explanation of this notion.

The Volume of an n-Box in $\mathbb{R}^m$

In Section 4.1, we saw that the area of the parallelogram (or *2-box*) in $\mathbb{R}^2$ determined by vectors $\mathbf{a}_1$ and $\mathbf{a}_2$ is the absolute value $|\det(A)|$ of the determinant of the 2×2 matrix A having $\mathbf{a}_1$ and $\mathbf{a}_2$ as column vectors.* We also saw that the volume of the parallelepiped (or *3-box*) in $\mathbb{R}^3$ determined by vectors $\mathbf{a}_1$, $\mathbf{a}_2$, and $\mathbf{a}_3$ is $|\det(A)|$ for the 3×3 matrix A whose column vectors are $\mathbf{a}_1$, $\mathbf{a}_2$, $\mathbf{a}_3$. We wish to extend these notions by defining an *n-box* in $\mathbb{R}^m$ for $m \geq n$ and finding its "volume."

DEFINITION 4.2   An n-Box in $\mathbb{R}^m$

Let $\mathbf{a}_1, \mathbf{a}_2, \ldots, \mathbf{a}_n$ be n independent vectors in $\mathbb{R}^m$ for $n \leq m$. The **n-box** in $\mathbb{R}^m$ determined by these vectors is the set of all vectors $\mathbf{x}$ satisfying

$$\mathbf{x} = t_1\mathbf{a}_1 + t_2\mathbf{a}_2 + \cdots + t_n\mathbf{a}_n$$

for $0 \leq t_i \leq 1$ and $i = 1, 2, \ldots, n$.

If the vectors $\mathbf{a}_1, \mathbf{a}_2, \ldots, \mathbf{a}_n$ in Definition 4.2 are dependent, the set described is a *degenerate n-box*.

EXAMPLE 1   Describe geometrically the 1-box determined by the "vector" 2 in $\mathbb{R}$ and the 1-box determined by a nonzero vector $\mathbf{a}$ in $\mathbb{R}^m$.

SOLUTION   The 1-box determined by the "vector" 2 in $\mathbb{R}$ consists of all numbers $t(2)$ for $0 \leq t \leq 1$, which is simply the closed interval $0 \leq x \leq 2$. Similarly, the 1-box in $\mathbb{R}^m$ determined by a nonzero vector $\mathbf{a}$ is the line segment joining the origin to the tip of $\mathbf{a}$.   ■

EXAMPLE 2   Draw symbolic sketches of a 2-box in $\mathbb{R}^m$ and a 3-box in $\mathbb{R}^m$.

SOLUTION   A 2-box in $\mathbb{R}^m$ is a parallelogram with a vertex at the origin, as shown in Figure 4.7. Similarly, a 3-box in $\mathbb{R}^m$ is a parallelepiped with a vertex at the origin, as illustrated in Figure 4.8.   ■

Notice that our boxes need not be rectangular boxes; perhaps we should have used the term *skew box* to make this clear.

---

*Anticipating our work in this section, we arrange the vectors $\mathbf{a}_i$ as columns rather than as rows of A. Recall that $\det(A) = \det(A^T)$.

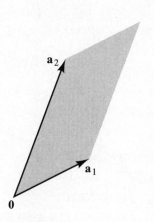

FIGURE 4.7
A 2-box in $\mathbb{R}^m$.

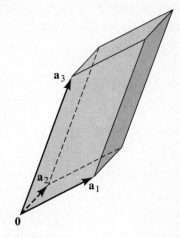

FIGURE 4.8
A 3-box in $\mathbb{R}^m$.

We are accustomed to speaking of the *length* of a line segment, of the *area* of a piece of the plane, and of the *volume* of a piece of space. To avoid introducing new terms when discussing a general *n*-box, we will use the three-dimensional term *volume* when speaking of its size. Notice that we already used the three-dimensional term *box* as the name for the object! Thus, by the *volume* of a 1-box, we simply mean its length; the *volume* of a 2-box is its area, and so on.

The volume of the 1-box determined by $\mathbf{a}_1$ in $\mathbb{R}^m$ is its length $\|\mathbf{a}_1\|$. Because the determinant of a 1×1 matrix is its sole entry, Exercise 36 shows that this length can be written as

$$\text{Length} = \|\mathbf{a}_1\| = \sqrt{\det([\mathbf{a}_1 \cdot \mathbf{a}_1])}. \tag{1}$$

Let us turn to a 2-box in $\mathbb{R}^m$ determined by nonzero and nonparallel vectors $\mathbf{a}_1$ and $\mathbf{a}_2$. We repeat an argument that we made in Section 4.1 for a 2-box in $\mathbb{R}^2$, using a slightly different notation for this general *m*. From Figure 4.9, we see that the area of this parallelogram is given by

$$\text{Area} = \|h\|\,\|\mathbf{a}_2\|,$$

where, for the angle θ between $\mathbf{a}_1$ and $\mathbf{a}_2$, we have $\|h\| = \|\mathbf{a}_1\| \sin \theta$. We then have

$$
\begin{aligned}
(\text{Area})^2 &= \|\mathbf{a}_1\|^2\|\mathbf{a}_2\|^2 \sin^2\theta \\
&= \|\mathbf{a}_1\|^2\|\mathbf{a}_2\|^2(1 - \cos^2\theta) \\
&= \|\mathbf{a}_1\|^2\|\mathbf{a}_2\|^2 - (\|\mathbf{a}_1\|\,\|\mathbf{a}_2\| \cos \theta)^2 \\
&= (\mathbf{a}_1 \cdot \mathbf{a}_1)(\mathbf{a}_2 \cdot \mathbf{a}_2) - (\mathbf{a}_1 \cdot \mathbf{a}_2)(\mathbf{a}_2 \cdot \mathbf{a}_1) \\
&= \begin{vmatrix} \mathbf{a}_1 \cdot \mathbf{a}_1 & \mathbf{a}_1 \cdot \mathbf{a}_2 \\ \mathbf{a}_2 \cdot \mathbf{a}_1 & \mathbf{a}_2 \cdot \mathbf{a}_2 \end{vmatrix} = \det([\mathbf{a}_i \cdot \mathbf{a}_j]).
\end{aligned}
\tag{2}
$$

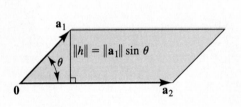

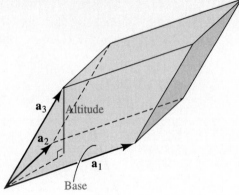

FIGURE 4.9

The volume of a 2-box in $\mathbb{R}^m$ is (Length of the base) × (Altitude).

FIGURE 4.10

The volume of a 3-box in $\mathbb{R}^m$ is (Area of the base) × (Altitude).

From Eqs. (1) and (2), we might guess that the square of the volume of an n-box in $\mathbb{R}^m$ is $\det([\mathbf{a}_i \cdot \mathbf{a}_j])$. Of course, we must define what we mean by the volume of such a box, but with the natural definition, this conjecture is true. If A is the matrix with jth column vector $\mathbf{a}_j$, then A^T is the matrix with ith row vector $\mathbf{a}_i$, and the $n \times n$ matrix $[\mathbf{a}_i \cdot \mathbf{a}_j]$ is A^TA, so Eq. (2) can be written as

$$(\text{Area}) = \sqrt{\det(A^TA)}.$$

We have an intuitive idea of the volume of an n-box in $\mathbb{R}^m$ for $n \le m$. For example, the 3-box in $\mathbb{R}^m$ determined by independent vectors $\mathbf{a}_1$, $\mathbf{a}_2$, $\mathbf{a}_3$ has a volume equal to the altitude of the box times the volume (that is, area) of the base, as shown in Figure 4.10. Roughly speaking, the volume of an n-box is equal to the altitude of the box times the volume of the base, which is an $(n - 1)$-box. This notion of the *altitude* of a box can be made precise after we develop projections in Chapter 6. The formal definition of the volume of an n-box appears in Appendix B, as does the proof of our main result on volumes (Theorem 4.7). For the remainder of this section, we are content to proceed with our intuitive notion of volume.

THEOREM 4.7 Volume of a Box

The volume of the n-box in $\mathbb{R}^m$ determined by independent vectors $\mathbf{a}_1, \mathbf{a}_2, \ldots, \mathbf{a}_n$ is given by

$$\text{Volume} = \sqrt{\det(A^TA)},$$

where A is the $m \times n$ matrix with $\mathbf{a}_j$ as jth column vector.

The volume of an n-box in $\mathbb{R}^n$ is of such importance that we restate this special case as a corollary.

COROLLARY   Volume of an n-Box in $\mathbb{R}^n$

If A is an $n \times n$ matrix with independent column vectors $\mathbf{a}_1, \mathbf{a}_2, \ldots, \mathbf{a}_n$, then $|\det(A)|$ is the volume of the n-box in $\mathbb{R}^n$ determined by these n vectors.

PROOF   By Theorem 4.7, the square of the volume of the n-box is $\det(A^TA)$. But because A is an $n \times n$ matrix, we have

$$\det(A^TA) = \det(A^T) \cdot \det(A) = (\det(A))^2.$$

The conclusion of the corollary then follows at once.   ▲

EXAMPLE 3   Find the area of the parallelogram in $\mathbb{R}^4$ determined by the vectors $[2, 1, -1, 3]$ and $[0, 2, 4, -1]$.

SOLUTION   If

$$A = \begin{bmatrix} 2 & 0 \\ 1 & 2 \\ -1 & 4 \\ 3 & -1 \end{bmatrix},$$

then

$$A^TA = \begin{bmatrix} 2 & 1 & -1 & 3 \\ 0 & 2 & 4 & -1 \end{bmatrix} \begin{bmatrix} 2 & 0 \\ 1 & 2 \\ -1 & 4 \\ 3 & -1 \end{bmatrix} = \begin{bmatrix} 15 & -5 \\ -5 & 21 \end{bmatrix}.$$

By Theorem 4.7, we have

$$(\text{Area})^2 = \begin{vmatrix} 15 & -5 \\ -5 & 21 \end{vmatrix} = 290.$$

Thus the area of the parallelogram is $\sqrt{290}$.   ■

EXAMPLE 4   Find the volume of the parallelepiped in $\mathbb{R}^3$ determined by the vectors $[1, 0, -1]$, $[-1, 1, 3]$, and $[2, 4, 1]$.

SOLUTION   We compute the determinant

$$\begin{vmatrix} 1 & -1 & 2 \\ 0 & 1 & 4 \\ -1 & 3 & 1 \end{vmatrix} = \begin{vmatrix} 1 & -1 & 2 \\ 0 & 1 & 4 \\ 0 & 2 & 3 \end{vmatrix} = \begin{vmatrix} 1 & 4 \\ 2 & 3 \end{vmatrix} = -5.$$

Applying the corollary of Theorem 4.7, the volume of the parallelepiped is therefore 5.   ■

Comparing Theorem 4.7 and its corollary, we see that the formula for the volume of an n-box in a space of larger dimension m involves a square root, whereas the formula for the volume of a box in a space of its own dimension does not involve a square root. The student of calculus discovers that the calculus formulas used to find the length of a curve (which is one-dimensional) in the plane or in space involve a square root. The same is true of the formulas used to find the area of a surface (two-dimensional) in space. However, the calculus formulas for finding the area of part of the plane or the volume of some part of space do not involve square roots. Theorem 4.7 and its corollary lie at the heart of this difference in the calculus formulas.

Volume-Change Factor of a Linear Transformation

Recall the multiplicative property of determinants: $\det(AB) = \det(A) \cdot \det(B)$, where A and B are $n \times n$ matrices. Suppose that both A and B have rank n. Then the column vectors $\mathbf{b}_1, \mathbf{b}_2, \ldots, \mathbf{b}_n$ of B determine an n-box having volume $|\det(B)| \neq 0$. From the definition of matrix multiplication, the jth column vector of AB is $A\mathbf{b}_j$, and we write

$$AB = \begin{bmatrix} | & | & & | \\ A\mathbf{b}_1 & A\mathbf{b}_2 & \cdots & A\mathbf{b}_n \\ | & | & & | \end{bmatrix}.$$

The linear transformation $T: \mathbb{R}^n \to \mathbb{R}^n$ defined by $T(\mathbf{x}) = A\mathbf{x}$ thus carries the original n-box determined by the column vectors of B into a new n-box determined by the column vectors of AB. The new n-box has volume $|\det(AB)| = |\det(A)| \cdot |\det(B)|$. That is, the volume of the new n-box, or *image box*, is equal to $|\det(A)|$ times the volume of the original box. Thus $|\det(A)|$ is referred to as the **volume-change factor** for the linear transformation T. We are interested in this concept only when $\det(A) \neq 0$, a requirement that ensures that A is invertible, that the n vectors $A\mathbf{b}_i$ are independent, and that T is an invertible transformation.

To illustrate this idea of the volume-change factor, consider the *n-cube* in $\mathbb{R}^n$ determined by the vectors $c\mathbf{e}_1, c\mathbf{e}_2, \ldots, c\mathbf{e}_n$ for $c > 0$. This n-cube with edges of length c has volume c^n. It is carried by T into an n-box having volume $|\det(A)| \cdot c^n$; the image n-box need not be a cube, nor even rectangular. We illustrate the image n-box in Figure 4.11 for the case $n = 1$ and in Figure 4.12 for the case $n = 3$.

EXAMPLE 5   Consider the linear transformation $T: \mathbb{R}^3 \to \mathbb{R}^3$ defined by

$$T([x_1, x_2, x_3]) = [x_1 + x_3, 2x_2, 2x_1 + 5x_3].$$

Find the volume of the image box when T acts on the cube determined by the vectors $c\mathbf{e}_1$, $c\mathbf{e}_2$, and $c\mathbf{e}_3$ for $c > 0$.

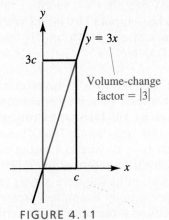

FIGURE 4.11
The volume-change factor of $T(x) = 3x$.

SOLUTION   The image box is determined by the vectors

$$T(c\mathbf{e}_1) = cT(\mathbf{e}_1) = c[1, 0, 2]$$
$$T(c\mathbf{e}_2) = cT(\mathbf{e}_2) = c[0, 2, 0]$$
$$T(c\mathbf{e}_3) = cT(\mathbf{e}_3) = c[1, 0, 5].$$

The standard matrix representation of T is

$$A = \begin{bmatrix} 1 & 0 & 1 \\ 0 & 2 & 0 \\ 2 & 0 & 5 \end{bmatrix},$$

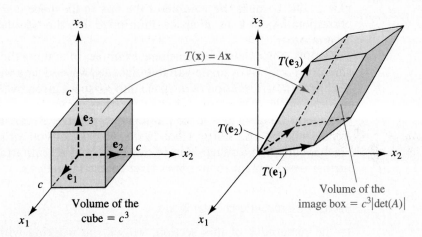

FIGURE 4.12
The volume-change factor of $T(x) = Ax$.

and the volume-change factor of T is $|\det(A)| = 6$. Therefore, the image box has volume $6c^3$. This volume can also be computed by taking the determinant of the matrix having as column vectors $T(c\mathbf{e}_1)$, $T(c\mathbf{e}_2)$, and $T(c\mathbf{e}_3)$. This matrix is cA, and has determinant $c^3 \cdot \det(A) = 6c^3$.   ∎

The sign of the determinant associated with an invertible linear transformation $T: \mathbb{R}^n \to \mathbb{R}^n$ depends on whether T preserves orientation. In the plane where $n = 2$, orientation is preserved by T if $T(\mathbf{e}_1)$ can be rotated counterclockwise through an angle of less than $180°$ to lie along $T(\mathbf{e}_2)$. It can be shown that this is the case if and only if $\det(A) > 0$, where A is the standard matrix representation of T. In general, a linear transformation $T: \mathbb{R}^n \to \mathbb{R}^n$ is said to *preserve orientation* if the determinant of its standard matrix representation A is positive. Thus the linear transformation in Example 5 preserves orientation because $\det(A) = 6$. Because this topic is more suitable for a course in analysis than for one in linear algebra, we do not pursue it further.

Application to Calculus

We can get an intuitive feel for the connection between the volume-change factor of T and integral calculus. The definition of an integral involves summing products of the form

$$\text{(Function value at some point of a box)(Volume of the box).} \qquad (3)$$
$$f(\mathbf{x}) \qquad\qquad\qquad\qquad\qquad\qquad d\mathbf{x}$$

Under a change of variables—say, from **x**-variables to **t**-variables—the boxes in the **dx**-space are replaced by boxes in **dt**-space via an invertible linear transformation—namely, the *differential* of the variable substitution function. Thus the second factor in product (3) must be expressed in terms of volumes of boxes in the **dt**-space. The determinant of the differential transformation must play a role, because the volume of the box in **dx**-space is the volume of the corresponding box in **dt**-space multiplied by the absolute value of the determinant.

Let us look at a one-dimensional example. In making the substitution $x = \sin t$ in an integral in single-variable calculus, we associate with each t-value t_0 the linear transformation of dt-space into dx-space given by the equation $dx = (\cos t_0)dt$. The determinant of this linear transformation is $\cos t_0$. A little 1-box of volume (length) dt and containing the point t_0 is carried by this linear transformation into a little 1-box in the dx-space of volume (length) $|\cos t_0|dt$. Having conveyed a rough idea of this topic and its importance, we leave its further development to an upper-level course in analysis.

Regions More General Than Boxes

In the remainder of this section, we will be working with the volume of *sufficiently nice regions* in $\mathbb{R}^n$. To define such a notion carefully would require an excursion into calculus of several variables. We will simply assume that we all have an intuitive notion of such regions having a well-defined volume.

Let $T: \mathbb{R}^n \to \mathbb{R}^n$ be a linear transformation of rank n, and let A be its standard matrix representation. Recall that the jth column vector of A is $T(\mathbf{e}_j)$ (see p. 146). We will show that, if a region G in $\mathbb{R}^n$ has volume V, the image of G under T has volume

$$|\det(A)| \cdot V.$$

That is, the volume of a region is multiplied by $|\det(A)|$ when the region is transformed by T. This result has important applications to integral calculus.

The volume of a sufficiently nice region G in $\mathbb{R}^n$ may be approximated by adding the volumes of small n-cubes contained in G and having edges of length c parallel to the coordinate axes. Figure 4.13(a) illustrates this situation for a plane region in $\mathbb{R}^2$, where a grid of small squares (2-cubes) is placed on the region. As the length c of the edges of the squares approaches zero, the sum of the areas of the colored squares inside the region approaches the area of the region. These squares inside G are mapped by T into parallelograms of area $c^2|\det(A)|$ inside the image of G under T. (See the colored parallelograms in Figure 4.13(b). As c approaches zero, the sum of the areas of these parallelograms approaches the area of the image of G under T, which thus must be the area of G multiplied by $|\det(A)|$. A similar construction can be made with a grid of n-cubes for a region G in $\mathbb{R}^n$. Each such cube is mapped by T into an n-box of volume $c^n|\det(A)|$. Adding the volumes of these n-boxes and taking the limiting value of the sum as c approaches zero, we see that the volume of the image under T of the region G is given by

$$\text{Volume of image of } G = |\det(A)| \cdot (\text{Volume of } G). \tag{4}$$

We summarize this work in a theorem on the following page.

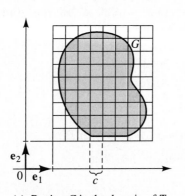

(a) Region G in the domain of T

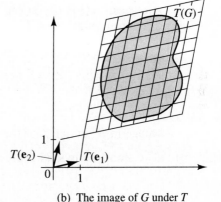

(b) The image of G under T

FIGURE 4.13

THEOREM 4.8 Volume-Change Factor for $T: \mathbb{R}^n \to \mathbb{R}^n$

Let G be a region in $\mathbb{R}^n$ of volume V, and let $T: \mathbb{R}^n \to \mathbb{R}^n$ be a linear transformation of rank n with standard matrix representation A. Then the volume of the image of G under T is $|\det(A)| \cdot V$.

EXAMPLE 6 Let $T: \mathbb{R}^2 \to \mathbb{R}^2$ be the linear transformation of the plane given by $T([x, y]) = [2x - y, x + 3y]$. Find the area of the image under T of the disk $x^2 + y^2 \le 4$ in the domain of T.

SOLUTION The disk $x^2 + y^2 \le 4$ has radius 2 and area 4π. It is mapped by T into a region (actually bounded by an ellipse) of area

$$|\det(A)| \cdot (4\pi) = \begin{vmatrix} 2 & -1 \\ 1 & 3 \end{vmatrix} \cdot (4\pi) = (6 + 1)(4\pi) = 28\pi. \qquad \blacksquare$$

We can generalize Theorem 4.8 to a linear transformation $T: \mathbb{R}^n \to \mathbb{R}^m$, where $m \ge n$ and T has rank n. This time, the standard matrix representation A is an $m \times n$ matrix. The image under T of the unit n-cube in $\mathbb{R}^n$ outlined by $\mathbf{e}_1, \mathbf{e}_2, \ldots, \mathbf{e}_n$ is the n-box in $\mathbb{R}^m$ outlined by $T(\mathbf{e}_1), T(\mathbf{e}_2), \ldots, T(\mathbf{e}_n)$. According to Theorem 4.7, the volume of this box in $\mathbb{R}^m$ is

$$\sqrt{\det(A^TA)}.$$

The same grid argument used earlier and illustrated in Figure 4.13 shows that a region G in $\mathbb{R}^n$ of volume V is mapped by T into a region of $\mathbb{R}^m$ of volume

$$\sqrt{\det(A^TA)} \cdot V.$$

We summarize this generalization in a theorem.

THEOREM 4.9 Volume-Change Factor for $T: \mathbb{R}^n \to \mathbb{R}^m$

Let G be a region in $\mathbb{R}^n$ of volume V. Let $m \ge n$ and let $T: \mathbb{R}^n \to \mathbb{R}^m$ be a linear transformation of rank n. Let A be the standard matrix representation of T. Then the volume of the image of G in $\mathbb{R}^m$ under the transformation T is

$$\sqrt{\det(A^TA)} \cdot V.$$

EXAMPLE 7 Let $T: \mathbb{R}^2 \to \mathbb{R}^3$ be given by $T([x, y]) = [2x + 3y, x - y, 2y]$. Find the area of the image in $\mathbb{R}^3$ under T of the disk $x^2 + y^2 \le 4$ in $\mathbb{R}^2$.

SOLUTION The standard matrix representation A of T is

$$A = \begin{bmatrix} 2 & 3 \\ 1 & -1 \\ 0 & 2 \end{bmatrix}$$

and

$$A^T A = \begin{bmatrix} 2 & 1 & 0 \\ 3 & -1 & 2 \end{bmatrix} \begin{bmatrix} 2 & 3 \\ 1 & -1 \\ 0 & 2 \end{bmatrix} = \begin{bmatrix} 5 & 5 \\ 5 & 14 \end{bmatrix}.$$

Thus,

$$\sqrt{\det(A^T A)} = \sqrt{70 - 25} = \sqrt{45} = 3\sqrt{5}.$$

A region G of $\mathbb{R}^2$ having area V is mapped by T into a plane region of area $3\sqrt{5} \cdot V$ in $\mathbb{R}^3$. Thus the disk $x^2 + y^2 \leq 4$ of area 4π is mapped into a plane region in $\mathbb{R}^3$ of area

$$(3\sqrt{5})(4\pi) = 12\pi\sqrt{5}. \quad \blacksquare$$

SUMMARY

1. An n-box in $\mathbb{R}^m$, where $m \geq n$, is determined by n independent vectors $\mathbf{a}_1, \mathbf{a}_2, \ldots, \mathbf{a}_n$ and consists of all vectors $\mathbf{x}$ in $\mathbb{R}^m$ such that

   $$\mathbf{x} = t_1\mathbf{a}_1 + t_2\mathbf{a}_2 + \cdots + t_n\mathbf{a}_n,$$

   where $0 \leq t_i \leq 1$ for $i = 1, 2, \ldots, n$.

2. A 1-box in $\mathbb{R}^m$ is a line segment, and its "volume" is its length.

3. A 2-box in $\mathbb{R}^m$ is a parallelogram determined by two independent vectors, and the "volume" of the 2-box is the area of the parallelogram.

4. A 3-box in $\mathbb{R}^m$ is a skewed box (parallelepiped) in the usual sense, and its volume is the usual volume.

5. Let $\mathbf{a}_1, \mathbf{a}_2, \ldots, \mathbf{a}_n$ be independent vectors in $\mathbb{R}^m$ for $m \geq n$, and let A be the $m \times n$ matrix with jth column vector $\mathbf{a}_j$. The volume of the n-box in $\mathbb{R}^m$ determined by the n vectors is $\sqrt{\det(A^T A)}$.

6. For the case of an n-box in the space $\mathbb{R}^n$ of the same dimension, the formula for its volume given in summary item 5 reduces to $|\det(A)|$.

7. If $T: \mathbb{R}^n \rightarrow \mathbb{R}^n$ is a linear transformation of rank n with standard matrix representation A, then T maps a region in its domain of volume V into a region of volume $|\det(A)|V$.

8. If $T: \mathbb{R}^n \rightarrow \mathbb{R}^m$ is a linear transformation of rank n with standard matrix representation A, then T maps a region in its domain of volume V into a region of $\mathbb{R}^m$ of volume $\sqrt{\det(A^T A)} \cdot V$.

EXERCISES

1. Find the area of the parallelogram in $\mathbb{R}^3$ determined by the vectors $[0, 1, 4]$ and $[-1, 3, -2]$.

2. Find the area of the parallelogram in $\mathbb{R}^5$ determined by the vectors $[1, 0, 1, 2, -1]$ and $[0, 1, -1, 1, 3]$.

3. Find the volume of the 3-box in $\mathbb{R}^4$ determined by the vectors $[-1, 2, 0, 1]$, $[0, 1, 3, 0]$, and $[0, 0, 2, -1]$.

4. Find the volume of the 4-box in $\mathbb{R}^5$ determined by the vectors $[1, 1, 1, 0, 1]$, $[0, 1, 1, 0, 0]$, $[3, 0, 1, 0, 0]$, and $[1, -1, 0, 0, 1]$.

In Exercises 5–10, find the volume of the n-box determined by the given vectors in $\mathbb{R}^n$.

5. $[-1, 4]$, $[2, 3]$ in $\mathbb{R}^2$
6. $[-5, 3]$, $[1, 7]$ in $\mathbb{R}^2$
7. $[1, 3, -5]$, $[2, 4, -1]$, $[3, 1, 2]$ in $\mathbb{R}^3$
8. $[-1, 4, 7]$, $[3, -2, -1]$, $[4, 0, 2]$ in $\mathbb{R}^3$
9. $[1, 0, 0, 1]$, $[2, -1, 3, 0]$, $[0, 1, 3, 4]$, $[-1, 1, -2, 1]$ in $\mathbb{R}^4$
10. $[1, -1, 0, 1]$, $[2, -1, 3, 1]$, $[-1, 4, 2, -1]$, $[0, 1, 0, 2]$ in $\mathbb{R}^4$

11. Find the area of the triangle in $\mathbb{R}^3$ with vertices $(-1, 2, 3)$, $(0, 1, 4)$, and $(2, 1, 5)$. [HINT: Think of vectors emanating from $(-1, 2, 3)$. The triangle may be viewed as half a parallelogram.]

12. Find the volume of the tetrahedron in $\mathbb{R}^3$ with vertices $(1, 0, 3)$, $(-1, 2, 4)$, $(3, -1, 2)$, and $(2, 0, -1)$. [HINT: Think of vectors emanating from $(1, 0, 3)$.]

13. Find the volume of the tetrahedron in $\mathbb{R}^4$ with vertices $(1, 0, 0, 1)$, $(-1, 2, 0, 1)$, $(3, 0, 1, 1)$, and $(-1, 4, 0, 1)$. [HINT: See the hint for Exercise 12.]

14. Give a geometric interpretation of the fact that an $n \times n$ matrix with two equal rows has determinant zero.

15. Using the results of this section, give a criterion that four points in $\mathbb{R}^n$ lie in a plane.

16. Determine whether the points $(1, 0, 1, 0)$, $(-1, 1, 0, 1)$, $(0, 1, -1, 1)$, and $(1, -1, 4, -1)$ lie in a plane in $\mathbb{R}^4$. (See Exercise 15.)

17. Determine whether the points $(2, 0, 1, 3)$, $(3, 1, 0, 1)$, $(-1, 2, 0, 4)$, and $(3, 1, 2, 4)$ lie in a plane in $\mathbb{R}^4$. (See Exercise 15.)

In Exercises 18–21, let $T: \mathbb{R}^2 \to \mathbb{R}^2$ be the linear transformation defined by $T([x, y]) =$ $[4x - 2y, 2x + 3y]$. Find the area of the image under T of each of the given regions in $\mathbb{R}^2$.

18. The square $0 \le x \le 1$, $0 \le y \le 1$
19. The rectangle $-1 \le x \le 1$, $1 \le y \le 2$
20. The parallelogram determined by $2\mathbf{e}_1 + 3\mathbf{e}_2$ and $4\mathbf{e}_1 - \mathbf{e}_2$
21. The disk $(x - 1)^2 + (y + 2)^2 \le 9$

In Exercises 22–25, let $T: \mathbb{R}^3 \to \mathbb{R}^3$ be defined by $T([x, y, z]) = [x - 2y, 3x + z, 4x + 3y]$. Find the volume of the image under T of each of the given regions in $\mathbb{R}^3$.

22. The cube $0 \le x \le 1$, $0 \le y \le 1$, $0 \le z \le 1$
23. The box $0 \le x \le 2$, $-1 \le y \le 3$, $2 \le z \le 5$
24. The box determined by $2\mathbf{e}_1 + 3\mathbf{e}_2 - \mathbf{e}_3$, $4\mathbf{e}_1 - 2\mathbf{e}_3$, and $\mathbf{e}_1 - \mathbf{e}_2 + 2\mathbf{e}_3$
25. The ball $x^2 + (y - 3)^2 + (z + 2)^2 \le 16$

In Exercises 26–29, let $T: \mathbb{R}^2 \to \mathbb{R}^3$ be the linear transformation defined by $T([x, y]) =$ $[y, x, x + y]$. Find the area of the image under T of each of the given regions in $\mathbb{R}^2$.

26. The square $0 \le x \le 1$, $0 \le y \le 1$
27. The rectangle $2 \le x \le 3$, $-1 \le y \le 4$
28. The triangle with vertices $(0, 0)$, $(6, 0)$, $(0, 3)$
29. The disk $x^2 + y^2 \le 25$

In Exercises 30–32, let $T: \mathbb{R}^2 \to \mathbb{R}^4$ be defined by $T([x, y]) = [x - y, x, -y, 2x + y]$. Find the area of the image under T of each of the given regions in $\mathbb{R}^2$.

30. The square $0 \le x \le 1$, $0 \le y \le 1$
31. The square $-1 \le x \le 3$, $-1 \le y \le 3$
32. The disk $x^2 + y^2 \le 9$

33. **a.** If one attempts to define an n-box in $\mathbb{R}^m$ for $n > m$, what will its volume as an n-box be?
    b. Let A be an $m \times n$ matrix with $n > m$. Find $\det(A^T A)$.

34. We have seen that, for $n \times n$ matrices A and B, we have $\det(AB) = \det(A) \cdot \det(B)$, but the proof was not intuitive. Give an intuitive

geometric argument showing that at least $|\det(AB)| = |\det(A)| \cdot |\det(B)|$. [HINT: Use the fact that, if A is the standard matrix representation of $T: \mathbb{R}^n \to \mathbb{R}^n$ and B is the standard matrix representation of $T': \mathbb{R}^n \to \mathbb{R}^n$, then AB is the standard matrix representation $T \circ T'$.]

35. Let $T: \mathbb{R}^n \to \mathbb{R}^n$ be a linear transformation of rank n with standard matrix representation A. Mark each of the following True or False.

___ **a.** The image under T of a box in $\mathbb{R}^n$ is again a box in $\mathbb{R}^n$.

___ **b.** The image under T of an n-box in $\mathbb{R}^n$ of volume V is a box in $\mathbb{R}^n$ of volume $\det(A) \cdot V$.

___ **c.** The image under T of an n-box in $\mathbb{R}^n$ of volume > 0 is a box in $\mathbb{R}^n$ of volume > 0.

___ **d.** If the image under T of an n-box B in $\mathbb{R}^n$ has volume 12, the box B has volume $|\det(A)| \cdot 12$.

___ **e.** If the image under T of an n-box B in $\mathbb{R}^n$ has volume 12, the box B has volume $12/|\det(A)|$.

___ **f.** If $n = 2$, the image under T of the unit disk $x^2 + y^2 \leq 1$ has area $|\det(A)|$.

___ **g.** The linear transformation T is an isomorphism.

___ **h.** The image under $T \circ T$ of an n-box in $\mathbb{R}^n$ of volume V is a box in $\mathbb{R}^n$ of volume $\det(A^2) \cdot V$.

___ **i.** The image under $T \circ T \circ T$ of an n-box in $\mathbb{R}^n$ of volume V is a box in $\mathbb{R}^n$ of volume $\det(A^3) \cdot V$.

___ **j.** The image under T of a nondegenerate 1-box is again nondegenerate.

36. Prove Eq. (1); that is, prove that the square of the length of the line segment determined by $\mathbf{a}_1$ in $\mathbb{R}^n$ is $\|\mathbf{a}_1\|^2 = \det([\mathbf{a}_1 \cdot \mathbf{a}_1])$.

EIGENVALUES AND EIGENVECTORS

This chapter introduces the important topic of eigenvalues and eigenvectors of a square matrix and the associated linear transformation. Determinants enable us to give illustrative computations and applications involving matrices of very small size. Eigenvectors and eigenvalues continue to appear at intervals throughout much of the rest of the text. In Section 8.4 we discuss some other methods for computing them.

5.1   EIGENVALUES AND EIGENVECTORS

Encounters with $A^k \mathbf{x}$

In Section 1.7 we studied Markov chains dealing with the distribution of a population among states, measured over evenly spaced time intervals. An $n \times n$ transition matrix T describes the movement of the population among the states during one time interval. The matrix T has the property that all entries are nonnegative and the sum of the entries in any column is 1. Suppose that $\mathbf{p}$ is the initial population distribution vector—that is, the column vector whose ith component is the proportion of the population in the ith state at the start of the process. Then $T\mathbf{p}$ is the corresponding population distribution vector after one time interval. Similarly, $T^2\mathbf{p}$ is the population distribution vector after two time intervals, and in general, $T^k\mathbf{p}$ gives the distribution of population among the states after k time intervals.

Markov chains provide one example in which we are interested in computing $A^k\mathbf{x}$ for an $n \times n$ matrix A and a column vector $\mathbf{x}$ of n components. We give a famous classical problem that provides another illustration.

EXAMPLE 1   (*Fibonacci's rabbits*) Suppose that newly born pairs of rabbits produce no offspring during the first month of their lives, but each pair produces one new pair each subsequent month. Starting with $F_1 = 1$ newly born pair in the first month, find the number F_k of pairs in the kth month, assuming that no rabbit dies.

SOLUTION   In the kth month, the number of pairs of rabbits is

F_k = (Number of pairs alive the preceding month)
$\quad$ + (Number of newly born pairs for the kth month).

Because our rabbits do not produce offspring during the first month of their lives, we see that the number of newly born pairs for the kth month is the number F_{k-2} of pairs alive two months before. Thus we can write the equation above as

$$F_k = F_{k-1} + F_{k-2}. \quad \text{Fibonacci's relation} \tag{1}$$

It is convenient to set $F_0 = 0$, denoting 0 pairs for month 0 before the arrival of the first newly born pair, which is presumably a gift. Thus the sequence

$$F_0, F_1, F_2, \ldots, F_k, \ldots$$

for the number of pairs of rabbits becomes the **Fibonacci sequence**

$$0, 1, 1, 2, 3, 5, 8, 13, 21, 34, \ldots, \tag{2}$$

where each term starting with $F_2 = 0 + 1 = 1$ is the sum of the two preceding terms. For any particular k, we could compute F_k by writing out the sequence far enough. ∎

Fibonacci published this problem early in the thirteenth century. The Fibonacci sequence (2) occurs naturally in a surprising number of places. For example, leaves appear in a spiral pattern along a branch. Some trees have five growths of leaves for every two turns, others have eight growths for every three turns, and still others have 13 growths for every five turns; note the appearance of these numbers in the sequence (2). A mathematical journal, the *Fibonacci Quarterly,* has published many papers relating to the Fibonacci sequence.

We said in Example 1 that F_k can be found by simply writing out enough terms of the sequence (2). That can be a tedious task, even if we want to compute only F_{30}. Linear algebra gives us another approach to this problem. The Fibonacci relation (1) can be expressed in matrix form. We see that

$$\begin{bmatrix} F_k \\ F_{k-1} \end{bmatrix} = \begin{bmatrix} 1 & 1 \\ 1 & 0 \end{bmatrix} \begin{bmatrix} F_{k-1} \\ F_{k-2} \end{bmatrix}.$$

Thus, if we set

$$\mathbf{x}_k = \begin{bmatrix} F_k \\ F_{k-1} \end{bmatrix} \quad \text{and} \quad A = \begin{bmatrix} 1 & 1 \\ 1 & 0 \end{bmatrix},$$

we find that

$$\mathbf{x}_k = A\mathbf{x}_{k-1}. \tag{3}$$

Applying Eq. (3) repeatedly, we see that

$$\mathbf{x}_2 = A\mathbf{x}_1, \qquad \mathbf{x}_3 = A\mathbf{x}_2 = A^2\mathbf{x}_1, \qquad \mathbf{x}_4 = A\mathbf{x}_3 = A^3\mathbf{x}_1,$$

and in general

$$\mathbf{x}_k = A^{k-1}\mathbf{x}_1 = \begin{bmatrix} 1 & 1 \\ 1 & 0 \end{bmatrix}^{k-1} \begin{bmatrix} 1 \\ 0 \end{bmatrix}. \tag{4}$$

Thus we can compute the kth Fibonacci number F_k by finding A^{k-1} and multiplying it on the right by the column vector $\mathbf{x}_1$. Raising a matrix to a power is also a bit of a job, but the routine MATCOMP (in LINTEK) or MATLAB can easily find F_{30} for us. (See Exercise 45.)

Both Markov chains and the Fibonacci sequence lead us to computations of the form $A^k\mathbf{x}$. Other examples leading to $A^k\mathbf{x}$ abound in the physical and social sciences.

> Computations of $A^k\mathbf{x}$ arise in any process in which information given by a column vector gives rise to analogous information at a later time by multiplying the vector by a matrix A.

Eigenvalues and Eigenvectors

Suppose that A is an $n \times n$ matrix and $\mathbf{v}$ is a column vector with n components such that

$$A\mathbf{v} = \lambda\mathbf{v} \tag{5}$$

for some scalar λ. Geometrically, Eq. (5) asserts that $A\mathbf{v}$ is a vector parallel to $\mathbf{v}$. From $A\mathbf{v} = \lambda\mathbf{v}$, we obtain

$$A^2\mathbf{v} = A(A\mathbf{v}) = A(\lambda\mathbf{v}) = \lambda(A\mathbf{v}) = \lambda(\lambda\mathbf{v}) = \lambda^2\mathbf{v}.$$

It is easy to show, in general, that $A^k\mathbf{v} = \lambda^k\mathbf{v}$. (See Exercise 27.) Thus, $A^k\mathbf{x}$ is easily computed if $\mathbf{x}$ is equal to *this vector* $\mathbf{v}$.

DEFINITION 5.1   Eigenvalues and Eigenvectors

Let A be an $n \times n$ matrix. A scalar λ is an **eigenvalue** of A if there is a *nonzero* column vector $\mathbf{v}$ in $\mathbb{R}^n$ such that $A\mathbf{v} = \lambda\mathbf{v}$. The vector $\mathbf{v}$ is then an **eigenvector** of A corresponding to λ. (The terms **characteristic vector** and **characteristic value** or **proper vector** and **proper value** are also used in place of *eigenvector* and *eigenvalue*, respectively.)

For example, the computations

$$\begin{bmatrix} 2 & 2 \\ 3 & 1 \end{bmatrix} \begin{bmatrix} 1 \\ 1 \end{bmatrix} = \begin{bmatrix} 4 \\ 4 \end{bmatrix} = 4 \begin{bmatrix} 1 \\ 1 \end{bmatrix}$$

show that the vector $\begin{bmatrix} 1 \\ 1 \end{bmatrix}$ is an eigenvector of the matrix $\begin{bmatrix} 2 & 2 \\ 3 & 1 \end{bmatrix}$ corresponding to the eigenvalue 4.

For many matrices A, the computation of $A^k\mathbf{x}$ for a *general vector* $\mathbf{x}$ is greatly simplified by first finding all eigenvalues and eigenvectors of A. Suppose that we can find a basis $\{\mathbf{b}_1, \mathbf{b}_2, \ldots, \mathbf{b}_n\}$ for $\mathbb{R}^n$ consisting of eigenvectors of A, so that for eigenvalues $\lambda_1, \lambda_2, \ldots, \lambda_n$ we have

$$A\mathbf{b}_i = \lambda_i\mathbf{b}_i \quad \text{for} \quad i = 1, 2, \ldots, n.$$

To compute $A^k\mathbf{x}$, we first express $\mathbf{x}$ as a linear combination of these basis eigenvectors:

$$\mathbf{x} = d_1\mathbf{b}_1 + d_2\mathbf{b}_2 + \cdots + d_n\mathbf{b}_n.$$

Because $A^k\mathbf{b}_i = \lambda_i^k\mathbf{b}_i$, we then obtain

$$A^k\mathbf{x} = \lambda_1^k d_1\mathbf{b}_1 + \lambda_2^k d_2\mathbf{b}_2 + \cdots + \lambda_n^k d_n\mathbf{b}_n. \tag{6}$$

It can be shown that if complex numbers are allowed as vector components and scalars, then such a basis consisting of eigenvectors usually exists, in the sense that if A is chosen at random, it exists with probability 1. When we limit ourselves to real-number components and scalars, such a basis often does not exist. However, it can be shown that if A is a real *symmetric* matrix, which is the case for some applications, then we can find a basis of eigenvectors. This idea is explored further in Section 5.2.

Equation (6) can give important information on the effect of repeatedly transforming $\mathbb{R}^n$ using multiplication by A. Suppose that at least one of the eigenvalues λ_i—we may as well assume that it is λ_1—has magnitude greater than 1. Then vectors of arbitrarily large magnitude can be obtained under repeated multiplication of $\mathbf{b}_1$ by A—that is, $\lim_{k\to\infty} |A^k\mathbf{b}_1| = \infty$. Multiplication by A is then called an **unstable transformation** of $\mathbb{R}^n$. On the other hand, if $|\lambda_i| < 1$ for $i = 1, 2, \ldots, n$, then $\lim_{k\to\infty} |A^k\mathbf{x}| = 0$ for all $\mathbf{x}$ in $\mathbb{R}^n$ and the transformation is **stable**. If the maximum of the $|\lambda_i|$ is 1, the transformation is **neutrally stable**.

Suppose now that in Eq. (6), one of the λ_i—again, we may as well assume that it is λ_1—is of magnitude greater than all of the other λ_i. Then λ_1^k *dominates* the other λ_i^k for large values of k. If $d_1 \neq 0$, this means that the term $\lambda_1^k d_1\mathbf{b}_1$ dominates the other terms in Eq. (6), so that for large values of k, the vector $A^k\mathbf{x}$ is nearly parallel to $\mathbf{b}_1$, and one more multiplication of $A^k\mathbf{x}$ by A amounts approximately to multiplying each component of $A^k\mathbf{x}$ by λ_1. Indeed, one way to compute such a dominant eigenvalue λ_1 is to use a computer and find the ratio of the components of $A^{k+1}\mathbf{x}$ to those of $A^k\mathbf{x}$ for a suitable value of k. Quite small values of k will serve if λ_1 dominates strongly—for instance, if its magnitude is several times the next largest magnitude of an eigenvalue. If the ratio of those magnitudes is closer to 1—say, 1.2—then we have to use a larger value for k. MATLAB can be used to illustrate this. (See the MATLAB exercises.) This method of computing the eigenvalue of maximum magnitude is called the *power method*, and is refined and treated in more detail in Section 8.4, where some additional methods for computing eigenvalues are described.

Computing Eigenvalues

In this section we show how a determinant can be used to find eigenvalues of an $n \times n$ matrix A; the computational technique is practical only for relatively small matrices.

We write the equation $A\mathbf{v} = \lambda\mathbf{v}$ as $A\mathbf{v} - \lambda\mathbf{v} = \mathbf{0}$, or as $A\mathbf{v} - \lambda I\mathbf{v} = \mathbf{0}$, where I is the $n \times n$ identity matrix. This last equation can be written as $(A - \lambda I)\mathbf{v} = \mathbf{0}$, so $\mathbf{v}$ must be a solution of the homogeneous linear system

$$(A - \lambda I)\mathbf{x} = \mathbf{0}. \tag{7}$$

An eigenvalue of A is thus a scalar λ for which system (7) has a nontrivial solution $\mathbf{v}$. (Recall that an eigenvector is nonzero by definition.) We know that system (7) has a nontrivial solution precisely when the determinant of the coefficient matrix is zero—that is, if and only if

$$\det(A - \lambda I) = 0. \tag{8}$$

HISTORICAL NOTE   THE CONCEPT OF AN EIGENVALUE, in its origins and its later development, was independent of matrix theory. In fact, its original context was that of the solution of systems of linear differential equations with constant coefficients (see Section 5.3). Jean Le Rond D'Alembert (1717–1783), in his work in the 1740s and 1750s on the motion of a string loaded with a finite number of masses, considered the system

$$\frac{d^2y_i}{dt^2} + \sum_{k=1}^{3} a_{ik}y_k = 0, \qquad i = 1, 2, 3.$$

(Here the number of masses is restricted to 3 for simplicity.) To solve this system, D'Alembert multiplied the ith equation by a constant v_i (to be determined) for each i and added the equations together to obtain

$$\sum_{i=1}^{3} v_i \frac{d^2y_i}{dt^2} + \sum_{i,k=1}^{3} v_i a_{ik}y_k = 0.$$

If the v_i are chosen so that $\Sigma_i \, v_i a_{ki} + \lambda v_k = 0$ for $k = 1, 2, 3$—that is, if the vector $[v_1, v_2, v_3]$ is an eigenvector corresponding to the eigenvalue $-\lambda$ for the matrix $A = [a_{ki}]$, then the substitution $u = v_1y_1 + v_2y_2 + v_3y_3$ converts the original system to the single differential equation

$$\frac{d^2u}{dt^2} + \lambda u = 0.$$

This equation can now easily be solved and leads to solutions for the y_k. It is not difficult to show that λ is determined by a cubic equation that has three roots.

Eigenvalues also appeared in other situations involving systems of differential equations, including physical situations studied by Euler and Lagrange.

If $A = [a_{ij}]$, then the previous equation can be written

$$\begin{vmatrix} a_{11} - \lambda & a_{12} & \cdots & a_{1n} \\ a_{21} & a_{22} - \lambda & \cdots & a_{2n} \\ \vdots & \vdots & & \vdots \\ a_{n1} & a_{n2} & \cdots & a_{nn} - \lambda \end{vmatrix} = 0. \qquad (9)$$

If we expand the determinant in Eq. (9), we obtain a polynomial expression $p(\lambda)$ of degree n with coefficients involving the a_{ij}. That is,

$$\det(A - \lambda I) = p(\lambda).$$

The polynomial $p(\lambda)$ is the **characteristic polynomial*** of the matrix A. The eigenvalues of A are precisely the solutions of the **characteristic equation** $p(\lambda) = 0$.

EXAMPLE 2   Find the eigenvalues of the matrix

$$A = \begin{bmatrix} 3 & 2 \\ 2 & 0 \end{bmatrix}.$$

SOLUTION   The characteristic polynomial of A is

$$\det(A - \lambda I) = \begin{vmatrix} 3 - \lambda & 2 \\ 2 & -\lambda \end{vmatrix} = \lambda^2 - 3\lambda - 4.$$

The characteristic equation is

$$\lambda^2 - 3\lambda - 4 = 0,$$

and we obtain $(\lambda - 4)(\lambda + 1) = 0$; therefore, $\lambda_1 = -1$ and $\lambda_2 = 4$ are eigenvalues of A.   ■

EXAMPLE 3   Show that $\lambda_1 = 1$ is an eigenvalue of the transition matrix for any Markov chain.

SOLUTION   Let T be an $n \times n$ transition matrix for a Markov chain; that is, all entries in T are nonnegative and the sum of the entries in each column of T is 1. We easily see that the sum of the entries in any column of $T - I$ must be zero. Thus the

---

*Many authors define the characteristic polynomial of A to be $p(\lambda) = \det(\lambda I - A)$ rather than $p(\lambda) = \det(A - \lambda I)$. Because $\lambda I - A = (-1)(A - \lambda I)$, we see that for an $n \times n$ matrix A, we have $\det(\lambda I - A) = (-1)^n \det(A - \lambda I)$. Thus the two definitions differ only when n is an odd integer, in which case the two polynomials differ only in sign. The definition $\det(\lambda I - A)$ has the advantage that the term of highest degree is always λ^n, rather than being $-\lambda^n$ when n is odd. We use the definition $p(\lambda) = \det(A - \lambda I)$ in this first course because for our pencil-and-paper computations, it is easier to write "$-\lambda$" after each diagonal entry than to change the sign of every entry and then write "$\lambda +$" before the diagonal entries. However, the command **poly(A)** in MATLAB will produce the coefficients of the polynomial $\det(\lambda I - A)$.

sum of the row vectors of $T - I$ is the zero vector, so the rows of $T - I$ are linearly dependent. Consequently, rank$(T - I) < n$, and $(T - I)\mathbf{x} = \mathbf{0}$ has a nontrivial solution, so $\lambda_1 = 1$ is an eigenvalue of T.   ∎

The characteristic equation of an $n \times n$ matrix is a polynomial equation of degree n. This equation has n solutions if we allow both real and complex numbers and if we count the possible multiplicities greater than 1 of some solutions. Linear algebra can be done using complex scalars as well as real scalars. Introducing complex scalars makes the theory simpler and illuminates behavior in the real-scalar case. However, pencil-and-paper computations involving complex numbers can be very laborious. We can expect an eigenvector corresponding to an eigenvalue $\lambda = a + bi$ with $b \neq 0$ to have some components that are not real numbers. Exercise 44 and some of the computer exercises require computations using complex numbers.

We have now run into complex numbers while attempting to do linear algebra within $\mathbb{R}^n$. With the exception of the exercises just mentioned, we leave the computation of complex eigenvalues and of eigenvectors with complex components to Chapter 9. It is only in the context of complex n-space $\mathbb{C}^n$, which consists of all n-tuples of complex numbers, that the complete eigenstory can be explained. For example, diagonalization of a square matrix with randomly chosen real number entries, which we will discuss in Section 5.2, often cannot be achieved if we restrict ourselves only to real numbers. Using complex numbers, such a matrix can be diagonalized with probability 1. We believe it is important that you be aware of the complete eigenstory, even if you do not study Chapter 9. For this reason, theorems and definitions in this chapter are often phrased using a parenthetical *possibly complex* or stated in terms of n-*space*, meaning either $\mathbb{R}^n$ or $\mathbb{C}^n$. Looking back at Definition 5.1, the same definitions of eigenvalues and eigenvectors apply if A has complex entries and we allow vectors $\mathbf{v}$ in $\mathbb{C}^n$ and complex scalars λ.

The characteristic polynomial of a matrix A may have multiple roots; perhaps it has -2 as a root of multiplicity 1, and 5 as a root of multiplicity 2, corresponding to factors $(\lambda + 2)(\lambda - 5)^2$ of the characteristic polynomial. Suppose that these are the only roots. We will say that the eigenvalues are $\lambda_1 = -2$ and $\lambda_2 = \lambda_3 = 5$. That is, if there are a total of k distinct roots of the characteristic polynomial of degree n, it is convenient for us to denote the roots by $\lambda_1, \lambda_2, \ldots, \lambda_n$. Our next example illustrates this.

EXAMPLE 4   Find the eigenvalues of the matrix

$$A = \begin{bmatrix} 2 & 1 & 0 \\ -1 & 0 & 1 \\ 1 & 3 & 1 \end{bmatrix}.$$

SOLUTION   The characteristic polynomial is

$$p(\lambda) = \begin{vmatrix} 2 - \lambda & 1 & 0 \\ -1 & -\lambda & 1 \\ 1 & 3 & 1-\lambda \end{vmatrix} = (2 - \lambda)\begin{vmatrix} -\lambda & 1 \\ 3 & 1-\lambda \end{vmatrix} - 1\begin{vmatrix} -1 & 1 \\ 1 & 1-\lambda \end{vmatrix}$$

$$= (2 - \lambda)(\lambda^2 - \lambda - 3) - (\lambda - 2) = -(\lambda - 2)(\lambda^2 - \lambda - 2)$$
$$= -(\lambda - 2)(\lambda - 2)(\lambda + 1).$$

Hence, the eigenvalues of A are $\lambda_1 = -1$ and $\lambda_2 = \lambda_3 = 2$. ∎

Computation of Eigenvectors

We turn to the computation of the eigenvectors corresponding to an eigenvalue λ of a matrix A. Having found the eigenvalue, we substitute it in homogeneous system (7) and solve to find the nontrivial solutions of the system. We will obtain an infinite number of nontrivial solutions, each of which is an eigenvector corresponding to the eigenvalue λ.

EXAMPLE 5   Find the eigenvectors corresponding to each eigenvalue found in Example 4 for the matrix

$$A = \begin{bmatrix} 2 & 1 & 0 \\ -1 & 0 & 1 \\ 1 & 3 & 1 \end{bmatrix}.$$

SOLUTION   The eigenvalues of A were found to be $\lambda_1 = -1$ and $\lambda_2 = \lambda_3 = 2$. We substitute each of these values in the homogeneous system (7). The eigenvectors are obtained by reducing the coefficient matrix $A - \lambda I$ in the augmented matrix for the system. For $\lambda_1 = -1$, we obtain

$$[A - \lambda_1 I \mid \mathbf{0}] = [A + I \mid \mathbf{0}]$$

$$= \begin{bmatrix} 3 & 1 & 0 & \mid & 0 \\ -1 & 1 & 1 & \mid & 0 \\ 1 & 3 & 2 & \mid & 0 \end{bmatrix} \sim \begin{bmatrix} 1 & 3 & 2 & \mid & 0 \\ 0 & 4 & 3 & \mid & 0 \\ 0 & -8 & -6 & \mid & 0 \end{bmatrix}$$

$$\sim \begin{bmatrix} 1 & 3 & 2 & \mid & 0 \\ 0 & 4 & 3 & \mid & 0 \\ 0 & 0 & 0 & \mid & 0 \end{bmatrix}.$$

(In the future, we will drop the column of zeros to the right of the partition when solving for eigenvectors.) The solution of the homogeneous system is given by

$$\begin{bmatrix} r/4 \\ -3r/4 \\ r \end{bmatrix} \quad \text{for any scalar } r.$$

Therefore,

$$\mathbf{v}_1 = \begin{bmatrix} r/4 \\ -3r/4 \\ r \end{bmatrix} = r \begin{bmatrix} \frac{1}{4} \\ -\frac{3}{4} \\ 1 \end{bmatrix} \quad \text{for any } \textit{nonzero} \text{ scalar } r$$

is an eigenvector corresponding to the eigenvalue $\lambda_1 = -1$. Replacing r by $4r$, we can express this result without fractions as

$$\mathbf{v}_1 = \begin{bmatrix} r \\ -3r \\ 4r \end{bmatrix} = r \begin{bmatrix} 1 \\ -3 \\ 4 \end{bmatrix} \quad \text{for any nonzero scalar } r.$$

For $\lambda_2 = 2$, we obtain

$$[A - \lambda_2 I] = [A - 2I]$$

$$= \begin{bmatrix} 0 & 1 & 0 \\ -1 & -2 & 1 \\ 1 & 3 & -1 \end{bmatrix} \sim \begin{bmatrix} 1 & 3 & -1 \\ 0 & 1 & 0 \\ 0 & 0 & 0 \end{bmatrix}.$$

This time we find that

$$\mathbf{v}_2 = \begin{bmatrix} s \\ 0 \\ s \end{bmatrix} = s \begin{bmatrix} 1 \\ 0 \\ 1 \end{bmatrix} \quad \text{for any nonzero scalar } s$$

is an eigenvector. As a check, we could compute $A\mathbf{v}_1$ and $A\mathbf{v}_2$. For example, we have

$$A\mathbf{v}_2 = \begin{bmatrix} 2 & 1 & 0 \\ -1 & 0 & 1 \\ 1 & 3 & 1 \end{bmatrix} \begin{bmatrix} s \\ 0 \\ s \end{bmatrix} = \begin{bmatrix} 2s \\ 0 \\ 2s \end{bmatrix} = 2 \begin{bmatrix} s \\ 0 \\ s \end{bmatrix} = 2\mathbf{v}_2 = \lambda_2 \mathbf{v}_2. \qquad \blacksquare$$

EXAMPLE 6   Find the eigenvalues and eigenvectors of the matrix

$$A = \begin{bmatrix} 1 & 0 & 0 \\ -8 & 4 & -6 \\ 8 & 1 & 9 \end{bmatrix}.$$

SOLUTION   The characteristic polynomial of A is

$$p(\lambda) = |A - \lambda I| = \begin{vmatrix} 1 - \lambda & 0 & 0 \\ -8 & 4 - \lambda & -6 \\ 8 & 1 & 9 - \lambda \end{vmatrix} = (1 - \lambda) \begin{vmatrix} 4 - \lambda & -6 \\ 1 & 9 - \lambda \end{vmatrix}$$

$$= (1 - \lambda)(\lambda^2 - 13\lambda + 42) = (1 - \lambda)(\lambda - 6)(\lambda - 7).$$

The eigenvalues of A are $\lambda_1 = 1$, $\lambda_2 = 6$, and $\lambda_3 = 7$. For $\lambda_1 = 1$, we have

$$A - \lambda_1 I = A - I = \begin{bmatrix} 0 & 0 & 0 \\ -8 & 3 & -6 \\ 8 & 1 & 8 \end{bmatrix} \sim \begin{bmatrix} 0 & 0 & 0 \\ 0 & 4 & 2 \\ 8 & 1 & 8 \end{bmatrix}$$

$$\sim \begin{bmatrix} 8 & 1 & 8 \\ 0 & 2 & 1 \\ 0 & 0 & 0 \end{bmatrix} \sim \begin{bmatrix} 1 & \frac{1}{8} & 1 \\ 0 & 1 & \frac{1}{2} \\ 0 & 0 & 0 \end{bmatrix} \sim \begin{bmatrix} 1 & 0 & \frac{15}{16} \\ 0 & 1 & \frac{1}{2} \\ 0 & 0 & 0 \end{bmatrix},$$

so

$$\mathbf{v}_1 = \begin{bmatrix} -15r/16 \\ -r/2 \\ r \end{bmatrix} = r \begin{bmatrix} -\frac{15}{16} \\ -\frac{1}{2} \\ 1 \end{bmatrix} \quad \text{for any nonzero scalar } r$$

is an eigenvector. Replacing r by $-16r$, we can express this result without the fractions as

$$\mathbf{v}_1 = \begin{bmatrix} 15r \\ 8r \\ -16r \end{bmatrix} = r \begin{bmatrix} 15 \\ 8 \\ -16 \end{bmatrix} \quad \text{for any nonzero scalar } r.$$

For $\lambda_2 = 6$, we have

$$A - \lambda_2 I = A - 6I = \begin{bmatrix} -5 & 0 & 0 \\ -8 & -2 & -6 \\ 8 & 1 & 3 \end{bmatrix} \sim \begin{bmatrix} 1 & 0 & 0 \\ 0 & -2 & -6 \\ 0 & 1 & 3 \end{bmatrix} \sim \begin{bmatrix} 1 & 0 & 0 \\ 0 & 1 & 3 \\ 0 & 0 & 0 \end{bmatrix},$$

so

$$\mathbf{v}_2 = \begin{bmatrix} 0 \\ -3s \\ s \end{bmatrix} = s \begin{bmatrix} 0 \\ -3 \\ 1 \end{bmatrix} \quad \text{for any nonzero scalar } s$$

is an eigenvector. Finally, for $\lambda_3 = 7$, we have

$$A - \lambda_3 I = A - 7I = \begin{bmatrix} -6 & 0 & 0 \\ -8 & -3 & -6 \\ 8 & 1 & 2 \end{bmatrix} \sim \begin{bmatrix} 1 & 0 & 0 \\ 0 & -3 & -6 \\ 0 & 1 & 2 \end{bmatrix} \sim \begin{bmatrix} 1 & 0 & 0 \\ 0 & 1 & 2 \\ 0 & 0 & 0 \end{bmatrix},$$

so

$$\mathbf{v}_3 = \begin{bmatrix} 0 \\ -2t \\ t \end{bmatrix} = t \begin{bmatrix} 0 \\ -2 \\ 1 \end{bmatrix} \quad \text{for any nonzero scalar } t$$

is an eigenvector. ∎

Properties of Eigenvalues and Eigenvectors

We turn now to algebraic properties of eigenvalues and eigenvectors. The properties given in Theorem 5.1 are so straightforward and instructive to prove that we leave the proofs as Exercises 27–29.

THEOREM 5.1   Properties of Eigenvalues and Eigenvectors

Let A be an $n \times n$ matrix.

1. If λ is an eigenvalue of A with $\mathbf{v}$ as a corresponding eigenvector, then λ^k is an eigenvalue of A^k, again with $\mathbf{v}$ as a corresponding eigenvector, for any positive integer k.

2. If λ is an eigenvalue of an invertible matrix A with $\mathbf{v}$ as a corresponding eigenvector, then $\lambda \neq 0$ and $1/\lambda$ is an eigenvalue of A^{-1}, again with $\mathbf{v}$ as a corresponding eigenvector.

3. If λ is an eigenvalue of A, then the set E_λ consisting of the zero vector together with all eigenvectors of A for this eigenvalue λ is a subspace of n-space, the **eigenspace** of λ.

Eigenvalues and Transformations

Let us consider the significance of an eigenvalue λ and corresponding eigenvector $\mathbf{v}$ of an $n \times n$ matrix A for the associated linear transformation $T(\mathbf{x}) = A\mathbf{x}$. The equation

$$A\mathbf{v} = \lambda\mathbf{v}$$

takes the form

$$T(\mathbf{v}) = \lambda\mathbf{v}.$$

Thus, the linear transformation T maps the vector $\mathbf{v}$ onto a vector that is parallel to $\mathbf{v}$. (See Fig. 5.1.) We present a definition of eigenvalues and eigenvectors for linear transformations that is more general than for matrices, in that we need not restrict ourselves to a finite-dimensional vector space.

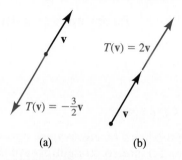

(a)   (b)

FIGURE 5.1

(a) T has eigenvalue $\lambda = -\frac{3}{2}$; (b) T has eigenvalue $\lambda = 2$.

DEFINITION 5.2 Eigenvalues and Eigenvectors

Let T be a linear transformation of a vector space V into itself. A scalar λ is an **eigenvalue** of T if there is a *nonzero* vector $\mathbf{v}$ in V such that $T(\mathbf{v}) = \lambda\mathbf{v}$. The vector $\mathbf{v}$ is then an **eigenvector** of T corresponding to λ.

It is significant that we can define eigenvalues and eigenvectors for a linear transformation $T: V \to V$ without any reference to a matrix representation and without even assuming that V is finite-dimensional. Example 8 will discuss the eigenvalues and eigenvectors of a linear transformation that lies at the heart of calculus and deals with an infinite-dimensional vector space. Students who have studied exponential growth problems in calculus will recognize the importance of this example.

ILLUSTRATION 1   Not every linear transformation has eigenvectors. Rotation of the plane counterclockwise through a positive angle θ is a linear transformation (see page 156). If $0 < \theta < 180°$, then no vector is mapped onto one parallel to it—that is, no vector is an eigenvector. If $\theta = 180°$, then every nonzero vector is an eigenvector and they all have the same associated eigenvalue $\lambda_1 = -1$. ∎

ILLUSTRATION 2   The linear transformation $T: \mathbb{R}^2 \to \mathbb{R}^2$ that reflects vectors in the line $x + 2y = 0$ maps the vector $[2, -1]$ onto itself and maps the vector $[1, 2]$ onto $[-1, -2]$, as indicated in Figure 5.2. The equations $T([2, -1]) = [2, -1]$ and $T([1, 2]) = [-1, -2]$ show that $[2, -1]$ and $[1, 2]$ are eigenvectors of T with corresponding eigenvalues 1 and -1, respectively. ∎

For a linear transformation $T: \mathbb{R}^n \to \mathbb{R}^n$, we can find the transformation's eigenvalues and eigenvectors by finding those of its standard matrix representation. The following example illustrates how this works.

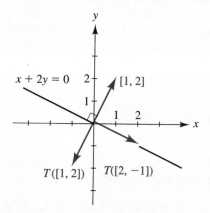

FIGURE 5.2
Reflection in the line $x + 2y = 0$.

EXAMPLE 7   Find the eigenvalues λ and eigenvectors $\mathbf{v}$ of the linear transformation:
$T: \mathbb{R}^3 \to \mathbb{R}^3$ defined by $T([x_1, x_2, x_3]) = [x_1, -8x_1 + 4x_2 - 6x_3, 8x_1 + x_2 + 9x_3]$.
Illustrate the equation $T(\mathbf{v}) = \lambda\mathbf{v}$ for each eigenvalue.

SOLUTION   Writing vectors as column vectors, we may express the linear transformation T as $T(\mathbf{x}) = A\mathbf{x}$, where A is the matrix given in Example 6. The rest of the solution proceeds precisely as in that example. Let us return to row notation and illustrate the action of T on the basic eigenvectors obtained by taking $r = s = t = 1$ in the expressions for $\mathbf{v}_1$, $\mathbf{v}_2$, and $\mathbf{v}_3$ in Example 6. We have

$$T([15, 8, -16]) = [15, -8(15) + 4(8) - 6(-16), 8(15) + 8 + 9(-16)]$$
$$= [15, 8, -16],$$
$$T([0, -3, 1]) = [0, -8(0) + 4(-3) - 6(1), 8(0) - 3 + 9(1)]$$
$$= [0, -18, 6] = 6[0, -3, 1],$$
$$T([0, -2, 1]) = [0, -8(0) + 4(-2) - 6(1), 8(0) - 2 + 9(1)]$$
$$= [0, -14, 7] = 7[0, -2, 1]. \qquad\blacksquare$$

We now give an example from calculus, involving a vector space that is not finite-dimensional.

EXAMPLE 8   Let D_∞ be the vector space of all functions mapping $\mathbb{R}$ into $\mathbb{R}$ and having derivatives of all orders. Let $T: D_\infty \to D_\infty$ be the differentiation map, so that $T(f) = f'$, the derivative of f. Describe all eigenvalues and eigenvectors of T. (We have seen in Section 3.4 that differentiation does give a linear transformation.)

SOLUTION   We must find scalars λ and nonzero functions f such that $T(f) = \lambda f$—that is, such that $f' = \lambda f$. We consider two cases: if $\lambda = 0$, and if $\lambda \neq 0$.

If $\lambda = 0$, we are trying to solve the differential equation $f' = 0$, or to use Leibniz notation, $dy/dx = 0$. We know from calculus that the only solutions of this equation are the constant functions. Thus the nonzero constant functions are eigenvectors corresponding to the eigenvalue 0.

If $\lambda \neq 0$, the differential equation becomes $f' = \lambda f$ or $dy/dx = \lambda y$. It is readily checked that $y = e^{\lambda x}$ is a solution of this equation, so $f(x) = ke^{\lambda x}$ is an eigenvector for every nonzero scalar k. To see that these are the only solutions, we can solve the differential equation by separating variables, which yields the equation

$$\frac{dy}{y} = \lambda\,dx.$$

Integrating both sides of the equation yields

$$\ln|y| = \lambda x + c.$$

Solving for y, we obtain $y = \pm e^c e^{\lambda x} = ke^{\lambda x}$, so the only solutions of the differential equation are indeed of the form $y = ke^{\lambda x}$. $\quad\blacksquare$

ILLUSTRATION 3   There are several physical situations in which a point of a vibrating body is subject to a restoring force proportional to its displacement from its position at rest. One such situation is the vibration of a body suspended by a spring, as

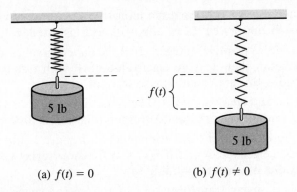

FIGURE 5.3
A vibrating body suspended by a spring.

illustrated in Figure 5.3. We let $f(t)$ be the displacement of the body at time t. The acceleration of the body at time t is then given by $f''(t)$. Using Newton's law $F = ma$, we see that we can express the relation between the restoring force and the displacement as $mf''(t) = -cf(t)$, where c is a positive constant. Dividing by m, we can rewrite this differential equation as

$$f''(t) = -k^2 f(t) \tag{10}$$

for some constant k. Now differentiating twice is a linear transformation of the vector space of all infinitely differentiable function into itself, and we see that the equation $f''(t) = -k^2 f(t)$ asserts that the function f is an eigenvector of this transformation with eigenvalue $-k^2$. We can easily check that the functions $\sin kt$ and $\cos kt$ are solutions of Eq. (10). By property 3 of Theorem 5.1, every linear combination

$$f(t) = a \sin kt + b \cos kt$$

is also a solution. Note that for this vibration situation, the eigenvalue $-k^2$ determines the *frequency* of the vibration. To illustrate, suppose in the case of the spring in Figure 5.3 we have $f(t) = 2$ and $f'(t) = 0$ when $t = 0$. It is readily determined that we then must have $b = 2$ and $a = 0$, so $f(t) = 2 \cos kt$. Thus the frequency of vibration, which is the reciprocal of the period, is $k/(2\pi)$. ∎

SUMMARY

Let A be an n × n matrix.

1. If $A\mathbf{v} = \lambda\mathbf{v}$, where $\mathbf{v}$ is a nonzero column vector and λ is a scalar, then λ is an eigenvalue of A and $\mathbf{v}$ is an eigenvector of A corresponding to λ.

2. The characteristic polynomial $p(\lambda)$ of A is obtained by expanding the determinant $|A - \lambda I|$, where I is the $n \times n$ identity matrix.

3. The eigenvalues λ of A can be found by solving the characteristic equation $p(\lambda) = |A - \lambda I| = 0$. There are at most n real solutions λ of this equation.

4. The eigenvectors of A corresponding to λ are the nontrivial solutions of the homogeneous system $(A - \lambda I)\mathbf{x} = \mathbf{0}$, as illustrated in Examples 5 and 6.

5. Let k be a positive integer. If λ is an eigenvalue of A having $\mathbf{v}$ as eigenvector, then λ^k is an eigenvalue of A^k with $\mathbf{v}$ as eigenvector. If A is invertible, this statement is also true for $k = -1$.

6. Let λ be an eigenvalue of A. The set E_λ in n-space consisting of the zero vector and all eigenvectors corresponding to λ is a subspace of n-space.

7. Let T be a linear transformation of a vector space V into itself. A nonzero $\mathbf{v}$ in V is an eigenvector of T if $T(\mathbf{v}) = \lambda\mathbf{v}$ for some scalar λ, which is called the eigenvalue of T corresponding to $\mathbf{v}$.

8. If T is a linear transformation of the vector space $\mathbb{R}^n$ into itself, the eigenvalues of T are the eigenvalues of the standard matrix representation of T.

EXERCISES

1. Consider the matrices

$$A_1 = \begin{bmatrix} 1 & -1 & -1 \\ -1 & 1 & -1 \\ -1 & -1 & 1 \end{bmatrix}, A_2 = \begin{bmatrix} 3 & -2 & 0 \\ -2 & 3 & 0 \\ 0 & 0 & 5 \end{bmatrix}$$

and the vectors

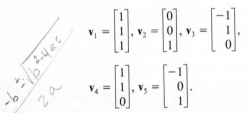

$$\mathbf{v}_1 = \begin{bmatrix} 1 \\ 1 \\ 1 \end{bmatrix}, \mathbf{v}_2 = \begin{bmatrix} 0 \\ 0 \\ 1 \end{bmatrix}, \mathbf{v}_3 = \begin{bmatrix} -1 \\ 1 \\ 0 \end{bmatrix},$$

$$\mathbf{v}_4 = \begin{bmatrix} 1 \\ 1 \\ 0 \end{bmatrix}, \mathbf{v}_5 = \begin{bmatrix} -1 \\ 0 \\ 1 \end{bmatrix}.$$

List the vectors that are eigenvectors of A_1 and the ones that are eigenvectors of A_2. Give the eigenvalue in each case.

In Exercises 2–16, find the characteristic polynomial, the real eigenvalues, and the corresponding eigenvectors of the given matrix.

2. $\begin{bmatrix} 7 & 5 \\ -10 & -8 \end{bmatrix}$

3. $\begin{bmatrix} -1 & -2 \\ 4 & 5 \end{bmatrix}$

4. $\begin{bmatrix} -7 & -5 \\ 16 & 17 \end{bmatrix}$

5. $\begin{bmatrix} 0 & -1 \\ 1 & 0 \end{bmatrix}$

6. $\begin{bmatrix} 1 & -2 \\ 1 & 2 \end{bmatrix}$

7. $\begin{bmatrix} 2 & 0 & 0 \\ 1 & -1 & -2 \\ -1 & 0 & 1 \end{bmatrix}$

8. $\begin{bmatrix} -1 & 0 & 0 \\ -4 & 2 & -1 \\ 4 & 0 & 3 \end{bmatrix}$

9. $\begin{bmatrix} 8 & 0 & 0 \\ 7 & -1 & -2 \\ -7 & 0 & 1 \end{bmatrix}$

10. $\begin{bmatrix} 1 & 0 & 0 \\ -8 & 4 & -5 \\ 8 & 0 & 9 \end{bmatrix}$

11. $\begin{bmatrix} -2 & 0 & 0 \\ -5 & -2 & -5 \\ 5 & 0 & 3 \end{bmatrix}$

12. $\begin{bmatrix} -4 & 0 & 0 \\ -7 & 2 & -1 \\ 7 & 0 & 3 \end{bmatrix}$

13. $\begin{bmatrix} -1 & 0 & 1 \\ -7 & 2 & 5 \\ 3 & 0 & 1 \end{bmatrix}$

14. $\begin{bmatrix} 4 & 0 & 0 \\ 8 & 4 & 8 \\ 0 & 0 & 4 \end{bmatrix}$

15. $\begin{bmatrix} 0 & 0 & 1 \\ -2 & -2 & 1 \\ 2 & 0 & -1 \end{bmatrix}$

16. $\begin{bmatrix} 2 & 0 & 1 \\ 6 & 4 & -3 \\ 2 & 0 & 3 \end{bmatrix}$

In Exercises 17–22, find the eigenvalues λ_i and the corresponding eigenvectors $\mathbf{v}_i$ of the linear transformation T.

17. T defined on $\mathbb{R}^2$ by $T([x, y]) = [2x - 3y, -3x + 2y]$

18. T defined on $\mathbb{R}^2$ by $T([x, y]) = [x - y, -x + y]$

19. T defined on $\mathbb{R}^3$ by $T([x_1, x_2, x_3]) = [x_1 + x_3, x_2, x_1 + x_3]$

20. T defined on $\mathbb{R}^3$ by $T([x_1, x_2, x_3]) =$ $[x_1, 4x_2 + 7x_3, 2x_2 - x_3]$

21. T defined on $\mathbb{R}^3$ by $T([x_1, x_2, x_3]) =$ $[x_1, -5x_1 + 3x_2 - 5x_3, -3x_1 - 2x_2]$

22. T defined on $\mathbb{R}^3$ by $T([x_1, x_2, x_3]) = [3x_1 - x_2 + x_3, -2x_1 + 2x_2 - x_3, 2x_1 + x_2 + 4x_3]$

23. Mark each of the following True or False.
____ a. Every square matrix has real eigenvalues.
____ b. Every $n \times n$ matrix has n distinct (possibly complex) eigenvalues.
____ c. Every $n \times n$ matrix has n not necessarily distinct and possibly complex eigenvalues.
____ d. There can be only one eigenvalue associated with an eigenvector of a linear transformation.
____ e. There can be only one eigenvector associated with an eigenvalue of a linear transformation.
____ f. If $\mathbf{v}$ is an eigenvector of a matrix A, then $\mathbf{v}$ is an eigenvector of $A + cI$ for all scalars c.
____ g. If λ is an eigenvalue of a matrix A, then λ is an eigenvalue of $A + cI$ for all scalars c.
____ h. If $\mathbf{v}$ is an eigenvector of an invertible matrix A, then $c\mathbf{v}$ is an eigenvector of A^{-1} for all nonzero scalars c.
____ i. Every vector in a vector space V is an eigenvector of the identity transformation of V into V.
____ j. Every nonzero vector in a vector space V is an eigenvector of the identity transformation of V into V.

24. Let $T: V \rightarrow V$ be a linear transformation of a vector space V into itself, and let λ be a scalar. Prove that $\{\mathbf{v} \in V \mid T(\mathbf{v}) = \lambda\mathbf{v}\}$ is a subspace of V.

25. Prove that if A is a square matrix, then AA^T and A^TA have the same eigenvalues.

26. Following the idea in Example 8, show that the functions e^{ax}, e^{-ax}, $\sin ax$, and $\cos ax$ are eigenvectors for the linear transformation $T: D_\infty \rightarrow D_\infty$ defined by $T(f) = f^{(4)}$, the fourth derivative of f. Indicate the eigenvalue in each case.

27. Prove property 1 of Theorem 5.1.

28. Prove property 2 of Theorem 5.1.

29. Prove property 3 of Theorem 5.1.

30. Prove that a square matrix is invertible if and only if no eigenvalue is zero.

31. Find the eigenvalues and eigenvectors of the matrix

$$A = \begin{bmatrix} 1 & 1 \\ 1 & 0 \end{bmatrix}$$

used to generate the Fibonacci sequence (2).

32. Let A be an $n \times n$ matrix and let I be the $n \times n$ identity matrix. Compare the eigenvectors and eigenvalues of A with those of $A + rI$ for a scalar r.

33. Let a square matrix A with real eigenvalues have a unique eigenvalue of greatest magnitude. Numerical computation of this eigenvalue by the *power method* (discussed in Section 8.4) can be difficult if there is another eigenvalue of almost equal magnitude, so the ratio of these magnitudes is close to 1.
   a. Suppose we know that a 4×4 matrix A has eigenvalues of approximately 20, 2, -3, and -19.5. Using Exercise 32, how might we modify A so that the aforementioned ratio is not so close to 1?
   b. Repeat part (a), given that the eigenvalues are known to be approximately 19.5, 2, -3, and -20.

34. Let A be an $n \times n$ real matrix. An eigenvector $\mathbf{w}$ in $\mathbb{R}^n$ and a corresponding eigenvalue α of A^T are also called a *left eigenvector and eigenvalue* of A. Explain the reason for this name.

35. (*Principle of biorthogonality*) Let A be an $n \times n$ real matrix. Let $\mathbf{v}$ in $\mathbb{R}^n$ be an eigenvector of A with corresponding eigenvalue λ, and let $\mathbf{w}$ in $\mathbb{R}^n$ be an eigenvector of A^T with corresponding eigenvalue α. Prove that if $\lambda \neq \alpha$, then $\mathbf{v}$ and $\mathbf{w}$ are perpendicular vectors. [HINT: Refer to Exercise 34, and compute $\mathbf{w}^T A \mathbf{v}$ in two ways, using associativity of matrix multiplication.]

36. a. Prove that the eigenvalues of an $n \times n$ real matrix A are the same as the eigenvalues of A^T.
   b. With reference to part (a), show by a counterexample that an eigenvector of A need not be an eigenvector of A^T.

37. The **trace** of an $n \times n$ matrix A is defined by

$$\text{tr}(A) = a_{11} + a_{22} + \cdots + a_{nn}.$$

Let the characteristic polynomial $p(\lambda)$ factor into linear factors, so that A has n (not necessarily distinct) eigenvalues $\lambda_1, \lambda_2, \ldots, \lambda_n$. Prove that

$$\text{tr}(A) = (-1)^{n-1}(\text{Coefficient of } \lambda^{n-1} \text{ in } p(\lambda))$$
$$= \lambda_1 + \lambda_2 + \cdots + \lambda_n.$$

38. Let A be an $n \times n$ matrix, and let C be an invertible $n \times n$ matrix. Show that the eigenvalues of A and of $C^{-1}AC$ are the same. [Hint: Show that the characteristic polynomials of the two matrices are the same.]

39. *Cayley–Hamilton theorem: Every square matrix A satisfies its characteristic equation. That is, if the characteristic equation is*
$p_n\lambda^n + p_{n-1}\lambda^{n-1} + \cdots + p_1\lambda + p_0 = 0$, *then*
$p_nA^n + p_{n-1}A^{n-1} + \cdots + p_1A + p_0I = O$, *the zero matrix.* Illustrate the Cayley–Hamilton theorem for the matrix $\begin{bmatrix} 2 & -1 \\ 1 & 3 \end{bmatrix}$.

40. Let A be an invertible $n \times n$ matrix. Using the Cayley–Hamilton theorem stated in the preceding exercise, show that A^{-1} can be computed as a linear combination of the powers $A, A^2, \ldots, A^n$ of A. Compute the inverse of the matrix A in the preceding exercise in this fashion.

41. Let $\mathbf{v}_1$ and $\mathbf{v}_2$ be eigenvectors of a linear transformation $T: V \rightarrow V$ with corresponding eigenvalues λ_1 and λ_2, respectively. Prove that, if $\lambda_1 \neq \lambda_2$, then $\mathbf{v}_1$ and $\mathbf{v}_2$ are independent vectors.

42. The analogue of Exercise 41 for a list of r eigenvectors in V having distinct eigenvalues is also true; that is, the vectors are independent. See if you can prove it. [Hint: Suppose that the vectors are dependent; consider the first vector in the list that is a linear combination of its predecessors, and apply T.]

43. State the result for matrices corresponding to Exercise 42. Explain why successful completion of Exercise 42 gives a proof of this statement for matrices.

44. Let

$$A = \begin{bmatrix} 0 & -1 \\ 1 & 0 \end{bmatrix}.$$

It can be shown that, if $\mathbf{x}$ is any column vector in $\mathbb{R}^2$, then $A\mathbf{x}$ can be obtained geometrically from $\mathbf{x}$ by rotating $\mathbf{x}$ counterclockwise through an angle of $90°$. For example, we find that $A\mathbf{e}_1 = \mathbf{e}_2$ and $A\mathbf{e}_2 = -\mathbf{e}_1$.
   a. Argue geometrically that A has no real eigenvalues.
   b. Find the complex eigenvalues and eigenvectors of A.
   c. Argue geometrically that A^2 should have real eigenvalues _____. (Fill in the blank.)
   d. Use part (b) and Theorem 5.1 to find the eigenvalues of A^2, and compare with the answer obtained for part (c). What are the real eigenvectors of A^2? the complex eigenvectors of A^2?
   e. Find eigenvalues and eigenvectors for A^3.
   f. Find eigenvalues and eigenvectors for A^4.

45. Use MATCOMP in LINTEK, or MATLAB, and relation (4) to find the following terms of the Fibonacci sequence (2) as accurately as possible. (Use double-precision printing if possible.)
   a. F_8 (Note that $F_8 = 21$; this part is to check procedure.)
   b. F_{30}
   c. F_{50}
   d. F_{77}
   e. F_{150}

46. The first two terms of a sequence are $a_0 = 0$ and $a_1 = 1$. Subsequent terms are generated using the relation

$$a_k = 2a_{k-1} + a_{k-2} \quad \text{for} \quad k \geq 2.$$

   a. Write the terms of the sequence through a_8.
   b. Find a matrix that can be used to generate the sequence, as the matrix A in Exercise 31 can be used to generate the Fibonacci sequence.
   c. Use MATCOMP in LINTEK, or MATLAB, to find a_{30}.

47. Repeat Exercise 46 for a sequence where $a_0 = 0$, $a_1 = 1$, $a_2 = 2$, and $a_k = 2a_{k-1} - 3a_{k-2} + a_{k-3}$ for $k \geq 3$.

For a square matrix A, the MATLAB command **eig(A)** produces the eigenvalues (both real and complex) of A. The command **[V, D] = eig(A)** produces a matrix V whose column vectors are (perhaps complex) eigenvectors of A and a diagonal matrix D whose entry d_{ii} is the eigenvalue for the eigenvector in the ith column of V. In Exercises 48–51, use either MATLAB or the routine MATCOMP in LINTEK to find the real eigenvalues and corresponding eigenvectors of the given matrix.

48. $\begin{bmatrix} 7 & 10 & 6 \\ 2 & -1 & -6 \\ -2 & -5 & 0 \end{bmatrix}$ **49.** $\begin{bmatrix} -1 & 0 & 0 & 0 \\ 3 & 2 & 0 & 0 \\ -3 & 0 & 2 & 0 \\ -3 & 1 & 0 & 3 \end{bmatrix}$

50. $\begin{bmatrix} 0 & 0 & -2 & 2 \\ -3 & 4 & -3 & 3 \\ -4 & 0 & 2 & 4 \\ -2 & 0 & 2 & 4 \end{bmatrix}$

51. $\begin{bmatrix} 4 & 0 & 0 & 0 \\ -6 & 16 & -6 & 6 \\ -16 & 0 & 20 & 16 \\ -16 & 0 & 16 & 20 \end{bmatrix}$

The routine ALLROOTS in LINTEK can be used to find both real and complex roots of a polynomial. The program uses Newton's method, which finds a solution by successive approximations of the polynomial function by a linear one. ALLROOTS is designed so that the user can watch the approximations approach a solution. Of course, a program such as MATLAB, which is designed for research, simply spits out the answers. In Exercises 52–55, either

a. use the command **eig(A)** in MATLAB to find all eigenvalues of the matrix or

b. first use MATCOMP in LINTEK to find the characteristic equation of the given matrix.

Copy down the equation, and then use ALLROOTS to find all eigenvalues of the matrix.

52. $\begin{bmatrix} -1 & 4 & 6 \\ 2 & 7 & 9 \\ -3 & 11 & 13 \end{bmatrix}$ **53.** $\begin{bmatrix} 10 & -13 & 8 \\ 3 & -20 & 5 \\ -11 & 7 & -6 \end{bmatrix}$

54. $\begin{bmatrix} -7 & 11 & -7 & 10 \\ 5 & 8 & -13 & 3 \\ -15 & 8 & -9 & 2 \\ 3 & -4 & 20 & -6 \end{bmatrix}$

55. $\begin{bmatrix} 21 & -8 & 0 & 32 \\ -14 & 17 & -6 & 9 \\ 15 & 11 & -13 & 16 \\ -18 & 30 & 43 & 31 \end{bmatrix}$

56. Use the routine MATCOMP in LINTEK, or MATLAB, to illustrate the Cayley–Hamilton theorem for the matrix

$$\begin{bmatrix} -2 & 4 & 6 & -1 \\ 5 & -8 & 3 & 2 \\ 11 & -3 & 7 & 1 \\ 0 & -5 & 9 & 10 \end{bmatrix}.$$

(See Exercise 39.)

57. The eigenvalue option of the routine VECTGRPH in LINTEK is designed to illustrate graphically, for a linear transformation of $\mathbb{R}^2$ into itself having real eigenvalues, how repeatedly transforming a vector makes it swing in the direction of an eigenvector having eigenvalue of maximum magnitude. By finding graphically a vector whose transform is parallel to it, one can estimate from the graph an eigenvector and the corresponding eigenvalue. Read the directions for this option, and then work with it until you can reliably achieve a score of 85% or better.

MATLAB

The command **v = poly(A)** in MATLAB produces a vector **v** whose components are the coefficients of the characteristic polynomial $p(\lambda)$ of A, appearing in order of decreasing powers of λ. The command **roots(v)** then produces the solutions of $p(\lambda) = 0$.

M1. Enter the command **format long** and then use the commands just explained to find the characteristic polynomial and both the real and the complex eigenvalues of

a. the matrix in Exercise 53

b. the matrix in Exercise 54.

[Recall that in MATLAB, the characteristic polynomial of an $n \times n$ matrix A is $\det(\lambda I - A)$, rather than $\det(A - \lambda I)$. Thus the coefficient of λ^n should be 1 rather than $(-1)^n$. We enter the long format because otherwise, with the displays in scientific notation, it might seem from MATLAB that the coefficient of λ^n is 0.]

Let A be an $n \times n$ matrix. In the introduction to this section, we considered the case where there is a basis $\{\mathbf{b}_1, \mathbf{b}_2, \ldots, \mathbf{b}_n\}$ for $\mathbb{R}^n$ consisting of eigenvectors of A, and where, for one of the eigenvectors, say $\mathbf{b}_1$, the corresponding eigenvalue λ_1 is of greater magnitude than each of $\lambda_2, \ldots, \lambda_n$. Also if $\mathbf{x} = d_1\mathbf{b}_1 + d_2\mathbf{b}_2 + \cdots + d_n\mathbf{b}_n$ with $d_1 \neq 0$, then for sufficiently large values of k, we saw that

$$A^{k+1}\mathbf{x} \approx \lambda_1(A^k\mathbf{x}),$$

so that this eigenvalue λ_1 of maximum magnitude might be estimated by taking the ratio of components of $A^{k+1}\mathbf{x}$ to components of $A^k\mathbf{x}$. Recall that a period before an arithmetic operation symbol, such as .* or ./, will cause that operation to be performed component by component for vectors or matrices. Thus the MATLAB command

$$\mathbf{r} = (A^\wedge(k+1)*\mathbf{x}) ./ (A^\wedge k*\mathbf{x})$$

will cause the desired ratios of components to be printed. If these ratios agree for all components, we expect we have computed the eigenvalue of maximum magnitude accurate to all figures shown.

M2. Enter **k = 1; x = [1 2 3]'; A = [3 5 −11; 5 −8 −3; −11 −3 14];** in MATLAB. (Recall that the final semicolon keeps the data from being displayed.) Then enter **r = (A^(k+1)∗x) ./ (A^k∗x)** to see the ratios displayed. Now enter **k = 5;** and use the up-arrow key to have the ratios computed for that k. Continue using the up-arrow key, setting k successively equal to 10, 15, 20, 25, . . . until the ratios agree to all places shown. Then change to **format long**, and continue until you have found the eigenvalue of maximum magnitude accurate to all figures shown. Copy it down as your answer to this problem. Check it using the command **eig(A)**.

M3. Property 2 of Theorem 5.1 indicates that the eigenvalue of minimum magnitude of A is the reciprocal of the eigenvalue of maximum magnitude of A^{-1}. Continuing the preceding exercise, enter **B = A;** to save A, and then enter **A = inv(A);**. Now enter **s = ones(3,1) ./ r.** (Enter **help ones** to learn the effect of the ones(3,1) statement.) Using the up-arrow key to access statements defining k and computing $\mathbf{r}$ and $\mathbf{s}$, compute the eigenvalue of A of minimum magnitude accurate to all figures shown in the long format. Copy down your answer, and check it using the **eig(B)** command.

M4. Raising a matrix A to a high power can generate error, and can cause overflow (numbers too large to handle) in a computer. Continuing the preceding two exercises, let us avoid this by only raising A to a low power, say the 5th power, and replace $\mathbf{x}$ by $A^5\mathbf{x}$ after each iteration. Repeating this m times should have the same effect as raising A to the power $5m$. Of course, now the entries in the vector $\mathbf{x}$ may get large. We compensate for this by *norming* $\mathbf{x}$ to be of magnitude 1 before the next iteration. This does not change the *direction* of **x**, which should swing to parallel the eigenvector $\mathbf{b}_1$ having eigenvalue λ_1 of maximum magnitude as the iterations are performed. To execute this procedure, enter **A = B**; to recover A as in Exercise M2, and then enter

$$\mathbf{x} = \mathbf{A^\wedge 5_* x}; \ \mathbf{x} = (\mathbf{1/norm(x))_* x}; \ \mathbf{r} = (\mathbf{A_* x}) ./ \mathbf{x}$$

to compute $A^5\mathbf{x}$, normalize **x**, and compute the ratios of components of $A\mathbf{x}$ to those of **x**. One repetition of the up-arrow key followed by the Enter key executes these iterations rapidly. Establish your result in Exercise M2 again, and then, replacing A by A^{-1}, establish your result in Exercise M3 again.

M5. Explain why the final vector **x** obtained when finding the eigenvalue of either maximum or of minimum magnitude in Exercise M4 should be an eigenvector corresponding to that eigenvalue. Check that this is so using the **[V, D] = eig(A)** command explained before Exercise 48.

In Exercises M6–M8, use the displayed command in Exercise M3 preceded by any necessary modifications in the initial vector **x** *to find the eigenvalues of maximum and minimum magnitude. Always reinitialize* **x** *to something like* **x = [1 2 3]** *before finding the eigenvalue of minimum magnitude. Give answers in short format. The matrices are in the file of matrices for this section, if it is available at your installation.*

M6. $\begin{bmatrix} -2 & 6 & 8 \\ 6 & -3 & 2 \\ 8 & 2 & -4 \end{bmatrix}$ **M7.** $\begin{bmatrix} 4 & 11 & 5 & -7 \\ 11 & -2 & -12 & 0 \\ 5 & 12 & -3 & 8 \\ -7 & -3 & 8 & 13 \end{bmatrix}$ **M8.** $\begin{bmatrix} -4 & 7 & 8 & -11 & 5 \\ 7 & -3 & 0 & 2 & 10 \\ 8 & 0 & 2 & 1 & 4 \\ -11 & 2 & 1 & 5 & -9 \\ 5 & 10 & 4 & -9 & 3 \end{bmatrix}$

5.2   DIAGONALIZATION

Recall that a square matrix is called **diagonal** if all entries not on the main diagonal are zero. In the preceding section we indicated the importance of being able to compute $A^k\mathbf{x}$ for an $n \times n$ matrix A and a column vector **x** in $\mathbb{R}^n$. In this section, we show that, if A has n distinct eigenvalues, then computation of A^k can be essentially replaced by computation of D^k, where D is a diagonal matrix with the eigenvalues of A as diagonal entries. Notice that D^k is the diagonal matrix obtained from D by raising each diagonal entry to the power k. For example,

$$\begin{bmatrix} 2 & 0 & 0 \\ 0 & -1 & 0 \\ 0 & 0 & -2 \end{bmatrix}^3 = \begin{bmatrix} 8 & 0 & 0 \\ 0 & -1 & 0 \\ 0 & 0 & -8 \end{bmatrix}.$$

The theorem that follows shows how to summarize the action of a square matrix on its eigenvectors in a single matrix equation. This theorem is the first step toward our goal of *diagonalizing* a matrix. Theorems stated are valid for matrices and vectors with complex entries and complex scalars unless we use the adjective *real*. We sometimes refer to *n-space* in this section, meaning either $\mathbb{R}^n$ or $\mathbb{C}^n$.

THEOREM 5.2   Matrix Summary of Eigenvalues of A

Let A be an $n \times n$ matrix and let $\lambda_1, \lambda_2, \ldots, \lambda_n$ be (possibly complex) scalars and $\mathbf{v}_1, \mathbf{v}_2, \ldots, \mathbf{v}_n$ be nonzero vectors in *n*-space. Let C be the $n \times n$ matrix having $\mathbf{v}_j$ as *j*th column vector, and let

$$D = \begin{bmatrix} \lambda_1 & & & \\ & \lambda_2 & & \mathbf{0} \\ & & \ddots & \\ \mathbf{0} & & & \lambda_n \end{bmatrix}.$$

Then $AC = CD$ if and only if $\lambda_1, \lambda_2, \ldots, \lambda_n$ are eigenvalues of A and $\mathbf{v}_j$ is an eigenvector of A corresponding to λ_j for $j = 1, 2, \ldots, n$.

PROOF   **We have**

$$CD = \begin{bmatrix} | & | & & | \\ \mathbf{v}_1 & \mathbf{v}_2 & \cdots & \mathbf{v}_n \\ | & | & & | \end{bmatrix} \begin{bmatrix} \lambda_1 & & & \\ & \lambda_2 & & \mathbf{0} \\ & & \ddots & \\ \mathbf{0} & & & \lambda_n \end{bmatrix}.$$

$$= \begin{bmatrix} | & | & & | \\ \lambda_1\mathbf{v}_1 & \lambda_2\mathbf{v}_2 & \cdots & \lambda_n\mathbf{v}_n \\ | & | & & | \end{bmatrix}.$$

On the other hand,

$$AC = A\begin{bmatrix} | & | & & | \\ \mathbf{v}_1 & \mathbf{v}_2 & \cdots & \mathbf{v}_n \\ | & | & & | \end{bmatrix} = \begin{bmatrix} | & | & & | \\ A\mathbf{v}_1 & A\mathbf{v}_2 & \cdots & A\mathbf{v}_n \\ | & | & & | \end{bmatrix}.$$

Thus, $AC = CD$ if and only if $A\mathbf{v}_j = \lambda_j\mathbf{v}_j$. ▲

The $n \times n$ matrix C is invertible if and only if $\text{rank}(C) = n$—that is, if and only if the column vectors of C form a basis for n-space. In this case, the criterion $AC = CD$ in Theorem 5.2 can be written as $D = C^{-1}AC$. The equation $D = C^{-1}AC$ transforms a matrix A into a diagonal matrix D that is much easier to work with, as we will see in a moment. We give a formal definition of such a transformation of A into D, followed by a corollary summarizing the results of this paragraph.

DEFINITION 5.3   Diagonalizable Matrix

> An $n \times n$ matrix A is **diagonalizable** if there exists an invertible matrix C such that $C^{-1}AC = D$, a diagonal matrix. The matrix C is said to **diagonalize** the matrix A.

COROLLARY 1   A Criterion for Diagonalization

An $n \times n$ matrix A is diagonalizable if and only if n-space has a basis consisting of eigenvectors of A.

If an $n \times n$ matrix A is diagonalizable as described in Definition 5.3, then we have $A = CDC^{-1}$ as well. From this we obtain

$$A^k = \underbrace{(CDC^{-1})(CDC^{-1})(CDC^{-1}) \ldots (CDC^{-1})}_{k \text{ factors}} \tag{1}$$

The adjacent terms $C^{-1}C$ cancel in Eq. (1) to give $A^k = CD^kC^{-1}$. Thus, the computation of A^k is essentially reduced to the computation of D^k. We summarize these observations as a corollary to Theorem 5.2.

COROLLARY 2   Computation of A^k

Let an $n \times n$ matrix A have n eigenvectors and eigenvalues, giving rise to the matrices C and D so that $AC = CD$, as described in Theorem 5.2. If the n eigenvectors are independent, then C is an invertible matrix and $C^{-1}AC = D$. Under these conditions, we have

$$A^k = CD^kC^{-1}$$

Diagonalization of square matrices plays a very important role in linear algebra. As we show in the next theorem, diagonalization of an $n \times n$ matrix A can always be achieved if the characteristic polynomial has n distinct roots.

THEOREM 5.3   Independence of Eigenvectors

> Let A be an $n \times n$ matrix. If $\mathbf{v}_1, \mathbf{v}_2, \ldots, \mathbf{v}_n$ are eigenvectors of A corresponding to *distinct* eigenvalues $\lambda_1, \lambda_2, \ldots, \lambda_n$, respectively, the set $\{\mathbf{v}_1, \mathbf{v}_2, \ldots, \mathbf{v}_n\}$ is linearly independent and A is diagonalizable.

PROOF   Suppose that the conclusion is false, so the eigenvectors $\mathbf{v}_1, \mathbf{v}_2, \ldots, \mathbf{v}_n$ are linearly dependent. Then one of them is a linear combination of its predecessors. (See Exercise 37, page 203.) Let $\mathbf{v}_k$ be the first such vector, so that

$$\mathbf{v}_k = d_1\mathbf{v}_1 + d_2\mathbf{v}_2 + \cdots + d_{k-1}\mathbf{v}_{k-1} \tag{2}$$

and $\{\mathbf{v}_1, \mathbf{v}_2, \ldots, \mathbf{v}_{k-1}\}$ is independent. Multiplying Eq. (2) by λ_k, we obtain

$$\lambda_k\mathbf{v}_k = d_1\lambda_k\mathbf{v}_1 + d_2\lambda_k\mathbf{v}_2 + \cdots + d_{k-1}\lambda_k\mathbf{v}_{k-1}. \tag{3}$$

On the other hand, multiplying both sides of Eq. (2) on the left by the matrix A yields

$$\lambda_k\mathbf{v}_k = d_1\lambda_1\mathbf{v}_1 + d_2\lambda_2\mathbf{v}_2 + \cdots + d_{k-1}\lambda_{k-1}\mathbf{v}_{k-1}, \tag{4}$$

because $A\mathbf{v}_i = \lambda_i\mathbf{v}_i$. Subtracting Eq. (4) from Eq. (3), we see that

$$\mathbf{0} = d_1(\lambda_k - \lambda_1)\mathbf{v}_1 + d_2(\lambda_k - \lambda_2)\mathbf{v}_2 + \cdots + d_{k-1}(\lambda_k - \lambda_{k-1})\mathbf{v}_{k-1}.$$

This last equation is a dependence relation because not all the coefficients are zero. (Not all d_i are zero because of Eq. (2) and because the λ_i are distinct.) But this contradicts the linear independence of the set $\{\mathbf{v}_1, \mathbf{v}_2, \ldots, \mathbf{v}_{k-1}\}$. We conclude that $\{\mathbf{v}_1, \mathbf{v}_2, \ldots, \mathbf{v}_n\}$ is independent. That A is diagonalizable follows at once from Corollary 1 of Theorem 5.2.   ▲

EXAMPLE 1   Diagonalize the matrix $A = \begin{bmatrix} -3 & 5 \\ -2 & 4 \end{bmatrix}$, and compute A^k in terms of k.

SOLUTION   We compute

$$\det(A - \lambda I) = \begin{vmatrix} -3 - \lambda & 5 \\ -2 & 4 - \lambda \end{vmatrix} = \lambda^2 - \lambda - 2 = (\lambda - 2)(\lambda + 1).$$

The eigenvalues of A are $\lambda_1 = 2$ and $\lambda_2 = -1$. For $\lambda_1 = 2$, we have

$$A - 2I = \begin{bmatrix} -5 & 5 \\ -2 & 2 \end{bmatrix} \sim \begin{bmatrix} 1 & -1 \\ 0 & 0 \end{bmatrix},$$

which yields an eigenvector

$$\mathbf{v}_1 = \begin{bmatrix} 1 \\ 1 \end{bmatrix}.$$

For $\lambda_2 = -1$, we have

$$A + I = \begin{bmatrix} -2 & 5 \\ -2 & 5 \end{bmatrix} \sim \begin{bmatrix} 2 & -5 \\ 0 & 0 \end{bmatrix},$$

which yields an eigenvector

$$\mathbf{v}_2 = \begin{bmatrix} 5 \\ 2 \end{bmatrix}.$$

A diagonalization of A is given by

$$A = CDC^{-1} = \begin{bmatrix} 1 & 5 \\ 1 & 2 \end{bmatrix} \begin{bmatrix} 2 & 0 \\ 0 & -1 \end{bmatrix} \begin{bmatrix} -\frac{2}{3} & \frac{5}{3} \\ \frac{1}{3} & -\frac{1}{3} \end{bmatrix}.$$

(We omit the computation of C^{-1}.) Thus,

$$A^k = \begin{bmatrix} 1 & 5 \\ 1 & 2 \end{bmatrix} \begin{bmatrix} 2^k & 0 \\ 0 & (-1)^k \end{bmatrix} \begin{bmatrix} -\frac{2}{3} & \frac{5}{3} \\ \frac{1}{3} & -\frac{1}{3} \end{bmatrix} = \frac{1}{3} \begin{bmatrix} -2^{k+1} \pm 5 & 5(2^k) \mp 5 \\ -2^{k+1} \pm 2 & 5(2^k) \mp 2 \end{bmatrix},$$

the colored sign being used only when k is odd. ∎

Diagonalization has uses other than computing A^k. The next section presents an application to differential equations. Section 8.1 gives another application. Here is an application to geometry in $\mathbb{R}^2$.

EXAMPLE 2   Find a formula for the linear transformation $T: \mathbb{R}^2 \to \mathbb{R}^2$ that reflects vectors in the line $x + 2y = 0$.

SOLUTION   We saw in Illustration 2 of Section 5.1 that T has eigenvectors $[1, 2]$ and $[2, -1]$ with corresponding eigenvalues -1 and 1, respectively. Let A be the standard matrix representation of T. Using column-vector notation, we have

$$T\left(\begin{bmatrix} 1 \\ 2 \end{bmatrix} \right) = A \begin{bmatrix} 1 \\ 2 \end{bmatrix} = \begin{bmatrix} -1 \\ -2 \end{bmatrix} \quad \text{and} \quad T\left(\begin{bmatrix} 2 \\ -1 \end{bmatrix} \right) = A \begin{bmatrix} 2 \\ -1 \end{bmatrix} = \begin{bmatrix} 2 \\ -1 \end{bmatrix}.$$

Letting $C = \begin{bmatrix} 1 & 2 \\ 2 & -1 \end{bmatrix}$ and $D = \begin{bmatrix} -1 & 0 \\ 0 & 1 \end{bmatrix}$, we see that $AC = CD$ as in Theorem 5.2. Because C is invertible,

$$A = CDC^{-1} = \begin{bmatrix} 1 & 2 \\ 2 & -1 \end{bmatrix} \begin{bmatrix} -1 & 0 \\ 0 & 1 \end{bmatrix} \begin{bmatrix} \frac{1}{5} & \frac{2}{5} \\ \frac{2}{5} & -\frac{1}{5} \end{bmatrix} = \frac{1}{5} \begin{bmatrix} 3 & -4 \\ -4 & -3 \end{bmatrix}.$$

Thus

$$T\left(\begin{bmatrix} x \\ y \end{bmatrix} \right) = A\left(\begin{bmatrix} x \\ y \end{bmatrix} \right) = \frac{1}{5} \begin{bmatrix} 3x - 4y \\ -4x - 3y \end{bmatrix}$$

or, in row notation, $T([x, y]) = \frac{1}{5}[3x - 4y, -4x - 3y]$. ∎

The preceding matrices A and D provide examples of *similar matrices*, a term we now define.

DEFINITION 5.4   Similar Matrices

An $n \times n$ matrix P is **similar** to an $n \times n$ matrix Q if there exists an invertible $n \times n$ matrix C such that $C^{-1}PC = Q$.

The relationship "P is similar to Q" satisfies the properties required for an *equivalence relation* (see Exercise 16). In particular, if P is similar to Q, then Q is also similar to P. This means that similarity need not be stated in a directional way; we can simply say that P and Q are similar matrices.

EXAMPLE 3   Find a diagonal matrix similar to the matrix

$$A = \begin{bmatrix} 1 & 0 & 0 \\ -8 & 4 & -6 \\ 8 & 1 & 9 \end{bmatrix}$$

of Example 6 in Section 5.1.

SOLUTION   Taking $r = s = t = 1$ in Example 6 of Section 5.1, we see that eigenvalues and corresponding eigenvectors of A are given by

$$\lambda_1 = 1, \qquad \lambda_2 = 6, \qquad \lambda_3 = 7,$$

$$\mathbf{v}_1 = \begin{bmatrix} 15 \\ 8 \\ -16 \end{bmatrix}, \qquad \mathbf{v}_2 = \begin{bmatrix} 0 \\ -3 \\ 1 \end{bmatrix}, \qquad \mathbf{v}_3 = \begin{bmatrix} 0 \\ -2 \\ 1 \end{bmatrix}.$$

If we let

$$C = \begin{bmatrix} 15 & 0 & 0 \\ 8 & -3 & -2 \\ -16 & 1 & 1 \end{bmatrix},$$

then Theorem 5.3 tells us that C is invertible. Theorem 5.2 then shows that

$$C^{-1}AC = D = \begin{bmatrix} 1 & 0 & 0 \\ 0 & 6 & 0 \\ 0 & 0 & 7 \end{bmatrix}.$$

HISTORICAL NOTE   THE IDEA OF SIMILARITY, like many matrix notions, appears without a definition in works from as early as the 1820s. In fact, in his 1826 work on quadratric forms (see note on page 409), Cauchy showed that if two quadratic forms (polynomials) are related by a change of variables—that is, if their matrices are similar—then their characteristic equations are the same. But like the concept of orthogonality, that of similarity was first formally defined and discussed by Georg Frobenius in 1878. Frobenius began by discussing the general case: he called two matrices A, D *equivalent* if there existed invertible matrices P, Q such that $D = PAQ$. The latter matrices were called the *substitutions* through which A was transformed into D.

Frobenius then dealt with the special cases where $P = Q^T$ (the two matrices were then called *congruent*) and where $P = Q^{-1}$ (the similarity case of this section). Frobenius went on to prove many results on similarity, including the useful theorem that, if A is similar to D, then $f(A)$ is similar to $f(D)$, where f is any polynomial matrix function.

Thus D is similar to A. We are not eager to check that $C^{-1}AC = D$; however, it is easy to check the equivalent statement:

$$AC = CD = \begin{bmatrix} 15 & 0 & 0 \\ 8 & -18 & -14 \\ -16 & 6 & 7 \end{bmatrix}.$$

■

It is not always essential that a matrix have distinct eigenvalues in order to be diagonalizable. As long as the $n \times n$ matrix A has n *independent* eigenvectors to form the column vectors of an invertible C, we have $C^{-1}AC = D$, the diagonal matrix of the eigenvalues corresponding to C.

EXAMPLE 4   Diagonalize the matrix

$$A = \begin{bmatrix} 1 & -3 & 3 \\ 0 & -5 & 6 \\ 0 & -3 & 4 \end{bmatrix}.$$

SOLUTION   We find that the characteristic equation of A is

$$(1 - \lambda)((-5 - \lambda)(4 - \lambda) + 18) = (1 - \lambda)(\lambda^2 + \lambda - 2)$$
$$= (1 - \lambda)(\lambda + 2)(\lambda - 1) = 0.$$

Thus, the eigenvalues of A are $\lambda_1 = 1$, $\lambda_2 = 1$, and $\lambda_3 = -2$. Notice that 1 is a root of multiplicity 2 of the characteristic equation; we say that the eigenvalue 1 has *algebraic multiplicity* 2.

Reducing $A - I$, we obtain

$$A - I = \begin{bmatrix} 0 & -3 & 3 \\ 0 & -6 & 6 \\ 0 & -3 & 3 \end{bmatrix} \sim \begin{bmatrix} 0 & 1 & -1 \\ 0 & 0 & 0 \\ 0 & 0 & 0 \end{bmatrix}.$$

We see that the eigenspace E_1 (that is, the nullspace of $A - I$) has dimension 2 and consists of vectors of the form

$$\begin{bmatrix} s \\ r \\ r \end{bmatrix} \text{ for any scalars } r \text{ and } s.$$

Taking $s = 1$ and $r = 0$, and then taking $s = 0$ and $r = 1$, we obtain the independent eigenvectors

$$\mathbf{v}_1 = \begin{bmatrix} 1 \\ 0 \\ 0 \end{bmatrix} \text{ and } \mathbf{v}_2 = \begin{bmatrix} 0 \\ 1 \\ 1 \end{bmatrix}$$

corresponding to the eigenvalues $\lambda_1 = \lambda_2 = 1$.

Reducing $A + 2I$, we find that

$$A + 2I = \begin{bmatrix} 3 & -3 & 3 \\ 0 & -3 & 6 \\ 0 & -3 & 6 \end{bmatrix} \sim \begin{bmatrix} 3 & 0 & -3 \\ 0 & 1 & -2 \\ 0 & 0 & 0 \end{bmatrix}.$$

Thus an eigenvector corresponding to $\lambda_3 = -2$ is

$$\mathbf{v}_3 = \begin{bmatrix} 1 \\ 2 \\ 1 \end{bmatrix}.$$

Therefore, if we take

$$C = \begin{bmatrix} 1 & 0 & 1 \\ 0 & 1 & 2 \\ 0 & 1 & 1 \end{bmatrix},$$

we should have

$$C^{-1}AC = D = \begin{bmatrix} 1 & 0 & 0 \\ 0 & 1 & 0 \\ 0 & 0 & -2 \end{bmatrix}.$$

A check shows that indeed

$$AC = CD = \begin{bmatrix} 1 & 0 & -2 \\ 0 & 1 & -4 \\ 0 & 1 & -2 \end{bmatrix}. \qquad \blacksquare$$

As we indicated in Example 4, the **algebraic multiplicity** of an eigenvalue λ_i of A is its multiplicity as a root of the characteristic equation of A. Its **geometric multiplicity** is the dimension of the eigenspace E_{λ_i}. Of course, the geometric multiplicity of each eigenvalue must be at least 1, because there always exists a nonzero eigenvector in the eigenspace. However, it is possible for the algebraic multiplicity to be greater than the geometric multiplicity.

EXAMPLE 5   Referring back to Examples 4 and 5 in Section 5.1, find the algebraic and geometric multiplicities of the eigenvalue 2 of the matrix

$$A = \begin{bmatrix} 2 & 1 & 0 \\ -1 & 0 & 1 \\ 1 & 3 & 1 \end{bmatrix}.$$

SOLUTION   Example 4 on page 292 shows that the characteristic equation of A is $-(\lambda - 2)^2(\lambda + 1) = 0$, so 2 is an eigenvalue of algebraic multiplicity 2. Example 5 on page 293 shows that the reduced form of $A - 2I$ is

$$\begin{bmatrix} 1 & 3 & -1 \\ 0 & 1 & 0 \\ 0 & 0 & 0 \end{bmatrix}.$$

Thus, the eigenspace E_2, which is the set of vectors of the form

$$\begin{bmatrix} r \\ 0 \\ r \end{bmatrix} \text{ for } r \in \mathbb{R},$$

has dimension 1, so the eigenvalue 2 has geometric multiplicity 1. $\blacksquare$

We state a relationship between the algebraic multiplicity and the geometric multiplicity of a (possibly complex) eigenvalue. (See Exercise 33 in Section 9.4.)

> The geometric multiplicity of an eigenvalue of a matrix A is less than or equal to its algebraic multiplicity.

Let $\lambda_1, \lambda_2, \ldots, \lambda_m$ be the distinct (possibly complex) eigenvalues of an $n \times n$ matrix A. Let B_i be a basis for the eigenspace of λ_i for $i = 1, 2, \ldots, m$. It can be shown by an argument similar to the proof of Theorem 5.3 that the union of these bases B_i is an independent set of vectors in n-space (see Exercise 24). Corollary 1 on page 307 shows that the matrix A is diagonalizable if this union of the B_i is a basis for n-space. This will occur precisely when the geometric multiplicity of each eigenvalue is equal to its algebraic multiplicity. (See the boxed statement.) Conversely, it can be shown that, if A is diagonalizable, the algebraic multiplicity of each eigenvalue is the same as its geometric multiplicity. We summarize this in a theorem.

THEOREM 5.4   A Criterion for Diagonalization

> An $n \times n$ matrix A is diagonalizable if and only if the algebraic multiplicity of each (possibly complex) eigenvalue is equal to its geometric multiplicity.

Thus the 3×3 matrix in Example 4 is diagonalizable, because its eigenvalue 1 has algebraic and geometric multiplicity 2, and its eigenvalue -2 has algebraic and geometric multiplicity 1. However, the matrix in Example 5 is not diagonalizable, because the eigenvalue 2 has algebraic multiplicity 2 but geometric multiplicity 1.

In Section 9.4, we show that every square matrix A is similar to a matrix J, its *Jordan canonical form*. If A is diagonalizable, then J is a diagonal matrix, found precisely as in the preceding examples. If A is not diagonalizable, then J again has the eigenvalues of A on its main diagonal, but it also has entries 1 immediately above some of the diagonal entries. The remaining entries are all zero. For example, the matrix

$$A = \begin{bmatrix} 2 & 1 & 0 \\ -1 & 0 & 1 \\ 1 & 3 & 1 \end{bmatrix}$$

of Example 5 has Jordan canonical form

$$J = \begin{bmatrix} -1 & 0 & 0 \\ 0 & 2 & 1 \\ 0 & 0 & 2 \end{bmatrix}.$$

Section 9.4 describes a technique for finding J. This Jordan canonical form is as close to a diagonalization of A as we can come. The Jordan canonical form has applications to the solution of systems of differential equations.

To conclude this discussion, we state a result whose proof requires an excursion into complex numbers. The proof is given in Chapter 9. Thus far, we have seen nothing to indicate that symmetric matrices play any significant role in linear algebra. The following theorem immediately elevates them into a position of prominence.

THEOREM 5.5   Diagonalization of Real Symmetric Matrices

Every real symmetric matrix is *real diagonalizable*. That is, if A is an $n \times n$ symmetric matrix with real-number entries, then each eigenvalue of A is a real number, and its algebraic multiplicity equals its geometric multiplicity.

A diagonalizing matrix C for a symmetric matrix A can be chosen to have some very nice properties, as we will show for real symmetric matrices in Chapter 6.

SUMMARY

Let A be an $n \times n$ matrix.

1. If A has n distinct eigenvalues $\lambda_1, \lambda_2, \ldots, \lambda_n$, and C is an $n \times n$ matrix having as jth column vector an eigenvector corresponding to λ_j, then $C^{-1}AC$ is the diagonal matrix having λ_j on the main diagonal in the jth column.

2. If $C^{-1}AC = D$, then $A^k = CD^kC^{-1}$.

3. A matrix P is similar to a matrix Q if there exists an invertible matrix C such that $C^{-1}PC = Q$.

4. The algebraic multiplicity of an eigenvalue λ of A is its multiplicity as a root of the characteristic equation; its geometric multiplicity is the dimension of the corresponding eigenspace E_λ.

5. Any eigenvalue's geometric multiplicity is less than or equal to its algebraic multiplicity.

6. The matrix A is diagonalizable if and only if the geometric multiplicity of each of its eigenvalues is the same as the algebraic multiplicity.

7. Every symmetric matrix is diagonalizable. All eigenvalues of a real symmetric matrix are real numbers.

EXERCISES

In Exercises 1–8, find the eigenvalues λ_i and the corresponding eigenvectors $\mathbf{v}_i$ of the given matrix A, and also find an invertible matrix C and a diagonal matrix D such that $D = C^{-1}AC$.

1. $A = \begin{bmatrix} -3 & 4 \\ 4 & 3 \end{bmatrix}$
    2. $A = \begin{bmatrix} 3 & 2 \\ 1 & 4 \end{bmatrix}$

3. $A = \begin{bmatrix} 7 & 8 \\ -4 & -5 \end{bmatrix}$
    4. $A = \begin{bmatrix} 6 & 3 & -3 \\ -2 & -1 & 2 \\ 16 & 8 & -7 \end{bmatrix}$

5. $A = \begin{bmatrix} -3 & 10 & -6 \\ 0 & 7 & -6 \\ 0 & 0 & 1 \end{bmatrix}$
    6. $A = \begin{bmatrix} -3 & 5 & -20 \\ 2 & 0 & 8 \\ 2 & 1 & 7 \end{bmatrix}$

7. $A = \begin{bmatrix} -2 & 0 & -1 \\ 0 & 2 & 0 \\ 3 & 0 & 2 \end{bmatrix}$
    8. $A = \begin{bmatrix} -4 & 6 & -12 \\ 3 & -1 & 6 \\ 3 & -3 & 8 \end{bmatrix}$

In Exercises 9–12, determine whether the given matrix is diagonalizable.

9. $\begin{bmatrix} 1 & 2 & 6 \\ 2 & 0 & -4 \\ 6 & -4 & 3 \end{bmatrix}$
    10. $\begin{bmatrix} 3 & 1 & 0 \\ 0 & 3 & 1 \\ 0 & 0 & 3 \end{bmatrix}$

11. $\begin{bmatrix} -1 & 4 & 2 & -7 \\ 0 & 5 & -3 & 6 \\ 0 & 0 & -5 & 1 \\ 0 & 0 & 0 & 11 \end{bmatrix}$
    12. $\begin{bmatrix} 3 & 2 & 5 & 1 \\ 2 & 0 & 2 & 6 \\ 5 & 2 & 7 & -1 \\ 1 & 6 & -1 & 3 \end{bmatrix}$

13. Mark each of the following True or False.
___ **a.** Every $n \times n$ matrix is diagonalizable.
___ **b.** If an $n \times n$ matrix has n distinct real eigenvalues, it is diagonalizable.
___ **c.** Every $n \times n$ real symmetric matrix is real diagonalizable.
___ **d.** An $n \times n$ matrix is diagonalizable if and only if it has n distinct eigenvalues.
___ **e.** An $n \times n$ matrix is diagonalizable if and only if the algebraic multiplicity of each of its eigenvalues equals the geometric multiplicity.
___ **f.** Every invertible matrix is diagonalizable.
___ **g.** Every triangular matrix is diagonalizable.
___ **h.** If A and B are similar square matrices and A is diagonalizable, then B is also diagonalizable.

___ **i.** If an $n \times n$ matrix A is diagonalizable, there is a unique diagonal matrix D that is similar to A.
___ **j.** If A and B are similar square matrices, then $\det(A) = \det(B)$.

14. Give two different diagonal matrices that are similar to the matrix $\begin{bmatrix} 1 & 4 \\ 0 & -3 \end{bmatrix}$.

15. Prove that, if a matrix is diagonalizable, so is its transpose.

16. Let P, Q, and R be $n \times n$ matrices. Recall that P is similar to Q if there exists an invertible $n \times n$ matrix C such that $C^{-1}PC = Q$. This exercise shows that similarity is an *equivalence relation*.
a. (*Reflexive.*) Show that P is similar to itself.
b. (*Symmetric.*) Show that, if P is similar to Q, then Q is similar to P.
c. (*Transitive.*) Show that, if P is similar to Q and Q is similar to R, then P is similar to R.

17. Prove that, for every square matrix A all of whose eigenvalues are real, the product of its eigenvalues is $\det(A)$.

18. Prove that similar square matrices have the same eigenvalues with the same algebraic multiplicities.

19. Let A be an $n \times n$ matrix.
a. Prove that if A is similar to rA where r is a real scalar other than 1 or -1, then all eigenvalues of A are zero. [HINT: See the preceding exercise.]
b. What can you say about A if it is diagonalizable and similar to rA for some r where $|r| \neq 1$?
c. Find a nonzero 2×2 matrix A which is similar to rA for every $r \neq 0$. (See part a.)
d. Show that $A = \begin{bmatrix} 1 & 1 \\ 0 & -1 \end{bmatrix}$ is similar to $-A$.
(Observe that the eigenvalues of A are not all zero.)

20. Let A be a real **tridiagonal matrix**—that is, an $n \times n$ matrix for $n > 2$ all of whose entries are zero except possibly those of the form $a_{i,i-1}$, a_{ii}, or $a_{i-1,i}$. Show that if $a_{i,i-1}$ and $a_{i-1,i}$ are both positive, both negative, or both zero for $i = 2, 3, \ldots, n$, then A has real eigenvalues. [HINT: Show that a diagonal matrix D can be found such that DAD^{-1} is real symmetric.]

21. Find a formula for the linear transformation $T: \mathbb{R}^2 \to \mathbb{R}^2$ that reflects vectors in the line $y = mx$. [HINT: Proceed as in Example 2.]

22. Let A and C be $n \times n$ matrices, and let C be invertible. Prove that, if $\mathbf{v}$ is an eigenvector of A with corresponding eigenvalue λ, then $C^{-1}\mathbf{v}$ is an eigenvector of $C^{-1}AC$ with corresponding eigenvalue λ. Then prove that all eigenvectors of $C^{-1}AC$ are of the form $C^{-1}\mathbf{v}$, where $\mathbf{v}$ is an eigenvector of A.

23. Explain how we can deduce from Exercise 22 that, if A and B are similar square matrices, each eigenvalue of A has the same geometric multiplicity for A that it has for B. (See Exercise 18 for the corresponding statement on algebraic multiplicities.)

24. Prove that, if $\lambda_1, \lambda_2, \ldots, \lambda_m$ are distinct real eigenvalues of an $n \times n$ real matrix A and if B_i is a basis for the eigenspace E_{λ_i}, then the union of the bases B_i is an independent set of vectors in $\mathbb{R}^n$. [HINT: Make use of Theorem 5.3.]

25. Let $T: V \to V$ be a linear transformation of a vector space V into itself. Prove that, if $\mathbf{v}_1, \mathbf{v}_2, \ldots, \mathbf{v}_k$ are eigenvectors of T corresponding to distinct nonzero eigenvalues $\lambda_1, \lambda_2, \ldots, \lambda_k$, then the set $\{T(\mathbf{v}_1), T(\mathbf{v}_2), \ldots, T(\mathbf{v}_k)\}$ is independent.

26. Show that the set $\{e^{\lambda_1 x}, e^{\lambda_2 x}, \ldots, e^{\lambda_k x}\}$, where the λ_i are distinct, is independent in the vector space W of all functions mapping $\mathbb{R}$ into $\mathbb{R}$ and having derivatives of all orders.

27. Using Exercise 26, show that the infinite set $\{e^{kx} \mid k \in \mathbb{R}\}$ is an independent set in the vector space W described in Exercise 26. [HINT: How many vectors are involved in any dependence relation?]

28. In Section 5.1, we stated that if we allow complex numbers, then an $n \times n$ matrix with entries chosen at random has eigenvectors that form a basis for n-space with probability 1. In light of our work in

this section, give the best justification you can for this statement.

29. The MATLAB command

$$\mathbf{A} = (2*\mathbf{r})*\mathbf{rand(n)} - \mathbf{r}*\mathbf{ones(n)}; \ \mathbf{eig(A)}$$

should exhibit the eigenvalues of an $n \times n$ matrix A with random entries between $-r$ and r. Illustrate your argument in the preceding exercise for seven matrices with values of r from 1 to 50 and values of n from 2 to 8. (Recall that we used **rand(n)** in the MATLAB exercises of Section 1.6.)

In Exercises 30–41, use the routines MATCOMP and ALLROOTS in LINTEK or use MATLAB to classify the given matrix as real diagonalizable, complex (but not real) diagonalizable, or not diagonalizable. (Load the matrices from a matrix file if it is accessible.)

30. $\begin{bmatrix} 18 & 25 & -25 \\ 1 & 6 & -1 \\ 18 & 34 & -25 \end{bmatrix}$

31. $\begin{bmatrix} 8.3 & 8.0 & -6.0 \\ -2.0 & 0.3 & 3.0 \\ 0.0 & 0.0 & 4.3 \end{bmatrix}$

32. $\begin{bmatrix} 24.55 & 46.60 & 46.60 \\ -4.66 & -8.07 & -9.32 \\ -9.32 & -18.64 & -17.39 \end{bmatrix}$

33. $\begin{bmatrix} 0.8 & -1.6 & 1.8 \\ -0.6 & -3.8 & 3.4 \\ -20.6 & -1.2 & 9.6 \end{bmatrix}$

34. $\begin{bmatrix} 7 & -20 & -5 & 5 \\ 5 & -13 & -5 & 0 \\ -5 & 10 & 7 & 5 \\ 5 & -10 & -5 & -3 \end{bmatrix}$

35. $\begin{bmatrix} 2 & 5 & -9 & 10 \\ 4 & 9 & 8 & -3 \\ 8 & 2 & 0 & 12 \\ 7 & -6 & 3 & 2 \end{bmatrix}$

36. $\begin{bmatrix} -22.7 & -26.9 & -6.3 & -46.5 \\ -59.7 & -40.9 & 20.9 & -99.5 \\ 15.9 & 9.6 & -8.4 & 26.5 \\ 43.8 & 36.5 & -7.3 & 78.2 \end{bmatrix}$

37. $\begin{bmatrix} 66.2 & 58.0 & -11.6 & 116.0 \\ 120.6 & 89.6 & -42.6 & 201.0 \\ -21.0 & -15.0 & 7.6 & -35.0 \\ -99.6 & -79.0 & 28.6 & -169.4 \end{bmatrix}$

38.
$$\begin{bmatrix} -253 & -232 & -96 & 1088 & 280 \\ 213 & 204 & 93 & -879 & -225 \\ -90 & -90 & -47 & 360 & 90 \\ -38 & -36 & -18 & 162 & 40 \\ 62 & 64 & 42 & -251 & -57 \end{bmatrix}$$

40.
$$\begin{bmatrix} -2513 & 596 & -414 & -2583 & 1937 \\ 127 & -32 & 33 & 132 & -81 \\ -421 & 94 & -83 & -434 & 306 \\ 2610 & -615 & 443 & 2684 & -1994 \\ 90 & -19 & 29 & 94 & -50 \end{bmatrix}$$

39.
$$\begin{bmatrix} 154 & -24 & -36 & -1608 & -336 \\ -126 & 16 & 18 & 1314 & 270 \\ 54 & 0 & 4 & -540 & -108 \\ 24 & 0 & 0 & -236 & -48 \\ -42 & -12 & -18 & 366 & 70 \end{bmatrix}$$

41.
$$\begin{bmatrix} 2 & -4 & 6 & 2 & 1 \\ 6 & 3 & -8 & 1 & 2 \\ -4 & 0 & 5 & 1 & 2 \\ 4 & 3 & 5 & -11 & 3 \\ 6 & 7 & 9 & 2 & 3 \end{bmatrix}$$

5.3   TWO APPLICATIONS

In this section, A will always denote a matrix with real-number entries.

In Section 5.1, we were motivated to introduce eigenvalues by our desire to compute $A^k\mathbf{x}$. Recall that we regard $\mathbf{x}$ as an initial information vector of some process, and A as a matrix that transforms, by left multiplication, an information vector at any stage of the process into the information vector at the next stage. In our first application, we examine this computation of $A^k\mathbf{x}$ and the significance of the eigenvalues of A. As illustration, we determine the behavior of the terms F_n of the Fibonacci sequence for large values of n. Our second application shows how diagonalization of matrices can be used to solve some linear systems of differential equations.

Application: Computing $A^k\mathbf{x}$

Let A be a real-diagonalizable $n \times n$ matrix, and let $\lambda_1, \lambda_2, \ldots, \lambda_n$ be the n not necessarily distinct eigenvalues of A. That is, each eigenvalue of A is repeated in this list in accord with its algebraic multiplicity. Let $B = (\mathbf{v}_1, \mathbf{v}_2, \ldots, \mathbf{v}_n)$ be an ordered basis for $\mathbb{R}^n$, where $\mathbf{v}_j$ is an eigenvector for λ_j. We have seen that, if C is the matrix having $\mathbf{v}_j$ as jth column vector, then

$$C^{-1}AC = D = \begin{bmatrix} \lambda_1 & & & \\ & \lambda_2 & & \text{\Large 0} \\ & & \ddots & \\ \text{\Large 0} & & & \lambda_n \end{bmatrix}.$$

For any vector $\mathbf{x}$ in $\mathbb{R}^n$, let $\mathbf{d} = [d_1, d_2, \ldots, d_n]$ be its coordinate vector relative to the basis B. Thus,

$$\mathbf{x} = d_1\mathbf{v}_1 + d_2\mathbf{v}_2 + \cdots + d_n\mathbf{v}_n.$$

Then

$$A^k\mathbf{x} = d_1\lambda_1{}^k\mathbf{v}_1 + d_2\lambda_2{}^k\mathbf{v}_2 + \cdots + d_n\lambda_n{}^k\mathbf{v}_n. \tag{1}$$

Equation (1) expresses $A^k\mathbf{x}$ as a linear combination of the eigenvectors $\mathbf{v}_j$.

Let us regard $\mathbf{x}$ as an initial *information vector* in a process in which the information vector at the next stage of the process is found by multiplying the present information vector on the left by a matrix A. Illustrations of this situation are provided by Markov chains in Section 1.7 and the generation of the Fibonacci sequence in Section 5.1. We are interested here in the long-term outcome of the process. That is, we wish to study $A^k\mathbf{x}$ for large values of k.

Let us number our eigenvalues and eigenvectors in Eq. (1) so that $|\lambda_i| \geq |\lambda_j|$ if $i < j$; that is, the eigenvalues are arranged in order of decreasing magnitude. Suppose that $|\lambda_1| > |\lambda_2|$ so that λ_1 is the unique eigenvalue of maximum magnitude. Equation (1) may be written

$$A^k\mathbf{x} = \lambda_1{}^k(d_1\mathbf{v}_1 + d_2(\lambda_2/\lambda_1)^k\mathbf{v}_2 + \cdots + d_n(\lambda_n/\lambda_1)^k\mathbf{v}_n).$$

Thus, if k is large and $d_1 \neq 0$, the vector $A^k\mathbf{x}$ is approximately equal to $d_1\lambda_1{}^k\mathbf{v}_1$ in the sense that $\|A^k\mathbf{x} - d_1\lambda_1{}^k\mathbf{v}_1\|$ is small compared with $\|A^k\mathbf{x}\|$.

EXAMPLE 1 Show that a diagonalizable transition matrix T for a Markov chain has no eigenvalues of magnitude > 1.

SOLUTION Example 3 in Section 5.1 shows that 1 is an eigenvalue for every transition matrix of a Markov chain. For every choice of population distribution vector $\mathbf{p}$, the vector $T^k\mathbf{p}$ is again a vector with nonnegative entries having sum 1. The preceding discussion shows that all eigenvalues of T must have magnitude ≤ 1; otherwise, entries in some $T^k\mathbf{p}$ would have very large magnitude as k increases. ∎

EXAMPLE 2 Find the order of magnitude of the term F_k of the Fibonacci sequence F_0, F_1, F_2, F_3, ... —that is, the sequence

$$0, 1, 1, 2, 3, 5, 8, 13, \ldots$$

for large values of k.

SOLUTION We saw in Section 5.1 that, if we let

$$\mathbf{x}_k = \begin{bmatrix} F_k \\ F_{k-1} \end{bmatrix},$$

then

$$\mathbf{x}_k = \begin{bmatrix} 1 & 1 \\ 1 & 0 \end{bmatrix}\mathbf{x}_{k-1}.$$

We compute relation (1) for

$$A = \begin{bmatrix} 1 & 1 \\ 1 & 0 \end{bmatrix} \quad \text{and} \quad \mathbf{x} = \mathbf{x}_1 = \begin{bmatrix} 1 \\ 0 \end{bmatrix}.$$

The characteristic equation of A is

$$(1 - \lambda)(-\lambda) - 1 = \lambda^2 - \lambda - 1 = 0.$$

Using the quadratic formula, we find the eigenvalues

$$\lambda_1 = \frac{1 + \sqrt{5}}{2} \quad \text{and} \quad \lambda_2 = \frac{1 - \sqrt{5}}{2}.$$

Reducing $A - \lambda_1 I$, we obtain

$$A - \lambda_1 I \sim \begin{bmatrix} \dfrac{1 - \sqrt{5}}{2} & 1 \\ 0 & 0 \end{bmatrix}, \quad \text{so} \quad \mathbf{v}_1 = \begin{bmatrix} 2 \\ \sqrt{5} - 1 \end{bmatrix}$$

is an eigenvector for λ_1. In an analogous fashion, we find that

$$\mathbf{v}_2 = \begin{bmatrix} -2 \\ \sqrt{5} + 1 \end{bmatrix}$$

is an eigenvector corresponding to λ_2. Thus we take

$$C = \begin{bmatrix} 2 & -2 \\ \sqrt{5} - 1 & \sqrt{5} + 1 \end{bmatrix}.$$

To find the coordinate vector $\mathbf{d}$ of $\mathbf{x}_1$ relative to the basis $(\mathbf{v}_1, \mathbf{v}_2)$, we observe that $\mathbf{x}_1 = C\mathbf{d}$. We find that

$$C^{-1} = \frac{1}{4\sqrt{5}} \begin{bmatrix} \sqrt{5} + 1 & 2 \\ 1 - \sqrt{5} & 2 \end{bmatrix};$$

thus,

$$\begin{bmatrix} d_1 \\ d_2 \end{bmatrix} = C^{-1}\mathbf{x}_1 = C^{-1}\begin{bmatrix} 1 \\ 0 \end{bmatrix} = \frac{1}{4\sqrt{5}} \begin{bmatrix} \sqrt{5} + 1 \\ 1 - \sqrt{5} \end{bmatrix}.$$

Equation 1 takes the form

$$A^k\mathbf{x}_1 = \begin{bmatrix} F_{k+1} \\ F_k \end{bmatrix} = \left(\frac{\sqrt{5} + 1}{4\sqrt{5}}\right)\left(\frac{1 + \sqrt{5}}{2}\right)^k \begin{bmatrix} 2 \\ \sqrt{5} - 1 \end{bmatrix}$$

$$- \left(\frac{\sqrt{5} - 1}{4\sqrt{5}}\right)\left(\frac{1 - \sqrt{5}}{2}\right)^k \begin{bmatrix} -2 \\ \sqrt{5} + 1 \end{bmatrix}. \quad (2)$$

For large k, the kth power of the eigenvalue $\lambda_1 = (1 + \sqrt{5})/2$ dominates, so $A^k\mathbf{x}_1$ is approximately equal to the shaded portion of Eq. (2). Computing the second component of $A^k\mathbf{x}_1$, we find that

$$F_k = \frac{1}{\sqrt{5}}\left(\left(\frac{1 + \sqrt{5}}{2}\right)^k - \left(\frac{1 - \sqrt{5}}{2}\right)^k\right). \quad (3)$$

Thus,

$$F_k \approx \frac{1}{\sqrt{5}}\left(\frac{1 + \sqrt{5}}{2}\right)^k \quad \text{for large } k. \tag{4}$$

Indeed, because $|\lambda_2| = |(1 - \sqrt{5})/2| < 1$, we see that the contribution from this second eigenvalue to the right-hand side of Eq. (3) approaches zero as k increases. Because $|\lambda_2^k/\sqrt{5}| < \frac{1}{2}$ for $k = 1$ and hence for all k, we see that F_k can be characterized as the closest integer to $(1/\sqrt{5})((1 + \sqrt{5})/2)^k$ for all k. Approximation (4) verifies that F_k increases exponentially with k, as expected for a population of rabbits. ■

Example 2 is a typical analysis of a process in which an information vector after the kth stage is equal to $A^k\mathbf{x}_1$ for an initial information vector $\mathbf{x}_1$ and a diagonalizable matrix A. An important consideration is whether any of the eigenvalues of A have magnitude greater than 1. When this is the case, the components of the information vector may grow exponentially in magnitude, as illustrated by the Fibonacci sequence in Example 2, where $|\lambda_1| > 1$. On the other hand, if all the eigenvalues have magnitude less than 1, the components of the information vector must approach zero as k increases.

The process just described is called *unstable* if A has an eigenvalue of magnitude greater than 1, *stable* if all eigenvalues have magnitude less than 1, and *neutrally stable* if the maximum magnitude of the eigenvalues is 1. Thus a Markov chain is a neutrally stable process, whereas generation of the Fibonacci sequence is unstable. The eigenvectors are called the **normal modes** of the process.

In the type of process just described, we study information at evenly spaced time intervals. If we study such a process as the number of time intervals increases and their duration approaches zero, we find ourselves in calculus. Eigenvalues and eigenvectors play an important role in applications of calculus, especially in studying any sort of vibration. In these applications of calculus, components of vectors are functions of time. Our second application illustrates this.

Application: Systems of Linear Differential Equations

In calculus, we see the importance of the differential equation

$$\frac{dx}{dt} = kx$$

in simple rate of growth problems involving the time derivative of the amount x present of a single quantity. We may also write this equation as $x' = kx$, where we understand that x is a function of the time variable t. In more complex growth situations, n quantities may be present in amounts $x_1, x_2, \ldots, x_n$. The rate of change of x_i may depend not only on the amount of x_i present, but also on the amounts of the other $n - 1$ quantities present at time t. We consider a situation in which the rate of growth of each x_i depends linearly on the amounts present of the n quantities. This leads to a system of linear differential equations

$$
\begin{aligned}
x_1' &= a_{11}x_1 + a_{12}x_2 + \cdots + a_{1n}x_n \\
x_2' &= a_{21}x_1 + a_{22}x_2 + \cdots + a_{2n}x_n \\
&\quad\vdots \\
x_n' &= a_{n1}x_1 + a_{n2}x_2 + \cdots + a_{nn}x_n,
\end{aligned}
\tag{5}
$$

where each x_i is a differentiable function of the real variable t and each a_{ij} is a scalar. The simplest such system is the single differential equation

$$
x' = ax,
\tag{6}
$$

which has the general solution

$$
x(t) = ke^{at},
\tag{7}
$$

where k is a scalar. (See Example 8 on page 298.) Direct computation verifies that function (7) is the general solution of Eq. (6).

Turning to the solution of system (5), we write it in matrix form as

$$
\mathbf{x}' = A\mathbf{x},
\tag{8}
$$

where

$$
\mathbf{x} = \begin{bmatrix} x_1(t) \\ x_2(t) \\ \vdots \\ x_n(t) \end{bmatrix}, \quad
\mathbf{x}' = \begin{bmatrix} x_1'(t) \\ x_2'(t) \\ \vdots \\ x_n'(t) \end{bmatrix}, \quad \text{and} \quad A = [a_{ij}].
$$

If the matrix A is a diagonal matrix so that $a_{ij} = 0$ for $i \neq j$, then system (8) reduces to a system of n equations, each like Eq. (6), namely:

$$
\begin{aligned}
x_1' &= a_{11}x_1 \\
x_2' &= a_{22}x_2 \\
&\quad\vdots \\
x_n' &= a_{nn}x_n.
\end{aligned}
\tag{9}
$$

The general solution is given by

$$
\mathbf{x} = \begin{bmatrix} x_1 \\ x_2 \\ \vdots \\ x_n \end{bmatrix} = \begin{bmatrix} k_1 e^{a_{11}t} \\ k_2 e^{a_{22}t} \\ \vdots \\ k_n e^{a_{nn}t} \end{bmatrix}.
$$

In the general case, we try to diagonalize A and reduce system (8) to a system like (9). If A is diagonalizable, we have

$$D = C^{-1}AC = \begin{bmatrix} \lambda_1 & & & \\ & \lambda_2 & & \mathbf{0} \\ & & \ddots & \\ \mathbf{0} & & & \lambda_n \end{bmatrix}$$

for some invertible $n \times n$ matrix C. If we substitute $A = CDC^{-1}$, Eq. (8) takes the form $C^{-1}\mathbf{x}' = D(C^{-1}\mathbf{x})$, or

$$\mathbf{y}' = D\mathbf{y}, \tag{10}$$

where

$$\mathbf{x} = C\mathbf{y}. \tag{11}$$

(If we let $\mathbf{x} = C\mathbf{y}$, we can confirm that $\mathbf{x}' = C\mathbf{y}'$.) The general solution of system (10) is

$$\mathbf{y} = \begin{bmatrix} y_1 \\ y_2 \\ \cdot \\ \cdot \\ \cdot \\ y_n \end{bmatrix} = \begin{bmatrix} k_1 e^{\lambda_1 t} \\ k_2 e^{\lambda_2 t} \\ \cdot \\ \cdot \\ \cdot \\ k_n e^{\lambda_n t} \end{bmatrix},$$

where the λ_j are eigenvalues of the diagonalizable matrix A. The general solution $\mathbf{x}$ of system (8) is then obtained from Eq. (11), using the corresponding eigenvectors $\mathbf{v}_j$ of A as columns of the matrix C.

We emphasize that we have described the general solution of system (8) only in the case where the matrix A is diagonalizable. The algebraic multiplicity of each eigenvalue of A must equal its geometric multiplicity. The following example illustrates that the eigenvalues need not be distinct, as long as their algebraic and geometric multiplicities are the same.

EXAMPLE 3   Solve the linear differential system

$$\begin{aligned} x_1' &= x_1 - x_2 - x_3 \\ x_2' &= -x_1 + x_2 - x_3 \\ x_3' &= -x_1 - x_2 + x_3. \end{aligned}$$

SOLUTION   The first step is to diagonalize the matrix

$$A = \begin{bmatrix} 1 & -1 & -1 \\ -1 & 1 & -1 \\ -1 & -1 & 1 \end{bmatrix}.$$

Expanding the determinant $|A - \lambda I|$ across the first row, we obtain

$$|A - \lambda I| = \begin{vmatrix} 1 - \lambda & -1 & -1 \\ -1 & 1 - \lambda & -1 \\ -1 & -1 & 1 - \lambda \end{vmatrix}$$

$$= (1 - \lambda)\begin{vmatrix} 1 - \lambda & -1 \\ -1 & 1 - \lambda \end{vmatrix} + \begin{vmatrix} -1 & -1 \\ -1 & 1 - \lambda \end{vmatrix} - \begin{vmatrix} -1 & 1 - \lambda \\ -1 & -1 \end{vmatrix}$$

$$= (1 - \lambda)(\lambda^2 - 2\lambda) + 2(\lambda - 2)$$

$$= (\lambda - 2)((1 - \lambda)\lambda + 2)$$

$$= -(\lambda + 1)(\lambda - 2)^2.$$

This yields eigenvalues $\lambda_1 = -1$ and $\lambda_2 = \lambda_3 = 2$.

Next we compute eigenvectors. For $\lambda_1 = -1$, we have

$$A + I = \begin{bmatrix} 2 & -1 & -1 \\ -1 & 2 & -1 \\ -1 & -1 & 2 \end{bmatrix} \sim \begin{bmatrix} 1 & 1 & -2 \\ 0 & 3 & -3 \\ 0 & 0 & 0 \end{bmatrix},$$

which gives the eigenvector

$$\mathbf{v}_1 = \begin{bmatrix} 1 \\ 1 \\ 1 \end{bmatrix}.$$

For $\lambda_2 = \lambda_3 = 2$, we have

$$A - 2I = \begin{bmatrix} -1 & -1 & -1 \\ -1 & -1 & -1 \\ -1 & -1 & -1 \end{bmatrix} \sim \begin{bmatrix} 1 & 1 & 1 \\ 0 & 0 & 0 \\ 0 & 0 & 0 \end{bmatrix},$$

which gives the independent eigenvectors

$$\mathbf{v}_2 = \begin{bmatrix} -1 \\ 1 \\ 0 \end{bmatrix} \quad \text{and} \quad \mathbf{v}_3 = \begin{bmatrix} -1 \\ 0 \\ 1 \end{bmatrix}.$$

A diagonalizing matrix is then

$$C = \begin{bmatrix} 1 & -1 & -1 \\ 1 & 1 & 0 \\ 1 & 0 & 1 \end{bmatrix},$$

so that

$$D = C^{-1}AC = \begin{bmatrix} -1 & 0 & 0 \\ 0 & 2 & 0 \\ 0 & 0 & 2 \end{bmatrix}.$$

Setting $\mathbf{x} = C\mathbf{y}$, we see that the system becomes

$$y_1' = -y_1$$
$$y_2' = 2y_2$$
$$y_3' = 2y_3,$$

whose solution is given by

$$\begin{bmatrix} y_1 \\ y_2 \\ y_3 \end{bmatrix} = \begin{bmatrix} k_1 e^{-t} \\ k_2 e^{2t} \\ k_3 e^{2t} \end{bmatrix}$$

Therefore, the solution to the original system is

$$\begin{bmatrix} x_1 \\ x_2 \\ x_3 \end{bmatrix} = \begin{bmatrix} 1 & -1 & -1 \\ 1 & 1 & 0 \\ 1 & 0 & 1 \end{bmatrix} \begin{bmatrix} y_1 \\ y_2 \\ y_3 \end{bmatrix} = \begin{bmatrix} k_1 e^{-t} - k_2 e^{2t} - k_3 e^{2t} \\ k_1 e^{-t} + k_2 e^{2t} \\ k_1 e^{-t} \qquad\quad + k_3 e^{2t} \end{bmatrix}. \qquad \blacksquare$$

In Section 9.4, we show that every square matrix A is similar to a matrix J in *Jordan canonical form*, so that $J = C^{-1}AC$. The Jordan canonical form J and the invertible matrix C may have complex entries. In the Jordan canonical form J, all nondiagonal entries are zero except for some entries 1 immediately above the diagonal entries. If the Jordan form $J = C^{-1}AC$ is known, the substitution $\mathbf{x} = C\mathbf{y}$ again reduces a system $\mathbf{x}' = A\mathbf{x}$ of linear differential equations to one that can be solved easily for $\mathbf{y}$. The solution of the original system is computed as $\mathbf{x} = C\mathbf{y}$, just as was the solution in Example 3. This is an extremely important technique in the study of differential equations.

SUMMARY

1. Let A be diagonalizable by a matrix C, let $\mathbf{x}$ be any column vector, and let $\mathbf{d} = C^{-1}\mathbf{x}$. Then

$$A^k\mathbf{x} = d_1\lambda_1{}^k\mathbf{v}_1 + d_2\lambda_2{}^k\mathbf{v}_2 + \cdots + d_n\lambda_n{}^k\mathbf{v}_n,$$

   where $\mathbf{v}_j$ is the jth column vector of C, as described on page 318.

2. Let multiplication of a column information vector by A give the information vector for the next stage of a process, as described on page 320. The process is stable, neutrally stable, or unstable, according as the maximum magnitude of the eigenvalues of A is less than 1, equal to 1, or greater than 1, respectively.

3. The system $\mathbf{x}' = A\mathbf{x}$ of linear differential equations can be solved, if A is diagonalizable, using the following three steps.

   Step 1. Find a matrix C so that $D = C^{-1}AC$ is a diagonal matrix.
   Step 2. Solve the simpler diagonal system $\mathbf{y}' = D\mathbf{y}$.
   Step 3. The solution of the original system is $\mathbf{x} = C\mathbf{y}$.

EXERCISES

1. Let the sequence a_0, a_1, a_2, ... be given by $a_0 = 0$, $a_1 = 1$, and $a_k = (a_{k-1} + a_{k-2})/2$ for $k \geq 2$.

   a. Find the matrix A that can be used to generate this sequence as we used a matrix to generate the Fibonacci sequence in Section 5.1.

   b. Classify this generation process as stable, neutrally stable, or unstable.

   c. Compute expression (1) for $\mathbf{x} = \begin{bmatrix} a_1 \\ a_0 \end{bmatrix}$ for this process. Check computations with the first few terms of the sequence.

   d. Use the answer to part (c) to estimate a_k for large k.

2. Repeat Exercise 1 if $a_k = a_{k-1} - \left(\frac{3}{16}\right)a_{k-2}$ for $k \geq 2$.

3. Repeat Exercise 1, but change the initial data to $a_0 = 1$, $a_1 = 0$.

4. Repeat Exercise 1 if $a_k = \left(\frac{1}{2}\right)a_{k-1} + \left(\frac{3}{16}\right)a_{k-2}$ for $k \geq 2$.

5. Repeat Exercise 1 if $a_k = a_{k-1} + \left(\frac{3}{4}\right)a_{k-2}$ for $k \geq 2$.

In Exercises 6–13, solve the given system of linear differential equations as outlined in the summary.

6. $\begin{aligned} x_1' &= 3x_1 - 5x_2 \\ x_2' &= \qquad 2x_2 \end{aligned}$

7. $\begin{aligned} x_1' &= x_1 + 4x_2 \\ x_2' &= 3x_1 \end{aligned}$

8. $\begin{aligned} x_1' &= x_1 + 2x_2 \\ x_2' &= 2x_1 + x_2 \end{aligned}$

9. $\begin{aligned} x_1' &= 2x_1 + 2x_2 \\ x_2' &= x_1 + 3x_2 \end{aligned}$

10. $\begin{aligned} x_1' &= 6x_1 + 3x_2 - 3x_3 \\ x_2' &= -2x_1 - x_2 + 2x_3 \\ x_3' &= 16x_1 + 8x_2 - 7x_3 \end{aligned}$

11. $\begin{aligned} x_1' &= -3x_1 + 10x_2 - 6x_3 \\ x_2' &= \qquad 7x_2 - 6x_3 \\ x_3' &= \qquad\qquad x_3 \end{aligned}$

12. $\begin{aligned} x_1' &= -3x_1 + 5x_2 - 20x_3 \\ x_2' &= 2x_1 \qquad + 8x_3 \\ x_3' &= 2x_1 + x_2 + 7x_3 \end{aligned}$

13. $\begin{aligned} x_1' &= -2x_1 \qquad - x_3 \\ x_2' &= \qquad 2x_2 \\ x_3' &= 3x_1 \qquad + 2x_3 \end{aligned}$

6

ORTHOGONALITY

We are accustomed to working in $\mathbb{R}^2$ with coordinates of vectors relative to the standard ordered basis $(\mathbf{e}_1, \mathbf{e}_2) = (\mathbf{i}, \mathbf{j})$. These basis vectors are orthogonal (perpendicular) and have length 1. The vectors in the standard ordered basis $(\mathbf{e}_1, \mathbf{e}_2, \ldots, \mathbf{e}_n)$ of $\mathbb{R}^n$ have these same two properties. It is precisely these two properties of the standard bases that make computations using coordinates relative to them quite easy. For example, computations of angles and of length are easy when we use coordinates relative to standard bases. Throughout linear algebra, computations using coordinates relative to bases consisting of orthogonal unit vectors are generally the easiest to perform, and they generate less error in work with a computer.

This chapter is devoted to orthogonality. Orthogonal projection, which is the main tool for finding a basis of orthogonal vectors, is developed in Section 6.1. Section 6.2 shows that every finite-dimensional inner-product space has a basis of orthogonal unit vectors, and shows how to construct such a basis from any given basis. Section 6.3 deals with orthogonal matrices and orthogonal linear transformations. In Section 6.4 we show that orthogonal projection can be achieved by matrix multiplication. To conclude the chapter, we give applications to the method of least squares and overdetermined linear systems in Section 6.5.

6.1   PROJECTIONS

The Projection of b on sp(a)

For convenience, we develop the ideas in this section for the vector space $\mathbb{R}^n$, using the dot product. The work is valid in general finite-dimensional inner-product spaces, using the more cumbersome notation $\langle \mathbf{u}, \mathbf{v} \rangle$ for the inner product of $\mathbf{u}$ and $\mathbf{v}$. An illustration in function spaces, using an inner product defined by an integral, appears at the end of this section.

ILLUSTRATION 1   A practical concern in vector applications involves determining what portion of a vector **b** can be considered to act in the direction given by another vector **a**. A force vector acting in a certain direction may be moving a body along a line having a different direction. For example, suppose that you are trying to roll your stalled car off the road by pushing on the door jamb at the side, so you can reach in and control the steering wheel when necessary. You are not applying the force in precisely the direction in which the car moves, as you would be if you could push from directly behind the car. Such considerations lead the physicist to consider the projection **p** of the force vector **F** on a direction vector **a**, as shown in Figure 6.1. In Figure 6.1, the vector **p** is found by dropping a perpendicular from the tip of **F** to the vector **a**.   ∎

Figure 6.2 shows the situation where the projection of **F** on **a** has a direction opposite to the direction of **a**; in terms of our car in Illustration 1, you would be applying the force to move the car backward rather than forward. Figure 6.2 suggests that it is preferable to speak of *projection on the subspace* sp(**a**) (which in this example is a line) than to speak of *projection on **a***.

We derive geometrically a formula for the projection **p** of the force vector **F** on sp(**a**), based on Figures 6.1 and 6.2. We see that **p** is a multiple of **a**. Now $(1/\|\mathbf{a}\|)\mathbf{a}$ is a unit vector having the same direction as **a**, so **p** is a scalar multiple of this unit vector. We need only find the appropriate scalar. Referring to Figure 6.1, we see that the appropriate scalar is $\|\mathbf{F}\| \cos \theta$, because it is the length of the leg labeled **p** of the right triangle. This same formula also gives the appropriate negative scalar for the case shown in Figure 6.2, because $\cos \theta$ is negative for this angle θ lying between 90° and 180°. Thus we obtain

$$\mathbf{p} = \frac{\|\mathbf{F}\| \cos \theta}{\|\mathbf{a}\|}\, \mathbf{a} = \frac{\|\mathbf{F}\|\, \|\mathbf{a}\| \cos \theta}{\|\mathbf{a}\|\, \|\mathbf{a}\|}\, \mathbf{a} = \frac{\mathbf{F} \cdot \mathbf{a}}{\mathbf{a} \cdot \mathbf{a}}\, \mathbf{a}.$$

Of course, we assume that $\mathbf{a} \neq \mathbf{0}$. We replace our force vector **F** by a general vector **b** and box this formula.

Projection p of b on sp(a) in $\mathbb{R}^n$

$$\mathbf{p} = \frac{\mathbf{b} \cdot \mathbf{a}}{\mathbf{a} \cdot \mathbf{a}}\, \mathbf{a}$$

(1)

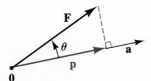

FIGURE 6.1
Projection p of F on a.

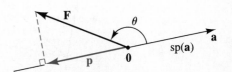

FIGURE 6.2
Projection p of F on sp(a).

EXAMPLE 1   Find the projection **p** of the vector [1, 2, 3] on sp([2, 4, 3]) in $\mathbb{R}^3$.

SOLUTION   We let **a** = [2, 4, 3] and **b** = [1, 2, 3] in formula (1), obtaining

$$\mathbf{p} = \frac{\mathbf{b} \cdot \mathbf{a}}{\mathbf{a} \cdot \mathbf{a}}\, \mathbf{a} = \frac{2 + 8 + 9}{4 + 16 + 9}\, \mathbf{a} = \frac{19}{29}\, [2,\, 4,\, 3].$$ ∎

The Concept of Projection

We now explain what is meant by the *projection* of a vector **b** in $\mathbb{R}^n$ on a general subspace W of $\mathbb{R}^n$. We will show that there are unique vectors $\mathbf{b}_W$ and $\mathbf{b}_{W^\perp}$ such that

1. $\mathbf{b}_W$ is in the subspace W;
2. $\mathbf{b}_{W^\perp}$ is orthogonal to every vector in W; and
3. $\mathbf{b} = \mathbf{b}_W + \mathbf{b}_{W^\perp}$.

Let $W^\perp$ be the set of all vectors in $\mathbb{R}^n$ that are perpendicular to every vector in W. Properties of $W^\perp$ will appear shortly in Theorem 6.1. Figure 6.3 gives a symbolic illustration of this decomposition of **b** into a sum of a vector in W and a vector orthogonal to W. Once we have demonstrated the existence of this decomposition, we will define the *projection of* **b** *on* W to be the vector $\mathbf{b}_W$.

The projection $\mathbf{b}_W$ of **b** on W is the vector **w** in W that is closest to **b**. That is, $\mathbf{w} = \mathbf{b}_W$ *minimizes* the distance $\|\mathbf{b} - \mathbf{w}\|$ from **b** to W for all **w** in W. This seems reasonable, because we have $\mathbf{b} = \mathbf{b}_W + \mathbf{b}_{W^\perp}$ and because $\mathbf{b}_{W^\perp}$ is orthogonal to every vector in W. We can demonstrate algebraically that for any $\mathbf{w} \in W$, we have $\|\mathbf{b} - \mathbf{w}\| \geq \|\mathbf{b} - \mathbf{b}_W\|$. We work with $\|\mathbf{b} - \mathbf{w}\|^2$ so we can use the dot product. Because the dot product of any vector in W and any vector in $W^\perp$ is 0, we obtain, for all $\mathbf{w} \in W$,

$$\|\mathbf{b} - \mathbf{w}\|^2 = (\mathbf{b} - \mathbf{w}) \cdot (\mathbf{b} - \mathbf{w})$$
$$= ((\mathbf{b} - \mathbf{b}_W) + (\mathbf{b}_W - \mathbf{w})) \cdot ((\mathbf{b} - \mathbf{b}_W) + (\mathbf{b}_W - \mathbf{w}))$$
$$= (\mathbf{b} - \mathbf{b}_W) \cdot (\mathbf{b} - \mathbf{b}_W) + 2\underbrace{(\mathbf{b} - \mathbf{b}_W)}_{\text{in } W^\perp} \cdot \underbrace{(\mathbf{b}_W - \mathbf{w})}_{\text{in } W} + (\mathbf{b}_W - \mathbf{w}) \cdot (\mathbf{b}_W - \mathbf{w})$$
$$= \|\mathbf{b} - \mathbf{b}_W\|^2 + \|\mathbf{b}_W - \mathbf{w}\|^2 \geq \|\mathbf{b} - \mathbf{b}_W\|^2.$$

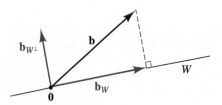

FIGURE 6.3
The decomposition b = b$_W$ + b$_{W^\perp}$

Of course, this minimum distance from **b** to W is just $\|\mathbf{b}_{W^\perp}\|$, as indicated by Figure 6.3.

ILLUSTRATION 2   Suppose that you are pushing a box across a floor by pushing forward and downward on the top edge. (See Figure 6.4.) We take as origin the point where the force is applied to the box and as W the plane through that origin parallel to the floor. If **F** is the force vector applied at the origin, then $\mathbf{F}_W$ is the portion of the force vector that actually moves the body along the floor, and $\mathbf{F}_{W^\perp}$ (which is directed straight down) is the portion of the force vector that attempts to push the box into the floor and thereby increases the friction between the box and the floor.   ■

Orthogonal Complement of a Subspace

As a preliminary to proving the existence and uniqueness of the vectors $\mathbf{b}_W$ and $\mathbf{b}_{W^\perp}$ just described, we consider all vectors in $\mathbb{R}^n$ that are orthogonal to every vector in a subspace W.

DEFINITION 6.1   Orthogonal Complement

Let W be a subspace of $\mathbb{R}^n$. The set of all vectors in $\mathbb{R}^n$ that are orthogonal to every vector in W is the **orthogonal complement** of W, and is denoted by $W^\perp$.

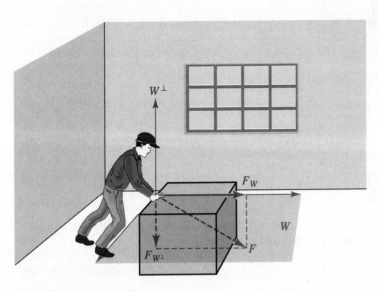

FIGURE 6.4
The decomposition of a force vector.

It is not difficult to find the orthogonal complement of W if we know a generating set for W. Let $\{v_1, v_2, \ldots, v_k\}$ be a generating set for W. Let A be the $k \times n$ matrix having v_i as its ith row vector. That is,

$$A = \begin{bmatrix} \underline{\quad\quad} v_1 \underline{\quad\quad} \\ \underline{\quad\quad} v_2 \underline{\quad\quad} \\ \vdots \\ \underline{\quad\quad} v_k \underline{\quad\quad} \end{bmatrix}.$$

Thus, W is the row space of A. Now the nullspace of A consists of all vectors x in $\mathbb{R}^n$ that are solutions of the homogeneous system $Ax = 0$. But $Ax = 0$ if and only if $v_i \cdot x = 0$ for $i = 1, 2, \ldots, k$. Therefore, the nullspace of A is the set of all vectors x in $\mathbb{R}^n$ that are orthogonal to each of the rows of A, and hence to the row space of A. In other words, the orthogonal complement of the row space of A is the nullspace of A. Thus we have found $W^\perp$. We summarize this procedure in a box. Notice that this procedure marks one of the rare occasions in this text when vectors must be placed as rows of a matrix.

Finding the Orthogonal Complement of a Subspace W of $\mathbb{R}^n$

1. Find a matrix A having as *row* vectors a generating set for W.
2. Find the nullspace of A—that is, the solution space of $Ax = 0$. This nullspace is $W^\perp$.

EXAMPLE 2   Find a basis for the orthogonal complement in $\mathbb{R}^4$ of the subspace

$$W = \text{sp}([1, 2, 2, 1], [3, 4, 2, 3]).$$

SOLUTION   We find the nullspace of the matrix

$$A = \begin{bmatrix} 1 & 2 & 2 & 1 \\ 3 & 4 & 2 & 3 \end{bmatrix}.$$

Reducing A, we have

$$\begin{bmatrix} 1 & 2 & 2 & 1 \\ 3 & 4 & 2 & 3 \end{bmatrix} \sim \begin{bmatrix} 1 & 2 & 2 & 1 \\ 0 & -2 & -4 & 0 \end{bmatrix} \sim \begin{bmatrix} 1 & 0 & -2 & 1 \\ 0 & 1 & 2 & 0 \end{bmatrix}.$$

Therefore, the nullspace of A, which is the orthogonal complement of W, is the set of vectors of the form

$$[2r - s, -2r, r, s] \quad \text{for any scalars } r \text{ and } s.$$

Thus $\{[2, -2, 1, 0], [-1, 0, 0, 1]\}$ is a basis for $W^\perp$.   ∎

ILLUSTRATION 3   Note that if W is the $(n - 1)$-dimensional solution space in $\mathbb{R}^n$ of a single linear equation $a_1x_1 + a_2x_2 + \cdots + a_nx_n = 0$, then the vector $[a_1, a_2, \ldots, a_n]$ of coefficients is orthogonal to W, because the equation can be written as

$$[a_1, a_2, \ldots, a_n] \cdot [x_1, x_2, \ldots, x_n] = 0.$$

Thus $W^\perp = \text{sp}([a_1, a_2, \ldots, a_n])$. The subspace W is a line if $n = 2$, a plane if $n = 3$, and is called a *hyperplane* for other values of n.) ∎

We now show that the orthogonal complement of a subspace W has some very nice properties. In particular, we exhibit the decomposition $\mathbf{b} = \mathbf{b}_W + \mathbf{b}_{W^\perp}$ described earlier and show that it is unique.

THEOREM 6.1   Properties of $W^\perp$

The orthogonal complement $W^\perp$ of a subspace W of $\mathbb{R}^n$ has the following properties:

1. $W^\perp$ is a subspace of $\mathbb{R}^n$.
2. $\dim(W^\perp) = n - \dim(W)$.
3. $(W^\perp)^\perp = W$; that is, the orthogonal complement of $W^\perp$ is W.
4. Each vector $\mathbf{b}$ in $\mathbb{R}^n$ can be expressed uniquely in the form $\mathbf{b} = \mathbf{b}_W + \mathbf{b}_{W^\perp}$ for $\mathbf{b}_W$ in W and $\mathbf{b}_{W^\perp}$ in $W^\perp$.

PROOF   We may assume $W \neq \{\mathbf{0}\}$. Let $\dim(W) = k$, and let $\{\mathbf{v}_1, \mathbf{v}_2, \ldots, \mathbf{v}_k\}$ be a basis for W. Let A be the $k \times n$ matrix having $\mathbf{v}_i$ as its ith row vector for $i = 1, \ldots, k$.

For property 1, we have seen that $W^\perp$ is the nullspace of the matrix A, so it is a subspace of $\mathbb{R}^n$.

For property 2, consider the rank equation of A:

$$\text{rank}(A) + \text{nullity}(A) = n.$$

Because $\dim(W) = \text{rank}(A)$ and because $W^\perp$ is the nullspace of A, we see that $\dim(W^\perp) = n - \dim(W)$.

For property 3, applying property 2 with the subspace $W^\perp$, we find that

$$\dim(W^\perp)^\perp = n - \dim(W^\perp) = n - (n - k) = k.$$

However, every vector in W is orthogonal to every vector in $W^\perp$, so W is a subspace of $(W^\perp)^\perp$. Because both W and $(W^\perp)^\perp$ have the same dimension k, we conclude that W must be equal to $(W^\perp)^\perp$. (See Exercise 38 in Section 2.1.)

For property 4, let $\{\mathbf{v}_{k+1}, \mathbf{v}_{k+2}, \ldots, \mathbf{v}_n\}$ be a basis for $W^\perp$. We claim that the set

$$\{\mathbf{v}_1, \mathbf{v}_2, \ldots, \mathbf{v}_n\} \tag{2}$$

is a basis for $\mathbb{R}^n$. Consider a relation

$$r_1\mathbf{v}_1 + r_2\mathbf{v}_2 + \cdots + r_k\mathbf{v}_k + s_{k+1}\mathbf{v}_{k+1} + s_{k+2}\mathbf{v}_{k+2} + \cdots + s_n\mathbf{v}_n = \mathbf{0}.$$

Rewrite this relation as

$$r_1\mathbf{v}_1 + r_2\mathbf{v}_2 + \cdots + r_k\mathbf{v}_k = -s_{k+1}\mathbf{v}_{k+1} - s_{k+2}\mathbf{v}_{k+2} - \cdots - s_n\mathbf{v}_n. \tag{3}$$

The sum on the left-hand side is in W, and the sum on the right-hand side is in $W^\perp$. Because these sums are equal, they represent a vector that is in both W and $W^\perp$ and must be orthogonal to itself. The only such vector is the zero vector, so both sides of Eq. (3) must equal $\mathbf{0}$. Because the $\mathbf{v}_i$ are independent for $1 \le i \le k$ and the $\mathbf{v}_j$ are independent for $k + 1 \le j \le n$, we see that all the scalars r_i and s_j are zero. This shows that set (2) is independent; because it contains n vectors, it must be a basis for $\mathbb{R}^n$. Therefore, for every vector $\mathbf{b}$ in $\mathbb{R}^n$ we can express $\mathbf{b}$ in the form

$$\mathbf{b} = \underbrace{r_1\mathbf{v}_1 + r_2\mathbf{v}_2 + \cdots + r_k\mathbf{v}_k}_{\mathbf{b}_W} + \underbrace{s_{k+1}\mathbf{v}_{k+1} + s_{k+2}\mathbf{v}_{k+2} + \cdots + s_n\mathbf{v}_n}_{\mathbf{b}_{W^\perp}}.$$

This shows that $\mathbf{b}$ can indeed be expressed as a sum of a vector in W and a vector in $W^\perp$, and this expression is unique because each vector in $\mathbb{R}^n$ is a unique combination of the vectors in the basis $\{\mathbf{v}_1, \mathbf{v}_2, \ldots, \mathbf{v}_n\}$. ▲

Projection of a Vector on a Subspace

Now that the groundwork is firmly established, we can define the projection of a vector in $\mathbb{R}^n$ on a subspace and then illustrate with some examples.

DEFINITION 6.2   Projection of b on W

Let $\mathbf{b}$ be a vector in $\mathbb{R}^n$, and let W be a subspace of $\mathbb{R}^n$. Let

$$\mathbf{b} = \mathbf{b}_W + \mathbf{b}_{W^\perp},$$

as described in Theorem 6.1. Then $\mathbf{b}_W$ is the **projection of b on** W.

Theorem 6.1 shows that the projection of $\mathbf{b}$ on W is unique. It also shows one way in which it can be computed.

Steps to Find the Projection of b on *W*

1. Select a basis $\{v_1, v_2, \ldots, v_k\}$ for the subspace W. (Often this is given.)
2. Find a basis $\{v_{k+1}, v_{k+2}, \ldots, v_n\}$ for $W^\perp$, as in Example 2.
3. Find the coordinate vector $\mathbf{r} = [r_1, r_2, \ldots, r_n]$ of **b** relative to the basis $(v_1, v_2, \ldots, v_n)$ so that $\mathbf{b} = r_1 v_1 + r_2 v_2 + \cdots + r_n v_n$. (See the box on page 207.)
4. Then $\mathbf{b}_W = r_1 v_1 + r_2 v_2 + \cdots + r_k v_k$.

EXAMPLE 3   Find the projection of $\mathbf{b} = [2, 1, 5]$ on the subspace $W = \text{sp}([1, 2, 1], [2, 1, -1])$.

SOLUTION   We follow the boxed procedure.

Step 1: Because $\mathbf{v}_1 = [1, 2, 1]$ and $\mathbf{v}_2 = [2, 1, -1]$ are independent, they form a basis for W.

Step 2: A basis for $W^\perp$ can be found by obtaining the nullspace of the matrix

$$A = \begin{bmatrix} 1 & 2 & 1 \\ 2 & 1 & -1 \end{bmatrix}.$$

An echelon form of A is

$$\begin{bmatrix} 1 & 2 & 1 \\ 0 & -3 & -3 \end{bmatrix},$$

and the nullspace of A is the set of vectors $[r, -r, r]$, where r is any scalar. Let us take $\mathbf{v}_3 = [1, -1, 1]$ to form the basis $\{v_3\}$ of $W^\perp$. (Alternatively, we could have computed $\mathbf{v}_1 \times \mathbf{v}_2$ to find a suitable $\mathbf{v}_3$.)

Step 3: To find the coordinate vector $\mathbf{r}$ of **b** relative to the ordered basis (v_1, v_2, v_3), we proceed as described in Section 3.3, and perform the reduction

$$\begin{bmatrix} 1 & 2 & 1 & | & 2 \\ 2 & 1 & -1 & | & 1 \\ 1 & -1 & 1 & | & 5 \end{bmatrix} \sim \begin{bmatrix} 1 & 2 & 1 & | & 2 \\ 0 & -3 & -3 & | & -3 \\ 0 & -3 & 0 & | & 3 \end{bmatrix} \sim \begin{bmatrix} 1 & 0 & 1 & | & 4 \\ 0 & 1 & 0 & | & -1 \\ 0 & 0 & -3 & | & -6 \end{bmatrix}$$

$$\quad\; \mathbf{v}_1 \;\; \mathbf{v}_2 \;\; \mathbf{v}_3 \;\;\; \mathbf{b}$$

$$\sim \begin{bmatrix} 1 & 0 & 0 & | & 2 \\ 0 & 1 & 0 & | & -1 \\ 0 & 0 & 1 & | & 2 \end{bmatrix}.$$

Thus, $\mathbf{r} = [2, -1, 2]$.

Step 4: The projection of **b** on W is

$$\mathbf{b}_W = 2\mathbf{v}_1 - \mathbf{v}_2 = 2[1, 2, 1] - [2, 1, -1] = [0, 3, 3].$$

As a check, notice that $2\mathbf{v}_3 = [2, -2, 2]$ is the projection of **b** on $W^\perp$, and $\mathbf{b} = \mathbf{b}_W + \mathbf{b}_{W^\perp} = [0, 3, 3] + [2, -2, 2] = [2, 1, 5]$.   ∎

Our next example shows that the procedure described in the box preceding Example 3 yields formula (1) for the projection of a vector **b** on sp(**a**).

EXAMPLE 4   Let $\mathbf{a} \neq \mathbf{0}$ and **b** be vectors in $\mathbb{R}^n$. Find the projection of **b** on sp(**a**), using the same boxed procedure we have been applying.

SOLUTION   We project **b** on the subspace $W = $ sp(**a**) of $\mathbb{R}^n$. Let $\{\mathbf{v}_2, \mathbf{v}_3, \ldots, \mathbf{v}_n\}$ be a basis for $W^{\perp}$, and let $\mathbf{r} = [r_1, r_2, \ldots, r_n]$ be the coordinate vector of **b** relative to the ordered basis $(\mathbf{a}, \mathbf{v}_2, \ldots, \mathbf{v}_n)$. Then

$$\mathbf{b} = r_1\mathbf{a} + r_2\mathbf{v}_2 + \cdots + r_n\mathbf{v}_n.$$

Because $\mathbf{v}_i \cdot \mathbf{a} = 0$ for $i = 2, \ldots, n$, we see that $\mathbf{b} \cdot \mathbf{a} = r_1\mathbf{a} \cdot \mathbf{a} = r_1(\mathbf{a} \cdot \mathbf{a})$. Because $\mathbf{a} \neq \mathbf{0}$, we can write

$$r_1 = \frac{\mathbf{b} \cdot \mathbf{a}}{\mathbf{a} \cdot \mathbf{a}}.$$

The projection of **b** on sp(**a**) is then $r_1\mathbf{a} = \dfrac{\mathbf{b} \cdot \mathbf{a}}{\mathbf{a} \cdot \mathbf{a}}\,\mathbf{a}$.   ∎

We have seen that if W is a one-dimensional subspace of $\mathbb{R}^n$, then we can readily find the projection $\mathbf{b}_W$ of $\mathbf{b} \in \mathbb{R}^n$ on W. We simply use formula (1). On the other hand, if $W^{\perp}$ is one-dimensional, we can use formula (1) to find the projection of **b** on $W^{\perp}$, and then find $\mathbf{b}_W$ using the relation $\mathbf{b}_W = \mathbf{b} - \mathbf{b}_{W^{\perp}}$. Our next example illustrates this.

EXAMPLE 5   Find the projection of the vector $[3, -1, 2]$ on the plane $x + y + z = 0$ through the origin in $\mathbb{R}^3$.

SOLUTION   Let W be the subspace of $\mathbb{R}^3$ given by the plane $x + y + z = 0$. Then $W^{\perp}$ is one-dimensional, and a generating vector for $W^{\perp}$ is $\mathbf{a} = [1, 1, 1]$, obtained by taking the coefficients of x, y, and z in this equation. (See Illustration 3.) Let $\mathbf{b} = [3, -1, 2]$. By formula (1), we have

$$\mathbf{b}_{W^{\perp}} = \frac{\mathbf{b} \cdot \mathbf{a}}{\mathbf{a} \cdot \mathbf{a}}\,\mathbf{a} = \frac{3 - 1 + 2}{1 + 1 + 1}[1, 1, 1] = \frac{4}{3}[1, 1, 1].$$

Thus $\mathbf{b}_W = \mathbf{b} - \mathbf{b}_{W^{\perp}} = [3, -1, 2] - \frac{4}{3}[1, 1, 1] = \left[\frac{5}{3}, -\frac{7}{3}, \frac{2}{3}\right].$ ∎

Projections in Inner-Product Spaces (Optional)

Everything we have done in this section is equally valid for any finite-dimensional inner-product space. For example, formula (1) for projecting one vector on a one-dimensional subspace takes the form

$$\mathbf{p} = \frac{\langle \mathbf{b}, \mathbf{a} \rangle}{\langle \mathbf{a}, \mathbf{a} \rangle}\,\mathbf{a}. \tag{4}$$

The fact that the projection $\mathbf{b}_W$ of a vector **b** on a subspace W is the vector **w** in W that minimizes $\|\mathbf{b} - \mathbf{w}\|$ is useful in many contexts.

For an example involving function spaces, suppose that f is a complicated function and that W consists of functions that are easily handled, such as polynomials or trigonometric functions. Using a suitable inner product and projection, we find that the function f_W in W becomes a best approximation to the function f by functions in W. Example 6 illustrates this by approximating the function $f(x) = x$ over the interval $0 \leq x \leq 1$ by a function in the space W of all constant functions on this interval. Visualizing their graphs, we are not surprised that the constant function $p(x) = \frac{1}{2}$ turns out to be the best approximation to x in W, using the inner product we defined on function spaces in Section 3.5. In Section 6.5, we will use this minimization feature again to find a best approximate solution of an inconsistent overdetermined linear system.

EXAMPLE 6   Let the inner product of two polynomials $p(x)$ and $q(x)$ in the space $P_{0,1}$ of polynomial functions with domain $0 \leq x \leq 1$ be defined by

$$\langle p(x), q(x) \rangle = \int_0^1 p(x)q(x) \, dx.$$

(See Example 3 in Section 3.5.) Find the projection of $f(x) = x$ on sp(1), using formula (4). Then find the projection of x on sp(1)$^\perp$.

SOLUTION   By formula (4), the projection of x on sp(1) is

$$\left(\frac{\langle x, 1 \rangle}{\langle 1, 1 \rangle} \right) 1 = \frac{\int_0^1 (x)(1) \, dx}{\int_0^1 (1)(1) \, dx} 1 = \left(\frac{1/2}{1} \right) 1 = \frac{1}{2}.$$

The projection of x on sp(1)$^\perp$ is obtained by subtracting the projection on sp(1) from x. We obtain $x - \frac{1}{2}$. As a check, we should then have $\langle \frac{1}{2}, x - \frac{1}{2} \rangle = 0$, and a computation of $\int_0^1 (\frac{1}{2})(x - \frac{1}{2}) \, dx$ shows that this is indeed so.   ∎

SUMMARY

1. The projection of $\mathbf{b}$ in $\mathbb{R}^n$ on sp($\mathbf{a}$) for a nonzero vector $\mathbf{a}$ in $\mathbb{R}^n$ is given by $((\mathbf{b} \cdot \mathbf{a})/(\mathbf{a} \cdot \mathbf{a}))\mathbf{a}$.

2. The orthogonal complement $W^\perp$ of a subspace W of $\mathbb{R}^n$ is the set of all vectors in $\mathbb{R}^n$ that are orthogonal to every vector in W. Further, $W^\perp$ is a subspace of $\mathbb{R}^n$ of dimension $n - \dim(W)$, and $(W^\perp)^\perp = W$.

3. The row space and the nullspace of an $m \times n$ matrix A are orthogonal complements of each other. In particular, $W^\perp$ can be computed as the nullspace of a matrix A having as its row vectors the vectors in a generating set for W.

4. Let W be a subspace of $\mathbb{R}^n$. Each vector $\mathbf{b}$ in $\mathbb{R}^n$ can be expressed uniquely in the form $\mathbf{b} = \mathbf{b}_W + \mathbf{b}_{W^\perp}$ for $\mathbf{b}_W$ in W and $\mathbf{b}_{W^\perp}$ in $W^\perp$.

5. The vectors $\mathbf{b}_W$ and $\mathbf{b}_{W^\perp}$ are the projections of $\mathbf{b}$ on W and on $W^\perp$, respectively. They can be computed by means of the boxed procedure on page 333.

EXERCISES

In Exercises 1–6, find the indicated projection.

1. The projection of [2, 1] on sp([3, 4]) in $\mathbb{R}^2$

2. The projection of [3, 4] on sp([2, 1]) in $\mathbb{R}^2$

3. The projection of [1, 2, 1] on each of the unit coordinate vectors in $\mathbb{R}^3$

4. The projection of [1, 2, 1] on the line with parametric equations $x = 3t$, $y = t$, $z = 2t$ in $\mathbb{R}^3$

5. The projection of [−1, 2, 0, 1] on sp([2, −3, 1, 2]) in $\mathbb{R}^4$

6. The projection of [2, −1, 3, −5] on the line sp([1, 0, −1, 2]) in $\mathbb{R}^4$

In Exercises 7–12, find the orthogonal complement of the given subspace.

7. The subspace sp([1, 2, −1]) in $\mathbb{R}^3$

8. The line sp([2, −1, 0, −3]) in $\mathbb{R}^4$

9. The subspace sp([1, 3, 0], [2, 1, 4]) in $\mathbb{R}^3$

10. The plane $2x + y + 3z = 0$ in $\mathbb{R}^3$

11. The subspace sp([2, 1, 3, 4], [1, 0, −2, 1]) in $\mathbb{R}^4$

12. The subspace (hyperplane) $ax_1 + bx_2 + cx_3 + dx_4 = 0$ in $\mathbb{R}^4$ [HINT: See Illustration 3.]

13. Find a nonzero vector in $\mathbb{R}^3$ perpendicular to [1, 1, 2] and [2, 3, 1] by
    a. the methods of this section,
    b. computing a determinant.

14. Find a nonzero vector in $\mathbb{R}^4$ perpendicular to [1, 0, −1, 1], [0, 0, −1, 1], and [2, −1, 2, 0] by
    a. the methods of this section,
    b. computing a determinant.

In Exercises 15–22, find the indicated projection.

15. The projection of [1, 2, 1] on the subspace sp([3, 1, 2], [1, 0, 1]) in $\mathbb{R}^3$

16. The projection of [1, 2, 1] on the plane $x + y + z = 0$ in $\mathbb{R}^3$

17. The projection of [1, 0, 0] on the subspace sp([2, 1, 1], [1, 0, 2]) in $\mathbb{R}^3$

18. The projection of [−1, 0, 1] on the plane $x + y = 0$ in $\mathbb{R}^3$

19. The projection of [0, 0, 1] on the plane $2x − y − z = 0$ in $\mathbb{R}^3$

20. The projection in $\mathbb{R}^4$ of [−2, 1, 3, −5] on
    a. the subspace sp(e_3)
    b. the subspace sp(e_1, e_4)
    c. the subspace sp(e_1, e_3, e_4)
    d. $\mathbb{R}^4$

21. The projection of [1, 0, −1, 1] on the subspace sp([1, 0, 0, 0], [0, 1, 1, 0], [0, 0, 1, 1]) in $\mathbb{R}^4$

22. The projection of [0, 1, −1, 0] on the subspace (hyperplane) $x_1 − x_2 + x_3 + x_4 = 0$ in $\mathbb{R}^4$ [HINT: See Example 5.]

23. Assume that **a**, **b**, and **c** are vectors in $\mathbb{R}^n$ and that W is a subspace of $\mathbb{R}^n$. Mark each of the following True or False.
    ____ a. The projection of **b** on sp(**a**) is a scalar multiple of **b**.
    ____ b. The projection of **b** on sp(**a**) is a scalar multiple of **a**.
    ____ c. The set of all vectors in $\mathbb{R}^n$ orthogonal to every vector in W is a subspace of $\mathbb{R}^n$.
    ____ d. The vector $\mathbf{w} \in W$ that minimizes $\|\mathbf{c} - \mathbf{w}\|$ is $\mathbf{c}_W$.
    ____ e. If the projection of **b** on W is **b** itself, then **b** is orthogonal to every vector in W.
    ____ f. If the projection of **b** on W is **b** itself, then **b** is in W.
    ____ g. The vector **b** is orthogonal to every vector in W if and only if $\mathbf{b}_W = \mathbf{0}$.
    ____ h. The intersection of W and $W^\perp$ is empty.
    ____ i. If **b** and **c** have the same projection on W, then $\mathbf{b} = \mathbf{c}$.
    ____ j. If **b** and **c** have the same projection on every subspace of $\mathbb{R}^n$, then $\mathbf{b} = \mathbf{c}$.

24. Let **a** and **b** be nonzero vectors in $\mathbb{R}^n$, and let θ be the angle between **a** and **b**. The scalar $\|\mathbf{b}\| \cos \theta$ is called the **scalar component** of **b** along **a**. Interpret this scalar graphically (see Figures 6.1 and 6.2), and give a formula for it in terms of the dot product.

25. Let W be a subspace of $\mathbb{R}^n$ and let **b** be a vector in $\mathbb{R}^n$. Prove that there is one and only one vector **p** in W such that $\mathbf{b} - \mathbf{p}$ is

perpendicular to every vector in W. [HINT: Suppose that $\mathbf{p}_1$ and $\mathbf{p}_2$ are two such vectors, and show that $\mathbf{p}_1 - \mathbf{p}_2$ is in $W^{\perp}$.]

26. Let A be an $m \times n$ matrix.
    a. Prove that the set W of row vectors $\mathbf{x}$ in $\mathbb{R}^m$ such that $\mathbf{x}A = \mathbf{0}$ is a subspace of $\mathbb{R}^m$.
    b. Prove that the subspace W in part (a) and the column space of A are orthogonal complements.

27. Subspaces U and W of $\mathbb{R}^n$ are **orthogonal** if $\mathbf{u} \cdot \mathbf{w} = 0$ for all $\mathbf{u}$ in U and all $\mathbf{w}$ in W. Let U and W be orthogonal subspaces of $\mathbb{R}^n$, and let $\dim(U) = n - \dim(W)$. Prove that each subspace is the orthogonal complement of the other.

28. Let W be a subspace of $\mathbb{R}^n$ with orthogonal complement $W^{\perp}$. Writing $\mathbf{a} = \mathbf{a}_W + \mathbf{a}_{W^{\perp}}$, as in Theorem 6.1, prove that

$$\|\mathbf{a}\| = \sqrt{\|\mathbf{a}_W\|^2 + \|\mathbf{a}_{W^{\perp}}\|^2}.$$

    [HINT: Use the formula $\|\mathbf{a}\|^2 = \mathbf{a} \cdot \mathbf{a}$.]

29. (*Distance from a point to a subspace*) Let W be a subspace of $\mathbb{R}^n$. Figure 6.5 suggests that the distance from the tip of $\mathbf{a}$ in $\mathbb{R}^n$ to the subspace W is equal to the magnitude of the projection of the vector $\mathbf{a}$ on the orthogonal complement of W. Find the distance from the point $(1, 2, 3)$ in $\mathbb{R}^3$ to the subspace (plane) $\mathrm{sp}([2, 2, 1], [1, 2, 1])$.

30. Find the distance from the point $(2, 1, 3, 1)$ in $\mathbb{R}^4$ to the plane $\mathrm{sp}([1, 0, 1, 0], [1, -1, 1, 1])$. [HINT: See Exercise 29.]

In Exercises 31–36, use the idea in Exercise 29 to find the distance from the tip of $\mathbf{a}$ to the given one-dimensional subspace (line). [NOTE: To calculate $\|\mathbf{a}_{W^{\perp}}\|$, first calculate $\|\mathbf{a}_W\|$ and then use Exercise 28.]

31. $\mathbf{a} = [1, 2, 1]$,
    $W = \mathrm{sp}([2, 1, 0])$ in $\mathbb{R}^3$

32. $\mathbf{a} = [2, -1, 3]$,
    $W = \mathrm{sp}([1, 2, 4])$ in $\mathbb{R}^3$

33. $\mathbf{a} = [1, 2, -1, 0]$,
    $W = \mathrm{sp}([3, 1, 4, -1])$ in $\mathbb{R}^4$

34. $\mathbf{a} = [2, 1, 1, 2]$,
    $W = \mathrm{sp}([1, 2, 1, 3])$ in $\mathbb{R}^4$

35. $\mathbf{a} = [1, 2, 3, 4, 5]$,
    $W = \mathrm{sp}([1, 1, 1, 1, 1])$ in $\mathbb{R}^5$

36. $\mathbf{a} = [1, 0, 1, 0, 1, 0, 1]$,
    $W = \mathrm{sp}([1, 2, 3, 4, 3, 2, 1])$ in $\mathbb{R}^7$

Exercises 37–39 involve inner-product spaces discussed in optional Section 3.5.

37. Referring to Example 6, find the projection of $f(x) = 1$ on $\mathrm{sp}(x)$ in P_2.

38. Referring to Example 6, find the projection of $f(x) = x$ on $\mathrm{sp}(1 + x)$.

39. Let S and T be nonempty subsets of an inner-product space V with the property that every vector in S is orthogonal to every vector in T. Prove that the span of S and the span of T are orthogonal subspaces of V.

40. Work with Topic 3 of the routine VECTGRPH in LINTEK until you are able to get a score of at least 80% most of the time.

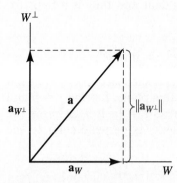

FIGURE 6.5
The distance from $\mathbf{a}$ to W is $\|\mathbf{a}_{W^{\perp}}\|$.

6.2   THE GRAM–SCHMIDT PROCESS

In the preceding section, we saw how to project a vector **b** on a subspace W of $\mathbb{R}^n$. The calculations can be somewhat tedious. We open this section by observing that, if we know a basis for W consisting of mutually perpendicular vectors, the computational burden can be eased. We then present the Gram–Schmidt algorithm, showing how such a nice basis for W can be found.

Orthogonal and Orthonormal Bases

A set $\{\mathbf{v}_1, \mathbf{v}_2, \ldots, \mathbf{v}_k\}$ of nonzero vectors in $\mathbb{R}^n$ is **orthogonal** if the vectors $\mathbf{v}_j$ are mutually perpendicular—that is, if $\mathbf{v}_i \cdot \mathbf{v}_j = 0$ for $i \neq j$. Our next theorem shows that an orthogonal generating set for a subspace of $\mathbb{R}^n$ is sure to be a basis.

THEOREM 6.2   Orthogonal Bases

Let $\{\mathbf{v}_1, \mathbf{v}_2, \ldots, \mathbf{v}_k\}$ be an orthogonal set of nonzero vectors in $\mathbb{R}^n$. Then this set is independent and consequently is a basis for the subspace $\text{sp}(\mathbf{v}_1, \mathbf{v}_2, \ldots, \mathbf{v}_k)$.

PROOF   To show that the orthogonal set $\{\mathbf{v}_1, \mathbf{v}_2, \ldots, \mathbf{v}_k\}$ is independent, let us suppose that

$$\mathbf{v}_j = s_1\mathbf{v}_1 + s_2\mathbf{v}_2 + \cdots + s_{j-1}\mathbf{v}_{j-1}.$$

Taking the dot product of both sides of this equation with $\mathbf{v}_j$ yields $\mathbf{v}_j \cdot \mathbf{v}_j = 0$, which contradicts the hypothesis that $\mathbf{v}_j \neq \mathbf{0}$. Thus, no $\mathbf{v}_j$ is a linear combination of its predecessors, so $\{\mathbf{v}_1, \mathbf{v}_2, \ldots, \mathbf{v}_k\}$ is independent and thus is a basis for $\text{sp}(\mathbf{v}_1, \mathbf{v}_2, \ldots, \mathbf{v}_k)$. (See Exercise 37 on page 203.)   ▲

EXAMPLE 1   Find an orthogonal basis for the plane $2x - y + z = 0$ in $\mathbb{R}^3$.

SOLUTION   The given plane contains the origin and hence is a subspace of $\mathbb{R}^3$. We need only find two perpendicular vectors $\mathbf{v}_1$ and $\mathbf{v}_2$ in this plane. Letting $y = 0$ and $z = 2$, we find that $x = -1$ in the given equation, so $\mathbf{v}_1 = [-1, 0, 2]$ lies in the plane. Because the vector $[2, -1, 1]$ of coefficients is perpendicular to the plane, we compute a cross product, and let

$$\mathbf{v}_2 = [-1, 0, 2] \times [2, -1, 1] = [2, 5, 1].$$

This vector is perpendicular to the coefficient vector $[2, -1, 1]$, so it lies in the plane; and of course, it is also perpendicular to the vector $[-1, 0, 2]$. Thus, $\{\mathbf{v}_1, \mathbf{v}_2\}$ is an orthogonal basis for the plane.   ■

Now we show how easy it is to project a vector $\mathbf{b}$ on a subspace W of $\mathbb{R}^n$ if we know an orthogonal basis $\{\mathbf{v}_1, \mathbf{v}_2, \ldots, \mathbf{v}_k\}$ for W. Recall from Section 6.1 that

$$\mathbf{b} = \mathbf{b}_W + \mathbf{b}_{W^\perp}, \tag{1}$$

where $\mathbf{b}_W$ is the projection of $\mathbf{b}$ on W, and $\mathbf{b}_{W^\perp}$ is the projection of $\mathbf{b}$ on $W^\perp$. Because $\mathbf{b}_W$ lies in W, we have

$$\mathbf{b}_W = r_1\mathbf{v}_1 + r_2\mathbf{v}_2 + \cdots + r_k\mathbf{v}_k \tag{2}$$

for some choice of scalars r_i. Computing the dot product of $\mathbf{b}$ with $\mathbf{v}_i$ and using Eqs. (1) and (2), we have

$$
\begin{aligned}
\mathbf{b} \cdot \mathbf{v}_i &= (\mathbf{b}_W \cdot \mathbf{v}_i) + (\mathbf{b}_{W^\perp} \cdot \mathbf{v}_i) \\
&= (r_1\mathbf{v}_1 \cdot \mathbf{v}_i + r_2\mathbf{v}_2 \cdot \mathbf{v}_i + \cdots + r_k\mathbf{v}_k \cdot \mathbf{v}_i) + 0 \qquad \mathbf{v}_i \text{ is in } W \\
&= r_i\mathbf{v}_i \cdot \mathbf{v}_i. \qquad\qquad\qquad\qquad\qquad\qquad \mathbf{v}_i \cdot \mathbf{v}_j = 0 \text{ for } i \ne j
\end{aligned}
$$

Therefore, $r_i = (\mathbf{b} \cdot \mathbf{v}_i)/(\mathbf{v}_i \cdot \mathbf{v}_i)$, so

$$r_i\mathbf{v}_i = \frac{\mathbf{b} \cdot \mathbf{v}_i}{\mathbf{v}_i \cdot \mathbf{v}_i}\mathbf{v}_i,$$

which is just the projection of $\mathbf{b}$ on $\mathbf{v}_i$. In other words, to project $\mathbf{b}$ on W, we need only project $\mathbf{b}$ on each of the orthogonal basis vectors, and then add! We summarize this in a theorem.

T H E O R E M   6.3   Projection Using an Orthogonal Basis

Let $\{\mathbf{v}_1, \mathbf{v}_2, \ldots, \mathbf{v}_k\}$ be an orthogonal basis for a subspace W of $\mathbb{R}^n$, and let $\mathbf{b}$ be any vector in $\mathbb{R}^n$. The projection of $\mathbf{b}$ on W is

$$\mathbf{b}_W = \frac{\mathbf{b} \cdot \mathbf{v}_1}{\mathbf{v}_1 \cdot \mathbf{v}_1}\mathbf{v}_1 + \frac{\mathbf{b} \cdot \mathbf{v}_2}{\mathbf{v}_2 \cdot \mathbf{v}_2}\mathbf{v}_2 + \cdots + \frac{\mathbf{b} \cdot \mathbf{v}_k}{\mathbf{v}_k \cdot \mathbf{v}_k}\mathbf{v}_k. \tag{3}$$

EXAMPLE 2   Find the projection of $\mathbf{b} = [3, -2, 2]$ on the plane $2x - y + z = 0$ in $\mathbb{R}^3$.

SOLUTION   In Example 1, we found an orthogonal basis for the given plane, consisting of the vectors $\mathbf{v}_1 = [-1, 0, 2]$ and $\mathbf{v}_2 = [2, 5, 1]$. Thus, the plane may be expressed as $W = \mathrm{sp}(\mathbf{v}_1, \mathbf{v}_2)$. Using Eq. (3), we have

$$
\begin{aligned}
\mathbf{b}_W &= \frac{\mathbf{b} \cdot \mathbf{v}_1}{\mathbf{v}_1 \cdot \mathbf{v}_1}\mathbf{v}_1 + \frac{\mathbf{b} \cdot \mathbf{v}_2}{\mathbf{v}_2 \cdot \mathbf{v}_2}\mathbf{v}_2 \\
&= \frac{1}{5}[-1, 0, 2] + \left(-\frac{2}{30}\right)[2, 5, 1] = \frac{3}{15}[-1, 0, 2] - \frac{1}{15}[2, 5, 1] \\
&= \left[-\frac{1}{3}, -\frac{1}{3}, \frac{1}{3}\right].
\end{aligned}
$$
■

It is sometimes desirable to *normalize* the vectors in an orthogonal basis, converting each basis vector to one parallel to it but of unit length. The result remains a basis for the same subspace. Notice that the standard basis in $\mathbb{R}^n$ consists of such *perpendicular unit vectors*. Such bases are extremely useful and merit a formal definition.

DEFINITION 6.3   Orthonormal Basis

Let W be a subspace of $\mathbb{R}^n$. A basis $\{\mathbf{q}_1, \mathbf{q}_2, \ldots, \mathbf{q}_k\}$ for W is **orthonormal** if

1. $\mathbf{q}_i \cdot \mathbf{q}_j = 0$ for $i \neq j$,   **Mutually perpendicular**
2. $\mathbf{q}_i \cdot \mathbf{q}_i = 1$.   **Length 1**

The standard basis for $\mathbb{R}^n$ is just one of many orthonormal bases for $\mathbb{R}^n$ if $n > 1$. For example, any two perpendicular vectors $\mathbf{v}_1$ and $\mathbf{v}_2$ on the unit circle (illustrated in Figure 6.6) form an orthonormal basis for $\mathbb{R}^2$.

For the projection of a vector on a subspace that has a known orthonormal basis, Eq. (3) in Theorem 6.3 assumes a simpler form:

Projection of b on *W* with Orthonormal Basis $\{q_1, q_2, \ldots, q_k\}$

$$\mathbf{b}_W = (\mathbf{b} \cdot \mathbf{q}_1)\mathbf{q}_1 + (\mathbf{b} \cdot \mathbf{q}_2)\mathbf{q}_2 + \cdots + (\mathbf{b} \cdot \mathbf{q}_k)\mathbf{q}_k \qquad (4)$$

EXAMPLE 3   Find an orthonormal basis for $W = \text{sp}(\mathbf{v}_1, \mathbf{v}_2, \mathbf{v}_3)$ in $\mathbb{R}^4$ if $\mathbf{v}_1 = [1, 1, 1, 1]$, $\mathbf{v}_2 = [-1, 1, -1, 1]$, and $\mathbf{v}_3 = [1, -1, -1, 1]$. Then find the projection of $\mathbf{b} = [1, 2, 3, 4]$ on W.

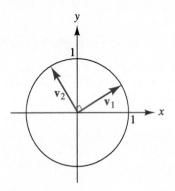

FIGURE 6.6
One of many orthonormal bases for $\mathbb{R}_2$.

SOLUTION   We see that $\mathbf{v}_1 \cdot \mathbf{v}_2 = \mathbf{v}_2 \cdot \mathbf{v}_3 = \mathbf{v}_1 \cdot \mathbf{v}_3 = 0$. Because $\|\mathbf{v}_1\| = \|\mathbf{v}_2\| = \|\mathbf{v}_3\| = 2$, let $\mathbf{q}_i = \frac{1}{2}\mathbf{v}_i$ for $i = 1, 2, 3$, to obtain an orthonormal basis $\{\mathbf{q}_1, \mathbf{q}_2, \mathbf{q}_3\}$ for W, so that

$$\mathbf{q}_1 = \left[\frac{1}{2}, \frac{1}{2}, \frac{1}{2}, \frac{1}{2}\right], \quad \mathbf{q}_2 = \left[-\frac{1}{2}, \frac{1}{2}, -\frac{1}{2}, \frac{1}{2}\right], \quad \text{and} \quad \mathbf{q}_3 = \left[\frac{1}{2}, -\frac{1}{2}, -\frac{1}{2}, \frac{1}{2}\right].$$

To find the projection of $\mathbf{b} = [1, 2, 3, 4]$ on W, we use Eq. (4) and obtain

$$\begin{aligned}\mathbf{b}_W &= (\mathbf{b} \cdot \mathbf{q}_1)\mathbf{q}_1 + (\mathbf{b} \cdot \mathbf{q}_2)\mathbf{q}_2 + (\mathbf{b} \cdot \mathbf{q}_3)\mathbf{q}_3 \\ &= 5\mathbf{q}_1 + \mathbf{q}_2 + 0\mathbf{q}_3 = [2, 3, 2, 3].\end{aligned}$$ ∎

The Gram–Schmidt Process

We now describe a computational technique for creating an orthonormal basis from a given basis of a subspace W of $\mathbb{R}^n$. The theorem that follows asserts the existence of such a basis; its proof is constructive. That is, the proof shows how an orthonormal basis can be constructed.

THEOREM 6.4   Orthonormal Basis (Gram–Schmidt) Theorem

> Let W be a subspace of $\mathbb{R}^n$, let $\{\mathbf{a}_1, \mathbf{a}_2, \dots, \mathbf{a}_k\}$ be any basis for W, and let
>
> $$W_j = \mathrm{sp}(\mathbf{a}_1, \mathbf{a}_2, \dots, \mathbf{a}_j) \quad \text{for} \quad j = 1, 2, \dots, k.$$
>
> Then there is an orthonormal basis $\{\mathbf{q}_1, \mathbf{q}_2, \dots, \mathbf{q}_k\}$ for W such that $W_j = \mathrm{sp}(\mathbf{q}_1, \mathbf{q}_2, \dots, \mathbf{q}_j)$.

PROOF   Let $\mathbf{v}_1 = \mathbf{a}_1$. For $j = 2, \dots, k$, let $\mathbf{p}_j$ be the projection of $\mathbf{a}_j$ on W_{j-1}, and let $\mathbf{v}_j = \mathbf{a}_j - \mathbf{p}_j$. That is, $\mathbf{v}_j$ is obtained by subtracting from $\mathbf{a}_j$ its projection on the subspace generated by its predecessors. Figure 6.7 gives a symbolic illustration. The decomposition

$$\mathbf{a}_j = \mathbf{p}_j + (\mathbf{a}_j - \mathbf{p}_j)$$

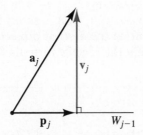

FIGURE 6.7
The vector $\mathbf{v}_j$ in the Gram–Schmidt construction.

is the unique expression for $\mathbf{a}_j$ as the sum of the vector $\mathbf{p}_j$ in W_{j-1} and the vector $\mathbf{a}_j - \mathbf{p}_j$ in $(W_{j-1})^\perp$, described in Theorem 6.1. Because $\mathbf{a}_j$ is in W_j and because $\mathbf{p}_j$ is in W_{j-1}, which is itself contained in W_j, we see that $\mathbf{v}_j = \mathbf{a}_j - \mathbf{p}_j$ lies in the subspace W_j. Now $\mathbf{v}_j$ is perpendicular to every vector in W_{j-1}. Consequently, $\mathbf{v}_j$ is perpendicular to $\mathbf{v}_1, \mathbf{v}_2, \ldots, \mathbf{v}_{j-1}$. We conclude that each vector in the set

$$\{\mathbf{v}_1, \mathbf{v}_2, \ldots, \mathbf{v}_j\} \tag{5}$$

is perpendicular to each of its predecessors. Thus the set (5) of vectors consists of j mutually perpendicular nonzero vectors in the j-dimensional subspace W_j, and so the set constitutes an orthogonal basis for W_j. It follows that, if we set $\mathbf{q}_i = (1/\|\mathbf{v}_i\|)\mathbf{v}_i$ for $i = 1, 2, \ldots, j$, then $W_j = \mathrm{sp}(\mathbf{q}_1, \mathbf{q}_2, \ldots, \mathbf{q}_j)$. Taking $j = k$, we see that

$$\{\mathbf{q}_1, \mathbf{q}_2, \ldots, \mathbf{q}_k\}$$

is an orthonormal basis for W. ▲

The proof of Theorem 6.4 was computational, providing us with a technique for constructing an orthonormal basis for a subspace W of $\mathbb{R}^n$. The technique is known as the *Gram–Schmidt process*, and we have boxed it for easy reference.

Gram–Schmidt Process

To find an orthonormal basis for a subspace W of $\mathbb{R}^n$:

1. Find a basis $\{\mathbf{a}_1, \mathbf{a}_2, \ldots, \mathbf{a}_k\}$ for W.
2. Let $\mathbf{v}_1 = \mathbf{a}_1$. For $j = 2, \ldots, k$, compute in succession the vector $\mathbf{v}_j$ given by subtracting from $\mathbf{a}_j$ its projection on the subspace generated by its predecessors.
3. The $\mathbf{v}_j$ so obtained form an orthogonal basis for W, and they may be normalized to yield an orthonormal basis.

When actually executing the Gram–Schmidt process, we project vectors on subspaces, as described in step 2 in the box. We know that it is best to work with an orthogonal or orthonormal basis for a subspace when projecting on it; and because the subspace $W_{j-1} = \mathrm{sp}(\mathbf{a}_1, \mathbf{a}_2, \ldots, \mathbf{a}_{j-1})$ is also the subspace generated by the orthogonal set $\{\mathbf{v}_1, \mathbf{v}_2, \ldots, \mathbf{v}_{j-1}\}$, it is surely best to work with the latter basis for W_{j-1} when computing the desired projection of $\mathbf{a}_j$ on W_{j-1}. Step 2 in the box and Eq. (3) show that the specific formula for $\mathbf{v}_j$ is as follows:

General Gram–Schmidt Formula

$$\mathbf{v}_j = \mathbf{a}_j - \left(\frac{\mathbf{a}_j \cdot \mathbf{v}_1}{\mathbf{v}_1 \cdot \mathbf{v}_1} \mathbf{v}_1 + \frac{\mathbf{a}_j \cdot \mathbf{v}_2}{\mathbf{v}_2 \cdot \mathbf{v}_2} \mathbf{v}_2 + \cdots + \frac{\mathbf{a}_j \cdot \mathbf{v}_{j-1}}{\mathbf{v}_{j-1} \cdot \mathbf{v}_{j-1}} \mathbf{v}_{j-1} \right). \tag{6}$$

One may normalize the $\mathbf{v}_j$, forming the vector $\mathbf{q}_j = (1/\|\mathbf{v}_j\|)\mathbf{v}_j$, to obtain a vector of length 1 at each step of the construction. In that case, formula (6) can be replaced by the following simple form:

Normalized Gram–Schmidt Formula

$$\mathbf{v}_j = \mathbf{a}_j - ((\mathbf{a}_j \cdot \mathbf{q}_1)\mathbf{q}_1 + (\mathbf{a}_j \cdot \mathbf{q}_2)\mathbf{q}_2 + \cdots + (\mathbf{a}_j \cdot \mathbf{q}_{j-1})\mathbf{q}_{j-1}) \qquad (7)$$

The arithmetic using formula (6) and that using formula (7) are similar, but formula (6) postpones the introduction of the radicals from normalizing until the entire orthogonal basis is obtained. We shall use formula (6) in our work. However, a computer will generate less error if it normalizes as it goes along. This is indicated in the next section.

EXAMPLE 4   Find an orthonormal basis for the subspace $W = \text{sp}([1, 0, 1], [1, 1, 1])$ of $\mathbb{R}^3$.

SOLUTION   We use the Gram–Schmidt process with formula (6), finding first an orthogonal basis for W. We take $\mathbf{v}_1 = [1, 0, 1]$. From formula (6) with $\mathbf{v}_1 = [1, 0, 1]$ and $\mathbf{a}_2 = [1, 1, 1]$, we have

$$\mathbf{v}_2 = \mathbf{a}_2 - \frac{\mathbf{a}_2 \cdot \mathbf{v}_1}{\mathbf{v}_1 \cdot \mathbf{v}_1}\mathbf{v}_1 = [1, 1, 1] - (2/2)[1, 0, 1] = [0, 1, 0].$$

An orthogonal basis for W is $\{[1, 0, 1], [0, 1, 0]\}$, and an orthonormal basis is $\{[1/\sqrt{2}, 0, 1/\sqrt{2}], [0, 1, 0]\}$. ∎

Referring to the proof of Theorem 6.4, we see that the sets $\{\mathbf{v}_1, \mathbf{v}_2, \ldots, \mathbf{v}_j\}$ and $\{\mathbf{q}_1, \mathbf{q}_2, \ldots, \mathbf{q}_j\}$ are both bases for the subspace $W_j = \text{sp}(\mathbf{a}_1, \mathbf{a}_2, \ldots, \mathbf{a}_j)$. Consequently, the vector $\mathbf{a}_j$ can be expressed as a linear combination

$$\mathbf{a}_j = r_{1j}\mathbf{q}_1 + r_{2j}\mathbf{q}_2 + \cdots + r_{jj}\mathbf{q}_j. \qquad (8)$$

HISTORICAL NOTE   The Gram–Schmidt process is named for the Danish mathematician Jorgen P. Gram (1850–1916) and the German Erhard Schmidt (1876–1959). It was first published by Gram in 1883 in a paper entitled "Series Development Using the Method of Least Squares." It was published again with a careful proof by Schmidt in 1907 in a work on integral equations. In fact, Schmidt even referred to Gram's result. For Schmidt, as for Gram, the vectors were continuous functions defined on an interval $[a, b]$ with the inner product of two such functions ϕ, ψ being given as $\int_a^b \phi(x)\psi(x)\,dx$. Schmidt was more explicit than Gram, however, writing out the process in great detail and proving that the set of functions ψ_i derived from his original set ϕ_i was in fact an orthonormal set.

Schmidt, who was at the University of Berlin from 1917 until his death, is best known for his definitive work on Hilbert spaces—spaces of square summable sequences of complex numbers. In fact, he applied the Gram–Schmidt process to sets of vectors in these spaces to help develop necessary and sufficient conditions for such sets to be linearly independent.

In particular, we see that $\mathbf{a}_1 = r_{11}\mathbf{q}_1$, $\mathbf{a}_2 = r_{12}\mathbf{q}_1 + r_{22}\mathbf{q}_2$, and so on. These equations arising from Eq. (8) for $j = 1, 2, \ldots, k$ can be written in matrix form as

$$\underbrace{\begin{bmatrix} | & | & & | \\ \mathbf{a}_1 & \mathbf{a}_2 & \cdots & \mathbf{a}_k \\ | & | & & | \end{bmatrix}}_{A} = \underbrace{\begin{bmatrix} | & | & & | \\ \mathbf{q}_1 & \mathbf{q}_2 & \cdots & \mathbf{q}_k \\ | & | & & | \end{bmatrix}}_{Q} \underbrace{\begin{bmatrix} r_1 & r_{12} & \cdots & r_{1k} \\ & r_{22} & \cdots & r_{2k} \\ & & \ddots & \\ \mathbf{0} & & & r_{kk} \end{bmatrix}}_{R},$$

so $A = QR$ for the indicated matrices. Because each $\mathbf{a}_j$ is in W_j but not in W_{j-1}, we see that no r_{jj} is zero, so R is an invertible $k \times k$ matrix. This factorization $A = QR$ is important in numerical linear algebra; we state it as a corollary. We will find use for it in Sections 6.5 and 8.4.

COROLLARY 1   QR-Factorization

Let A be an $n \times k$ matrix with independent column vectors in $\mathbb{R}^n$. There exists an $n \times k$ matrix Q with orthonormal column vectors and an upper-triangular invertible $k \times k$ matrix R such that $A = QR$.

EXAMPLE 5   Let

$$A = \begin{bmatrix} 1 & 1 \\ 0 & 1 \\ 1 & 1 \end{bmatrix}.$$

Factor A in the form $A = QR$ described in Corollary 1 of Theorem 6.4, using the computations in Example 4.

SOLUTION   From Example 4, we see that we can take

$$Q = \begin{bmatrix} 1/\sqrt{2} & 0 \\ 0 & 1 \\ 1/\sqrt{2} & 0 \end{bmatrix}$$

and solve $QR = A$ for the matrix R. That is, we solve the matrix equation

$$\begin{bmatrix} 1/\sqrt{2} & 0 \\ 0 & 1 \\ 1/\sqrt{2} & 0 \end{bmatrix} \begin{bmatrix} r_{11} & r_{12} \\ 0 & r_{22} \end{bmatrix} = \begin{bmatrix} 1 & 1 \\ 0 & 1 \\ 1 & 1 \end{bmatrix}$$

for the entries r_{11}, r_{12}, and r_{22}. This corresponds to two linear systems of three equations each, but by inspection we see that $r_{11} = \sqrt{2}$, $r_{12} = \sqrt{2}$, and $r_{22} = 1$. Thus,

$$A = \begin{bmatrix} 1 & 1 \\ 0 & 1 \\ 1 & 1 \end{bmatrix} = \begin{bmatrix} 1/\sqrt{2} & 0 \\ 0 & 1 \\ 1/\sqrt{2} & 0 \end{bmatrix} \begin{bmatrix} \sqrt{2} & \sqrt{2} \\ 0 & 1 \end{bmatrix} = QR.$$
∎

We give another illustration of the Gram–Schmidt process, this time requiring two applications of formula (6).

EXAMPLE 6   Find an orthonormal basis for the subspace

$$W = \text{sp}([1, 2, 0, 2], [2, 1, 1, 1], [1, 0, 1, 1])$$

of $\mathbb{R}^4$.

SOLUTION   First we find an orthogonal basis, using formula (6). We take $\mathbf{v}_1 = [1, 2, 0, 2]$ and compute $\mathbf{v}_2$ by subtracting from $\mathbf{a}_2 = [2, 1, 1, 1]$ its projection on $\mathbf{v}_1$:

$$\mathbf{v}_2 = \mathbf{a}_2 - \frac{\mathbf{a}_2 \cdot \mathbf{v}_1}{\mathbf{v}_1 \cdot \mathbf{v}_1} \mathbf{v}_1 = [2, 1, 1, 1] - \frac{6}{9}[1, 2, 0, 2] = \left[\frac{4}{3}, -\frac{1}{3}, 1, -\frac{1}{3}\right].$$

To ease computations, we replace $\mathbf{v}_2$ by the parallel vector $3\mathbf{v}_2$, which serves just as well, obtaining $\mathbf{v}_2 = [4, -1, 3, -1]$. Finally, we subtract from $\mathbf{a}_3 = [1, 0, 1, 1]$ its projection on the subspace $\text{sp}(\mathbf{v}_1, \mathbf{v}_2)$, obtaining

$$\mathbf{v}_3 = \mathbf{a}_3 - \frac{\mathbf{a}_3 \cdot \mathbf{v}_1}{\mathbf{v}_1 \cdot \mathbf{v}_1} \mathbf{v}_1 - \frac{\mathbf{a}_3 \cdot \mathbf{v}_2}{\mathbf{v}_2 \cdot \mathbf{v}_2} \mathbf{v}_2$$

$$= [1, 0, 1, 1] - \frac{3}{9}[1, 2, 0, 2] - \frac{6}{27}[4, -1, 3, -1]$$

$$= \left[-\frac{2}{9}, -\frac{4}{9}, \frac{3}{9}, \frac{5}{9}\right].$$

Replacing $\mathbf{v}_3$ by $9\mathbf{v}_3$, we see that

$$\{[1, 2, 0, 2], [4, -1, 3, -1], [-2, -4, 3, 5]\}$$

is an orthogonal basis for W. Normalizing each vector to length 1, we obtain

$$\left\{\frac{1}{3}[1, 2, 0, 2], \frac{1}{3\sqrt{3}}[4, -1, 3, -1], \frac{1}{3\sqrt{6}}[-2, -4, 3, 5]\right\}$$

as an orthonormal basis for W.   ∎

As you can see, the arithmetic involved in the Gram–Schmidt process can be a bit tedious with pencil and paper, but it is very easy to implement the process on a computer.

We know that any independent set of vectors in $\mathbb{R}^n$ can be extended to a basis for $\mathbb{R}^n$. Using Theorem 6.4, we can prove a similar result for orthogonal sets.

COROLLARY 2   Expansion of an Orthogonal Set to an Orthogonal Basis

Every orthogonal set of vectors in a subspace W of $\mathbb{R}^n$ can be expanded if necessary to an orthogonal basis for W.

PROOF   An orthogonal set $\{\mathbf{v}_1, \mathbf{v}_2, \ldots, \mathbf{v}_r\}$ of vectors in W is an independent set by Theorem 6.2, and can be expanded to a basis $\{\mathbf{v}_1, \ldots, \mathbf{v}_r, \mathbf{a}_1, \ldots, \mathbf{a}_s\}$ of W by Theorem 2.3. We apply the Gram–Schmidt process to this basis for W. Because the $\mathbf{v}_j$ are already mutually perpendicular, none of them will be changed by the Gram–Schmidt process, which thus yields an orthogonal basis containing the given vectors $\mathbf{v}_j$ for $j = 1, \ldots, r$.   ▲

EXAMPLE 7   Expand $\{[1, 1, 0], [1, -1, 1]\}$ to an orthogonal basis for $\mathbb{R}^3$, and then transform this to an orthonormal basis for $\mathbb{R}^3$.

SOLUTION   First we expand the given set to a basis $\{\mathbf{a}_1, \mathbf{a}_2, \mathbf{a}_3\}$ for $\mathbb{R}^3$. We take $\mathbf{a}_1 = [1, 1, 0]$, $\mathbf{a}_2 = [1, -1, 1]$, and $\mathbf{a}_3 = [1, 0, 0]$, which we can see form a basis for $\mathbb{R}^3$. (See Theorem 2.3.)

Now we use the Gram–Schmidt process with formula (6). Because $\mathbf{a}_1$ and $\mathbf{a}_2$ are perpendicular, we let $\mathbf{v}_1 = \mathbf{a}_1 = [1, 1, 0]$ and $\mathbf{v}_2 = \mathbf{a}_2 = [1, -1, 1]$. From formula (6), we have

$$\mathbf{v}_3 = \mathbf{a}_3 - \left(\frac{\mathbf{a}_3 \cdot \mathbf{v}_1}{\mathbf{v}_1 \cdot \mathbf{v}_1} \mathbf{v}_1 + \frac{\mathbf{a}_3 \cdot \mathbf{v}_2}{\mathbf{v}_2 \cdot \mathbf{v}_2} \mathbf{v}_2 \right)$$

$$= [1, 0, 0] - \left(\frac{1}{2} [1, 1, 0] + \frac{1}{3} [1, -1, 1] \right)$$

$$= [1, 0, 0] - \left[\frac{5}{6}, \frac{1}{6}, \frac{1}{3} \right] = \left[\frac{1}{6}, -\frac{1}{6}, -\frac{1}{3} \right].$$

Multiplying this vector by -6, we replace $\mathbf{v}_3$ by $[-1, 1, 2]$. Thus we have expanded the given set to an orthogonal basis

$$\{[1, 1, 0], [1, -1, 1], [-1, 1, 2]\}$$

of $\mathbb{R}^3$. Normalizing these vectors to unit length, we obtain

$$\{[1/\sqrt{2}, 1/\sqrt{2}, 0], [1/\sqrt{3}, -1/\sqrt{3}, 1/\sqrt{3}], [-1/\sqrt{6}, 1/\sqrt{6}, 2/\sqrt{6}]\}$$

as an orthonormal basis.   ■

The Gram–Schmidt Process in Inner-Product Spaces (Optional)

The results in this section easily extend to any inner-product space. We have the notions of an orthogonal set, an orthogonal basis, and an orthonormal basis, with essentially the same definitions given earlier. The Gram–Schmidt process is still valid. We conclude with an example.

EXAMPLE 8   Find an orthogonal basis for the subspace $\text{sp}(1, \sqrt{x}, x)$ of the vector space $C_{0,1}$ of continuous functions with domain $0 \leq x \leq 1$, where $\langle f, g \rangle = \int_0^1 f(x)g(x) \, dx$.

SOLUTION   We let $\mathbf{v}_1 = 1$ and compute

$$\mathbf{v}_2 = \sqrt{x} - \frac{\langle \sqrt{x}, 1 \rangle}{\langle 1, 1 \rangle} 1 = \sqrt{x} - \frac{\int_0^1 \sqrt{x} \, dx}{\int_0^1 1 \, dx} = \sqrt{x} - \frac{2/3}{1} = \sqrt{x} - \frac{2}{3}.$$

We replace $\mathbf{v}_2$ by $3\mathbf{v}_2$, obtaining $\mathbf{v}_2 = 3\sqrt{x} - 2$, and compute $\mathbf{v}_3$ as

$$\mathbf{v}_3 = x - \frac{\int_0^1 x\, dx}{\int_0^1 1\, dx}\, 1 - \frac{\int_0^1 x(3\sqrt{x} - 2)\, dx}{\int_0^1 (3\sqrt{x} - 2)^2\, dx}\, (3\sqrt{x} - 2)$$

$$= x - \frac{1/2}{1} - \frac{6/5 - 1}{9/2 - 8 + 4}\, (3\sqrt{x} - 2) = x - \frac{1}{2} - \frac{2}{5}\, (3\sqrt{x} - 2)$$

$$= x - \frac{6}{5}\sqrt{x} + \frac{3}{10}.$$

Replacing $\mathbf{v}_3$ by $10\mathbf{v}_3$, we obtain the orthogonal basis

$$\{1,\ 3\sqrt{x} - 2,\ 10x - 12\sqrt{x} + 3\}. \qquad \blacksquare$$

SUMMARY

1. A basis for a subspace W is orthogonal if the basis vectors are mutually perpendicular, and it is orthonormal if the vectors also have length 1.

2. Any orthogonal set of vectors in $\mathbb{R}^n$ is a basis for the subspace it generates.

3. Let W be a subspace of $\mathbb{R}^n$ with an orthogonal basis. The projection of a vector $\mathbf{b}$ in $\mathbb{R}^n$ on W is equal to the sum of the projections of $\mathbf{b}$ on each basis vector.

4. Every nonzero subspace W of $\mathbb{R}^n$ has an orthonormal basis. Any basis can be transformed into an orthogonal basis by means of the Gram–Schmidt process, in which each vector $\mathbf{a}_j$ of the given basis is replaced by the vector $\mathbf{v}_j$ obtained by subtracting from $\mathbf{a}_j$ its projection on the subspace generated by its predecessors.

5. Any orthogonal set of vectors in a subspace W of $\mathbb{R}^n$ can be expanded, if necessary, to an orthogonal basis for W.

6. Let A be an $n \times k$ matrix of rank k. Then A can be factored as QR, where Q is an $n \times k$ matrix with orthonormal column vectors and R is a $k \times k$ upper-triangular invertible matrix.

EXERCISES

In Exercises 1–4, verify that the generating set of the given subspace W is orthogonal, and find the projection of the given vector $\mathbf{b}$ on W.

1. $W = \mathrm{sp}([2, 3, 1], [-1, 1, -1])$; $\mathbf{b} = [2, 1, 4]$
2. $W = \mathrm{sp}([-1, 0, 1], [1, 1, 1])$; $\mathbf{b} = [1, 2, 3]$

3. $W = \mathrm{sp}([1, -1, -1, 1], [1, 1, 1, 1],$ $[-1, 0, 0, 1])$; $\mathbf{b} = [2, 1, 3, 1]$
4. $W = \mathrm{sp}([1, -1, 1, 1], [-1, 1, 1, 1],$ $[1, 1, -1, 1])$; $\mathbf{b} = [1, 4, 1, 2]$
5. Find an orthonormal basis for the plane $2x + 3y + z = 0$.

6. Find an orthonormal basis for the subspace

$$W = \{[x_1, x_2, x_3, x_4] \mid x_1 = x_2 + 2x_3,$$
$$x_4 = -x_2 + x_3\}$$

of $\mathbb{R}^4$.

7. Find an orthonormal basis for the subspace sp([0, 1, 0], [1, 1, 1]) of $\mathbb{R}^3$.

8. Find an orthonormal basis for the subspace sp([1, 1, 0], [−1, 2, 1]) of $\mathbb{R}^3$.

9. Transform the basis {[1, 0, 1], [0, 1, 2], [2, 1, 0]} for $\mathbb{R}^3$ into an orthonormal basis, using the Gram–Schmidt process.

10. Repeat Exercise 9, using the basis {[1, 1, 1], [1, 0, 1], [0, 1, 1]} for $\mathbb{R}^3$.

11. Find an orthonormal basis for the subspace of $\mathbb{R}^4$ spanned by [1, 0, 1, 0], [1, 1, 1, 0], and [1, −1, 0, 1].

12. Find an orthonormal basis for the subspace sp([1, −1, 1, 0, 0], [−1, 0, 0, 0, 1], [0, 0, 1, 0, 1], [1, 0, 0, 1, 1]) of $\mathbb{R}^5$.

13. Find the projection of [5, −3, 4] on the subspace in Exercise 7, using the orthonormal basis found there.

14. Repeat Exercise 13, but use the subspace in Exercise 8.

15. Find the projection of [2, 0, −1, 1] on the subspace in Exercise 11, using the orthonormal basis found there.

16. Find the projection of [−1, 0, 0, 1, −1] on the subspace in Exercise 12, using the orthonormal basis found there.

17. Find an orthonormal basis for $\mathbb{R}^4$ that contains an orthonormal basis for the subspace sp([1, 0, 1, 0], [0, 1, 1, 0]).

18. Find an orthogonal basis for the orthogonal complement of sp([1, −1, 3]) in $\mathbb{R}^3$.

19. Find an orthogonal basis for the nullspace of the matrix

$$\begin{bmatrix} 1 & 2 & 1 & 1 \\ 0 & 1 & -1 & 2 \\ 2 & 5 & 1 & 4 \\ 1 & 1 & 2 & -1 \end{bmatrix}.$$

20. Find an orthonormal basis for $\mathbb{R}^3$ that contains the vector $(1/\sqrt{3})[1, 1, 1]$.

21. Find an orthonormal basis for sp([2, 1, 1], [1, −1, 2]) that contains $(1/\sqrt{6})[2, 1, 1]$.

22. Find an orthogonal basis for sp([1, 2, 1, 2], [2, 1, 2, 0]) that contains [1, 2, 1, 2].

23. Find an orthogonal basis for sp([2, 1, −1, 1], [1, 1, 3, 0], [1, 1, 1, 1]) that contains [2, 1, −1, 1] and [1, 1, 3, 0].

24. Let B be the ordered orthonormal basis

$$\left(\left[\tfrac{1}{3}, \tfrac{2}{3}, -\tfrac{2}{3} \right], \left[\tfrac{2}{3}, \tfrac{1}{3}, \tfrac{2}{3} \right], \left[-\tfrac{2}{3}, \tfrac{2}{3}, \tfrac{1}{3} \right] \right) \text{ for } \mathbb{R}^3.$$

a. Find the coordinate vectors $[c_1, c_2, c_3]$ for [1, 2, −4] and $[d_1, d_2, d_3]$ for [5, −3, 2], relative to the ordered basis B.

b. Compute [1, 2, −4] · [5 − 3, 2], and then compute $[c_1, c_2, c_3] \cdot [d_1, d_2, d_3]$. What do you notice?

25. Mark each of the following True or False.

____ a. All vectors in an orthogonal basis have length 1.

____ b. All vectors in an orthonormal basis have length 1.

____ c. Every nontrivial subspace of $\mathbb{R}^n$ has an orthonormal basis.

____ d. Every vector in $\mathbb{R}^n$ is in some orthonormal basis for $\mathbb{R}^n$.

____ e. Every nonzero vector in $\mathbb{R}^n$ is in some orthonormal basis for $\mathbb{R}^n$.

____ f. Every unit vector in $\mathbb{R}^n$ is in some orthonormal basis for $\mathbb{R}^n$.

____ g. Every $n \times k$ matrix A has a factorization $A = QR$, where the column vectors of Q form an orthonormal set and R is an invertible $k \times k$ matrix.

____ h. Every $n \times k$ matrix A of rank k has a factorization $A = QR$, where the column vectors of Q form an orthonormal set and R is an invertible $k \times k$ matrix.

____ i. It is advantageous to work with an orthogonal basis for W when projecting a vector **b** in $\mathbb{R}^n$ on a subspace W of $\mathbb{R}^n$.

____ j. It is even more advantageous to work with an orthonormal basis for W when performing the projection in part (i).

In Exercises 26–28, use the text answers for the indicated earlier exercise to find a QR-factorization of the matrix having as column vectors the transposes of the row vectors given in that exercise.

26. Exercise 7   27. Exercise 9   28. Exercise 11

29. Let A be an $n \times k$ matrix. Prove that the column vectors of A are orthonormal if and only if $A^T A = I$.

30. Let A be an $n \times n$ matrix. Prove that A has orthonormal column vectors if and only if A is invertible with inverse $A^{-1} = A^T$.

31. Let A be an $n \times n$ matrix. Prove that the column vectors of A are orthonormal if and only if the row vectors of A are orthonormal. [HINT: Use Exercise 30 and the fact that A commutes with its inverse.]

Exercises 32–35 involve inner-product spaces.

32. Let V be an inner-product space of dimension n, and let B be an ordered orthonormal basis for V. Prove that, for any vectors $\mathbf{a}$ and $\mathbf{b}$ in V, the inner product $\langle \mathbf{a}, \mathbf{b} \rangle$ is equal to the dot product of the coordinate vectors of $\mathbf{a}$ and $\mathbf{b}$ relative to B. (See Exercise 24 for an illustration.)

33. Find an orthonormal basis for sp(sin x, cos x) for $0 \leq x \leq \pi$ if the inner product is defined by $\langle f, g \rangle = \int_0^\pi f(x)g(x)\, dx$.

34. Find an orthonormal basis for sp(1, x, x^2) for $-1 \leq x \leq 1$ if the inner product is defined by $\langle f, g \rangle = \int_{-1}^1 f(x)g(x)\, dx$.

35. Find an orthonormal basis for sp(1, e^x) for $0 \leq x \leq 1$ if the inner product is defined by $\langle f, g \rangle = \int_0^1 f(x)g(x)\, dx$.

The routine QRFACTOR in LINTEK allows the user to enter k independent row vectors in $\mathbb{R}^n$ for n and k at most 10. The program can then be used to find an orthonormal set of vectors spanning the same subspace. It will also exhibit a QR-factorization of the n × k matrix A having the entered vectors as column vectors.

*For an n × k matrix A of rank k, the command **[Q,R] = qr(A)** in MATLAB produces an n × n matrix Q whose columns form an orthonormal basis for $\mathbb{R}^n$ and an n × k upper-triangular matrix R (that is, with $r_{ij} = 0$ for $i > j$) such that A = QR. The first k columns of Q comprise the n × k matrix Q described in Corollary 1 of Theorem 6.4, and R is the k × k matrix R described in Corollary 1 with $n - k$ rows of zeros supplied at the bottom to make it the same size as A.*

In Exercises 36–38, use MATLAB or LINTEK as just described to check the answers you gave for the indicated preceding exercise. (Note that the order you took for the vectors in the Gram–Schmidt process in those exercises must be the same as the order in which you supply them in the software to be able to check your answers.)

36. Exercises 7–12   **37.** Exercises 17, 20–23
38. Exercises 26–28

6.3   ## ORTHOGONAL MATRICES

Let A be the $n \times n$ matrix with column vectors $\mathbf{a}_1, \mathbf{a}_2, \ldots, \mathbf{a}_n$. Recall that these vectors form an orthonormal basis for $\mathbb{R}^n$ if and only if

$$\mathbf{a}_i \cdot \mathbf{a}_j = \begin{cases} 0 & \text{if} \quad i \neq j, \quad \mathbf{a}_i \perp \mathbf{a}_j \\ 1 & \text{if} \quad i = j. \quad \|\mathbf{a}_j\| = 1 \end{cases}$$

Because

$$A^T A = \begin{bmatrix} \rule{1.5em}{0.4pt} & \mathbf{a}_1 & \rule{1.5em}{0.4pt} \\ \rule{1.5em}{0.4pt} & \mathbf{a}_2 & \rule{1.5em}{0.4pt} \\ & \vdots & \\ \rule{1.5em}{0.4pt} & \mathbf{a}_n & \rule{1.5em}{0.4pt} \end{bmatrix} \begin{bmatrix} \vert & \vert & & \vert \\ \mathbf{a}_1 & \mathbf{a}_2 & \cdots & \mathbf{a}_n \\ \vert & \vert & & \vert \end{bmatrix}$$

has $\mathbf{a}_i \cdot \mathbf{a}_j$ in the ith row and jth column, we see that the columns of A form an orthonormal basis of $\mathbb{R}^n$ if and only if

$$A^T A = I. \tag{1}$$

In computations with matrices using a computer, it is desirable to use matrices satisfying Eq. (1) as much as possible, as we discuss later in this section.

DEFINITION 6.4 Orthogonal Matrix

An $n \times n$ matrix A is **orthogonal** if $A^T A = I$.

The term *orthogonal* applied to a matrix is just a bit misleading. For example, $\begin{bmatrix} 1 & -2 \\ 2 & 1 \end{bmatrix}$ is not an orthogonal matrix. For an orthogonal matrix, not only must the columns be mutually orthogonal, they must also be unit vectors; that is, they must have length 1. This is not indicated by the name. It is unfortunate that the conventional name is *orthogonal matrix* rather than *orthonormal matrix*, but it is very difficult to change established terminology.

From Definition 6.4 we see that an $n \times n$ matrix A is orthogonal if and only if A is invertible and $A^{-1} = A^T$. Because every invertible matrix commutes with its inverse, it follows that $AA^T = I$, too; that is, $(A^T)^T A^T = I$. This means that the column vectors of A^T, which are the row vectors of A, also form an orthonormal basis for $\mathbb{R}^n$. Conversely, if the row vectors of an $n \times n$ matrix A form an orthonormal basis for $\mathbb{R}^n$, then so do the column vectors. We summarize these remarks in a theorem.

THEOREM 6.5 Characterizing Properties of an Orthogonal Matrix

Let A be an $n \times n$ matrix. The following conditions are equivalent:

1. The rows of A form an orthonormal basis for $\mathbb{R}^n$.
2. The columns of A form an orthonormal basis for $\mathbb{R}^n$.
3. The matrix A is orthogonal—that is, invertible with $A^{-1} = A^T$.

EXAMPLE 1   Verify that the matrix

$$A = \frac{1}{7} \begin{bmatrix} 2 & 3 & 6 \\ 3 & -6 & 2 \\ 6 & 2 & -3 \end{bmatrix}$$

is an orthogonal matrix, and find A^{-1}.

SOLUTION   We have

$$A^T A = \frac{1}{49} \begin{bmatrix} 2 & 3 & 6 \\ 3 & -6 & 2 \\ 6 & 2 & -3 \end{bmatrix} \begin{bmatrix} 2 & 3 & 6 \\ 3 & -6 & 2 \\ 6 & 2 & -3 \end{bmatrix} = \begin{bmatrix} 1 & 0 & 0 \\ 0 & 1 & 0 \\ 0 & 0 & 1 \end{bmatrix}.$$

In this example, A is symmetric, so

$$A^{-1} = A^T = A = \frac{1}{7}\begin{bmatrix} 2 & 3 & 6 \\ 3 & -6 & 2 \\ 6 & 2 & -3 \end{bmatrix}.$$

We now present the properties of orthogonal matrices that make them especially desirable for use in matrix computations.

THEOREM 6.6 Properties of $A\mathbf{x}$ for an Orthogonal Matrix A

Let A be an orthogonal $n \times n$ matrix and let $\mathbf{x}$ and $\mathbf{y}$ be any column vectors in $\mathbb{R}^n$.

1. $(A\mathbf{x}) \cdot (A\mathbf{y}) = \mathbf{x} \cdot \mathbf{y}$.   Preservation of dot product
2. $\|A\mathbf{x}\| = \|\mathbf{x}\|$.   Preservation of length
3. The angle between nonzero vectors   Preservation of angle
   $\mathbf{x}$ and $\mathbf{y}$ equals the angle between
   $A\mathbf{x}$ and $A\mathbf{y}$.

PROOF   For property 1, we need only recall that the dot product $\mathbf{x} \cdot \mathbf{y}$ of two *column* vectors can be found by using the matrix multiplication $(\mathbf{x}^T)\mathbf{y}$. Because A is orthogonal, we know that $A^T A = I$, so

$$[(A\mathbf{x}) \cdot (A\mathbf{y})] = (A\mathbf{x})^T A\mathbf{y} = \mathbf{x}^T A^T A\mathbf{y} = \mathbf{x}^T I\mathbf{y} = \mathbf{x}^T \mathbf{y} = [\mathbf{x} \cdot \mathbf{y}].$$

HISTORICAL NOTE   PROPERTIES OF ORTHOGONAL MATRICES for square systems of coefficients appear in various works of the early nineteenth century. For example, in 1833 Carl Gustav Jacob Jacobi (1804–1851) sought to find a linear substitution

$$y_1 = \Sigma \; \alpha_{1i}x_i, \qquad y_2 = \Sigma \; \alpha_{2i}x_i, \dots, \qquad y_n = \Sigma \; \alpha_{ni}x_i$$

such that $\Sigma y_i^2 = \Sigma x_i^2$. He found that the coefficients of the substitution must satisfy the orthogonality property

$$\sum_i \alpha_{ij}\alpha_{ik} = \begin{cases} 0, & j \neq k, \\ 1, & j = k. \end{cases}$$

One can even trace orthogonal systems of coefficients back to seventeenth- and eighteenth-century works in analytic geometry, when rotations of the plane or of 3-space are given in order to transform the equations of curves or surfaces. Expressed as matrices, these rotations would give orthogonal ones.

The formal definition of an orthogonal matrix, however, and a comprehensive discussion appeared in an 1878 paper of Georg Ferdinand Frobenius (1849–1917) entitled "On Linear Substitutions and Bilinear Forms." In particular, Frobenius dealt with the eigenvalues of such a matrix.

Frobenius, who was a full professor in Zurich and later in Berlin, made his major mathematical contribution in the area of group theory. He was instrumental in developing the concept of an abstract group, as well as in investigating the theory of finite matrix groups and group characters.

For property 2, the length of a vector can be defined in terms of the dot product—namely, $\|\mathbf{x}\| = \sqrt{\mathbf{x} \cdot \mathbf{x}}$. Because multiplication by A preserves the dot product, it must preserve the length.

For property 3, the angle between nonzero vectors $\mathbf{x}$ and $\mathbf{y}$ can be defined in terms of the dot product—namely, as

$$\cos^{-1}\left(\frac{\mathbf{x} \cdot \mathbf{y}}{\sqrt{\mathbf{x} \cdot \mathbf{x}}\sqrt{\mathbf{y} \cdot \mathbf{y}}}\right).$$

Because multiplication by A preserves the dot product, it must preserve angles. ▲

EXAMPLE 2   Let $\mathbf{v}$ be a vector in $\mathbb{R}^3$ with coordinate vector $[2, 3, 5]$ relative to some ordered orthonormal basis $(\mathbf{a}_1, \mathbf{a}_2, \mathbf{a}_3)$ of $\mathbb{R}^3$. Find $\|\mathbf{v}\|$.

SOLUTION   We have $\mathbf{v} = 2\mathbf{a}_1 + 3\mathbf{a}_2 + 5\mathbf{a}_3$, which can be expressed as

$$\mathbf{v} = A\mathbf{x}, \quad \text{where} \quad A = \begin{bmatrix} | & | & | \\ \mathbf{a}_1 & \mathbf{a}_2 & \mathbf{a}_3 \\ | & | & | \end{bmatrix} \quad \text{and} \quad \mathbf{x} = \begin{bmatrix} 2 \\ 3 \\ 5 \end{bmatrix}.$$

Using property (ii) of Theorem 6.6, we obtain

$$\|\mathbf{v}\| = \|A\mathbf{x}\| = \|\mathbf{x}\| = \sqrt{4 + 9 + 25} = \sqrt{38}. \qquad \blacksquare$$

Property 2 of Theorem 6.6 is the reason that it is desirable to use orthogonal matrices in matrix computations on a computer. Suppose, for example, that we have occasion to perform a multiplication $A\mathbf{x}$ for a square matrix A and a column vector $\mathbf{x}$ whose components are quantities we have to measure. Our measurements are apt to have some error, so rather than using the true vector $\mathbf{x}$ for these measured quantities, we probably work with $\mathbf{x} + \mathbf{e}$, where $\mathbf{e}$ is a nonzero error vector. Upon multiplication by A, we then obtain

$$A(\mathbf{x} + \mathbf{e}) = A\mathbf{x} + A\mathbf{e}.$$

The new error vector is $A\mathbf{e}$. If the matrix A is orthogonal, we know that $\|A\mathbf{e}\| = \|\mathbf{e}\|$, so the magnitude of the error vector remains the same under multiplication by A. We express this important fact as follows:

> Multiplication by orthogonal matrices is a *stable* operation.

If A is not orthogonal, $\|A\mathbf{e}\|$ can be a great deal larger than $\|\mathbf{e}\|$. Repeated multiplication by nonorthogonal matrices can cause the error vector to blow up to such an extent that the final answer is meaningless. To take advantage of this stability, scientists try to *orthogonalize* computational algorithms with matrices, in order to produce reliable results.

Orthogonal Diagonalization of Real Symmetric Matrices

Recall that a symmetric matrix is an $n \times n$ square matrix A in which the kth row vector is equal to the kth column vector for each $k = 1, 2, \ldots, n$. Equivalently, A is symmetric if and only if it is equal to its transpose A^T. The problem of diagonalizing a real symmetric matrix arises in many applications. (See Section 8.1, for example.) As we stated in Theorem 5.5, this diagonalization can always be achieved. That is, there is an invertible matrix C such that $C^{-1}AC = D$ is a diagonal matrix. In fact, C can be chosen to be an *orthogonal matrix*, as we will show. This means that diagonalization of real symmetric matrices is a computationally stable process. We begin by proving the perpendicularity of eigenvectors of a real symmetric matrix that correspond to distinct eigenvalues.

THEOREM 6.7   Orthogonality of Eigenspaces of a Real Symmetric Matrix

> Eigenvectors of a real symmetric matrix that correspond to different eigenvalues are orthogonal. That is, the eigenspaces of a real symmetric matrix are orthogonal.

PROOF   Let A be an $n \times n$ symmetric matrix, and let $\mathbf{v}_1$ and $\mathbf{v}_2$ be eigenvectors corresponding to distinct eigenvalues λ_1 and λ_2, respectively. Writing vectors as column vectors, we have

$$A\mathbf{v}_1 = \lambda_1\mathbf{v}_1 \quad \text{and} \quad A\mathbf{v}_2 = \lambda_2\mathbf{v}_2.$$

We want to show that $\mathbf{v}_1$ and $\mathbf{v}_2$ are orthogonal. We begin by showing that $\lambda_1(\mathbf{v}_1 \cdot \mathbf{v}_2) = \lambda_2(\mathbf{v}_1 \cdot \mathbf{v}_2)$. We compute

$$\lambda_1(\mathbf{v}_1 \cdot \mathbf{v}_2) = (\lambda_1\mathbf{v}_1) \cdot \mathbf{v}_2 = (A\mathbf{v}_1) \cdot \mathbf{v}_2.$$

The final dot product can be written in matrix form as

$$(A\mathbf{v}_1)^T\mathbf{v}_2 = (\mathbf{v}_1{}^T A^T)\mathbf{v}_2.$$

Therefore,

$$[\lambda_1(\mathbf{v}_1 \cdot \mathbf{v}_2)] = \mathbf{v}_1{}^T A^T \mathbf{v}_2. \tag{2}$$

On the other hand,

$$[\lambda_2(\mathbf{v}_1 \cdot \mathbf{v}_2)] = [\mathbf{v}_1 \cdot (\lambda_2\mathbf{v}_2)] = \mathbf{v}_1{}^T A \mathbf{v}_2. \tag{3}$$

Because $A = A^T$, Eqs. (2) and (3) show that

$$\lambda_1(\mathbf{v}_1 \cdot \mathbf{v}_2) = \lambda_2(\mathbf{v}_1 \cdot \mathbf{v}_2) \quad \text{or} \quad (\lambda_1 - \lambda_2)(\mathbf{v}_1 \cdot \mathbf{v}_2) = 0.$$

Because $\lambda_1 - \lambda_2 \neq 0$, we conclude that $\mathbf{v}_1 \cdot \mathbf{v}_2 = 0$, which shows that $\mathbf{v}_1$ is orthogonal to $\mathbf{v}_2$. ▲

The results stated for real symmetric matrices in Section 5.2 tell us that an $n \times n$ real symmetric matrix A has only real numbers as the roots of its characteristic polynomial, and that the algebraic multiplicity of each eigenvalue is equal to its geometric multiplicity; therefore, we can find a basis for $\mathbb{R}^n$ consisting of eigenvectors of A. Using the Gram–Schmidt process, we can modify the vectors of the basis in each *eigenspace* to be an orthonormal set. Theorem 6.7 then tells us that the basis vectors from different eigenspaces are also perpendicular, so we obtain a basis of mutually perpendicular real eigenvectors of unit length. We can take as the diagonalizing matrix C, such that $C^{-1}AC = D$, an orthogonal matrix whose column vectors consist of the vectors in this orthonormal basis for $\mathbb{R}^n$. We summarize our discussion in a theorem.

THEOREM 6.8 Fundamental Theorem of Real Symmetric Matrices

> Every real symmetric matrix A is diagonalizable. The diagonalization $C^{-1}AC = D$ can be achieved by using a real orthogonal matrix C.

The converse of Theorem 6.8 is also true. If $D = C^{-1}AC$ is a diagonal matrix and C is an orthogonal matrix, then A is symmetric. (See Exercise 24.) The equation $D = C^{-1}AC$ is said to be an **orthogonal diagonalization** of A.

EXAMPLE 3 Find an orthogonal diagonalization of the matrix

$$A = \begin{bmatrix} 1 & 2 \\ 2 & 4 \end{bmatrix}.$$

SOLUTION The eigenvalues of A are the roots of the characteristic equation

$$\begin{vmatrix} 1 - \lambda & 2 \\ 2 & 4 - \lambda \end{vmatrix} = \lambda^2 - 5\lambda = 0.$$

They are $\lambda_1 = 0$ and $\lambda_2 = 5$. We proceed to find the corresponding eigenvectors.
For $\lambda_1 = 0$, we have

$$A - \lambda_1 I = A = \begin{bmatrix} 1 & 2 \\ 2 & 4 \end{bmatrix} \sim \begin{bmatrix} 1 & 2 \\ 0 & 0 \end{bmatrix},$$

which yields the eigenspace

$$E_0 = \mathrm{sp}\left(\begin{bmatrix} -2 \\ 1 \end{bmatrix}\right).$$

For $\lambda_2 = 5$, we have

$$A - \lambda_2 I = A - 5I = \begin{bmatrix} -4 & 2 \\ 2 & -1 \end{bmatrix} \sim \begin{bmatrix} 2 & -1 \\ 0 & 0 \end{bmatrix},$$

which yields the eigenspace

$$E_5 = \text{sp}\left(\begin{bmatrix} 1 \\ 2 \end{bmatrix}\right).$$

Thus,

$$\begin{bmatrix} 0 & 0 \\ 0 & 5 \end{bmatrix} = \begin{bmatrix} -2 & 1 \\ 1 & 2 \end{bmatrix}^{-1} \begin{bmatrix} 1 & 2 \\ 2 & 4 \end{bmatrix} \begin{bmatrix} -2 & 1 \\ 1 & 2 \end{bmatrix}$$

is a diagonalization of A, and

$$\begin{bmatrix} 0 & 0 \\ 0 & 5 \end{bmatrix} = \begin{bmatrix} -2/\sqrt{5} & 1/\sqrt{5} \\ 1/\sqrt{5} & 2/\sqrt{5} \end{bmatrix}^{-1} \begin{bmatrix} 1 & 2 \\ 2 & 4 \end{bmatrix} \begin{bmatrix} -2/\sqrt{5} & 1/\sqrt{5} \\ 1/\sqrt{5} & 2/\sqrt{5} \end{bmatrix}$$

is an orthogonal diagonalization of A.   ∎

EXAMPLE 4   Find an orthogonal diagonalization of the matrix

$$A = \begin{bmatrix} 1 & -1 & -1 \\ -1 & 1 & -1 \\ -1 & -1 & 1 \end{bmatrix},$$

that is, find an orthogonal matrix C such that $C^{-1}AC$ is a diagonal matrix D.

SOLUTION   Example 3 of Section 5.3 shows that the eigenvalues and associated eigenspaces of A are

$$\lambda_1 = -1, \qquad E_1 = \text{sp}\left(\underbrace{\begin{bmatrix} 1 \\ 1 \\ 1 \end{bmatrix}}_{\mathbf{v}_1}\right),$$

$$\lambda_2 = \lambda_3 = 2, \qquad E_2 = \text{sp}\left(\underbrace{\begin{bmatrix} -1 \\ 1 \\ 0 \end{bmatrix}}_{\mathbf{v}_2}, \underbrace{\begin{bmatrix} -1 \\ 0 \\ 1 \end{bmatrix}}_{\mathbf{v}_3}\right).$$

Notice that vectors $\mathbf{v}_2$ and $\mathbf{v}_3$ in E_2 are orthogonal to the vector $\mathbf{v}_1$ in E_1, as must be the case for this symmetric matrix A. The vectors $\mathbf{v}_2$ and $\mathbf{v}_3$ in E_2 are not orthogonal, but we can use the Gram–Schmidt process to find an orthogonal basis for E_2. We replace $\mathbf{v}_3$ by

$$\mathbf{v}_3 - \frac{\mathbf{v}_3 \cdot \mathbf{v}_2}{\mathbf{v}_2 \cdot \mathbf{v}_2} \mathbf{v}_2 = \begin{bmatrix} -1 \\ 0 \\ 1 \end{bmatrix} - \frac{1}{2} \begin{bmatrix} -1 \\ 1 \\ 0 \end{bmatrix} = \begin{bmatrix} -1/2 \\ -1/2 \\ 1 \end{bmatrix}, \quad \text{or by} \quad \begin{bmatrix} -1 \\ -1 \\ 2 \end{bmatrix}.$$

Thus $\{[1, 1, 1], [-1, 1, 0], [-1, -1, 2]\}$ is an orthogonal basis for $\mathbb{R}^3$ of eigenvectors of A. An orthogonal diagonalizing matrix C is obtained by normalizing these vectors and taking the vectors in the resulting orthonormal basis as column vectors in C. We obtain

$$C = \begin{bmatrix} 1/\sqrt{3} & -1/\sqrt{2} & -1/\sqrt{6} \\ 1/\sqrt{3} & 1/\sqrt{2} & -1/\sqrt{6} \\ 1/\sqrt{3} & 0 & 2/\sqrt{6} \end{bmatrix}.$$ ∎

Orthogonal Linear Transformations

Every theorem about matrices has an interpretation for linear transforma-tions. Let A be an orthogonal $n \times n$ matrix, and let $T: \mathbb{R}^n \to \mathbb{R}^n$ be defined by $T(\mathbf{x}) = A\mathbf{x}$. Theorem 6.6 immediately establishes the following properties of T:

1. $T(\mathbf{x}) \cdot T(\mathbf{y}) = \mathbf{x} \cdot \mathbf{y};$ **Preservation of dot product**
2. $\|T(\mathbf{x})\| = \|\mathbf{x}\|;$ **Preservation of length**
3. The angle between $T(\mathbf{x})$ and $T(\mathbf{y})$ equals   **Preservation of angle** the angle between $\mathbf{x}$ and $\mathbf{y}$.

The first property is commonly used to define an orthogonal linear transfor-mation $T: \mathbb{R}^n \to \mathbb{R}^n$.

DEFINITION 6.5 Orthogonal Linear Transformation

> A linear transformation $T: \mathbb{R}^n \to \mathbb{R}^n$ is **orthogonal** if it satisfies $T(\mathbf{v}) \cdot T(\mathbf{w}) = \mathbf{v} \cdot \mathbf{w}$ for all vectors $\mathbf{v}$ and $\mathbf{w}$ in $\mathbb{R}^n$.

For example, the linear transformation that reflects the plane $\mathbb{R}^2$ in a line containing the origin clearly preserves both the angle θ between vectors $\mathbf{u}$ and $\mathbf{v}$ and the magnitudes of the vectors. Because the dot product in $\mathbb{R}^2$ satisfies

$$\mathbf{u} \cdot \mathbf{v} = \|\mathbf{u}\| \, \|\mathbf{v}\| \cos \theta,$$

it follows that dot products are also preserved. Therefore, this reflection of the plane is an orthogonal linear transformation.

We just showed that every orthogonal matrix gives rise to an orthogonal linear transformation of $\mathbb{R}^n$ into itself. The converse is also true.

THEOREM 6.9 Orthogonal Transformations vis-à-vis Matrices

> A linear transformation T of $\mathbb{R}^n$ into itself is orthogonal if and only if its standard matrix representation A is an orthogonal matrix.

PROOF It remains for us to show that A is an orthogonal matrix if T preserves the dot product. The columns of A are the vectors $T(\mathbf{e}_1), T(\mathbf{e}_2), \ldots, T(\mathbf{e}_n)$, where $\mathbf{e}_j$ is the jth unit coordinate vector of $\mathbb{R}^n$. We have

$$T(\mathbf{e}_i) \cdot T(\mathbf{e}_j) = \mathbf{e}_i \cdot \mathbf{e}_j = \begin{cases} 0 & \text{if } i \neq j, \\ 1 & \text{if } i = j, \end{cases}$$

showing that the columns of A form an orthonormal basis of $\mathbb{R}^n$. Thus, A is an orthogonal matrix. ▲

EXAMPLE 5   Show that the linear transformation $T: \mathbb{R}^3 \to \mathbb{R}^3$ defined by $T([x_1, x_2, x_3]) = [x_1/\sqrt{2} + x_3/\sqrt{2}, x_2, -x_1/\sqrt{2} + x_3/\sqrt{2}]$ is orthogonal.

SOLUTION   The orthogonality of the transformation follows from the fact that the standard matrix representation

$$A = \begin{bmatrix} 1/\sqrt{2} & 0 & 1/\sqrt{2} \\ 0 & 1 & 0 \\ -1/\sqrt{2} & 0 & 1/\sqrt{2} \end{bmatrix}$$

is an orthogonal matrix.   ∎

In Exercise 40, we ask you to show that a linear transformation $T: \mathbb{R}^n \to \mathbb{R}^n$ is orthogonal if and only if T maps unit vectors into unit vectors. Sometimes, this is an easy condition to verify, as our next example illustrates.

EXAMPLE 6   Show that the linear transformation that rotates the plane counterclockwise through any angle is an orthogonal linear transformation.

SOLUTION   A rotation of the plane preserves the lengths of all vectors—in particular, unit vectors. Example 1 in Section 2.4 shows that a rotation is a linear transformation, and Exercise 40 then shows that the transformation is orthogonal.   ∎

You might conjecture that a linear transformation $T: \mathbb{R}^n \to \mathbb{R}^n$ is orthogonal if and only if it preserves the angle between vectors. Exercise 38 asks you to give a counterexample, showing that this is *not* the case.

For students who studied inner-product spaces in Section 3.5, we mention that a linear transformation $T: V \to V$ of an inner-product space V is defined to be **orthogonal** if $\langle T(\mathbf{v}), T(\mathbf{w}) \rangle = \langle \mathbf{v}, \mathbf{w} \rangle$ for all vectors $\mathbf{v}, \mathbf{w} \in V$. This is the natural extension of Definition 6.5.

SUMMARY

1. A square $n \times n$ matrix A is orthogonal if it satisfies any one (and hence all) of these three equivalent conditions:
    a. The rows of A form an orthonormal basis for $\mathbb{R}^n$.
    b. The columns of A form an orthonormal basis for $\mathbb{R}^n$.
    c. The matrix A is invertible, and $A^{-1} = A^T$.

2. Multiplication of column vectors in $\mathbb{R}^n$ on the left by an $n \times n$ orthogonal matrix preserves length, dot product, and the angle between vectors. Such multiplication is computationally stable.

3. A linear transformation of $\mathbb{R}^n$ into itself is orthogonal if and only if it preserves the dot product, or (equivalently) if and only if its standard matrix representation is orthogonal, or (equivalently) if and only if it maps unit vectors into unit vectors.

4. The eigenspaces of a symmetric matrix A are mutually orthogonal, and A has n mutually perpendicular eigenvectors.

5. A symmetric matrix A is diagonalizable by an orthogonal matrix C. That is, there exists an orthogonal matrix C such that $D = C^{-1}AC$ is a diagonal matrix.

EXERCISES

In Exercises 1–4, verify that the given matrix is orthogonal, and find its inverse.

1. $(1/\sqrt{2})\begin{bmatrix} 1 & 1 \\ -1 & 1 \end{bmatrix}$ **2.** $\begin{bmatrix} \frac{3}{5} & 0 & \frac{4}{5} \\ -\frac{4}{5} & 0 & \frac{3}{5} \\ 0 & 1 & 0 \end{bmatrix}$

3. $\frac{1}{7}\begin{bmatrix} 2 & -3 & 6 \\ 3 & 6 & 2 \\ -6 & 2 & 3 \end{bmatrix}$ **4.** $\frac{1}{2}\begin{bmatrix} 1 & -1 & 1 & 1 \\ -1 & 1 & 1 & 1 \\ 1 & 1 & -1 & 1 \\ 1 & 1 & 1 & -1 \end{bmatrix}$

If A and D are square matrices, D is diagonal, and AD is orthogonal, then $(AD)^{-1} = (AD)^T$ and $D^{-1}A^{-1} = D^T A^T$ so that $A^{-1} = DD^T A^T = D^2 A^T$. In Exercises 5–8, find the inverse of each matrix A by first finding a diagonal matrix D so that AD has column vectors of length 1, and then applying the formula $A^{-1} = D^2 A^T$.

5. $\begin{bmatrix} 1 & 3 \\ -1 & 3 \end{bmatrix}$ **6.** $\begin{bmatrix} 3 & 0 & 8 \\ -4 & 0 & 6 \\ 0 & 1 & 0 \end{bmatrix}$

7. $\begin{bmatrix} 4 & -3 & 6 \\ 6 & 6 & 2 \\ -12 & 2 & 3 \end{bmatrix}$ **8.** $\begin{bmatrix} 2 & -1 & 3 & 1 \\ -2 & 1 & 3 & 1 \\ 2 & 1 & -3 & 1 \\ 2 & 1 & 3 & -1 \end{bmatrix}$

9. Supply a third column vector so that the matrix

$$\begin{bmatrix} 1/\sqrt{3} & 1/\sqrt{2} & \\ 1/\sqrt{3} & 0 & \\ 1/\sqrt{3} & -1/\sqrt{2} & \end{bmatrix}$$ is orthogonal.

10. Repeat Exercise 9 for the matrix

$$\begin{bmatrix} \frac{2}{7} & 3/\sqrt{13} & \\ \frac{3}{7} & -2/\sqrt{13} & \\ \frac{6}{7} & 0 & \end{bmatrix}.$$

11. Let $(\mathbf{a}_1, \mathbf{a}_2, \mathbf{a}_3)$ be an ordered orthonormal basis for $\mathbb{R}^3$, and let **b** be a unit vector with coordinate vector $\left[\frac{1}{2}, \frac{1}{3}, c\right]$ relative to this basis. Find all possible values for c.

12. Let $(\mathbf{a}_1, \mathbf{a}_2, \mathbf{a}_3, \mathbf{a}_4)$ be an ordered orthonormal basis for $\mathbb{R}^4$, and let $[2, 1, 4, -3]$ be the coordinate vector of a vector **b** in $\mathbb{R}^4$ relative to this basis. Find $\|\mathbf{b}\|$.

In Exercises 13–18, find a matrix C such that $D = C^{-1}AC$ is an orthogonal diagonalization of the given symmetric matrix A.

13. $\begin{bmatrix} 1 & 2 \\ 2 & 1 \end{bmatrix}$ **14.** $\begin{bmatrix} 3 & 2 \\ 2 & 0 \end{bmatrix}$

15. $\begin{bmatrix} 2 & 1 & 0 \\ 1 & 2 & 1 \\ 0 & 1 & 2 \end{bmatrix}$ **16.** $\begin{bmatrix} 0 & 1 & 1 \\ 1 & 2 & 1 \\ 1 & 1 & 0 \end{bmatrix}$

17. $\begin{bmatrix} 0 & 1 & 1 & 0 \\ 1 & -2 & 2 & 1 \\ 1 & 2 & -2 & 1 \\ 0 & 1 & 1 & 0 \end{bmatrix}$ **18.** $\begin{bmatrix} 1 & -1 & 0 & 0 \\ -1 & 1 & 0 & 0 \\ 0 & 0 & 1 & 3 \\ 0 & 0 & 3 & 1 \end{bmatrix}$

19. Mark each of the following True or False.

____ **a.** A square matrix is orthogonal if its column vectors are orthogonal.

____ **b.** Every orthogonal matrix has nullspace $\{\mathbf{0}\}$.

____ **c.** If A^T is orthogonal, then A is orthogonal.

____ **d.** If A is an $n \times n$ symmetric orthogonal matrix, then $A^2 = I$.

____ **e.** If A is an $n \times n$ symmetric matrix such that $A^2 = I$, then A is orthogonal.

____ **f.** If A and B are orthogonal $n \times n$ matrices, then AB is orthogonal.

____ **g.** Every orthogonal linear transformation carries every unit vector into a unit vector.

____ **h.** Every linear transformation that carries each unit vector into a unit vector is orthogonal.

____ **i.** Every map of the plane into itself that is an *isometry* (that is, preserves distance between points) is given by an orthogonal linear transformation.

____ **j.** Every map of the plane into itself that is an *isometry* and that leaves the origin fixed is given by an orthogonal linear transformation.

20. Let A be an orthogonal $n \times n$ matrix. Show that $\|A\mathbf{x}\| = \|A^{-1}\mathbf{x}\|$ for any vector $\mathbf{x}$ in $\mathbb{R}^n$.

21. Let A be an orthogonal matrix. Show that A^2 is an orthogonal matrix, too.

22. Show that, if A is an orthogonal matrix, then $\det(A) = \pm 1$.

23. Find a 2×2 matrix with determinant 1 that is not an orthogonal matrix.

24. Let $D = C^{-1}AC$ be a diagonal matrix, where C is an orthogonal matrix. Show that A is symmetric.

25. Let A be an $n \times n$ matrix such that $A\mathbf{x} \cdot A\mathbf{y} = \mathbf{x} \cdot \mathbf{y}$ for all vectors $\mathbf{x}$ and $\mathbf{y}$ in $\mathbb{R}^n$. Show that A is an orthogonal matrix.

26. Let A be an $n \times n$ matrix such that $\|A\mathbf{x}\| = \|\mathbf{x}\|$ for all vectors $\mathbf{x}$ in $\mathbb{R}^n$. Show that A is an orthogonal matrix. [HINT: Show that $\mathbf{x} \cdot \mathbf{y} = \frac{1}{4}(\|x + y\|^2 - \|x - y\|^2)$, and then use Exercise 25.]

27. Show that the real eigenvalues of an orthogonal matrix must be equal to 1 or -1. [HINT: Think in terms of linear transformations.]

28. Describe all real diagonal orthogonal matrices.

29. **a.** Show that a row-interchange elementary matrix is orthogonal.
    b. Let A be a matrix obtained by permuting (that is, changing the order of) the rows of the $n \times n$ identity matrix. Show that A is an orthogonal matrix.

30. Let $\{\mathbf{a}_1, \mathbf{a}_2, \dots, \mathbf{a}_n\}$ be an orthonormal basis of column vectors for $\mathbb{R}^n$, and let C be an orthogonal $n \times n$ matrix. Show that

$$\{C\mathbf{a}_1, C\mathbf{a}_2, \dots, C\mathbf{a}_n\}$$

is also an orthonormal basis for $\mathbb{R}^n$.

31. Let A and C be orthogonal $n \times n$ matrices. Show that $C^{-1}AC$ is orthogonal.

In Exercises 32–37, determine whether the given linear transformation is orthogonal.

32. $T: \mathbb{R}^2 \to \mathbb{R}^2$ defined by $T([x, y]) = [y, x]$

33. $T: \mathbb{R}^3 \to \mathbb{R}^3$ defined by $T([x, y, z]) = [x, y, 0]$

34. $T: \mathbb{R}^2 \to \mathbb{R}^2$ defined by $T([x, y]) = [2x, y]$

35. $T: \mathbb{R}^2 \to \mathbb{R}^2$ defined by $T([x, y]) = [x, -y]$

36. $T: \mathbb{R}^2 \to \mathbb{R}^2$ defined by $T([x, y]) = [x/2 - \sqrt{3}y/2, -\sqrt{3}x/2 + y/2]$

37. $T: \mathbb{R}^3 \to \mathbb{R}^3$ defined by $T([x, y, z]) = [x/3 + 2y/3 + 2z/3, -2x/3 - y/3 + 2z/3, -2x/3 + 2y/3 - z/3]$

38. Find a linear transformation $T: \mathbb{R}^n \to \mathbb{R}^n$ that preserves the angle between vectors but is not an orthogonal transformation.

39. Show that every 2×2 orthogonal matrix is of one of two forms: either

$$\begin{bmatrix} \cos\theta & \sin\theta \\ \sin\theta & -\cos\theta \end{bmatrix} \quad \text{or} \quad \begin{bmatrix} \cos\theta & -\sin\theta \\ \sin\theta & \cos\theta \end{bmatrix},$$

for some angle θ.

40. Let $T: \mathbb{R}^n \to \mathbb{R}^n$ be a linear transformation. Show that T is orthogonal if and only if T maps unit vectors to unit vectors. [HINT: Use Exercise 26.]

41. (*Real Householder matrix*) Let $\mathbf{v}$ be a nonzero column vector in $\mathbb{R}^n$. Show that $C_\mathbf{v} = I - \frac{2}{\mathbf{v} \cdot \mathbf{v}}(\mathbf{v}\mathbf{v}^T)$ is an orthogonal matrix. (These Householder matrices can be used to perform certain important stable reductions of matrices.)

6.4 THE PROJECTION MATRIX

Let W be a subspace of $\mathbb{R}^n$. Projection of vectors in $\mathbb{R}^n$ on W gives a mapping T of $\mathbb{R}^n$ into itself. Figure 6.8 illustrates the projection $T(\mathbf{a} + \mathbf{b})$ of $\mathbf{a} + \mathbf{b}$ on W, and Figure 6.9 illustrates the projection $T(r\mathbf{a})$ of $r\mathbf{a}$ on W. These figures suggest that

$$T(\mathbf{a} + \mathbf{b}) = T(\mathbf{a}) + T(\mathbf{b})$$

and

$$T(r\mathbf{a}) = rT(\mathbf{a}).$$

Thus we expect that T is a linear transformation. If this is the case, there must be a matrix P such that $T(\mathbf{x}) = P\mathbf{x}$—namely, the standard matrix representation of T. Recall that in Section 6.1 we showed how to find the projection of a vector $\mathbf{b}$ on W by finding a basis for W, then finding a basis for its orthogonal complement $W^{\perp}$, then finding coordinates of $\mathbf{b}$ with respect to the resulting basis of $\mathbb{R}^n$, and so on. It is a somewhat involved process. It would be nice to have a matrix P such that projection of $\mathbf{b}$ on W could be found by computing $P\mathbf{b}$.

In this section, we start by showing that there is indeed a matrix P such that the projection of $\mathbf{b}$ on W is equal to $P\mathbf{b}$. Of course, this then shows that projection is a linear transformation, because we know that multiplication of column vectors in $\mathbb{R}^n$ on the left by any $n \times n$ matrix A gives a linear transformation of $\mathbb{R}^n$ into itself. We derive a formula for this *projection matrix* P corresponding to a subspace W. The formula involves choosing a basis for W, but the final matrix P obtained is independent of the basis chosen.

We have seen that projection on a subspace W is less difficult to compute when an orthonormal basis for W is used. The formula for the matrix P also becomes quite simple when an orthonormal basis is used, and consequently projection becomes an easy computation. Section 6.5 gives an application of these ideas to data analysis.

We will need the following result concerning the rank of $A^T A$ for our work in this section.

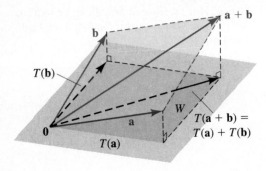

FIGURE 6.8
Projection of **a** + **b** on *W*.

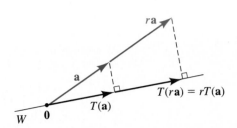

FIGURE 6.9
Projection of *r***a** on *W*.

THEOREM 6.10   The Rank of $(A^T)A$

Let A be an $m \times n$ matrix of rank r. Then the $n \times n$ symmetric matrix $(A^T)A$ also has rank r.

PROOF   We will work with nullspaces. If $\mathbf{v}$ is any solution vector of the system $A\mathbf{x} = \mathbf{0}$, so that $A\mathbf{v} = \mathbf{0}$, then upon multiplying the sides of this last equation on the left by A^T, we see that $\mathbf{v}$ is also a solution of the system $(A^T)A\mathbf{x} = \mathbf{0}$. Conversely, assuming that $(A^T)A\mathbf{w} = \mathbf{0}$ for an $n \times 1$ vector $\mathbf{w}$, then

$$[0] = \mathbf{w}^T[(A^T)A\mathbf{w}] = (A\mathbf{w})^T(A\mathbf{w}),$$

which may be written $\|A\mathbf{w}\|^2 = 0$. That is, $A\mathbf{w}$ is a vector of magnitude 0, and consequently $A\mathbf{w} = \mathbf{0}$. It follows that A and $(A^T)A$ have the same nullspace. Because they have the same number of columns, we apply the rank equation (Theorem 2.5) and conclude that $\text{rank}(A) = \text{rank}((A^T)A)$.   ▲

The Formula for the Projection Matrix

Let $W = \text{sp}(\mathbf{a}_1, \mathbf{a}_2, \ldots, \mathbf{a}_k)$ be a subspace of $\mathbb{R}^n$, where the vectors $\mathbf{a}_1, \mathbf{a}_2, \ldots, \mathbf{a}_k$ are independent. Let $\mathbf{b}$ be a vector in $\mathbb{R}^n$, and let $\mathbf{p} = \mathbf{b}_W$ be its projection on W. From the unique decomposition

$$\underset{\substack{\quad \\ \mathbf{b}_W \qquad \mathbf{b}_{W^\perp}}}{\mathbf{b} = \mathbf{p} + (\mathbf{b} - \mathbf{p})}$$

explained in Section 6.1, we see that the projection vector $\mathbf{p}$ is the unique vector satisfying the following two properties. (See Figure 6.10 for an illustration when $\dim(W) = 2$.)

Properties of the Projection $\mathbf{p}$ of Vector $\mathbf{b}$ on the Subspace W

1. The vector $\mathbf{p}$ must lie in the subspace W.
2. The vector $\mathbf{b} - \mathbf{p}$ must be perpendicular to every vector in W.

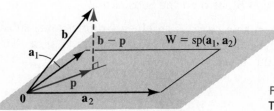

FIGURE 6.10
The projection of $\mathbf{b}$ on $\text{sp}(\mathbf{a}_1, \mathbf{a}_2)$.

If we write our vectors in $\mathbb{R}^n$ as column vectors, the subspace W is the column space of the $n \times k$ matrix A, whose columns are $\mathbf{a}_1, \mathbf{a}_2, \ldots, \mathbf{a}_k$. All vectors in $W = \mathrm{sp}(\mathbf{a}_1, \mathbf{a}_2, \ldots, \mathbf{a}_k)$ have the form $A\mathbf{x}$, where

$$\mathbf{x} = \begin{bmatrix} x_1 \\ x_2 \\ \cdot \\ \cdot \\ \cdot \\ x_k \end{bmatrix} \quad \text{for any scalars } x_1, x_2, \ldots, x_k.$$

Because $\mathbf{p}$ lies in the space W, we see that $\mathbf{p} = A\mathbf{r}$, where

$$\mathbf{r} = \begin{bmatrix} r_1 \\ r_2 \\ \cdot \\ \cdot \\ \cdot \\ r_k \end{bmatrix} \quad \text{for some scalars } r_1, r_2, \ldots, r_k.$$

Because $\mathbf{b} - A\mathbf{r}$ must be perpendicular to each vector in W, the dot product of $\mathbf{b} - A\mathbf{r}$ and $A\mathbf{x}$ must be zero for all vectors $\mathbf{x}$. This dot-product condition can be written as the matrix equation

$$(A\mathbf{x})^T(\mathbf{b} - A\mathbf{r}) = \mathbf{x}^T(A^T\mathbf{b} - A^TA\mathbf{r}) = [0].$$

In other words, the dot product of the vectors $\mathbf{x}$ and $A^T\mathbf{b} - A^TA\mathbf{r}$ must be zero *for all vectors* $\mathbf{x}$. This can happen only if the vector $A^T\mathbf{b} - A^TA\mathbf{r}$ is itself the zero vector. (See Exercise 41 in Section 1.3.) Therefore, we have

$$A^T\mathbf{b} - A^TA\mathbf{r} = \mathbf{0}. \tag{1}$$

Now the $k \times k$ matrix A^TA appearing in Eq. (1) is invertible, because it has the same rank as A (see Theorem 6.10); and A has rank k, because its columns are independent. Solving Eq. (1) for $\mathbf{r}$, we obtain

$$\mathbf{r} = (A^TA)^{-1}A^T\mathbf{b}. \tag{2}$$

Denoting the projection $\mathbf{p} = A\mathbf{r}$ of $\mathbf{b}$ on W by $\mathbf{p} = \mathbf{b}_W$ and writing $\mathbf{b}$ as a column vector, we obtain the formula in the following box.

Projection $\mathbf{b}_W$ of $\mathbf{b}$ on the Subspace W

Let $W = \mathrm{sp}(\mathbf{a}_1, \mathbf{a}_2, \ldots, \mathbf{a}_k)$ be a k-dimensional subspace of $\mathbb{R}^n$, and let A have as columns the vectors $\mathbf{a}_1, \mathbf{a}_2, \ldots, \mathbf{a}_k$. The projection of $\mathbf{b}$ in $\mathbb{R}^n$ on W is given by

$$\mathbf{b}_W = A(A^TA)^{-1}A^T\mathbf{b}. \tag{3}$$

We leave as Exercise 12 the demonstration that the formula in Section 6.1 for projecting **b** on sp(**a**) can be written in the form of formula (3), using matrices of appropriate size.

Our first example reworks Example 1 in Section 6.1, using formula (3).

EXAMPLE 1   Using formula (3), find the projection of the vector **b** on the subspace $W =$ sp(**a**), where

$$\mathbf{b} = \begin{bmatrix} 1 \\ 2 \\ 3 \end{bmatrix} \quad \text{and} \quad \mathbf{a} = \begin{bmatrix} 2 \\ 4 \\ 3 \end{bmatrix}.$$

SOLUTION   Let A be the matrix whose single column is **a**. Then

$$A^T A = [2 \quad 4 \quad 3] \begin{bmatrix} 2 \\ 4 \\ 3 \end{bmatrix} = [29].$$

Putting this into formula (3), we have

$$\mathbf{b}_W = A(A^T A)^{-1} A^T \mathbf{b} = \begin{bmatrix} 2 \\ 4 \\ 3 \end{bmatrix} \left(\frac{1}{29} \right) [2 \quad 4 \quad 3] \begin{bmatrix} 1 \\ 2 \\ 3 \end{bmatrix} = \left(\frac{1}{29} \right) \begin{bmatrix} 4 & 8 & 6 \\ 8 & 16 & 12 \\ 6 & 12 & 9 \end{bmatrix} \begin{bmatrix} 1 \\ 2 \\ 3 \end{bmatrix} = \left(\frac{19}{29} \right) \begin{bmatrix} 2 \\ 4 \\ 3 \end{bmatrix}. \quad \blacksquare$$

We refer to the matrix $A(A^T A)^{-1} A^T$ in formula (3) as the **projection matrix** for the subspace W. It takes any vector **b** in $\mathbb{R}^n$ and, by left multiplication, projects it onto the vector $\mathbf{b}_W$, which lies in W. We will show shortly that this matrix is uniquely determined by W, which allows us to use the definite article *the* when talking about it. Before demonstrating this uniqueness, we box the formula for this matrix and present two more examples.

The Projection Matrix P for the Subspace W

Let $W = \text{sp}(\mathbf{a}_1, \mathbf{a}_2, \ldots, \mathbf{a}_k)$ be a k-dimensional subspace of $\mathbb{R}^n$, and let A have as columns the vectors $\mathbf{a}_1, \mathbf{a}_2, \ldots, \mathbf{a}_k$. The projection matrix for the subspace W is given by

$$P = A(A^T A)^{-1} A^T.$$

EXAMPLE 2   Find the projection matrix for the x_2, x_3-plane in $\mathbb{R}^3$.

SOLUTION   The x_2, x_3-plane is the subspace $W = \text{sp}(\mathbf{e}_2, \mathbf{e}_3)$, where $\mathbf{e}_2$ and $\mathbf{e}_3$ are the column vectors of the matrix

$$A = \begin{bmatrix} 0 & 0 \\ 1 & 0 \\ 0 & 1 \end{bmatrix}.$$

We find that $A^TA = I$, the 2×2 identity matrix, and that the projection matrix is

$$P = A(A^TA)^{-1}A^T = \begin{bmatrix} 0 & 0 \\ 1 & 0 \\ 0 & 1 \end{bmatrix}\begin{bmatrix} 0 & 1 & 0 \\ 0 & 0 & 1 \end{bmatrix} = \begin{bmatrix} 0 & 0 & 0 \\ 0 & 1 & 0 \\ 0 & 0 & 1 \end{bmatrix}.$$

Notice that P projects each vector

$$\begin{bmatrix} b_1 \\ b_2 \\ b_3 \end{bmatrix} \text{ in } \mathbb{R}^3 \text{ onto } \quad P\begin{bmatrix} b_1 \\ b_2 \\ b_3 \end{bmatrix} = \begin{bmatrix} 0 \\ b_2 \\ b_3 \end{bmatrix}$$

in the x_2,x_3-plane, as expected. ∎

EXAMPLE 3   Find the matrix that projects vectors in $\mathbb{R}^3$ on the plane $2x - y - 3z = 0$. Also, find the projection of a general vector $\mathbf{b}$ in $\mathbb{R}^3$ on this plane.

SOLUTION   We observe that the given plane contains the zero vector and can therefore be written as the subspace $W = \text{sp}(\mathbf{a}_1, \mathbf{a}_2)$, where $\mathbf{a}_1$ and $\mathbf{a}_2$ are any two nonzero and nonparallel vectors in the plane. We choose

$$\mathbf{a}_1 = \begin{bmatrix} 0 \\ 3 \\ -1 \end{bmatrix} \text{ and } \mathbf{a}_2 = \begin{bmatrix} 1 \\ 2 \\ 0 \end{bmatrix},$$

so that

$$A = \begin{bmatrix} 0 & 1 \\ 3 & 2 \\ -1 & 0 \end{bmatrix}.$$

Then

$$(A^TA)^{-1} = \begin{bmatrix} 10 & 6 \\ 6 & 5 \end{bmatrix}^{-1} = \frac{1}{14}\begin{bmatrix} 5 & -6 \\ -6 & 10 \end{bmatrix}$$

and the desired matrix is

$$P = \frac{1}{14}\begin{bmatrix} 0 & 1 \\ 3 & 2 \\ -1 & 0 \end{bmatrix}\begin{bmatrix} 5 & -6 \\ -6 & 10 \end{bmatrix}\begin{bmatrix} 0 & 3 & -1 \\ 1 & 2 & 0 \end{bmatrix}$$

$$= \frac{1}{14}\begin{bmatrix} -6 & 10 \\ 3 & 2 \\ -5 & 6 \end{bmatrix}\begin{bmatrix} 0 & 3 & -1 \\ 1 & 2 & 0 \end{bmatrix} = \frac{1}{14}\begin{bmatrix} 10 & 2 & 6 \\ 2 & 13 & -3 \\ 6 & -3 & 5 \end{bmatrix}.$$

Each vector $\mathbf{b} = [b_1, b_2, b_3]$ in $\mathbb{R}^3$ projects onto the vector

$$\mathbf{b}_W = P\mathbf{b} = \frac{1}{14}\begin{bmatrix} 10 & 2 & 6 \\ 2 & 13 & -3 \\ 6 & -3 & 5 \end{bmatrix}\begin{bmatrix} b_1 \\ b_2 \\ b_3 \end{bmatrix} = \frac{1}{14}\begin{bmatrix} 10b_1 + 2b_2 + 6b_3 \\ 2b_1 + 13b_2 - 3b_3 \\ 6b_1 - 3b_2 + 5b_3 \end{bmatrix}.$$

∎

Uniqueness and Properties of the Projection Matrix

It might appear from the formula $P = A(A^TA)^{-1}A^T$ that a projection matrix P for the subspace W of $\mathbb{R}^n$ depends on the particular choice of basis for W that is used for the column vectors of A. However, this is not so. Because the projection map of $\mathbb{R}^n$ onto W can be computed for $\mathbf{x} \in \mathbb{R}^n$ by the multiplication $P\mathbf{x}$ of the vector $\mathbf{x}$ on the left by the matrix P, we see that this projection map is a linear transformation, and P must be its *unique* standard matrix representation. Thus we may refer to the matrix $P = A(A^TA)^{-1}A^T$ as *the* projection matrix for the subspace W. We summarize our work in a theorem.

THEOREM 6.11   Projection Matrix

Let W be a subspace of $\mathbb{R}^n$. There is a unique $n \times n$ matrix P such that, for each column vector $\mathbf{b}$ in $\mathbb{R}^n$, the vector $P\mathbf{b}$ is the projection of $\mathbf{b}$ on W. This projection matrix P can be found by selecting any basis $\{\mathbf{a}_1, \mathbf{a}_2, \ldots, \mathbf{a}_k\}$ for W and computing $P = A(A^TA)^{-1}A^T$, where A is the $n \times k$ matrix having column vectors $\mathbf{a}_1, \mathbf{a}_2, \ldots, \mathbf{a}_k$.

Exercise 16 indicates that the projection matrix P given in Theorem 6.11 satisfies two properties:

Properties of a Projection Matrix P

1. $P^2 = P$. P is **idempotent**.
2. $P^T = P$. P is symmetric.

We can use property 2 as a partial check for errors in long computations that lead to P, as in Example 3. These two properties completely characterize the projection matrices, as we now show.

THEOREM 6.12   Characterization of Projection Matrices

The projection matrix P for a subspace W of $\mathbb{R}^n$ is both idempotent and symmetric. Conversely, every $n \times n$ matrix that is both idempotent and symmetric is a projection matrix: specifically, it is the projection matrix for its column space.

PROOF   Exercise 16 indicates that a projection matrix is both idempotent and symmetric.

To establish the converse, let P be an $n \times n$ matrix that is both symmetric and idempotent. We show that P is the projection matrix for its own column space W. Let $\mathbf{b}$ be any vector in $\mathbb{R}^n$. By Theorem 6.11, we need show only that

$P\mathbf{b}$ satisfies the characterizing properties of the projection of $\mathbf{b}$ on W given in the box on page 361. Now $P\mathbf{b}$ surely lies in the column space W of P, because W consists of all vectors $P\mathbf{x}$ for any vector $\mathbf{x}$ in $\mathbb{R}^n$. The second requirement is that $\mathbf{b} - P\mathbf{b}$ must be perpendicular to each vector $P\mathbf{x}$ in the column space of P. Writing the dot product of $\mathbf{b} - P\mathbf{b}$ and $P\mathbf{x}$ in matrix form, and using the hypotheses $P^2 = P = P^T$, we have

$$
\begin{aligned}
(\mathbf{b} - P\mathbf{b})^T P\mathbf{x} &= ((I - P)\mathbf{b})^T P\mathbf{x} = \mathbf{b}^T(I - P)^T P\mathbf{x} \\
&= \mathbf{b}^T(I - P)P\mathbf{x} = \mathbf{b}^T(P - P^2)\mathbf{x} \\
&= \mathbf{b}^T(P - P)\mathbf{x} = \mathbf{b}^T O\mathbf{x} = [0].
\end{aligned}
$$

Because their dot product is zero, we see that $\mathbf{b} - P\mathbf{b}$ and $P\mathbf{x}$ are indeed perpendicular, and our proof is complete. ▲

The Orthonormal Case

In Example 2, we saw that the projection matrix for the x_2,x_3-plane in $\mathbb{R}^3$ has the very simple description

$$
P = \begin{bmatrix} 0 & 0 & 0 \\ 0 & 1 & 0 \\ 0 & 0 & 1 \end{bmatrix}.
$$

The usually complicated formula $P = A(A^TA)^{-1}A^T$ was simplified when we used the standard unit coordinate vectors in our basis for W. Namely, we had

$$
A = \begin{bmatrix} 0 & 0 \\ 1 & 0 \\ 0 & 1 \end{bmatrix} \quad \text{so} \quad A^TA = \begin{bmatrix} 0 & 1 & 0 \\ 0 & 0 & 1 \end{bmatrix}\begin{bmatrix} 0 & 0 \\ 1 & 0 \\ 0 & 1 \end{bmatrix} = \begin{bmatrix} 1 & 0 \\ 0 & 1 \end{bmatrix} = I.
$$

Thus, $(A^TA)^{-1} = I$, too, which simplifies what is normally the worst part of the computation in our formula for P. This simplification can be made in computing P for any subspace W of $\mathbb{R}^n$, provided that we know an *orthonormal basis* $\{\mathbf{a}_1, \mathbf{a}_2, \ldots, \mathbf{a}_k\}$ for W. If A is the $n \times k$ matrix whose columns are $\mathbf{a}_1, \mathbf{a}_2, \ldots, \mathbf{a}_k$, then we know that $A^TA = I$. The formula for the projection matrix becomes

$$
P = A(A^TA)^{-1}A^T = AIA^T = AA^T.
$$

We box this result for easy reference.

Projection Matrix: Orthonormal Case

Let $\{\mathbf{a}_1, \mathbf{a}_2, \ldots, \mathbf{a}_k\}$ be an orthonormal basis for a subspace W of $\mathbb{R}^n$. The projection matrix for W is

$$
P = AA^T, \tag{4}
$$

where A is the $n \times k$ matrix having column vectors $\mathbf{a}_1, \mathbf{a}_2, \ldots, \mathbf{a}_k$.

EXAMPLE 4   Find the projection matrix for the subspace $W = \text{sp}(\mathbf{a}_1, \mathbf{a}_2)$ of $\mathbb{R}^3$ if

$$\mathbf{a}_1 = \begin{bmatrix} 1/\sqrt{3} \\ -1/\sqrt{3} \\ 1/\sqrt{3} \end{bmatrix} \quad \text{and} \quad \mathbf{a}_2 = \begin{bmatrix} 1/\sqrt{2} \\ 1/\sqrt{2} \\ 0 \end{bmatrix}.$$

Find the projection of each vector $\mathbf{b}$ in $\mathbb{R}^3$ on W.

SOLUTION   Let

$$A = \begin{bmatrix} 1/\sqrt{3} & 1/\sqrt{2} \\ -1/\sqrt{3} & 1/\sqrt{2} \\ 1/\sqrt{3} & 0 \end{bmatrix}.$$

Then

$$A^T A = \begin{bmatrix} 1 & 0 \\ 0 & 1 \end{bmatrix},$$

so

$$P = AA^T = \begin{bmatrix} \frac{5}{6} & \frac{1}{6} & \frac{1}{3} \\ \frac{1}{6} & \frac{5}{6} & -\frac{1}{3} \\ \frac{1}{3} & -\frac{1}{3} & \frac{1}{3} \end{bmatrix}.$$

Each column vector $\mathbf{b}$ in $\mathbb{R}^3$ projects onto

$$\mathbf{b}_W = P\mathbf{b} = \begin{bmatrix} \frac{5}{6} & \frac{1}{6} & \frac{1}{3} \\ \frac{1}{6} & \frac{5}{6} & -\frac{1}{3} \\ \frac{1}{3} & -\frac{1}{3} & \frac{1}{3} \end{bmatrix}\begin{bmatrix} b_1 \\ b_2 \\ b_3 \end{bmatrix} = \frac{1}{6}\begin{bmatrix} 5b_1 + b_2 + 2b_3 \\ b_1 + 5b_2 - 2b_3 \\ 2b_1 - 2b_2 + 2b_3 \end{bmatrix}.$$

Alternatively, we can compute $\mathbf{b}_W$ directly, using boxed Eq. (4) in Section 6.2. We obtain

$$\mathbf{b}_W = (\mathbf{b} \cdot \mathbf{a}_1)\mathbf{a}_1 + (\mathbf{b} \cdot \mathbf{a}_2)\mathbf{a}_2 = \frac{1}{3}\begin{bmatrix} b_1 - b_2 + b_3 \\ -b_1 + b_2 - b_3 \\ b_1 - b_2 + b_3 \end{bmatrix} + \frac{1}{2}\begin{bmatrix} b_1 + b_2 \\ b_1 + b_2 \\ 0 \end{bmatrix}$$

$$= \frac{1}{6}\begin{bmatrix} 5b_1 + b_2 + 2b_3 \\ b_1 + 5b_2 - 2b_3 \\ 2b_1 - 2b_2 + 2b_3 \end{bmatrix}. \quad \blacksquare$$

SUMMARY

Let $\{\mathbf{a}_1, \mathbf{a}_2, \ldots, \mathbf{a}_k\}$ be a basis for a subspace W of $\mathbb{R}^n$, and let A be the $n \times k$ matrix having $\mathbf{a}_j$ as jth column vector, so that W is the column space of A.

1.   The projection of a column vector $\mathbf{b}$ in $\mathbb{R}^n$ on W is $\mathbf{b}_W = A(A^TA)^{-1}A^T\mathbf{b}$.
2.   The matrix $P = A(A^TA)^{-1}A^T$ is the projection matrix for the subspace W. It is the unique matrix P such that, for every vector $\mathbf{b}$ in $\mathbb{R}^n$, the vector $P\mathbf{b}$ lies in W and the vector $\mathbf{b} - P\mathbf{b}$ is perpendicular to every vector in W.
3.   If the basis $\{\mathbf{a}_1, \mathbf{a}_2, \ldots, \mathbf{a}_k\}$ of the subspace W is orthonormal, the projection matrix for W is $P = AA^T$.

4. The projection matrix P of a subspace W is idempotent and symmetric. Every symmetric idempotent matrix is the projection matrix for its column space.

EXERCISES

In Exercises 1–8, find the projection matrix for the given subspace, and find the projection of the indicated vector on the subspace.

1. $[1, 2, 1]$ on $\text{sp}([2, 1, -1])$ in $\mathbb{R}^3$
2. $[1, 3, 4]$ on $\text{sp}([1, -1, 2])$ in $\mathbb{R}^3$
3. $[2, -1, 3]$ on $\text{sp}([2, 1, 1], [-1, 2, 1])$ in $\mathbb{R}^3$
4. $[1, 2, 1]$ on $\text{sp}([3, 0, 1], [1, 1, 1])$ in $\mathbb{R}^3$
5. $[1, 3, 1]$ on the plane $x + y - 2z = 0$ in $\mathbb{R}^3$
6. $[4, 2, -1]$ on the plane $3x + 2y + z = 0$ in $\mathbb{R}^3$
7. $[1, 2, 1, 3]$ on $\text{sp}([1, 2, 1, 1], [-1, 1, 0, -1])$ in $\mathbb{R}^4$
8. $[1, 1, 2, 1]$ on $\text{sp}([1, 1, 1, 1], [1, -1, 1, -1], [-1, 1, 1, -1])$ in $\mathbb{R}^4$
9. Find the projection matrix for the x_1, x_2-plane in $\mathbb{R}^3$.
10. Find the projection matrix for the x_1, x_3-coordinate subspace of $\mathbb{R}^4$.
11. Find the projection matrix for the x_1, x_2, x_4-coordinate subspace of $\mathbb{R}^4$.
12. Show that boxed Eq. (3) of this section reduces to Eq. (1) of Section 6.1 for projecting $\mathbf{b}$ on $\text{sp}(\mathbf{a})$.
13. Give a geometric argument indicating that every projection matrix is idempotent.
14. Let $\mathbf{a}$ be a unit column vector in $\mathbb{R}^n$. Show that $\mathbf{a}\mathbf{a}^T$ is the projection matrix for the subspace $\text{sp}(\mathbf{a})$.
15. Mark each of the following True or False.
   ___ a. A subspace W of dimension k in $\mathbb{R}^n$ has associated with it a $k \times k$ projection matrix.
   ___ b. Every subspace W of $\mathbb{R}^n$ has associated with it an $n \times n$ projection matrix.
   ___ c. Projection of $\mathbb{R}^n$ on a subspace W is a linear transformation of $\mathbb{R}^n$ into itself.
   ___ d. Two different subspaces of $\mathbb{R}^n$ may have the same projection matrix.
   ___ e. Two different matrices may be projection matrices for the same subspace of $\mathbb{R}^n$.
   ___ f. Every projection matrix is symmetric.
   ___ g. Every symmetric matrix is a projection matrix.
   ___ h. An $n \times n$ symmetric matrix A is a projection matrix if and only if $A^2 = I$.
   ___ i. Every symmetric idempotent matrix is the projection matrix for its column space.
   ___ j. Every symmetric idempotent matrix is the projection matrix for its row space.
16. Show that the projection matrix $P = A(A^TA)^{-1}A^T$ given in Theorem 6.11 satisfies the following two conditions:
   a. $P^2 = P$,
   b. $P^T = P$.
17. What is the projection matrix for the subspace $\mathbb{R}^n$ of $\mathbb{R}^n$?
18. Let U be a subspace of W, which is a subspace of $\mathbb{R}^n$. Let P be the projection matrix for W, and let R be the projection matrix for U. Find PR and RP. [HINT: Argue geometrically.]
19. Let P be the projection matrix for a k-dimensional subspace of $\mathbb{R}^n$.
   a. Find all eigenvalues of P.
   b. Find the algebraic multiplicity and the geometric multiplicity of each eigenvalue found in part (a).
   c. Explain how we can deduce that P is diagonalizable, without using the fact that P is a symmetric matrix.
20. Show that every symmetric matrix whose only eigenvalues are 0 and 1 is a projection matrix.
21. Find all invertible projection matrices.

In Exercises 22–28, find the projection matrix for the subspace W having the given orthonormal basis. The vectors are given in row notation to save space in printing.

22. $W = \text{sp}(\mathbf{a}_1, \mathbf{a}_2)$ in $\mathbb{R}^3$, where
   $\mathbf{a}_1 = [1/\sqrt{2}, 0, -1/\sqrt{2}]$ and $\mathbf{a}_2 = [1/\sqrt{3}, -1/\sqrt{3}, 1/\sqrt{3}]$

23. $W = \text{sp}(\mathbf{a}_1, \mathbf{a}_2)$ in $\mathbb{R}^3$, where
   $\mathbf{a}_1 = [\frac{3}{5}, \frac{4}{5}, 0]$ and $\mathbf{a}_2 = [0, 0, 1]$

24. $W = \mathrm{sp}(\mathbf{a}_1, \mathbf{a}_2)$ in $\mathbb{R}^4$, where
$\mathbf{a}_1 = \left[3/(5\sqrt{2}), 4/(5\sqrt{2}), \frac{1}{2}, -\frac{1}{2}\right]$ and
$\mathbf{a}_2 = \left[4/(5\sqrt{2}), -3/(5\sqrt{2}), \frac{1}{2}, \frac{1}{2}\right]$

25. $W = \mathrm{sp}(\mathbf{a}_1, \mathbf{a}_2)$ in $\mathbb{R}^4$, where $\mathbf{a}_1 = \left[\frac{2}{7}, 0, \frac{3}{7}, -\frac{6}{7}\right]$
and $\mathbf{a}_2 = \left[-\frac{3}{7}, \frac{6}{7}, \frac{2}{7}, 0\right]$

26. $W = \mathrm{sp}(\mathbf{a}_1, \mathbf{a}_2)$ in $\mathbb{R}^4$, where
$\mathbf{a}_1 = \left[0, -\frac{2}{3}, \frac{2}{3}, \frac{1}{3}\right]$ and $\mathbf{a}_2 = \left[\frac{2}{3}, 0, -\frac{1}{3}, \frac{2}{3}\right]$

27. $W = \mathrm{sp}(\mathbf{a}_1, \mathbf{a}_2, \mathbf{a}_3)$ in $\mathbb{R}^4$, where
$\mathbf{a}_1 = [1/\sqrt{3}, 0, 1/\sqrt{3}, 1/\sqrt{3}]$,
$\mathbf{a}_2 = [1/\sqrt{3}, 1/\sqrt{3}, -1/\sqrt{3}, 0]$, and
$\mathbf{a}_3 = [1/\sqrt{3}, -1/\sqrt{3}, 0, -1/\sqrt{3}]$

28. $W = \mathrm{sp}(\mathbf{a}_1, \mathbf{a}_2, \mathbf{a}_3)$ in $\mathbb{R}^4$, where
$\mathbf{a}_1 = \left[\frac{1}{2}, \frac{1}{2}, \frac{1}{2}, \frac{1}{2}\right]$, $\mathbf{a}_2 = \left[-\frac{1}{2}, \frac{1}{2}, -\frac{1}{2}, \frac{1}{2}\right]$, and
$\mathbf{a}_3 = \left[\frac{1}{2}, \frac{1}{2}, -\frac{1}{2}, -\frac{1}{2}\right]$

In Exercises 29–32, find the projection of **b** *on W.*

29. The subspace W in Exercise 22;
$\mathbf{b} = [6, -12, -6]$

30. The subspace W in Exercise 23;
$\mathbf{b} = [20, -15, 5]$

31. The subspace W in Exercise 26;
$\mathbf{b} = [9, 0, -9, 18]$

32. The subspace W in Exercise 28;
$\mathbf{b} = [4, -12, -4, 0]$

33. Let W be a subspace of $\mathbb{R}^n$, and let P be the projection matrix for W. **Reflection of $\mathbb{R}^n$ in W** is the mapping of $\mathbb{R}^n$ into itself that carries each vector **b** in $\mathbb{R}^n$ into its reflection $\mathbf{b}_r$, according to the following geometric description:

Let **p** be the projection of **b** on W. Starting at the tip of **b**, travel in a straight line to the tip of **p**, and then continue in the same

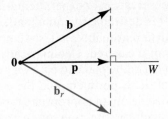

FIGURE 6.11
Reflection of $\mathbb{R}^n$ through W.

direction an equal distance to arrive at $\mathbf{b}_r$. (See Figure 6.11.)

Show that $\mathbf{b}_r = (2P - I)\mathbf{b}$. (Notice that, because reflection can be accomplished by matrix multiplication, this reflection must be a linear transformation of $\mathbb{R}^n$ into itself.)

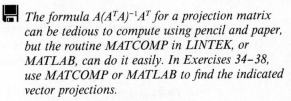 *The formula $A(A^TA)^{-1}A^T$ for a projection matrix can be tedious to compute using pencil and paper, but the routine MATCOMP in LINTEK, or MATLAB, can do it easily. In Exercises 34–38, use MATCOMP or MATLAB to find the indicated vector projections.*

34. The projections in $\mathbb{R}^6$ of $[-1, 2, 3, 1, 6, 2]$ and $[2, 0, 3, -1, 4, 5]$ on
$\mathrm{sp}([1, -2, 3, 1, 4, 0])$

35. The projections in $\mathbb{R}^3$ of $[1, -1, 4]$, $[3, 3, -1]$, and $[-2, 4, 7]$ on
$\mathrm{sp}([1, 3, -4], [2, 0, 3])$

36. The projections in $\mathbb{R}^4$ of $[-1, 3, 2, 0]$ and $[4, -1, 1, 5]$ on $\mathrm{sp}([0, 1, 2, 1], [-1, 2, 1, 4])$

37. The projections in $\mathbb{R}^4$ of $[2, 1, 0, 3]$, $[1, 1, -1, 2]$, and $[4, 3, 1, 3]$ on
$\mathrm{sp}([1, 0, -1, 0], [1, 2, -1, 4], [2, 1, 3, -1])$

38. The projections in $\mathbb{R}^5$ of $[2, 1, -3, 2, 4]$ and $[1, -4, 0, 1, 5]$ on
$\mathrm{sp}([3, 1, 4, 0, 1], [2, 1, 3, -5, 1])$

6.5 THE METHOD OF LEAST SQUARES

The Nature of the Problem

In this section we apply our work on projections to problems of data analysis. Suppose that data measurements of the form (a_i, b_i) are obtained from observation or experimentation and are plotted as data points in the x,y-plane. It is desirable to find a mathematical relationship $y = f(x)$ that represents the data reasonably well, so that we can make predictions of data values that were not measured. Geometrically, this means that we would like the graph of $y = f(x)$ in the plane to pass very close to our data points. Depending on the nature of the experiment and the configuration of the plotted data points, we might decide on an appropriate type of function $y = f(x)$ such as a linear function, a quadratic function, or an exponential function. We illustrate with three types of problems.

PROBLEM 1   According to Hooke's law, the distance that a spring stretches is proportional to the force applied. Suppose that we attach four different weights a_1, a_2, a_3, and a_4 in turn to the bottom of a spring suspended vertically. We measure the four lengths b_1, b_2, b_3, and b_4 of the stretched spring, and the data in Table 6.1 are obtained. Because of Hooke's law, we expect the data points (a_i, b_i) to be close to some line with equation

$$y = f(x) = r_0 + r_1x,$$

where r_0 is the unstretched length of the spring and r_1 is the *spring constant*. That is, if our measurements were exact and the spring ideal, we would have $b_i = r_0 + r_1a_i$ for specific values r_0 and r_1.   ■

TABLE 6.1

a_i = Weight in ounces	2.0	4.0	5.0	6.0
b_i = Length in inches	6.5	8.5	11.0	12.5

In Problem 1, we have only the two unknowns r_0 and r_1; and in theory, just two measurements should suffice to find them. In practice, however, we expect to have some error in physical measurements. It is standard procedure to make more measurements than are theoretically necessary in the hope that the errors will roughly cancel each other out, in accordance with the laws of probability. Substitution of each data point (a_i, b_i) from Problem 1 into the equation $y = r_0 + r_1x$ gives a single linear equation in the two unknowns r_0 and r_1. The four data points of Problem 1 thus give rise to a linear system of four equations in only two unknowns. Such a linear system with more equations than unknowns is called **overdetermined**, and one expects to find that the system is *inconsistent*, having no actual solution. It will be our task to find values for the unknowns r_0 and r_1 that will come as close as possible, in some sense, to satisfying all four of the equations.

We have used the illustration presented in Problem 1 to introduce our goal in this section, and we will solve the problem in a moment. We first present two more hypothetical problems, which we will also solve later in the section.

PROBLEM 2   At a recent boat show, the observations listed in Table 6.2 were made relating the prices b_i of sailboats and their weights a_i. Plotting the data points (a_i, b_i), as shown in Figure 6.12, we might expect a quadratic function of the form

$$y = f(x) = r_0 + r_1 x + r_2 x^2$$

to fit the data fairly well.   ∎

TABLE 6.2

a_i = Weight in tons	2	4	5	8
b_i = Price in units of \$10,000	1	3	5	12

PROBLEM 3   A population of rabbits on a large island was estimated each year from 1991 to 1994, giving the data in Table 6.3. Knowing that population growth is exponential in the absence of disease, predators, famine, and so on, we expect an exponential function

$$y = f(x) = re^{sx}$$

to provide the best representation of these data. Notice that, by using logarithms, we can convert this exponential function into a form linear in x:

$$\ln y = \ln r + sx.$$

TABLE 6.3

a_i = (Year observed) – 1990	1	2	3	4
b_i = Number of rabbits in units of 1000	3	4.5	8	17

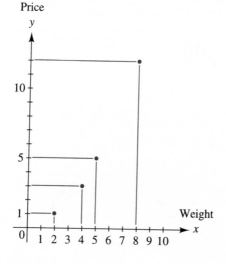

FIGURE 6.12
Problem 2 data.

Least-Squares Solution of $Ar \approx b$

Let A be a matrix with independent column vectors. The least-squares solution $\bar{r}$ of $Ar \approx b$ can be computed in either of the following ways:

1. Compute $r = (A^T A)^{-1} A^T b$.
2. Solve $(A^T A) r = A^T b$.

When a computer is being used, the second method is more efficient.

EXAMPLE 1   Find the least-squares fit to the data points in Problem 1 by a straight line—that is, by a linear function $y = r_0 + r_1 x$.

SOLUTION   We form the system $b \approx Ar$ in system (1) for the data in Table 6.1:

$$\begin{bmatrix} 6.5 \\ 8.5 \\ 11.0 \\ 12.5 \end{bmatrix} \approx \begin{bmatrix} 1 & 2 \\ 1 & 4 \\ 1 & 5 \\ 1 & 6 \end{bmatrix} \begin{bmatrix} r_0 \\ r_1 \end{bmatrix}.$$
$$\quad b \qquad\qquad A \qquad r$$

We have

$$A^T A = \begin{bmatrix} 1 & 1 & 1 & 1 \\ 2 & 4 & 5 & 6 \end{bmatrix} \begin{bmatrix} 1 & 2 \\ 1 & 4 \\ 1 & 5 \\ 1 & 6 \end{bmatrix} = \begin{bmatrix} 4 & 17 \\ 17 & 81 \end{bmatrix}$$

and

$$(A^T A)^{-1} = \frac{1}{35} \begin{bmatrix} 81 & -17 \\ -17 & 4 \end{bmatrix}$$

From Eq. (3), we obtain the least-squares solution

$$\bar{r} = (A^T A)^{-1} A^T b = \frac{1}{35} \begin{bmatrix} 81 & -17 \\ -17 & 4 \end{bmatrix} \begin{bmatrix} 1 & 1 & 1 & 1 \\ 2 & 4 & 5 & 6 \end{bmatrix} \begin{bmatrix} 6.5 \\ 8.5 \\ 11.0 \\ 12.5 \end{bmatrix}$$

$$= \frac{1}{35} \begin{bmatrix} 47 & 13 & -4 & -21 \\ -9 & -1 & 3 & 7 \end{bmatrix} \begin{bmatrix} 6.5 \\ 8.5 \\ 11.0 \\ 12.5 \end{bmatrix} = \frac{1}{35} \begin{bmatrix} 109.5 \\ 53.5 \end{bmatrix} \approx \begin{bmatrix} 3.1 \\ 1.5 \end{bmatrix}.$$

Therefore, the linear function that best fits the data in the least-squares sense is $y \approx f(x) = 1.5x + 3.1$. This line and the data points are shown in Figure 6.15. ∎

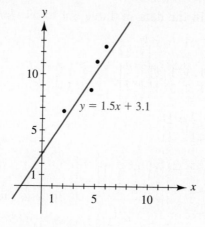

FIGURE 6.15
The least-squares fit of data points.

EXAMPLE 2   Use the method of least squares to fit the data in Problem 3 by an exponential function $y = f(x) = re^{sx}$.

SOLUTION   We use logarithms and convert the exponential equation to an equation that is linear in x:

$$\ln y = \ln r + sx.$$

TABLE 6.4

$x = a_i$	1	2	3	4
$y = b_i$	3	4.5	8	17
$z = \ln(b_i)$	1.10	1.50	2.08	2.83

HISTORICAL NOTE   A TECHNIQUE VERY CLOSE TO THAT OF LEAST SQUARES was developed by Roger Cotes (1682–1716), the gifted mathematician who edited the second edition of Isaac Newton's *Principia*, in a work dealing with errors in astronomical observations, written around 1715.

The complete principle, however, was first formulated by Carl Gauss at around the age of 16 while he was adjusting approximations in dealing with the distribution of prime numbers. Gauss later stated that he used the method frequently over the years—for example, when he did calculations concerning the orbits of asteroids. Gauss published the method in 1809 and gave a definitive exposition 14 years later.

On the other hand, it was Adrien-Marie Legendre (1752–1833), founder of the theory of elliptic functions, who first published the method of least squares, in an 1806 work on determining the orbits of comets. After Gauss's 1809 publication, Legendre wrote to him, censuring him for claiming the method as his own. Even as late as 1827, Legendre was still berating Gauss for "appropriating the discoveries of others." In fact, the problem lay in Gauss's failure to publish many of his discoveries promptly; he mentioned them only after they were published by others.

From Table 6.3, we obtain the data in Table 6.4 to use for our logarithmic equation. We obtain

$$A^T A = \begin{bmatrix} 1 & 1 & 1 & 1 \\ 1 & 2 & 3 & 4 \end{bmatrix} \begin{bmatrix} 1 & 1 \\ 1 & 2 \\ 1 & 3 \\ 1 & 4 \end{bmatrix} = \begin{bmatrix} 4 & 10 \\ 10 & 30 \end{bmatrix}$$

$$(A^T A)^{-1} = \begin{bmatrix} \frac{3}{2} & -\frac{1}{2} \\ -\frac{1}{2} & \frac{1}{5} \end{bmatrix}$$

$$(A^T A)^{-1} A^T = \begin{bmatrix} \frac{3}{2} & -\frac{1}{2} \\ -\frac{1}{2} & \frac{1}{5} \end{bmatrix} \begin{bmatrix} 1 & 1 & 1 & 1 \\ 1 & 2 & 3 & 4 \end{bmatrix} = \begin{bmatrix} 1 & \frac{1}{2} & 0 & -\frac{1}{2} \\ -\frac{3}{10} & -\frac{1}{10} & \frac{1}{10} & \frac{3}{10} \end{bmatrix}.$$

Multiplying this last matrix on the right by

$$\begin{bmatrix} 1.10 \\ 1.50 \\ 2.08 \\ 2.83 \end{bmatrix},$$

we obtain, from Eq. (3),

$$\bar{\mathbf{r}} = \begin{bmatrix} \ln r \\ s \end{bmatrix} \approx \begin{bmatrix} .435 \\ .577 \end{bmatrix}.$$

Thus $r = e^{.435} \approx 1.54$, and we obtain $y = f(x) = 1.54 e^{.577x}$ as a fitting exponential function.

The graph of the function and of the data points in Table 6.5 is shown in Figure 6.16. On the basis of the function $f(x)$ obtained, we can project the population of rabbits on the island in the year 2000 to be about

$$f(10) \cdot 1000 \approx 494{,}000 \text{ rabbits},$$

unless predators, disease, or lack of food interferes. ∎

TABLE 6.5

a_i	b_i	$f(a_i)$
1	3	2.7
2	4.5	4.9
3	8	8.7
4	17	15.5

In Example 2, we used the least-squares method with the *logarithm* of our original y-coordinate data. Thus, it is the equation $\ln y = r + sx$ that is the least-squares fit of the logarithmic data. Using the least-squares method to fit the logarithm of the y-coordinate data produces an exponential function that approximates the smaller y-values in the data better than it does the larger y-values, as illustrated by Table 6.5. It can be shown that the exponential

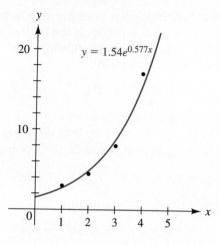

FIGURE 6.16
Data points and the exponential fit.

fit $y = f(x) = re^{sx}$ obtained from the least-squares fit of $\ln y = \ln r + sx$ amounts roughly to minimizing the *percent of error* in taking $f(a_i)$ for b_i. The least-squares fit of logarithmic data does not yield the least-squares fit of the original data. The routine YOUFIT in LINTEK can be used to illustrate this. (See Exercises 29 and 30.)

Least-Squares Solutions for Larger Systems

Data points of more than two components may lead to overdetermined linear systems in more than two unknowns. Suppose that an experiment is repeated m times, and data values

$$b_i, a_{i1}, \ldots, a_{in}$$

are obtained from measurements on the ith experiment. For example, the data values a_{ij} might be ones that can be controlled and the value b_i measured. Of course, there may be errors in the controlled as well as in the measured values. Suppose that we have reason to believe that the data obtained for each experiment should *theoretically* satisfy the same linear relation

$$y = r_0 + r_1 x_1 + r_2 x_2 + \cdots + r_n x_n, \tag{4}$$

with $y = b_i$ and $x_j = a_{ij}$. We then obtain a system of m linear approximations in $n + 1$ unknowns $r_0, r_1, r_2, \ldots, r_n$:

$$
\underbrace{\begin{bmatrix} b_1 \\ b_2 \\ \cdot \\ \cdot \\ \cdot \\ b_m \end{bmatrix}}_{\mathbf{b}} \approx \underbrace{\begin{bmatrix} 1 & a_{11} & a_{12} & \cdots & a_{1n} \\ 1 & a_{21} & a_{22} & \cdots & a_{2n} \\ \cdot & \cdot & \cdot & & \cdot \\ \cdot & \cdot & \cdot & & \cdot \\ \cdot & \cdot & \cdot & & \cdot \\ 1 & a_{m1} & a_{m2} & \cdots & a_{mn} \end{bmatrix}}_{A} \underbrace{\begin{bmatrix} r_0 \\ r_1 \\ r_2 \\ \cdot \\ \cdot \\ \cdot \\ r_n \end{bmatrix}}_{\mathbf{r}}. \tag{5}
$$

If $m > n + 1$, then system (5) corresponds to an overdetermined linear system $\mathbf{b} = A\mathbf{r}$, which probably has no exact solution. If the rank of A is $n + 1$, then a repetition of our geometric argument above indicates that the least-squares solution $\bar{\mathbf{r}}$ for the system $\mathbf{b} \approx A\mathbf{r}$ is given by

$$\bar{\mathbf{r}} = (A^TA)^{-1}A^T\mathbf{b}. \tag{6}$$

Again, it is more efficient computationally to solve $(A^TA)\mathbf{r} = A^T\mathbf{b}$.

A linear system of the form shown in system (5) arises if m data points (a_i, b_i) are found and a least-squares fit is sought for a polynomial function

$$y = r_0 + r_1x + \cdots + r_{n-1}x^{n-1} + r_nx^n.$$

The data point (a_i, b_i) leads to the approximation

$$b_i \approx r_0 + r_1a_i + \cdots + r_{n-1}a_i^{n-1} + r_na_i^n.$$

The m data points thus give rise to a linear system of the form of system (5), where the matrix A is given by

$$A = \begin{bmatrix} 1 & a_1 & a_1^2 & \cdots & a_1^n \\ 1 & a_2 & a_2^2 & \cdots & a_2^n \\ \cdot & \cdot & \cdot & & \cdot \\ \cdot & \cdot & \cdot & & \cdot \\ \cdot & \cdot & \cdot & & \cdot \\ 1 & a_m & a_m^2 & \cdots & a_m^n \end{bmatrix}. \tag{7}$$

EXAMPLE 3   Use a computer to find the least-squares fit to the data in Problem 2 by a parabola—that is, by a quadratic function

$$y = r_0 + r_1x + r_2x^2.$$

SOLUTION   We write the data in the form $\mathbf{b} \approx A\mathbf{r}$, where A has the form of matrix (7):

$$\begin{bmatrix} 1 \\ 3 \\ 5 \\ 12 \end{bmatrix} \approx \begin{bmatrix} 1 & 2 & 4 \\ 1 & 4 & 16 \\ 1 & 5 & 25 \\ 1 & 8 & 64 \end{bmatrix} \begin{bmatrix} r_0 \\ r_1 \\ r_2 \end{bmatrix}.$$

Entering A and $\mathbf{b}$ in either LINTEK or MATLAB, we find that

$$A^TA = \begin{bmatrix} 4 & 19 & 109 \\ 19 & 109 & 709 \\ 109 & 709 & 4993 \end{bmatrix} \quad \text{and} \quad A^T\mathbf{b} = \begin{bmatrix} 21 \\ 135 \\ 945 \end{bmatrix}.$$

Solving the linear system $(A^TA)\mathbf{r} = A^T\mathbf{b}$ using either package then yields

$$\bar{\mathbf{r}} \approx \begin{bmatrix} .207 \\ .010 \\ .183 \end{bmatrix}.$$

Thus, the quadratic function that best approximates the data in the least-squares sense is

$$y \approx f(x) = .207 + .01x + .183x^2.$$

In Figure 6.17, we show the graph of this quadratic function and plot the data points. Data are given in Table 6.6. On the basis of the least-squares fit, we might project that the price of a 10-ton sailing yacht would be about 18.6 times $10,000, or $186,000, and that the price of a 20-ton yacht would be about $f(20)$ times $10,000, or about $736,000. However, one should be wary of using a function $f(x)$ that seems to fit data well for measured values of x to project data for values of x far from those measured values. Our quadratic function seems to fit our data quite well, but we should not expect the cost we might project for a 100-ton ship to be at all accurate. ∎

TABLE 6.6

a_i	b_i	$f(a_i)$
2	1	.959
4	3	3.17
5	5	4.83
8	12	12.0

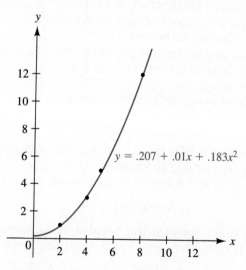

FIGURE 6.17
The graph and data points for Example 3.

EXAMPLE 4   Show that the least-squares linear fit to data points (a_i, b_i) for $i = 1, 2, \ldots, m$ by a constant function $f(x) = r_0$ is achieved when r_0 is the average y-value

$$(b_1 + b_2 + \cdots + b_m)/m$$

of the data values b_i.

SOLUTION   Such a constant function $y = f(x) = r_0$ has a horizontal line as its graph. In this situation, we have $n = 1$ as the number of unknowns and system (5) becomes

$$\begin{bmatrix} b_1 \\ b_2 \\ \cdot \\ \cdot \\ \cdot \\ b_m \end{bmatrix} \approx \begin{bmatrix} 1 \\ 1 \\ \cdot \\ \cdot \\ \cdot \\ 1 \end{bmatrix} [r_0].$$

$$\quad \mathbf{b} \qquad A \quad \mathbf{r}$$

Thus A is the $m \times 1$ matrix with all entries 1. We easily find that $A^T A = [m]$, so $(A^T A)^{-1} = [1/m]$. We also find that $A^T \mathbf{b} = b_1 + b_2 + \cdots + b_m$. Thus, the least-squares solution [Eq. (6)] is

$$\bar{\mathbf{r}} = (b_1 + b_2 + \cdots + b_m)/m,$$

as asserted.   ∎

Overdetermined Systems of Linear Equations

We have presented examples in which least-squares solutions were found to systems of linear approximations arising from applications. We used $\mathbf{r}$ as the column vector of unknowns in those examples, because we were using x as the independent variable in the formula for a fitting function $f(x)$. We now discuss the mathematics of overdetermined systems, and we use $\mathbf{x}$ as the column vector of unknowns again.

Consider a general overdetermined system of linear equations

$$A\mathbf{x} = \mathbf{b},$$

where A is an $m \times n$ matrix of rank n and $m > n$. We expect such a system to be inconsistent—that is, to have no solution. As we have seen, the system of linear approximations $A\mathbf{x} \approx \mathbf{b}$ has least-squares solution

$$\bar{\mathbf{x}} = (A^T A)^{-1} A^T \mathbf{b}. \tag{8}$$

The vector $\bar{\mathbf{x}}$ is thus the **least-squares solution of the overdetermined system** $A\mathbf{x} = \mathbf{b}$. Equation (8) can be written in the equivalent form

$$A^T A \bar{\mathbf{x}} = A^T \mathbf{b}, \tag{9}$$

which exhibits $\bar{\mathbf{x}}$ as the solution of the consistent linear system $A^T A \mathbf{x} = A^T \mathbf{b}$.

EXAMPLE 5   Find the least-squares approximate solution of the overdetermined linear system

$$\begin{bmatrix} 1 & 0 & 1 \\ 2 & 1 & 1 \\ 1 & 3 & 0 \\ 0 & 2 & 1 \\ -1 & 2 & -1 \end{bmatrix} \begin{bmatrix} x_1 \\ x_2 \\ x_3 \end{bmatrix} = \begin{bmatrix} 1 \\ 0 \\ 1 \\ 2 \\ 0 \end{bmatrix}$$

by transforming it into the consistent form [system (9)].

SOLUTION   The consistent system [system (9)] is

$$\begin{bmatrix} 1 & 2 & 1 & 0 & -1 \\ 0 & 1 & 3 & 2 & 2 \\ 1 & 1 & 0 & 1 & -1 \end{bmatrix} \begin{bmatrix} 1 & 0 & 1 \\ 2 & 1 & 1 \\ 1 & 3 & 0 \\ 0 & 2 & 1 \\ -1 & 2 & -1 \end{bmatrix} \begin{bmatrix} x_1 \\ x_2 \\ x_3 \end{bmatrix} = \begin{bmatrix} 1 & 2 & 1 & 0 & -1 \\ 0 & 1 & 3 & 2 & 2 \\ 1 & 1 & 0 & 1 & -1 \end{bmatrix} \begin{bmatrix} 1 \\ 0 \\ 1 \\ 2 \\ 0 \end{bmatrix}$$

or

$$\begin{bmatrix} 7 & 3 & 4 \\ 3 & 18 & 1 \\ 4 & 1 & 4 \end{bmatrix} \begin{bmatrix} x_1 \\ x_2 \\ x_3 \end{bmatrix} = \begin{bmatrix} 2 \\ 7 \\ 3 \end{bmatrix},$$

whose solution is found, using MATLAB or LINTEK, to be

$$\begin{bmatrix} x_1 \\ x_2 \\ x_3 \end{bmatrix} \approx \begin{bmatrix} -.614 \\ .421 \\ 1.259 \end{bmatrix},$$

accurate to three decimal places. This is the least-squares approximate solution $\bar{x}$ to the given overdetermined system.   ∎

The Orthogonal Case

To obtain the least-squares linear fit $y = r_0 + r_1 x$ of k data points (a_i, b_i), we form the matrix

$$A = \begin{bmatrix} 1 & a_1 \\ 1 & a_2 \\ & \cdot \\ & \cdot \\ & \cdot \\ 1 & a_k \end{bmatrix} \tag{10}$$

and compute the least-squares solution vector $\bar{r} = (A^T A)^{-1} A^T \mathbf{b}$. If the set of column vectors in A were orthonormal, then $A^T A$ would be the 2×2 identity matrix I. Of course, this is never the case, because the first column vector of A

has length $\sqrt{k}$. However, it may happen that the column vectors in A are mutually perpendicular; we can see then that

$$A^T A = \begin{bmatrix} k & 0 \\ 0 & \mathbf{a} \cdot \mathbf{a} \end{bmatrix}, \quad \text{so} \quad (A^T A)^{-1} = \begin{bmatrix} 1/k & 0 \\ 0 & 1/(\mathbf{a} \cdot \mathbf{a}) \end{bmatrix}.$$

The matrix A has this property if the x-values $a_1, a_2, \ldots, a_k$ for the data are symmetrically positioned about zero. We illustrate with an example.

EXAMPLE 6   Find the least-squares linear fit of the data points $(-3, 8)$, $(-1, 5)$, $(1, 3)$, and $(3, 0)$.

SOLUTION   The matrix A is given by

$$A = \begin{bmatrix} 1 & -3 \\ 1 & -1 \\ 1 & 1 \\ 1 & 3 \end{bmatrix}.$$

We can see why the symmetry of the x-values about zero causes the column vectors of this matrix to be orthogonal. We find that

$$A^T A = \begin{bmatrix} 1 & 1 & 1 & 1 \\ -3 & -1 & 1 & 3 \end{bmatrix} \begin{bmatrix} 1 & -3 \\ 1 & -1 \\ 1 & 1 \\ 1 & 3 \end{bmatrix} = \begin{bmatrix} 4 & 0 \\ 0 & 20 \end{bmatrix}.$$

Then

$$\bar{\mathbf{r}} = \begin{bmatrix} r_0 \\ r_1 \end{bmatrix} = (A^T A)^{-1} A^T \mathbf{b} = \begin{bmatrix} \frac{1}{4} & 0 \\ 0 & \frac{1}{20} \end{bmatrix} \begin{bmatrix} 1 & 1 & 1 & 1 \\ -3 & -1 & 1 & 3 \end{bmatrix} \begin{bmatrix} 8 \\ 5 \\ 3 \\ 0 \end{bmatrix}$$

$$= \begin{bmatrix} \frac{1}{4} & 0 \\ 0 & \frac{1}{20} \end{bmatrix} \begin{bmatrix} 16 \\ -26 \end{bmatrix} = \begin{bmatrix} 4 \\ -1.3 \end{bmatrix}.$$

Thus, the least-squares linear fit is given by $y = 4 - 1.3x$. ∎

Suppose now that the x-values in the data are not symmetrically positioned about zero, but that there is a number $x = c$ about which they are symmetrically located. If we make the variable transformation $t = x - c$, then the t-values for the data are symmetrically positioned about $t = 0$. We can find the t,y-equation for the least-squares linear fit, and then replace t by $x - c$ to obtain the x,y-equation. (Exercises 14 and 15 refine this idea still further.)

EXAMPLE 7   The number of sales of a particular item by a manufacturer in each of the first four months of 1995 is given in Table 6.7. Find the least-squares linear fit of these data, and use it to project sales in the fifth month.

TABLE 6.7

$\mathbf{a}_i$ = Month	1	2	3	4
$\mathbf{b}_i$ = Thousands sold	2.5	3	3.8	4.5

SOLUTION   Our x-values a_i are symmetrically located about $c = 2.5$. If we take $t = x - 2.5$, the data points (t, y) become

$$(-1.5, 2.5), \qquad (-.5, 3), \qquad (.5, 3.8), \qquad (1.5, 4.5).$$

We find that

$$A^TA = \begin{bmatrix} 1 & 1 & 1 & 1 \\ -1.5 & -.5 & .5 & 1.5 \end{bmatrix} \begin{bmatrix} 1 & -1.5 \\ 1 & -.5 \\ 1 & .5 \\ 1 & 1.5 \end{bmatrix} = \begin{bmatrix} 4 & 0 \\ 0 & 5 \end{bmatrix}.$$

Thus,

$$(A^TA)^{-1}A^T\mathbf{b} = \begin{bmatrix} \frac{1}{4} & 0 \\ 0 & \frac{1}{5} \end{bmatrix} \begin{bmatrix} 1 & 1 & 1 & 1 \\ -1.5 & -.5 & .5 & 1.5 \end{bmatrix} \begin{bmatrix} 2.5 \\ 3 \\ 3.8 \\ 4.5 \end{bmatrix}$$

$$= \begin{bmatrix} \frac{1}{4} & 0 \\ 0 & \frac{1}{5} \end{bmatrix} \begin{bmatrix} 13.8 \\ 3.4 \end{bmatrix} = \begin{bmatrix} 3.45 \\ .68 \end{bmatrix}.$$

Thus, the t,y-equation is $y = 3.45 + .68t$. Replacing t by $x - 2.5$, we obtain

$$y = 3.45 + .68(x - 2.5) = 1.75 + .68x.$$

Setting $x = 5$, we estimate sales in the fifth month to be 5.15 units, or 5150 items. ∎

Using the *QR*-Factorization

The QR-factorization discussed at the end of Section 6.2 has application to the method of least squares. Consider an overdetermined system $A\mathbf{x} = \mathbf{b}$, where the matrix A has independent column vectors. We factor A as QR and remember that $Q^TQ = I$, because Q has orthonormal column vectors. Then we obtain

$$(A^TA)^{-1}A^T\mathbf{b} = ((QR)^TQR)^{-1}(QR)^T\mathbf{b} = (R^TQ^TQR)^{-1}R^TQ^T\mathbf{b}$$
$$= (R^TR)^{-1}R^TQ^T\mathbf{b} = R^{-1}(R^T)^{-1}R^TQ^T\mathbf{b} = R^{-1}Q^T\mathbf{b}. \qquad (11)$$

To find the least-squares solution vector $R^{-1}Q^T\mathbf{b}$ once Q and R have been found, we first multiply $\mathbf{b}$ by Q^T, which is a stable computation because Q is orthogonal. We then solve the upper-triangular system

$$R\mathbf{x} = Q^T\mathbf{b} \qquad (12)$$

by back substitution. Because R is already upper triangular, no matrix reduction is needed! The routine QRFACTOR in LINTEK has an option for this least-squares computation. The program demonstrates how quickly the least-squares solution vector can be found once a QR-factorization of A is known.

Suppose we believe that the output y of a process should be a linear function of time t. We measure the y-values at specified times during the process. We can then find the least-squares linear fit $y = r_1 + s_1 t$ from the measured data by finding the QR-factorization of an appropriate matrix A and subsequently finding the solution vector [Eq. (11)] in the way we just described. If we repeat the process later and make measurements again at the same time intervals, the matrix A will not change, so we can find another least-squares linear fit using the solution vector with the same matrices Q and R. We can repeat this often, obtaining linear fits $r_1 + s_1 t, r_2 + s_2 t, \ldots, r_m + s_m t$. For our final estimate of the proper linear fit for this process, we might take $r + st$, where r is the average of the m values r_i and s is the average of the m values s_i. We expect that errors made in measurements will roughly cancel each other out over numerous repetitions of the process, so that our estimated linear fit $r + st$ will be fairly accurate.

SUMMARY

1. A set of k data points (a_i, b_i) whose first coordinates are not all equal can be fitted with a polynomial function or with an exponential function, using the method of least squares illustrated in Examples 1 through 3.

2. Suppose that the x-values a_i in a set of k data points (a_i, b_i) are symmetrically positioned about zero. Let A be the $k \times 2$ matrix with first column vector having all entries 1 and second column vector $\mathbf{a}$. The columns of A are orthogonal, and

$$A^T A = \begin{bmatrix} k & 0 \\ 0 & \mathbf{a} \cdot \mathbf{a} \end{bmatrix}.$$

Computation of the least-squares linear fit for the data points is then simplified. If the x-values of the data points are symmetrically positioned about $x = c$, then the substitution $t = x - c$ gives data points with t-values symmetrically positioned about zero, and the above simplification applies. See Example 7.

Let $A\mathbf{x} = \mathbf{b}$ be a linear system of m equations in n unknowns, where $m > n$ (an overdetermined system) and where the rank of A is n.

3. The least-squares solution of the corresponding system $A\mathbf{x} \approx \mathbf{b}$ of linear approximations is the vector $\mathbf{x} = \overline{\mathbf{x}}$, which minimizes the magnitude of the error vectors $A\mathbf{x} - \mathbf{b}$ for $\mathbf{x} \in \mathbb{R}^n$.

4. The least-squares solution $\bar{\mathbf{x}}$ of $A\mathbf{x} \approx \mathbf{b}$ and of $A\mathbf{x} = \mathbf{b}$ is given by the formula

$$\bar{\mathbf{x}} = (A^T A)^{-1} A^T \mathbf{b}.$$

Geometrically, $A\bar{\mathbf{x}}$ is the projection of $\mathbf{b}$ on the column space of A.

5. An alternative to using the formula for $\bar{\mathbf{x}}$ in summary item 4 is to convert the overdetermined system $A\mathbf{x} = \mathbf{b}$ to the consistent system $(A^T A)\mathbf{x} = A^T \mathbf{b}$ by multiplying both sides of $A\mathbf{x} = \mathbf{b}$ by A^T, and then to find its unique solution, which is the least-squares solution $\bar{\mathbf{x}}$ of $A\mathbf{x} = \mathbf{b}$.

EXERCISES

1. Let the length b_i of a spring with an attached weight a_i be determined by measurements, as shown in Table 6.8.
   a. Find the least-squares linear fit, in accordance with Hooke's law.
   b. Use the answer to part (a) to estimate the length of the spring if a weight of 5 ounces is attached.

TABLE 6.8

a_i = Weight in ounces	1	2	4	6
b_i = Length in inches	3	4.1	5.9	8.2

2. A company had profits (in units of $10,000) of 0.5 in 1989, 1 in 1991, and 2 in 1994. Let time t be measured in years, with $t = 0$ in 1989.
   a. Find the least-squares linear fit of the data.
   b. Using the answer to part (a), estimate the profit in 1995.

3. Repeat Exercise 2, but find an exponential fit of the data, working with logarithms as explained in Example 2.

4. A publishing company specializing in college texts starts with a field sales force of ten people, and it has profits of $100,000. On increasing this sales force to 20, it has profits of $300,000; and increasing its sales force to 30 produces profits of $400,000.
   a. Find the least-squares linear fit for these data. [HINT: Express the numbers of salespeople in multiples of 10 and the profit in multiples of $100,000.]

   b. Use the answer to part (a) to estimate the profit if the sales force is reduced to 25.
   c. Does the profit obtained using the answer to part (a) for a sales force of 0 people seem in any way plausible?

In Exercises 5–7, find the least-squares fit to the given data by a linear function $f(x) = r_0 + r_1 x$. Graph the linear function and the data points.

5. (0, 1), (2, 6), (3, 11), (4, 12)
6. (1, 1), (2, 4), (3, 6), (4, 9)
7. (0, 0), (1, 1), (2, 3), (3, 8)
8. Find the least-squares fit to the data in Exercise 6 by a parabola (a quadratic polynomial function).
9. Repeat Exercise 8, but use the data in Exercise 7 instead.

In Exercises 10–13, use the technique illustrated in Examples 6 and 7 to solve the least-squares problem.

10. Find the least-squares linear fit to the data points $(-4, -2)$, $(-2, 0)$, $(0, 1)$, $(2, 4)$, $(4, 5)$.
11. Find the least-squares linear fit to the data points (0, 1), (1, 4), (2, 6), (3, 8), (4, 9).
12. The gallons of maple syrup made from the sugar bush of a Vermont farmer over the past five years were:

    80 gallons five years ago,
    70 gallons four years ago,
    75 gallons three years ago,
    65 gallons two years ago,
    60 gallons last year.

Use a least-squares linear fit of these data to project the number of gallons that will be produced this year. (Does this problem make practical sense? Why?)

13. The minutes required for a rat to find its way out of a maze on each repeated attempt were 8, 8, 6, 5, and 6 on its first five tries. Use a least-squares linear fit of these data to project the time the rat will take on its sixth try.

14. Let $(a_1, b_1), (a_2, b_2), \ldots, (a_m, b_m)$ be data points. If $\sum_{i=1}^{m} a_i = 0$, show that the line that best fits the data in the least-squares sense is given by $r_0 + r_1 x$, where

$$r_0 = \left(\sum_{i=1}^{m} b_i \right) \bigg/ m$$

and

$$r_1 = \left(\sum_{i=1}^{m} a_i b_i \right) \bigg/ \left(\sum_{i=1}^{m} a_i^2 \right).$$

15. Repeat Exercise 14, but do not assume that $\sum_{i=1}^{m} a_i = 0$. Show that the least-squares linear fit of the data is given by $y = r_0 + r_1(x - c)$, where $c = (\sum_{i=1}^{m} a_i)/m$ and r_0 and r_1 have the values given in Exercise 14.

In Exercises 16–20, find the least-squares solution of the given overdetermined system $A\mathbf{x} = \mathbf{b}$ by converting it to a consistent system and then solving, as illustrated in Example 5.

16. $\begin{bmatrix} 1 & 2 \\ 1 & 1 \\ 2 & 3 \end{bmatrix} \begin{bmatrix} x_1 \\ x_2 \end{bmatrix} = \begin{bmatrix} 3 \\ 1 \\ 3 \end{bmatrix}$

17. $\begin{bmatrix} 3 & 1 \\ 1 & 2 \\ 2 & -1 \end{bmatrix} \begin{bmatrix} x_1 \\ x_2 \end{bmatrix} = \begin{bmatrix} 1 \\ 0 \\ -2 \end{bmatrix}$

18. $\begin{bmatrix} 1 & 1 & 1 \\ -1 & 0 & 1 \\ 1 & -1 & 0 \\ 0 & 1 & -1 \end{bmatrix} \begin{bmatrix} x_1 \\ x_2 \\ x_3 \end{bmatrix} = \begin{bmatrix} 0 \\ 1 \\ -1 \\ -2 \end{bmatrix}$

19. $\begin{bmatrix} 1 & 1 & 3 \\ -1 & 0 & 5 \\ 0 & 1 & -2 \\ 1 & -1 & 1 \\ 1 & 0 & 1 \end{bmatrix} \begin{bmatrix} x_1 \\ x_2 \\ x_3 \end{bmatrix} = \begin{bmatrix} 1 \\ -1 \\ 3 \\ -2 \\ 0 \end{bmatrix}$

20. $\begin{bmatrix} 2 & -1 & 1 \\ 1 & 0 & 1 \\ -1 & 0 & 1 \\ 0 & -5 & -1 \\ 1 & 2 & -2 \end{bmatrix} \begin{bmatrix} x_1 \\ x_2 \\ x_3 \end{bmatrix} = \begin{bmatrix} 2 \\ 1 \\ 2 \\ 5 \\ 0 \end{bmatrix}$

21. Mark each of the following True or False.
___ a. There is a unique polynomial function of degree k with graph passing through any given k points in $\mathbb{R}^2$.
___ b. There is a unique polynomial function of degree k with graph passing through any k points in $\mathbb{R}^2$ having distinct first coordinates.
___ c. There is a unique polynomial function of degree at most $k - 1$ with graph passing through any k points in $\mathbb{R}^2$ having distinct first coordinates.
___ d. The least-squares solution of $A\mathbf{x} = \mathbf{b}$ is unique.
___ e. The least-squares solution of $A\mathbf{x} = \mathbf{b}$ can be an actual solution only if A is a square matrix.
___ f. The least-squares solution vector of $A\mathbf{x} = \mathbf{b}$ is the projection of $\mathbf{b}$ on the column space of A.
___ g. The least-squares solution vector of $A\mathbf{x} = \mathbf{b}$ is the vector $\overline{\mathbf{x}}$ such that $A\overline{\mathbf{x}}$ is the projection of $\mathbf{b}$ on the column space of A.
___ h. The least-squares solution vector of $A\mathbf{x} = \mathbf{b}$ is the vector $\mathbf{x} = \overline{\mathbf{x}}$ that minimizes the magnitude of $A\mathbf{x} - \mathbf{b}$.
___ i. A linear system has a least-squares solution only if the number of equations is greater than the number of unknowns.
___ j. Every linear system has a least-squares solution.

In Exercises 22–26, use LINTEK or MATLAB to find the least-squares solution of the linear system. If matrix files are available, the matrix A and the vector $\mathbf{b}$ for Exercise 22 are denoted in the file by A22 and b22, and similarly for Exercises 23–26. Recall that in MATLAB, the command **rref([A'∗A A'∗b])** *will row-reduce the augmented matrix $[A^T A \mid A^T \mathbf{b}]$.*

22. Exercise 16   25. Exercise 19

23. Exercise 17   26. Exercise 20

24. Exercise 18

The routine YOUFIT in LINTEK can be used to illustrate graphically the fitting of data points by linear, quadratic, or exponential functions. In Exercises 27–31, use YOUFIT to try visually to fit the given data with the indicated type of graph. When this is done, enter the zero data suggested to see the computer's fit. Run twice more with the same data points but without trying to fit the data visually, and determine whether the data are best fitted by a linear, quadratic, or (logarithmically fitted) exponential function, by comparing the least-squares sums for the three cases.

27. Fit (1, 2), (4, 6), (7, 10), (10, 14), (14, 19) by a linear function.

28. Fit (2, 2), (6, 10), (10, 12), (16, 2) by a quadratic function.

29. Fit (1, 1), (10, 8), (14, 12), (16, 20) by an exponential function. Try to achieve a lower squares sum than the computer obtains with its least-squares fit that uses logarithms of y-values.

30. Repeat Exercise 29 with data (1, 9), (5, 1), (6, .5), (9, .01).

31. Fit (2, 9), (4, 6), (7, 1), (8, .1) by a linear function.

The routine QRFACTOR in LINTEK has an option to use a QR-factorization of A to find the least-squares solution of a linear system A$\mathbf{x}$ = $\mathbf{b}$,

executing back substitution on R$\mathbf{x}$ = $Q^T\mathbf{b}$ as in Eq. (12).

Recall that in MATLAB, if A is $n \times k$, then Q is $n \times n$ and R is $n \times k$. Cutting Q and R down to the text sizes $n \times k$ and $k \times k$, respectively, we can use the command lines

$$[Q \; R] = qr(A); \; [n, k] = size(A);$$
$$rref([R(1:k,1:k) \; Q(:,1:k)'*b])$$

to compute the solution of R$\mathbf{x}$ = $Q^T\mathbf{b}$.

Use LINTEK or MATLAB in this fashion for Exercises 32–37. You must supply the matrix A and the vector $\mathbf{b}$.

32. Find the least-squares linear fit for the data points (−3, 10), (−2, 8), (−1, 7), (0, 6), (1, 4), (2, 5), (3, 6).

33. Find the least-squares quadratic fit for the data points in Exercise 32.

34. Find the least-squares cubic fit for the data points in Exercise 32.

35. Find the least-squares quartic fit for the data points in Exercise 32.

36. Find the quadratic polynomial function whose graph passes through the points (1, 4), (2, 15), (3, 32).

37. Find the cubic polynomial function whose graph passes through the points (−1, 13), (0, −5), (2, 7), (3, 25).

CHANGE OF BASIS

Recall from Section 3.4 that a linear transformation is a mapping of one vector space into another that preserves addition and scalar multiplication. Two vector spaces V and V' are isomorphic if there exists a linear transformation of V onto V' that is one-to-one. Of particular importance are the *coordinatization isomorphisms* of an n-dimensional vector space V with $\mathbb{R}^n$. One chooses an ordered basis B for V and defines $T: V \to \mathbb{R}^n$ by taking $T(\mathbf{v}) = \mathbf{v}_B$, the coordinate vector of $\mathbf{v}$ relative to B, as described in Section 3.3. Such an isomorphism describes V and $\mathbb{R}^n$ as being virtually the same as vector spaces. Thus much of the work in finite-dimensional vector spaces can be reduced to computations in $\mathbb{R}^n$. We take full advantage of this feature as we continue our study of linear transformations in this chapter.

We have seen that bases other than the standard basis E of $\mathbb{R}^n$ can be useful. For example, suppose A is an $n \times n$ matrix. Changing from the standard basis of $\mathbb{R}^n$ to a basis of eigenvectors (when this is possible) facilitates computation of the powers A^k of A using the more easily computed powers of a diagonal matrix, as described in Section 5.2. Our work in the preceding chapter gives another illustration of a desirable change of basis: recall the convenience that an orthonormal basis can provide. We expect that changing to a new basis will change coordinate vectors and matrix representations of linear transformations.

In Section 7.1 we review the notion of the coordinate vector $\mathbf{v}_B$ relative to an ordered basis B, and consider the effect of changing the ordered basis from B to B'. With this backdrop we turn to matrix representations of a linear transformation in Section 7.2, examining the relationship between matrix representations of a linear transformation $T: V \to V'$ relative to different ordered bases of V and of V'.

7.1

COORDINATIZATION AND CHANGE OF BASIS

Let $B = (\mathbf{b}_1, \mathbf{b}_2, \ldots, \mathbf{b}_n)$ be an ordered basis for a vector space V. Recall that if $\mathbf{v}$ is a vector in V and $\mathbf{v} = r_1\mathbf{b}_1 + r_2\mathbf{b}_2 + \cdots + r_n\mathbf{b}_n$, then the *coordinate vector of* $\mathbf{v}$ *relative to B* is

$$\mathbf{v}_B = [r_1, r_2, \ldots, r_n].$$

For example, the coordinate vector of $\cos^2 x$ relative to the ordered basis $(1, \sin^2 x)$ in the vector space $\mathrm{sp}(1, \sin^2 x)$ is $[1, -1]$. If this ordered basis is changed to some other ordered basis, the coordinate vector may change. In this section we consider the relationship between two coordinate vectors $\mathbf{v}_B$ and $\mathbf{v}_{B'}$ of the same vector $\mathbf{v}$.

We begin our discussion with $\mathbb{R}^n$, which we will view as a space of column vectors. Let $\mathbf{v}$ be a vector in $\mathbb{R}^n$ with coordinate vector

$$\mathbf{v}_B = \begin{bmatrix} r_1 \\ r_2 \\ \cdot \\ \cdot \\ \cdot \\ r_n \end{bmatrix}$$

relative to an ordered basis $B = (\mathbf{b}_1, \mathbf{b}_2, \ldots, \mathbf{b}_n)$, so that

$$r_1\mathbf{b}_1 + r_2\mathbf{b}_2 + \cdots + r_n\mathbf{b}_n = \mathbf{v}. \tag{1}$$

Let M_B be the matrix having the vectors in the ordered basis B as column vectors; this is the **basis matrix** for B, which we display as

$$M_B = \begin{bmatrix} | & | & & | \\ \mathbf{b}_1 & \mathbf{b}_2 & \cdots & \mathbf{b}_n \\ | & | & & | \end{bmatrix}. \tag{2}$$

Equation (1), which expresses $\mathbf{v}$ as a vector in the column space of the matrix M_B, can be written in the form

$$M_B\mathbf{v}_B = \mathbf{v}. \tag{3}$$

If B' is another ordered basis for $\mathbb{R}^n$, we can similarly obtain

$$M_{B'}\mathbf{v}_{B'} = \mathbf{v}. \tag{4}$$

Equations (3) and (4) together yield

$$M_{B'}\mathbf{v}_{B'} = M_B\mathbf{v}_B. \tag{5}$$

Equation (5) is easy to remember, and it turns out to be very useful. Both M_B and $M_{B'}$ are invertible, because they are square matrices whose column vectors are independent. Thus, Eq. (5) yields

$$\mathbf{v}_{B'} = M_{B'}^{-1} M_B \mathbf{v}_B. \tag{6}$$

Equation (6) shows that, given any two ordered bases B and B' of $\mathbb{R}^n$, there exists an invertible matrix C—namely,

$$C = M_{B'}^{-1} M_B, \tag{7}$$

such that, for all $\mathbf{v}$ in $\mathbb{R}^n$,

$$\mathbf{v}_{B'} = C\mathbf{v}_B. \tag{8}$$

If we know this matrix C, we can conveniently convert coordinates relative to B into coordinates relative to B'. The matrix C in Eq. (7) is the unique matrix satisfying Eq. (8). This can be seen by assuming that D is also a matrix satisfying Eq. (8), so that $\mathbf{v}_{B'} = D\mathbf{v}_B$; then $C\mathbf{v}_B = D\mathbf{v}_B$ for all vectors $\mathbf{v}_B$ in $\mathbb{R}^n$. Exercise 41 in Section 1.3 shows that we must have $C = D$.

The matrix C in Eq. (7) and Eq. (8) is computed in terms of the ordered bases B and B', and we will now change to the subscripted notation $C_{B,B'}$ to suggest this dependency. Thus Eq. (8) becomes

$$\mathbf{v}_{B'} = C_{B,B'}\mathbf{v}_B,$$

and the subscripts on C, read from left to right, indicate that we are changing from coordinates relative to B to coordinates relative to B'.

Because every n-dimensional vector space is isomorphic to $\mathbb{R}^n$, all our results here are valid for coordinates with respect to ordered bases B and B' in any finite-dimensional vector space. We phrase the definition that follows in these terms.

DEFINITION 7.1   Change-of-Coordinates Matrix

Let B and B' be ordered bases for a finite-dimensional vector space V. The **change-of-coordinates matrix** from B to B' is the unique matrix $C_{B,B'}$ such that $C_{B,B'}\mathbf{v}_B = \mathbf{v}_{B'}$ for all vectors $\mathbf{v}$ in V.

The term *transition matrix* is used in some texts in place of *change-of-coordinates matrix*. But we used *transition matrix* with reference to Markov chains in Section 1.7, so we avoid duplicate terminology here by using the more descriptive term *change-of-coordinates matrix*.

Equation (8), written in the form $C^{-1}\mathbf{v}_{B'} = \mathbf{v}_B$, shows that the inverse of the change-of-coordinates matrix from B to B' is the change-of-coordinates matrix from B to B'—that is, $C_{B',B} = C_{B,B'}^{-1}$.

Equation (7) shows us exactly what the change-of-coordinates matrix must be for any two ordered bases B and B' in $\mathbb{R}^n$. A direct way to compute this product $M_{B'}^{-1}M_B$, if $M_{B'}^{-1}$ is not already available, is to form the partitioned matrix

$$[M_{B'} \mid M_B]$$

and reduce it (by using elementary row operations) to the form $[I \mid C]$. We can regard this reduction as solving n linear systems, one for each column vector of M_B and all having the same coefficient matrix $M_{B'}$. From this perspective, the matrix $M_{B'}$ times the jth column vector of C (that is, the jth "solution vector") must yield the jth column vector of M_B. This shows that $M_{B'}C = M_B$. Consequently, we must have $C = M_{B'}^{-1}M_B = C_{B,B'}$. We now have a convenient procedure for finding a change-of-coordinates matrix.

Finding the Change-of-Coordinates Matrix from B to B' in $\mathbb{R}^n$

Let $B = (\mathbf{b}_1, \mathbf{b}_2, \ldots, \mathbf{b}_n)$ and $B' = (\mathbf{b}'_1, \mathbf{b}'_2, \ldots, \mathbf{b}'_n)$ be ordered bases of $\mathbb{R}^n$. The change-of-coordinates matrix from B to B' is the matrix $C_{B,B'}$ obtained by the row reduction

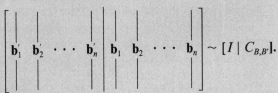

New basis matrix   Old basis matrix

EXAMPLE 1   Let $B = ([1, 1, 0], [2, 0, 1], [1, -1, 0])$ and let $E = (\mathbf{e}_1, \mathbf{e}_2, \mathbf{e}_3)$ be the standard ordered basis of $\mathbb{R}^3$. Find the change-of-coordinates matrix $C_{E,B}$ from E to B, and use it to find the coordinate vector $\mathbf{v}_B$ of

$$\mathbf{v} = \begin{bmatrix} 2 \\ 3 \\ 4 \end{bmatrix}$$

relative to B.

SOLUTION   Following the boxed procedure, we place the "new" basis matrix to the left and the "old" basis matrix to the right in an augmented matrix, and then we proceed with the row reduction:

$$\begin{bmatrix} 1 & 2 & 1 & 1 & 0 & 0 \\ 1 & 0 & -1 & 0 & 1 & 0 \\ 0 & 1 & 0 & 0 & 0 & 1 \end{bmatrix} \sim \begin{bmatrix} 1 & 2 & 1 & 1 & 0 & 0 \\ 0 & -2 & -2 & -1 & 1 & 0 \\ 0 & 1 & 0 & 0 & 0 & 1 \end{bmatrix}$$
$$\quad\ M_B \qquad\qquad M_E$$

$$\sim \begin{bmatrix} 1 & 0 & -1 \\ 0 & 1 & 1 \\ 0 & 0 & -1 \end{bmatrix} \begin{array}{|ccc} 0 & 1 & 0 \\ \frac{1}{2} & -\frac{1}{2} & 0 \\ -\frac{1}{2} & \frac{1}{2} & 1 \end{array} \sim \begin{bmatrix} 1 & 0 & 0 \\ 0 & 1 & 0 \\ 0 & 0 & 1 \end{bmatrix} \begin{array}{|ccc} \frac{1}{2} & \frac{1}{2} & -1 \\ 0 & 0 & 1 \\ \frac{1}{2} & -\frac{1}{2} & -1 \end{array}.$$

Thus the change-of-coordinates matrix from E to B is

$$C_{E,B} = \begin{bmatrix} \frac{1}{2} & \frac{1}{2} & -1 \\ 0 & 0 & 1 \\ \frac{1}{2} & -\frac{1}{2} & -1 \end{bmatrix}.$$

The coordinate vector of $v = \begin{bmatrix} 2 \\ 3 \\ 4 \end{bmatrix}$ relative to B is

$$v_B = C_{E,B}v = \begin{bmatrix} \frac{1}{2} & \frac{1}{2} & -1 \\ 0 & 0 & 1 \\ \frac{1}{2} & -\frac{1}{2} & -1 \end{bmatrix} \begin{bmatrix} 2 \\ 3 \\ 4 \end{bmatrix} = \begin{bmatrix} -\frac{3}{2} \\ 4 \\ -\frac{9}{2} \end{bmatrix}. \qquad \blacksquare$$

In the solution to Example 1, notice that $C_{E,B}$ is the inverse of the matrix appearing to the left of the partition in the original augmented matrix; that is, $C_{E,B} = M_B^{-1}$. Taking the inverse of both matrices, we see that $C_{B,E} = M_B$. Of course, this is true in general, because by our boxed procedure, we find $C_{B,E}$ by reducing the matrix $[M_E \mid M_B] = [I \mid M_B]$, and this augmented matrix is already in reduced row-echelon form.

In order to find a change-of-coordinates matrix $C_{B,B'}$ for bases B and B' in an n-dimensional vector space V other than $\mathbb{R}^n$, we choose a convenient ordered basis for V and coordinatize the vectors in V relative to that basis, making V look just like $\mathbb{R}^n$. We illustrate this procedure with the vector space P_2 of polynomials of degree at most 2, showing how the work can be carried out with coordinate vectors in $\mathbb{R}^3$.

EXAMPLE 2   Let $B = (x^2, x, 1)$ and $B' = (x^2 - x, 2x^2 - 2x + 1, x^2 - 2x)$ be ordered bases of the vector space P_2. Find the change-of-coordinates matrix from B to B', and use it to find the coordinate vector of $2x^2 + 3x - 1$ relative to B'.

SOLUTION   Let us use the ordered basis $B = (x^2, x, 1)$ to coordinatize polynomials in P_2. Identifying each polynomial $a_2x^2 + a_1x + a_0$ with its coordinate vector $[a_2, a_1, a_0]$ relative to the basis B, we obtain the following correspondence from P_2 to $\mathbb{R}^3$:

	Polynomial Basis in P_2	Coordinate Basis in $\mathbb{R}^3$
Old	$B = (x^2, x, 1)$	$([1, 0, 0], [0, 1, 0], [0, 0,1])$
New	$B' = (x^2 - x, 2x^2 - 2x + 1, x^2 - 2x)$	$([1, -1, 0], [2, -2, 1], [1, -2, 0])$.

Working with coordinate vectors in $\mathbb{R}^3$, we compute the desired change-of-coordinates matrix, as described in the box preceding Example 1.

$$\left[\begin{array}{rrr|rrr} 1 & 2 & 1 & 1 & 0 & 0 \\ -1 & -2 & -2 & 0 & 1 & 0 \\ 0 & 1 & 0 & 0 & 0 & 1 \end{array}\right] \sim \left[\begin{array}{rrr|rrr} 1 & 2 & 1 & 1 & 0 & 0 \\ 0 & 0 & -1 & 1 & 1 & 0 \\ 0 & 1 & 0 & 0 & 0 & 1 \end{array}\right]$$

$$\underbrace{\text{New basis}}\quad\underbrace{\text{Old basis}}$$

$$\sim \left[\begin{array}{rrr|rrr} 1 & 0 & 1 & 1 & 0 & -2 \\ 0 & 1 & 0 & 0 & 0 & 1 \\ 0 & 0 & -1 & 1 & 1 & 0 \end{array}\right] \sim \left[\begin{array}{rrr|rrr} 1 & 0 & 0 & 2 & 1 & -2 \\ 0 & 1 & 0 & 0 & 0 & 1 \\ 0 & 0 & 1 & -1 & -1 & 0 \end{array}\right].$$

The change-of-coordinates matrix is

$$C_{B,B'} = \left[\begin{array}{rrr} 2 & 1 & -2 \\ 0 & 0 & 1 \\ -1 & -1 & 0 \end{array}\right].$$

We compute the coordinate vector of $\mathbf{v} = 2x^2 + 3x - 1$ relative to B', using $C_{B,B'}$ and the coordinate vector of $\mathbf{v}$ relative to B, as follows:

$$\underbrace{\left[\begin{array}{rrr} 2 & 1 & -2 \\ 0 & 0 & 1 \\ -1 & -1 & 0 \end{array}\right]}_{C_{B,B'}} \underbrace{\left[\begin{array}{r} 2 \\ 3 \\ -1 \end{array}\right]}_{\mathbf{v}_B} = \underbrace{\left[\begin{array}{r} 9 \\ -1 \\ -5 \end{array}\right]}_{\mathbf{v}_{B'}}.$$

As a check, we have

$$2x^2 + 3x - 1 = 9(x^2 - x) - 1(2x^2 - 2x + 1) - 5(x^2 - 2x).$$

$$\underset{\mathbf{v}}{}\qquad\underset{\mathbf{b}_1'}{}\qquad\underset{\mathbf{b}_2'}{}\qquad\underset{\mathbf{b}_3'}{} \quad \blacksquare$$

There is another way besides $M_{B'}^{-1}M_B$ to express the change-of-coordinates matrix $C_{B,B'}$. Recall that the coordinate vector $(\mathbf{b}_j)_{B'}$ of $\mathbf{b}_j$ relative to B' is found by reducing the augmented matrix $[M_{B'} \mid \mathbf{b}_j]$. Thus all n coordinate vectors $(\mathbf{b}_j)_{B'}$ can be found at once by reducing the augmented matrix $[M_{B'} \mid M_B]$. But this is precisely what our boxed procedure for finding $C_{B,B'}$ calls for. We conclude that

$$C_{B,B'} = \left[\begin{array}{cccc} | & | & & | \\ (\mathbf{b}_1)_{B'} & (\mathbf{b}_2)_{B'} & \cdots & (\mathbf{b}_n)_{B'} \\ | & | & & | \end{array}\right]. \tag{9}$$

Because every n-dimensional vector space V is isomorphic to $\mathbb{R}^n$, Eq. (9) continues to hold for ordered bases B and B' in V. To see explicitly why this is true, we use B' to coordinatize V. Relative to B' the coordinate vectors of vectors in B' are the vectors in the standard ordered basis E of $\mathbb{R}^n$, whereas the coordinate vectors of basis vectors in B are the vectors in the basis $B* = ((\mathbf{b}_1)_{B'}, (\mathbf{b}_2)_{B'}, \ldots, (\mathbf{b}_n)_{B'})$ of $\mathbb{R}^n$. Now the change-of-coordinates matrix from $B*$ to E is precisely the basis matrix M_{B*} on the right-hand side of Eq. (9). Because

V and $\mathbb{R}^n$ are isomorphic, we see (as in Example 2) that this matrix is also the change-of-coordinates matrix from B to B'.

There are times when it is feasible to use Eq. (9) to find the change-of-coordinates matrix—just noticing how the vector $\mathbf{b}_j$ in B can be expressed as a linear combination of vectors in B' to determine the jth column vector of $C_{B,B'}$ rather than actually coordinatizing V and working within $\mathbb{R}^n$. The next example illustrates this.

EXAMPLE 3   Use Eq. (9) to find the change-of-coordinates matrix $C_{B,B'}$ from the basis $B = (x^2 - 1, x^2 + 1, x^2 + 2x + 1)$ to the basis $B' = (x^2, x, 1)$ in the vector space P_2 of polynomials of degree at most 2.

SOLUTION   Relative to $B' = (x^2, x, 1)$, we see at once that the coordinate vectors of $x^2 - 1$, $x^2 + 1$, and $x^2 + 2x + 1$ are $[1, 0, -1]$, $[1, 0, 1]$, and $[1, 2, 1]$, respectively. Changing these to column vectors, we obtain

$$C_{B,B'} = \begin{bmatrix} 1 & 1 & 1 \\ 0 & 0 & 2 \\ -1 & 1 & 1 \end{bmatrix}.$$ ∎

SUMMARY

Let V be a vector space with ordered basis $B = (\mathbf{b}_1, \mathbf{b}_2, \ldots, \mathbf{b}_n)$.

1. Each vector $\mathbf{v}$ in V has a unique expression as a linear combination

$$\mathbf{v} = r_1\mathbf{b}_1 + r_2\mathbf{b}_2 + \cdots + r_n\mathbf{b}_n.$$

The vector $\mathbf{v}_B = [r_1, r_2, \ldots, r_n]$ is the *coordinate vector* of $\mathbf{v}$ relative to B. Associating with each vector $\mathbf{v}$ its coordinate vector $\mathbf{v}_B$ *coordinatizes* V, so that V is *isomorphic* to $\mathbb{R}^n$.

2. Let B and B' be ordered bases of V. There exists a unique $n \times n$ matrix $C_{B,B'}$ such that $C_{B,B'}\mathbf{v}_B = \mathbf{v}_{B'}$ for all vectors $\mathbf{v}$ in V. This is the *change-of-coordinates* matrix from B to B'. It can be computed by coordinatizing V and then applying the boxed procedure on page 391. Alternatively, it can be computed using Eq. (9).

EXERCISES

Exercises 1–7 are a review of Section 3.3. In Exercises 1–6, find the coordinate vector of the given vector relative to the given ordered basis.

1. $\cos 2x$ in $\mathrm{sp}(\sin^2 x, \cos^2 x)$ relative to $(\sin^2 x, \cos^2 x)$

2. $x^3 + x^2 - 2x + 4$ in P_3 relative to $(1, x^2, x, x^3)$.

3. $x^3 + 3x^2 - 4x + 2$ in P_3 relative to $(x, x^2 - 1, x^3, 2x^2)$

4. $x + x^4$ in P_4 relative to $(1, 2x - 1, x^3 + x^4, 2x^3, x^2 + 2)$

5. $\begin{bmatrix} 1 & -2 \\ 3 & 4 \end{bmatrix}$ in M_2 relative to

$$\left(\begin{bmatrix} 0 & 1 \\ 1 & 0 \end{bmatrix}, \begin{bmatrix} 0 & -1 \\ 0 & 0 \end{bmatrix}, \begin{bmatrix} 1 & -1 \\ 0 & 3 \end{bmatrix}, \begin{bmatrix} 0 & 1 \\ 0 & 1 \end{bmatrix} \right)$$

6. $\sinh x$ in $sp(e^x, e^{-x})$ relative to (e^x, e^{-x})

7. Find the polynomial in P_2 whose coordinate vector relative to the ordered basis $B = (x + x^2, x - x^2, 1 + x)$ is [3, 1, 2].

8. Let B be an ordered basis for $\mathbb{R}^3$. If

$$C_{E,B} = \begin{bmatrix} 3 & 1 & 2 \\ 4 & 1 & 2 \\ -1 & 2 & 1 \end{bmatrix} \quad \text{and} \quad \mathbf{v} = \begin{bmatrix} 2 \\ 5 \\ -1 \end{bmatrix},$$

find the coordinate vector $\mathbf{v}_B$.

9. Let V be a vector space with ordered bases B and B'. If

$$C_{B,B'} = \begin{bmatrix} 1 & 2 & 0 \\ 0 & 1 & -2 \\ -1 & 0 & 1 \end{bmatrix} \quad \text{and}$$

$$\mathbf{v} = 3\mathbf{b}_1 - 2\mathbf{b}_2 + \mathbf{b}_3,$$

find the coordinate vector $\mathbf{v}_{B'}$.

In Exercises 10–14, find the change-of-coordinates matrix (a) from B to B', and (b) from B' to B. Verify that these matrices are inverses of each other.

10. $B = ([1, 1], [1, 0])$ and $B' = ([0, 1], [1, 1])$ in $\mathbb{R}^2$

11. $B = ([2, 3, 1], [1, 2, 0], [2, 0, 3])$ and $B' = ([1, 0, 0], [0, 1, 0], [0, 0, 1])$ in $\mathbb{R}^3$

12. $B = ([1, 0, 1], [1, 1, 0], [0, 1, 1])$ and $B' = ([0, 1, 1], [1, 1, 0], [1, 0, 1])$ in $\mathbb{R}^3$

13. $B = ([1, 1, 1, 1], [1, 1, 1, 0], [1, 1, 0, 0], [1, 0, 0, 0])$ and the standard ordered basis $B' = E$ for $\mathbb{R}^4$

14. $B = (\sinh x, \cosh x)$ and $B' = (e^x, e^{-x})$ in $sp(\sinh x, \cosh x)$

15. Find the change-of-coordinates matrix from B' to B for the bases $B = (x^2, x, 1)$ and $B' = (x^2 - x, 2x^2 - 2x + 1, x^2 - 2x)$ of P_2 in Example 2. Verify that this matrix is the inverse of the change-of-coordinates matrix from B to B' found in that example.

16. Proceeding as in Example 2, find the change-of-coordinates matrix from $B = (x^3, x^2, x, 1)$ to $B' = (x^3 - x^2, x^2 - x, x - 1, x^3 + 1)$ in the vector space P_3 of polynomials of degree at most 3.

17. Find the change-of-coordinates matrix from B' to B for the vector space and ordered bases given in Exercise 16. Show that this matrix is the inverse of the matrix found in Exercise 16.

In Exercises 18–21, use Eq. (9) in the text to find the change-of-coordinates matrix from B to B' for the given vector space V with ordered bases B and B'.

18. $V = P_3$, B is the basis in Exercise 3, $B' = (x_3, x_2, x, 1)$

19. $V = P_3$, $B = (x^3 + x^2 + 1, 2x^3 - x^2 + x + 1, x^3 - x + 1, x^2 + x + 1)$, $B' = (x^3, x^2, x, 1)$

20. V is the space M_2 of all 2×2 matrices, B is the basis in Exercise 5,

$$B' = \left(\begin{bmatrix} 1 & 0 \\ 0 & 0 \end{bmatrix}, \begin{bmatrix} 0 & 1 \\ 0 & 0 \end{bmatrix}, \begin{bmatrix} 0 & 0 \\ 1 & 0 \end{bmatrix}, \begin{bmatrix} 0 & 0 \\ 0 & 1 \end{bmatrix} \right)$$

21. V is the subspace $sp(\sin^2 x, \cos^2 x)$ of the vector space F of all real-valued functions with domain $\mathbb{R}$, $B = (\cos 2x, 1)$, $B' = (\sin^2 x, \cos^2 x)$

22. Find the change-of-coordinates matrix from B' to B for the vector space V and ordered bases B and B' of Exercise 21.

23. Let B and B' be ordered bases for $\mathbb{R}^n$. Mark each of the following True or False.

___ a. Every change-of-coordinates matrix is square.

___ b. Every $n \times n$ matrix is a change-of-coordinates matrix relative to some bases of $\mathbb{R}^n$.

___ c. If B and B' are orthonormal bases, then $C_{B,B'}$ is an orthogonal matrix.

___ d. If $C_{B,B'}$ is an orthogonal matrix, then B and B' are orthonormal bases.

___ e. If $C_{B,B'}$ is an orthogonal matrix and B is an orthonormal basis, then B' is an orthonormal basis.

___ f. For all choices of B and B', we have $\det(C_{B,B'}) = 1$.

____ g. For all choices of B and B', we have $\det(C_{B,B'}) \neq 0$.

____ h. $\det(C_{B,B'}) = 1$ if and only if $B = B'$.

____ i. $C_{B,B'} = I$ if and only if $B = B'$.

____ j. Every invertible $n \times n$ matrix is the change-of-coordinates matrix $C_{B,B'}$ for some ordered bases B and B' of $\mathbb{R}^n$.

24. For ordered bases B and B' in $\mathbb{R}^n$, explain how the change-of-coordinates matrix from B to B' is related to the change-of-coordinates matrices from B to E and from E to B'.

25. Let B, B', and B'' be ordered bases for $\mathbb{R}^n$. Find the change-of-coordinates matrix from B to B'' in terms of $C_{B,B'}$ and $C_{B',B''}$. [HINT: For a vector $\mathbf{v}$ in $\mathbb{R}^n$, what matrix times $\mathbf{v}_B$ gives $\mathbf{v}_{B'}$?]

7.2   MATRIX REPRESENTATIONS AND SIMILARITY

Let V and V' be vector spaces with ordered bases $B = (\mathbf{b}_1, \mathbf{b}_2, \ldots, \mathbf{b}_n)$ and $B' = (\mathbf{b}_1', \mathbf{b}_2', \ldots, \mathbf{b}_m')$, respectively. If $T: V \to V'$ is a linear transformation, then Theorem 3.10 shows that the *matrix representation of T* relative to B, B', which we now denote as $R_{B,B'}$ is given by

$$R_{B,B'} = \begin{bmatrix} | & | & & | \\ T(\mathbf{b}_1)_{B'} & T(\mathbf{b}_2)_{B'} & \cdots & T(\mathbf{b}_n)_{B'} \\ | & | & & | \end{bmatrix}, \tag{1}$$

where $T(\mathbf{b}_j)_{B'}$ is the coordinate vector of $T(\mathbf{b}_j)$ relative to B'. Furthermore, $R_{B,B'}$ is the *unique* matrix satisfying

$$T(\mathbf{v})_{B'} = R_{B,B'}\mathbf{v}_B \quad \text{for all } \mathbf{v} \text{ in } V. \tag{2}$$

Let us denote by $M_{T(B)}$ the $m \times n$ matrix whose jth column vector is $T(\mathbf{b}_j)$. From Eq. (1), we see that we can compute $R_{B,B'}$ by row-reducing the augmented matrix $[M_{B'} \mid M_{T(B)}]$ to obtain $[I \mid R_{B,B'}]$. The discussion surrounding Theorem 3.10 shows how computations involving $T: V \to V'$ can be carried out in $\mathbb{R}^n$ and $\mathbb{R}^m$ using $R_{B,B'}$ and coordinates of vectors relative to the basis B of V and B' of V', in view of the isomorphisms of V with $\mathbb{R}^n$ and of V' with $\mathbb{R}^m$ that these coordinatizations provide.

It is the purpose of this section to study the effect that choosing different bases for coordinatization has on the matrix representations of a linear transformation. For simplicity, we shall derive our results in terms of the vector spaces $\mathbb{R}^n$. They can then be carried over to other finite-dimensional vector spaces using coordinatization isomorphisms.

The Multiplicative Property of Matrix Representations

Let $T: \mathbb{R}^n \to \mathbb{R}^m$ and $T': \mathbb{R}^m \to \mathbb{R}^s$ be linear transformations. Section 2.3 showed that composition of linear transformations corresponds to multiplication of their standard matrix representations; that is, for standard bases, we have

$$\text{Matrix for } (T' \circ T) = (\text{Matrix for } T')(\text{Matrix for } T).$$

The analogous property holds for representations relative to any bases. Consider the linear transformations and ordered bases shown by the diagram

The following vector and matrix counterpart diagram shows the action of the transformations on vectors:

$$(3)$$

The matrix $R_{B,B'}$ under the first arrow transforms the vector on the left of the arrow into the vector on the right by left multiplication, as indicated by Eq. (2). The matrix under the second arrow acts in a similar way. Thus the matrix product $R_{B',B''}R_{B,B'}$ transforms $\mathbf{v}_B$ into $T'(T(\mathbf{v}))_{B''}$. However, the matrix representation $R_{B,B''}$ of $T'' \circ T$ relative to B,B'' is the *unique* matrix that carries $\mathbf{v}_B$ into $T'(T(\mathbf{v}))_{B''}$ for all $\mathbf{v}_B$ in $\mathbb{R}^n$. Thus we must have

$$R_{B,B''} = R_{B',B''}R_{B,B'}. \tag{4}$$

Notice that the order of the matrices in this product is *opposite* to the order in which they appear in diagram (3). (See the footnote on page 160 to see why this is so.)

Equation (4) holds equally well for matrix representations of linear transformations of general finite-dimensional vector spaces, as shown in the diagram

EXAMPLE 1   Let $B = (x^4, x^3, x^2, x, 1)$, which is an ordered basis for P_4, and let $T: P_4 \rightarrow P_4$ be the differentiation transformation. Find the matrix representation $R_{B,B}$ of T relative to B,B, and illustrate how it can be used to differentiate the polynomial function

$$3x^4 - 5x^3 + 7x^2 - 8x + 2.$$

Then use calculus to show that $(R_{B,B})^5 = O$.

SOLUTION   Because $T(x^k) = kx^{k-1}$, we see that the matrix representation $R_{B,B}$ in Eq. (1) is

$$R_{B,B} = \begin{bmatrix} 0 & 0 & 0 & 0 & 0 \\ 4 & 0 & 0 & 0 & 0 \\ 0 & 3 & 0 & 0 & 0 \\ 0 & 0 & 2 & 0 & 0 \\ 0 & 0 & 0 & 1 & 0 \end{bmatrix}.$$

Now the coordinate vector of $p(x) = 3x^4 - 5x^3 + 7x^2 - 8x + 2$ relative to B is

$$p(x)_B = \begin{bmatrix} 3 \\ -5 \\ 7 \\ -8 \\ 2 \end{bmatrix}.$$

Applying Eq. (2) with $B' = B$, we find that

$$T(p(x))_B = R_{B,B}p(x)_B = \begin{bmatrix} 0 & 0 & 0 & 0 & 0 \\ 4 & 0 & 0 & 0 & 0 \\ 0 & 3 & 0 & 0 & 0 \\ 0 & 0 & 2 & 0 & 0 \\ 0 & 0 & 0 & 1 & 0 \end{bmatrix} \begin{bmatrix} 3 \\ -5 \\ 7 \\ -8 \\ 2 \end{bmatrix} = \begin{bmatrix} 0 \\ 12 \\ -15 \\ 14 \\ -8 \end{bmatrix}.$$

Thus, $p'(x) = T(p(x)) = 12x^3 - 15x^2 + 14x - 8$.

The discussion surrounding Eq. (4) shows that the linear transformation of P_4 that computes the fifth derivative has matrix representation $(R_{B,B})^5$. Because the fifth derivative of any polynomial in P_4 is zero, we see that $(R_{B,B})^5 = O$. ∎

Similarity of Representations Relative to Different Bases

Let B and B' be ordered bases for $\mathbb{R}^n$. In the preceding section we saw that the change-of-coordinates matrix $C_{B,B'}$ from B to B' is the unique matrix satisfying

$$\mathbf{v}_{B'} = C_{B,B'}\mathbf{v}_B \quad \text{for all } \mathbf{v} \text{ in } \mathbb{R}^n.$$

Comparison of this equation with Eq. (2) shows that $C_{B,B'}$ is the matrix representation, relative to B,B', of the identity transformation $\iota\colon \mathbb{R}^n \to \mathbb{R}^n$, where $\iota(\mathbf{v}) = \mathbf{v}$ for all $\mathbf{v} \in \mathbb{R}^n$.

The relationship between the matrix representations R_B and $R_{B'}$ for the linear transformation $T\colon \mathbb{R}^n \to \mathbb{R}^n$ can be derived from the following diagram:

$$
\begin{array}{ccccccc}
 & & & T & & & \\
 & & \overbrace{\qquad\qquad\qquad\qquad} & & & & \\
\mathbb{R}^n & \xrightarrow{\ \iota\ } & \mathbb{R}^n & \xrightarrow{\ T\ } & \mathbb{R}^n & \xrightarrow{\ \iota\ } & \mathbb{R}^n. \\
B' & & B & & B & & B'
\end{array}
$$

The vector and matrix counterpart diagram similar to diagram (3) is

$$R_{B'}$$

$$\mathbf{v}_{B'} \xrightarrow{\quad} \mathbf{v}_B \xrightarrow{\quad} T(\mathbf{v})_B \xrightarrow{\quad} T(\mathbf{v})_{B'}.$$
$$\qquad C_{B',B} \qquad\qquad R_B \qquad\qquad C_{B,B'}$$

Remembering to reverse the order, we find that

$$R_{B'} = C_{B,B'} R_B C_{B',B}. \tag{5}$$

Reading the matrix product in Eq. (5) from right to left, we see that in order to compute $T(\mathbf{v})_{B'}$ from $\mathbf{v}_{B'}$ if we know R_B, we:

1. change from B' to B coordinates,
2. compute the transformation relative to B coordinates,
3. change back to B' coordinates.

Equation (5) makes this procedure easy to remember.

The matrices $C_{B,B'}$ and $C_{B',B}$ are inverses of each other. If we drop subscripts for a moment and let $C = C_{B',B}$, then Eq. (5) becomes

$$R_{B'} = C^{-1} R_B C. \tag{6}$$

Recall that two $n \times n$ matrices A and R are *similar* if there exists an invertible $n \times n$ matrix C such that $R = C^{-1}AC$. (See Definition 5.4.) We have shown that matrix representations of the same transformation relative to different bases are similar. We state this as a theorem in the context of a general finite-dimensional vector space.

THEOREM 7.1 Similarity of Matrix Representations of *T*

Let T be a linear transformation of a finite-dimensional vector space V into itself, and let B and B' be ordered bases of V. Let R_B and $R_{B'}$ be the matrix representations of T relative to B and B', respectively. Then

$$R_{B'} = C^{-1} R_B C,$$

where $C = C_{B',B}$ is the change-of-coordinates matrix from B' to B. Consequently, $R_{B'}$ and R_B are similar matrices.

EXAMPLE 2   Consider the linear transformation $T: \mathbb{R}^3 \to \mathbb{R}^3$ defined by

$$T(x_1, x_2, x_3) = (x_1 + x_2 + x_3, x_1 + x_2, x_3).$$

Find the standard matrix representation A of T, and find the matrix representation R_B of T relative to B, where

$$B = ([1, 1, 0], [1, 0, 1], [0, 1, 1]).$$

In addition, find an invertible matrix C such that $R_B = C^{-1}AC$.

SOLUTION Here the standard ordered basis E plays the role that the basis B played in Theorem 7.1, and the basis B here plays the role played by B' in that theorem. Of course, the standard matrix representation of T is

$$A = \begin{bmatrix} 1 & 1 & 1 \\ 1 & 1 & 0 \\ 0 & 0 & 1 \end{bmatrix}.$$

We compute the matrix representation R_B by reducing $[M_B \mid M_{T(B)}]$ to $[I \mid R_B]$ as follows:

$$\begin{bmatrix} 1 & 1 & 0 & \vline & 2 & 2 & 2 \\ 1 & 0 & 1 & \vline & 2 & 1 & 1 \\ 0 & 1 & 1 & \vline & 0 & 1 & 1 \end{bmatrix} \sim \begin{bmatrix} 1 & 1 & 0 & \vline & 2 & 2 & 2 \\ 0 & -1 & 1 & \vline & 0 & -1 & -1 \\ 0 & 1 & 1 & \vline & 0 & 1 & 1 \end{bmatrix}$$

$$\quad \mathbf{b_1} \ \ \mathbf{b_2} \ \ \mathbf{b_3} \quad T(\mathbf{b_1}) \ \ T(\mathbf{b_2}) \ \ T(\mathbf{b_3})$$

$$\sim \begin{bmatrix} 1 & 0 & 1 & \vline & 2 & 1 & 1 \\ 0 & 1 & -1 & \vline & 0 & 1 & 1 \\ 0 & 0 & 2 & \vline & 0 & 0 & 0 \end{bmatrix} \sim \begin{bmatrix} 1 & 0 & 0 & \vline & 2 & 1 & 1 \\ 0 & 1 & 0 & \vline & 0 & 1 & 1 \\ 0 & 0 & 1 & \vline & 0 & 0 & 0 \end{bmatrix}.$$

$$\qquad\qquad\qquad\qquad\qquad\qquad I \qquad\qquad R_B$$

Thus,

$$R_B = \begin{bmatrix} 2 & 1 & 1 \\ 0 & 1 & 1 \\ 0 & 0 & 0 \end{bmatrix}.$$

An invertible matrix C such that $R_B = C^{-1}AC$ is $C = C_{B,E}$. We find $C = C_{B,E}$ by reducing the augmented matrix $[M_E \mid M_B] = [I \mid M_B]$. Because this matrix is already reduced, we see that

$$C = M_B = \begin{bmatrix} 1 & 1 & 0 \\ 1 & 0 & 1 \\ 0 & 1 & 1 \end{bmatrix}$$

in this case. The matrices A and R_B are similar matrices, both representing the given linear transformation. As a check, we could compute that $AC = CR_B$. ∎

We give an example illustrating Theorem 7.1 in the case of a finite-dimensional vector space other than $\mathbb{R}^n$.

EXAMPLE 3 For the space P_2 of polynomials of degree at most 2, let $T: P_2 \to P_2$ be defined by $T(p(x)) = p(x - 1)$. Consider the two ordered bases $B = (x^2, x, 1)$ and $B' = (x, x + 1, x^2 - 1)$. Find the matrix representations R_B and $R_{B'}$ of T and a matrix C such that $R_{B'} = C^{-1}R_B C$.

SOLUTION Because

$$T(x^2) = (x - 1)^2 = x^2 - 2x + 1,$$
$$T(x) = x - 1,$$
$$T(1) = 1,$$

the matrix representation of T relative to $B = (x^2, x, 1)$ is

$$R_B = \begin{bmatrix} 1 & 0 & 0 \\ -2 & 1 & 0 \\ 1 & -1 & 1 \end{bmatrix}.$$

Next we compute the change-of-coordinates matrices $C_{B,B'}$ and $C_{B',B}$. We see that

$$C = C_{B',B} = \begin{bmatrix} 0 & 0 & 1 \\ 1 & 1 & 0 \\ 0 & 1 & -1 \end{bmatrix}.$$

Moreover

$$\begin{aligned} x^2 &= -1(x) + 1(x + 1) + 1(x^2 - 1), \\ x &= 1(x) + 0(x + 1) + 0(x^2 - 1), \\ 1 &= -1(x) + 1(x + 1) + 0(x^2 - 1), \end{aligned}$$

so

$$C_{B,B'} = \begin{bmatrix} -1 & 1 & -1 \\ 1 & 0 & 1 \\ 1 & 0 & 0 \end{bmatrix}.$$

Notice that $C_{B,B'}$ can be computed as $(C_{B',B})^{-1}$. We now have

$$R_{B'} = \underbrace{\begin{bmatrix} -1 & 1 & -1 \\ 1 & 0 & 1 \\ 1 & 0 & 0 \end{bmatrix}}_{C_{B,B'}} \underbrace{\begin{bmatrix} 1 & 0 & 0 \\ -2 & 1 & 0 \\ 1 & -1 & 1 \end{bmatrix}}_{R_B} \underbrace{\begin{bmatrix} 0 & 0 & 1 \\ 1 & 1 & 0 \\ 0 & 1 & -1 \end{bmatrix}}_{C_{B',B}}$$

$$= \begin{bmatrix} -1 & 1 & -1 \\ 1 & 0 & 1 \\ 1 & 0 & 0 \end{bmatrix} \begin{bmatrix} 0 & 0 & 1 \\ 1 & 1 & -2 \\ -1 & 0 & 0 \end{bmatrix} = \begin{bmatrix} 2 & 1 & -3 \\ -1 & 0 & 1 \\ 0 & 0 & 1 \end{bmatrix}.$$

Alternatively, $R_{B'}$ can be computed directly as

$$R_{B'} = \begin{bmatrix} \Big| & \Big| & \Big| \\ T(\mathbf{b}_1')_{B'} & T(\mathbf{b}_2')_{B'} & T(\mathbf{b}_3')_{B'} \\ \Big| & \Big| & \Big| \end{bmatrix}. \qquad \blacksquare$$

We have seen that matrix representations of the same transformation relative to different bases are similar. Conversely, any two similar matrices can be viewed as representations of the same transformation relative to different bases. To see this, let A be an $n \times n$ matrix, and let C be any invertible $n \times n$ matrix. Let $T: \mathbb{R}^n \to \mathbb{R}^n$ be defined by $T(\mathbf{x}) = A\mathbf{x}$ so that A is the standard matrix representation of T. Because C is invertible, its column vectors are independent and form a basis for $\mathbb{R}^n$. Let B be the ordered basis having as jth vector the *j*th column vector of C. Then C is precisely the change-of-coordinates matrix from B to the standard ordered basis E. That is, $C = C_{B,E}$. Consequently, $C^{-1}AC = C_{E,B}AC_{B,E}$ is the matrix representation of T relative to B.

Significance of the Similarity Relationship for Matrices

Two $n \times n$ matrices are similar if and only if they are matrix representations of the same linear transformation T relative to suitable ordered bases.

The Interplay of Matrices and Linear Transformations

Let V be an n-dimensional vector space, and let $T: V \to V$ be a linear transformation. Suppose that T has a property that can be characterized in terms of the mapping, without reference to coordinates relative to any basis. Assertions about T of the form

There is a basis of eigenvectors of T,

The nullspace of T has dimension 4,

The eigenspace E_5 of T has dimension 2,

are *coordinate-independent* assertions. For another example, T has λ as an eigenvalue if and only if

$$T(\mathbf{v}) = \lambda\mathbf{v} \quad \text{for some nonzero vector } \mathbf{v} \text{ in } V. \tag{7}$$

This statement makes no reference to coordinates relative to a basis. The existence and value of an eigenvalue λ is coordinate independent. Of course, the existence and the characterization of eigenvectors $\mathbf{v}$ corresponding to λ as given in Eq. (7) are also coordinate independent. However, an eigenvector $\mathbf{v}$ can have different coordinates relative to different bases B and B' of V, so the matrix representations R_B and $R_{B'}$ of V may have different eigenvectors. Equation (7) expressed in terms of the coordinatizations of V by B and B' becomes

$$R_B(\mathbf{v}_B) = \lambda\mathbf{v}_B \quad \text{and} \quad R_{B'}(\mathbf{v}_{B'}) = \lambda\mathbf{v}_{B'},$$

respectively. While the coordinates of the eigenvector $\mathbf{v}$ change, the value λ doesn't change. That is, if λ is an eigenvalue of T, then λ is an eigenvalue of *every* matrix representation of T.

Now any two similar matrices can be viewed as matrix representations of the same linear transformation T, relative to suitable bases. Consequently, similar matrices must share any properties that can be described for the transformation in a coordinate-free fashion. In particular we obtain, with no additional work, this nice result:

Similar matrices have the same eigenvalues.

We take a moment to expand on the ideas we have just introduced. Let V be a vector space of finite dimension n. The study of linear transformations $T: V \to V$ is essentially the same as the study of products $A\mathbf{x}$ of vectors $\mathbf{x}$ in $\mathbb{R}^n$ by $n \times n$ matrices A. We should understand this relationship thoroughly and be able to bounce back and forth at will from $T(\mathbf{v})$ for $\mathbf{v}$ in V to $A\mathbf{x}$ for $\mathbf{x}$ in $\mathbb{R}^n$. Sometimes a theorem that is not immediately obvious from one point of view is quite easy from another. For example, from the matrix point of view, it is not immediately apparent that similar matrices have the same eigenvalues; we ask for a matrix proof in Exercise 27. However, the truth of this statement becomes obvious from the linear transformation point of view. On the other hand, it is not obvious from Eq. (7) concerning an eigenvalue of a linear transformation $T: \mathbb{R}^n \to \mathbb{R}^n$ that a linear transformation of V can have at most n eigenvalues. However, this is easy to establish from the matrix point of view: $A\mathbf{x} = \lambda\mathbf{x}$ has a nontrivial solution if and only if the coefficient matrix $A - \lambda I$ of the system $(A - \lambda I)\mathbf{x} = \mathbf{0}$ has determinant zero, and $\det(A - \lambda I) = 0$ is a polynomial equation of degree n.

We now state a theorem relating eigenvalues and eigenvectors of similar matrices. Exercises 24 and 25 ask for proofs of the second and third statements in the theorem. We have already proved the first statement.

THEOREM 7.2   Eigenvalues and Eigenvectors of Similar Matrices

Let A and R be similar $n \times n$ matrices, so that $R = C^{-1}AC$ for some invertible $n \times n$ matrix C. Let the eigenvalues of A be the (not necessarily distinct) numbers $\lambda_1, \lambda_2, \ldots, \lambda_n$.

1. The eigenvalues of R are also $\lambda_1, \lambda_2, \ldots, \lambda_n$.
2. The algebraic and geometric multiplicity of each λ_i as an eigenvalue of A remains the same as when it is viewed as an eigenvalue of R.
3. If $\mathbf{v}_i$ in $\mathbb{R}^n$ is an eigenvector of the matrix A corresponding to λ_i, then $C^{-1}\mathbf{v}_i$ is an eigenvector of the matrix R corresponding to λ_i.

We give one more illustration of the usefulness of this interplay between linear transformations and matrices, and of the significance of similarity. Let V be a finite-dimensional vector space, and let $T: V \to V$ be a linear transformation. It was easy to define the notions of *eigenvalue* and *eigenvector* in terms of T. We can now define the **geometric multiplicity** of an eigenvalue λ to be the dimension of the eigenspace $E_\lambda = \{\mathbf{v} \in V \mid T(\mathbf{v}) = \lambda\mathbf{v}\}$. However, at this time there is no obvious way to characterize the algebraic multiplicity of λ exclusively in terms of the mapping T, without coordinatization. Consequently, we define the **algebraic multiplicity** of λ to be its algebraic multiplicity as an eigenvalue of a matrix representation of λ. This makes sense because this algebraic multiplicity of λ is the same for *all* matrix representations of T. Statement 2 of Theorem 7.2 assures us that this is the case.

Diagonalization

Let $T: V \rightarrow V$ be a linear transformation of an n-dimensional vector space into itself. Suppose that there exists an ordered basis $B = (\mathbf{b}_1, \mathbf{b}_2, \ldots, \mathbf{b}_n)$ of V composed of eigenvectors of T. Let the eigenvalue corresponding to $\mathbf{b}_i$ be λ_i. Then the matrix representation of T relative to B has the simple diagonal form

$$D = \begin{bmatrix} \lambda_1 & & & \\ & \lambda_2 & & \mathbf{0} \\ & & \ddots & \\ \mathbf{0} & & & \lambda_n \end{bmatrix}.$$

We give a definition of diagonalization for linear transformations that clearly parallels one for matrices.

DEFINITION 7.2 Diagonalizable Transformation

> A linear transformation T of a finite-dimensional vector space V into itself is **diagonalizable** if V has an ordered basis consisting of eigenvectors of T.

EXAMPLE 4   Show that reflection of the plane $\mathbb{R}^2$ in the line $y = mx$ is a diagonalizable transformation and find a diagonal matrix representation for it.

SOLUTION   Reflection of the plane $\mathbb{R}^2$ in a line through the origin is a linear transformation. (See Illustration 2 on page 297.) Figure 7.1 shows that $\mathbf{b}_1 = [1, m]$ is carried into itself and $\mathbf{b}_2 = [-m, 1]$ is carried into $-\mathbf{b}_2$. Thus $\mathbf{b}_1$ is an eigenvector of this transformation with corresponding eigenvalue 1 whereas $\mathbf{b}_2$

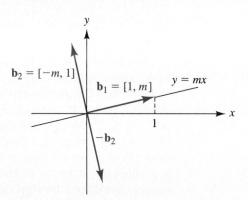

FIGURE 7.1
Reflection in the line $y = mx$.

is an eigenvector with eigenvalue -1. Because $\{\mathbf{b}_1, \mathbf{b}_2\}$ is a basis of eigenvectors for $\mathbb{R}^2$, the reflection transformation is diagonalizable. Relative to the ordered basis $B = (\mathbf{b}_1, \mathbf{b}_2)$,

$$R_B = \begin{bmatrix} 1 & 0 \\ 0 & -1 \end{bmatrix}.$$ ∎

If V has a known ordered basis B of eigenvectors, it becomes easy to compute the k-fold composition $T^k(\mathbf{v})$ for any positive integer k and for any vector $\mathbf{v}$ in V. We need only find the coordinate vector $\mathbf{d}$ in $\mathbb{R}^n$ of $\mathbf{v}$ relative to B, so that

$$\mathbf{v} = d_1\mathbf{b}_1 + d_2\mathbf{b}_2 + \cdots + d_n\mathbf{b}_n.$$

Then

$$T^k(\mathbf{v}) = d_1\lambda_1{}^k\mathbf{b}_1 + d_2\lambda_2{}^k\mathbf{b}_2 + \cdots + d_n\lambda_n{}^k\mathbf{b}_n. \tag{8}$$

Of course, this is the transformation analogue of the computation of $A^k\mathbf{x}$ discussed in Section 5.1. We illustrate with an example.

EXAMPLE 5   Consider the vector space P_2 of polynomials of degree at most 2, and let B' be the ordered basis $(1, x, x^2)$ for P_2. Let $T: P_2 \to P_2$ be the linear transformation such that

$$T(1) = 3 + 2x + x^2, \qquad T(x) = 2, \qquad T(x^2) = 2x^2.$$

Find $T^4(x + 2)$.

SOLUTION   The matrix representation of T relative to B' is

$$R_{B'} = \begin{bmatrix} 3 & 2 & 0 \\ 2 & 0 & 0 \\ 1 & 0 & 2 \end{bmatrix}.$$

Using the methods of Chapter 5, we easily find the eigenvalues and eigenvectors of $R_{B'}$ and of T given in Table 7.1.

TABLE 7.1

Eigenvalues	Eigenvectors of $R_{B'}$	Eigenvectors of T
$\lambda_1 = -1$	$\mathbf{w}_1 = [-3, 6, 1]$	$p_1(x) = -3 + 6x + x^2$
$\lambda_2 = 2$	$\mathbf{w}_2 = [0, 0, 1]$	$p_2(x) = x^2$
$\lambda_3 = 4$	$\mathbf{w}_3 = [2, 1, 1]$	$p_3(x) = 2 + x + x^2$

Let B be the ordered basis $(-3 + 6x + x^2, x^2, 2 + x + x^2)$ consisting of these eigenvectors. We can find the coordinate vector $\mathbf{d}$ of $x + 2$ relative to the basis B by inspection. Because

$$x + 2 = 0(x^2 + 6x - 3) + (-1)x^2 + 1(x^2 + x + 2),$$

we see that

$$d_1 = 0, \quad d_2 = -1, \quad \text{and} \quad d_3 = 1.$$

Thus, Eq. (8) has the form

$$T^k(x + 2) = 2^k(-1)x^2 + 4^k(1)(x^2 + x + 2).$$

In particular,

$$T^4(x + 2) = -16x^2 + 256(x^2 + x + 2) = 240x^2 + 256x + 512. \qquad \blacksquare$$

SUMMARY

Let $T: V \to V$ be a linear transformation of a finite-dimensional vector space into itself.

1. If B and B' are ordered bases of V, then the matrix representations R_B and $R_{B'}$ of T relative to B and to B' are similar. That is, there is an invertible matrix C—namely, $C = C_{B',B}$—such that

   $$R_{B'} = C^{-1}R_B C.$$

2. Conversely, two similar $n \times n$ matrices represent the same linear transformation of $\mathbb{R}^n$ into $\mathbb{R}^n$ relative to two suitably chosen ordered bases.

3. Similar matrices have the same eigenvalues with the same algebraic and geometric multiplicities.

4. If A and R are similar matrices with $R = C^{-1}AC$, and if $\mathbf{v}$ is an eigenvector of A, then $C^{-1}\mathbf{v}$ is an eigenvector of R corresponding to the same eigenvalue.

5. The transformation T is diagonalizable if V has a basis B consisting of eigenvectors of T. In this case, the matrix representation R_B is a diagonal matrix and computation of $T^k(\mathbf{v})$ by $R_B{}^k(\mathbf{v}_B)$ becomes relatively easy.

EXERCISES

In Exercises 1–14, find the matrix representations R_B and $R_{B'}$ and an invertible matrix C such that $R_{B'} = C^{-1}R_B C$ for the linear transformation T of the given vector space with the indicated ordered bases B and B'.

1. $T: \mathbb{R}^2 \to \mathbb{R}^2$ defined by $T([x, y]) = [x - y, x + 2y]$; $B = ([1, 1], [2, 1])$, $B' = E$

2. $T: \mathbb{R}^2 \to \mathbb{R}^2$ defined by $T([x, y]) = [2x + 3y, x + 2y]$; $B = ([1, -1], [1, 1])$, $B' = ([2, 3], [1, 2])$

3. $T: \mathbb{R}^3 \to \mathbb{R}^3$ defined by $T([x, y, z]) = [x + y, x + z, y - z]$; $B = ([1, 1, 1], [1, 1, 0], [1, 0, 0])$, $B' = E$

4. $T: \mathbb{R}^3 \to \mathbb{R}^3$ defined by $T([x, y, z]) = [5x, 2y, 3z]$; B and B' as in Exercise 3

5. $T: \mathbb{R}^3 \to \mathbb{R}^3$ defined by $T([x, y, z]) = [z, 0, x]$; $B = ([3, 1, 2], [1, 2, 1], [2, -1, 0])$, $B' = ([1, 2, 1], [2, 1, -1], [5, 4, 1])$

6. $T: \mathbb{R}^2 \to \mathbb{R}^2$ defined as reflection of the plane through the line $5x = 3y$; $B = ([3, 5], [5, -3])$, $B' = E$

7. $T: \mathbb{R}^3 \rightarrow \mathbb{R}^3$ defined as reflection of $\mathbb{R}^3$ through the plane $x + y + z = 0$; $B = ([1, 0, -1], [1, -1, 0], [1, 1, 1])$, $B' = E$

8. $T: \mathbb{R}^3 \rightarrow \mathbb{R}^3$ defined as reflection of $\mathbb{R}^3$ through the plane $2x + 3y + z = 0$; $B = ([2, 3, 1], [0, 1, -3], [1, 0, -2])$, $B' = E$

9. $T: \mathbb{R}^3 \rightarrow \mathbb{R}^3$ defined as projection on the plane $x_2 = 0$; $B = E$, $B' = ([1, 0, 1], [1, 0, -1], [0, 1, 0])$

10. $T: \mathbb{R}^3 \rightarrow \mathbb{R}^3$ defined as projection on the plane $x + y + z = 0$; B and B' as in Exercise 7.

11. $T: P_2 \rightarrow P_2$ defined by $T(p(x)) = p(x + 1) + p(x)$; $B = (x^2, x, 1)$, $B' = (1, x, x^2)$

12. $T: P_2 \rightarrow P_2$ as in Exercise 11, but using $B = (x^2, x, 1)$ and $B' = (x^2 + 1, x + 1, 2)$

13. $T: P_3 \rightarrow P_3$ defined by $T(p(x)) = p'(x)$, the derivative of $p(x)$; $B = (x^3, x^2, x, 1)$, $B' = (1, x + 1, x^2 + 1, x^3 + 1)$

14. $T: W \rightarrow W$, where $W = \text{sp}(e^x, xe^x)$ and T is the derivative transformation; $B = (e^x, xe^x)$, $B' = (2xe^x, 3e^x)$

15. Let $T: P_2 \rightarrow P_2$ be the linear transformation and B' the ordered basis of P_2 given in Example 3. Find the matrix representation $R_{B'}$ of T by computing the matrix with column vectors $T(\mathbf{b}_1')_{B'}$, $T(\mathbf{b}_2')_{B'}$, $T(\mathbf{b}_3')_{B'}$.

16. Repeat Exercise 15 for the transformation in Example 5 and the basis $B' = (x + 1, -1, x^2 + x)$ of P_2.

In Exercises 17–22, find the eigenvalues λ_i and the corresponding eigenspaces of the linear transformation T. Determine whether the linear transformation is diagonalizable.

17. T defined on $\mathbb{R}^2$ by $T([x, y]) = [2x - 3y, -3x + 2y]$

18. T defined on $\mathbb{R}^2$ by $T([x, y]) = [x - y, -x + y]$

19. T defined on $\mathbb{R}^3$ by $T([x_1, x_2, x_3]) = [x_1 + x_3, x_2, x_1 + x_3]$

20. T defined on $\mathbb{R}^3$ by $T([x_1, x_2, x_3]) = [x_1, 4x_2 + 7x_3, 2x_2 - x_3]$

21. T defined on $\mathbb{R}^3$ by $T([x_1, x_2, x_3]) = [5x_1, -5x_1 + 3x_2 - 5x_3, -3x_1 - 2x_2]$

22. T defined on $\mathbb{R}^3$ by $T([x_1, x_2, x_3]) = [3x_1 - x_2 + x_3, -2x_1 + 2x_2 - x_3, 2x_1 + x_2 + 4x_3]$

23. Mark each of the following True or False

___ a. Two similar $n \times n$ matrices represent the same linear transformation of $\mathbb{R}^n$ into itself relative to the standard basis.

___ b. Two different $n \times n$ matrices represent different linear transformations of $\mathbb{R}^n$ into itself relative to the standard basis.

___ c. Two similar $n \times n$ matrices represent the same linear transformation of $\mathbb{R}^n$ into itself relative to two suitably chosen bases for $\mathbb{R}^n$.

___ d. Similar matrices have the same eigenvalues and eigenvectors.

___ e. Similar matrices have the same eigenvalues with the same algebraic and geometric multiplicities.

___ f. If A and C are $n \times n$ matrices and C is invertible and $\mathbf{v}$ is an eigenvector of A, then $C^{-1}\mathbf{v}$ is an eigenvector of $C^{-1}AC$.

___ g. If A and C are $n \times n$ matrices and C is invertible and $\mathbf{v}$ is an eigenvector of A, then $C\mathbf{v}$ is an eigenvector of CAC^{-1}.

___ h. Any two $n \times n$ diagonal matrices are similar.

___ i. Any two $n \times n$ diagonalizable matrices having the same eigenvectors are similar.

___ j. Any two $n \times n$ diagonalizable matrices having the same eigenvalues of the same algebraic multiplicities are similar.

24. Prove statement 2 of Theorem 7.2.

25. Prove statement 3 of Theorem 7.2.

26. Let A and R be similar matrices. Prove in two ways that A^2 and R^2 are similar matrices: using a matrix argument, and using a linear transformation argument.

27. Give a determinant proof that similar matrices have the same eigenvalues.

EIGENVALUES: FURTHER APPLICATIONS AND COMPUTATIONS

This chapter deals with further applications of eigenvalues and with the computation of eigenvalues. In Section 8.1, we discuss quadratic forms and their diagonalization. The principal axis theorem (Theorem 8.1) asserts that every quadratic form can be diagonalized. This is probably the most important result in the chapter, having applications to the vibration of elastic bodies, to quantum mechanics, and to electric circuits. Presentations of such applications are beyond the scope of this text, and we have chosen to present a more accessible application in Section 8.2: classification of conic sections and quadric surfaces. Although the facts about conic sections may be familiar to you, their easy derivation from the principal axis theorem should be seen and enjoyed.

Section 8.3 applies the principal axis theorem to local extrema of functions and discusses maximization and minimization of quadratic forms on unit spheres. The latter topic is again important in vibration problems; it indicates that eigenvalues of maximum and of minimum magnitudes can be found for symmetric matrices by using techniques from advanced calculus for finding extrema of quadratic forms on unit spheres.

In Section 8.4, we sketch three methods for computing eigenvalues: the power method, Jacobi's method, and the QR method. We attempt to make as intuitively clear as we can how and why each method works, but proofs and discussions of efficiency are omitted.

This chapter contains some applications to geometry and analysis that are usually phrased in terms of points in $\mathbb{R}^n$ rather than in terms of vectors. We will be studying these applications using our work with vectors. To permit a natural and convenient use of the classical terminology of points while working with vectors, we relax for this chapter our convention that the boldface $\mathbf{x}$ always be a vector $[x_1, x_2, \ldots, x_n]$, and allow the same notation to represent the point $(x_1, x_2, \ldots, x_n)$ as well.

DIAGONALIZATION OF QUADRATIC FORMS

Quadratic Forms

A quadratic form in one variable x is a polynomial $f(x) = ax^2$, where $a \neq 0$. A quadratic form in two variables x and y is a polynomial $f(x, y) = ax^2 + bxy + cy^2$, where at least one of a, b, or c is nonzero. The term *quadratic* means *degree* 2. The term *form* means *homogeneous*; that is, each summand contains a product of the *same number* of variables—namely, 2 for a quadratic form. Thus, $3x^2 - 4xy$ is a quadratic form in x and y, but $x^2 + y^2 - 4$ and $x - 3y^2$ are not quadratic forms.

Turning to the general case, a **quadratic form** $f(\mathbf{x}) = f(x_1, x_2, \ldots, x_n)$ in n variables is a polynomial that can be written, using summation notation, as

$$f(\mathbf{x}) = \sum_{\substack{i \leq j \\ i,j=1}}^{n} u_{ij} x_i x_j, \tag{1}$$

where not all u_{ij} are zero. To illustrate, the general quadratic form in x_1, x_2, and x_3 is

$$u_{11} x_1^2 + u_{12} x_1 x_2 + u_{13} x_1 x_3 + u_{22} x_2^2 + u_{23} x_2 x_3 + u_{33} x_3^2. \tag{2}$$

HISTORICAL NOTE   IN 1826, CAUCHY DISCUSSED QUADRATIC FORMS IN THREE VARIABLES—forms of the type $Ax^2 + By^2 + Cz^2 + 2Dxy + 2Exz + 2Fyz$. He showed that the characteristic equation formed from the determinant

$$\begin{vmatrix} A & D & E \\ D & B & F \\ E & F & C \end{vmatrix}$$

remains the same under any change of rectangular axes, what we would call an *orthogonal coordinate change*. Furthermore, he demonstrated that one could always find axes such that the new form has only the square terms. Three years later, Cauchy generalized the result to quadratic forms in n variables. (The matrices of such forms are $n \times n$ symmetric matrices.) He showed that the roots $\lambda_1, \lambda_2, \ldots, \lambda_n$ of the characteristic equation are all real, and he showed how to find the linear substitution that converts the original form to the form $\lambda_1 x_1^2 + \lambda_2 x_2^2 + \cdots + \lambda_n x_n^2$. In modern terminology, Cauchy had proved Theorem 5.5, that every real symmetric matrix is diagonalizable. In the two-variable case, Cauchy's proof amounts to finding the maximum and minimum of the quadratic form $f(x,y) = Ax^2 + 2Bxy + Cy^2$ subject to the condition that $x^2 + y^2 = 1$. In geometric terms, the point at which the extreme value occurs is that point on the unit circle which also lies on the end of one axis of one of the family of ellipses (or hyperbolas) described by the quadratic form. If one then takes the line from the origin to that point as one of the axes and the perpendicular to that line as the other, the equation in relation to those axes will have only the squares of the variables, as desired. To determine an extreme point subject to a condition, Cauchy uses what is today called the principle of Lagrange multipliers.

A computation shows that

$$[x_1, x_2, x_3] \begin{bmatrix} u_{11} & u_{12} & u_{13} \\ 0 & u_{22} & u_{23} \\ 0 & 0 & u_{33} \end{bmatrix} \begin{bmatrix} x_1 \\ x_2 \\ x_3 \end{bmatrix}$$

$$= [x_1, x_2, x_3] \begin{bmatrix} u_{11}x_1 + u_{12}x_2 + u_{13}x_3 \\ u_{22}x_2 + u_{23}x_3 \\ u_{33}x_3 \end{bmatrix}$$

$$= x_1(u_{11}x_1 + u_{12}x_2 + u_{13}x_3) + x_2(u_{22}x_2 + u_{23}x_3) + x_3(u_{33}x_3)$$

$$= u_{11}x_1^2 + u_{12}x_1x_2 + u_{13}x_1x_3 + u_{22}x_2^2 + u_{23}x_2x_3 + u_{33}x_3^2,$$

which is again form (2). We can verify that the term involving u_{ij} for $i \le j$ in the expansion of

$$[x_1, x_2, \ldots, x_n] \begin{bmatrix} u_{11} & u_{12} & \cdots & u_{1n} \\ 0 & u_{22} & \cdots & u_{2n} \\ \cdot & \cdot & & \cdot \\ \cdot & \cdot & & \cdot \\ \cdot & \cdot & & \cdot \\ 0 & 0 & \cdots & u_{nn} \end{bmatrix} \begin{bmatrix} x_1 \\ x_2 \\ \cdot \\ \cdot \\ \cdot \\ x_n \end{bmatrix} \tag{3}$$

is precisely $u_{ij}x_ix_j$. Thus, matrix product (3) is a 1×1 matrix whose sole entry is equal to sum (1).

> Every quadratic form in n variables x_i can be written as $\mathbf{x}^T U \mathbf{x}$, where $\mathbf{x}$ is the column vector of variables and U is a nonzero upper-triangular matrix.

We will call the matrix U the **upper-triangular coefficient matrix** of the quadratic form.

EXAMPLE 1   Write $x^2 - 2xy + 6xz + z^2$ in the form of matrix product (3).

SOLUTION   We obtain the matrix expression

$$[x, y, z] \begin{bmatrix} 1 & -2 & 6 \\ 0 & 0 & 0 \\ 0 & 0 & 1 \end{bmatrix} \begin{bmatrix} x \\ y \\ z \end{bmatrix}.$$

Just think of x as the first variable, y as the second, and z as the third. The summand $-2xy$, for example, gives the coefficient -2 in the row 1, column 2 position.   ∎

A matrix expression for a given quadratic form is by no means unique: $\mathbf{x}^T A \mathbf{x}$ gives a quadratic form for any nonzero $n \times n$ matrix A, and the form can be rewritten as $\mathbf{x}^T U \mathbf{x}$, where U is upper triangular.

EXAMPLE 2   Expand

$$[x, y, z] \begin{bmatrix} -1 & 3 & 1 \\ 2 & 1 & 0 \\ -2 & 2 & 4 \end{bmatrix} \begin{bmatrix} x \\ y \\ z \end{bmatrix},$$

and find the upper-triangular coefficient matrix for the quadratic form.

SOLUTION   We find that

$$[x, y, z] \begin{bmatrix} -1 & 3 & 1 \\ 2 & 1 & 0 \\ -2 & 2 & 4 \end{bmatrix} \begin{bmatrix} x \\ y \\ z \end{bmatrix} = [x, y, z] \begin{bmatrix} -x + 3y + z \\ 2x + y \\ -2x + 2y + 4z \end{bmatrix}$$

$$= x(-x + 3y + z) + y(2x + y) + z(-2x + 2y + 4z)$$
$$= -x^2 + 3xy + xz + 2xy + y^2 - 2xz + 2yz + 4z^2$$
$$= -x^2 + 5xy + y^2 - xz + 2yz + 4z^2.$$

The upper-triangular coefficient matrix is

$$U = \begin{bmatrix} -1 & 5 & -1 \\ 0 & 1 & 2 \\ 0 & 0 & 4 \end{bmatrix}. \qquad ■$$

All the nice things that we will prove about quadratic forms come from the fact that any quadratic form $f(\mathbf{x})$ can be expressed as $f(\mathbf{x}) = \mathbf{x}^T A \mathbf{x}$, where A is a *symmetric* matrix.

EXAMPLE 3   Find the symmetric coefficient matrix of the form $x^2 - 2xy + 6xz + z^2$ discussed in Example 1.

SOLUTION   Rewriting the form as

$$x^2 - xy - yx + 3xz + 3zx + z^2,$$

we obtain the symmetric coefficient matrix

$$A = \begin{bmatrix} 1 & -1 & 3 \\ -1 & 0 & 0 \\ 3 & 0 & 1 \end{bmatrix}. \qquad ■$$

As illustrated in Example 3, we obtain the symmetric matrix A of the form

$$f(\mathbf{x}) = \sum_{\substack{i \leq j \\ i,j=1}}^{n} u_{ij} x_i x_j \qquad (4)$$

by writing each *cross term* $u_{ij} x_i x_j$ as $(u_{ij}/2) x_i x_j + (u_{ij}/2) x_j x_i$, and by taking $a_{ij} = a_{ji} = u_{ij}/2$. The resulting matrix A is symmetric. We call it the **symmetric coefficient matrix** of the form.

> Every quadratic form in n variables x_i can be written as $\mathbf{x}^T A \mathbf{x}$, where $\mathbf{x}$ is the column vector of variables and A is a symmetric matrix.

Diagonalization of Quadratic Forms

We saw in Section 6.3 that a symmetric matrix can be diagonalized by an orthogonal matrix. That is, if A is an $n \times n$ symmetric matrix, there exists an $n \times n$ orthogonal change-of-coordinates matrix C such that $C^{-1}AC = D$, where D is diagonal. Recall that the diagonal entries of D are $\lambda_1, \lambda_2, \ldots, \lambda_n$, where the λ_j are the (not necessarily distinct) eigenvalues of the matrix A, the jth column of C is a unit eigenvector corresponding to λ_j, and the column vectors of C form an orthonormal basis for $\mathbb{R}^n$.

Consider now a quadratic form $\mathbf{x}^T A \mathbf{x}$, where A is symmetric, and let C be an orthogonal diagonalizing matrix for A. Because $C^{-1} = C^T$, the substitution

$$\mathbf{x} = C\mathbf{t} \quad \text{Diagonalizing substitution}$$

changes our quadratic form as follows:

$$\mathbf{x}^T A \mathbf{x} = (C\mathbf{t})^T A (C\mathbf{t}) = \mathbf{t}^T C^T A C \mathbf{t} = \mathbf{t}^T C^{-1} A C \mathbf{t}$$
$$= \mathbf{t}^T D \mathbf{t} = \lambda_1 t_1^2 + \lambda_2 t_2^2 + \cdots + \lambda_n t_n^2,$$

where the λ_j are the eigenvalues of A. We have thus diagonalized the quadratic form. The value of $\mathbf{x}^T A \mathbf{x}$ for any $\mathbf{x}$ in $\mathbb{R}^n$ is the same as the value of $\mathbf{t}^T D \mathbf{t}$ for $\mathbf{t} = C^{-1}\mathbf{x}$.

EXAMPLE 4   Find a substitution $\mathbf{x} = C\mathbf{t}$ that diagonalizes the form $3x_1^2 + 10x_1x_2 + 3x_2^2$, and find the corresponding diagonalized form.

SOLUTION   The symmetric coefficient matrix for the quadratic form is

$$A = \begin{bmatrix} 3 & 5 \\ 5 & 3 \end{bmatrix}.$$

We need to find the eigenvalues and eigenvectors for A. We have

$$|A - \lambda I| = \begin{vmatrix} 3 - \lambda & 5 \\ 5 & 3 - \lambda \end{vmatrix} = \lambda^2 - 6\lambda - 16 = (\lambda + 2)(\lambda - 8).$$

Thus, $\lambda_1 = -2$ and $\lambda_2 = 8$ are eigenvalues of A. Finding the eigenvectors that are to become the columns of the substitution matrix C, we compute

$$A - \lambda_1 I = A + 2I = \begin{bmatrix} 5 & 5 \\ 5 & 5 \end{bmatrix} \sim \begin{bmatrix} 1 & 1 \\ 0 & 0 \end{bmatrix}$$

and

$$A - \lambda_2 I = A - 8I = \begin{bmatrix} -5 & 5 \\ 5 & -5 \end{bmatrix} \sim \begin{bmatrix} 1 & -1 \\ 0 & 0 \end{bmatrix}.$$

Thus eigenvectors are $\mathbf{v}_1 = [-1, 1]$ and $\mathbf{v}_2 = [1, 1]$. Normalizing them to length 1 and placing them in the columns of the substitution matrix C, we obtain the diagonalizing substitution

$$\begin{bmatrix} x_1 \\ x_2 \end{bmatrix} = C \begin{bmatrix} t_1 \\ t_2 \end{bmatrix} = \begin{bmatrix} -1/\sqrt{2} & 1/\sqrt{2} \\ 1/\sqrt{2} & 1/\sqrt{2} \end{bmatrix} \begin{bmatrix} t_1 \\ t_2 \end{bmatrix}.$$

Our theory then tells us that making the variable substitution

$$x_1 = \left(\frac{1}{\sqrt{2}}\right)(-t_1 + t_2),$$

$$x_2 = \left(\frac{1}{\sqrt{2}}\right)(t_1 + t_2),$$

(5)

in the form $3x_1^2 + 10x_1x_2 + 3x_2^2$ will give the diagonal form

$$\lambda_1 t_1^2 + \lambda_2 t_2^2 = -2t_1^2 + 8t_2^2.$$

This can, of course, be checked by actually substituting the expressions for x_1 and x_2 in Eqs. (5) into the form $3x_1^2 + 10x_1x_2 + 3x_2^2$. ∎

In the preceding example, we might like to clear denominators in our substitution given in Eqs. (5) by using $\mathbf{x} = (\sqrt{2}C)\mathbf{t}$, so that $x_1 = -t_1 + t_2$ and $x_2 = t_1 + t_2$. Using the substitution $\mathbf{x} = kC\mathbf{t}$ with a quadratic form $\mathbf{x}^T A\mathbf{x}$, we obtain

$$\mathbf{x}^T A\mathbf{x} = (kC\mathbf{t})^T A(kC\mathbf{t}) = k^2(\mathbf{t}^T C^T A C\mathbf{t}) = k^2(\mathbf{t}^T D\mathbf{t}).$$

Thus, with $k = \sqrt{2}$, the substitution $x_1 = -t_1 + t_2$, $x_2 = t_1 + t_2$ in Example 1 results in the diagonal form

$$2(-2t_1^2 + 8t_2^2) = -4t_1^2 + 16t_2^2.$$

In any particular situation, we would have to balance the desire for arithmetic simplicity against the desire to have the new orthogonal basis actually be orthonormal.

As an orthogonal matrix, C has determinant ± 1. (See Exercise 22 in Section 6.3.) We state without proof the significance of the sign of $\det(C)$. If $\det(C) = 1$, then the new ordered orthonormal basis B given by the column vectors of C has the **same orientation** in $\mathbb{R}^n$ as the standard ordered basis E; while if $\det(C) = -1$, then B has the **opposite orientation** to E. In order for $B = (\mathbf{b}_1, \mathbf{b}_2)$ in $\mathbb{R}^2$ to have the same orientation as $(\mathbf{e}_1, \mathbf{e}_2)$, there must be a *rotation* of the plane, given by a matrix transformation $\mathbf{x} = C\mathbf{t}$, which carries both $\mathbf{e}_1$ to $\mathbf{b}_1$ and $\mathbf{e}_2$ to $\mathbf{b}_2$. The same interpretation in terms of rotation is true in $\mathbb{R}^3$, with the

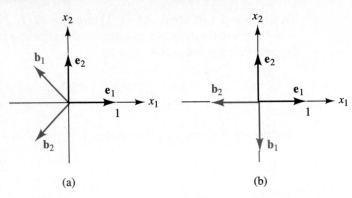

FIGURE 8.1
(a) Rotation of axes, hence (**b**$_1$, **b**$_2$) and (**e**$_1$, **e**$_2$) have the same orientation;
(b) not a rotation of axes, hence, (**b**$_1$, **b**$_2$) and (**e**$_1$, **e**$_2$) have opposite orientations.

additional condition that **e**$_3$ be carried to **b**$_3$. To illustrate for the plane $\mathbb{R}^2$, Figure 8.1(a) shows an ordered orthonormal basis $B = (\mathbf{b}_1, \mathbf{b}_2)$, where $\mathbf{b}_1 = [-1/\sqrt{2}, 1/\sqrt{2}]$ and $\mathbf{b}_2 = [-1/\sqrt{2}, -1/\sqrt{2}]$, having the same orientation as E. Notice that

$$\det(C) = \begin{vmatrix} -1/\sqrt{2} & -1/\sqrt{2} \\ 1/\sqrt{2} & -1/\sqrt{2} \end{vmatrix} = \frac{1}{2} + \frac{1}{2} = 1.$$
$$\quad\quad\quad \mathbf{b}_1 \quad\quad \mathbf{b}_2$$

Counterclockwise rotation of the plane through an angle of 135° carries E into B, preserving the order of the vectors. However, we see that the basis $(\mathbf{b}_1, \mathbf{b}_2) = ([0, -1], [-1, 0])$ shown in Figure 8.1(b) does not have the same orientation as E, and this time

$$\det(C) = \begin{vmatrix} 0 & -1 \\ -1 & 0 \end{vmatrix} = -1.$$
$$\quad\quad \mathbf{b}_1 \quad \mathbf{b}_2$$

For any orthogonal matrix C having determinant -1, multiplication of any *single* column of C by -1 gives an orthogonal matrix with determinant 1. For the diagonalization we are considering, where the columns are normalized eigenvectors, multiplication by -1 still gives a normalized eigenvector. Although there will be some sign changes in the diagonalizing substitution, the final diagonal form remains the same, because the eigenvalues are the same and their order has not been changed.

We summarize all our work in one main theorem, and then conclude with a final example. Section 8.2 will give an application of the theorem to geometry, and Section 8.3 will give an application to optimization.

THEOREM 8.1 Principal Axis Theorem

Every quadratic form $f(\mathbf{x})$ in n variables $x_1, x_2, \ldots, x_n$ can be diagonalized by a substitution $\mathbf{x} = C\mathbf{t}$, where C is an $n \times n$ orthogonal matrix. The diagonalized form appears as

$$\lambda_1 t_1^2 + \lambda_2 t_2^2 + \cdots + \lambda_n t_n^2,$$

where the λ_j are the eigenvalues of the symmetric coefficient matrix A of $f(\mathbf{x})$. The jth column vector of C is a normalized eigenvector $\mathbf{v}_j$ of A corresponding to λ_j. Moreover, C can be chosen so that $\det(C) = 1$.

We box a step-by-step outline for diagonalizing a quadratic form.

Diagonalizing a Quadratic Form $f(\mathbf{x})$

Step 1 Find the symmetric coefficient matrix A of the form $f(\mathbf{x})$.

Step 2 Find the (not necessarily distinct) eigenvalues $\lambda_1, \lambda_2, \ldots, \lambda_n$ of A.

Step 3 Find an orthonormal basis for $\mathbb{R}^n$ consisting of normalized eigenvectors of A.

Step 4 Form the matrix C, whose columns are the basis vectors found in step 3, in the order corresponding to the listing of eigenvalues in step 2. The transformation $\mathbf{x} = C\mathbf{t}$ is a **rotation** if $\det(C) = 1$. If a rotation is desired and $\det(C) = -1$, change the sign of all components of just one column vector in C.

Step 5 The substitution $\mathbf{x} = C\mathbf{t}$ transforms $f(\mathbf{x})$ to the diagonal form $\lambda_1 t_1^2 + \lambda_2 t_2^2 + \cdots + \lambda_n t_n^2$.

EXAMPLE 5 Find a variable substitution that diagonalizes the form $2xy + 2xz$, and give the resulting diagonal form.

SOLUTION The symmetric coefficient matrix for the given quadratic form is

$$A = \begin{bmatrix} 0 & 1 & 1 \\ 1 & 0 & 0 \\ 1 & 0 & 0 \end{bmatrix}.$$

Expanding the determinant on the last column, we find that

$$|A - \lambda I| = \begin{vmatrix} -\lambda & 1 & 1 \\ 1 & -\lambda & 0 \\ 1 & 0 & -\lambda \end{vmatrix} = 1(\lambda) - \lambda(\lambda^2 - 1)$$

$$= -\lambda(-1 + \lambda^2 - 1) = -\lambda(\lambda^2 - 2).$$

The eigenvalues of A are then $\lambda_1 = 0$, $\lambda_2 = \sqrt{2}$, and $\lambda_3 = -\sqrt{2}$.
Computing eigenvectors, we have

$$A - \lambda_1 I = A = \begin{bmatrix} 0 & 1 & 1 \\ 1 & 0 & 0 \\ 1 & 0 & 0 \end{bmatrix} \sim \begin{bmatrix} 1 & 0 & 0 \\ 0 & 1 & 1 \\ 0 & 0 & 0 \end{bmatrix},$$

so $\mathbf{v}_1 = [0, -1, 1]$ is an eigenvector for $\lambda_1 = 0$. Moreover,

$$A - \lambda_2 I = A - \sqrt{2}I = \begin{bmatrix} -\sqrt{2} & 1 & 1 \\ 1 & -\sqrt{2} & 0 \\ 1 & 0 & -\sqrt{2} \end{bmatrix} \sim \begin{bmatrix} 1 & -\sqrt{2} & 0 \\ -\sqrt{2} & 1 & 1 \\ 1 & 0 & -\sqrt{2} \end{bmatrix}$$

$$\sim \begin{bmatrix} 1 & -\sqrt{2} & 0 \\ 0 & -1 & 1 \\ 0 & \sqrt{2} & -\sqrt{2} \end{bmatrix} \sim \begin{bmatrix} 1 & -\sqrt{2} & 0 \\ 0 & 1 & -1 \\ 0 & 0 & 0 \end{bmatrix},$$

so $\mathbf{v}_2 = [\sqrt{2}, 1, 1]$ is an eigenvector for $\lambda_2 = \sqrt{2}$. In a similar fashion, we find
that $\mathbf{v}_3 = [-\sqrt{2}, 1, 1]$ is an eigenvector for $\lambda_3 = -\sqrt{2}$. Thus,

$$C = \begin{bmatrix} 0 & \sqrt{2}/2 & -\sqrt{2}/2 \\ -1/\sqrt{2} & \frac{1}{2} & \frac{1}{2} \\ 1/\sqrt{2} & \frac{1}{2} & \frac{1}{2} \end{bmatrix}$$

is an orthogonal diagonalizing matrix for A. The substitution

$$x = \left(\frac{1}{\sqrt{2}}\right)(t_2 - t_3)$$

$$y = \frac{1}{2}(-\sqrt{2}t_1 + t_2 + t_3)$$

$$z = \frac{1}{2}(\sqrt{2}t_1 + t_2 + t_3)$$

will diagonalize $2xy + 2xz$ to become $\sqrt{2}t_2^2 - \sqrt{2}t_3^2$. ∎

Let $\mathbf{x}$ *and* $\mathbf{t}$ *be* $n \times 1$ *column vectors with entries* x_i *and* t_i, *respectively.*

1. For every nonzero $n \times n$ matrix A, the product $f(\mathbf{x}) = \mathbf{x}^T A \mathbf{x}$ gives a quadratic form in the variables x_i.
2. Given any quadratic form $f(\mathbf{x})$ in the variables x_i, there exist an upper-triangular coefficient matrix U such that $f(\mathbf{x}) = \mathbf{x}^T U \mathbf{x}$ and symmetric coefficient matrix A such that $f(\mathbf{x}) = \mathbf{x}^T A \mathbf{x}$.
3. Any quadratic form $f(\mathbf{x})$ can be orthogonally diagonalized by using a variable substitution $\mathbf{x} = C\mathbf{t}$ as described in the boxed steps preceding Example 5.

EXERCISES

In Exercises 1–8, find the upper-triangular coefficient matrix U and the symmetric coefficient matrix A of the given quadratic form.

1. $3x^2 - 6xy + y^2$ **2.** $8x^2 + 9xy - 3y^2$

3. $x^2 - y^2 - 4xy + 3xz - 8yz$

4. $x_1^2 - 2x_2^2 + x_3^2 + 6x_4^2 - 2x_1x_4 + 6x_2x_4 - 8x_1x_3$

5. $[x, y] \begin{bmatrix} -2 & 1 \\ 7 & 3 \end{bmatrix} \begin{bmatrix} x \\ y \end{bmatrix}$

6. $[x, y] \begin{bmatrix} 7 & -10 \\ 15 & 2 \end{bmatrix} \begin{bmatrix} x \\ y \end{bmatrix}$

7. $[x, y, z] \begin{bmatrix} 8 & 3 & 1 \\ 2 & 1 & -4 \\ -5 & 2 & 10 \end{bmatrix} \begin{bmatrix} x \\ y \\ z \end{bmatrix}$

8. $[x_1, x_2, x_3, x_4] \begin{bmatrix} 2 & -1 & 3 & 0 \\ 4 & 2 & -1 & 3 \\ -6 & 3 & 0 & 7 \\ 10 & 2 & 1 & 5 \end{bmatrix} \begin{bmatrix} x_1 \\ x_2 \\ x_3 \\ x_4 \end{bmatrix}$

In Exercises 9–16, find an orthogonal substitution that diagonalizes the given quadratic form, and find the diagonalized form.

9. $2xy$

10. $3x^2 + 4xy$

11. $-6xy + 8y^2$

12. $x^2 + 2xy + y^2$

13. $3x^2 - 4xy + 3y^2$

14. $x_2^2 + 2x_1x_3$

15. $x_1^2 + x_2^2 + x_3^2 - 2x_1x_2 - 2x_1x_3 - 2x_2x_3$
[SUGGESTION: Use Example 4 in Section 6.3.]

16. $x_1^2 + 2x_2^2 - 6x_1x_3 + x_3^2 - 4x_4^2$

17. Find a necessary and sufficient condition on a, b, and c such that the quadratic form $ax^2 + 2bxy + cy^2$ can be orthogonally diagonalized to kt^2.

18. Repeat Exercise 17, but require also that $k = 1$.

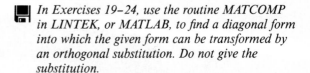 *In Exercises 19–24, use the routine MATCOMP in LINTEK, or MATLAB, to find a diagonal form into which the given form can be transformed by an orthogonal substitution. Do not give the substitution.*

19. $3x^2 + 4xy - 5y^2$ **20.** $x^2 - 8xy + y^2$

21. $3x^2 + y^2 - 2z^2 - 4xy + 6yz$

22. $y^2 - 8z^2 + 3xy - 4xz + 7yz$

23. $x_1^2 - 3x_1x_4 + 5x_4^2 - 8x_2x_3$

24. $x_1^2 - 8x_1x_2 + 6x_2x_3 - 4x_3x_4$

APPLICATIONS TO GEOMETRY

Conic Sections in $\mathbb{R}^2$

Figure 8.2 shows three different types of plane curves obtained when a double right-circular cone is cut by a plane. These *conic sections* are the ellipse, the hyperbola, and the parabola. Figure 8.3(a) shows an ellipse in *standard position*, with center at the origin. The equation of this ellipse is

$$\frac{x^2}{a^2} + \frac{y^2}{b^2} = 1. \tag{1}$$

The ellipse in Figure 8.3(b) with center (h, k) has the equation

$$\frac{(x - h)^2}{a^2} + \frac{(y - k)^2}{b^2} = 1. \tag{2}$$

If we make the translation substitution $\bar{x} = x - h$, $\bar{y} = y - k$, then Eq. (2) becomes $\bar{x}^2/a^2 + \bar{y}^2/b^2 = 1$, which resembles Eq. (1).

By completing the square, we can put any quadratic polynomial equation in x and y with no xy-term but with x^2 and y^2 having coefficients of the same sign into the form of Eq. (2), possibly with 0 or -1 on the right-hand side. The procedure should be familiar to you from work with circles. We give one illustration for an ellipse.

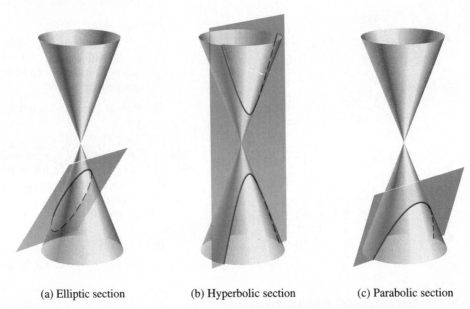

(a) Elliptic section   (b) Hyperbolic section   (c) Parabolic section

FIGURE 8.2
Sections of a cone: (a) an elliptic section; (b) a hyperbolic section; (c) a parabolic section.

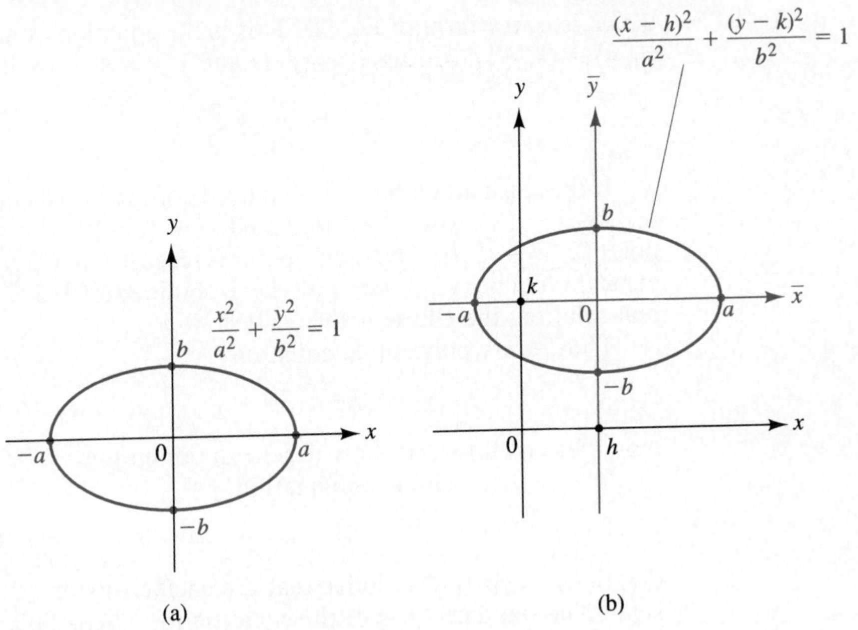

$$\frac{(x-h)^2}{a^2} + \frac{(y-k)^2}{b^2} = 1$$

(a)

(b)

FIGURE 8.3
(a) Ellipse in standard position; (b) ellipse centered at (h, k).

EXAMPLE 1   Complete the square in the equation

$$x^2 + 3y^2 - 4x + 6y = -1.$$

SOLUTION   Completing the square, we obtain

$$(x^2 - 4x) + 3(y^2 + 2y) = -1$$
$$(x - 2)^2 + 3(y + 1)^2 = 4 + 3 - 1 = 6$$
$$\frac{(x-2)^2}{6} + \frac{(y+1)^2}{2} = 1,$$

HISTORICAL NOTE ANALYTIC GEOMETRY is generally considered to have been founded by René Descartes (1596–1650) and Pierre Fermat (1601–1665) in the first half of the seventeenth century. But it was not until the appearance of a Latin version of Descartes' *Geometry* in 1661 by Frans van Schooten (1615–1660) that its influence began to be felt. This Latin version was published along with many commentaries; in particular, the *Elements of Curves* by Jan de Witt (1625–1672) gave a systematic treatment of conic sections. De Witt gave canonical forms of these equations similar to those in use today; for example, $y^2 = ax$, $by^2 + x^2 = f^2$, and $x^2 - by^2 = f^2$ represented the parabola, ellipse, and hyperbola, respectively. He then showed how, given an arbitrary second-degree equation in x and y, to find a transformation of axes that reduces the given equation to one of the canonical forms. This is, of course, equivalent to diagonalizing a particular symmetric matrix.

De Witt was a talented mathematician who, because of his family background, could devote but little time to mathematics. In 1653 he became in effect the Prime Minister of the Netherlands. Over the next decades, he guided the fortunes of the country through a most difficult period, including three wars with England. In 1672 the hostility of one of the Dutch factions culminated in his murder.

which is of the form of Eq. (2). This is the equation of an ellipse with center $(h, k) = (2, -1)$. Setting $\bar{x} = x - 2$ and $\bar{y} = y + 1$, we obtain the equation

$$\frac{\bar{x}^2}{6} + \frac{\bar{y}^2}{2} = 1.$$ ∎

If the constant on the right-hand side of the initial equation in Example 1 had been -7, we would have obtained $\bar{x}^2 + 3\bar{y}^2 = 0$, which describes the single point $(\bar{x}, \bar{y}) = (0, 0)$. This single point is regarded as a *degenerate* ellipse. If the constant had been -8, we would have obtained $\bar{x}^2 + 3\bar{y}^2 = -1$, which has no real solution; the ellipse is then called *empty*.

Thus, every polynomial equation

$$c_1x^2 + c_2y^2 + c_3x + c_4y = d, \qquad c_1c_2 > 0, \tag{3}$$

describes an ellipse, possibly degenerate or empty.

In a similar fashion, an equation

$$c_1x^2 + c_2y^2 + c_3x + c_4y = d, \qquad c_1c_2 < 0, \tag{4}$$

describes a hyperbola. Notice that the coefficients of x^2 and y^2 have *opposite* signs. The standard form of the equation for a hyperbola centered at $(0, 0)$ is

$$\frac{x^2}{a^2} - \frac{y^2}{b^2} = 1 \quad \text{or} \quad -\frac{x^2}{a^2} + \frac{y^2}{b^2} = 1,$$

as shown in Figure 8.4. The dashed diagonal lines $y = \pm(b/a)x$ shown in the figure are the *asymptotes* of the hyperbola. By completing squares and

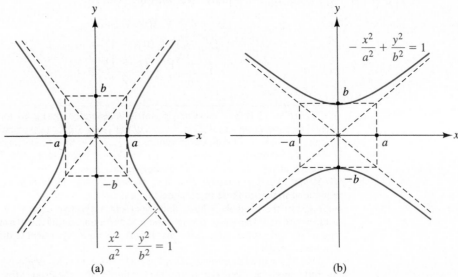

(a)   (b)

FIGURE 8.4

(a) The hyperbola $\dfrac{x^2}{a^2} - \dfrac{y^2}{b^2} = 1$; (b) the hyperbola $-\dfrac{x^2}{a^2} + \dfrac{y^2}{b^2} = 1$.

translating axes, we can reduce Eq. (4) to one of the standard forms shown in Figure 8.4, in variables $\bar{x}$ and $\bar{y}$, unless the constant in the final equation reduces to zero. In that case we obtain

$$\frac{\bar{x}^2}{a^2} - \frac{\bar{y}^2}{b^2} = 0 \quad \text{or} \quad \bar{y} = \pm\frac{b}{a}\bar{x}.$$

These equations represent two lines that can be considered a degenerate hyperbola. Thus any equation of the form of Eq. (4) describes a (possibly degenerate) hyperbola.

Finally, the equations

$$c_1x^2 + c_2x + c_3y = d \quad \text{and} \quad c_1y^2 + c_2x + c_3y = d, \qquad c_1 \neq 0, \qquad (5)$$

describe parabolas. If $c_3 \neq 0$ in the first equation in Eqs. (5) and $c_2 \neq 0$ in the second, these equations can be reduced to the form

$$\bar{x}^2 = a\bar{y} \quad \text{and} \quad \bar{y}^2 = a\bar{x} \qquad (6)$$

by completing the square and translating axes. Figure 8.5 shows two parabolas in standard position. If Eqs. (5) reduce to $c_1x^2 + c_2x = d$ and $c_1y^2 + c_3y = d$, each describes two parallel lines that can be considered degenerate parabolas.

In summary, every equation of the form

$$c_1x^2 + c_2y^2 + c_3x + c_4y = d \qquad (7)$$

with at least one of c_1 or c_2 nonzero describes a (possibly degenerate or empty) ellipse, hyperbola, or parabola.

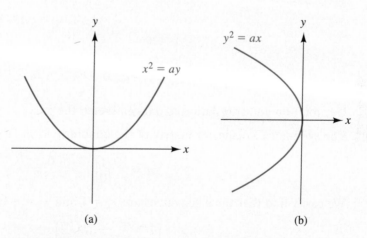

(a)   (b)

FIGURE 8.5
(a) The parabola $x^2 = ay$, $a > 0$; (b) the parabola $y^2 = ax$, $a < 0$.

Classification of Second-Degree Curves

We can now apply our work in Section 8.1 on diagonalizing quadratic forms to classification of the plane curves described by an equation of the type

$$ax^2 + bxy + cy^2 + dx + ey + f = 0 \quad \text{for } a, b, c \text{ not all zero.} \tag{8}$$

We make a substitution

$$\begin{bmatrix} x \\ y \end{bmatrix} = C \begin{bmatrix} t_1 \\ t_2 \end{bmatrix}, \quad \text{where} \quad \det(C) = 1, \tag{9}$$

which orthogonally diagonalizes the quadratic-form part of Eq. (8), which is in color, and we obtain an equation of the form

$$\lambda_1 t_1^2 + \lambda_2 t_2^2 + g t_1 + h t_2 + k = 0. \tag{10}$$

This equation has the form of Eq. (7) and describes an ellipse, hyperbola, or parabola. Remember that Eq. (9) corresponds to a rotation that carries the vector $\mathbf{e}_1$ to the first column vector $\mathbf{b}_1$ of C and carries $\mathbf{e}_2$ to the second column vector $\mathbf{b}_2$. We think of $\mathbf{b}_1$ as pointing out the t_1-axis and $\mathbf{b}_2$ as pointing out the t_2-axis. We summarize our work in a theorem and give just one illustration, leaving others to the exercises.

THEOREM 8.2 Classification of Second-Degree Plane Curves

Every equation of the form of Eq. (8) can be reduced to an equation of the form of Eq. (10) by means of an orthogonal substitution corresponding to a rotation of the plane. The coefficients λ_1 and λ_2 in Eq. (10) are the eigenvalues of the symmetric coefficient matrix of the quadratic-form portion of Eq. (8). The curve describes a (possibly degenerate or empty)

$$
\begin{array}{ll}
\text{ellipse} & \text{if } \lambda_1 \lambda_2 > 0, \\
\text{hyperbola} & \text{if } \lambda_1 \lambda_2 < 0, \\
\text{parabola} & \text{if } \lambda_1 \lambda_2 = 0.
\end{array}
$$

EXAMPLE 2   Use rotation and translation of axes to sketch the curve $2xy + 2\sqrt{2}x = 1$.

SOLUTION   The symmetric coefficient matrix of the quadratic form $2xy$ is

$$A = \begin{bmatrix} 0 & 1 \\ 1 & 0 \end{bmatrix}.$$

We easily find that the eigenvalues are $\lambda_1 = 1$ and $\lambda_2 = -1$, and that

$$C = \begin{bmatrix} 1/\sqrt{2} & -1/\sqrt{2} \\ 1/\sqrt{2} & 1/\sqrt{2} \end{bmatrix}$$

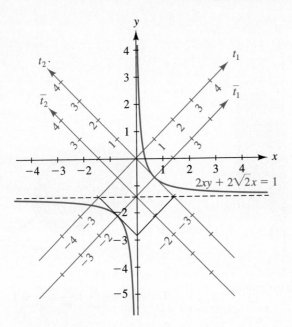

FIGURE 8.6
The hyperbola $2xy + 2\sqrt{2}x = 1$.

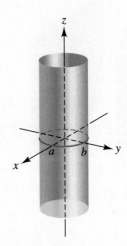

FIGURE 8.7
The elliptic cylinder $\dfrac{x^2}{a^2} + \dfrac{y^2}{b^2} = 1$.

is an orthogonal diagonalizing matrix with determinant 1. The substitution

$$x = \frac{1}{\sqrt{2}}(t_1 - t_2)$$

$$y = \frac{1}{\sqrt{2}}(t_1 + t_2)$$

then yields

$$t_1^2 - t_2^2 + 2t_1 - 2t_2 = 1.$$

Completing the square, we obtain

$$(t_1 + 1)^2 - (t_2 + 1)^2 = 1,$$

which describes the hyperbola shown in Figure 8.6. ∎

Quadric Surfaces

An equation in three variables of the form

$$c_1x^2 + c_2y^2 + c_3z^2 + c_4x + c_5y + c_6z = d, \tag{11}$$

where at least one of $c_1, c_2,$ or c_3 is nonzero, describes a *quadric surface* in space, which again might be degenerate or empty. Figures 8.7 through 8.15 show some of the quadric surfaces in standard position.

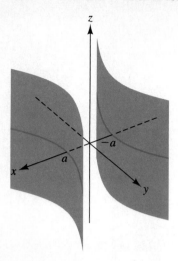

FIGURE 8.8
The hyperbolic cylinder
$\dfrac{x^2}{a^2} - \dfrac{y^2}{b^2} = 1$.

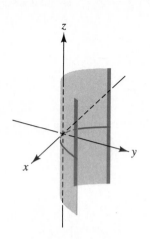

FIGURE 8.9
The parabolic cylinder
$ay = x^2$.

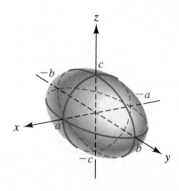

FIGURE 8.10
The ellipsoid $\dfrac{x^2}{a^2} + \dfrac{y^2}{b^2} + \dfrac{z^2}{c^2} = 1$.

By completing the square in Eq. (11), which corresponds to translating axes, we see that Eq. (11) can be reduced to an equation involving $\bar{x}$, $\bar{y}$, and $\bar{z}$ in which a variable appearing to the second power does not appear to the first power. Notice that this is true for all the equations in Figures 8.7 through 8.15.

HISTORICAL NOTE THE EARLIEST CLASSIFICATION OF QUADRIC SURFACES was given by Leonhard Euler, in his precalculus text *Introduction to Infinitesimal Analysis* (1748). Euler's classification was similar to the conic-section classification of De Witt. Euler considered the second-degree equation in three variables $Ap^2 + Bq^2 + Cr^2 + Dpq + Epr + Fqr + Gp + Hq + Ir + K = 0$ as representing a surface in 3-space. As did De Witt, he gave canonical forms for these surfaces and showed how to rotate and translate the axes to reduce any given equation to a standard form such as $Ap^2 + Bq^2 + Cr^2 + K = 0$. An analysis of the signs of the new coefficients determined whether the given equation represented an ellipsoid, a hyperboloid of one or two sheets, an elliptic or hyperbolic paraboloid, or one of the degenerate cases. Euler did not, however, make explicit use of eigenvalues. He gave a general formula for rotation of axes in 3-space as functions of certain angles and then showed how to choose the angles to make the coefficients D, E, and F all zero.

Euler was the most prolific mathematician of all time. His collected works fill over 70 large volumes. Euler was born in Switzerland, but spent his professional life in St. Petersburg and Berlin. His texts on precalculus, differential calculus, and integral calculus (1748, 1755, 1768) had immense influence and became the bases for such texts up to the present. He standardized much of our current notation, introducing the numbers e, π, and i, as well as giving our current definitions for the trigonometric functions. Even though he was blind for the last 17 years of his life, he continued to produce mathematical papers almost up to the day he died.

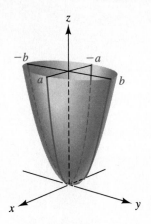

FIGURE 8.11

The elliptic paraboloid $z = \dfrac{x^2}{a^2} + \dfrac{y^2}{b^2}$.

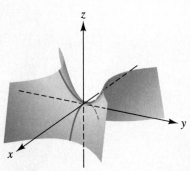

FIGURE 8.12

The hyperbolic paraboloid
$z = \dfrac{y^2}{b^2} - \dfrac{x^2}{a^2}$.

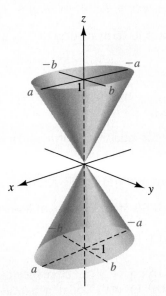

FIGURE 8.13

The elliptic cone $z^2 = \dfrac{x^2}{a^2} + \dfrac{y^2}{b^2}$.

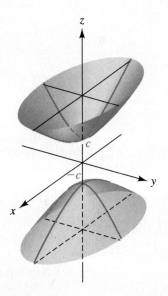

FIGURE 8.14

The hyperboloid of two sheets
$\dfrac{z^2}{c^2} - 1 = \dfrac{x^2}{a^2} + \dfrac{y^2}{b^2}$.

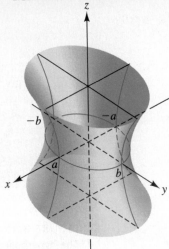

FIGURE 8.15

The hyperboloid of one sheet $\dfrac{z^2}{c^2} + 1 = \dfrac{x^2}{a^2} + \dfrac{y^2}{b^2}$.

Again, degenerate and empty cases are possible. For example, the equation $x^2 + 2y^2 + z^2 = -4$ gives an empty ellipsoid. The elliptic cone, hyperboloid of two sheets, and hyperboloid of one sheet in Figures 8.13 through 8.15 differ only in whether a constant in their equations is zero, negative, or positive.

Now consider a general second-degree polynomial equation

$$ax^2 + by^2 + cz^2 + dxy + exz + fyz + px + qy + rz + s = 0, \qquad (12)$$

where the coefficient of at least one term of degree 2 is nonzero. Making a substitution

$$\begin{bmatrix} x \\ y \\ z \end{bmatrix} = C \begin{bmatrix} t_1 \\ t_2 \\ t_3 \end{bmatrix}, \qquad \text{where} \quad \det(C) = 1, \qquad (13)$$

that orthogonally diagonalizes the quadratic-form portion of Eq. (12) (the portion in color), we obtain

$$\lambda_1 t_1^2 + \lambda_2 t_2^2 + \lambda_3 t_3^2 + p't_1 + q't_2 + r't_3 + s' = 0, \qquad (14)$$

which is of the form of Eq. (11). We state this as a theorem.

THEOREM 8.3 Principal Axis Theorem for $\mathbb{R}^3$

Every equation of the form of Eq. (12) can be reduced to an equation of the form of Eq. (14) by an orthogonal substitution (13) that corresponds to a rotation of axes.

Again, computation of the eigenvalues λ_1, λ_2, and λ_3 of the symmetric coefficient matrix of the quadratic-form portion of Eq. (12) may give us quite a bit of information as to the type of quadric surface. However, actually executing the substitution and completing squares, which can be tedious, may be necessary to distinguish among certain surfaces. We give a rough classification scheme in Table 8.1. Remember that empty or degenerate cases are possible, even where we do not explicitly give them.

TABLE 8.1

Eigenvalues λ_1, λ_2, λ_3	Quadric Surface
All of the same sign	Ellipsoid
Two of one sign and one of the other sign	Elliptic cone, hyperboloid of two sheets, or hyperboloid of one sheet
One zero, two of the same sign	Elliptic paraboloid or elliptic cylinder (degenerate case)
One zero, two of opposite signs	Hyperbolic paraboloid or hyperbolic cylinder (degenerate case)
Two zero, one nonzero	Parabolic cylinder or two parallel planes (degenerate case)

If you try to verify Table 8.1, you will wonder about an equation of the form $ax^2 + by + cz = d$ in the last case given. Exercise 11 indicates that, by means of a rotation of axes, this equation can be reduced to $at_1^2 + rt_2 = d$, which can be written as $at_1^2 + r(t_2 - d/r) = 0$, and consequently describes a parabolic cylinder. We conclude with four examples.

EXAMPLE 3   Classify the quadric surface $2xy + 2xz = 1$.

SOLUTION   Example 5 in Section 8.1 shows that the orthogonal substitution

$$x = \frac{1}{\sqrt{2}}(t_2 - t_3)$$

$$y = \frac{1}{2}\left(-\sqrt{2}t_1 + t_2 + t_3\right)$$

$$z = \frac{1}{2}\left(\sqrt{2}t_1 + t_2 + t_3\right)$$

transforms $2xy + 2xz$ into $\sqrt{2}t_2^2 - \sqrt{2}t_3^2$. It can be checked that the matrix C corresponding to this substitution has determinant 1. Thus, the substitution corresponds to a rotation of axes and transforms the equation $2xy + 2xz = 1$ into $\sqrt{2}t_2^2 - \sqrt{2}t_3^2 = 1$, which we recognize as a hyperbolic cylinder.   ∎

EXAMPLE 4   Classify the quadric surface $2xy + 2xz = y + 1$.

SOLUTION Using the same substitution as we did in Example 3, we obtain the equation

$$\sqrt{2}t_2^2 - \sqrt{2}t_3^2 = \frac{1}{2}\left(-\sqrt{2}t_1 + t_2 + t_3\right) + 1.$$

Translation of axes by completing squares yields an equation of the form

$$\sqrt{2}(t_2 - h)^2 - \sqrt{2}(t_3 - k)^2 = -\frac{1}{\sqrt{2}}(t_1 - r),$$

which we recognize as a hyperbolic paraboloid. ∎

EXAMPLE 5 Classify the quadric surface $2xy + 2xz = x + 1$.

SOLUTION Using again the substitution in Example 3, we obtain

$$\sqrt{2}t_2^2 - \sqrt{2}t_3^2 = \frac{1}{\sqrt{2}}(t_2 - t_3) + 1$$

or

$$2t_2^2 - t_2 - 2t_3^2 + t_3 = \sqrt{2}.$$

Completing squares yields

$$2\left(t_2 - \frac{1}{4}\right)^2 - 2\left(t_3 - \frac{1}{4}\right)^2 = \sqrt{2},$$

which represents a hyperbolic cylinder. ∎

EXAMPLE 6 Classify the quadric surface

$$2x^2 - 3y^2 + z^2 - 2xy + 4yz - 6x + 8y - 8z = 17$$

as far as possible, by finding just the eigenvalues of the symmetric coefficient matrix of the quadratic-form portion of the equation.

SOLUTION The symmetric matrix of the quadratic-form portion is

$$A = \begin{bmatrix} 2 & -1 & 0 \\ -1 & -3 & 2 \\ 0 & 2 & 1 \end{bmatrix}.$$

We find that

$$\det(A - \lambda I) = \begin{vmatrix} 2 - \lambda & -1 & 0 \\ -1 & -3 - \lambda & 2 \\ 0 & 2 & 1 - \lambda \end{vmatrix}$$

$$= (2 - \lambda)\begin{vmatrix} -3 - \lambda & 2 \\ 2 & 1 - \lambda \end{vmatrix} + \begin{vmatrix} -1 & 2 \\ 0 & 1 - \lambda \end{vmatrix}$$

$$= (2 - \lambda)(\lambda^2 + 2\lambda - 7) + \lambda - 1$$

$$= -\lambda^3 + 12\lambda - 15.$$

We can see that $\lambda = 0$ is not a solution of the characteristic equation $-\lambda^3 + 12\lambda - 15 = 0$. We could plot a rough sketch of the graph of $y = -\lambda^3 + 12\lambda - 15$, just to determine the signs of the eigenvalues. However, we prefer to use the routine MATCOMP in LINTEK, or MATLAB, with the matrix A. We quickly find that the eigenvalues are approximately

$$\lambda_1 = -3.9720, \qquad \lambda_2 = 1.5765, \qquad \lambda_3 = 2.3954.$$

According to Table 8.1, we have an elliptic cone, a hyperboloid of two sheets, or a hyperboloid of one sheet. ■

SUMMARY

1. Given an equation $ax^2 + bxy + cy^2 + dx + ey + f = 0$, let A be the symmetric coefficient matrix of the quadratic-form portion of the equation (the portion in color), let λ_1 and λ_2 be the eigenvalues of A, and let C be an orthogonal matrix of determinant 1 with eigenvectors of A for columns. The substitution corresponding to matrix C followed by translation of axes reduces the given equation to a standard form for the equation of a conic section. In particular, the equation describes a (possibly degenerate or empty)

$$
\begin{array}{ll}
\text{ellipse} & \text{if } \lambda_1 \lambda_2 > 0, \\
\text{hyperbola} & \text{if } \lambda_1 \lambda_2 < 0, \\
\text{parabola} & \text{if } \lambda_1 \lambda_2 = 0.
\end{array}
$$

2. Proceeding in a way analogous to that described in summary item (1) but for the equation

$$ax^2 + by^2 + cz^2 + dxy + exz + fyz + px + qy + rz + s = 0,$$

one obtains a standard form for the equation of a (possibly degenerate or empty) quadric surface in space. Table 8.1 lists the information that can be obtained from the three eigenvalues λ_1, λ_2, and λ_3 alone.

EXERCISES

In Exercises 1–8, rotate axes, using a substitution

$$\begin{bmatrix} x \\ y \end{bmatrix} = C \begin{bmatrix} t_1 \\ t_2 \end{bmatrix},$$

complete squares if necessary, and sketch the graph (if it is not empty) of the given conic section.

1. $2xy = 1$

2. $2xy - 2\sqrt{2}y = 1$

3. $x^2 + 2xy + y^2 = 4$

4. $x^2 - 2xy + y^2 + 4\sqrt{2}x = 4$

5. $10x^2 + 6xy + 2y^2 = 4$

6. $5x^2 + 4xy + 2y^2 = -1$

7. $3x^2 + 4xy + 6y^2 = 8$

8. $x^2 + 8xy + 7y^2 + 18\sqrt{5}x = -9$

9. Show that the plane curve $ax^2 + bxy + cy^2 + dx + ey + f = 0$ is a (possibly degenerate or empty)

$$\begin{array}{ll} \text{ellipse} & \text{if } b^2 - 4ac < 0, \\ \text{hyperbola} & \text{if } b^2 - 4ac > 0, \\ \text{parabola} & \text{if } b^2 - 4ac = 0. \end{array}$$

[HINT: Diagonalize $ax^2 + bxy + cy^2$, using the quadratic formula, and check the signs of the eigenvalues.]

10. Use Exercise 9 to classify the conic section with the given equation.

a. $2x^2 + 8xy + 8y^2 - 3x + 2y = 13$
b. $y^2 + 4xy - 5x^2 - 3x = 12$
c. $-x^2 + 5xy - 7y^2 - 4y + 11 = 0$
d. $xy + 4x - 3y = 8$
e. $2x^2 - 3xy + y^2 - 8x + 5y = 30$
f. $x^2 + 6xy + 9y^2 - 2x + 14y = 10$
g. $4x^2 - 2xy - 3y^2 + 8x - 5y = 17$
h. $8x^2 + 6xy + 2y^2 - 5x = 25$
i. $x^2 - 2xy + 4x - 5y = 6$
j. $2x^2 - 3xy + 2y^2 - 8y = 15$

11. Show that the equation $ax^2 + by + cz = d$ can be transformed into an equation of the form $at_1^2 + rt_2 = d$ by a rotation of axes in space. [HINT:

If $\begin{bmatrix} x \\ y \\ z \end{bmatrix} = C \begin{bmatrix} t_1 \\ t_2 \\ t_3 \end{bmatrix}$, then

$$\begin{bmatrix} t_1 \\ t_2 \\ t_3 \end{bmatrix} = C^{-1} \begin{bmatrix} x \\ y \\ z \end{bmatrix} = C^T \begin{bmatrix} x \\ y \\ z \end{bmatrix}.$$

Find an orthogonal matrix C^T such that $\det(C^T) = 1$, $t_1 = x$, and $t_2 = (by + cz)/r$ for some r.]

In Exercises 12–20, classify the quadric surface with the given equation as one of the (possibly empty or degenerate) types illustrated in Figures 8.7–8.15.

12. $2x^2 + 2y^2 + 6yz + 10z^2 = 9$

13. $2xy + z^2 + 1 = 0$

14. $2xz + y^2 + 2y + 1 = 0$

15. $x^2 + 2yz + 4x + 1 = 0$

16. $3x^2 + 2y^2 + 6xz + 3z^2 = 1$

17. $x^2 - 8xy + 16y^2 - 3z^2 = 8$

18. $x^2 + 4y^2 - 4xz + 4z^2 = 8$

19. $-3x^2 + 2y^2 + 8yz + 16z^2 = 10$

20. $x^2 + y^2 + z^2 - 2xy + 2xz - 2yz = 9$

In Exercises 21–27, use the routine MATCOMP in LINTEK, or MATLAB, to classify the quadric surface according to Table 8.1.

21. $x^2 + y^2 + z^2 - 2xy - 2xz - 2yz + 3x - 3z = 8$

22. $3x^2 + 2y^2 + 5z^2 + 4xy + 2yz - 3x + 10y = 4$

23. $2x^2 - 8y^2 + 3z^2 - 4xy + yz + 6xz - 3x - 8z = 3$

24. $3x^2 + 7y^2 + 4z^2 + 6xy - 8yz + 16x = 20$

25. $x^2 + 4y^2 + 16z^2 + 4xy + 8xz + 16yz - 8x + 3y = 8$

26. $x^2 + 6y^2 + 4z^2 - xy - 2xz - 3yz - 9x = 20$

27. $4x^2 - 3y^2 + z^2 + 8xz + 6yz + 2x - 3y = 8$

8.3   APPLICATIONS TO EXTREMA

Finding Extrema of Functions

We turn now to one of the applications of linear algebra to calculus. We simply state the facts from calculus that we need, trying to make them seem reasonable where possible.

There are many situations in which one desires to maximize or minimize a function of one or more variables. Applications involve maximizing profit, minimizing costs, maximizing speed, minimizing time, and so on. Such problems are of great practical importance.

Polynomial functions are especially easy to work with. Many important functions, such as trigonometric, exponential, and logarithmic functions, can't be expressed by polynomial formulas. However, each of these functions, and many others, can be expressed near a point in the domain of the function as a "polynomial" of infinite degree—an *infinite series*, as it is called. For example, it is shown in calculus that

$$\cos x = 1 - \frac{1}{2!}x^2 + \frac{1}{4!}x^4 - \frac{1}{6!}x^6 + \cdots \tag{1}$$

for any number x. [Recall that $2! = 2 \cdot 1$, $3! = 3 \cdot 2 \cdot 1$, and in general, $n! = n(n-1)(n-2) \cdots 3 \cdot 2 \cdot 1$.] We leave to calculus the discussion of such things as the interpretation of the infinite sum

$$1 - \frac{1}{2!} + \frac{1}{4!} - \frac{1}{6!} + \cdots$$

as $\cos 1$. From Eq. (1), it would seem that we should have

$$\cos(2x - y) = 1 - \frac{1}{2!}(2x - y)^2 + \frac{1}{4!}(2x - y)^4 - \frac{1}{6!}(2x - y)^6 + \cdots. \tag{2}$$

Notice that the term

$$-\frac{1}{2!}(2x - y)^2 = -\frac{1}{2}(4x^2 - 4xy + y^2)$$

is a quadratic form—that is, a form (homogeneous polynomial) of degree 2. Similarly,

$$\frac{1}{4!}(2x - y)^4 = \frac{1}{24}(16x^4 - 32x^3y + 24x^2y^2 - 8xy^3 + y^4)$$

is a form of degree 4.

Consider a function $g(x_1, x_2, \ldots, x_n)$ of n variables, which we denote as usual by $g(\mathbf{x})$. For many of the most common functions $g(\mathbf{x})$, it can be shown that, if $\mathbf{x}$ is near $\mathbf{0}$, so that all x_i are small in magnitude, then

$$g(\mathbf{x}) = c + f_1(\mathbf{x}) + f_2(\mathbf{x}) + f_3(\mathbf{x}) + \cdots + f_n(\mathbf{x}) + \cdots, \tag{3}$$

where each $f_i(\mathbf{x})$ is a form of degree i or is zero. Equation (2) illustrates this.

We wish to determine whether $g(\mathbf{x})$ in Eq. (3) has a *local extremum*—that is, a *local maximum* or *local minimum* at the origin $\mathbf{x} = \mathbf{0}$. A function $g(\mathbf{x})$ has a **local maximum at 0** if $g(\mathbf{x}) \leq g(\mathbf{0})$ for all $\mathbf{x}$ where $\|\mathbf{x}\|$ is sufficiently small. The notion of a **local minimum at 0** for $g(\mathbf{x})$ is analogously defined. For example, the function of one variable $g(x) = x^2 + 1$, whose graph is shown in Figure 8.16, has a local minimum of 1 at $x = 0$.

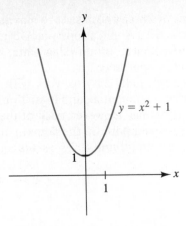

FIGURE 8.16
The graph of $y = g(x) = x^2 + 1$.

Now if $\mathbf{x}$ is really close to $\mathbf{0}$, all the forms $f_1(\mathbf{x})$, $f_2(\mathbf{x})$, $f_3(\mathbf{x})$, ... in Eq. (3) have very small values, so the constant c, if $c \neq 0$, is the dominant term of Eq. (3) near zero. Notice that, for $g(x) = 1 + x^2$ in Figure 8.16, the function has values close to the constant 1 for x close to zero.

After a nonzero constant, the form in Eq. (3) that contributes the most to $g(\mathbf{x})$ for $\|\mathbf{x}\|$ close to $\mathbf{0}$ is the nonzero form $f_i(\mathbf{x})$ of lowest degree, because the *lower* the degree of a term of a polynomial, the *more* it contributes when the variables are all near zero. For example, x is greater than x^2 near zero; if $x = \frac{1}{100}$, then x^2 is only $\frac{1}{10,000}$. If $f_1(\mathbf{x}) = 0$ and $f_2(\mathbf{x}) \neq 0$, then $f_2(\mathbf{x})$ is the dominant form of Eq. (3) after a nonzero constant c, and so on.

We claim that, if $f_1(\mathbf{x}) \neq 0$ in Eq. (3), then $g(\mathbf{x})$ does not have a local maximum or minimum at $\mathbf{x} = \mathbf{0}$. Suppose that

$$f_1(\mathbf{x}) = d_1 x_1 + d_2 x_2 + \cdots + d_n x_n,$$

with some coefficient—say, d_1—nonzero. If k is small but of the same sign as d_1, then near the point $(k, 0, 0, \ldots, 0)$ the function $g(\mathbf{x}) \approx c + d_1 k > c$. On the other hand, if k is small but of opposite sign to d_1, then near this point $g(\mathbf{x}) < c$. Thus, $g(\mathbf{x})$ can have a local maximum or minimum of c at $\mathbf{0}$ only if $f_1(\mathbf{x}) = 0$ for all $\mathbf{x}$.

If $f_1(\mathbf{x}) = 0$ and $f_2(\mathbf{x}) \neq 0$ in Eq. (3), then, for $\mathbf{x}$ near $\mathbf{0}$, $f_2(\mathbf{x})$ is the dominant form in the equation; it can be expected to dominate the terms of higher degree for $\mathbf{x}$ close to $\mathbf{0}$. It seems reasonable that, if $f_2(\mathbf{x}) > 0$ for *all* $\mathbf{x} \neq \mathbf{0}$ but near $\mathbf{0}$, then for such $\mathbf{x}$

$$g(\mathbf{x}) = c + (\text{Little bit}),$$

and $g(\mathbf{x})$ has a local minimum of c at $\mathbf{0}$. On the other hand, if $f_2(\mathbf{x}) < 0$ for *all* such $\mathbf{x}$, then we expect that

$$g(\mathbf{x}) = c - (\text{Little bit})$$

for these values $\mathbf{x}$, and $g(\mathbf{x})$ has a local maximum of c at $\mathbf{0}$. This is proved in an advanced calculus course.

We know that we can orthogonally diagonalize the form $f_2(\mathbf{x})$ with a substitution $\mathbf{x} = C\mathbf{t}$ to become

$$\lambda_1 t_1^2 + \lambda_2 t_2^2 + \cdots + \lambda_n t_n^2, \tag{4}$$

where the λ_i are the eigenvalues of the symmetric coefficient matrix of $f_2(\mathbf{x})$. Form (4) is > 0 for *all* nonzero $\mathbf{t}$, and hence $f_2(\mathbf{x}) > 0$ for *all* nonzero $\mathbf{x}$, if and only if we have *all* $\lambda_i > 0$. Similarly, $f_2(\mathbf{x}) < 0$ for *all* nonzero $\mathbf{x}$ if and only if *all* $\lambda_i < 0$. It is also clear that, if some λ_i are positive and some are negative, then form (4) and hence $f_2(\mathbf{x})$ assume both positive and negative values arbitrarily close to zero.

DEFINITION 8.1   Definite Quadratic Forms

A quadratic form $f(\mathbf{x})$ is **positive definite** if $f(\mathbf{x}) > 0$ for all nonzero $\mathbf{x}$ in $\mathbb{R}^n$, and it is **negative definite** if $f(\mathbf{x}) < 0$ for all such nonzero $\mathbf{x}$.

Our work in Section 8.1 and the preceding statement give us the following theorem and corollary.

THEOREM 8.4   Criteria for Definite Quadratic Forms

A quadratic form is positive definite if and only if all the eigenvalues of its symmetric coefficient matrix are positive, and it is negative definite if and only if all those eigenvalues are negative.

COROLLARY   A Test for Local Extrema

Let $g(\mathbf{x})$ be a function of n variables given by Eq. (3). Suppose that the function $f_1(\mathbf{x})$ in Eq. (3) is zero. If $f_2(\mathbf{x})$ is positive definite, then $g(\mathbf{x})$ has a local minimum of c at $\mathbf{x} = \mathbf{0}$, whereas if $f_2(\mathbf{x})$ is negative definite, then $g(\mathbf{x})$ has a local maximum of c at $\mathbf{x} = \mathbf{0}$. If $f_2(\mathbf{x})$ assumes both positive and negative values, then $g(\mathbf{x})$ has no local extremum at $\mathbf{x} = \mathbf{0}$.

We have discussed local extrema only at the origin. To do similar work at another point $(h_1, h_2, \ldots, h_n)$, we need only translate axes to this point, letting $\overline{x}_i = x_i - h_i$ so that Eq. (3) becomes

$$g(\overline{\mathbf{x}}) = c + f_1(\overline{\mathbf{x}}) + f_2(\overline{\mathbf{x}}) + \cdots + f_n(\overline{\mathbf{x}}) + \cdots . \tag{5}$$

Our discussion indicates a method for attempting to find local extrema of a function, which we box.

Finding an Extremum of g(x)

Step 1 Find a point **h** where the function $f_1(\overline{\mathbf{x}}) = f_1(\mathbf{x} - \mathbf{h})$ in Eq. (5) becomes the zero function. (This is the province of calculus.)

Step 2 Find the quadratic form $f_2(\overline{\mathbf{x}})$ at the point **h**. (This also requires calculus.)

Step 3 Find the eigenvalues of the symmetric coefficient matrix of the quadratic form.

Step 4 If all eigenvalues are positive, then $g(\overline{\mathbf{x}})$ has a local minimum of c at **h**. If all are negative, then $g(\overline{\mathbf{x}})$ has a local maximum of c at **h**. If eigenvalues of both signs occur, then $g(\overline{\mathbf{x}})$ has no local extremum at **h**.

Step 5 If any of the eigenvalues is zero, further study is necessary.

Exercises 17 through 22 illustrate some of the things that can occur if step 5 is the case. We shall not tackle steps 1 or 2, because they require calculus, but will simply start with equations of the form of Eqs. (3) or (5) that meet the requirements in steps 1 and 2.

EXAMPLE 1   Let

$$g(x, y) = 3 + (2x^2 - 4xy + 4y^2) + (x^3 + 4xy^2 + y^3).$$

Determine whether $g(x, y)$ has a local extremum at the origin.

SOLUTION   The symmetric coefficient matrix for the quadratic-form portion $2x^2 - 4xy + 4y^2$ of $g(x, y)$ is

$$A = \begin{bmatrix} 2 & -2 \\ -2 & 4 \end{bmatrix}.$$

We find that

$$|A - \lambda I| = \begin{vmatrix} 2 - \lambda & -2 \\ -2 & 4 - \lambda \end{vmatrix} = \lambda^2 - 6\lambda + 4.$$

The solutions of the characteristic equation $\lambda^2 - 6\lambda + 4 = 0$ are found by the quadratic formula to be

$$\lambda = \frac{6 \pm \sqrt{36 - 16}}{2} = 3 \pm \sqrt{5}.$$

We see that $\lambda_1 = 3 + \sqrt{5}$ and $\lambda_2 = 3 - \sqrt{5}$ are both positive, so our form is positive definite. Thus, $g(x, y)$ has a local minimum of 3 at $(0, 0)$.   ■

EXAMPLE 2 Suppose that

$$g(x, y, z) = 7 + (2x^2 - 8y^2 + 3z^2 - 4xy + 2yz + 6xz) + (xz^2 - 5y^3)$$
$$+ \text{(higher-degree terms)}.$$

Determine whether $g(x, y, z)$ has a local extremum at the origin.

SOLUTION The symmetric coefficient matrix of the quadratic-form portion is

$$A = \begin{bmatrix} 2 & -2 & 3 \\ -2 & -8 & 1 \\ 3 & 1 & 3 \end{bmatrix}.$$

We could find $p(\lambda) = |A - \lambda I|$ and attempt to solve the characteristic equation, or we could try to sketch the graph of $p(\lambda)$ well enough to determine the signs of the eigenvalues. However, we prefer to use the routine MATCOMP in LINTEK, or MATLAB, and we find that the eigenvalues are approximately

$$\lambda_1 = -8.605, \qquad \lambda_2 = 0.042, \qquad \lambda_3 = 5.563.$$

Because both positive and negative eigenvalues appear, there is no local extremum at the origin. ■

Maximizing or Minimizing a Quadratic Form on the Unit Sphere

Let $f(\mathbf{x})$ be a quadratic form in the variables $x_1, x_2, \ldots, x_n$. We consider the problem of finding the maximum and minimum values of $f(\mathbf{x})$ for $\mathbf{x}$ on the *unit sphere*, where $\|\mathbf{x}\| = 1$—that is, where $x_1^2 + x_2^2 + \cdots + x_n^2 = 1$. It is shown in advanced calculus that such extrema of $f(\mathbf{x})$ on the unit sphere always exist.

We need only orthogonally diagonalize the form $f(\mathbf{x})$, using an orthogonal transformation $\mathbf{x} = C\mathbf{t}$, obtaining as usual

$$\lambda_1 t_1^2 + \lambda_2 t_2^2 + \cdots + \lambda_n t_n^2. \tag{6}$$

Because our new basis for $\mathbb{R}^n$ is again orthonormal, the unit sphere has $\mathbf{t}$-equation

$$t_1^2 + t_2^2 + \cdots + t_n^2 = 1; \tag{7}$$

this is a very important point. Suppose that the λ_i are arranged so that

$$\lambda_1 \geq \lambda_2 \geq \cdots \geq \lambda_n.$$

On the unit sphere with Eq. (7), we see that formula (6) can be written as

$$\lambda_1(1 - t_2^2 - \cdots - t_n^2) + \lambda_2 t_2^2 + \cdots + \lambda_n t_n^2$$
$$= \lambda_1 - (\lambda_1 - \lambda_2)t_2^2 - \cdots - (\lambda_1 - \lambda_n)t_n^2. \tag{8}$$

Because $\lambda_1 - \lambda_i \geq 0$ for $i > 1$, we see that the maximum value assumed by formula (8) is λ_1 when $t_2 = t_3 = \cdots = t_n = 0$, and $t_1 = \pm 1$. Exercise 32 indicates similarly that the minimum value assumed by form (6) is λ_n, when $t_n = \pm 1$ and all other $t_i = 0$. We state this as a theorem.

THEOREM 8.5   Extrema of a Quadratic Form on the Unit Sphere

Let $f(\mathbf{x})$ be a quadratic form, and let $\lambda_1, \lambda_2, \ldots, \lambda_n$ be the eigenvalues of the symmetric coefficient matrix of $f(\mathbf{x})$. The maximum value assumed by $f(\mathbf{x})$ on the unit sphere $\|\mathbf{x}\| = 1$ is the maximum of the λ_i, and the minimum value assumed is the minimum of the λ_i. Each extremum is assumed at any eigenvector of length 1 corresponding to the eigenvalue that gives the extremum.

The preceding theorem is very important in vibration applications ranging from aerodynamics to particle physics. In such applications, one often needs to know the eigenvalue of maximum or of minimum magnitude for a symmetric matrix. These eigenvalues are frequently found by using advanced-calculus techniques to maximize or minimize the value of a quadratic form on a unit sphere, rather than using algebraic techniques such as those presented in this text for finding eigenvalues. The principal axis theorem (Theorem 8.1) and the preceding theorem are the algebraic foundation for such an analytic approach. We illustrate Theorem 8.5 with an example that maximizes a quadratic form on a unit sphere by finding the eigenvalues, rather than illustrating the more important reverse procedure.

EXAMPLE 3   Find the maximum and minimum values assumed by $2xy + 2xz$ on the unit sphere $x^2 + y^2 + z^2 = 1$, and find all points where these extrema are assumed.

SOLUTION   From Example 5 in Section 8.1, we see that the eigenvalues of the symmetric coefficient matrix A and an associated eigenvector are given by

$$\begin{aligned}
\lambda_1 &= 0, & \mathbf{v}_1 &= [0, -1, 1], \\
\lambda_2 &= \sqrt{2}, & \mathbf{v}_2 &= [\sqrt{2}, 1, 1], \\
\lambda_3 &= -\sqrt{2}, & \mathbf{v}_3 &= [-\sqrt{2}, 1, 1].
\end{aligned}$$

We see that the maximum value assumed by $2xy + 2xz$ on the unit sphere is $\sqrt{2}$, and it is assumed at the points $\pm(\sqrt{2}/2, \frac{1}{2}, \frac{1}{2})$. The minimum value assumed is $-\sqrt{2}$, and it is assumed at the points $\pm(-\sqrt{2}/2, \frac{1}{2}, \frac{1}{2})$. Notice that we normalized our eigenvectors to length 1 so that they extend to the unit sphere.   ∎

The extension of Theorem 8.5 to a sphere centered at the origin, with radius other than 1, is left as Exercise 33.

SUMMARY

1. Let $g(\mathbf{x}) = c + f_1(\mathbf{x}) + f_2(\mathbf{x}) + \cdots + f_n(\mathbf{x}) + \cdots$ near $\mathbf{x} = \mathbf{0}$, where $f_i(\mathbf{x})$ is a form of degree i or is zero.

    a. If $f_1(\mathbf{x}) = 0$ and $f_2(\mathbf{x})$ is positive definite, then $g(\mathbf{x})$ has a local minimum of c at $\mathbf{x} = \mathbf{0}$.

    b. If $f_1(\mathbf{x}) = 0$ and $f_2(\mathbf{x})$ is negative definite, then $g(\mathbf{x})$ has a local maximum of c at $\mathbf{x} = \mathbf{0}$.

    c. If $f_1(\mathbf{x}) \neq 0$ or if the symmetric coefficient matrix of $f_2(\mathbf{x})$ has both positive and negative eigenvalues, then $g(\mathbf{x})$ has no local extremum at $\mathbf{x} = \mathbf{0}$.

2. The natural analogue of summary item 1 holds at $\mathbf{x} = \mathbf{h}$; just translate the axes to the point $\mathbf{h}$, and replace x_i by $\bar{x}_i = x_i - h_i$.

3. A quadratic form in n variables has as maximum (minimum) value on the unit sphere $\|\mathbf{x}\| = 1$ in $\mathbb{R}^n$ the maximum (minimum) of the eigenvalues of the symmetric coefficient matrix of the form. The maximum (minimum) is assumed at each corresponding eigenvector of length 1.

EXERCISES

In Exercises 1–15, assume that $g(\mathbf{x})$, or $g(\bar{\mathbf{x}})$, is described by the given formula for values of $\mathbf{x}$, or $\bar{\mathbf{x}}$, near zero. Draw whatever conclusions are possible concerning local extrema of the function g.

1. $g(x, y) = -7 + (3x^2 - 6xy + 4y^2) + (x^3 - 4y^3)$

2. $g(x, y) = 8 - (2x^2 - 8xy + 3y^2) + (2x^2y - y^3)$

3. $g(x, y) = 4 - 3x + (2x^2 - 2xy + y^2) + (2x^2y + y^3) + \cdots$

4. $g(x, y) = 5 - (8x^2 - 6xy + 2y^2) + (4x^3 - xy^2) + \cdots$

5. $g(\bar{x}, \bar{y}) = 3 - (4\bar{x}^2 - 8\bar{x}\bar{y} + 5\bar{y}^2) + (2\bar{x}^2\bar{y} - \bar{y}^3) + \cdots, \bar{x} = x + 5, \bar{y} = y$

6. $g(x, y) = -2 + (8x^2 + 4xy + y^2) + (x^3 + 5x^2y)$

7. $g(x, y) = 5 + (3x^2 + 10xy + 7y^2) + (7xy^2 - y^3)$

8. $g(x, y) = 4 - (x^2 - 6xy + 9y^2) + (x^3 - y^3) + \cdots$

9. $g(\bar{x}, \bar{y}) = 3 + (2\bar{x}^2 + 8\bar{x}\bar{y} + 8\bar{y}^2) + (4\bar{x}^3 - \bar{x}\bar{y}^2) + \cdots, \bar{x} = x - 3, \bar{y} = y - 1$

10. $g(\bar{x}, \bar{y}) = 4 - (\bar{x}^2 + 3\bar{x}\bar{y} - \bar{y}^2) + (4\bar{x}^2\bar{y} - 5\bar{x}\bar{y}^2), \bar{x} = x + 1, \bar{y} = y - 7$

11. $g(x, y, z) = 4 - (x^2 + 4xy + 5y^2 + 3z^2) + (x^3 - xyz)$

12. $g(x, y, z) = 3 + (2x^2 + 6xz - y^2 + 5z^2) + (x^2z - y^2z) + \cdots$

13. $g(x, y, z) = 5 + (4x^2 + 2xy + y^2 - z^2) + (xy^2 - 4xyz) + \cdots$

14. $g(\bar{x}, \bar{y}, \bar{z}) = 4 + (\bar{x}^2 + \bar{y}^2 + \bar{z}^2 - 2\bar{x}\bar{z} - 2\bar{x}\bar{y} - 2\bar{y}\bar{z}) + (3\bar{x}^3 - \bar{z}^3) + \cdots, \bar{x} = x + 1, \bar{y} = y - 2, \bar{z} = z + 5$

15. $g(\bar{x}, \bar{y}, \bar{z}) = 4 - 3\bar{z} + (2\bar{x}^2 - 2\bar{x}\bar{y} + 3\bar{x}\bar{z} + 5\bar{y}^2 + \bar{z}^2) + (2\bar{x}\bar{y}\bar{z} - \bar{z}^3) + \cdots, \bar{x} = x - 7, \bar{y} = y + 6, \bar{z} = z$

16. Define the notion of a local minimum c of a function $f(\mathbf{x})$ of n variables at a point $\mathbf{h}$.

In Exercises 17–22, let

$$g(x, y) = c + f_1(x, y) + f_2(x, y) + f_3(x, y) + \cdots$$

in accordance with the notation we have been using. Let λ_1 and λ_2 be the eigenvalues of the symmetric coefficient matrix of $f_2(x, y)$.

17. Give an example of a polynomial function $g(x, y)$ for which $f_1(x, y) = 0$, $\lambda_1 = 0$, $|\lambda_2| = 1$, and $g(x, y)$ has a local minimum of 10 at the origin.

18. Repeat Exercise 17, but make $g(x, y)$ have a local maximum of -5 at the origin.

19. Repeat Exercise 17, but make $g(x, y)$ have no local extremum at the origin.

20. Give an example of a polynomial function $g(x, y)$ such that $f_1(x, y) = f_2(x, y) = 0$, having a local maximum of 1 at the origin.

21. Repeat Exercise 20, but make $g(x, y)$ have a local minimum of 40 at the origin.

22. Repeat Exercise 20, but make $g(x, y)$ have no local extremum at the origin.

In Exercises 23–31, find the maximum and minimum values of the quadratic form on the unit circle in $\mathbb{R}^2$, or unit sphere in $\mathbb{R}^n$, and find all points on the unit circle or sphere where the extrema are assumed.

23. $xy, n = 2$

24. $3x^2 + 4xy, n = 2$

25. $-6xy + 8y^2, n = 2$

26. $x^2 + 2xy + y^2, n = 3$

27. $3x^2 - 6xy + 3y^2, n = 2$

28. $y^2 + 2xz, n = 3$

29. $x^2 + y^2 + z^2 - 2xy + 2xz - 2yz, n = 3$

30. $x^2 + y^2 + z^2 - 2xy - 2xz - 2yz, n = 3$
    [SUGGESTION: Use Example 4 in Section 6.3.]

31. $x^2 + w^2 + 4yz - 2xw, n = 4$

32. Prove that the minimum value assumed by a quadratic form in n variables on the unit sphere in $\mathbb{R}^n$ is the minimum eigenvalue of the symmetric coefficient matrix of the form, and prove that this minimum value is assumed at any corresponding eigenvector of length 1.

33. Let $f(\mathbf{x})$ be a quadratic form in n variables. Describe the maximum and minimum values assumed by $f(\mathbf{x})$ on the sphere $\|\mathbf{x}\| = a^2$ for any $a > 0$. Describe the points of the sphere at which these extrema are assumed. [HINT: For any k in $\mathbb{R}$, how does $f(k\mathbf{x})$ compare with $f(\mathbf{x})$?]

In Exercises 34–38, use the routine MATCOMP in LINTEK, or MATLAB, and follow the directions for Exercises 1–15.

34. $g(x, y, z) = 7 + (3x^2 + 2y^2 + 5z^2 + 4xy + 2yz) + (xyz - x^2y)$

35. $g(x, y, z) = 5 - (x^2 + 6y^2 + 4z^2 - xy - 2xz - 3yz) + (xz^2 - 5z^3) + \cdots$

36. $g(x, y, z) = -1 - (2x^2 - 8y^2 + 3z^2 - 4xy + yz + 6xz) + (8x^3 - 4y^2z + 3z^3)$

37. $g(x, y, z) = 7 + (x^2 + 2y^2 + z^2 - 3xy - 4xz - 3yz) + (x^2y - 4yz^2)$

38. $g(x, y, z) = 5 + (x^2 + 4y^2 + 16z^2 + 4xy + 8xz + 16yz) + (x^3 + y^3 + 7xyz) + \cdots$

8.4   COMPUTING EIGENVALUES AND EIGENVECTORS

Computing the eigenvalues of a matrix is one of the toughest jobs in linear algebra. Many algorithms have been developed, but no one method can be considered the best for all cases. We have used the characteristic equation for computation of eigenvalues in most of our examples and exercises. The routine MATCOMP in LINTEK also uses it. Professionals frown on the use of this method in practical applications because small errors made in computing coefficients of the characteristic polynomial can lead to significant errors in eigenvalues. We describe three other methods in this section.

1. The *power method* is especially useful if one wants only the eigenvalue of largest (or of smallest) magnitude, as in many vibration problems.

2. *Jacobi's method* for symmetric matrices is presented without proof. This method is chiefly of historical interest, because the third and most recent of the methods we describe is more general and usually more efficient.

3. The *QR method* was developed by H. Rutishauser in 1958 and J. G. F. Francis in 1961 and is probably the most widely used method today. It finds all eigenvalues, both real and complex, of a real matrix. Details are beyond the scope of this text. We give only a rough idea of the method, with no proofs.

The routines POWER, JACOBI, and QRFACTOR in LINTEK can be used to illustrate the three methods.

The Power Method

Let A be an $n \times n$ diagonalizable matrix with real eigenvalues. Suppose that one eigenvalue—say, λ_1—has greater magnitude than all the others. That is, $|\lambda_1| > |\lambda_i|$ for $i > 1$. We call λ_1 the **dominant eigenvalue** of A. In many vibration problems, we are interested only in computing this dominant eigenvalue and the eigenvalue of minimum absolute value.

Because A is diagonalizable, there exists a basis

$$\{\mathbf{b}_1, \mathbf{b}_2, \ldots, \mathbf{b}_n\}$$

for $\mathbb{R}^n$ composed of eigenvectors of A. We assume that $\mathbf{b}_i$ is the eigenvector corresponding to λ_i, and that the numbering is such that $|\lambda_1| > |\lambda_2| \geq |\lambda_3| \geq \cdots \geq |\lambda_n|$.

Let $\mathbf{w}_1$ be any nonzero vector in $\mathbb{R}^n$. Then

$$\mathbf{w}_1 = c_1\mathbf{b}_1 + c_2\mathbf{b}_2 + \cdots + c_n\mathbf{b}_n \tag{1}$$

for some constants c_i in $\mathbb{R}$. Applying A^s to both sides of Eq. (1) and remembering that $A^s\mathbf{b}_i = \lambda_i^s\mathbf{b}_i$, we see that

$$A^s\mathbf{w}_1 = \lambda_1^s c_1\mathbf{b}_1 + \lambda_2^s c_2\mathbf{b}_2 + \cdots + \lambda_n^s c_n\mathbf{b}_n. \tag{2}$$

Because λ_1 is dominant, we see that, for large s, the summand $\lambda_1^s c_1\mathbf{b}_1$ dominates the right-hand side of Eq. (2), as long as $c_1 \neq 0$. This is even more evident if we rewrite Eq. (2) in the form

$$A^s\mathbf{w}_1 = \lambda_1^s(c_1\mathbf{b}_1 + (\lambda_2/\lambda_1)^s c_2\mathbf{b}_2 + \cdots + (\lambda_n/\lambda_1)^s c_n\mathbf{b}_n). \tag{3}$$

If s is large, the quotients $(\lambda_i/\lambda_1)^s$ for $i > 1$ are close to zero, because $|\lambda_i/\lambda_1| < 1$. Thus, if $c_1 \neq 0$ and s is large enough, $A^s\mathbf{w}_1$ is very nearly parallel to $\lambda_1^s c_1\mathbf{b}_1$, which is an eigenvector of A corresponding to the eigenvalue λ_1. This suggests that we can approximate an eigenvector of A corresponding to the dominant eigenvalue λ_1 by multiplying an appropriate initial approximation vector $\mathbf{w}_1$ repeatedly by A.

A few comments are in order. In a practical application, we may have a rough idea of an eigenvector for λ_1 and be able to choose a reasonably good first approximation $\mathbf{w}_1$. In any case, $\mathbf{w}_1$ *should not* be in the subspace of $\mathbb{R}^n$

generated by the eigenvectors corresponding to the λ_j for $j > 1$. This corresponds to the requirement that $c_1 \neq 0$.

Repeated multiplication of $\mathbf{w}_1$ by A may produce very large (or very small) numbers. It is customary to scale after each multiplication, to keep the components of the vectors at a reasonable size. After the first multiplication, we find the maximum d_1 of the magnitudes of all the components of $A\mathbf{w}_1$ and apply A the next time to the vector $\mathbf{w}_2 = (1/d_1)A\mathbf{w}_1$. Similarly, we let $\mathbf{w}_3 = (1/d_2)A\mathbf{w}_2$, where d_2 is the maximum of the magnitudes of components of $A\mathbf{w}_2$, and so on. Thus we are always multiplying A times a vector $\mathbf{w}_j$ with components of maximum magnitude 1. This scaling also aids us in estimating the number-of-significant-figures accuracy we have attained in the components of our approximations to an eigenvector.

If $\mathbf{x}$ is an eigenvector corresponding to λ_1, then

$$\frac{A\mathbf{x} \cdot \mathbf{x}}{\mathbf{x} \cdot \mathbf{x}} = \frac{\lambda_1 \mathbf{x} \cdot \mathbf{x}}{\mathbf{x} \cdot \mathbf{x}} = \lambda_1. \tag{4}$$

The quotient $(A\mathbf{x} \cdot \mathbf{x})/(\mathbf{x} \cdot \mathbf{x})$ is called a **Rayleigh quotient**. As we compute the $\mathbf{w}_j$, the Rayleigh quotients $(A\mathbf{w}_j \cdot \mathbf{w}_j)/(\mathbf{w}_j \cdot \mathbf{w}_j)$ should approach λ_1.

This *power method* for finding the dominant eigenvector should, mathematically, break down if we choose the initial approximation $\mathbf{w}_1$ in Eq. (1) in such a way that the coefficient c_1 of $\mathbf{b}_1$ is zero. However, due to roundoff error, it often happens that a nonzero component of $\mathbf{b}_1$ creeps into the $\mathbf{w}_j$ as they are computed, and the $\mathbf{w}_j$ then start swinging toward an eigenvector for λ_1 as desired. This is one case where roundoff error is helpful!

Equation (3) indicates that the ratio $|\lambda_2/\lambda_1|$, which is the maximum of the magnitudes $|\lambda_i/\lambda_1|$ for $i > 1$, should control the speed of convergence of the $\mathbf{w}_j$ to an eigenvector. If $|\lambda_2/\lambda_1|$ is close to 1, convergence may be quite slow.

We summarize the steps of the power method in the following box.

HISTORICAL NOTE   THE RAYLEIGH QUOTIENT is named for John William Strutt, the third Baron Rayleigh (1842–1919). Rayleigh was a hereditary peer who surprised his family by pursuing a scientific career instead of contenting himself with the life of a country gentleman. He set up a laboratory at the family seat in Terling Place, Essex, and spent most of his life there pursuing his research into many aspects of physics—in particular, sound and optics. He is especially famous for his resolution of the long-standing question in optics as to why the sky is blue, as well as for his codiscovery of the element argon, for which he won the Nobel prize in 1904. When he received the British Order of Merit in 1902 he said that "the only merit of which he personally was conscious was that of having pleased himself by his studies, and any results that may have been due to his researches were owing to the fact that it had been a pleasure to him to become a physicist."

Rayleigh used the Rayleigh quotient early in his career in an 1873 work in which he needed to evaluate approximately the normal modes of a complex vibrating system. He subsequently used it and related methods in his classic text *The Theory of Sound* (1877).

> **The Power Method for Finding the Dominant Eigenvalue λ_1 of A**
>
> **Step 1** Choose an appropriate vector $\mathbf{w}_1$ in $\mathbb{R}^n$ as first approximation to an eigenvector corresponding to λ_1.
>
> **Step 2** Compute $A\mathbf{w}_1$ and the Rayleigh quotient $(A\mathbf{w}_1 \cdot \mathbf{w}_1)/(\mathbf{w}_1 \cdot \mathbf{w}_1)$.
>
> **Step 3** Let $\mathbf{w}_2 = (1/d_1)A\mathbf{w}_1$, where d_1 is the maximum of the magnitudes of components of $A\mathbf{w}_1$.
>
> **Step 4** Repeat step 2, with all subscripts increased by 1. The Rayleigh quotients should approach λ_1, and the $\mathbf{w}_j$ should approach an eigenvector of A corresponding to λ_1.

EXAMPLE 1   Illustrate the power method for the matrix

$$A = \begin{bmatrix} 3 & -2 \\ -2 & 0 \end{bmatrix} \quad \text{starting with} \quad \mathbf{w}_1 = \begin{bmatrix} 1 \\ -1 \end{bmatrix},$$

finding $\mathbf{w}_2$, $\mathbf{w}_3$, and the first two Rayleigh quotients.

SOLUTION   We have

$$A\mathbf{w}_1 = \begin{bmatrix} 3 & -2 \\ -2 & 0 \end{bmatrix}\begin{bmatrix} 1 \\ -1 \end{bmatrix} = \begin{bmatrix} 5 \\ -2 \end{bmatrix}.$$

We find as first Rayleigh quotient

$$\frac{A\mathbf{w}_1 \cdot \mathbf{w}_1}{\mathbf{w}_1 \cdot \mathbf{w}_1} = \frac{7}{2} = 3.5.$$

Because 5 is the maximum magnitude of a component of $A\mathbf{w}_1$, we have

$$\mathbf{w}_2 = \frac{1}{5}A\mathbf{w}_1 = \begin{bmatrix} 1 \\ -\frac{2}{5} \end{bmatrix}.$$

Then

$$A\mathbf{w}_2 = \begin{bmatrix} 3 & -2 \\ -2 & 0 \end{bmatrix}\begin{bmatrix} 1 \\ -\frac{2}{5} \end{bmatrix} = \begin{bmatrix} \frac{19}{5} \\ -2 \end{bmatrix}.$$

The next Rayleigh quotient is

$$\frac{A\mathbf{w}_2 \cdot \mathbf{w}_2}{\mathbf{w}_2 \cdot \mathbf{w}_2} = \frac{\dfrac{23}{5}}{\dfrac{29}{25}} = \frac{115}{29} \approx 3.966.$$

Finally,

$$\mathbf{w}_3 = \frac{5}{19}\begin{bmatrix} \frac{19}{5} \\ -2 \end{bmatrix} = \begin{bmatrix} 1 \\ -\frac{10}{19} \end{bmatrix}.$$

∎

EXAMPLE 2   Use the routine POWER in LINTEK with the data in Example 1, and give the vectors $\mathbf{w}_j$ and the Rayleigh quotients until stabilization to all decimal places printed occurs.

SOLUTION   POWER prints fewer significant figures for the vectors to save space. The data obtained using POWER are shown in Table 8.2. It is easy to solve the characteristic equation $\lambda^2 - 3\lambda - 4 = 0$ of A, and to see that the eigenvalues are really $\lambda_1 = 4$ and $\lambda_2 = -1$. POWER found the dominant eigenvalue 4, and it shows that $[1, -0.5]$ is an eigenvector for this eigenvalue.   ∎

If A has an eigenvalue of nonzero magnitude *smaller* than that of any other eigenvalue, the power method can be used with A^{-1} to find this smallest eigenvalue. The eigenvalues of A^{-1} are the reciprocals of the eigenvalues of A, and the eigenvectors are the same. To illustrate, if 5 is the dominant eigenvalue of A^{-1} and $\mathbf{v}$ is an associated eigenvector, then $\frac{1}{5}$ is the eigenvalue of A of smallest magnitude, and $\mathbf{v}$ is still an associated eigenvector.

TABLE 8.2

Power Method for $A = \begin{bmatrix} 3 & -2 \\ -2 & 0 \end{bmatrix}$

Vector Approximations	Rayleigh Quotients
$[1, -1]$	
$[1, -.4]$	3.5
$[1, -.5263158]$	3.96551724137931
$[1, -.4935065]$	3.997830802603037
$[1, -.5016286]$	3.999864369998644
$[1, -.4995932]$	3.999991522909338
$[1, -.5001018]$	3.999999470180991
$[1, -.4999746]$	3.999999966886309
$[1, -.5000064]$	3.999999997930394
$[1, -.4999984]$	3.99999999987065
$[1, -.5000004]$	3.999999999991916
$[1, -.4999999]$	3.999999999999495
$[1, -.5]$	3.999999999999968
$[1, -.5]$	3.999999999999998
$[1, -.5]$	4
$[1, -.5]$	4

Deflation for Symmetric Matrices

The method of deflation gives a way to compute eigenvalues of intermediate magnitude of a *symmetric* matrix by the power method. It is based on an interesting decomposition of a matrix product AB. Let A be an $m \times n$ matrix, and let B be an $n \times s$ matrix. We write AB symbolically as

$$AB = \begin{bmatrix} | & | & & | \\ \mathbf{c}_1 & \mathbf{c}_2 & \cdots & \mathbf{c}_n \\ | & | & & | \end{bmatrix} \begin{bmatrix} \rule[0.5ex]{2em}{0.4pt} \mathbf{r}_1 \rule[0.5ex]{2em}{0.4pt} \\ \rule[0.5ex]{2em}{0.4pt} \mathbf{r}_2 \rule[0.5ex]{2em}{0.4pt} \\ \vdots \\ \rule[0.5ex]{2em}{0.4pt} \mathbf{r}_n \rule[0.5ex]{2em}{0.4pt} \end{bmatrix},$$

where $\mathbf{c}_j$ is the jth column vector of A and $\mathbf{r}_i$ is the ith row vector of B. We claim that

$$AB = \mathbf{c}_1\mathbf{r}_1 + \mathbf{c}_2\mathbf{r}_2 + \cdots + \mathbf{c}_n\mathbf{r}_n, \tag{5}$$

where each $\mathbf{c}_i\mathbf{r}_i$ is the product of an $m \times 1$ matrix with a $1 \times s$ matrix. To see this, remember that the entry in the ith row and jth column of AB is

$$a_{i1}b_{1j} + a_{i2}b_{2j} + \cdots + a_{in}b_{nj}.$$

But $\mathbf{c}_1\mathbf{r}_1$ contributes precisely $a_{i1}b_{1j}$ to the ith row and jth column of the sum in Eq. (5), while $\mathbf{c}_2\mathbf{r}_2$ contributes $a_{i2}b_{2j}$, and so on. This establishes Eq. (5).

Let A be an $n \times n$ *symmetric* matrix with eigenvalues $\lambda_1, \lambda_2, \ldots, \lambda_n$, where

$$|\lambda_1| \geq |\lambda_2| \geq \cdots \geq |\lambda_n|.$$

We know from Section 6.3 that there exists an orthogonal matrix C such that $C^{-1}AC = D$, where D is a diagonal matrix, with $d_{ii} = \lambda_i$. Then

$$A = CDC^{-1} = CDC^T$$

$$= \begin{bmatrix} | & | & & | \\ \mathbf{b}_1 & \mathbf{b}_2 & \cdots & \mathbf{b}_n \\ | & | & & | \end{bmatrix} \begin{bmatrix} \lambda_1 & & & \mathbf{0} \\ & \lambda_2 & & \\ & & \ddots & \\ \mathbf{0} & & & \lambda_n \end{bmatrix} \begin{bmatrix} \rule[0.5ex]{2em}{0.4pt} \mathbf{b}_1^T \rule[0.5ex]{2em}{0.4pt} \\ \rule[0.5ex]{2em}{0.4pt} \mathbf{b}_2^T \rule[0.5ex]{2em}{0.4pt} \\ \vdots \\ \rule[0.5ex]{2em}{0.4pt} \mathbf{b}_n^T \rule[0.5ex]{2em}{0.4pt} \end{bmatrix}, \tag{6}$$

where $\{\mathbf{b}_1, \mathbf{b}_2, \ldots, \mathbf{b}_n\}$ is an orthonormal basis for $\mathbb{R}^n$ consisting of eigenvectors of A and forming the column vectors of C. From Eq. (6), we see that

$$A = \begin{bmatrix} | & | & & | \\ \mathbf{b}_1 & \mathbf{b}_2 & \cdots & \mathbf{b}_n \\ | & | & & | \end{bmatrix} \begin{bmatrix} \rule[0.5ex]{2em}{0.4pt} \lambda_1\mathbf{b}_1^T \rule[0.5ex]{2em}{0.4pt} \\ \rule[0.5ex]{2em}{0.4pt} \lambda_2\mathbf{b}_2^T \rule[0.5ex]{2em}{0.4pt} \\ \vdots \\ \rule[0.5ex]{2em}{0.4pt} \lambda_n\mathbf{b}_n^T \rule[0.5ex]{2em}{0.4pt} \end{bmatrix}. \tag{7}$$

Using Eq. (5) and writing the $\mathbf{b}_i$ as column vectors, we see that A can be written in the following form:

Spectral Decomposition of A

$$A = \lambda_1 \mathbf{b}_1 \mathbf{b}_1{}^T + \lambda_2 \mathbf{b}_2 \mathbf{b}_2{}^T + \cdots + \lambda_n \mathbf{b}_n \mathbf{b}_n{}^T. \tag{8}$$

Equation (8) is known as the **spectral theorem** for symmetric matrices.

EXAMPLE 3   Illustrate the spectral theorem for the matrix

$$A = \begin{bmatrix} 3 & -2 \\ -2 & 0 \end{bmatrix}$$

of Example 1.

SOLUTION   Computation shows that the matrix A has eigenvalues $\lambda_1 = 4$ and $\lambda_2 = -1$, with corresponding eigenvectors

$$\mathbf{v}_1 = \begin{bmatrix} 1 \\ -\frac{1}{2} \end{bmatrix} \quad \text{and} \quad \mathbf{v}_2 = \begin{bmatrix} \frac{1}{2} \\ 1 \end{bmatrix}.$$

Eigenvectors $\mathbf{b}_1$ and $\mathbf{b}_2$ forming an orthonormal basis are

$$\mathbf{b}_1 = \begin{bmatrix} 2/\sqrt{5} \\ -1/\sqrt{5} \end{bmatrix} \quad \text{and} \quad \mathbf{b}_2 = \begin{bmatrix} 1/\sqrt{5} \\ 2/\sqrt{5} \end{bmatrix}.$$

We have

$$\lambda_1 \mathbf{b}_1 \mathbf{b}_1{}^T + \lambda_2 \mathbf{b}_2 \mathbf{b}_2{}^T = 4 \begin{bmatrix} 2/\sqrt{5} \\ -1/\sqrt{5} \end{bmatrix} [2/\sqrt{5},\ -1/\sqrt{5}] - \begin{bmatrix} 1/\sqrt{5} \\ 2/\sqrt{5} \end{bmatrix} [1/\sqrt{5},\ 2/\sqrt{5}]$$

$$= 4 \begin{bmatrix} \frac{4}{5} & -\frac{2}{5} \\ -\frac{2}{5} & \frac{1}{5} \end{bmatrix} - 1 \begin{bmatrix} \frac{1}{5} & \frac{2}{5} \\ \frac{2}{5} & \frac{4}{5} \end{bmatrix}$$

$$= \begin{bmatrix} \frac{16}{5} & -\frac{8}{5} \\ -\frac{8}{5} & \frac{4}{5} \end{bmatrix} - \begin{bmatrix} \frac{1}{5} & \frac{2}{5} \\ \frac{2}{5} & \frac{4}{5} \end{bmatrix} = \begin{bmatrix} 3 & -2 \\ -2 & 0 \end{bmatrix} = A. \qquad \blacksquare$$

HISTORICAL NOTE   THE TERM *SPECTRUM* was coined around 1905 by David Hilbert (1862–1943) for use in dealing with the eigenvalues of quadratic forms in infinitely many variables. The notion of such forms came out of his study of certain linear operators in spaces of functions. Hilbert was struck by the analogy one could make between these operators and quadratic forms in finitely many variables. Hilbert's approach was greatly expanded and generalized during the next decade by Erhard Schmidt, Frigyes Riesz (1880–1956), and Hermann Weyl. Interestingly enough, in the 1920s physicists called on spectra of certain linear operators to explain optical spectra.

David Hilbert was the most influential mathematician of the early twentieth century. He made major contributions in many fields, including algebraic forms, algebraic number theory, integral equations, the foundations of geometry, theoretical physics, and the foundations of mathematics. His speech at the 2nd International Congress of Mathematicians in Paris in 1900, outlining the important mathematical problems of the day, proved extremely significant in providing the direction for twentieth-century mathematics.

Suppose now that we have found the eigenvalue λ_1 of maximum magnitude of A, and a corresponding eigenvector $\mathbf{v}_1$ by the power method. We compute the unit vector $\mathbf{b}_1 = \mathbf{v}_1/\|\mathbf{v}_1\|$. From Eq. (8), we see that

$$A - \lambda_1\mathbf{b}_1\mathbf{b}_1{}^T = 0\mathbf{b}_1\mathbf{b}_1{}^T + \lambda_2\mathbf{b}_2\mathbf{b}_2{}^T + \cdots + \lambda_n\mathbf{b}_n\mathbf{b}_n{}^T \tag{9}$$

is a matrix with eigenvalues λ_2, λ_3, $\ldots$, λ_n, 0 in order of descending magnitude, and with corresponding eigenvectors $\mathbf{b}_2$, $\mathbf{b}_3$, ..., $\mathbf{b}_n$, $\mathbf{b}_1$. We can now use the power method on this matrix to compute λ_2 and $\mathbf{b}_2$. We then execute this *deflation* again, forming $A - \lambda_1\mathbf{b}_1\mathbf{b}_1{}^T - \lambda_2\mathbf{b}_2\mathbf{b}_2{}^T$ to find λ_3 and $\mathbf{b}_3$, and so on.

The routine POWER in LINTEK has an option to use this method of deflation. For the symmetric matrices of the small size that we use in our examples and exercises, POWER handles deflation well, provided that we compute each eigenvalue until stabilization to all places shown on the screen is achieved. In practice, scientists are wary of using deflation to find more than one or two further eigenvalues, because any error made in computation of an eigenvalue or eigenvector will propagate errors in the computation of subsequent ones.

EXAMPLE 4   Illustrate Eq. (9) for deflation with the matrix

$$A = \begin{bmatrix} 3 & -2 \\ -2 & 0 \end{bmatrix}$$

of Example 3.

SOLUTION   From Example 3, we have

$$A = \lambda_1\mathbf{b}_1\mathbf{b}_1{}^T + \lambda_2\mathbf{b}_2\mathbf{b}_2{}^T = 4\begin{bmatrix} \frac{4}{5} & -\frac{2}{5} \\ -\frac{2}{5} & \frac{1}{5} \end{bmatrix} - \begin{bmatrix} \frac{1}{5} & \frac{2}{5} \\ \frac{2}{5} & \frac{4}{5} \end{bmatrix}.$$

Thus,

$$A - \lambda_1\mathbf{b}_1\mathbf{b}_1{}^T = -\begin{bmatrix} \frac{1}{5} & \frac{2}{5} \\ \frac{2}{5} & \frac{4}{5} \end{bmatrix} = \begin{bmatrix} -\frac{1}{5} & -\frac{2}{5} \\ -\frac{2}{5} & -\frac{4}{5} \end{bmatrix}.$$

The characteristic polynomial for this matrix is

$$\begin{vmatrix} -\frac{1}{5} - \lambda & -\frac{2}{5} \\ -\frac{2}{5} & -\frac{4}{5} - \lambda \end{vmatrix} = (\lambda^2 + \lambda) = \lambda(\lambda + 1),$$

and the eigenvalues are indeed $\lambda_2 = -1$ together with 0, as indicated following Eq. (9).   ∎

EXAMPLE 5   Use the routine POWER in LINTEK to find the eigenvalues and eigenvectors, using deflation, for the symmetric matrix

$$\begin{bmatrix} 2 & -8 & 5 \\ -8 & 0 & 10 \\ 5 & 10 & -6 \end{bmatrix}.$$

SOLUTION   Using POWER with deflation and finding eigenvalues as accurately as the printing on the screen permits, we obtain the eigenvectors and eigenvalues

$$\mathbf{v}_1 = [-.6042427, -.8477272, 1], \qquad \lambda_1 = -17.49848531152027,$$
$$\mathbf{v}_2 = [-.760632, 1, .3881208], \qquad \lambda_2 = 9.96626448890372,$$
$$\mathbf{v}_3 = [1, .3958651, .9398283], \qquad \lambda_3 = 3.532220822616553.$$

Using MATCOMP as a check, we find the same eigenvalues and eigenvectors.   ∎

Jacobi's Method for Symmetric Matrices

We present Jacobi's method for diagonalizing a symmetric matrix, omitting proofs. Let $A = [a_{ij}]$ be an $n \times n$ symmetric matrix, and suppose that a_{pq} is an entry of maximum magnitude among the entries of A that lie above the main diagonal. For example, in the matrix

$$A = \begin{bmatrix} 2 & -8 & 5 \\ -8 & 0 & 10 \\ 5 & 10 & -6 \end{bmatrix}, \tag{10}$$

the entry above the diagonal having maximum magnitude is 10. Then form the 2×2 matrix

$$\begin{bmatrix} a_{pp} & a_{pq} \\ a_{qp} & a_{qq} \end{bmatrix}. \tag{11}$$

From matrix (10), we would form the matrix consisting of the portion shown in color—namely,

$$\begin{bmatrix} 0 & 10 \\ 10 & -6 \end{bmatrix}.$$

Let $C = [c_{ij}]$ be a 2×2 orthogonal matrix that diagonalizes matrix (11). (Recall that C can always be chosen to correspond to a rotation of the plane, although this is not essential in order for Jacobi's method to work.) Now form an $n \times n$ matrix R, the same size as the matrix A, which looks like the identity matrix except that $r_{pp} = c_{11}$, $r_{pq} = c_{12}$, $r_{qp} = c_{21}$, and $r_{qq} = c_{22}$. For matrix (10), where 10 has maximum magnitude above the diagonal, we would have

$$R = \begin{bmatrix} 1 & 0 & 0 \\ 0 & c_{11} & c_{12} \\ 0 & c_{21} & c_{22} \end{bmatrix}.$$

This matrix R will be an orthogonal matrix, with $\det(R) = \det(C)$. Now form the new symmetric matrix $B_1 = R^T A R$, which has zero entries in the row p, column q position and in the row q, column p position. Other entries in B_1 can also be changed from those in A, but it can be shown that the maximum magnitude of off-diagonal entries has been reduced, assuming that no other above-diagonal entry in A had the magnitude of a_{pq}. Then repeat this process, starting with B_1 instead of A, to obtain another symmetric matrix B_2, and so on. It can be shown that the maximum magnitude of off-diagonal entries in the matrices B_i approaches zero as i increases. Thus the sequence of matrices

$$B_1, B_2, B_3, \ldots$$

will approach a diagonal matrix D whose eigenvalues $d_{11}, d_{22}, \ldots, d_{nn}$ are the same as those of A.

If one is going to use Jacobi's method much, one should find the 2×2 matrix C that diagonalizes a general 2×2 symmetric matrix

$$\begin{bmatrix} a & b \\ b & c \end{bmatrix};$$

that is, one should find formulas for computing C in terms of the entries a, b, and c. Exercise 23 develops such formulas.

Rather than give a tedious pencil-and-paper example of Jacobi's method, we choose to present data generated by the routine JACOBI in LINTEK for matrix (10)—which is the same matrix as that in Example 5, where we used the power method. Observe how, in each step, the colored entries of maximum magnitude off the diagonal are reduced to "zero." Although they may not remain zero in the next step, they never return to their original size.

EXAMPLE 6   Use the routine JACOBI in LINTEK to diagonalize the matrix

$$\begin{bmatrix} 2 & -8 & 5 \\ -8 & 0 & 10 \\ 5 & 10 & -6 \end{bmatrix}.$$

SOLUTION   The routine JACOBI gives the following matrices:

$$\begin{bmatrix} 2 & -8 & 5 \\ -8 & 0 & 10 \\ 5 & 10 & -6 \end{bmatrix}$$

$$\begin{bmatrix} 2 & 8.786909 & 3.433691 \\ 8.786909 & -13.44031 & 0 \\ 3.433691 & 0 & 7.440307 \end{bmatrix}$$

$$\begin{bmatrix} -17.41676 & 0 & -1.415677 \\ 0 & 5.97645 & -3.128273 \\ -1.415677 & -3.128273 & 7.440307 \end{bmatrix}$$

$$\begin{bmatrix} -17.41676 & -.8796473 & -1.109216 \\ -.8796473 & 3.495621 & 0 \\ -1.109216 & 0 & 9.921136 \end{bmatrix}$$

$$\begin{bmatrix} -17.46169 & -.8789265 & 0 \\ -.8789265 & 3.495621 & 3.560333\text{E-}02 \\ 0 & 3.560333\text{E-}02 & 9.966067 \end{bmatrix}$$

$$\begin{bmatrix} -17.49849 & 0 & 1.489243\text{E-}03 \\ 0 & 3.532418 & 3.557217\text{E-}02 \\ 1.489243\text{E-}03 & 3.557217\text{E-}02 & 9.966067 \end{bmatrix}$$

$$\begin{bmatrix} -17.49849 & 8.233765\text{E-}06 & -1.48922\text{E-}03 \\ 8.233765\text{E-}06 & 3.532221 & 0 \\ -1.48922\text{E-}03 & 0 & 9.966265 \end{bmatrix}$$

$$\begin{bmatrix} -17.49849 & 8.233765\text{E-}06 & 0 \\ 8.233765\text{E-}06 & 3.532221 & -4.464509\text{E-}10 \\ 0 & -4.464508\text{E-}10 & 9.966265 \end{bmatrix}$$

The off-diagonal entries are now quite small, and we obtain the same eigenvalues from the diagonal that we did in Example 5 using the power method. ∎

QR Algorithm

At the present time, an algorithm based on the QR factorization of an *invertible* matrix, discussed in Section 6.2, is often used by professionals to find eigenvalues of a matrix. A full treatment of the QR algorithm is beyond the scope of this text, but we give a brief description of the method. As with the Jacobi method, pencil-and-paper computations of eigenvalues using the QR algorithm are too cumbersome to include in this text. The routine QRFACTOR in LINTEK can be used to illustrate the features of the QR algorithm that we now describe.

Let A be a nonsingular matrix. The QR algorithm generates a sequence of matrices $A_1, A_2, A_3, A_4, \ldots,$ all having the same eigenvalues as A. To generate this sequence, let $A_1 = A$, and factor $A_1 = Q_1 R_1$, where Q_1 is the orthogonal matrix and R_1 is the upper-triangular matrix described in Section 6.2. Then let $A_2 = R_1 Q_1$, factor A_2 into $Q_2 R_2$, and set $A_3 = R_2 Q_2$. Continue in this fashion, factoring A_n into $Q_n R_n$ and setting $A_{n+1} = R_n Q_n$. Under fairly general conditions, the matrices A_i will approach an almost upper-triangular matrix of the form

$$\begin{bmatrix} X & X & X & X & X & \cdots & X & X \\ X & X & X & X & X & \cdots & X & X \\ 0 & 0 & X & X & X & \cdots & X & X \\ 0 & 0 & X & X & X & \cdots & X & X \\ 0 & 0 & 0 & 0 & X & \cdots & X & X \\ 0 & 0 & 0 & 0 & X & \cdots & X & X \\ & & & & \vdots & & & \\ 0 & 0 & 0 & 0 & & \cdots & X & X \end{bmatrix}.$$

The colored entries just below the main diagonal may or may not be zero. If one of these entries is nonzero, the 2×2 submatrix having the entry in its lower left-hand corner, like the matrix shaded above, has a pair of complex conjugate numbers $a \pm bi$ as eigenvalues that are also eigenvalues of the large matrix, and of A. Entries on the diagonal that do not lie in such a 2×2 block are real eigenvalues of the matrix and of A.

The routine QRFACTOR in LINTEK can be used to illustrate this procedure. A few comments about the procedure and the program are in order.

From $A_1 = Q_1 R_1$, we have $R_1 = Q_1^{-1} A_1$. Then $A_2 = R_1 Q_1 = Q_1^{-1} A_1 Q_1$, so we see that A_2 is similar to $A_1 = A$ and therefore has the same eigenvalues as A. Continuing in this fashion, we see that each matrix A_i is similar to A. This explains why the eigenvalues don't change as the matrices of the sequence are generated.

Notice, too, that Q_1 and Q_1^{-1} are orthogonal matrices, so $Q_1^{-1}\mathbf{x}$ and $\mathbf{y}Q_1$ have the same magnitudes as the vectors $\mathbf{x}$ and $\mathbf{y}$, respectively. It follows that, if E is the matrix of errors in the entries of A, then the error matrix $Q_1^{-1}EQ_1$ arising in the computation of $Q_1^{-1}AQ_1$ is of magnitude comparable to that of E. That is, the generation of the sequence of matrices A_i is *stable*. This is highly desirable in numerical computations.

Finally, it is often useful to perform a **shift**, adding a scalar multiple rI of the identity matrix to A_i before generating the next matrix A_{i+1}. Such a shift increases all eigenvalues by r (see Exercise 32 in Section 5.1), but we can keep track of the total change due to such shifts and adjust the eigenvalues of the final matrix found to obtain those of A. To illustrate one use of shifts, suppose that we wish to find eigenvalues of a singular matrix B. We can form an initial shift, perhaps taking $A = B + (.001)I$, to obtain an invertible matrix A to start the algorithm. For another illustration, we can find by using the routine QRFACTOR that the QR algorithm applied to the matrix

$$\begin{bmatrix} 0 & 1 \\ 1 & 0 \end{bmatrix}$$

generates this same matrix repeatedly, even though the eigenvalues are 1 and -1 rather than complex numbers. This is an example of a matrix A for which the sequence of matrices A_i does not approach a form described earlier. However, a shift that adds $(.9)I$ produces the matrix

$$\begin{bmatrix} .9 & 1 \\ 1 & .9 \end{bmatrix},$$

which generates a sequence that quickly converges to

$$\begin{bmatrix} 1.9 & 0 \\ 0 & -.1 \end{bmatrix}.$$

Subtracting the scalar .9 from the eigenvalues 1.9 and $-.1$ of this last matrix, we obtain the eigenvalues 1 and -1 of the original matrix.

Shifts can also be used to speed convergence, which is quite fast when the ratios of magnitudes of eigenvalues are large. The routine QRFACTOR displays the matrices A_i as they are generated, and it allows shifts. If we notice that we are going to obtain an eigenvalue whose decimal expansion starts with 17.52, then we can speed convergence greatly by adding the shift $(-17.52)I$. The resulting eigenvalue will be near zero, and the ratios of the magnitudes of other eigenvalues to its magnitude will probably be large. Using this technique with QRFACTOR, it is quite easy to find all eigenvalues, both real and complex, of most matrices of reasonable size that can be displayed conveniently.

Professional programs make many further improvements in the algorithm we have presented, in order to speed the creation of zeros in the lower part of the matrix.

SUMMARY

1. Let A be an $n \times n$ diagonalizable matrix with real eigenvalues λ_j and with a dominant eigenvalue λ_1 of algebraic multiplicity 1, so that $|\lambda_1| > |\lambda_j|$ for $j = 2, 3, \ldots, n$. Start with any vector $\mathbf{w}_1$ in $\mathbb{R}^n$ that is not in the subspace generated by eigenvectors corresponding to the eigenvalues λ_j for $j > 1$. Form the vectors $\mathbf{w}_2 = (A\mathbf{w}_1)/d_1$, $\mathbf{w}_3 = (A\mathbf{w}_2)/d_2, \ldots$, where d_i is the maximum of the magnitudes of the components of $A\mathbf{w}_i$. The sequence of vectors

$$\mathbf{w}_1, \mathbf{w}_2, \mathbf{w}_3, \ldots$$

approaches an eigenvector of A corresponding to λ_1, and the associated Rayleigh quotients $(A\mathbf{w}_i \cdot \mathbf{w}_i)/(\mathbf{w}_i \cdot \mathbf{w}_i)$ approach λ_1. This is the foundation for the power method, which is summarized in the box before Example 1.

2. If A is diagonalizable and invertible, and if $|\lambda_n| < |\lambda_i|$ for $i < n$ with λ_n of algebraic multiplicity 1, then the power method may be used with A^{-1} to find λ_n.

3. Let A be an $n \times n$ symmetric matrix with eigenvalues λ_i such that $|\lambda_1| > |\lambda_2| \geq \cdots \geq |\lambda_n|$. If $\mathbf{b}_1$ is a unit eigenvector corresponding to λ_1, then $A - \lambda_1 \mathbf{b}_1 \mathbf{b}_1^T$ has eigenvalues $\lambda_2, \lambda_3, \ldots, \lambda_n, 0$, and if $|\lambda_2| > |\lambda_3|$, then λ_2 can be found by applying the power method to $A - \lambda_1 \mathbf{b}_1 \mathbf{b}_1^T$. This deflation can be continued with $A - \lambda_1 \mathbf{b}_1 \mathbf{b}_1^T - \lambda_2 \mathbf{b}_2 \mathbf{b}_2^T$ if $|\lambda_3| > |\lambda_4|$, and so on, to find more eigenvalues.

4. In the Jacobi method for diagonalizing a symmetric matrix A, one generates a sequence of symmetric matrices, starting with A. Each matrix of the sequence is obtained from the preceding one by multiplying it on the left by R^T and on the right by R, where R is an orthogonal "rotation" matrix designed to annihilate the two (symmetrically located) entries of maximum magnitude off the diagonal. The matrices in the sequence approach a diagonal matrix D having the same eigenvalues as A.

5. In the QR method, one begins with an invertible matrix A and generates a sequence of matrices A_i by setting $A_1 = A$ and $A_{n+1} = R_n Q_n$, where the QR factorization of A_n is $Q_n R_n$. The matrices A_i approach almost upper-triangular matrices having the real eigenvalues of A on the diagonal and pairs of complex conjugate eigenvalues of A as eigenvalues of 2×2 blocks appearing along the diagonal. Shifts may be used to speed convergence or to find eigenvalues of a singular matrix.

EXERCISES

In Exercises 1–4, use the power method to estimate the eigenvalue of maximum magnitude and a corresponding eigenvector for the given matrix. Start with first estimate

$$\mathbf{w}_1 = \begin{bmatrix} 1 \\ 1 \end{bmatrix},$$

and compute $\mathbf{w}_2$, $\mathbf{w}_3$, and $\mathbf{w}_4$. Also find the three Rayleigh quotients. Then find the exact eigenvalues, for comparison, using the characteristic equation.

1. $\begin{bmatrix} 3 & -3 \\ -5 & 1 \end{bmatrix}$ 2. $\begin{bmatrix} 3 & -3 \\ 4 & -5 \end{bmatrix}$

3. $\begin{bmatrix} -3 & 10 \\ -3 & 8 \end{bmatrix}$ 4. $\begin{bmatrix} -4 & 9 \\ -2 & 5 \end{bmatrix}$

In Exercises 5–8, find the spectral decomposition (8) of the given symmetric matrix.

5. $\begin{bmatrix} 2 & 3 \\ 3 & 2 \end{bmatrix}$ 6. $\begin{bmatrix} 3 & 5 \\ 5 & 3 \end{bmatrix}$

7. $\begin{bmatrix} 1 & 1 & 1 \\ 1 & 0 & 0 \\ 1 & 0 & 0 \end{bmatrix}$

8. $\begin{bmatrix} 1 & -1 & -1 \\ -1 & 1 & -1 \\ -1 & -1 & 1 \end{bmatrix}$

[HINT: Use Example 4 in Section 6.3.]

In Exercises 9–12, find the matrix obtained by deflation after the (exact) eigenvalue of maximum magnitude and a corresponding eigenvector are found.

9. The matrix in Exercise 5

10. The matrix in Exercise 6

11. The matrix in Exercise 7

12. The matrix in Exercise 8

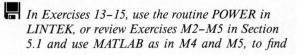 In Exercises 13–15, use the routine POWER in LINTEK, or review Exercises M2–M5 in Section 5.1 and use MATLAB as in M4 and M5, to find

the eigenvalue of maximum magnitude and a corresponding eigenvector of the given matrix.

13. $\begin{bmatrix} 1 & -44 & -88 \\ -5 & 55 & 113 \\ 1 & -24 & -48 \end{bmatrix}$

14. $\begin{bmatrix} 57 & -2 & 31 \\ -205 & 8 & -113 \\ -130 & 4 & -70 \end{bmatrix}$

15. $\begin{bmatrix} 3 & -22 & -46 \\ -3 & 23 & 47 \\ 1 & -10 & -20 \end{bmatrix}$

16. Use the software described for Exercises 13–15 with matrix inversion to find the eigenvalue of minimum magnitude and the corresponding eigenvector for the matrix in Exercise 13.

17. Repeat Exercise 16 for the matrix

$$\begin{bmatrix} -1 & 3 & 1 \\ 3 & 2 & -11 \\ 1 & -11 & 7 \end{bmatrix}.$$

In Exercises 18–21, use the routine POWER in LINTEK and deflation to find all eigenvalues and corresponding eigenvectors for the given symmetric matrix. Always continue the method before deflating until as much stabilization as possible is achieved. Note the relationship between ratios of magnitudes of eigenvalues and the speed of convergence.

18. $\begin{bmatrix} 3 & 5 & -7 \\ 5 & 10 & 11 \\ -7 & 11 & 0 \end{bmatrix}$

19. $\begin{bmatrix} 0 & -1 & 4 \\ -1 & 2 & -1 \\ 4 & -1 & 0 \end{bmatrix}$

20. $\begin{bmatrix} 3 & 1 & -2 & 1 \\ 1 & 4 & 0 & -3 \\ -2 & 0 & 0 & 5 \\ 1 & -3 & 5 & 2 \end{bmatrix}$

21. $\begin{bmatrix} 5 & 7 & -2 & 6 \\ 7 & 4 & 11 & 3 \\ -2 & 11 & -8 & 0 \\ 6 & 3 & 0 & 6 \end{bmatrix}$

22. The eigenvalue option of the routine VECTGRPH in LINTEK is designed to

strengthen your geometric understanding of the power method. There is no scaling, but you have the option of starting again if the numbers get so large that the vectors are off the screen. Read the directions in the program, and then run this eigenvalue option until you can reliably achieve a score of 85% or better.

23. Consider the matrix

$$A = \begin{bmatrix} a & b \\ b & c \end{bmatrix}.$$

The following steps will form an orthogonal diagonalizing matrix C for A.

Step 1 Let $g = (a - c)/2$.
Step 2 Let $h = \sqrt{g^2 + b^2}$.
Step 3 Let $r = \sqrt{b^2 + (g + h)^2}$.
Step 4 Let $s = \sqrt{b^2 + (g - h)^2}$.
Step 5 Let $C = \begin{bmatrix} -b/r & -b/s \\ (g + h)/r & (g - h)/s \end{bmatrix}$.

(If $b < 0$ and a rotation matrix is desired, change the sign of a column vector in C.) Prove this algorithm by proving the following.
a. The eigenvalues of A are

$$\lambda = \frac{a + c \pm \sqrt{(a - c)^2 + 4b^2}}{2}.$$

[HINT: Use the quadratic formula.]
b. The first row vector of $A - \lambda I$ is $(g \pm \sqrt{g^2 + b^2}, b)$.
c. The columns of the orthogonal matrix C in step 5 can be found using part b.
d. The parenthetical statement following step 5 is valid.

24. Use the algorithm in Exercise 23 to find an orthogonal diagonalizing matrix with determinant 1 for each of the following.

a. $\begin{bmatrix} 0 & 1 \\ 1 & 0 \end{bmatrix}$

b. $\begin{bmatrix} 3 & -2 \\ -2 & 1 \end{bmatrix}$

In Exercises 25–28, use the routine JACOBI in LINTEK to find the eigenvalues of the matrix by Jacobi's method.

25. The matrix in Exercise 18

26. The matrix in Exercise 19

27. The matrix in Exercise 20

28. The matrix in Exercise 21

In Exercises 29–32, use the routine QRFACTOR in LINTEK to find all eigenvalues, both real and complex, of the given matrix. Be sure to make use of shifts to speed convergence whenever you can see approximately what an eigenvalue will be.

29. $\begin{bmatrix} -1 & 5 & 7 \\ 3 & -6 & 8 \\ 9 & 0 & -11 \end{bmatrix}$

30. $\begin{bmatrix} -3 & 6 & -7 & 2 \\ 13 & -18 & 4 & 6 \\ 21 & 32 & -16 & 9 \\ -4 & 8 & 7 & 11 \end{bmatrix}$

31. $\begin{bmatrix} -12 & 3 & 15 & 2 & -21 \\ 47 & -34 & 87 & 24 & 7 \\ 35 & 72 & 33 & -57 & 82 \\ -145 & 67 & 32 & 10 & 46 \\ -9 & 22 & 21 & -45 & 8 \end{bmatrix}$

32. $\begin{bmatrix} -3 & 6 & 4 & -2 & 16 \\ 21 & -33 & 5 & 8 & -12 \\ 15 & -21 & 13 & 4 & 20 \\ -18 & 12 & 4 & 8 & 3 \\ -22 & 31 & 14 & 9 & 10 \end{bmatrix}$

COMPLEX SCALARS

Much of the linear algebra we have discussed is equally valid in applications using complex numbers as scalars. In fact, the use of complex scalars actually extends and clarifies many of the results we have developed. For example, when we are dealing with complex scalars, every $n \times n$ matrix has n eigenvalues, counted with their algebraic multiplicities.

In Section 9.1, we review the algebra of complex numbers. Section 9.2 summarizes properties of the complex vector space $\mathbb{C}^n$ and of matrices with complex entries. Diagonalization of these matrices and Schur's unitary triangularization theorem are discussed in Section 9.3. Section 9.4 is devoted to Jordan canonical forms.

9.1 ALGEBRA OF COMPLEX NUMBERS

The Number i

Numbers exist only in our minds. There is no physical entity that *is* the number 1. If there were, 1 would be in a place of high honor in some great museum of science, and past it would file a steady stream of mathematicians, gazing at 1 in wonder and awe. As mathematics developed, new numbers were *invented* (some prefer to say *discovered*) to fulfill algebraic needs that arose. If we start just with the positive integers, which are the most *natural numbers,* inventing *zero* and *negative integers* enables us not merely to add any two integers, but also to subtract any two integers. Inventing *rational numbers* (fractions) enables us to divide an integer by any nonzero integer. It can be shown that no rational number squared is equal to 2, so *irrational numbers* have to be invented to solve the equation $x^2 = 2$, and our familiar decimal notation comes into use. This decimal notation provides us with other numbers, such as π and e, that are of practical use. The numbers (positive, negative, and zero) that we customarily write using decimal notation have unfortunately become known as the *real numbers*; they are no more real than any other numbers we may invent, because all numbers exist only in our minds.

The real numbers are still inadequate to provide solutions of even certain polynomial equations. The simple equation $x^2 + 1 = 0$ has no real number as a solution. In terms of our needs in this text, the matrix

$$\begin{bmatrix} 0 & -1 \\ 1 & 0 \end{bmatrix}$$

has no eigenvalues. Mathematicians have had to invent a solution i of the equation $x^2 + 1 = 0$ to fill this need. Of course, once a new number is invented, it must take its place in the algebra of numbers; that is, we would like to multiply it and add it to all the numbers we already have. Thus, we also allow numbers bi and finally $a + bi$ for all of our old real numbers a and b. As we will subsequently show, it then becomes possible to add, subtract, multiply, and divide any of these new *complex numbers*, except that division by zero is still not possible. It is unfortunate that i has become known as an *imaginary number*. The purpose of this introduction is to point out that i is no less real and no more imaginary than any other number.

With the invention of i and the consequent enlargement of our number system to the set $\mathbb{C}$ of all the complex numbers $a + bi$, a marvelous thing happens. Not only does the equation $x^2 + a = 0$ have a solution for all real numbers a, but *every* polynomial equation has a solution in the complex numbers. We state without proof the famous *Fundamental Theorem of Algebra*, which shows this.

Fundamental Theorem of Algebra

Every polynomial equation with coefficients in $\mathbb{C}$ has n solutions in $\mathbb{C}$, where n is the degree of the polynomial and the solutions are counted with their algebraic multiplicity.

Review of Complex Arithmetic

A **complex number** z is any expression $z = a + bi$, where a and b are real numbers and $i = \sqrt{-1}$. The scalar a is the **real part** of z, and b is the **imaginary part** of z. It is useful to represent the complex number $z = a + bi$ as the vector $[a, b]$ in the x,y-plane $\mathbb{R}^2$, as shown in Figure 9.1. The x-axis is the **real axis**, and the y-axis is the **imaginary axis**. We let $\mathbb{C} = \{a + bi \mid a, b \in \mathbb{R}\}$ be the set of all complex numbers.

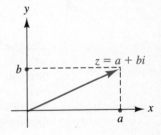

FIGURE 9.1
The complex number z.

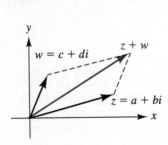

FIGURE 9.2
Representation of *z* + *w*.

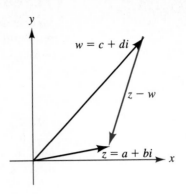

FIGURE 9.3
Representation of *z* − *w*.

Complex numbers are added, subtracted, and multiplied by real scalars in the natural way:

$$(a + bi) \pm (c + di) = (a \pm c) + (b \pm d)i,$$
$$\underset{\mathbf{z}}{} \qquad \underset{\mathbf{w}}{}$$

$$r(a + bi) = ra + (rb)i.$$

These operations have the same geometric representations as those for vectors in $\mathbb{R}^2$ and are illustrated in Figures 9.2, 9.3, and 9.4.

It is clear that the set $\mathbb{C}$ of complex numbers is a *real* vector space of dimension 2, isomorphic to $\mathbb{R}^2$.

The **modulus** (or **magnitude**) of the complex number $z = a + bi$ is $|z| = \sqrt{a^2 + b^2}$, which is the length of the vector in Figure 9.1. Notice that $|z| = 0$ if and only if $z = 0$—that is, if and only if $a = b = 0$.

Multiplication of two complex numbers is defined in the way it has to be to make the distributive laws $z_1(z_2 + z_3) = z_1z_2 + z_1z_3$ and $(z_1 + z_2)z_3 = z_1z_3 + z_2z_3$ valid. Namely, remembering that $i^2 = -1$, we have

$$(a + bi)(c + di) = (ac - bd) + (ad + bc)i.$$

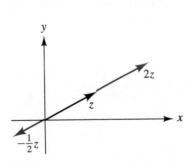

FIGURE 9.4
Representation of *rz*.

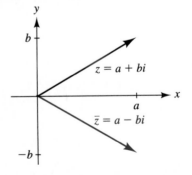

FIGURE 9.5
The conjugate of *z*.

Multiplication of complex numbers is commutative and associative. To divide one complex number by a nonzero complex number, we make use of the notion of a *complex conjugate*. The **conjugate** of $z = a + bi$ is $\bar{z} = a - bi$. Figure 9.5 illustrates the geometric relationship between $\bar{z}$ and z. Computing $z\bar{z} = (a + bi)(a - bi)$, we find that

$$z\bar{z} = a^2 + b^2 = |z|^2. \tag{1}$$

If $z \neq 0$, then Eq. (1) can be written as $z(\bar{z}/|z|^2) = 1$, so $1/z = \bar{z}/|z|^2$. More generally, if $z \neq 0$, then w/z is computed as

$$\frac{w}{z} = \frac{w\bar{z}}{z\bar{z}} = \frac{w\bar{z}}{|z|^2} = \frac{1}{|z|^2}(w\bar{z}). \tag{2}$$

We will see geometric representations of multiplication and division in a moment.

EXAMPLE 1   Find $(3 + 4i)^{-1} = 1/(3 + 4i)$.

SOLUTION   Using Eq. (2), we have

$$\frac{1}{3 + 4i} = \left(\frac{1}{3 + 4i}\right)\left(\frac{3 - 4i}{3 - 4i}\right) = \frac{3 - 4i}{25} = \frac{1}{25}(3 - 4i) = \frac{3}{25} - \frac{4}{25}i. \quad \blacksquare$$

EXAMPLE 2   Compute $(2 + 3i)/(1 + 2i)$.

SOLUTION   Using the technique of Eq. (2), we obtain

$$\frac{2 + 3i}{1 + 2i} = \left(\frac{2 + 3i}{1 + 2i}\right)\left(\frac{1 - 2i}{1 - 2i}\right) = \frac{1}{5}(2 + 3i)(1 - 2i) = \frac{8}{5} - \frac{1}{5}i. \quad \blacksquare$$

All of the familiar rules of arithmetic apply to the algebra of complex numbers. However, there is no notion of *order* for complex numbers; the notion $z_1 < z_2$ is defined only if z_1 and z_2 are real. The conjugation operation turns out to be very important; we summarize its properties in a theorem.

THEOREM 9.1   Properties of Conjugation in $\mathbb{C}$

Let $z = a + bi$ and $w = c + di$ be complex numbers. Then

1. $\overline{z + w} = \bar{z} + \bar{w}$,
2. $\overline{z - w} = \bar{z} - \bar{w}$,
3. $\overline{zw} = \bar{z}\,\bar{w}$,
4. $\overline{z/w} = \bar{z}/\bar{w}$,
5. $\overline{\bar{z}} = z$.

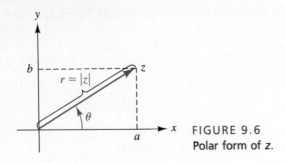

FIGURE 9.6
Polar form of z.

The proofs of the properties in Theorem 9.1 are very easy. We prove property 3 as an example and leave the rest as exercises. (See Exercises 12 and 13.)

EXAMPLE 3   Prove property 3 of Theorem 9.1.

SOLUTION   Let $z = a + bi$ and $w = c + di$. Then

$$\overline{z}\,\overline{w} = (\overline{a + bi})(\overline{c + di}) = (a - bi)(c - di)$$
$$= (ac - bd) - (ad + bc)i = \overline{zw}.$$ ∎

Polar Form of Complex Numbers

Let us return to the vector representation of $z = a + bi$ in $\mathbb{R}^2$. Figure 9.6 indicates that, for $z \neq 0$, if θ is an angle from the positive x-axis to the vector representation of z, and if $r = |z|$, then $a = r \cos \theta$ and $b = r \sin \theta$. Thus,

$$z = r(\cos \theta + i \sin \theta). \quad \textbf{A polar form of } z \tag{3}$$

HISTORICAL NOTE   Complex numbers made their initial appearance on the mathematical scene in *The Great Art, or On the Rules of Algebra* (1545) by the 16th century Italian mathematician and physician, Gerolamo Cardano (1501–1576). It was in this work that Cardano presented an algebraic method of solution of cubic and quartic equations. But it was the quadratic problem of dividing 10 into two parts such that their product is 40 to which Cardano found the solution $5 + \sqrt{-15}$ and $5 - \sqrt{-15}$ by standard techniques. Cardano was not entirely happy with this answer, as he wrote, "So progresses arithmetic subtlety the end of which, as is said, is as refined as it is useless."

Twenty-seven years later the engineer Raphael Bombelli (1526–1572) published an algebra text in which he dealt systematically with complex numbers. Bombelli wanted to clarify Cardano's cubic formula, which under certain circumstances could express a correct real solution of a cubic equation as the sum of two expressions each involving the square root of a negative number. Thus he developed our modern rules for operating with expressions of the form $a + b\sqrt{-1}$, including methods for determining cube roots of such numbers. Thus, for example, he showed that $\sqrt[3]{2 + \sqrt{-121}} = 2 + \sqrt{-1}$ and therefore that the solution to the cubic equation $x^3 = 15x + 4$, which Cardano's formula gave as $x = \sqrt[3]{2 + \sqrt{-121}} + \sqrt[3]{2 - \sqrt{-121}}$, could be written simply as $x = (2 + \sqrt{-1}) + (2 - \sqrt{-1})$, or as $x = 4$, the obvious answer.

The angle θ is an **argument** of z. Of particular importance is the restricted value

$$-\pi < \theta \le \pi \quad \text{The principal argument of } z$$

denoted by Arg(z). We usually use this **principal argument** of z and refer to the corresponding form (3) as **the polar form** of z.

EXAMPLE 4   Find the principal argument and the polar form of the complex number $z = -\sqrt{3} - i$.

SOLUTION   Because $|z| = \sqrt{(-\sqrt{3})^2 + (-1)^2} = \sqrt{4} = 2$, we have $\cos\theta = -\sqrt{3}/2$ and $\sin\theta = -\frac{1}{2}$, as indicated in Figure 9.7. The principal argument of z is the angle θ between $-\pi$ and π satisfying these two conditions—that is, $\theta = -5\pi/6$. The required polar form is $z = 2(\cos(-5\pi/6) + i\sin(-5\pi/6))$. ∎

If we multiply two complex numbers in their polar form, we quickly discover the geometric representation of multiplication. Suppose that we let $z_1 = r_1(\cos\theta_1 + i\sin\theta_1)$ and $z_2 = r_2(\cos\theta_2 + i\sin\theta_2)$. Then

$$z_1z_2 = r_1r_2((\cos\theta_1\cos\theta_2 - \sin\theta_1\sin\theta_2) + i(\sin\theta_1\cos\theta_2 + \cos\theta_1\sin\theta_2))$$
$$= r_1r_2(\cos(\theta_1 + \theta_2) + i\sin(\theta_1 + \theta_2)).$$

The last equation arises from the familiar trigonometric identities for the cosine and sine of a sum. This computation shows that, when two complex numbers are multiplied, the moduli are multiplied and the arguments are added. Notice, however, that the principal argument of a product may not be the sum of the principal arguments of the factors; the sum of the principal arguments may have to be adjusted to lie between $-\pi$ and π. (See Figure 9.8.)

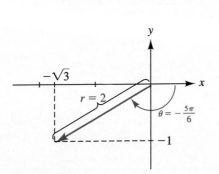

FIGURE 9.7
The polar form of $z = -\sqrt{3} - i$.

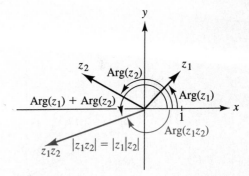

FIGURE 9.8
Representation of z_1z_2.

Geometric Representation of $z_1 z_2$

1. $|z_1 z_2| = |z_1||z_2|$.
2. $\text{Arg}(z_1) + \text{Arg}(z_2)$ is an argument of $z_1 z_2$.

(4)

Because z_1/z_2 is the complex number w such that $z_2 w = z_1$, we see that, in division, one divides the moduli and subtracts the arguments. (See Figure 9.9.)

Geometric Representation of z_1/z_2

1. $|z_1/z_2| = |z_1|/|z_2|$.
2. $\text{Arg}(z_1) - \text{Arg}(z_2)$ is an argument of z_1/z_2.

(5)

EXAMPLE 5   Illustrate relations (4) and (5) for the complex numbers $z_1 = \sqrt{3} + i$ and $z_2 = 1 + i$.

SOLUTION   For z_1, we have $r_1 = |z_1| = \sqrt{4} = 2$, $\cos \theta_1 = \sqrt{3}/2$, and $\sin \theta_1 = \frac{1}{2}$, so

$$z_1 = 2(\cos \frac{\pi}{6} + i \sin \frac{\pi}{6}).$$

For z_2, we have $r_2 = |z_2| = \sqrt{2}$ and $\cos \theta_2 = \sin \theta_2 = 1/\sqrt{2}$, so

$$z_2 = \sqrt{2}(\cos \frac{\pi}{4} + i \sin \frac{\pi}{4}).$$

Thus,

$$z_1 z_2 = 2\sqrt{2}(\cos(\frac{\pi}{6} + \frac{\pi}{4}) + i \sin(\frac{\pi}{6} + \frac{\pi}{4}))$$

$$= 2\sqrt{2}(\cos \frac{5\pi}{12} + i \sin \frac{5\pi}{12}),$$

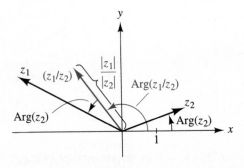

FIGURE 9.9
Representation of z_1/z_2.

and

$$\frac{z_1}{z_2} = \frac{2}{\sqrt{2}}(\cos(\frac{\pi}{6} - \frac{\pi}{4}) + i\sin(\frac{\pi}{6} - \frac{\pi}{4}))$$

$$= \frac{2}{\sqrt{2}}(\cos(-\frac{\pi}{12}) + i\sin(-\frac{\pi}{12})). \qquad \blacksquare$$

Relations (4) can be used repeatedly with $z_1 = z_2 = z = r(\cos\theta + i\sin\theta)$ to compute z^n for a positive integer n. We obtain the formula

$$\boxed{z^n = r^n(\cos n\theta + i\sin n\theta).} \qquad (6)$$

We can find nth roots of z by solving the equation $w^n = z$. Writing $z = r(\cos\theta + i\sin\theta)$ and $w = s(\cos\phi + i\sin\phi)$ and using Eq. (6), the equation $z = w^n$ becomes

$$r(\cos\theta + i\sin\theta) = s^n(\cos n\phi + i\sin n\phi).$$

Therefore, $r = s^n$ and $n\phi = \theta \pm 2k\pi$ for $k = 0, 1, 2, \ldots$, so

$$w = s(\cos\phi + i\sin\phi),$$

where

$$s = r^{1/n} \quad \text{and} \quad \phi = \frac{\theta}{n} \pm \frac{2k\pi}{n}.$$

Exercise 27 indicates that these values for ϕ represent precisely n distinct complex numbers, as indicated in the following box.

---

nth roots of $z = r(\cos\theta + i\sin\theta)$

The nth roots of z are

$$r^{1/n}\left(\cos\left(\frac{\theta}{n} + \frac{2k\pi}{n}\right) + i\sin\left(\frac{\theta}{n} + \frac{2k\pi}{n}\right)\right) \qquad (7)$$

for $k = 0, 1, 2, \ldots, n - 1$.

This illustrates the Fundamental Theorem of Algebra for the equation $w^n = z$ of degree n that has n solutions in $\mathbb{C}$.

EXAMPLE 6   Find the fourth roots of 16, and draw their vector representations in $\mathbb{R}^2$.

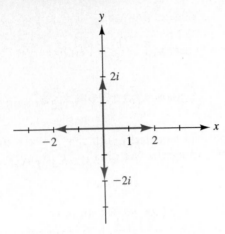

FIGURE 9.10
The fourth roots of 16.

SOLUTION   The polar form of $z = 16$ is $z = 16(\cos 0 + i \sin 0)$, where $\text{Arg}(z) = 0$. Applying formula (7), with $n = 4$, we find the following fourth roots of 16 (see Figure 9.10):

k	$16^{1/4}\left(\cos\dfrac{2k\pi}{4} + i\sin\dfrac{2k\pi}{4}\right)$
0	$2(\cos 0 + i \sin 0) = 2$
1	$2\left(\cos\dfrac{\pi}{2} + i\sin\dfrac{\pi}{2}\right) = 2i$
2	$2(\cos \pi + i \sin \pi) = -2$
3	$2\left(\cos\dfrac{3\pi}{2} + i\sin\dfrac{3\pi}{2}\right) = -2i$

■

SUMMARY

1. A complex number is a number of the form $z = a + bi$, where a and b are real numbers and $i = \sqrt{-1}$.

2. The modulus of the complex number $z = a + bi$ is
$$|z| = \sqrt{a^2 + b^2}.$$

3. The complex number $z = a + bi$ has the polar form $z = |z|(\cos \theta + i \sin \theta)$, and $\theta = \text{Arg}(z)$ is the principal argument of z if $-\pi < \theta \leq \pi$.

4. Arithmetic computations in the set $\mathbb{C}$ of complex numbers, including division and extraction of roots, can be accomplished as illustrated in this section and can be represented geometrically as shown in the figures of this section.

EXERCISES

1. Find the sum $z + w$ and the product zw if
   a. $z = 1 + 2i, w = 3 - i$,
   b. $z = 3 + i, w = i$.

2. Find the sum $z + w$ and the product zw if
   a. $z = 2 + 3i, w = 5 - i$,
   b. $z = 1 + 2i, w = 2 - i$.

3. Find $|z|$ and $\bar{z}$, and verify that $z\bar{z} = |z|^2$, if
   a. $z = 3 + 2i$,
   b. $z = 4 - i$.

4. Find $|z|$ and $\bar{z}$, and verify that $z\bar{z} = |z|^2$, if
   a. $z = 2 + i$,
   b. $z = 3 - 4i$.

5. Show that z is a real number if and only if $z = \bar{z}$.

6. Express z^{-1} in the form $a + bi$, for a and b real numbers, if
   a. $z = -1 + i$,
   b. $z = 3 + 4i$.

7. Express z/w in the form $a + bi$, for a and b real numbers, if
   a. $z = 1 + 2i, w = 1 + i$,
   b. $z = 3 + i, w = 3 + 4i$.

8. Find the modulus and principal argument for
   a. $\sqrt{3} - i$,
   b. $-\sqrt{3} - i$.

9. Find the modulus and principal argument for
   a. $-2 + 2i$,
   b. $-2 - 2i$.

10. Express $(\sqrt{3} + i)^6$ in the form $a + bi$ for a and b real numbers. [HINT: Write the given number in polar form.]

11. Express $(1 + i)^8$ in the form $a + bi$ for a and b real numbers. [HINT: Write the given number in polar form.]

12. Prove properties 1, 2, and 5 of Theorem 9.1.

13. Prove property 4 of Theorem 9.1.

14. Illustrate Eqs. (5) in the text for $z_1 = \sqrt{3} + i$ and $z_2 = -1 + \sqrt{3}i$.

15. Illustrate Eqs. (5) in the text for $z_1 = 2 + 2i$ and $z_2 = 1 + \sqrt{3}i$.

16. If $z^8 = 16$, find $|z|$.

17. Mark each of the following True or False.

_____ a. The existence of complex numbers is more doubtful than the existence of real numbers.

_____ b. Pencil-and-paper computations with complex numbers are more cumbersome than with real numbers.

_____ c. The square of every complex number is a positive real number.

_____ d. Every complex number has two distinct square roots in $\mathbb{C}$.

_____ e. Every nonzero complex number has two distinct square roots in $\mathbb{C}$.

_____ f. The Fundamental Theorem of Algebra asserts that the algebraic operations of addition, subtraction, multiplication, and division are possible with any two complex numbers, as long as we do not divide by zero.

_____ g. The product of two complex numbers cannot be a real number unless both numbers are themselves real or unless both are of the form bi, where b is a real number.

_____ h. If $(a + bi)^3 = 8$, then $a^2 + b^2 = 4$.

_____ i. If $\text{Arg}(z) = 3\pi/4$ and $\text{Arg}(w) = -\pi/2$, then $\text{Arg}(z/w) = 5\pi/4$.

_____ j. If $z + \bar{z} = 2z$, then z is a real number.

18. Find the three cube roots of 8.

19. Find the four fourth roots of -16.

20. Find the three cube roots of -27.

21. Find the four fourth roots of 1.

22. Find the six sixth roots of 1.

23. Find the eight eighth roots of 256.

24. A **primitive nth root of unity** is a complex number z such that $z^n = 1$ but $z^m \neq 1$ for $m < n$.

    a. Give a formula for one primitive nth root of unity.
    b. Find the primitive fourth roots of unity.
    c. How many primitive eighth roots of unity are there? [HINT: Argue geometrically in terms of polar forms.]

25. Let $z, w \in \mathbb{C}$. Show that $|z + w| \leq |z| + |w|$. [HINT: Remember that $\mathbb{C}$ is a real vector space of dimension 2, naturally isomorphic to $\mathbb{R}^2$.]

26. Show that the nth roots of $z \in \mathbb{C}$ can be represented geometrically as n equally spaced points on the circle $x^2 + y^2 = |z|^2$.

27. Show that the infinite list of values $\phi = \frac{\theta}{n} \pm \frac{2k\pi}{n}$ for $k = 0, 1, 2, \ldots$ yields just n distinct complex numbers $w = s(\cos \phi + i \sin \phi)$ of modulus s.

28. In calculus it is shown that

$$e^x = 1 + x + \frac{x^2}{2!} + \frac{x^3}{3!} + \frac{x^4}{4!} + \cdots$$

$$\sin x = x - \frac{x^3}{3!} + \frac{x^5}{5!} - \frac{x^7}{7!} + \frac{x^9}{9!} - \cdots$$

$$\cos x = 1 - \frac{x^2}{2!} + \frac{x^4}{4!} - \frac{x^6}{6!} + \frac{x^8}{8!} - \cdots.$$

    a. Proceeding formally, show that $e^{i\theta} = \cos \theta + i \sin \theta$. (This is **Euler's formula**.)
    b. Show that every complex number z may be written in the form $z = re^{i\theta}$, where $r = |z|$ and $\theta = \mathrm{Arg}(z)$.
    c. Assuming that the usual laws of exponents hold for exponents that are complex numbers, use part b to derive again Eqs. (4) and (5) describing multiplication and division of complex numbers in polar form.

9.2   MATRICES AND VECTOR SPACES WITH COMPLEX SCALARS

Complex Matrices and Linear Systems

Both the real number system $\mathbb{R}$ and the complex number system $\mathbb{C}$ are algebraic structures known as *fields*. In a field, we can add any two elements and multiply any two elements to produce an element of the field. Addition and multiplication are commutative and associative operations, and multiplication is distributive over addition. The field contains an additive identity 0 and a multiplicative identity 1. Every element c in the field has an additive inverse $-c$ in the field—that is, an element that, when added to c, produces the additive identity. Similarly, every nonzero element d in the field has a multiplicative inverse $1/d$ in the field.

The part of our work in Chapters 1, 2, and 3 that rests only on the field axioms of $\mathbb{R}$ is equally valid if we allow complex scalars. In particular, we can work with *complex matrices*—that is, with matrices having complex entries: adding matrices of the same size, multiplying matrices of appropriate sizes, and multiplying a matrix by a complex scalar. We can solve linear systems by using the same Gauss or Gauss–Jordan methods that we used in Chapter 1. All of our work in Chapter 1, Sections 3 through 6, makes perfectly good sense when applied to complex scalars. Pencil-and-paper computations are more tedious, however.

EXAMPLE 1   Solve the linear system

$$
\begin{aligned}
z_1 - z_2 + (1 + i)z_3 &= i \\
iz_1 - 2iz_2 + iz_3 &= 2 - i \\
iz_2 - (1 + i)z_3 &= 1 + 2i.
\end{aligned}
$$

SOLUTION   We use the Gauss–Jordan method as follows:

$$
\begin{bmatrix}
1 & -1 & 1+i & | & i \\
i & -2i & i & | & 2-i \\
0 & i & -1-i & | & 1+2i
\end{bmatrix}
\sim
\begin{bmatrix}
1 & -1 & 1+i & | & i \\
0 & -i & 1 & | & 3-i \\
0 & i & -1-i & | & 1+2i
\end{bmatrix}
$$

$$
\sim
\begin{bmatrix}
1 & 0 & 1+2i & | & 1+4i \\
0 & 1 & i & | & 1+3i \\
0 & 0 & -i & | & 4+i
\end{bmatrix}
\sim
\begin{bmatrix}
1 & 0 & 0 & | & 10+2i \\
0 & 1 & 0 & | & 5+4i \\
0 & 0 & 1 & | & -1+4i
\end{bmatrix}.
$$

Thus we obtain the solution $z_1 = 10 + 2i$, $z_2 = 5 + 4i$, $z_3 = -1 + 4i$. ∎

EXAMPLE 2   Find the inverse of the matrix

$$
A = \begin{bmatrix}
1 & 2i & 1+i \\
1 & 3i & i \\
0 & 1+i & -1
\end{bmatrix}.
$$

SOLUTION   We proceed precisely as in Chapter 1:

$$
\begin{bmatrix}
1 & 2i & 1+i & | & 1 & 0 & 0 \\
1 & 3i & i & | & 0 & 1 & 0 \\
0 & 1+i & -1 & | & 0 & 0 & 1
\end{bmatrix}
\sim
\begin{bmatrix}
1 & 2i & 1+i & | & 1 & 0 & 0 \\
0 & i & -1 & | & -1 & 1 & 0 \\
0 & 1+i & -1 & | & 0 & 0 & 1
\end{bmatrix}
$$

$$
\sim
\begin{bmatrix}
1 & 0 & 3+i & | & 3 & -2 & 0 \\
0 & 1 & i & | & i & -i & 0 \\
0 & 0 & -i & | & 1-i & -1+i & 1
\end{bmatrix}
$$

$$
\sim
\begin{bmatrix}
1 & 0 & 0 & | & 1-4i & 4i & 1-3i \\
0 & 1 & 0 & | & 1 & -1 & 1 \\
0 & 0 & 1 & | & 1+i & -1-i & i
\end{bmatrix}.
$$

Thus,

$$
A^{-1} = \begin{bmatrix}
1-4i & 4i & 1-3i \\
1 & -1 & 1 \\
1+i & -1-i & i
\end{bmatrix}.
$$
∎

Complex Vector Spaces

The definition of a *complex vector space* is identical with that of a real vector space, except that the field of scalars used is $\mathbb{C}$ rather than $\mathbb{R}$. The set $\mathbb{C}^n$ of all n-tuples having entries in $\mathbb{C}$ is an example of a complex vector space. Another example is the set of all $m \times n$ matrices with complex entries. The vector space

$\mathbb{C}^n$ has the same standard basis $\{\mathbf{e}_1, \mathbf{e}_2, \ldots, \mathbf{e}_n\}$ as $\mathbb{R}^n$, but of course now the field of scalars is $\mathbb{C}$. Thus, $\mathbb{C}^n$ is an *n*-dimensional *complex* vector space—that is, if we use *complex scalars*. In particular, $\mathbb{C} = \mathbb{C}^1$ is a one-dimensional complex vector space, even though we can regard it geometrically as a plane. That plane is a two-dimensional *real* vector space, but is a one-dimensional *complex* vector space. In general, $\mathbb{C}^n$ can be regarded geometrically in a natural way as a $2n$-dimensional real vector space, as we ask you to explain in Exercise 1.

All of our work in Chapters 1, 2, and 3 regarding subspaces, generating sets, independence, and bases carries over to complex vector spaces, and the proofs are identical.

EXAMPLE 3   Determine whether the set

$$S = \{[1, 2i, 1 + i], [1, 3i, i], [0, 1 + i, -1]\}$$

is independent and is a basis for $\mathbb{C}^3$.

SOLUTION   The vectors given in S are the row vectors in matrix A of Example 2. Row reduction of the matrix in that example shows that the matrix has rank 3, so the vectors in S are independent and hence form a basis for the three-dimensional space $\mathbb{C}^3$.   ∎

EXAMPLE 4   Find the coordinate vector $\mathbf{v}_B$ in $\mathbb{C}^3$ of the vector $\mathbf{v} = [i, 2 - i, 1 + 2i]$ relative to the ordered basis

$$B = ([1, i, 0], [-1, -2i, i], [1 + i, i, -1 - i]).$$

SOLUTION   To find the coordinate vector of $\mathbf{v}$ relative to B, we reduce the augmented matrix having the vectors in B as column vectors and having the vector $\mathbf{v}$ as the column to the right of the partition. This is precisely the augmented matrix that we reduced in Example 1, so we see that the coordinate vector $\mathbf{v}_B$, written as usual as a column vector, is

$$\mathbf{v}_B = \begin{bmatrix} 10 + 2i \\ 5 + 4i \\ -1 + 4i \end{bmatrix}.$$

∎

Euclidean Inner Product in $\mathbb{C}^n$

We now come to an essential difference in the structures of $\mathbb{C}$ and of $\mathbb{R}$. We have a natural idea of order for the elements of $\mathbb{R}$. We know what it means to say $x_1 < x_2$, and we have often used the fact that $x^2 \geq 0$ for all $x \in \mathbb{R}$. There is no idea of order in $\mathbb{C}$, extending the ordering of $\mathbb{R}$, for $i^2 = -1$ in $\mathbb{C}$. The nonzero numbers in $\mathbb{C}$ cannot be classified as either positive or negative on the basis of whether or not they are squares, because *all* numbers in $\mathbb{C}$ are squares. This is a very important difference between $\mathbb{R}$ and $\mathbb{C}$.

Let us see what problems this causes as we try to extend some more of the ideas in Chapter 3. We multiply matrices with complex entries by taking dot products of row vectors of the first with column vectors of the second, just as we do for matrices with real entries. Recall that, in $\mathbb{R}^n$, the length of a vector $\mathbf{v}$ is $\|\mathbf{v}\| = \sqrt{\mathbf{v} \cdot \mathbf{v}}$. However, this dot product cannot be used as an inner product to define length of vectors in $\mathbb{C}$, because, if $\mathbf{v} = [1, i]$, we would have $\mathbf{v} \cdot \mathbf{v} = 0$. The fix for this problem is simple. Recalling that $|a + bi|^2 = (a - bi)(a + bi)$, we make an adjustment and define the *Euclidean inner product* in $\mathbb{C}^n$.

DEFINITION 9.1 Euclidean Inner Product

Let $\mathbf{u} = [u_1, u_2, \ldots, u_n]$ and $\mathbf{v} = [v_1, v_2, \ldots, v_n]$ be vectors in $\mathbb{C}^n$. The **Euclidean inner product** of $\mathbf{u}$ and $\mathbf{v}$ is

$$\langle \mathbf{u}, \mathbf{v} \rangle = \overline{u_1} v_1 + \overline{u_2} v_2 + \cdots + \overline{u_n} v_n.$$

Notice that Definition 9.1 gives $\langle [1, i], [1, i] \rangle = (1)(1) + (-i)(i) = 1 + 1 = 2$. Because $|a + bi|^2 = (a - bi)(a + bi)$, we see at once that, for $\mathbf{v} \in \mathbb{C}^n$, we have

$$\langle \mathbf{v}, \mathbf{v} \rangle = |v_1|^2 + |v_2|^2 + \cdots + |v_n|^2, \tag{1}$$

just as in the $\mathbb{R}^n$ case.

We list properties of the Euclidean inner product in $\mathbb{C}^n$ as a theorem, leaving the proofs as exercises. (See Exercises 16 through 19.)

THEOREM 9.2 Properties of the Euclidean Inner Product

Let $\mathbf{u}$, $\mathbf{v}$, and $\mathbf{w}$ be vectors in $\mathbb{C}^n$, and let z be a complex scalar. Then

1. $\langle \mathbf{u}, \mathbf{u} \rangle \geq 0$, and $\langle \mathbf{u}, \mathbf{u} \rangle = 0$ if and only if $\mathbf{u} = \mathbf{0}$,
2. $\langle \mathbf{u}, \mathbf{v} \rangle = \overline{\langle \mathbf{v}, \mathbf{u} \rangle}$,
3. $\langle (\mathbf{u} + \mathbf{v}), \mathbf{w} \rangle = \langle \mathbf{u}, \mathbf{w} \rangle + \langle \mathbf{v}, \mathbf{w} \rangle$,
4. $\langle \mathbf{w}, (\mathbf{u} + \mathbf{v}) \rangle = \langle \mathbf{w}, \mathbf{u} \rangle + \langle \mathbf{w}, \mathbf{v} \rangle$,
5. $\langle z\mathbf{u}, \mathbf{v} \rangle = \overline{z}\langle \mathbf{u}, \mathbf{v} \rangle$, and $\langle \mathbf{u}, z\mathbf{v} \rangle = z\langle \mathbf{u}, \mathbf{v} \rangle$.

Property 1 of Theorem 9.2 and Eq. (1) preceding the theorem suggest that we define the **magnitude** or **norm** of a vector $\mathbf{v}$ in $\mathbb{C}^n$ as

$$\|\mathbf{v}\| = \sqrt{\langle \mathbf{v}, \mathbf{v} \rangle} = \sqrt{\overline{v_1} v_1 + \overline{v_2} v_2 + \cdots + \overline{v_n} v_n}. \quad \text{Magnitude of } \mathbf{v}$$

Vectors $\mathbf{u}$ and $\mathbf{v}$ in $\mathbb{C}^n$ are **perpendicular** (or **orthogonal**) if $\langle \mathbf{u}, \mathbf{v} \rangle = 0$; property 2 of Theorem 9.2 tells us that $\langle \mathbf{u}, \mathbf{v} \rangle = 0$ if and only if $\langle \mathbf{v}, \mathbf{u} \rangle = 0$. The vectors are **parallel** if $\mathbf{u} = z\mathbf{v}$ for some scalar z in $\mathbb{C}$. Notice that $\|z\mathbf{v}\| = |z| \, \|\mathbf{v}\|$, just as in the real case. Vectors of magnitude 1 are **unit vectors**. Having made these definitions, we will feel free to consider without further definition such things as orthogonal subspaces and orthonormal bases.

EXAMPLE 5 Find a unit vector in $\mathbb{C}^3$ parallel to $\mathbf{v} = [1, i, 1 + i]$.

SOLUTION   Because $\|\mathbf{v}\| = \sqrt{\langle \mathbf{v}, \mathbf{v} \rangle}$, we have

$$\|\mathbf{v}\| = \sqrt{1(1) + (-i)i + (1 - i)(1 + i)} = \sqrt{1 + 1 + 2} = 2.$$

Either of the vectors $\pm\frac{1}{2}[1, i, 1 + i]$ satisfies the requirement.   ∎

The Euclidean inner product given in Definition 9.1 reduces to the usual dot product of vectors if the vectors are in $\mathbb{R}^n$. However, we have to watch one feature very carefully:

> *The Euclidean inner product in $\mathbb{C}^n$ is not commutative.*

(See Property 2 in Theorem 9.2.) For example,

$$\langle [1, i], [0, 1] \rangle = -i, \qquad \text{but} \quad \langle [0, 1], [1, i] \rangle = i.$$

To illustrate the care that must be taken as a result of the noncommutativity of the inner product, we consider the Gram–Schmidt process applied in the vector space $\mathbb{C}^n$. Let $\mathbf{u}$ and $\mathbf{v}$ be vectors in $\mathbb{C}^n$. Then the vector

$$\mathbf{w} = \mathbf{u} - \frac{\langle \mathbf{v}, \mathbf{u} \rangle}{\langle \mathbf{v}, \mathbf{v} \rangle}\mathbf{v}$$

is orthogonal to $\mathbf{v}$, because

$$\langle \mathbf{w}, \mathbf{v} \rangle = \langle \mathbf{u}, \mathbf{v} \rangle - \overline{\langle \mathbf{v}, \mathbf{u} \rangle} = \langle \mathbf{u}, \mathbf{v} \rangle - \langle \mathbf{u}, \mathbf{v} \rangle = 0.$$

We must not use $\mathbf{u} - \dfrac{\langle \mathbf{u}, \mathbf{v} \rangle}{\langle \mathbf{v}, \mathbf{v} \rangle}\mathbf{v}$, whose inner product with $\mathbf{v}$ is generally not zero.

EXAMPLE 6   Transform the basis

$$\{[1, i, i], [1, 0, -i], [1, 0, 1]\}$$

of $\mathbb{C}^3$ to an orthonormal one, using the Gram–Schmidt process.

SOLUTION   First we transform the given basis to an orthogonal one. Because $\mathbf{v}_1 = [1, i, i]$ and $\mathbf{v}_2 = [1, 0, -i]$ are orthogonal, we work with $\mathbf{v}_3 = [1, 0, 1]$ and replace it by

$$\mathbf{v}_3 - \frac{\langle \mathbf{v}_1, \mathbf{v}_3 \rangle}{\langle \mathbf{v}_1, \mathbf{v}_1 \rangle}\mathbf{v}_1 - \frac{\langle \mathbf{v}_2, \mathbf{v}_3 \rangle}{\langle \mathbf{v}_2, \mathbf{v}_2 \rangle}\mathbf{v}_2$$

$$= \mathbf{v}_3 - \frac{1 - i}{3}\mathbf{v}_1 - \frac{1 + i}{2}\mathbf{v}_2$$

$$= [1, 0, 1] - \frac{1}{3}[1 - i, 1 + i, 1 + i] - \frac{1}{2}[1 + i, 0, 1 - i]$$

$$= \frac{1}{6}[1 - i, -2 - 2i, 1 + i].$$

In fact, we prefer to replace $\mathbf{v}_3$ by $[1 - i, -2 - 2i, 1 + i]$, which is just as good. An orthogonal basis is

$$\{[1, i, i], [1, 0, -i], [1 - i, -2 - 2i, 1 + i]\},$$

and an orthonormal basis is

$$\left\{\frac{1}{3}[1, i, i], \frac{1}{2}[1, 0, -i], \frac{1}{2\sqrt{3}}[1 - i, -2 - 2i, 1 + i]\right\}.$$ ∎

The Conjugate Transpose

In fixing up our old inner product to serve in $\mathbb{C}^n$ in Definition 9.1, we had to decide whether to take conjugates of the components u_i of the first vector in $\langle \mathbf{u}, \mathbf{v} \rangle$ or to take conjugates of the components v_i of the second vector. It is more convenient to take conjugates of components of the first vector for the following reason. The bulk of our work in $\mathbb{R}^n$ has been formulated in terms of column vectors, although we often use row notation to save space. For example, when working with matrix representations of linear transformations, we always write $A\mathbf{x}$, where $\mathbf{x}$ is a column vector. You may recall several instances where it has been convenient for us to use the fact that the inner product (dot product) of two vectors in $\mathbb{R}^n$ appears as sole entry in the matrix product of the first vector as a row vector and of the second vector as a column vector. That is, for column vectors $\mathbf{x}, \mathbf{y} \in \mathbb{R}^n$, we can write $\mathbf{x} \cdot \mathbf{y} = \mathbf{x}^T\mathbf{y}$, where $\mathbf{x}^T$ is a row vector and where no distinction is made between a 1×1 matrix and a scalar. Thinking in these terms, we can express the condition that the column vectors of a real square matrix A form an orthonormal basis as $A^TA = I$. In order to preserve these convenient algebraic formulas with as little change as possible, we choose to take conjugates of the components of the first vector in Definition 9.1, and we continue with a definition that allows us to recover these formulas.

DEFINITION 9.2   Conjugate Transpose, or Hermitian Adjoint

Let $A = [a_{ij}]$ be an $m \times n$ matrix with complex scalar entries.

1. The **conjugate** of A is the $m \times n$ matrix $\overline{A} = [\overline{a_{ij}}]$.
2. The **conjugate transpose** (or **Hermitian adjoint**) of A is the matrix $A^* = [\overline{a_{ij}}]^T$.

For column vectors $\mathbf{v}, \mathbf{w} \in \mathbb{C}^n$, we have

$$\langle \mathbf{v}, \mathbf{w} \rangle = \mathbf{v}^*\mathbf{w}. \tag{2}$$

Moreover, the condition for an $n \times n$ complex matrix A to have orthogonal unit column vectors can be written as

$$A^*A = I.$$

EXAMPLE 7 Find the conjugate transpose A^* of the matrix

$$A = \begin{bmatrix} 1 & i & 1+i \\ 2 & 0 & i \\ 2i & 1 & 1-i \end{bmatrix}.$$

SOLUTION We form the transpose while taking the conjugate of each element, obtaining

$$A^* = \begin{bmatrix} 1 & 2 & -2i \\ -i & 0 & 1 \\ 1-i & -i & 1+i \end{bmatrix}. \qquad \blacksquare$$

Following are some properties of the conjugate transpose that can easily be verified. (See Exercise 32.)

THEOREM 9.3 Properties of the Conjugate Transpose

Let A and B be $m \times n$ matrices. Then

1. $(A^*)^* = A$,
2. $(A + B)^* = A^* + B^*$,
3. $(zA)^* = \bar{z}A^*$ for any scalar $z \in \mathbb{C}$,
4. If A and B are square matrices, $(AB)^* = B^*A^*$.

EXAMPLE 8 Using the properties in Theorem 9.3, show that, for any $n \times n$ matrix A, we have $(A + A^*)^* = A + A^*$.

SOLUTION Using properties 1 and 2 of Theorem 9.3, we have

$$(A + A^*)^* = A^* + (A^*)^* = A^* + A = A + A^*,$$

which is what we wished to show. $\blacksquare$

Unitary and Hermitian Matrices

Recall that a *real orthogonal matrix* is a square matrix having orthogonal unit column vectors. The complex analogue of such a matrix is known by another name.

DEFINITION 9.3 Unitary Matrix

A square matrix U with complex entries is **unitary** if its column vectors are orthogonal unit vectors—that is, if $U^*U = I$.

Our next example gives an important property of unitary matrices that is familiar to us from the real orthogonal case. Notice in this example how handy the notation $\mathbf{v}^*\mathbf{v}$ is for $\langle \mathbf{v}, \mathbf{v} \rangle$.

EXAMPLE 9   Let $T: \mathbb{C}^n \to \mathbb{C}^n$ be a linear transformation having a unitary matrix U as matrix representation with respect to the standard basis. Show that $\|T(\mathbf{z})\| = \|\mathbf{z}\|$ for all $\mathbf{z} \in \mathbb{C}^n$.

SOLUTION   We know that $T(\mathbf{z}) = U\mathbf{z}$, because U is the standard matrix representation of T. Because $\|\mathbf{v}\|^2 = \langle \mathbf{v}, \mathbf{v} \rangle = \mathbf{v}^*\mathbf{v}$, we find by using property 4 of Theorem 9.3 that

$$\|U\mathbf{z}\|^2 = (U\mathbf{z})^*U\mathbf{z} = \mathbf{z}^*(U^*U)\mathbf{z} = \mathbf{z}^*I\mathbf{z} = \mathbf{z}^*\mathbf{z} = \|\mathbf{z}\|^2.$$

Taking square roots, we find that $\|U\mathbf{z}\| = \|\mathbf{z}\|$, so $\|T(\mathbf{z})\| = \|\mathbf{z}\|$.   ∎

Real symmetric matrices play an important role in linear algebra. We saw in Chapter 8 that they are very handy in work with quadratic forms. We also stated without proof in Theorem 5.5 that every real symmetric matrix is diagonalizable. Symmetric matrices are defined in terms of the transpose operation. The useful analogue of symmetry for matrices with complex scalars involves the conjugate transpose operation.

DEFINITION 9.4   Hermitian Matrix

A square matrix H is **Hermitian** if $H^* = H$—that is, if H is equal to its conjugate transpose.

If a square matrix H actually has real entries, it is Hermitian if and only if it is symmetric. Notice that the condition $H = H^* = (\overline{H})^T$ implies that the entries on the diagonal of any Hermitian matrix are real numbers.

EXAMPLE 10   Example 8 shows that, for any square matrix A, the matrix $A + A^*$ is Hermitian. Illustrate that the entries on the diagonal of $A + A^*$ are real, using the matrix A in Example 7.

SOLUTION   Example 7 shows that, for the matrix

$$A = \begin{bmatrix} 1 & i & 1 + i \\ 2 & 0 & i \\ 2i & 1 & 1 - i \end{bmatrix},$$

we have

$$A^* = \begin{bmatrix} 1 & 2 & -2i \\ -i & 0 & 1 \\ 1 - i & -i & 1 + i \end{bmatrix}.$$

Thus,

$$A + A^* = \begin{bmatrix} 2 & 2 + i & 1 - i \\ 2 - i & 0 & 1 + i \\ 1 + i & 1 - i & 2 \end{bmatrix},$$

which has real diagonal entries.   ∎

Hermitian matrices provide the proper generalization of real symmetric matrices to enable us to prove in the next section that every Hermitian matrix is diagonalizable. The special case of this theorem for real matrices thus finally provides us with a proof of the fundamental theorem that every real symmetric matrix is diagonalizable. The fundamental theorem for real symmetric matrices is tough to prove if we stay within the real number system, but it is a corollary of a fairly easy theorem for complex matrices. This illustrates how we can obtain true insight into theorems in real analysis and linear algebra by studying analogous concepts that use complex numbers.

SUMMARY

1. $\mathbb{C}^n$ is an n-dimensional complex vector space.

2. If $\mathbf{u} = [u_1, u_2, \ldots, u_n]$ and $\mathbf{v} = [v_1, v_2, \ldots, v_n]$ are vectors in $\mathbb{C}^n$, then $\langle \mathbf{u}, \mathbf{v} \rangle = \overline{u_1} v_1 + \overline{u_2} v_2 + \cdots + \overline{u_n} v_n$ is the Euclidean inner product of $\mathbf{u}$ and $\mathbf{v}$ and satisfies the properties in Theorem 9.2. In general, $\langle \mathbf{u}, \mathbf{v} \rangle \neq \langle \mathbf{v}, \mathbf{u} \rangle$.

3. For $\mathbf{u}, \mathbf{v} \in \mathbb{C}^n$, the vector $\mathbf{u} - \dfrac{\langle \mathbf{v}, \mathbf{u} \rangle}{\langle \mathbf{v}, \mathbf{v} \rangle} \mathbf{v}$ is perpendicular to $\mathbf{v}$.

4. The conjugate transpose of an $m \times n$ matrix A is the $n \times m$ matrix $A^* = (\overline{A})^T$. The conjugate transpose operation satisfies the properties in Theorem 9.3.

5. A square matrix U is unitary if $U^*U = I$. A real unitary matrix is an orthogonal matrix.

6. A square matrix H is Hermitian if $H = H^*$. A real Hermitian matrix is a symmetric matrix.

EXERCISES

1. Explain how $\mathbb{C}^n$ can be viewed as a $2n$-dimensional real vector space and as an n-dimensional complex vector space. Give a basis in each case.

2. Is it appropriate to view $\mathbb{R}^n$ as a subspace of $\mathbb{C}^n$? Explain.

3. Find AB and BA if

$$A = \begin{bmatrix} 1 & i & i \\ 1+i & 1 & -i \\ i & 1+i & 1-i \end{bmatrix} \text{ and}$$

$$B = \begin{bmatrix} -1 & 1+i & i \\ 2+i & i & 1-i \\ i & 1 & i \end{bmatrix}.$$

4. Find A^2 and A^4 if $A = \begin{bmatrix} 1 & 1+i \\ -1+i & i \end{bmatrix}$.

5. Find A^{-1} if $A = \begin{bmatrix} 1 & i \\ 1+i & 2+i \end{bmatrix}$.

6. Find A^{-1} if $A = \begin{bmatrix} i & 1+i \\ 1 & i \end{bmatrix}$.

7. Find A^{-1} if $A = \begin{bmatrix} 1 & i & 1-i \\ 1 & 1 & 1+i \\ 0 & -1+i & 1 \end{bmatrix}$.

8. Find A^{-1} if $A = \begin{bmatrix} i & 1-i & 1+i \\ 0 & 1 & i \\ 1-i & -i & 1-i \end{bmatrix}$.

9. Solve the linear system $A\mathbf{z} = \begin{bmatrix} i \\ 1+i \\ i \end{bmatrix}$ if A is the matrix in Exercise 7.

10. Solve the linear system $A\mathbf{z} = \begin{bmatrix} -1 + i \\ 2 + i \\ 1 \end{bmatrix}$ if A is

the matrix in Exercise 8.

11. Find the solution space of the homogeneous system

$$z_1 + iz_2 + (1 - i)z_3 = 0$$
$$(1 + 2i)z_1 - z_2 + z_3 = 0$$
$$(1 + i)z_1 + 2iz_2 + (3 - 2i)z_3 = 0.$$

12. Find the rank of the matrix

$$\begin{bmatrix} 1 & 1 + i & i & 1 - i \\ 1 + i & 2 + i & 1 - i & 1 \\ 1 + i & 1 + 4i & 1 + 3i & 2 - i \end{bmatrix}.$$

13. Find the rank of the matrix

$$\begin{bmatrix} 2 - i & i & 1 - i & 1 + 3i \\ i & 1 + i & -1 + 2i & i \\ 1 - i & 1 + 2i & 1 - 3i & 2 + 3i \end{bmatrix}.$$

14. Find each of the indicated inner products.
  a. $\langle [2 + i, 2, i], [1, 1 + i, 2 - i] \rangle$ in $\mathbb{C}^3$
  b. $\langle [1 - i, 1 + i], [1 + i, 1 - i] \rangle$ in $\mathbb{C}^2$
  c. $\langle [1 + i, 1, 1 - i, 1], [i, 1 + i, i, 1 - i] \rangle$ in $\mathbb{C}^4$
  d. $\langle [2 - i, 3 + i, 1 + i], [1 + i, 2 - i, 1 + i] \rangle$ in $\mathbb{C}^3$

15. Compute $\langle \mathbf{u}, \mathbf{v} \rangle$ and $\langle \mathbf{v}, \mathbf{u} \rangle$ for each of the following.
  a. $\mathbf{u} = [1, i], \mathbf{v} = [1 + i, 1 - i]$
  b. $\mathbf{u} = [1 + i, 2 - i, i], \mathbf{v} = [1, 1 - i, 1 + i]$

16. Verify property 1 of Theorem 9.2.

17. Verify property 2 of Theorem 9.2.

18. Verify properties 3 and 4 of Theorem 9.2.

19. Verify property 5 of Theorem 9.2.

20. Find the magnitude of each of the following vectors.
  a. $[1, i, i]$
  b. $[1 + i, 1 - i, 1 + i]$
  c. $[1 + i, 2 + i, 3 + i]$
  d. $[i, 1 + i, 1 - i, i]$
  e. $[1 + i, 1 - i, i, 1 - i]$

21. Determine whether the given pairs of vectors are parallel, perpendicular, or neither.
  a. $[1, i], [i, 1]$
  b. $[1 + i, i, 1 - i], [1 - i, 1, -1 - i]$
  c. $[1 + i, 2 - i], [3i, 3 + i]$
  d. $[1, i, 1 - i], [1 - i, 1 + i, 2]$
  e. $[1 + i, 1 - i, 1], [i, 1 - i, -3 - i]$

22. Find a unit vector parallel to $[1 + i, 1 - i, i]$.

23. Find a vector of length 2 parallel to $[i, 1 - i, 1 + i, 1 - i]$.

24. Find a unit vector perpendicular to $[2 - i, 1 + i]$.

25. Find a vector perpendicular to both $[1, i, 1 - i]$ and $[1 + i, 1 - i, 1]$.

In Exercises 26–29, transform the given basis of $\mathbb{C}^n$ into an orthogonal basis, using the Gram–Schmidt process.

26. $\{[1 + i, 1 - i], [1, 1]\}$ in $\mathbb{C}^2$

27. $\{[2 + i, 1 + i], [1 + i, i]\}$ in $\mathbb{C}^2$

28. $\{[1 - i, 1 + i, 1 + i], [1, 1, -1 - i], [1, i, -i]\}$ in $\mathbb{C}^3$

29. $\{[1, i, i], [1 + i, 1, i], [i, 1 + i, 0]\}$ in $\mathbb{C}^3$

30. Find the conjugate transpose of each of the following matrices.
  a. $\begin{bmatrix} 1 & 1 + i & 2 \\ 1 - i & 1 + i & 2 \end{bmatrix}$
  b. $\begin{bmatrix} i & 1 + i \\ 2 + i & 1 - i \end{bmatrix}$
  c. $\begin{bmatrix} 1 + i & 1 + i \\ 2 - i & 1 - i \\ 1 & 1 - 2i \end{bmatrix}$
  d. $\begin{bmatrix} 1 + 2i & i & 1 - i \\ i & 1 - i & 1 + i \\ 1 + i & 2 - i & 1 + 3i \end{bmatrix}$

31. Label each of the following matrices as Hermitian, unitary, both, or neither.
  a. $\dfrac{1}{\sqrt{3}} \begin{bmatrix} 1 & 1 + i \\ 1 - i & -1 \end{bmatrix}$
  b. $\dfrac{1}{2} \begin{bmatrix} 1 & i & 1 - i \\ -i & 2 & 1 \\ 1 + i & 1 & 1 \end{bmatrix}$
  c. $\dfrac{1}{\sqrt{6}} \begin{bmatrix} \sqrt{2} & \sqrt{2}i & \sqrt{2}i \\ \sqrt{3}i & 0 & \sqrt{3} \\ -i & -2 & 1 \end{bmatrix}$
  d. $\dfrac{1}{4} \begin{bmatrix} 1 & i & 1 - i \\ i & 3 & 1 \\ 1 - i & 1 & 2 \end{bmatrix}$

32. Prove the properties of the conjugate transpose operation given in Theorem 9.3. [HINT: From Section 1.3, we know that analogous properties of the *transpose* operation hold for real matrices and real scalars and can be derived using just field properties of $\mathbb{R}$, so they are also true for matrices with complex entries. Thus we can focus on the effect of the *conjugation*. From Theorem 9.1, we know that $\overline{z + w} = \overline{z} + \overline{w}$, $\overline{zw} = \overline{z}\,\overline{w}$, and $\overline{\overline{z}} = z$, for $z, w \in \mathbb{C}$. Use these properties of conjugation to complete the proof of Theorem 9.3.]

33. Mark each of the following True or False. Assume that all matrices and scalars are complex.

____ **a.** The definition of a determinant, properties of determinants (the transpose property, the row-interchange property, and so on), and techniques for computing them are developed using only field properties of $\mathbb{R}$ in Chapter 4, and thus they remain equally valid for square complex matrices.

____ **b.** Cramer's rule is valid for square linear systems with complex coefficients.

____ **c.** If A is any square matrix and $\det(A) \neq 0$, then $\det(iA) \neq \det(A)$.

____ **d.** If U is unitary, then $U^{-1} = U^T$.

____ **e.** If U is unitary, then $(\overline{U})^{-1} = U^T$.

____ **f.** The Euclidean inner product in $\mathbb{C}^n$ is not commutative.

____ **g.** For $\mathbf{u}, \mathbf{v} \in \mathbb{C}^n$, we have $\langle \mathbf{u}, \mathbf{v} \rangle = \langle \mathbf{v}, \mathbf{u} \rangle$ if and only if $\langle \mathbf{u}, \mathbf{v} \rangle$ is a real number.

____ **h.** For a square matrix A, we have $\det(\overline{A}) = \overline{\det(A)}$.

____ **i.** For a square matrix A, we have $\det(A^*) = \det(A)$.

____ **j.** If U is a unitary matrix, then $\det(U^*) = \pm 1$.

34. Prove that, for vectors $\mathbf{v}_1, \mathbf{v}_2, \ldots, \mathbf{v}_n$ in $\mathbb{C}^n$, $\{\mathbf{v}_1, \mathbf{v}_2, \ldots, \mathbf{v}_n\}$ is a basis for $\mathbb{C}^n$ if and only if $\{\overline{\mathbf{v}}_1, \overline{\mathbf{v}}_2, \ldots, \overline{\mathbf{v}}_n\}$ is a basis for $\mathbb{C}^n$.

35. Prove that an $n \times n$ matrix U is unitary if and only if the rows of U form an orthonormal basis for $\mathbb{C}^n$.

36. Prove that, if A is a square matrix, then AA^* is a Hermitian matrix.

37. Prove that the product of two commuting $n \times n$ Hermitian matrices is also a Hermitian matrix. What can you say about the sum of two Hermitian matrices?

38. Prove that the product of two $n \times n$ unitary matrices is also a unitary matrix. What about the sum?

39. Let $T: \mathbb{C}^n \to \mathbb{C}^n$ be a linear transformation whose standard matrix representation is a unitary matrix U. Show that $\langle T(\mathbf{u}), T(\mathbf{v}) \rangle = \langle \mathbf{u}, \mathbf{v} \rangle$ for all $\mathbf{u}, \mathbf{v} \in \mathbb{C}^n$. [HINT: Remember that $\langle \mathbf{u}, \mathbf{v} \rangle = \mathbf{u}^*\mathbf{v}$.]

40. Prove that for $\mathbf{u}, \mathbf{v} \in \mathbb{C}^n$, we have $(\mathbf{u}^*\mathbf{v})^* = \overline{\mathbf{u}^*\mathbf{v}} = \mathbf{v}^*\mathbf{u} = \mathbf{u}^T\overline{\mathbf{v}}$.

41. Describe the unitary diagonal matrices.

42. Prove that, if U is unitary, then $\overline{U}$, U^T, and U^* are unitary matrices also.

43. A square matrix A is **normal** if $A^*A = AA^*$.

   a. Show that every Hermitian matrix is normal.

   b. Show that every unitary matrix is normal.

   c. Show that, if $A^* = -A$, then A is normal.

44. Let A be an $n \times n$ matrix. Referring to Exercise 43, prove that, if A is normal, then $\|A\mathbf{z}\| = \|A^*\mathbf{z}\|$ for all $\mathbf{z} \in \mathbb{C}^n$.

45. Prove the converse of the statement in Exercise 44.

MATLAB

MATLAB can work with complex numbers. When typing a complex number $a + bi$ as a component of a vector or as an entry of a matrix, be sure to type **a+b∗i** with the ∗ denoting multiplication and with *no spaces* before or after the + or the ∗. A space before the + would cause MATLAB to interpret **a +b∗i** as two numbers, the real number a followed by another entry containing the complex number bi.

For a matrix A, MATLAB interprets $\mathbf{A'}$ as the *conjugate* transpose of A—that is, as the transpose of A with every entry replaced by its conjugate. Here are three more MATLAB functions:

real(A) is the matrix of real parts of the entries in A,

imag(A) is the matrix of complex parts of the entries in A,

conj(A) is the matrix of conjugates of the entries in A.

Matrices for the exercises that follow are in a file. Some of the exercises ask you to check your answers to some of the more tedious pencil-and-paper computations in the exercises for this section.

M1. Check Exercise 3. (The file matrices are E3A and E3B.)

M2. Check Exercise 7.

M3. Check Exercise 8.

M4. Check Exercise 9.

M5. Check Exercise 10.

M6. Check Exercise 11.

M7. Check Exercise 12.

M8. Check Exercise 13.

M9. Check Exercise 28.

M10. Check Exercise 29.

M11. Consider the matrix

$$\begin{bmatrix} 2-3i & 3+7i & -5+2i & 7-3i & -10+4i \\ 8-i & 2+5i & 11-3i & 6+2i & 14-4i \\ 13+2i & 3-4i & 9+9i & 3-2i & 7+6i \\ 5+i & 8-4i & 12+8i & -3+2i & 1-5i \end{bmatrix}.$$

a. Find the norms of the four row vectors by multiplying appropriate matrices.

b. Find the norms of the five column vectors by multiplying appropriate matrices.

c. Find the inner product $\langle \mathbf{r}_1, \mathbf{r}_3 \rangle$ where $\mathbf{r}_1$ is the first row vector and $\mathbf{r}_3$ is the third row vector.

d. Find the inner product $\langle \mathbf{c}_2, \mathbf{c}_5 \rangle$ where $\mathbf{c}_2$ is the second column vector and $\mathbf{c}_5$ is the fifth column vector.

9.3   EIGENVALUES AND DIAGONALIZATION

Recall the fundamental theorem of real symmetric matrices that we stated without proof as Theorem 6.8:

> Every real symmetric matrix is diagonalizable by a real orthogonal matrix.

Our main goal in this section is to extend this result to complex matrices, as follows:

Every Hermitian matrix is diagonalizable by a unitary matrix.

We will prove this theorem, which has the theorem for real symmetric matrices as an easy corollary.

Eigenvalues for Complex Matrices

We begin by extending the notions of eigenvalues and eigenvectors to complex matrices. The definitions are identical to those for real matrices. If A is an $n \times n$ complex matrix and if $A\mathbf{v} = \lambda\mathbf{v}$, where $\lambda \in \mathbb{C}$ and $\mathbf{v} \in \mathbb{C}^n$, $\mathbf{v} \neq \mathbf{0}$, then λ is an **eigenvalue** of A and $\mathbf{v}$ is a corresponding **eigenvector**. The zero vector and the set of all eigenvectors of A corresponding to λ constitute the **eigenspace** E_λ. Computation of eigenvalues and eigenspaces of a complex matrix is the same as for real matrices, except that the arithmetic involves complex numbers and consequently is more laborious to do with pencil and paper. Every $n \times n$ complex matrix has n not necessarily distinct eigenvalues. This is a consequence of the Fundamental Theorem of Algebra, which we stated in Section 9.1. Recall that, for real matrices, there may exist no real eigenvalues.

EXAMPLE 1   Find the eigenvalues and eigenspaces of the matrix

$$A = \begin{bmatrix} 1 & 0 & i \\ 0 & 2 & 0 \\ -i & 0 & 1 \end{bmatrix}.$$

SOLUTION   The characteristic polynomial of A is

$$\det(A - \lambda I) = \begin{vmatrix} 1 - \lambda & 0 & i \\ 0 & 2 - \lambda & 0 \\ -i & 0 & 1 - \lambda \end{vmatrix} = (2 - \lambda)((1 - \lambda)^2 + i^2)$$

$$= (2 - \lambda)(1 - 2\lambda + \lambda^2 - 1) = -\lambda(2 - \lambda)^2.$$

The three roots of $-\lambda(2 - \lambda)^2 = 0$ are $\lambda_1 = 0$, $\lambda_2 = \lambda_3 = 2$.
For the eigenvalue $\lambda_1 = 0$, we have

$$A - \lambda_1 I = \begin{bmatrix} 1 & 0 & i \\ 0 & 2 & 0 \\ -i & 0 & 1 \end{bmatrix} \sim \begin{bmatrix} 1 & 0 & i \\ 0 & 1 & 0 \\ 0 & 0 & 0 \end{bmatrix},$$

which gives the eigenspace $E_0 = \text{sp}\left(\begin{bmatrix} -i \\ 0 \\ 1 \end{bmatrix}\right)$. For the double root $\lambda_2 = \lambda_3 = 2$, we have

$$A - 2I = \begin{bmatrix} -1 & 0 & i \\ 0 & 0 & 0 \\ -i & 0 & -1 \end{bmatrix} \sim \begin{bmatrix} 1 & 0 & -i \\ 0 & 0 & 0 \\ 0 & 0 & 0 \end{bmatrix},$$

which gives the two-dimensional eigenspace $E_2 = \text{sp}\left(\begin{bmatrix} i \\ 0 \\ 1 \end{bmatrix}, \begin{bmatrix} 0 \\ 1 \\ 0 \end{bmatrix}\right)$.   ∎

Definitions and theorems concerning eigenvalues and eigenvectors that depend only on the field axioms discussed at the beginning of Section 9.2 continue to make sense and to hold for complex matrices. In particular, an $n \times n$ complex matrix A is **diagonalizable** if and only if there exist an invertible matrix C and a diagonal matrix D such that $D = C^{-1}AC$. Just as for real matrices, two complex $n \times n$ matrices A and B are **similar** if there exists an invertible $n \times n$ matrix C such that $B = C^{-1}AC$. Similarity is an equivalence relation. Thus, A is similar to A; if A is similar to B, then B is similar to A; and if furthermore B is similar to D, then A is similar to D. All of these things are defined and proved using just field properties.

Consider again the equation $D = C^{-1}AC$, where D is a diagonal matrix. The equivalent equation, $CD = AC$, for an invertible matrix C shows that A is diagonalizable if and only if $\mathbb{C}^n$ has a basis of eigenvectors of A, and it shows that the matrix C must have such a basis of eigenvectors as its column vectors, whereas D has the corresponding eigenvalues on its diagonal. We obtain all of this from $CD = AC$ by considering the jth column vector of CD and comparing it with the jth column vector of AC, just as we did for the real case in Section 5.2. Such a basis for $\mathbb{C}^n$ of eigenvectors of A exists if and only if the algebraic multiplicity of each eigenvalue is equal to its geometric multiplicity (the dimension of the corresponding eigenspace).

EXAMPLE 2   Let $A = \begin{bmatrix} 1 & 0 & i \\ 0 & 2 & 0 \\ -i & 0 & 1 \end{bmatrix}$. Find a matrix C such that $C^{-1}AC$ is a diagonal matrix.

SOLUTION   From the preceding example, we see that A has an eigenvalue $\lambda_1 = 0$ of algebraic multiplicity 1 with eigenspace

$$E_0 = \text{sp}\left(\begin{bmatrix} -i \\ 0 \\ 1 \end{bmatrix}\right)$$

and that it has the double eigenvalue $\lambda_2 = \lambda_3 = 2$ with eigenspace

$$E_2 = \text{sp}\left(\begin{bmatrix} i \\ 0 \\ 1 \end{bmatrix}, \begin{bmatrix} 0 \\ 1 \\ 0 \end{bmatrix}\right).$$

Thus, the algebraic multiplicity of each eigenvalue is equal to its geometric multiplicity, and the vectors shown form a basis for $\mathbb{C}^n$. Therefore, the matrix

$$C = \begin{bmatrix} -i & i & 0 \\ 0 & 0 & 1 \\ 1 & 1 & 0 \end{bmatrix}$$

is invertible and diagonalizes A; and we must also have $C^{-1}AC = D$, where

$$D = \begin{bmatrix} 0 & 0 & 0 \\ 0 & 2 & 0 \\ 0 & 0 & 2 \end{bmatrix}. \qquad \blacksquare$$

The proof of Theorem 5.3, which asserts that eigenvectors corresponding to distinct eigenvalues are independent, depends only on field properties and thus is valid in the complex case. Consequently, every matrix having only eigenvalues of algebraic multiplicity 1 is diagonalizable. We focus our attention on the geometric multiplicity of any eigenvalues of algebraic multiplicity greater than 1, when determining whether a matrix is diagonalizable.

EXAMPLE 3   Find all values of c for which the matrix

$$A = \begin{bmatrix} i & c & 1 \\ 0 & i & 2i \\ 0 & 0 & 1 \end{bmatrix}$$

is diagonalizable.

SOLUTION   The eigenvalues of the upper-triangular matrix A are $\lambda_1 = \lambda_2 = i$ and $\lambda_3 = 1$. We focus our attention on the eigenvalue i of algebraic multiplicity 2. For A to be diagonalizable, its eigenspace E_i must have geometric multiplicity 2. The eigenspace is the nullspace of the matrix

$$A - \lambda_1 I = \begin{bmatrix} 0 & c & 1 \\ 0 & 0 & 2i \\ 0 & 0 & 1-i \end{bmatrix},$$

and the nullspace has dimension 2 if and only if the matrix has rank 1, which is the case if and only if $c = 0$. $\blacksquare$

Diagonalization of Hermitian Matrices

Diagonalization via a unitary matrix is of special importance, as we saw in the real case, where it becomes diagonalization by an orthogonal matrix. We call $n \times n$ matrices A and B **unitarily equivalent** if there is a unitary matrix U such that $B = U^{-1}AU$. Because the inverse of a unitary matrix is unitary and because a product of unitary matrices is unitary, we can show that unitary equivalence is an equivalence relation. Thus, A is unitarily equivalent to itself; and if A is unitarily equivalent to B, then B is unitarily equivalent to A; if furthermore B is unitarily equivalent to C, then A is unitarily equivalent to C.

Now we achieve the main goal of this section: to prove that Hermitian matrices are unitarily equivalent to a diagonal matrix. That is, a Hermitian matrix can be diagonalized using a unitary matrix. This follows from a very important result known as *Schur's lemma* (or *Schur's unitary triangularization theorem*), which we state, deferring the proof until the end of this section.

THEOREM 9.4   Schur's Lemma

Let A be an $n \times n$ (complex) matrix. There is a unitary matrix U such that $U^{-1}AU$ is upper triangular.

Using Schur's lemma, we can prove that every Hermitian matrix is diagonalizable, and that the diagonalizing matrix can be chosen to be unitary. We express this by saying that every Hermitian matrix is **unitarily diagonalizable**.

THEOREM 9.5   Spectral Theorem for Hermitian Matrices

If A is a Hermitian matrix, there exists a unitary matrix U such that $U^{-1}AU$ is a diagonal matrix. Furthermore, all eigenvalues of A are real.

PROOF   By Schur's lemma, there exists a unitary matrix U such that $U^{-1}AU$ is an upper-triangular matrix. Because U is unitary, we have $U^*U = I$, so $U^{-1} = U^*$; and because A is Hermitian, we also know that $A^* = A$. Thus, we have

$$(U^{-1}AU)^* = (U^*AU)^* = U^*A^*(U^*)^* = U^*AU = U^{-1}AU,$$

which shows that the upper-triangular matrix $U^{-1}AU$ is also Hermitian. Because the conjugate transpose of an upper-triangular matrix is a lower-triangular matrix, we see that the entries above the diagonal in $U^{-1}AU$ must all be zero; therefore, $U^{-1}AU = D$, where D is a diagonal matrix. Thus, A is unitarily diagonalizable.

It remains to be shown that each eigenvalue of A is a real number. From the theory of diagonalization, we know that the entries on the diagonal of D are the eigenvalues of A. Now we showed in the preceding paragraph that the matrix $D = U^{-1}AU$ is Hermitian, so $D^* = D$. Forming the conjugate transpose of a diagonal matrix amounts simply to taking the conjugates of the entries on the diagonal. Because $D^* = D$, the entries on the diagonal of D remain unchanged under conjugation, so they must be real numbers.   ▲

COROLLARY Fundamental Theorem of Real Symmetric Matrices

Every $n \times n$ real symmetric matrix has n real eigenvalues, counted with their algebraic multiplicity, and is diagonalizable by a real orthogonal matrix.

PROOF Because every real $n \times n$ symmetric matrix A is also Hermitian, Theorem 9.5 establishes that all of its eigenvalues in $\mathbb{C}$ actually lie in $\mathbb{R}$; therefore, the matrix has n real eigenvalues, counting them with their algebraic multiplicity. Furthermore, Theorem 9.5 asserts that A can be diagonalized by a unitary matrix U. We know that the column vectors of U are eigenvectors of A. Now the eigenvectors of A can be computed by row reductions of $A - \lambda_i I$, where the λ_i are eigenvalues of A. Because all the λ_i are real, the row reductions all take place in the field $\mathbb{R}$ of real numbers. The reduced echelon form of $A - \lambda_i I$ is thus a *real* matrix; it must have a nullspace of dimension (geometric multiplicity) equal to the algebraic multiplicity of λ_i, because A is diagonalizable. Thus, we can find bases for the eigenspaces E_{λ_i} consisting of vectors in $\mathbb{R}^n$. Using the Gram–Schmidt process, we can assume that the basis of each eigenspace is orthonormal. The matrix C having as column vectors the vectors in these orthonormal bases of eigenspaces E_{λ_i} is thus a real orthogonal matrix that diagonalizes A.   ▲

EXAMPLE 4   Find a unitary matrix that diagonalizes the matrix A in Example 2.

SOLUTION   We found in Example 2 that the matrix

$$C = \begin{bmatrix} -i & i & 0 \\ 0 & 0 & 1 \\ 1 & 1 & 0 \end{bmatrix}$$

diagonalizes A. Notice that the inner product of any two distinct column vectors of this matrix is zero, so the column vectors are orthogonal. We need only normalize them to length 1 in order to obtain a unitary matrix that diagonalizes A. Thus, such a matrix is

$$U = \frac{1}{\sqrt{2}} \begin{bmatrix} -i & i & 0 \\ 0 & 0 & \sqrt{2} \\ 1 & 1 & 0 \end{bmatrix}.$$ ■

Recall that in Theorem 6.7 we showed that real eigenvectors of a symmetric matrix corresponding to distinct eigenvalues are orthogonal. Generalizing this, we can show that the eigenvectors of a Hermitian matrix corresponding to distinct eigenvalues are orthogonal. We ask you to show this in Exercise 21, using the fact that a Hermitian matrix can be unitarily diagonalized; but it is easy to demonstrate this orthogonality by using properties of matrices and the fact that the eigenvalues must be real.

THEOREM 9.6   Orthogonality of Eigenspaces of a Hermitian Matrix

The eigenvectors of a Hermitian matrix corresponding to distinct eigenvalues are orthogonal.

PROOF   Let $\mathbf{v}$ and $\mathbf{w}$ be eigenvectors of a Hermitian matrix A corresponding to distinct eigenvalues λ_1 and λ_2, respectively. Using the facts that $A = A^*$ and that the eigenvalues are real, so that $\overline{\lambda}_2 = \lambda_2$, we have

$$\lambda_1(\mathbf{w}^*\mathbf{v}) = \mathbf{w}^*(\lambda_1\mathbf{v}) = \mathbf{w}^*(A\mathbf{v}) = \mathbf{w}^*(A^*\mathbf{v}) = (\mathbf{w}^*A^*)\mathbf{v}$$
$$= (A\mathbf{w})^*\mathbf{v} = (\lambda_2\mathbf{w})^*\mathbf{v} = \lambda_2(\mathbf{w}^*\mathbf{v}).$$

Therefore, $(\lambda_1 - \lambda_2)(\mathbf{w}^*\mathbf{v}) = 0$. Because $\lambda_1 \neq \lambda_2$, we must have $\mathbf{w}^*\mathbf{v} = 0$, so $\mathbf{w}$ and $\mathbf{v}$ are orthogonal.   ▲

EXAMPLE 5   Find a unitary matrix C that diagonalizes the Hermitian matrix

$$A = \begin{bmatrix} -1 & i & 1+i \\ -i & 1 & 0 \\ 1-i & 0 & 1 \end{bmatrix}.$$

SOLUTION   We find that

$$|A - \lambda I| = \begin{vmatrix} -1-\lambda & i & 1+i \\ -i & 1-\lambda & 0 \\ 1-i & 0 & 1-\lambda \end{vmatrix}$$

$$= (-1-\lambda)\begin{vmatrix} 1-\lambda & 0 \\ 0 & 1-\lambda \end{vmatrix} - i\begin{vmatrix} -i & 0 \\ 1-i & 1-\lambda \end{vmatrix} +$$

$$(1+i)\begin{vmatrix} -i & 1-\lambda \\ 1-i & 0 \end{vmatrix}$$

$$= (-1-\lambda)(1-\lambda)^2 - i(-i)(1-\lambda) + (1+i)(-1)(1-i)(1-\lambda)$$

$$= (1-\lambda)(\lambda^2 - 1 - 1 - 2) = (1-\lambda)(\lambda^2 - 4).$$

Thus, the eigenvalues are $\lambda_1 = 1$, $\lambda_2 = 2$, and $\lambda_3 = -2$. To find U, we need only compute one eigenvector of length 1 for each of these three distinct eigenvalues. The three eigenvectors we obtain must form an orthonormal set, according to Theorem 9.6.

For $\lambda_1 = 1$, we find that

$$A - I = \begin{bmatrix} -2 & i & 1+i \\ -i & 0 & 0 \\ 1-i & 0 & 0 \end{bmatrix},$$

so an eigenvector is

$$\mathbf{v}_1 = \begin{bmatrix} 0 \\ 1+i \\ -i \end{bmatrix}.$$

For $\lambda_2 = 2$, we find that

$$A - 2I = \begin{bmatrix} -3 & i & 1+i \\ -i & -1 & 0 \\ 1-i & 0 & -1 \end{bmatrix} \sim \begin{bmatrix} 1 & -i & 0 \\ 0 & -2i & 1+i \\ 0 & 1+i & -1 \end{bmatrix} \sim \begin{bmatrix} 1 & -i & 0 \\ 0 & 1 & (i-1)/2 \\ 0 & 0 & 0 \end{bmatrix},$$

so a corresponding eigenvector is

$$\mathbf{v}_2 = \begin{bmatrix} 1+i \\ 1-i \\ 2 \end{bmatrix}.$$

Finally, for $\lambda_3 = -2$, we have

$$A + 2I = \begin{bmatrix} 1 & i & 1+i \\ -i & 3 & 0 \\ 1-i & 0 & 3 \end{bmatrix} \sim \begin{bmatrix} 1 & i & 1+i \\ 0 & 2 & -1+i \\ 0 & -1-i & 1 \end{bmatrix} \sim \begin{bmatrix} 1 & i & 1+i \\ 0 & 1 & (-1+i)/2 \\ 0 & 0 & 0 \end{bmatrix},$$

and a corresponding eigenvector is

$$\mathbf{v}_3 = \begin{bmatrix} -3-3i \\ 1-i \\ 2 \end{bmatrix}.$$

We normalize the vectors $\mathbf{v}_1$, $\mathbf{v}_2$, and $\mathbf{v}_3$ and form the column vectors in U from the resulting vectors of magnitude 1, obtaining

$$U = \begin{bmatrix} 0 & (1+i)/(2\sqrt{2}) & (-3-3i)/(2\sqrt{6}) \\ (1+i)/\sqrt{3} & (1-i)/(2\sqrt{2}) & (1-i)/(2\sqrt{6}) \\ -i/\sqrt{3} & 1/\sqrt{2} & 1/\sqrt{6} \end{bmatrix}. \qquad \blacksquare$$

A Criterion for Unitary Diagonalization

We have seen that every Hermitian matrix is unitarily diagonalizable, but of course a unitarily diagonalizable matrix need not be Hermitian. For example, the 1×1 matrix $[i]$ is already diagonal, so it is diagonalizable by the identity matrix. However, it is not Hermitian because $[i]^* = [-i]$. There is actually a way to determine whether a square matrix is unitarily diagonalizable, without having to find its eigenvalues and eigenvectors.

DEFINITION 9.5   Normal Matrix

A square matrix A is **normal** if it commutes with its conjugate transpose, so that $AA^* = A^*A$.

Exercises 25 and 26 ask you to prove the following theorem, which gives a criterion for A to be unitarily diagonalizable in terms of matrix multiplication!

THEOREM 9.7   Spectral Theorem for Normal Matrices

A square matrix A is unitarily diagonalizable if and only if it is a normal matrix.

EXAMPLE 6   Determine all values of a such that the matrix

$$A = \begin{bmatrix} i & a \\ 2 & i \end{bmatrix}$$

is unitarily diagonalizable.

SOLUTION   In order for the matrix to be unitarily diagonalizable, we must have $AA^* = A^*A$ so that

$$\begin{bmatrix} i & a \\ 2 & i \end{bmatrix}\begin{bmatrix} -i & 2 \\ \bar{a} & -i \end{bmatrix} = \begin{bmatrix} -i & 2 \\ \bar{a} & -i \end{bmatrix}\begin{bmatrix} i & a \\ 2 & i \end{bmatrix}.$$

Equating entries, we obtain

    row 1, column 1: $1 + |a|^2 = 1 + 4$,   so $|a| = 2$,

    row 1, column 2: $2i - ai = -ai + 2i$,

    row 2, column 1: $-2i + \bar{a}i = \bar{a}i - 2i$,

    row 2, column 2: $4 + 1 = |a|^2 + 1$,   so $|a| = 2$.

Clearly these conditions are satisfied as long as $|a| = 2$, so a can be any number of the form $x + yi$, where $x^2 + y^2 = 4$.   ∎

Proof of Shur's Lemma

We now prove by induction that, if A is an $n \times n$ matrix, there exists a unitary matrix U such that $U^{-1}AU = U^*AU$ is upper triangular. If $n = 1$, the lemma is trivial. We assume as induction hypothesis that the lemma is true for all matrices of size at most $(n - 1) \times (n - 1)$, and we proceed to show that it must hold for an $n \times n$ matrix A.

Let λ_1 be an eigenvalue of A, and let $\mathbf{v}_1$ be a corresponding eigenvector of norm 1. We can expand $\{\mathbf{v}_1\}$ to a basis for $\mathbb{C}^n$, and by the Gram–Schmidt process we can transform it into an orthonormal basis $\{\mathbf{v}_1, \mathbf{v}_2, \ldots, \mathbf{v}_n\}$. Let U_1 be the unitary matrix whose jth column vector is $\mathbf{v}_j$. Now the first column vector of AU_1 is $A\mathbf{v}_1 = \lambda_1\mathbf{v}_1$. Because the ith row vector of U_1^* is $\mathbf{v}_i^*$, and because the vectors $\mathbf{v}_i$ are mutually orthogonal, we see that the first column vector of the matrix $U_1^*AU_1$ is

$$U_1^*(\lambda_1\mathbf{v}_1) = \begin{bmatrix} \lambda_1 \\ 0 \\ \cdot \\ \cdot \\ \cdot \\ 0 \end{bmatrix}.$$

This shows that we can write $U_1^*AU_1$ symbolically as

$$U_1^*AU_1 = \begin{bmatrix} \lambda_1 & \times & \times & \cdots & \times \\ 0 & & & & \\ \cdot & & & & \\ \cdot & & & A_1 & \\ \cdot & & & & \\ 0 & & & & \end{bmatrix}, \tag{1}$$

where we have denoted the $(n-1) \times (n-1)$ submatrix in the lower right-hand corner of $U_1^*AU_1$ by A_1. By our induction hypothesis, there exists an $(n-1) \times (n-1)$ unitary matrix C such that $C^*A_1C = B$, where B is upper triangular. Let

$$U_2 = \begin{bmatrix} 1 & 0 & 0 & \cdots & 0 \\ 0 & & & & \\ \cdot & & & & \\ \cdot & & & C & \\ \cdot & & & & \\ 0 & & & & \end{bmatrix}, \tag{2}$$

where we have used a symbolic notation similar to that in Eq. (1). Because C is unitary, it is clear that U_2 is a unitary matrix. Now let $U = U_1U_2$. Because $U^*U = U_2^*(U_1^*U_1)U_2 = U_2^*IU_2 = U_2^*U_2 = I$, we see that U is a unitary matrix. Now

$$U^*AU = U_2^*(U_1^*AU_1)U_2. \tag{3}$$

The matrix in parentheses in Eq. (3) is the matrix displayed in Eq. (1). From our definition of U_2 in Eq. (2), we see that the $(n-1) \times (n-1)$ block in the lower right-hand corner of U^*AU is $C^*A_1C = B$. Thus, we have

$$U^*AU = \begin{bmatrix} \lambda_1 & \times & \times & \cdots & \times \\ 0 & & & & \\ \cdot & & & & \\ \cdot & & & B & \\ \cdot & & & & \\ 0 & & & & \end{bmatrix},$$

which is upper triangular because B is an upper-triangular matrix. This completes our induction argument. ▲

SUMMARY

1. An $n \times n$ matrix A is diagonalizable if and only if $\mathbb{C}^n$ has a basis consisting of eigenvectors of A. Equivalently, each eigenvalue has algebraic multiplicity equal to its geometric multiplicity.
2. Every Hermitian matrix is diagonalizable by a unitary matrix.
3. Every Hermitian matrix has real eigenvalues.

4. A square matrix A is unitarily diagonalizable if and only if it is normal, so that $AA^* = A^*A$.

5. Schur's lemma states that every square matrix is unitarily equivalent to an upper-triangular matrix.

EXERCISES

In Exercises 1–12, find a unitary matrix U and a diagonal matrix D such that $D = U^{-1}AU$ for the given matrix A.

1. $A = \begin{bmatrix} 1 & i \\ -i & 1 \end{bmatrix}$

2. $A = \begin{bmatrix} 1 & 2i \\ -2i & 1 \end{bmatrix}$

3. $A = \begin{bmatrix} 1 & 1+i \\ 1-i & 2 \end{bmatrix}$

4. $A = \begin{bmatrix} 9 & 3-i \\ 3+i & 0 \end{bmatrix}$

5. $A = \begin{bmatrix} 0 & i & 0 \\ -i & 0 & 0 \\ 0 & 0 & 1 \end{bmatrix}$

6. $A = \begin{bmatrix} 1 & -i & 0 \\ i & 1 & 0 \\ 0 & 0 & 2 \end{bmatrix}$

7. $A = \begin{bmatrix} 1 & 2-2i & 0 \\ 2+2i & -1 & 0 \\ 0 & 0 & 3 \end{bmatrix}$

8. $A = \begin{bmatrix} 0 & 0 & 1+2i \\ 0 & 5 & 0 \\ 1-2i & 0 & 4 \end{bmatrix}$

9. $A = \begin{bmatrix} 1 & 0 & 2+2i \\ 0 & -3 & 0 \\ 2-2i & 0 & -1 \end{bmatrix}$

10. $A = \begin{bmatrix} 2 & 0 & 1-i \\ 0 & 3 & 0 \\ 1+i & 0 & 1 \end{bmatrix}$

11. $A = \begin{bmatrix} 3 & i & 1+i \\ -i & 1 & 0 \\ 1-i & 0 & 1 \end{bmatrix}$

12. $A = \begin{bmatrix} -3 & 5i & 1+i \\ -5i & 3 & 0 \\ 1-i & 0 & 3 \end{bmatrix}$

13. Find all $a \in \mathbb{C}$ such that the matrix $\begin{bmatrix} i & 4 \\ a & i \end{bmatrix}$ is unitarily diagonalizable.

14. Find all $a, b \in \mathbb{C}$ such that the matrix $\begin{bmatrix} i & a \\ b & i \end{bmatrix}$ is unitarily diagonalizable.

15. Find all $a \in \mathbb{C}$ such that the matrix $\begin{bmatrix} i & a \\ 1 & 3i \end{bmatrix}$ is unitarily diagonalizable.

16. Find all $a, b \in \mathbb{C}$ such that the matrix $\begin{bmatrix} a & -i \\ i & b \end{bmatrix}$ is unitarily diagonalizable.

17. Prove that every 2×2 real matrix that is unitarily diagonalizable has one of the following forms: $\begin{bmatrix} a & b \\ b & d \end{bmatrix}, \begin{bmatrix} a & b \\ -b & a \end{bmatrix}$, for $a, b, d \in \mathbb{R}$.

18. Determine whether the matrix $\begin{bmatrix} i & -1 & 1 \\ 1 & -i & -1 \\ -1 & 1 & i \end{bmatrix}$ is unitarily diagonalizable.

19. Mark each of the following True or False.
___ a. Every square matrix is unitarily equivalent to a diagonal matrix.
___ b. Every square matrix is unitarily equivalent to an upper-triangular matrix.
___ c. Every Hermitian matrix is unitarily equivalent to a diagonal matrix.
___ d. Every unitarily diagonalizable matrix is Hermitian.
___ e. Every real symmetric matrix is Hermitian.
___ f. Every diagonalizable matrix is normal.
___ g. Every unitarily diagonalizable matrix is normal.
___ h. Every real symmetric matrix is normal.
___ i. Every square matrix is diagonalizable, although perhaps not by a unitary matrix.
___ j. Every square matrix with eigenvalues of algebraic multiplicity 1 is diagonalizable by a unitary matrix.

20. Prove that the eigenvalues of a Hermitian matrix are real, without using Theorem 9.5 or Schur's lemma. [HINT: Let $A\mathbf{v} = \lambda\mathbf{v}$, and use the fact that $\mathbf{v}^*A\mathbf{v} = \mathbf{v}^*A^*\mathbf{v}$.]

21. Argue directly from Theorem 9.5 that eigenvectors from different eigenspaces of a Hermitian matrix are orthogonal.

22. Suppose that A is an $n \times n$ matrix such that $A^* = -A$. Show that
    a. A has eigenvalues of the form ri, where $r \in \mathbb{R}$,
    b. A is diagonalizable by a unitary matrix. [HINT FOR BOTH PARTS: Work with iA.]

23. Prove that an $n \times n$ matrix A is unitarily diagonalizable if and only if $\|A\mathbf{v}\| = \|A^*\mathbf{v}\|$ for all $\mathbf{v} \in \mathbb{C}^n$.

24. Prove that a normal matrix is Hermitian if and only if all its eigenvalues are in $\mathbb{R}$.

25. a. Prove that a diagonal matrix is normal.
    b. Prove that, if A is normal and B is unitarily equivalent to A, then B is normal.
    c. Deduce from parts a and b that a unitarily diagonalizable matrix is normal.

26. a. Prove that every normal matrix A is unitarily equivalent to a normal upper-triangular matrix B. (Use Schur's lemma and part b of Exercise 25.)
    b. Prove that an $n \times n$ normal upper-triangular matrix B must be diagonal. [HINT: Let $C = B^*B = BB^*$. Equating the computations of c_{11} from B^*B and from BB^*, show that $b_{1j} = 0$ for $1 < j \le n$. Then equate the computations of c_{22} from B^*B and from BB^* to show that $b_{2j} = 0$ for $2 < j \le n$. Continue this process to show that B is lower triangular.]
    c. Deduce from parts a and b that a normal matrix is unitarily diagonalizable.

In Exercises 27 and 28, use the command **[U, D] = eig(A)** in MATLAB to work the indicated exercise. If your MATLAB answer U for the unitary matrix differs from the U that we found using pencil and paper and put in the answers at the end of our text, explain how you can get from one answer to the other.

27. Exercise 9
28. Exercise 11

9.4   JORDAN CANONICAL FORM

Jordan Blocks

We have spent considerable time on diagonalization of matrices. The preceding section was concerned primarily with unitary diagonalization. As we have seen, diagonal matrices are easily handled. Unfortunately, not every $n \times n$ matrix A can be diagonalized, because we cannot always find a basis for $\mathbb{C}^n$ consisting of eigenvectors of A. We remind you of this with an example that is well worth studying.

EXAMPLE 1   Show that the matrix

$$J = \begin{bmatrix} 5 & 1 & 0 \\ 0 & 5 & 0 \\ 0 & 0 & 5 \end{bmatrix}$$

is not diagonalizable.

SOLUTION   We see that 5 is the only eigenvalue of J, and that it has algebraic multiplicity 3. However,

$$J - 5I = \begin{bmatrix} 0 & 1 & 0 \\ 0 & 0 & 0 \\ 0 & 0 & 0 \end{bmatrix},$$

which shows that the eigenspace E_5 has dimension 2 and has basis $\{e_1, e_3\}$. Thus, the geometric multiplicity of the eigenvalue 5 is only 2, and J is not diagonalizable. We cannot find a basis for $\mathbb{R}^3$ (or even $\mathbb{C}^3$) consisting entirely of eigenvectors of J. ■

Let us examine the matrix J in Example 1 a bit more. Notice that, although $J - 5I$ has a nullspace of dimension 2, the matrix $(J - 5I)^2$ is the zero matrix and has all of $\mathbb{C}^3$ as nullspace. Moreover, multiplication on the left by $J - 5I$ carries e_2 into e_1 and carries both e_1 and e_3 into $\mathbf{0}$. We say that $J - 5I$ **annihilates** e_1 and e_3. The action of $J - 5I$ on these standard basis vectors is denoted schematically by the two *strings*

$$J - 5I: \quad \begin{matrix} e_2 \to e_1 \to \mathbf{0}, \\ e_3 \to \mathbf{0}. \end{matrix} \tag{1}$$

Diagram (1) also shows that $(J - 5I)^2$ maps each of these basis vectors into $\mathbf{0}$. Because $(J - 5I)e_2 = e_1$, we have $Je_2 = 5e_2 + e_1$, whereas $Je_1 = 5e_1$ and $Je_3 = 5e_3$.

EXAMPLE 2   Let

$$J = \begin{bmatrix} \lambda & 1 & 0 & 0 & 0 \\ 0 & \lambda & 1 & 0 & 0 \\ 0 & 0 & \lambda & 1 & 0 \\ 0 & 0 & 0 & \lambda & 1 \\ 0 & 0 & 0 & 0 & \lambda \end{bmatrix}. \tag{2}$$

Discuss the action of $J - \lambda I$ on the standard basis vectors, drawing a schematic diagram similar to diagram (1). Describe also the action of J on the vectors in the standard basis.

SOLUTION   We find that

$$J - \lambda I = \begin{bmatrix} 0 & 1 & 0 & 0 & 0 \\ 0 & 0 & 1 & 0 & 0 \\ 0 & 0 & 0 & 1 & 0 \\ 0 & 0 & 0 & 0 & 1 \\ 0 & 0 & 0 & 0 & 0 \end{bmatrix}.$$

We see that multiplication of e_i on the left by $J - \lambda I$ yields e_{i-1} for $2 \leq i \leq 5$, whereas $(J - \lambda I)e_1 = \mathbf{0}$. Schematically, we have just one *string*,

$$J - \lambda I: \quad e_5 \to e_4 \to e_3 \to e_2 \to e_1 \to \mathbf{0}. \tag{3}$$

Left-multiplication by J yields

$$Je_5 = \lambda e_5 + e_4, \qquad Je_4 = \lambda e_4 + e_3, \qquad Je_3 = \lambda e_3 + e_2,$$
$$Je_2 = \lambda e_2 + e_1, \qquad Je_1 = 0.$$ ∎

The matrix J in Example 2 is an example of a *Jordan block* matrix.

DEFINITION 9.6   Jordan Block

An $m \times m$ matrix is a **Jordan block** if it is structured as follows:

1. All diagonal entries are equal.
2. Each entry immediately above a diagonal entry is 1.
3. All other entries are zero.

Thus, the matrix J in Example 2 is a Jordan block. However, the matrix in Example 1 is not a Jordan block, since the entry 5 at the bottom of the diagonal does not have a 1 just above it. A Jordan block has the properties described in the next theorem. These properties were illustrated in Example 2, and we leave a formal proof to you if you desire one. Notice that, for an $m \times m$ Jordan block

$$J = \begin{bmatrix} \lambda & 1 & 0 & \cdots & 0 & 0 \\ 0 & \lambda & 1 & \cdots & 0 & 0 \\ & & & \vdots & & \\ 0 & 0 & 0 & \cdots & \lambda & 1 \\ 0 & 0 & 0 & \cdots & 0 & \lambda \end{bmatrix},$$

we have just one string:

$$J - \lambda I: \qquad e_m \to e_{m-1} \to \cdots \to e_2 \to e_1 \to 0.$$

THEOREM 9.8   Properties of a Jordan Block

Let J be an $m \times m$ Jordan block with diagonal entries all equal to λ. Then the following properties hold:

1. $(J - \lambda I)e_i = e_{i-1}$ for $1 < i \leq m$, and $(J - \lambda I)e_1 = 0$.
2. $(J - \lambda I)^m = O$, but $(J - \lambda I)^i \neq O$ for $i < m$.
3. $Je_i = \lambda e_i + e_{i-1}$ for $1 < i \leq m$, whereas $Je_1 = \lambda e_1$.

Jordan Canonical Forms

We have seen that not every $n \times n$ matrix is diagonalizable. It is our purpose in this section to show that every $n \times n$ matrix is similar to a matrix having all entries 0 except for those on the diagonal and entries 1 immediately above some diagonal entries; each 1 above a diagonal entry must have the same number on its left as below it on the diagonal. An example of such a matrix is

$$
J = \begin{bmatrix}
-i & 1 & 0 & 0 & 0 & 0 & 0 & 0 \\
0 & -i & 1 & 0 & 0 & 0 & 0 & 0 \\
0 & 0 & -i & 0 & 0 & 0 & 0 & 0 \\
0 & 0 & 0 & -i & 1 & 0 & 0 & 0 \\
0 & 0 & 0 & 0 & -i & 0 & 0 & 0 \\
0 & 0 & 0 & 0 & 0 & 2 & 0 & 0 \\
0 & 0 & 0 & 0 & 0 & 0 & 5 & 1 \\
0 & 0 & 0 & 0 & 0 & 0 & 0 & 5
\end{bmatrix}. \tag{4}
$$

As the shading indicates, this matrix J is comprised of four Jordan blocks, placed corner-to-corner along the diagonal.

DEFINITION 9.7 Jordan Canonical Form

> An $n \times n$ matrix J is a **Jordan canonical form** if it consists of Jordan blocks, placed corner-to-corner along the main diagonal, as in matrix (4), with only zero entries outside these Jordan blocks.

Every diagonal matrix is a Jordan canonical form, because each diagonal entry can be viewed as being the sole entry in a 1×1 Jordan block. Notice that matrix (4) contains the 1×1 Jordan block [2]. Notice, too, that the breaks between the Jordan blocks in matrix (4) occur where some diagonal entry has a 0 rather than 1 immediately above it.

EXAMPLE 3 Is the matrix

$$
\begin{bmatrix}
2 & 1 & 0 \\
0 & 2 & 1 \\
0 & 0 & 7
\end{bmatrix}
$$

a Jordan canonical form? Why?

SOLUTION This matrix is not a Jordan canonical form. Because not all diagonal entries are equal, there should be at least two Jordan blocks present in order for the matrix to be a Jordan canonical form, and [7] should be a 1×1 Jordan block. However, the entry immediately above 7 is not 0. Consequently, this matrix is not a Jordan canonical form. ∎

EXAMPLE 4 Describe the effect of matrix J in Eq. (4) on each of the standard basis vectors in $\mathbb{C}^8$. Then give the eigenvalues and eigenspaces of J. Finally, find the dimension of the nullspace of $(J - \lambda I)^k$ for each eigenvalue λ of J and for each positive integer k.

SOLUTION   We find that

$$Je_3 = -ie_3 + e_2, \quad Je_2 = -ie_2 + e_1, \quad Je_1 = -ie_1,$$
$$Je_5 = -ie_5 + e_4, \quad Je_4 = -ie_4,$$
$$Je_6 = 2e_6,$$
$$Je_8 = 5e_8 + e_7, \quad Je_7 = 5e_7.$$

The eigenvalues of J are $-i$, 2, and 5, which have algebraic multiplicities of 5, 1, and 2, respectively. The eigenspaces of J are $E_{-i} = \mathrm{sp}(e_1, e_4)$, $E_2 = \mathrm{sp}(e_6)$, and $E_5 = \mathrm{sp}(e_7)$, as you can easily check.

The effect of $J - (-i)I$ on the first five standard basis vectors is given by the two strings

$$J + iI: \quad \begin{aligned} e_3 \to e_2 \to e_1 \to 0, \\ e_5 \to e_4 \to 0. \end{aligned} \tag{5}$$

The 3×3 lower right-hand corner of $J + iI$ describes the action of $J + iI$ on e_6, e_7, and e_8. Because this 3×3 matrix has a nonzero determinant, it causes $J + iI$ to carry these three vectors into three independent vectors, and the same is true of all powers of $J + iI$. Thus we can determine the dimension of the nullspace of $J + iI$ by diagram (5), and we find that

$J + iI$ has nullspace $\mathrm{sp}(e_1, e_4)$ of dimension 2,

$(J + iI)^2$ has nullspace $\mathrm{sp}(e_1, e_2, e_4, e_5)$ of dimension 4,

$(J + iI)^3$ has nullspace $\mathrm{sp}(e_1, e_2, e_3, e_4, e_5)$ of dimension 5,

$(J + iI)^k$ has the same nullspace as that of $(J + iI)^3$ for $k > 3$.

By a similar argument, we find that

$(J - 2I)^k$ has nullspace $\mathrm{sp}(e_6)$ of dimension 1 for $k \geq 1$,

$J - 5I$ has nullspace $\mathrm{sp}(e_7)$ of dimension 1,

$(J - 5I)^k$ has nullspace $\mathrm{sp}(e_7, e_8)$ of dimension 2 for $k > 1$.   ∎

HISTORICAL NOTE   THE JORDAN CANONICAL FORM appears in the *Treatise on Substitutions and Algebraic Equations*, the chief work of the French algebraist Camille Jordan (1838–1921). This text, which appeared in 1870, incorporated the author's group-theory work over the preceding decade and became the bible of the field for the remainder of the nineteenth century. The theorem containing the canonical form actually deals not with matrices over the real numbers, but with matrices with entries from the finite field of order p. And as the title of the book indicates, Jordan was not considering matrices as such, but the linear substitutions that they represented.

Camille Jordan, a brilliant student, entered the Ecole Polytechnique in Paris at the age of 17 and practiced engineering from the time of his graduation until 1885. He thus had ample time for mathematical research. From 1873 until 1912, he taught at both the Ecole Polytechnique and the Collège de France. Besides doing seminal work on group theory, he is known for important discoveries in modern analysis and topology.

EXAMPLE 5   Suppose a 9×9 Jordan canonical form J has the following properties:

    1. $(J - 3iI)^k$ has rank 7 for $k = 1$, rank 5 for $k = 2$, and rank 4 for $k \geq 3$,

    2. $(J + I)^j$ has rank 6 for $j = 1$ and rank 5 for $j \geq 2$.

Find the Jordan blocks that appear in J.

SOLUTION   Because the rank of $J - 3iI$ is 7, the dimension of its nullspace is $9 - 7 = 2$, so $3i$ is an eigenvalue of geometric multiplicity 2. It must give rise to two Jordan blocks. In addition, $J - 3iI$ must annihilate two eigenvectors $\mathbf{e}_r$ and $\mathbf{e}_s$ in the standard basis. Because the rank of $(J - 3iI)^2$ is 5, its nullspace must have dimension 4, so in a diagram of the effect of $J - 3iI$ on the standard basis, we must have $(J - 3iI)\mathbf{e}_{r+1} = \mathbf{e}_r$ and $(J - 3iI)\mathbf{e}_{s+1} = \mathbf{e}_s$. Because $(J - 3iI)^k$ has rank 4 for $k \geq 3$, its nullity is 5, and we have just one more standard basis vector—either $\mathbf{e}_{r+2}$ or $\mathbf{e}_{s+2}$—that is annihilated by $(J - 3iI)^3$. Thus, the two Jordan blocks in J that have $3i$ on the diagonal are

$$J_1 = \begin{bmatrix} 3i & 1 & 0 \\ 0 & 3i & 1 \\ 0 & 0 & 3i \end{bmatrix} \quad \text{and} \quad J_2 = \begin{bmatrix} 3i & 1 \\ 0 & 3i \end{bmatrix}.$$

Because $J + 1I$ has rank 6, its nullspace has dimension $9 - 6 = 3$, so -1 is an eigenvalue of geometric multiplicity 3 and gives rise to three Jordan blocks. Because $(J + I)^j$ has rank 5 for $j \geq 2$, its nullspace has dimension 4, so $(J + I)^2$ annihilates a total of four standard basis vectors. Thus, just one of these Jordan blocks is 2×2, and the other two are 1×1. The Jordan blocks arising from the eigenvalue -1 are then

$$J_3 = \begin{bmatrix} -1 & 1 \\ 0 & -1 \end{bmatrix} \quad \text{and} \quad J_4 = J_5 = [-1].$$

The matrix J might have these blocks in any order down its diagonal. Symbolically, we might have

$$J = \begin{bmatrix} J_3 & & & & \\ & J_1 & & \mathbf{0} & \\ & & J_4 & & \\ & & & J_2 & \\ & & & & J_5 \end{bmatrix}, \quad J = \begin{bmatrix} J_4 & & & & \\ & J_5 & & \mathbf{0} & \\ & & J_2 & & \\ & & & J_1 & \\ & & & & J_3 \end{bmatrix},$$

or any other order.   ∎

Jordan Bases

If an $n \times n$ matrix A is similar to a Jordan canonical form J, we call J a **Jordan canonical form** of A. When this is the case, there exists an invertible matrix C such that $C^{-1}AC = J$. We know that similar matrices represent the same linear transformation, but with respect to different bases. Thus, if A is similar to J, there must exist a basis $\{\mathbf{b}_1, \mathbf{b}_2, \ldots, \mathbf{b}_n\}$ of $\mathbb{C}^n$ with the same schematic string properties relative to A that the standard ordered basis has relative to the matrix J. We proceed to define such a *Jordan basis*.

DEFINITION 9.8   Jordan Basis

Let A be an $n \times n$ matrix. An ordered basis $B = (\mathbf{b}_1, \mathbf{b}_2, \ldots, \mathbf{b}_n)$ of $\mathbb{C}^n$ is a **Jordan basis for** A if, for $1 \le j \le n$, we have either $A\mathbf{b}_j = \lambda\mathbf{b}_j$ or $A\mathbf{b}_j = \lambda\mathbf{b}_j + \mathbf{b}_{j-1}$, where λ is an eigenvalue of A that we say is **associated with** $\mathbf{b}_j$. If $A\mathbf{b}_j = \lambda\mathbf{b}_j + \mathbf{b}_{j-1}$, we require that the eigenvalue associated with $\mathbf{b}_{j-1}$ also be λ.

If an $n \times n$ matrix A has a Jordan basis B, then the matrix representation of the linear transformation $T(\mathbf{z}) = A\mathbf{z}$ relative to B must be a Jordan canonical form. We know then that $J = C^{-1}AC$, where C is the $n \times n$ matrix whose jth column vector is the jth vector $\mathbf{b}_j$ in B. In a moment we will prove that, for every square matrix, there is an associated Jordan basis, and consequently that every square matrix is similar to a Jordan canonical form. First, though, we outline a method for the computation of a Jordan canonical form of A.

Finding a Jordan Canonical Form of A

1. Find the eigenvalues of A.
2. For each eigenvalue λ, compute the rank of $(A - \lambda I)^k$ for consecutive values of k, starting with $k = 1$, until the same rank is obtained for two consecutive values of k.
3. From the data generated, find a Jordan canonical form for A, as in Example 5.

We now illustrate this technique.

EXAMPLE 6   Find a Jordan canonical form of the matrix

$$A = \begin{bmatrix} 2 & 5 & 0 & 0 & 1 \\ 0 & 2 & 0 & 0 & 0 \\ 0 & 0 & -1 & 0 & 0 \\ 0 & 0 & 0 & -1 & 0 \\ 0 & 0 & 0 & 0 & -1 \end{bmatrix}.$$

SOLUTION   Because A is an upper-triangular matrix, we see that the eigenvalues of A are $\lambda_1 = \lambda_2 = 2$ and $\lambda_3 = \lambda_4 = \lambda_5 = -1$. Now

$$A - \lambda_1 I = A - 2I = \begin{bmatrix} 0 & 5 & 0 & 0 & 1 \\ 0 & 0 & 0 & 0 & 0 \\ 0 & 0 & -3 & 0 & 0 \\ 0 & 0 & 0 & -3 & 0 \\ 0 & 0 & 0 & 0 & -3 \end{bmatrix}$$

has rank 4 and consequently has a nullspace of dimension 1. We find that

$$(A - 2I)^2 = \begin{bmatrix} 0 & 0 & 0 & 0 & -3 \\ 0 & 0 & 0 & 0 & 0 \\ 0 & 0 & 9 & 0 & 0 \\ 0 & 0 & 0 & 9 & 0 \\ 0 & 0 & 0 & 0 & 9 \end{bmatrix},$$

which has rank 3 and therefore has a nullspace of dimension 2. Furthermore,

$$(A - 2I)^3 = \begin{bmatrix} 0 & 0 & 0 & 0 & 9 \\ 0 & 0 & 0 & 0 & 0 \\ 0 & 0 & -27 & 0 & 0 \\ 0 & 0 & 0 & -27 & 0 \\ 0 & 0 & 0 & 0 & -27 \end{bmatrix}$$

has the same rank and nullity as $(A - 2I)^2$. Thus we have $A\mathbf{b}_1 = 2\mathbf{b}_1$ and $A\mathbf{b}_2 = 2\mathbf{b}_2 + \mathbf{b}_1$ for some Jordan basis $B = (\mathbf{b}_1, \mathbf{b}_2, \mathbf{b}_3, \mathbf{b}_4, \mathbf{b}_5)$ for A. There is just one Jordan block associated with $\lambda_1 = 2$—namely,

$$J_1 = \begin{bmatrix} 2 & 1 \\ 0 & 2 \end{bmatrix}.$$

For the eigenvalue $\lambda_3 = -1$, we find that

$$A - \lambda_3 I = A + I = \begin{bmatrix} 3 & 5 & 0 & 0 & 1 \\ 0 & 3 & 0 & 0 & 0 \\ 0 & 0 & 0 & 0 & 0 \\ 0 & 0 & 0 & 0 & 0 \\ 0 & 0 & 0 & 0 & 0 \end{bmatrix},$$

which has rank 2 and therefore has a nullspace of dimension 3. Because -1 is an eigenvalue of both algebraic multiplicity and geometric multiplicity 3, we realize that $J_2 = J_3 = J_4 = [-1]$ are the remaining Jordan blocks. This is confirmed by the fact that

$$(A + I)^2 = \begin{bmatrix} 9 & 30 & 0 & 0 & 3 \\ 0 & 9 & 0 & 0 & 0 \\ 0 & 0 & 0 & 0 & 0 \\ 0 & 0 & 0 & 0 & 0 \\ 0 & 0 & 0 & 0 & 0 \end{bmatrix}$$

again has rank 2 and nullity 3. Thus, a Jordan canonical form for A is

$$J = \begin{bmatrix} 2 & 1 & 0 & 0 & 0 \\ 0 & 2 & 0 & 0 & 0 \\ 0 & 0 & -1 & 0 & 0 \\ 0 & 0 & 0 & -1 & 0 \\ 0 & 0 & 0 & 0 & -1 \end{bmatrix}.$$ ∎

EXAMPLE 7   Find a Jordan basis for matrix A in Example 6.

SOLUTION   For the part of a Jordan basis associated with the eigenvalue 2, we need to find a vector $\mathbf{b}_2$ in the nullspace of $(A - 2I)^2$ that is not in the nullspace of $A - 2I$; then we may take $\mathbf{b}_1 = (A - 2I)\mathbf{b}_2$. From the computation of $A - 2I$ and $(A - 2I)^2$ in Example 6, we see that we can take

$$\mathbf{b}_2 = \begin{bmatrix} 0 \\ 1 \\ 0 \\ 0 \\ 0 \end{bmatrix}, \quad \text{and then} \quad \mathbf{b}_1 = (A - 2I)\mathbf{b}_2 = \begin{bmatrix} 5 \\ 0 \\ 0 \\ 0 \\ 0 \end{bmatrix}.$$

For $\mathbf{b}_3$, $\mathbf{b}_4$, and $\mathbf{b}_5$, we need only take a basis for the nullspace of $A + I$. We see that we can take

$$\mathbf{b}_3 = \begin{bmatrix} 0 \\ 0 \\ 1 \\ 0 \\ 0 \end{bmatrix}, \quad \mathbf{b}_4 = \begin{bmatrix} 0 \\ 0 \\ 0 \\ 1 \\ 0 \end{bmatrix}, \quad \text{and} \quad \mathbf{b}_5 = \begin{bmatrix} -1 \\ 0 \\ 0 \\ 0 \\ 3 \end{bmatrix}.$$ ∎

In Example 7, it was easy to find vectors in a Jordan basis corresponding to the eigenvalue 2 whose geometric multiplicity is less than its algebraic multiplicity, because only one Jordan block corresponds to the eigenvalue 2. We now indicate how a Jordan basis can be constructed when more than one such block corresponds to a single eigenvalue λ. Let N_r be the nullspace of $(A - \lambda I)^r$ for $r \geq 1$, and suppose (for example) that $\dim(N_1) = 4$, $\dim(N_2) = 7$, and $\dim(N_r) = 8$ for $r \geq 3$. Then a Jordan basis for A contains four strings corresponding to λ, which we may represent as

$$\mathbf{b}_3 \rightarrow \mathbf{b}_2 \rightarrow \mathbf{b}_1 \rightarrow \mathbf{0},$$

$$\mathbf{b}_5 \rightarrow \mathbf{b}_4 \rightarrow \mathbf{0},$$

$$\mathbf{b}_7 \rightarrow \mathbf{b}_6 \rightarrow \mathbf{0},$$

$$\mathbf{b}_8 \rightarrow \mathbf{0}.$$

To find the first and longest of these strings, we *compute* a basis $\{\mathbf{v}_1, \mathbf{v}_2, \dots, \mathbf{v}_8\}$ for the nullspace N_3 of $(A - \lambda I)^3$. The preceding strings show that multiplication of all of the vectors in N_3 on the left by $(A - \lambda I)^2$ yields a space of dimension 1, so at least one of the vectors $\mathbf{v}_i$ has the property that $(A - \lambda I)^2 \mathbf{v}_i \neq \mathbf{0}$.

Let $\mathbf{b}_3$ be such a vector, and set $\mathbf{b}_2 = (A - \lambda I)\mathbf{b}_3$ and $\mathbf{b}_1 = (A - \lambda I)\mathbf{b}_2$. It is not difficult to show that $\mathbf{b}_1$, $\mathbf{b}_2$, and $\mathbf{b}_3$ must be independent. Thus we have found the first string.

Now $\mathbf{b}_1$ and $\mathbf{b}_2$ lie in N_2, and we can expand the independent set $\{\mathbf{b}_1, \mathbf{b}_2\}$ to a basis $\{\mathbf{b}_1, \mathbf{b}_2, \mathbf{w}_1, \ldots, \mathbf{w}_5\}$ of N_2. Again, the strings displayed earlier show that multiplication of the vectors in N_2 on the left by $A - \lambda I$ must yield a space of dimension 3, so there exist two vectors $\mathbf{w}_i$ and $\mathbf{w}_j$ such that the vectors $\mathbf{b}_1$, $(A - \lambda I)\mathbf{w}_i$, and $(A - \lambda I)\mathbf{w}_j$ are independent. Let $\mathbf{b}_5 = \mathbf{w}_i$ and $\mathbf{b}_4 = (A - \lambda I)\mathbf{b}_5$, while $\mathbf{b}_7 = \mathbf{w}_j$ and $\mathbf{b}_6 = (A - \lambda I)\mathbf{b}_7$. It can be shown that the vectors $\mathbf{b}_1$, $\mathbf{b}_2$, $\ldots$, $\mathbf{b}_7$ are independent. Finally, we expand the set $\{\mathbf{b}_1, \mathbf{b}_4, \mathbf{b}_6\}$ to a basis $\{\mathbf{b}_1, \mathbf{b}_4, \mathbf{b}_6, \mathbf{b}_8\}$ for N_1 to complete the portion of the Jordan basis corresponding to λ.

Although we know the techniques for finding bases for the nullspaces N_i and for expanding a given set of independent vectors to a basis, significant pencil-and-paper illustrations of this construction would be cumbersome, so we do not include them here. Any Jordan bases requested in the exercises can be found as in Example 7.

An application of the Jordan canonical form to differential equations is indicated in Exercise 32. We mention that computer-aided computation of a Jordan canonical form for a square matrix is not a stable process. Consider, for example, the matrix

$$A = \begin{bmatrix} 2 & c \\ 0 & 2 \end{bmatrix}.$$

If $c = 10^{-100}$, then the Jordan canonical form of A has 1 as its entry in the upper right-hand corner; but if $c = 0$, that entry is 0.

Existence of a Jordan Form for a Square Matrix

To demonstrate the existence of a Jordan canonical form similar to an $n \times n$ matrix A, we need only show that we have a Jordan basis B for A. Let us formalize the concept of a *string* in a Jordan basis $B = (\mathbf{b}_1, \mathbf{b}_2, \ldots, \mathbf{b}_n)$. Let λ be an eigenvalue of A. If $A\mathbf{b}_i = \lambda\mathbf{b}_i$ and $A\mathbf{b}_k = \lambda\mathbf{b}_k + \mathbf{b}_{k-1}$ for $i < k < j$, while $A\mathbf{b}_j \neq \lambda\mathbf{b}_j + \mathbf{b}_{j-1}$, we refer to the sequence $\mathbf{b}_i$, $\mathbf{b}_{i+1}$, $\ldots$, $\mathbf{b}_{j-1}$ as a **string** of basis vectors **starting** at $\mathbf{b}_{j-1}$, **ending** at $\mathbf{b}_i$, and **associated with** λ. This string is represented by the diagram

$$A - \lambda I: \qquad \mathbf{b}_{j-1} \to \cdots \to \mathbf{b}_{i+1} \to \mathbf{b}_i \to 0.$$

THEOREM 9.9 Jordan Canonical Form of a Square Matrix

Let A be a square matrix. There exists an invertible matrix C such that the matrix $J = C^{-1}AC$ is a Jordan canonical form. This Jordan canonical form is unique, except for the order of the Jordan blocks of which it is composed.

PROOF   We use a proof due to Filippov. First we note that it suffices to prove the theorem for matrices A having 0 as an eigenvalue. Observe that, if λ is an eigenvalue of A, then 0 is an eigenvalue of $A - \lambda I$. Now if we can find C such that $C^{-1}(A - \lambda I)C = J$ is a Jordan canonical form, then $C^{-1}AC = J + \lambda I$ is also a Jordan canonical form. Thus, we restrict ourselves to the case where A has an eigenvalue of 0.

In order to find a Jordan canonical form for A, it is useful to consider also the linear transformation $T: \mathbb{C}^n \to \mathbb{C}^n$, where $T(\mathbf{z}) = A\mathbf{z}$; a Jordan basis for A is considered to be a Jordan basis for T. We will prove the existence of a Jordan basis for any such linear transformation by induction on the dimension of the domain of the transformation.

If T is a linear transformation of a one-dimensional vector space sp($\mathbf{z}$), then $T(\mathbf{z}) = \lambda\mathbf{z}$ for some $\lambda \in \mathbb{C}$, and $\{\mathbf{z}\}$ is the required Jordan basis. (The matrix of T with respect to this ordered basis is the 1×1 matrix $[\lambda]$, which is already a Jordan canonical form.)

Now suppose that there exist Jordan bases for linear transformations on subspaces of $\mathbb{C}^n$ of dimension less than n, and let $T(\mathbf{z}) = A\mathbf{z}$ for $\mathbf{z} \in \mathbb{C}^n$ and an $n \times n$ matrix A. As noted, we can assume that zero is an eigenvalue of A. Then rank(A) < n; let r = rank(A). Now T maps $\mathbb{C}^n$ onto the column space of A that is of dimension $r < n$. Let T' be the induced linear transformation of the column space of A into itself, defined by $T'(\mathbf{v}) = T(\mathbf{v})$ for $\mathbf{v}$ in the column space of A. By our induction hypothesis, there is a Jordan basis

$$B' = (\mathbf{u}_1, \mathbf{u}_2, \ldots, \mathbf{u}_r)$$

for this column space of A.

Let S be the intersection of the column space and the nullspace of A. We wish to separate the vectors in B' that are in S from those that are not. The nonzero vectors in S are precisely the eigenvectors in the column space of A with corresponding eigenvalue 0; that is, they are the eigenvectors of T' with eigenvalue 0. In other words, S is the nullspace of T'. Let J' be the matrix representation of T' relative to B'. Because J' is a Jordan canonical form, we see that the nullity of T' (and of J') is precisely the number of zero rows in J'. This is true because J' is an upper-triangular square matrix; it can be brought to echelon form by means of row exchanges that place the zero rows at the bottom while sliding the nonzero rows up. Thus, if dim(S) = s, there are s zero rows in J'. Now in J' we have exactly one zero row for each Jordan block corresponding to the eigenvalue 0—namely, the row containing the bottom row of the block. Because the number of such blocks is equal to the number of strings in B' ending in S, we conclude that there are s such strings. Some of these strings may be of length 1 whereas others may be longer.

Figure 9.11 shows one possible situation when $s = 2$, where two vectors in S—namely, $\mathbf{u}_1$ and $\mathbf{u}_4$—are ending points of strings

$$\mathbf{u}_3 \to \mathbf{u}_2 \to \mathbf{u}_1 \to \mathbf{0} \quad \text{and} \quad \mathbf{u}_5 \to \mathbf{u}_4 \to \mathbf{0}$$

lying in the column space of A. These s strings of B' that end in S start at s vectors in the column space of A; these are the vectors $\mathbf{u}_3$ and $\mathbf{u}_5$ in Figure 9.11.

Because the vector at the beginning of the jth string is in the column space of A, it must have the form $A\mathbf{w}_j$ for some vector $\mathbf{w}_j$ in $\mathbb{C}^n$. Thus we obtain the vectors $\mathbf{w}_1, \mathbf{w}_2, \ldots, \mathbf{w}_s$ illustrated in Figure 9.11 for $s = 2$.

Finally, the nullspace of A has dimension $n - r$, and we can expand the set of s independent vectors in S to a basis for this nullspace. This gives rise to $n - r - s$ more vectors $\mathbf{v}_1, \mathbf{v}_2, \ldots, \mathbf{v}_{n-r-s}$. Of course, each $\mathbf{v}_i$ is an eigenvector with corresponding eigenvalue 0.

We claim that

$$(\mathbf{u}_1, \ldots, \mathbf{u}_r, \mathbf{w}_1, \ldots, \mathbf{w}_s, \mathbf{v}_1, \ldots, \mathbf{v}_{n-r-s})$$

can be reordered to become a Jordan basis B for A (and of course for T). We reorder it by moving the vectors $\mathbf{w}_i$, tucking each one in so that it starts the appropriate string in B' that was used to define it. For the situation in Figure 9.11, we obtain

$$(\mathbf{u}_1, \mathbf{u}_2, \mathbf{u}_3, \mathbf{w}_1, \mathbf{u}_4, \mathbf{u}_5, \mathbf{w}_2, \mathbf{u}_6, \ldots, \mathbf{u}_r, \mathbf{v}_1, \ldots, \mathbf{v}_{n-r-2})$$

as Jordan basis. From our construction, we see that B is a Jordan basis for A if it is a basis for $\mathbb{C}^n$. Because there are $r + s + (n - r - s) = n$ vectors in all, we need only show that they are independent.

Suppose that

$$\sum_{i=1}^{r} a_i \mathbf{u}_i + \sum_{j=1}^{s} c_j \mathbf{w}_j + \sum_{k=1}^{n-r-s} d_k \mathbf{v}_k = \mathbf{0}. \tag{6}$$

Because the vectors $\mathbf{v}_k$ lie in the nullspace of A, if we apply A to both sides of this equation, we obtain

$$\sum_{i=1}^{r} a_i A\mathbf{u}_i + \sum_{j=1}^{s} c_j A\mathbf{w}_j = \mathbf{0}. \tag{7}$$

Because each $A\mathbf{u}_i$ is either of the form $\lambda \mathbf{u}_i$ or of the form $\lambda \mathbf{u}_i + \mathbf{u}_{i-1}$, we see that the first sum is a linear combination of vectors $\mathbf{u}_i$. Moreover, these vectors

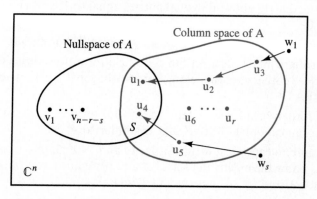

FIGURE 9.11
Construction of a Jordan basis for A ($s = 2$).

$A\mathbf{u}_i$ *do not begin* any string in B'. Now the vectors $A\mathbf{w}_j$ in the second sum are vectors $\mathbf{u}_i$ that appear at the *start* of the s strings in B' that end in S. Thus they do not appear in the first sum. Because B' is an independent set, all the coefficients c_j in Eq. (7) must be zero. Equation (6) can then be written as

$$\sum_{i=1}^{r} a_i \mathbf{u}_i = \sum_{k=1}^{n-r-s} -d_k \mathbf{v}_k. \tag{8}$$

Now the vector on the left-hand side of this equation lies in the column space of A, whereas the vector on the right-hand side is in the nullspace of A. Consequently, this vector lies in S and is a linear combination of the s basis vectors $\mathbf{u}_i$ in S. Because the $\mathbf{v}_k$ were obtained by extending these s vectors to a basis for the nullspace of A, the vector $\mathbf{0}$ is the only linear combination of the $\mathbf{v}_k$ that lies in S. Thus, the vector on both sides of Eq. (8) is $\mathbf{0}$. Because the $\mathbf{v}_k$ are independent, we see that all d_k are zero. Because the $\mathbf{u}_i$ are independent, it follows that the a_i are all zero. Therefore, B is an independent set of n vectors and is thus a basis for $\mathbb{C}^n$. We have seen that, by our construction, it must be a Jordan basis. This completes the induction part of our proof, demonstrating the existence of a Jordan canonical form for every square matrix A.

Our work prior to this theorem makes clear that the Jordan blocks constituting a Jordan canonical form for A are completely determined by the ranks of the matrices $(A - \lambda I)^k$ for all eigenvalues λ of A and for all positive integers k. Thus, a Jordan canonical form J for A is unique except as to the order in which these blocks appear along the diagonal of J. ▲

SUMMARY

1. A Jordan block is a square matrix with all diagonal entries equal, all entries immediately above diagonal entries equal to 1, and all other entries equal to 0.

2. Properties of a Jordan block are given in Theorem 9.8.

3. A square matrix is a Jordan canonical form if it consists of Jordan blocks placed corner to corner along its main diagonal, with entries elsewhere equal to 0.

4. A Jordan basis (see Definition 9.8) for an $n \times n$ matrix A gives rise to a Jordan canonical form J that is similar to A.

5. A Jordan canonical form similar to an $n \times n$ matrix A can be computed if we know the eigenvalues λ_i of A and if we know the rank of $(A - \lambda_i I)^k$ for each λ_i and for all positive integers k.

6. Every square matrix has a Jordan canonical form; that is, it is similar to a Jordan canonical form.

EXERCISES

In Exercises 1–6, determine whether the given matrix is a Jordan canonical form.

1. $\begin{bmatrix} 0 & 0 & 0 \\ 0 & 0 & 0 \\ 0 & 0 & 0 \end{bmatrix}$

2. $\begin{bmatrix} 3 & 0 & 0 \\ 0 & 3 & 1 \\ 0 & 0 & 3 \end{bmatrix}$

3. $\begin{bmatrix} 3 & 1 & 0 & 0 \\ 0 & 3 & 1 & 0 \\ 0 & 0 & 2 & 1 \\ 0 & 0 & 0 & 2 \end{bmatrix}$

4. $\begin{bmatrix} 1 & 0 & 0 & 0 \\ 0 & 2 & 0 & 0 \\ 0 & 0 & 3 & 0 \\ 0 & 0 & 0 & 4 \end{bmatrix}$

5. $\begin{bmatrix} i & 1 & 0 & 0 \\ 0 & -i & 0 & 0 \\ 0 & 0 & 3 & 1 \\ 0 & 0 & 0 & 3 \end{bmatrix}$

6. $\begin{bmatrix} 2 & 1 & 0 & 0 \\ 0 & 2 & 0 & 0 \\ 0 & 0 & i & 0 \\ 0 & 0 & 0 & -1 \end{bmatrix}$

In Exercises 7–10:

a) Find the eigenvalues of the given matrix J.

b) Give the rank and nullity of $(J - \lambda)^k$ for each eigenvalue λ of J and for every positive integer k.

c) Draw schemata of the strings of vectors in the standard basis arising from the Jordan blocks in J.

d) For each standard basis vector $\mathbf{e}_k$, express $J\mathbf{e}_k$ as a linear combination of vectors in the standard basis.

7. $\begin{bmatrix} -2 & 1 & 0 & 0 \\ 0 & -2 & 1 & 0 \\ 0 & 0 & -2 & 1 \\ 0 & 0 & 0 & -2 \end{bmatrix}$

8. $\begin{bmatrix} i & 0 & 0 & 0 & 0 \\ 0 & i & 1 & 0 & 0 \\ 0 & 0 & i & 0 & 0 \\ 0 & 0 & 0 & -2 & 0 \\ 0 & 0 & 0 & 0 & -2 \end{bmatrix}$

9. $\begin{bmatrix} -1 & 0 & 0 & 0 & 0 \\ 0 & 2 & 1 & 0 & 0 \\ 0 & 0 & 2 & 0 & 0 \\ 0 & 0 & 0 & 2 & 1 \\ 0 & 0 & 0 & 0 & 2 \end{bmatrix}$

10. $\begin{bmatrix} i & 1 & 0 & 0 & 0 & 0 & 0 & 0 \\ 0 & i & 0 & 0 & 0 & 0 & 0 & 0 \\ 0 & 0 & i & 0 & 0 & 0 & 0 & 0 \\ 0 & 0 & 0 & i & 1 & 0 & 0 & 0 \\ 0 & 0 & 0 & 0 & i & 0 & 0 & 0 \\ 0 & 0 & 0 & 0 & 0 & 2 & 1 & 0 \\ 0 & 0 & 0 & 0 & 0 & 0 & 2 & 1 \\ 0 & 0 & 0 & 0 & 0 & 0 & 0 & 2 \end{bmatrix}$

In Exercises 11–14, find a Jordan canonical form for A from the given data.

11. A is 5×5, $A - 3I$ has nullity 2, $(A - 3I)^2$ has nullity 3, $(A - 3I)^3$ has nullity 4, $(A - 3I)^k$ has nullity 5 for $k \geq 4$.

12. A is 7×7, $A + I$ has nullity 3, $(A + I)^k$ has nullity 5 for $k \geq 2$; $A + iI$ has nullity 1, $(A + iI)^j$ has nullity 2 for $j \geq 2$.

13. A is 8×8, $A - I$ has nullity 2, $(A - I)^2$ has nullity 4, $(A - I)^k$ has nullity 5 for $k \geq 3$; $(A + 2I)^j$ has nullity 3 for $j \geq 1$.

14. A is 8×8; $A + iI$ has rank 4, $(A + iI)^2$ has rank 2, $(A + iI)^3$ has rank 1, $(A + iI)^k = O$ for $k \geq 4$.

In Exercises 15–22, find a Jordan canonical form and a Jordan basis for the given matrix.

15. $\begin{bmatrix} -10 & 4 \\ -25 & 10 \end{bmatrix}$

16. $\begin{bmatrix} 5 & -4 \\ 9 & -7 \end{bmatrix}$

17. $\begin{bmatrix} 4 & 0 & 0 \\ 2 & 1 & 3 \\ 5 & 0 & 4 \end{bmatrix}$

18. $\begin{bmatrix} -3 & 0 & 1 \\ 2 & -2 & 1 \\ -1 & 0 & -1 \end{bmatrix}$

19. $\begin{bmatrix} 2 & 5 & 0 & 0 & 0 \\ 0 & 2 & 0 & 0 & 0 \\ 0 & 0 & -1 & 0 & -1 \\ 0 & 0 & 0 & -1 & 0 \\ 0 & 0 & 0 & 0 & -1 \end{bmatrix}$

20. $\begin{bmatrix} i & 0 & 0 & 0 & 0 \\ 0 & i & 0 & 0 & 0 \\ 0 & 0 & 2 & 0 & 0 \\ 0 & 0 & 0 & 2 & 0 \\ 2 & 0 & -1 & 0 & 2 \end{bmatrix}$

21. $\begin{bmatrix} 2 & 0 & 0 & 0 & 1 \\ 0 & 2 & 0 & 0 & 0 \\ 0 & 0 & 2 & 0 & 1 \\ 0 & 0 & 0 & 2 & 0 \\ 0 & 0 & 0 & 0 & 2 \end{bmatrix}$

22. $\begin{bmatrix} 1 & 2 & 0 & 0 & 1 \\ 0 & 1 & 0 & 0 & 0 \\ 0 & 0 & 1 & 0 & 1 \\ 0 & 0 & 0 & 1 & 2 \\ 0 & 0 & 0 & 0 & 1 \end{bmatrix}$

23. Mark each of the following True or False.

____ **a.** Every Jordan block matrix has just one eigenvalue.

____ **b.** Every matrix having a unique eigenvalue is a Jordan block.

____ **c.** Every diagonal matrix is a Jordan canonical form.

____ **d.** Every square matrix is similar to a Jordan canonical form.

____ **e.** Every square matrix is similar to a unique Jordan canonical form.

____ **f.** Every 1×1 matrix is similar to a unique Jordan canonical form.

____ **g.** There is a Jordan basis for every square matrix A.

____ **h.** There is a unique Jordan basis for every square matrix A.

____ **i.** Every 3×3 diagonalizable matrix is similar to exactly six Jordan canonical forms.

____ **j.** Every 3×3 matrix is similar to exactly six Jordan canonical forms.

24. Let $A = \begin{bmatrix} 0 & 1 & 0 & 0 \\ 0 & 0 & 1 & 0 \\ 0 & 0 & 0 & 1 \\ 0 & 0 & 0 & 0 \end{bmatrix}$. Compute

A^2, A^3, and A^4.

25. Let A be an $n \times n$ upper-triangular matrix with all diagonal entries 0. Compute A^m for all positive integers $m \geq n$. (See Exercise 24.) Prove that your answer is correct.

26. Let $A = \begin{bmatrix} 2 & 1 & 0 & 0 & 0 \\ 0 & 2 & 1 & 0 & 0 \\ 0 & 0 & 2 & 0 & 0 \\ 0 & 0 & 0 & 3 & 1 \\ 0 & 0 & 0 & 0 & 3 \end{bmatrix}$. Compute

$(A - 2I)^3(A - 3I)^2$.

27. Let $A = \begin{bmatrix} 2 & 0 & 0 & 0 & 0 \\ 0 & 2 & 1 & 0 & 0 \\ 0 & 0 & 2 & 0 & 0 \\ 0 & 0 & 0 & 3 & 1 \\ 0 & 0 & 0 & 0 & 3 \end{bmatrix}$. Compute

$(A - 2I)^2(A - 3I)^2$. Compare with Exercise 26.

28. Let $A = \begin{bmatrix} i & 0 & 0 & 0 & 0 \\ 0 & i & 1 & 0 & 0 \\ 0 & 0 & i & 0 & 0 \\ 0 & 0 & 0 & 2 & 0 \\ 0 & 0 & 0 & 0 & 2 \end{bmatrix}$. Find a polynomial in

A (that is, a sum of terms $a_j A^j$ with a term $a_0 I$) that gives the zero matrix. (See Exercises 24–27.)

29. Repeat Exercise 28 for the matrix $A = \begin{bmatrix} -1 & 1 & 0 & 0 & 0 \\ 0 & -1 & 1 & 0 & 0 \\ 0 & 0 & -1 & 0 & 0 \\ 0 & 0 & 0 & i & 0 \\ 0 & 0 & 0 & 0 & i \end{bmatrix}$.

30. The Cayley–Hamilton theorem states that, if $p(\lambda) = a_n \lambda^n + \cdots + a_1 \lambda + a_0$ is the characteristic polynomial of a matrix A, then $p(A) = a_n A^n + \cdots + a_1 A + a_0 I = O$, the zero matrix. Prove it. [HINT: Consider $(A - \lambda_i I)^{n_i} \mathbf{b}$, where $\mathbf{b}$ is a vector in a Jordan basis corresponding to λ_i.] In view of Exercises 24–29, explain why you expect $p(J)$ to be O, where J is a Jordan canonical form for A. Deduce that $p(A) = O$.

31. Let $T: \mathbb{C}^n \to \mathbb{C}^n$ be a linear transformation. A subspace W of $\mathbb{C}^n$ is **invariant under** T if $T(\mathbf{w}) \in W$ for all $\mathbf{w} \in W$. Let A be the standard matrix representation of T.

a. Describe the one-dimensional invariant subspaces of T.

b. Show that every eigenspace E_λ of T is invariant under T.

c. Show that the vectors in any string in a Jordan basis for A generate an invariant subspace of T.

d. Is it true that, if S is a subspace of a subspace W that is invariant under T, then S is also invariant under T? If not, give a counterexample.

e. Is it true that every subspace of $\mathbb{R}^n$ invariant under T is contained in the nullspace of $(A - \lambda I)^n$, where λ is some eigenvalue of T? If not, give a counterexample.

32. In Section 5.3, we considered systems
$\mathbf{x}' = A\mathbf{x}$ of differential equations, and we saw
that, if $A = CJC^{-1}$, then the system takes the
form $\mathbf{y}' = J\mathbf{y}$, where $\mathbf{x} = C\mathbf{y}$. (We used D in
place of J in Section 5.3, because we were
concerned only with diagonalization.) Let
$\lambda_1, \lambda_2, \ldots, \lambda_n$ be the (not necessarily
distinct) eigenvalues of an $n \times n$ matrix A,
and let J be a Jordan canonical form for A.

a. Show that the system $\mathbf{y}' = J\mathbf{y}$ is of the
form

$$y_1' = \lambda_1 y_1 + c_1 y_2,$$
$$y_2' = \lambda_2 y_2 + c_2 y_3,$$
$$\vdots$$
$$y_{n-1}' = \lambda_{n-1} y_{n-1} + c_{n-1} y_n,$$
$$y_n' = \lambda_n y_n,$$

where each c_i is either 0 or 1.

b. How can the system in part a be solved?
[HINT: Start with the last equation.]

c. Given that, for

$$A = \begin{bmatrix} 2 & 2 & 3 \\ 0 & -1 & 0 \\ 2 & 2 & 1 \end{bmatrix},$$

$$C = \begin{bmatrix} \frac{1}{2} & 1 & \frac{3}{2} \\ 0 & -\frac{5}{4} & 0 \\ -\frac{1}{2} & 0 & 1 \end{bmatrix},$$

$$J = \begin{bmatrix} -1 & 1 & 0 \\ 0 & -1 & 0 \\ 0 & 0 & 4 \end{bmatrix},$$

we have $C^{-1}AC = J$, find the solution of
the differential system $\mathbf{x}' = A\mathbf{x}$.

33. Let A be an $n \times n$ matrix with eigenvalue λ.
Prove that the algebraic multiplicity of
λ is at least as large as its geometric multiplicity.

SOLVING LARGE LINEAR SYSTEMS

The Gauss and Gauss–Jordan methods presented in Chapter 1 are fine for solving very small linear systems with pencil and paper. Some applied problems—in particular, those requiring numerical solutions of differential equations—can lead to very large linear systems, involving thousands of equations in thousands of unknowns. Of course, such large linear systems must be solved by the use of computers. That is the primary concern of this chapter. Although a computer can work tremendously faster than we can with pencil and paper, each individual arithmetic operation does take some time, and additional time is used whenever the value of a variable is stored or retrieved. In addition, indexing in subscripted arrays requires time. When solving a large linear system with a computer, we must use as efficient a computational algorithm as we can, so that the number of operations required is as small as practically possible.

We begin this chapter with a discussion of the time required for a computer to execute operations and a comparison of the efficiency of the Gauss method including back substitution with that of the Gauss–Jordan method.

Section 10.2 presents the LU (lower- and upper-triangular) factorization of the coefficient matrix of a square linear system. This factorization will appear as we develop an efficient algorithm for solving, by repeated computer runs, many systems all having the same coefficient matrix.

Section 10.3 deals with problems of roundoff and discusses ill-conditioned matrices. We will see that there are actually very small linear systems, consisting of only two equations in two unknowns, for which good computer programs may give incorrect solutions.

10.1   CONSIDERATIONS OF TIME

Timing Data for One PC

The computation involved in solving a linear system is just a lot of arithmetic. Arithmetic takes time to execute even with a computer, although the computer is obviously much faster than an individual working with pencil and paper. In reducing a matrix by using elementary row operations, we spend most of our time adding a multiple of a row vector to another row vector. A typical step in multiplying row k of a matrix A by r and adding it to row i consists of

$$\text{replacing} \quad a_{ij} \quad \text{by} \quad a_{ij} + ra_{kj}. \tag{1}$$

In a computer language such as BASIC, operation (1) might become

$$A(I,J) = A(I,J) + R*A(K,J). \tag{2}$$

Computer instruction (2) involves operations of addition and multiplication, as well as indexing, retrieving values, and storing values. We use terminology of C. B. Moler and call such an operation a **flop**. These flops require time to execute on any computer, although the times vary widely depending on the computer hardware and the language of the computer program. When we wrote the first edition of this text in 1986, we experimented with our personal computer, which was very slow by today's standards, and created the data in Table 10.1. Although computers are much faster today, the indication in the table that computation does require time remains as valid now as it was then. We left the original 1986 data in the table so that you can see, if you work Exercise 22, how much faster computers are today. To form the table, we generated two random numbers, R and S, and then found the time required to add them 10,000 times, using the BASIC loop

$$\text{FOR I} = 1 \text{ TO N: C} = \text{R} + \text{S: NEXT} \tag{3}$$

where we had earlier set N = 10,000. We obtained the data shown in the top row of Table 10.1. We also had the computer execute loop (3) with + replaced by − (subtraction), then by * (multiplication), and finally by / (division). We similarly timed the execution of 10,000 flops, using

$$\text{FOR I} = 1 \text{ TO N: A(K,J)} = \text{A(K,J)} + \text{R} * \text{A(M,J): NEXT,} \tag{4}$$

where we had set N = 10,000 and had also assigned values to all other variables except I. All our data are shown in Table 10.1 on the next page.

Table 10.1 gives us quite a bit of insight into the PC that generated the data. Here are some things we noticed immediately.

Point 1 Multiplication took a bit, but surprisingly little, more time than addition.

TABLE 10.1
Time (in Seconds) for Executing 10,000 Operations

Routine	Interpretive BASIC		Compiled BASIC	
	Single-Precision	Double-Precision	Single-Precision	Double-Precision
Addition [using (3)]	37	44	8	9
Subtraction [using (3) with −]	37	48	8	9
Multiplication [using (3) with *]	39	53	9	11
Division [using (3) with /]	47	223	9	15
Flops [using (4)]	143	162	15	18

Point 2   Division took the most time of the four arithmetic operations. Indeed, our PC found double-precision division in interpretive BASIC very time-consuming. We should try to minimize divisions as much as possible. For example, when computing

$$\left(\frac{4}{3}\right)(3, 2, 5, 7, 8) + (-4, 11, 2, 1, 5)$$

to obtain a row vector with first entry zero, we should *not* compute the remaining entries as

$$\left(\frac{4}{3}\right)2 + 11, \qquad \left(\frac{4}{3}\right)5 + 2, \qquad \left(\frac{4}{3}\right)7 + 1, \qquad \left(\frac{4}{3}\right)8 + 5,$$

which requires four divisions. Rather, we should do a single division, finding $r = \frac{4}{3}$, and then computing

$$2r + 11, \qquad 5r + 2, \qquad 7r + 1, \qquad 8r + 5.$$

Point 3   Our program ran much faster in compiled BASIC than in interpretive BASIC. In the compiled version on our PC, the time for double-precision division was not so far out of line with the time for other operations.

Point 4   The indexing in the flops required significant time in interpretive BASIC on our PC.

The routine TIMING in LINTEK was used to generate the data in Table 10.1 on our PC. Exercise 22 asks students to obtain analogous data with their PCs, using this program.

Counting Operations

We turn to counting the flops required to solve a square linear system $A\mathbf{x} = \mathbf{b}$ that has a unique solution. We assume that no row interchanges are necessary, which is frequently the case.

Suppose that we solve the system $A\mathbf{x} = \mathbf{b}$, using the Gauss method with back substitution. Form the augmented matrix

$$[A \mid \mathbf{b}] = \begin{bmatrix} a_{11} & a_{12} & \cdots & a_{1n} & b_1 \\ a_{21} & a_{22} & \cdots & a_{2n} & b_2 \\ & & \vdots & & \vdots \\ a_{n1} & a_{n2} & \cdots & a_{nn} & b_n \end{bmatrix}.$$

For the moment let us neglect the flops performed on $\mathbf{b}$ and count just the flops performed on A. In reducing the $n \times n$ matrix A to upper-triangular form U, we execute $n - 1$ flops in adding a multiple of the first row of $[A \mid \mathbf{b}]$ to the second row. (We do not need to compute the zero entry at the beginning of our new second row; we know it will be zero.) We similarly use $n - 1$ flops to obtain the new row 3, and so on. This gives a total of $(n - 1)^2$ flops executed using the pivot in row 1. The pivot in row 2 is then used for the $(n - 1) \times (n - 1)$ matrix obtained by crossing off the first row and column of the modified coefficient matrix. By the count just made for the $n \times n$ matrix, we realize that using this pivot in the second row will result in execution of $(n - 2)^2$ flops. Continuing in this fashion, we see that reducing A to upper-triangular form U will require

$$(n - 1)^2 + (n - 2)^2 + (n - 3)^2 + \cdots + 1 \tag{5}$$

flops, together with some divisions. Let's count the divisions. We expect to use just one division each time a row is multiplied by a constant and added to another row (see point 2). There will be $n - 1$ divisions involving the pivot in row 1, $n - 2$ involving the pivot in row 2, and so on, for a total of

$$(n - 1) + (n - 2) + (n - 3) + \cdots + 1 \tag{6}$$

divisions.

There are some handy formulas for finding a sum of consecutive integers and a sum of squares of consecutive integers. It can be shown by induction (see Appendix A) that

$$1 + 2 + 3 + \cdots + n = \frac{n(n + 1)}{2} \tag{7}$$

and

$$1^2 + 2^2 + 3^2 + \cdots + n^2 = \frac{n(n + 1)(2n + 1)}{6}. \tag{8}$$

Replacing n by $n - 1$ in formula (8), we see that the number of flops given by sum (5) is equal to

$$\frac{(n - 1)n(2n - 1)}{6} = \frac{(n^2 - n)(2n - 1)}{6} = \frac{1}{6}(2n^3 - 3n^2 + n)$$

$$= \frac{n^3}{3} - \frac{n^2}{2} + \frac{n}{6}. \tag{9}$$

We assume that we are using a computer to solve a *large* linear system, involving hundreds or thousands of equations. With $n = 1000$, the value of Eq. (9) becomes

$$\frac{1{,}000{,}000{,}000}{3} - \frac{1{,}000{,}000}{2} + \frac{1000}{6}. \tag{10}$$

The largest term in expression (10) is the first one, $1{,}000{,}000{,}000/3$, corresponding to the $n^3/3$ term in Eq. (9). The lower powers of n in the $n^2/2$ and $n/6$ terms contribute little in comparison with the $n^3/3$ term for large n. It is customary to keep just the $n^3/3$ term as a measure of the **order of magnitude** of the expression in Eq. (9) for large values of n.

Turning to the count of the divisions in sum (6), we see from Eq. (7) with n replaced by $n - 1$ that we are using

$$\frac{(n - 1)n}{2} = \frac{n^2}{2} - \frac{n}{2} \tag{11}$$

divisions. For large n, this is of order of magnitude $n^2/2$, which is inconsequential in comparison with the order of magnitude $n^3/3$ for the flops. Exercise 1 shows that the number of flops performed on the column vector **b** in reducing $[A \mid \mathbf{b}]$ is again given by Eq. (11), so it can be neglected for large n in view of the order of magnitude $n^3/3$. The result is shown in the following box.

Flop Count for Reducing $[A \mid \mathbf{b}]$ to $[U \mid \mathbf{c}]$

If $A\mathbf{x} = \mathbf{b}$ is a square linear system in which A is an $n \times n$ matrix, the number of flops executed in reducing $[A \mid \mathbf{b}]$ to the form $[U \mid \mathbf{c}]$ is of order of magnitude $n^3/3$ for large n.

We turn now to finding the number of flops used in back substitution to solve the upper-triangular linear system $U\mathbf{x} = \mathbf{c}$. This system can be written out as

$$u_{11}x_1 + \cdots + u_{1,n-1}x_{n-1} + u_{1n}x_n = c_1$$

$$\vdots$$

$$u_{n-1,n-1}x_{n-1} + u_{n-1,n}x_n = c_{n-1}$$

$$u_{nn}x_n = c_n.$$

Solving for x_n requires one indexed division, which we consider to be a flop. Solving then for x_{n-1} requires an indexed multiplication and subtraction, followed by an indexed division, which we consider to be two flops. Solving for x_{n-2} requires two flops, each consisting of a multiplication combined with a subtraction, followed by an indexed division, and we consider this to contribute three more flops, and so on. We obtain the count

$$1 + 2 + 3 + \cdots + n = \frac{(n + 1)n}{2} = \frac{n^2}{2} + \frac{n}{2} \tag{12}$$

for the flops in this back substitution. Again, this is of lower order of magnitude than the order of magnitude $n^3/3$ for the flops required to reduce $[A \mid \mathbf{b}]$ to $[U \mid \mathbf{c}]$.

Flop Count for Back Substitution

If U is an $n \times n$ upper-triangular matrix, then back substitution to solve $U\mathbf{x} = \mathbf{c}$ requires the order of magnitude $n^2/2$ flops for large n.

Combining the results shown in the last two boxes, we arrive at the following flop count.

Flop Count for Solving $A\mathbf{x} = \mathbf{b}$, Using the Gauss Method with Back Substitution

If A is an $n \times n$ matrix, the number of flops required to solve the system $A\mathbf{x} = \mathbf{b}$ using the Gauss method with back substitution is of order of magnitude $n^3/3$ for large n.

EXAMPLE 1   For the computer that produced the execution times shown in Table 10.1, find the approximate time required to solve a system of 100 equations in 100 unknowns, using single-precision arithmetic and (a) interpretive BASIC, (b) compiled BASIC.

SOLUTION   From the flop count for the Gauss method, we see that solving such a system with $n = 100$ requires about $100^3/3 = 1{,}000{,}000/3$ flops. In interpretive BASIC, the time required for 10,000 flops in single precision was about 143 seconds. Thus the $1{,}000{,}000/3$ flops require about $(1{,}000{,}000/30{,}000)143 = 14{,}300/3$ seconds, or about 1 hour and 20 minutes. In compiled BASIC, where 10,000 flops took about 15 seconds, we find that the time is approximately $(1{,}000{,}000/30{,}000)15 = 1500/3 = 500$ seconds, or about 8 minutes. The PC used for Table 10.1 is regarded today as terribly slow.   ∎

It is interesting to compare the efficiency of the Gauss method with back substitution to that of the Gauss–Jordan method. Recall that, in the Gauss–Jordan method, $[A \mid \mathbf{b}]$ is reduced to a form $[I \mid \mathbf{c}]$, where I is the identity matrix. Exercise 2 shows that Gauss–Jordan flop count is of order of magnitude $n^3/2$ for large n. Thus the Gauss–Jordan method is not as efficient as the Gauss method with back substitution if n is large. One expects a Gauss–Jordan program to take about one and a half times as long to execute. The routine TIMING in LINTEK can be used to illustrate this. (See Exercises 23 and 24.)

Counting Flops for Matrix Operations

The exercises ask us to count the flops involved in matrix addition, multiplication, and exponentiation. Recall that a matrix product $C = AB$ is obtained by taking dot products of row vectors of an $m \times n$ matrix A with column vectors of an $n \times s$ matrix B. We indicate the usual way that a computer finds the dot product c_{ij} in C. First the computer sets $c_{ij} = 0$. Then it replaces c_{ij} by $c_{ij} + a_{i1}b_{1j}$, which gives c_{ij} the value $a_{i1}b_{1j}$. Then the computer replaces c_{ij} by $c_{ij} + a_{i2}b_{2j}$, giving c_{ij} the value $a_{i1}b_{1j} + a_{i2}b_{2j}$. This process continues in the obvious way until finally we have accumulated the desired value,

$$c_{ij} = a_{i1}b_{1j} + a_{i2}b_{2j} + \cdots + a_{in}b_{nj}.$$

Each of these replacements of c_{ij} is accomplished by means of a flop, typically expressed by

$$C(I,J) = C(I,J) + A(I,K) * B(K,J) \tag{13}$$

in BASIC.

EXAMPLE 2   Find the number of flops required to compute the dot product of two vectors, each with n components.

SOLUTION   The preceding discussion shows that the dot product uses n flops of form (13), corresponding to the values $1, 2, 3, \ldots , n$ for K.   ∎

SUMMARY

1. A flop is a rather vaguely defined execution by a computer, consisting typically of a bit of indexing, the retrieval and storage of a couple of values, and one or two arithmetic operations. Typical flops might appear in a computer program in instruction lines such as

$$C(I,J) = A(I,J) + B(I,J)$$

or

$$A(I,J) = A(I,J) + R*A(K,J).$$

2. A computer takes time to execute a flop, and it is desirable to use as few flops as possible when performing an extensive computation.

3. If the number of flops used to solve a problem is given by a polynomial expression in n, it is customary to keep only the term of highest degree in the polynomial as a measure of the *order of magnitude* of the number of flops when n is large.

4. The order of magnitude of the number of flops used in solving a system $A\mathbf{x} = \mathbf{b}$ for an $n \times n$ matrix A is

$$\frac{n^3}{3} \quad \text{for the Gauss method with back substitution}$$

and

$$\frac{n^3}{2} \quad \text{for the Gauss–Jordan method.}$$

5. The formulas

$$1 + 2 + 3 + \cdots + n = \frac{n(n + 1)}{2}$$

and

$$1^2 + 2^2 + 3^2 + \cdots + n^2 = \frac{n(n + 1)(2n + 1)}{6}$$

are handy for determining the number of flops performed by a computer in matrix computations.

EXERCISES

In all of these exercises, assume that no row vectors are interchanged in a matrix.

1. Let A be an $n \times n$ matrix. Show that, in reducing $[A \mid \mathbf{b}]$ to $[U \mid \mathbf{c}]$ using the Gauss method, the number of flops performed on $\mathbf{b}$ is of order of magnitude $n^2/2$ for large n.

2. Let A be an $n \times n$ matrix. Show that, in solving $A\mathbf{x} = \mathbf{b}$ using the Gauss–Jordan method, the number of flops has order of magnitude $n^3/2$ for large n.

In Exercises 3–11, let A be an $n \times n$ matrix, and let B and C be $m \times n$ matrices. For each matrix, find the number of flops required for efficient computation of the matrix.

3. $B + C$ 4. A^2 5. BA

6. A^3 7. A^4 8. A^5

9. A^6 10. A^{63} 11. A^{64}

12. Which of the following is more efficient with a computer?

a. Solving a square system $A\mathbf{x} = \mathbf{b}$ by the Gauss–Jordan method, which makes each pivot 1 before creating the zeros in the column containing it.

b. Using a similar technique to reduce the system to a diagonal system $D\mathbf{x} = \mathbf{c}$, where the entries in D are not necessarily all 1, and then dividing by these entries to obtain the solution.

13. Let A be an $n \times n$ matrix, where n is large. Find the order of magnitude for the number of flops if A^{-1} is computed using the Gauss–Jordan method on the augmented matrix $[A \mid I]$ without trying to reduce the number of flops used on I in response to the zeros that appear in it.

14. Repeat Exercise 13, using the Gauss method with back substitution rather than the Gauss–Jordan method.

15. Repeat Exercise 14, but this time cut down the number of flops performed on I during the Gauss reduction by taking into account the zeros in I.

Exercises 16–20 concern band matrices. In a number of situations, square linear systems $A\mathbf{x} = \mathbf{b}$ occur in which the nonzero entries in the $n \times n$ matrix A are all concentrated near the main diagonal, running from the upper left-hand corner to the lower right-hand corner of A. Such a matrix is called a **band matrix.** *For example, the matrix*

$$w = 2$$
$$\begin{bmatrix} 2 & 1 & 0 & 0 & 0 & 0 \\ 1 & 3 & 4 & 0 & 0 & 0 \\ 0 & 4 & 1 & 2 & 0 & 0 \\ 0 & 0 & 2 & 2 & 7 & 0 \\ 0 & 0 & 0 & 7 & 1 & 3 \\ 0 & 0 & 0 & 0 & 3 & 5 \end{bmatrix} \quad \textbf{(14)}$$
$$w = 2$$

is a symmetric 6×6 band matrix. We say that the **band width** *of a band matrix $[a_{ij}]$ is w if w is the smallest integer such that $a_{ij} = 0$ for $|i - j| \geq w$. Thus matrix (14) has band width $w = 2$, as indicated. Such a matrix of band width 2 is also called* **tridiagonal.** *We usually assume that the band width is small compared with the size of n. As the band width approaches n, the matrix becomes* **full.**

In Exercises 16–20, assume that A is an $n \times n$ band matrix with band width w that is small in comparison with n.

16. What can be said concerning the band width of A^2? of A^3? of A^m?

17. Let A be tridiagonal, so $w = 2$. Find the order of magnitude of the number of flops required to reduce the partitioned matrix

$[A \mid \mathbf{b}]$ to a form $[U \mid \mathbf{c}]$, where U is an upper-triangular matrix, taking into account the banded character of A.

18. Repeat Exercise 17 for band width w, expressing the result in terms of w.

19. Repeat Exercise 18, but include the flops used in back substitution to solve $A\mathbf{x} = \mathbf{b}$.

20. Explain why the Gauss method with back substitution is much more efficient than the Gauss–Jordan method for a banded matrix where w is very small compared with n.

21. Mark each of the following True or False.
___ a. A flop is a very precisely defined entity.
___ b. Computers can work so fast that it is not worthwhile to try to minimize the number of computations a computer makes to solve a problem, provided that the number of computations is only a few hundred.
___ c. Computers can work so fast that it is not worthwhile to try to minimize the number of computations required to solve a problem.
___ d. The Gauss method with back substitution and the Gauss–Jordan method for solving a large linear system both take about the same amount of computer time.
___ e. The Gauss–Jordan method for solving a large linear system takes about half again as long to execute as does the Gauss method on a computer.
___ f. Multiplying two $n \times n$ matrices requires more flops than does solving a linear system with an $n \times n$ coefficient matrix.
___ g. About n^2 flops are required to execute the back substitution in solving a linear system with an $n \times n$ coefficient matrix by using the Gauss method.
___ h. Executing the Gauss method with back substitution for a large linear system with an $n \times n$ coefficient matrix requires about n^2 flops.
___ i. Executing the Gauss method with back substitution for a large linear system with an $n \times n$ coefficient matrix requires about $n^3/3$ flops.
___ j. Executing the Gauss–Jordan method for a large linear system with an $n \times n$ coefficient matrix requires about $n^2/2$ flops.

🖫 *LINTEK contains a routine TIMING that can be used to time algebraic operations and flops. The program can also be used to time the solution of square systems $A\mathbf{x} = \mathbf{b}$ by the Gauss method with back substitution and by the Gauss–Jordan method. For a user-specified integer $n \le 80$, the program generates the $n \times n$ matrix A and column vector $\mathbf{b}$, where all entries are in the interval $[-20, 20]$. Use TIMING for Exercises 22–24.*

22. Experiment with TIMING to see *roughly* how many of the indicated operations your computer can do in 5 seconds.
    a. Additions
    b. Subtractions
    c. Multiplications
    d. Divisions
    e. Flops

23. Run the routine TIMING in LINTEK to time the Gauss method and the Gauss–Jordan method, starting with small values of n and increasing them until a few seconds' difference in times for the two methods is obtained. Does the time for the Gauss–Jordan method seem to be about $\frac{3}{2}$ the time for the Gauss method with back substitution? If not, why not?

24. Continuing Exercise 23, increase the size of n until the solutions take 2 or 3 minutes. Now does the Gauss–Jordan method seem to take about $\frac{3}{2}$ times as long? If this ratio is significantly different from that obtained in Exercise 23, explain why. (The two ratios may or may not appear to be approximately the same, depending on the speed of the computer used and on whether time in fractions of seconds is displayed.)

MATLAB

The command **clock** *in MATLAB returns a row vector*

[year, month, day, hour, minute, second]

that gives the date and time in decimal form. To calculate the elapsed time for a computation, we can

set **t0 = clock**,

execute the computation,

give the command **etime(clock,t0)**, *which returns elapsed time since* **t0**.

The student edition of MATLAB will not accept a value of i *greater than 1024 in a* "FOR i = 1 to n" *loop, so to calculate the time for more than 1024 operations in a loop, we use a double*

FOR g = 1 to h, FOR i = 1 to n

loop with which we can find the time required for up to 1024^2 operations. Note below that the syntax of the MATLAB loops is a bit different from that just displayed. Recall that the colon can be used with the meaning "through." The first two lines below define data to be used in the last line. The last line is jammed together so that it will all fit on one screen line when you modify the **c=a+b** *portion in Exercise M5 to time flops. As shown below, the last line returns the elapsed time for* h · n *repeated additions of two numbers* a *and* b *between 0 and 1:*

```
A = rand(6,6); a = rand; b = rand; r = rand; j = 2; k = 4; m = 3;
h = 100; n = 50;
t0=clock;for g=1:h,for i=1:n,c=a+b;end;end;etime(clock,t0)
```

M1. Enter in the three lines shown above. *Put spaces in the last line only after* "**for**". Using the up-arrow key and changing the value of n (and h if necessary), find roughly how many additions MATLAB can perform in 5 seconds on your computer. (Remember, the time given is for $h \cdot n$ additions.)

M2. Modify the **c=a+b** portion and repeat M1 for subtractions.

M3. Modify the **c=a+b** portion and repeat M1 for multiplications.

M4. Modify the **c=a+b** portion and repeat M1 for divisions.

M5. Modify the **c=a+b** portion and repeat M1 for flops **A(k,j)=A(k,j)+r*A(m,j)**. *Delete* **c=a+b** *first, and don't insert any spaces!*

10.2   THE *LU*-FACTORIZATION

Keeping a Record of Row Operations

We continue to work with a square linear system $A\mathbf{x} = \mathbf{b}$ having a unique solution that can be found by using Gauss elimination with back substitution, without having to interchange any rows. That is, the matrix A can be row-reduced to an upper-triangular matrix U, without making any row interchanges.

Situations occur in which it is necessary to solve many such systems, all having the same coefficient matrix A but different column vectors $\mathbf{b}$. We solve such multiple systems by row-reducing a single augmented matrix

$$[A \mid \mathbf{b}_1, \mathbf{b}_2, \ldots, \mathbf{b}_s]$$

in which we line up all the different column vectors $\mathbf{b}_1, \mathbf{b}_2, \ldots, \mathbf{b}_s$ on the right of the partition. In practice, it may be impossible to solve all of these systems by using this single augmentation in a single computer run. Here are some possibilities in which a sequence of computer runs may be needed to solve all the systems:

1. Remember that we are concerned with *large* systems. If the number s of vectors $\mathbf{b}_j$ is large, there may not be room in the computer memory to accommodate all of these data at one time. Indeed, it might even be necessary to reduce the $n \times n$ matrix A in segments, if n is very large. If we can handle the s vectors $\mathbf{b}_j$ only in groups of m at a time, we must use at least s/m computer runs to solve all the systems.

2. Perhaps the vectors $\mathbf{b}_j$ are generated over a period of time, and we need to solve systems involving groups of the vectors $\mathbf{b}_j$ as they are generated. For example, we may want to solve $A\mathbf{x} = \mathbf{b}_j$ with r different vectors $\mathbf{b}_j$ each day.

3. Perhaps the vector $\mathbf{b}_{j+1}$ depends on the solution of $A\mathbf{x} = \mathbf{b}_j$. We would then have to solve $A\mathbf{x} = \mathbf{b}_1$, determine $\mathbf{b}_2$, solve $A\mathbf{x} = \mathbf{b}_2$, determine $\mathbf{b}_3$, and so on, until we finally solved $A\mathbf{x} = \mathbf{b}_s$.

From Section 10.1, we know that the magnitude of the number of flops required to reduce A to an upper-triangular matrix U is $n^3/3$ for large n. We want to avoid having to repeat all this work done in reducing A after the first computer run.

We assume that A can be reduced to U without interchanging rows—that is, that (nonzero) pivots always appear where we want them as we reduce A to U. This means that only elementary row-addition operations (those that add a multiple of a row vector to another row vector) are used. Recall that a row-addition operation can be accomplished by multiplying on the left by an $n \times n$ elementary matrix E, where E is obtained by applying the same row-addition operation to the identity matrix. There is a sequence $E_1, E_2, \ldots,$ E_h of such elementary matrices such that

$$E_h E_{h-1} \cdots E_2 E_1 A = U. \tag{1}$$

Once the matrix U has been found by the computer, the data in it can be stored on a disk or on tape, and simply read in for future computer runs. However, when we are solving $A\mathbf{x} = \mathbf{b}$ by reducing the augmented matrix $[A \mid \mathbf{b}]$, the sequence of elementary row operations described by $E_h E_{h-1} \cdots E_2 E_1$ in Eq. (1) must be applied to the *entire* augmented matrix, not merely to A. Thus we need to keep a *record* of this sequence of row operations to perform on column vectors $\mathbf{b}_j$ when they are used in subsequent computer runs.

Here is a way of recording the row-addition operations that is both efficient and algebraically interesting. As we make the reduction of A to U, we create a *lower-triangular matrix L*, which is a record of the row-addition operations performed. We start with the $n \times n$ identity matrix I, and as each row-addition operation on A is performed, we change one of the zero entries below the diagonal in I to produce a record of that operation. For example, if during the reduction we add 4 times row 2 to row 6, we place -4 in the *second* column position in the *sixth* row of the matrix L that we are creating as a record. The general formulation is shown in the following box.

Creation of the Matrix *L*

Start with the $n \times n$ identity matrix I. If during the reduction of A to U, r times row i is added to row k, replace the zero in row k and column i of the identity matrix by $-r$. The final result obtained from the identity matrix is L.

EXAMPLE 1   Reduce the matrix

$$A = \begin{bmatrix} 1 & 3 & -1 \\ 2 & 8 & 4 \\ -1 & 3 & 4 \end{bmatrix}$$

to upper-triangular form U, and create the matrix L described in the preceding box.

SOLUTION   We proceed in two columns, as follows:

Reduction of A to U   **Creation of L from I**

$$A = \begin{bmatrix} 1 & 3 & -1 \\ 2 & 8 & 4 \\ -1 & 3 & 4 \end{bmatrix}$$ $$I = \begin{bmatrix} 1 & 0 & 0 \\ 0 & 1 & 0 \\ 0 & 0 & 1 \end{bmatrix}$$

Add -2 times
row 1 to row 2.

$$\sim \begin{bmatrix} 1 & 3 & -1 \\ 0 & 2 & 6 \\ -1 & 3 & 4 \end{bmatrix}$$ $$\begin{bmatrix} 1 & 0 & 0 \\ 2 & 1 & 0 \\ 0 & 0 & 1 \end{bmatrix}$$

Add 1 times
row 1 to row 3.

$$\sim \begin{bmatrix} 1 & 3 & -1 \\ 0 & 2 & 6 \\ 0 & 6 & 3 \end{bmatrix}$$ $$\begin{bmatrix} 1 & 0 & 0 \\ 2 & 1 & 0 \\ -1 & 0 & 1 \end{bmatrix}$$

Add -3 times
row 2 to row 3.

$$\sim \begin{bmatrix} 1 & 3 & -1 \\ 0 & 2 & 6 \\ 0 & 0 & -15 \end{bmatrix} = U$$ $$L = \begin{bmatrix} 1 & 0 & 0 \\ 2 & 1 & 0 \\ -1 & 3 & 1 \end{bmatrix}.$$ ∎

We now illustrate how the record kept in L in Example 1 can be used to solve a linear system $A\mathbf{x} = \mathbf{b}$ having the matrix A in Example 1 as coefficient matrix.

EXAMPLE 2   Use the record in L in Example 1 to solve the linear system $A\mathbf{x} = \mathbf{b}$ given by

$$\begin{aligned} x_1 + 3x_2 - x_3 &= -4 \\ 2x_1 + 8x_2 + 4x_3 &= 2 \\ -x_1 + 3x_2 + 4x_3 &= 4. \end{aligned}$$

SOLUTION   We use the record in L to find the column vector $\mathbf{c}$ that would occur if we were to reduce $[A \mid \mathbf{b}]$ to $[U \mid \mathbf{c}]$, using these same row operations:

Entry ℓ_{ij} in L	Meaning of the Entry	Reduction of $\mathbf{b}$ to $\mathbf{c}$
		$\mathbf{b} = \begin{bmatrix} -4 \\ 2 \\ 4 \end{bmatrix}$
$\ell_{21} = 2$	Add -2 times row 1 to row 2.	$\sim \begin{bmatrix} -4 \\ 10 \\ 4 \end{bmatrix}$
$\ell_{31} = -1$	Add $-(-1) = 1$ times row 1 to row 3.	$\sim \begin{bmatrix} -4 \\ 10 \\ 0 \end{bmatrix}$
$\ell_{32} = 3$	Add -3 times row 2 to row 3.	$\sim \begin{bmatrix} -4 \\ 10 \\ -30 \end{bmatrix} = \mathbf{c}.$

If we put this result together with the matrix U obtained in Example 1, the reduced augmented matrix for the linear system becomes

$$[U \mid \mathbf{c}] = \begin{bmatrix} 1 & 3 & -1 & -4 \\ 0 & 2 & 6 & 10 \\ 0 & 0 & -15 & -30 \end{bmatrix}.$$

Back substitution then yields

$$x_3 = \frac{-30}{-15} = 2,$$

$$x_2 = \frac{10 - 6(2)}{2} = \frac{-2}{2} = -1,$$

$$x_1 = -4 + 3 + 2 = 1. \qquad \blacksquare$$

We give another example.

EXAMPLE 3   Let

$$A = \begin{bmatrix} 1 & -2 & 0 & 3 \\ -2 & 3 & 1 & -6 \\ -1 & 4 & -4 & 3 \\ 5 & -8 & 4 & 0 \end{bmatrix} \quad \text{and} \quad \mathbf{b} = \begin{bmatrix} 11 \\ -21 \\ -1 \\ 23 \end{bmatrix}.$$

Generate the matrix L while reducing the matrix A to U. Then use U and the record in L to solve $A\mathbf{x} = \mathbf{b}$.

SOLUTION   We work in two columns again. This time we fix up a whole column of A in each step:

Reduction of A \qquad Generation of L

$$A = \begin{bmatrix} 1 & -2 & 0 & 3 \\ -2 & 3 & 1 & -6 \\ -1 & 4 & -4 & 3 \\ 5 & -8 & 4 & 0 \end{bmatrix} \qquad I = \begin{bmatrix} 1 & 0 & 0 & 0 \\ 0 & 1 & 0 & 0 \\ 0 & 0 & 1 & 0 \\ 0 & 0 & 0 & 1 \end{bmatrix}$$

$$\sim \begin{bmatrix} 1 & -2 & 0 & 3 \\ 0 & -1 & 1 & 0 \\ 0 & 2 & -4 & 6 \\ 0 & 2 & 4 & -15 \end{bmatrix} \qquad \begin{bmatrix} 1 & 0 & 0 & 0 \\ -2 & 1 & 0 & 0 \\ -1 & 0 & 1 & 0 \\ 5 & 0 & 0 & 1 \end{bmatrix}$$

$$\sim \begin{bmatrix} 1 & -2 & 0 & 3 \\ 0 & -1 & 1 & 0 \\ 0 & 0 & -2 & 6 \\ 0 & 0 & 6 & -15 \end{bmatrix} \qquad \begin{bmatrix} 1 & 0 & 0 & 0 \\ -2 & 1 & 0 & 0 \\ -1 & -2 & 1 & 0 \\ 5 & -2 & 0 & 1 \end{bmatrix}$$

$$\sim \begin{bmatrix} 1 & -2 & 0 & 3 \\ 0 & -1 & 1 & 0 \\ 0 & 0 & -2 & 6 \\ 0 & 0 & 0 & 3 \end{bmatrix} = U \quad L = \begin{bmatrix} 1 & 0 & 0 & 0 \\ -2 & 1 & 0 & 0 \\ -1 & -2 & 1 & 0 \\ 5 & -2 & -3 & 1 \end{bmatrix}.$$

We now apply the record below the diagonal in L to the vector $\mathbf{b}$, working under the main diagonal down each column of the record in L in turn.

First column of L: $\mathbf{b} = \begin{bmatrix} 11 \\ -21 \\ -1 \\ 23 \end{bmatrix} \sim \begin{bmatrix} 11 \\ 1 \\ -1 \\ 23 \end{bmatrix} \sim \begin{bmatrix} 11 \\ 1 \\ 10 \\ 23 \end{bmatrix} \sim \begin{bmatrix} 11 \\ 1 \\ 10 \\ -32 \end{bmatrix}.$

Second column of L: $\begin{bmatrix} 11 \\ 1 \\ 10 \\ -32 \end{bmatrix} \sim \begin{bmatrix} 11 \\ 1 \\ 12 \\ -32 \end{bmatrix} \sim \begin{bmatrix} 11 \\ 1 \\ 12 \\ -30 \end{bmatrix}.$

Third column of L: $\begin{bmatrix} 11 \\ 1 \\ 12 \\ -30 \end{bmatrix} \sim \begin{bmatrix} 11 \\ 1 \\ 12 \\ 6 \end{bmatrix}.$

The augmented matrix

$$\begin{bmatrix} 1 & -2 & 0 & 3 & | & 11 \\ 0 & -1 & 1 & 0 & | & 1 \\ 0 & 0 & -2 & 6 & | & 12 \\ 0 & 0 & 0 & 3 & | & 6 \end{bmatrix}$$

yields, upon back substitution,

$$x_4 = \frac{6}{3} = 2,$$

$$x_3 = \frac{12 - 12}{-2} = 0,$$

$$x_2 = \frac{1 - 0}{-1} = -1,$$

$$x_1 = 11 - 2 - 6 = 3. \qquad \blacksquare$$

Two questions come to mind:

1. Why bother to put the entries 1 down the main diagonal in L when they are not used?
2. If we add r times a row to another row while reducing A, why do we put $-r$ rather than r in the record in L, and then change back to r again when performing the operations on the column vector $\mathbf{b}$?

We do these two things only because the matrix L formed in this way has an interesting algebraic property, which we will discuss in a moment. In fact, when solving a large system using a computer, we certainly would not fuss with the entries 1 down the diagonal. Indeed, we can save memory space by not even generating a matrix L separate from the one being reduced. When creating a zero below the main diagonal, place the record $-r$ or r desired, as described in question 2 above, directly in the matrix being reduced at the position where a zero is being created! The computer already has space reserved for an entry there. Just remember that the final matrix contains the desired entries of U on and above the diagonal and the record for L or $-L$ below the diagonal. *We will always use $-r$ rather than r as record entry in this text.* Thus, for the 4×4 matrix in Example 3, we obtain

$$\begin{bmatrix} 1 & -2 & 0 & 3 \\ -2 & -1 & 1 & 0 \\ -1 & -2 & -2 & 6 \\ 5 & -2 & -3 & 3 \end{bmatrix}, \quad \textbf{Combined } L\backslash U \textbf{ display} \quad (2)$$

where the black entries on or above the main diagonal give the essential data for U, and the color entries are the essential data for L.

Let us examine the efficiency of solving a system $A\mathbf{x} = \mathbf{b}$ if U and L are already known. Each entry in the record in L requires one flop to execute when applying this record to reduce a column vector $\mathbf{b}$. The number of entries is

$$1 + 2 + \cdots + n - 1 = \frac{(n-1)n}{2} = \frac{n^2}{2} - \frac{n}{2},$$

which is of order of magnitude $n^2/2$ for large n. We saw in Section 10.1 that back substitution requires about $n^2/2$ flops, too, giving a total of n^2 flops for large n to solve $A\mathbf{x} = \mathbf{b}$, once U and L are known. If instead we computed A^{-1} and found $\mathbf{x} = A^{-1}\mathbf{b}$, the product $A^{-1}\mathbf{b}$ would also require n^2 flops. But there are at least two advantages in using the LU-technique. First, finding U requires about $n^3/3$ flops for large n, whereas finding A^{-1} requires n^3 flops. (See Exercise 15 in Section 10.1.) If $n = 1000$, the difference in computer time is considerable. Second, more computer memory is used in reducing $[A \mid I]$ to $[I \mid A^{-1}]$ than is used in the efficient way we record L as we find U, illustrated in the combined $L\backslash U$ display (2).

We give a specific illustration in which keeping the record L is useful.

EXAMPLE 4   Let

$$A = \begin{bmatrix} 1 & 2 & -1 \\ -2 & -5 & 3 \\ -1 & -3 & 0 \end{bmatrix} \quad \text{and} \quad \mathbf{b} = \begin{bmatrix} 9 \\ -17 \\ -44 \end{bmatrix}.$$

Solve the linear system $A^3\mathbf{x} = \mathbf{b}$.

SOLUTION   We view $A^3\mathbf{x} = \mathbf{b}$ as $A(A^2\mathbf{x}) = \mathbf{b}$, and substitute $\mathbf{y} = A^2\mathbf{x}$ to obtain $A\mathbf{y} = \mathbf{b}$. We can solve this equation for $\mathbf{y}$. Then we write $A^2\mathbf{x} = \mathbf{y}$ as $A(A\mathbf{x}) = \mathbf{y}$ or $A\mathbf{z} = \mathbf{y}$, where $\mathbf{z} = A\mathbf{x}$. We then solve $A\mathbf{z} = \mathbf{y}$ for $\mathbf{z}$. Finally, we solve $A\mathbf{x} = \mathbf{z}$ for the desired $\mathbf{x}$. Because we are using the same coefficient matrix A each time, it is efficient to find the matrices L and U and then to proceed as in Example 3 to find $\mathbf{y}$, $\mathbf{z}$, and $\mathbf{x}$ in turn.

We find that the matrices U and L are given by

$$U = \begin{bmatrix} 1 & 2 & -1 \\ 0 & -1 & 1 \\ 0 & 0 & -2 \end{bmatrix} \quad \text{and} \quad L = \begin{bmatrix} 1 & 0 & 0 \\ -2 & 1 & 0 \\ -1 & 1 & 1 \end{bmatrix}.$$

Applying the record in L to $\mathbf{b}$, we obtain

$$\mathbf{b} = \begin{bmatrix} 9 \\ -17 \\ -44 \end{bmatrix} \sim \begin{bmatrix} 9 \\ 1 \\ -35 \end{bmatrix} \sim \begin{bmatrix} 9 \\ 1 \\ -36 \end{bmatrix}.$$

From the matrix U, we find that

$$\mathbf{y} = \begin{bmatrix} -7 \\ 17 \\ 18 \end{bmatrix}.$$

To solve $A\mathbf{z} = \mathbf{y}$, we apply the record in L to $\mathbf{y}$:

$$\mathbf{y} = \begin{bmatrix} -7 \\ 17 \\ 18 \end{bmatrix} \sim \begin{bmatrix} -7 \\ 3 \\ 11 \end{bmatrix} \sim \begin{bmatrix} -7 \\ 3 \\ 8 \end{bmatrix}.$$

Using U, we obtain

$$\mathbf{z} = \begin{bmatrix} 3 \\ -7 \\ -4 \end{bmatrix}.$$

Finally, to solve $A\mathbf{x} = \mathbf{z}$, we apply the record in L to $\mathbf{z}$:

$$\mathbf{z} = \begin{bmatrix} 3 \\ -7 \\ -4 \end{bmatrix} \sim \begin{bmatrix} 3 \\ -1 \\ -1 \end{bmatrix} \sim \begin{bmatrix} 3 \\ -1 \\ 0 \end{bmatrix}.$$

Using U, we find that

$$\mathbf{x} = \begin{bmatrix} 1 \\ 1 \\ 0 \end{bmatrix}.$$

■

The routine LUFACTOR in LINTEK can be used to find the matrices L and U; it has an option for iteration to solve a system $A^m\mathbf{x} = \mathbf{b}$, as we did in Example 4.

The Factorization $A = LU$

This heading shows why the matrix L we described is algebraically interesting: we have $A = LU$.

EXAMPLE 5   Illustrate $A = LU$ for the matrices obtained in Example 3.

SOLUTION   From Example 3, we have

$$LU = \begin{bmatrix} 1 & 0 & 0 & 0 \\ -2 & 1 & 0 & 0 \\ -1 & -2 & 1 & 0 \\ 5 & -2 & -3 & 1 \end{bmatrix} \begin{bmatrix} 1 & -2 & 0 & 3 \\ 0 & -1 & 1 & 0 \\ 0 & 0 & -2 & 6 \\ 0 & 0 & 0 & 3 \end{bmatrix} = \begin{bmatrix} 1 & -2 & 0 & 3 \\ -2 & 3 & 1 & -6 \\ -1 & 4 & -4 & 3 \\ 5 & -8 & 4 & 0 \end{bmatrix} = A.$$

■

It is not difficult to establish that we always have $A = LU$. Recall from Eq. (1) that

$$E_h E_{h-1} \cdots E_2 E_1 A = U, \tag{3}$$

for elementary matrices E_i corresponding to elementary row operations consisting of adding a scalar multiple of a row to a lower row. From Eq. (3), we obtain

$$A = E_1^{-1} E_2^{-1} \cdots E_{h-1}^{-1} E_h^{-1} U. \tag{4}$$

We proceed to show that the matrix L is equal to the product $E_1^{-1} E_2^{-1} \cdots E_{h-1}^{-1} E_h^{-1}$ appearing in Eq. (4). Now if E_i is obtained from the identity matrix by adding r (possibly $r = 0$) times a row to another row, then E_i^{-1} is obtained from the identity matrix by adding $-r$ times the same row to the other row. Because E_h corresponds to the last row operation performed, $E_h^{-1} I$ places the negative of the last multiplier in the nth row and $(n - 1)$st column of this product precisely where it should appear in L. Similarly,

$$E_{h-1}^{-1}(E_h^{-1} I)$$

has the appropriate entry desired in L in the nth row and $(n - 2)$nd column;

$$E_{h-2}^{-1}(E_{h-1}^{-1} E_h^{-1} I)$$

has the appropriate entry desired in L in the $(n - 1)$st row and $(n - 2)$nd column; and so on. That is, the record entries of L may be created from the expression

$$E_1^{-1} E_2^{-1} \cdots E_{h-1}^{-1} E_h^{-1} I,$$

starting at the right with I and working to the left. This is the order shown by the color numbers below the main diagonal in the matrix

$$
\begin{bmatrix}
1 & & & & & & \\
 & 1 & & & & 0 & \\
 & & 1 & & & & \\
 & & & \cdot & & & \\
 & & & & \cdot & & \\
 & & & & & 1 & \\
 & & \text{etc.} & & & 10 & 1 \\
 & & & & & 9 & 6 & 1 \\
 & & & & & 8 & 5 & 3 & 1 \\
 & & & & & 7 & 4 & 2 & 1 & 1
\end{bmatrix}
$$

As we perform elementary row operations that correspond to adding scalar multiples of the shaded row to lower rows, none of the entries already created below the main diagonal will be changed, because the zeros to the right of the main diagonal lie over those entries. Thus,

$$E_1^{-1} E_2^{-1} \cdots E_{h-1}^{-1} E_h^{-1} I = L, \tag{5}$$

and $A = LU$ follows at once from Eq. (4).

We have shown how to find a solution of $A\mathbf{x} = \mathbf{b}$ from a factorization $A = LU$ by using the record in L to modify $\mathbf{b}$ to a vector $\mathbf{c}$ and then solving $U\mathbf{x} = \mathbf{c}$ by back substitution. Approximately n^2 flops are required. An alternative method of determining $\mathbf{x}$ from $LU\mathbf{x} = \mathbf{b}$ is to view the equation as $L(U\mathbf{x}) = \mathbf{b}$, letting $\mathbf{c} = U\mathbf{x}$. We first solve $L\mathbf{c} = \mathbf{b}$ for $\mathbf{c}$ by *forward substitution*, and then we solve $U\mathbf{x} = \mathbf{c}$ by back substitution. Because each of the forward and back substitutions takes approximately $n^2/2$ flops, the total number required is again approximately n^2 flops. We illustrate this alternative method.

EXAMPLE 6   In Example 1, we found the factorization

$$\begin{bmatrix} 1 & 3 & -1 \\ 2 & 8 & 4 \\ -1 & 3 & 4 \end{bmatrix} = \begin{bmatrix} 1 & 0 & 0 \\ 2 & 1 & 0 \\ -1 & 3 & 1 \end{bmatrix}\begin{bmatrix} 1 & 3 & -1 \\ 0 & 2 & 6 \\ 0 & 0 & -15 \end{bmatrix}.$$
$$\quad\quad A \quad\quad\quad\quad\quad\quad L \quad\quad\quad\quad\quad U$$

Use the method of forward substitution and back substitution to solve the linear system $A\mathbf{x} = \begin{bmatrix} -4 \\ 2 \\ 4 \end{bmatrix}$, which we solved in Example 2.

SOLUTION   First, we solve $L\mathbf{c} = \mathbf{b}$ by forward substitution:

$$\left[\begin{array}{ccc|c} 1 & 0 & 0 & -4 \\ 2 & 1 & 0 & 2 \\ -1 & 3 & 1 & 4 \end{array}\right]$$

$$c_1 = -4,$$

$$2c_1 + c_2 = 2, \quad\quad c_2 = 2 + 8 = 10.$$

$$-c_1 + 3c_2 + c_3 = 4,$$

$$c_3 = c_1 - 3c_3 + 4 = -4 - 30 + 4 = -30.$$

Notice that this is the same $\mathbf{c}$ as was obtained in Example 2. The back substitution with $U\mathbf{x} = \mathbf{c}$ of Example 2 then yields the same solution $\mathbf{x}$.   ∎

Factorization of an invertible square matrix A into LU, with L lower triangular and U upper triangular, is not unique. For example, if r is a nonzero scalar, then rL is lower triangular and $(1/r)U$ is upper triangular, and if $A = LU$, then we also have $A = (rL)((1/r)U)$. But let D be the $n \times n$ diagonal matrix having the same main diagonal as U; that is,

$$D = \begin{bmatrix} u_{11} & & & \\ & u_{22} & & \mathbf{0} \\ & & \ddots & \\ \mathbf{0} & & & u_{nn} \end{bmatrix}.$$

Let U^* be the upper-triangular matrix obtained from U by multiplying the ith row by $1/u_{ii}$ for $i = 1, 2, \ldots, n$. Then $U = DU^*$, and we have

$$A = LDU^*.$$

Now both L and U^* have all entries 1 on their main diagonals. This type of factorization is unique, as we now show.

THEOREM 10.1 Unique Factorization

Let A be an invertible $n \times n$ matrix. A factorization $A = LDU$, where

    L is lower triangular with all main diagonal entries 1,

    U is upper triangular with all main diagonal entries 1,

    D is diagonal with all main diagonal entries nonzero,

is unique.

PROOF Suppose that $A = L_1 D_1 U_1$ and $A = L_2 D_2 U_2$ are two such factorizations. Observe that both L_1^{-1} and L_2^{-1} are also lower triangular, D_1^{-1} and D_2^{-1} are both diagonal, and U_1^{-1} and U_2^{-1} are both still upper triangular. Just think how the matrix reductions of $[L_1 \mid I]$ or $[D_1 \mid I]$ or $[U_1 \mid I]$ to find the inverses look. Furthermore, $L_1^{-1}, L_2^{-1}, U_1^{-1}$, and U_2^{-1} have all their main diagonal entries equal to 1.

Now from $L_1 D_1 U_1 = L_2 D_2 U_2$, we obtain

$$L_2^{-1} L_1 = D_2 U_2 U_1^{-1} D_1^{-1}. \tag{6}$$

We see that $L_2^{-1} L_1$ is again lower triangular with entries 1 on its main diagonal, whereas $D_2 U_2 U_1^{-1} D_1^{-1}$ is upper triangular. Equation (6) then shows that both matrices must be I, so $L_1 L_2^{-1} = I$ and $L_1 = L_2$. A similar argument starting over with $L_1 D_1 U_1 = L_2 D_2 U_2$ rewritten as

$$U_1 U_2^{-1} = D_1^{-1} L_1^{-1} L_2 D_2 \tag{7}$$

shows that $U_1 = U_2$. We then have $L_1 D_1 U_1 = L_1 D_2 U_1$, and multiplication on the left by L_1^{-1} and on the right by U_1^{-1} yields $D_1 = D_2$. ▲

Systems Requiring Row Interchanges

Let A be an invertible square matrix whose row reduction to an upper-triangular matrix U requires at least one row interchange. Then not all elementary matrices corresponding to the necessary row operations add multiples of row vectors to other row vectors. We can still write

$$E_h E_{h-1} \cdots E_2 E_1 A = U,$$

so

$$A = E_1^{-1} E_2^{-1} \cdots E_{h-1}^{-1} E_h^{-1} U,$$

but now that some of these E_i^{-1} interchange rows, their product may not be lower triangular. However, after we discover which row interchanges are necessary, we could start over, and make these necessary row interchanges in the matrix A before we start creating zeros below the diagonal. We can see that the upper-triangular matrix U obtained would still be the same. Suppose, for example, that to obtain a nonzero element in pivot position in the ith row we interchange this ith row with a kth row farther down in the matrix. The new ith row will be the same as though it had been put in the ith row position before the start of the reduction; in either case, it has been modified during the reduction only by the addition of multiples of rows *above* the ith row position. As multiples of it are now added to rows below the ith row position, the same rows (except possibly for order) below the ith row position are created, whether row interchange is performed during reduction or is completed before reduction starts.

Interchanging some rows before the start of the reduction amounts to multiplying A on the left by a sequence of elementary row-interchange matrices. Any product of elementary row-interchange matrices is called a **permutation matrix**. Thus we can form PA for a permutation matrix P, and PA will then admit a factorization $PA = LU$. We state this as a theorem.

THEOREM 10.2 *LU*-Factorization

Let A be an invertible square matrix. Then there exists a permutation matrix P, a lower-triangular matrix L, and an upper-triangular matrix U such that $PA = LU$.

EXAMPLE 7   Illustrate Theorem 10.2 for the matrix

$$A = \begin{bmatrix} 1 & 3 & 2 \\ -2 & -6 & 1 \\ 2 & 5 & 7 \end{bmatrix}.$$

SOLUTION   Starting to reduce A to upper-triangular form, we have

$$\begin{bmatrix} 1 & 3 & 2 \\ -2 & -6 & 1 \\ 2 & 5 & 7 \end{bmatrix} \sim \begin{bmatrix} 1 & 3 & 2 \\ 0 & 0 & 5 \\ 0 & -1 & 3 \end{bmatrix},$$

and we now find it necessary to interchange rows 2 and 3, which will then produce the desired U. Thus we take

$$P = \begin{bmatrix} 1 & 0 & 0 \\ 0 & 0 & 1 \\ 0 & 1 & 0 \end{bmatrix},$$

and we have

$$PA = \begin{bmatrix} 1 & 3 & 2 \\ 2 & 5 & 7 \\ -2 & -6 & 1 \end{bmatrix} \sim \begin{bmatrix} 1 & 3 & 2 \\ 0 & -1 & 3 \\ 0 & 0 & 5 \end{bmatrix} = U.$$

The record matrix L for reduction of PA becomes

$$L = \begin{bmatrix} 1 & 0 & 0 \\ 2 & 1 & 0 \\ -2 & 0 & 1 \end{bmatrix},$$

and we confirm that

$$PA = \begin{bmatrix} 1 & 3 & 2 \\ 2 & 5 & 7 \\ -2 & -6 & 1 \end{bmatrix} = \begin{bmatrix} 1 & 0 & 0 \\ 2 & 1 & 0 \\ -2 & 0 & 1 \end{bmatrix} \begin{bmatrix} 1 & 3 & 2 \\ 0 & -1 & 3 \\ 0 & 0 & 5 \end{bmatrix} = LU.$$ ∎

Suppose that, when solving a large square linear system $A\mathbf{x} = \mathbf{b}$ with a computer, keeping the record L, we find that row interchanges are advisable. (See the discussion of partial pivoting in Section 10.3.) We could keep going and make the row interchanges to find an upper-triangular matrix; we could then start over, make those row interchanges first, just as in Example 7, to obtain a matrix PA, and then obtain U and L for the matrix PA instead. Of course, we would first have to compute $P\mathbf{b}$ when solving $A\mathbf{x} = \mathbf{b}$, because the system would then become $PA\mathbf{x} = P\mathbf{b}$, before we could proceed with the record L and back substitution. This is an undesirable procedure, because it requires reduction of both A and PA, taking a total of about $2n^3/3$ flops, assuming that A is $n \times n$ and n is large. Surely it would be better to devise a method of record-keeping that would also keep track of any row interchanges as they occurred, and would then apply this improved record to $\mathbf{b}$ and use back substitution to find the solution. We toss this suggestion out for enthusiastic programmers to consider on their own.

SUMMARY

1. If A is an $n \times n$ invertible matrix that can be row-reduced to an upper-triangular matrix U without row interchanges, there exists a lower-triangular $n \times n$ matrix L such that $A = LU$.

2. The matrix L in summary item 1 can be found as follows. Start with the $n \times n$ identity matrix I. If during the reduction of A to U, r times row i is added to row k, replace the zero in row k and column i of the identity matrix by $-r$. The final result obtained from the identity matrix is the matrix L.

3. Once A has been reduced to U and L has been found, a computer can find the solution of $A\mathbf{x} = \mathbf{b}$ for a new column vector $\mathbf{b}$, using about n^2 flops for large n.

4. If A is as described in summary item 1, then A has a unique factorization of the form $A = LDU$, where

L is lower triangular with all diagonal entries 1,

U is upper triangular with all diagonal entries 1, and

D is a diagonal matrix with all diagonal entries nonzero.

5. For any invertible matrix A, there exists a permutation matrix P such that PA can be row-reduced to an upper-triangular matrix U and has the properties described for A in summary items 1, 2, 3, and 4.

EXERCISES

1. Discuss briefly the need to worry about the time required to create the record matrix L when solving

$$Ax = b_1, b_2, \ldots, b_s$$

for a large $n \times n$ matrix A.

2. Is there any practical value in creating the record matrix L when one needs to solve only a *single* square linear system $Ax = b$, as we did in Example 3?

In Exercises 3–7, find the solution of $Ax = b$ from the given combined $L\backslash U$ display of the matrix A and the given vector b.

3. $L\backslash U = \begin{bmatrix} 1 & 4 \\ 3 & -2 \end{bmatrix}$, $b = \begin{bmatrix} 2 \\ -4 \end{bmatrix}$

4. $L\backslash U = \begin{bmatrix} -2 & 5 \\ -2 & 1 \end{bmatrix}$, $b = \begin{bmatrix} 3 \\ -7 \end{bmatrix}$

5. $L\backslash U = \begin{bmatrix} 1 & -3 & 4 \\ -2 & -1 & 9 \\ 0 & 1 & -6 \end{bmatrix}$, $b = \begin{bmatrix} 2 \\ -2 \\ 8 \end{bmatrix}$

6. $L\backslash U = \begin{bmatrix} 1 & -4 & 2 \\ 0 & 2 & -1 \\ 3 & 2 & -2 \end{bmatrix}$, $b = \begin{bmatrix} -3 \\ 2 \\ 3 \end{bmatrix}$

7. $L\backslash U = \begin{bmatrix} 1 & 0 & 0 & 1 \\ 0 & -1 & 2 & 1 \\ -1 & -2 & 1 & 3 \\ 2 & 1 & -1 & 3 \end{bmatrix}$, $b = \begin{bmatrix} 4 \\ 7 \\ -8 \\ 14 \end{bmatrix}$

In Exercises 8–13, let A be the given matrix. Find a permutation matrix P, if necessary, and matrices L and U such that $PA = LU$. Check the

answer, using matrix multiplication. Then solve the system $Ax = b$, using P, L, and U.

8. $A = \begin{bmatrix} 2 & -1 \\ 6 & -5 \end{bmatrix}$, $b = \begin{bmatrix} 8 \\ 32 \end{bmatrix}$

9. $A = \begin{bmatrix} 2 & 1 & -3 \\ 6 & 3 & -8 \\ 2 & -1 & 5 \end{bmatrix}$, $b = \begin{bmatrix} 6 \\ 17 \\ 0 \end{bmatrix}$

10. $A = \begin{bmatrix} 1 & 1 & -3 \\ 0 & 1 & 1 \\ 3 & -1 & 1 \end{bmatrix}$, $b = \begin{bmatrix} -13 \\ 6 \\ -7 \end{bmatrix}$

11. $A = \begin{bmatrix} 1 & 2 & -1 \\ 3 & 7 & 2 \\ 4 & -2 & 1 \end{bmatrix}$, $b = \begin{bmatrix} -3 \\ 1 \\ -2 \end{bmatrix}$

12. $A = \begin{bmatrix} 1 & -4 & 1 & -2 \\ 0 & 2 & -1 & 1 \\ 2 & -7 & -2 & 1 \\ 0 & 3 & 0 & -4 \end{bmatrix}$, $b = \begin{bmatrix} -9 \\ 6 \\ 10 \\ -16 \end{bmatrix}$

13. $A = \begin{bmatrix} 1 & 2 & -3 & 0 \\ 2 & 5 & -6 & 0 \\ -1 & -2 & 1 & -1 \\ 4 & 10 & -9 & 1 \end{bmatrix}$, $b = \begin{bmatrix} 8 \\ 17 \\ -8 \\ 33 \end{bmatrix}$

In Exercises 14–17, proceed as in Example 4 to solve the given system.

14. Solve $A^2x = b$ if $A = \begin{bmatrix} 1 & 1 \\ 3 & 0 \end{bmatrix}$, $b = \begin{bmatrix} -11 \\ -6 \end{bmatrix}$.

15. Solve $A^4x = b$ if $A = \begin{bmatrix} -1 & 2 \\ 2 & -3 \end{bmatrix}$,

$b = \begin{bmatrix} 144 \\ -233 \end{bmatrix}$.

16. Solve $A^2\mathbf{x} = \mathbf{b}$ if $A = \begin{bmatrix} 2 & -1 & 0 \\ 4 & -1 & 2 \\ -6 & 2 & 0 \end{bmatrix}$,

$\mathbf{b} = \begin{bmatrix} -2 \\ 14 \\ 12 \end{bmatrix}$.

17. Solve $A^3\mathbf{x} = \mathbf{b}$ if $A = \begin{bmatrix} -1 & 0 & 1 \\ 3 & 1 & 0 \\ -2 & 0 & 4 \end{bmatrix}$,

$\mathbf{b} = \begin{bmatrix} 27 \\ 29 \\ 122 \end{bmatrix}$.

In Exercises 18–20, find the unique factorization LDU for the given matrix A.

18. The matrix in Exercise 8

19. The matrix in Exercise 10

20. The matrix in Exercise 11

21. Mark each of the following True or False.

____ **a.** Every matrix A has an *LU*-factorization.

____ **b.** Every square matrix A has an *LU*-factorization.

____ **c.** Every square matrix A has a factorization $P^{-1}LU$ for some permutation matrix P.

____ **d.** If an *LU*-factorization of an $n \times n$ matrix A is known, using this factorization to solve a large linear system with coefficient matrix A requires about $n^2/2$ flops.

____ **e.** If an *LU*-factorization of an $n \times n$ matrix A is known, using this factorization to solve a large linear system with coefficient matrix A requires about n^2 flops.

____ **f.** If $A = LU$, then this is the only factorization of this type for A.

____ **g.** If $A = LU$, then the matrix L can be regarded as a record of the steps in the row reduction of A to U.

____ **h.** All three types of elementary row operations used to reduce A to U can be recorded in the matrix L in an *LU*-factorization of A.

____ **i.** Two of the three types of elementary row operations used to reduce A to U can be recorded in the matrix L in an *LU*-factorization of A.

____ **j.** One can solve a linear system $LU\mathbf{x} = \mathbf{b}$ by means of a forward substitution followed by a back substitution.

In Exercises 22–24, use the routine LUFACTOR in LINTEK to find the combined L\U display (2) for the matrix A. Then solve $A\mathbf{x} = \mathbf{b}_i$ for each of the column vectors $\mathbf{b}_i$.

22. $A = \begin{bmatrix} 2 & 1 & -3 & 4 & 0 \\ 1 & 4 & 6 & -1 & 5 \\ -2 & 0 & 3 & 1 & 4 \\ 6 & 1 & 2 & 8 & -3 \\ 4 & 1 & 3 & 2 & 1 \end{bmatrix}$;

$\mathbf{b}_i = \begin{bmatrix} 13 \\ 7 \\ 4 \\ 27 \\ 17 \end{bmatrix}, \begin{bmatrix} 10 \\ 0 \\ 10 \\ 60 \\ 20 \end{bmatrix}$, and $\begin{bmatrix} -20 \\ 96 \\ 53 \\ 5 \\ 26 \end{bmatrix}$ for $i = 1$, 2, and 3, respectively.

23. $A = \begin{bmatrix} 1 & 3 & -5 & 2 & 1 \\ 4 & -6 & 10 & 8 & 3 \\ 3 & 6 & -1 & 4 & 7 \\ 2 & 1 & 11 & -3 & 13 \\ -6 & 3 & 1 & 4 & 21 \end{bmatrix}$;

$\mathbf{b}_i = \begin{bmatrix} 0 \\ 20 \\ -9 \\ -31 \\ -43 \end{bmatrix}, \begin{bmatrix} -48 \\ 218 \\ 0 \\ 100 \\ 143 \end{bmatrix}$, and $\begin{bmatrix} 87 \\ -151 \\ 102 \\ -46 \\ 223 \end{bmatrix}$

for $i = 1$, 2, and 3, respectively.

24. $A = \begin{bmatrix} -1 & 3 & 2 & 3 & 4 & 6 \\ 3 & 1 & 0 & -2 & 1 & 7 \\ 6 & 8 & -2 & 1 & 4 & 6 \\ 4 & -2 & 3 & 1 & 7 & 6 \\ 3 & -2 & 1 & 10 & 8 & 0 \\ 4 & -5 & 8 & -3 & 2 & 10 \end{bmatrix}$;

$\mathbf{b}_i = \begin{bmatrix} 19 \\ 1 \\ -8 \\ 82 \\ 76 \\ 103 \end{bmatrix}, \begin{bmatrix} 82 \\ 2 \\ 82 \\ 31 \\ 92 \\ -61 \end{bmatrix}$, and $\begin{bmatrix} 25 \\ -3 \\ -24 \\ -73 \\ -115 \\ 20 \end{bmatrix}$

for $i = 1$, 2, and 3, respectively.

In Exercises 25–29, use the routine LUFACTOR in LINTEK to solve the indicated system.

25. $A^5\mathbf{x} = \mathbf{b}$ for A and $\mathbf{b}$ in Exercise 14

26. $A^7\mathbf{x} = \mathbf{b}$ for A and $\mathbf{b}$ in Exercise 15

27. $A^6\mathbf{x} = \mathbf{b}$ for A and $\mathbf{b}$ in Exercise 16

28. $A^8\mathbf{x} = \mathbf{b}$ for A and $\mathbf{b}$ in Exercise 17

29. $A^3\mathbf{x} = \mathbf{b}_1$ for A and $\mathbf{b}_1$ in Exercise 24

The routine TIMING in LINTEK has an option to time the formation of the combined L\U display from a matrix A. The user specifies the size n for an n × n linear system Ax = b. The program then generates an n × n matrix A with entries in the interval [−20, 20] and creates the L\U display shown in matrix (2) in the text. The time required for the reduction is indicated. The user may then specify a number s for solution of Ax = b₁, b₂, . . . , bₛ. The computer generates a column vector and solves s systems, using the record in L and back substitution; the time to find these s solutions is also indicated. Recall that the reduction of A should require about n³/3 flops for

large n, and the solution for each column vector should require about n² flops. In Exercises 30–34, use the routine TIMING and see if the ratios of times obtained seem to conform to the ratios of the numbers of flops required, at least as n increases. Compare the time required to solve one system, including the computation of L\U, with the time required to solve a system of the same size using the Gauss method with back substitution. The times should be about the same.

30. $n = 4$; $s = 1, 3, 12$

31. $n = 10$; $s = 1, 5, 10$

32. $n = 16$; $s = 1, 8, 16$

33. $n = 24$; $s = 1, 12, 24$

34. $n = 30$; $s = 1, 15, 30$

MATLAB

MATLAB uses *LU*-factorization, for example, in its computation of $A \backslash B$. (Recall that $A \backslash B = A^{-1}B$ if A is a square matrix.) In finding an *LU*-factorization of A, MATLAB uses partial pivoting (row swapping to make pivots as large as possible) for increased accuracy (see Section 10.3). It obtains a lower-triangular matrix L and an upper-triangular matrix U such that $LU = PA$, where P is a permutation matrix, as illustrated in Example 7. The MATLAB command

$$[\mathbf{L},\mathbf{U},\mathbf{P}] = \mathbf{lu}(\mathbf{A})$$

produces these matrices L, U, and P. In pencil-and-paper work, we like small pivots rather than large ones, and we swapped rows in Exercises 8–13 only if we encountered a zero pivot. Thus this MATLAB command cannot be used to check our answers to Exercises 8–13.

| 10.3 | PIVOTING, SCALING, AND ILL-CONDITIONED MATRICES |

Some Problems Encountered with Computers

A computer can't do absolutely precise arithmetic with real numbers. For example, any computer can only compute using an approximation of the number $\sqrt{2}$, never with the number $\sqrt{2}$ itself. Computers work using base-2 arithmetic, representing numbers as strings of zeros and ones. But they encounter the same problems that we encounter if we pretend that we are a computer using base-10 arithmetic and are capable of handling only some fixed finite number of significant digits. We will illustrate computer problems in base-10 notation, because it is more familiar to all of us.

Suppose that our base-10 computer is asked to compute the quotient $\frac{1}{3}$, and suppose that it can represent a number in floating-point arithmetic using eight significant figures. It represents $\frac{1}{3}$ in **truncated** form as 0.33333333. It may represent $\frac{2}{3}$ as 0.66666667, **rounding off** to eight significant figures. For convenience, we will refer to all errors generated by either truncation or roundoff as **roundoff errors**.

Most people realize that in an extended arithmetic computation by a computer, the roundoff error can accumulate to such an extent that the final result of the computation becomes meaningless. We will see that there are at least two situations in which a computer cannot meaningfully execute even a single arithmetic operation. In order to avoid working with long strings of digits, we assume that our computer can compute only to three significant figures in its work. Thus, our three-figure, base-10 computer computes the quotient $\frac{2}{3}$ as 0.666. We box the first computer problem that concerns us, and give an example.

Addition of two numbers of very different magnitude may result in the loss of some or even all of the significant figures of the smaller number.

EXAMPLE 1   Evaluate our three-figure computer's computation of

$$45.1 + .0725.$$

SOLUTION   Because it can handle only three significant figures, our computer represents the actual sum 45.1725 as 45.1. In other words, the second summand .0725 might as well be zero, so far as our computer is concerned. The datum .0725 is completely lost.   ■

The difficulty illustrated in Example 1 can cause serious problems in attempting to solve a linear system $A\mathbf{x} = \mathbf{b}$ with a computer, as we will illustrate in a moment. First we box another problem a computer may have.

Subtraction of nearly equal numbers can result in a loss of significant figures.

EXAMPLE 2   Evaluate our three-figure computer's computation of

$$\frac{2}{3} - \frac{665}{1000}.$$

SOLUTION   The actual difference is

$$.666666\cdots - .665 = .00166666\cdots.$$

However, our three-figure computer obtains

$$.666 - .665 = .001.$$

Two of the three significant figures with which the computer can work have been lost. ∎

The difficulty illustrated in Example 2 is encountered if one tries to use a computer to do differential calculus.

The cure for both difficulties is to have the computer work with more significant figures. But using more significant figures requires the computer to take more time to execute a program, and of course the same errors can occur "farther out." For example, the typical microcomputer of this decade, with software designed for routine computations, will compute $10^{10} + 10^{-10}$ as 10^{10}, losing the second summand 10^{-10} entirely.

Partial Pivoting

In row reduction of a matrix to echelon form, a technique called *partial pivoting* is often used. In **partial pivoting,** one interchanges the row in which the pivot is to occur with a row farther down, if necessary, so that the pivot becomes as large in absolute value as possible. To illustrate, suppose that row reduction of a matrix to echelon form leads to an intermediate matrix

$$\begin{bmatrix} 2 & 8 & -1 & 3 & 4 \\ 0 & -2 & 3 & -5 & 6 \\ 0 & 4 & 1 & 2 & 0 \\ 0 & -7 & 3 & 1 & 4 \end{bmatrix}.$$

Using partial pivoting, we would then interchange rows 2 and 4 to use the entry -7 of maximum magnitude among the possibilities -2, 4, and -7 for pivots in the second column. That is, we would form the matrix

$$\begin{bmatrix} 2 & 8 & -1 & 3 & 4 \\ 0 & -7 & 3 & 1 & 4 \\ 0 & 4 & 1 & 2 & 0 \\ 0 & -2 & 3 & -5 & 6 \end{bmatrix}$$

and continue with the reduction of this matrix.

We show by example the advantage of partial pivoting. Let us consider the linear system

$$\begin{aligned} .01x_1 + 100x_2 &= 100 \\ -100x_1 + 200x_2 &= 100. \end{aligned} \qquad \text{(1)}$$

EXAMPLE 3   Find the actual solution of linear system (1). Then compare the result with that obtained by a three-figure computer using the Gauss method with back substitution, but without partial pivoting.

SOLUTION   First we find the actual solution:

$$\begin{bmatrix} .01 & 100 & | & 100 \\ -100 & 200 & | & 100 \end{bmatrix} \sim \begin{bmatrix} .01 & 100 & | & 100 \\ 0 & 1{,}000{,}200 & | & 1{,}000{,}100 \end{bmatrix},$$

and back substitution yields

$$x_2 = \frac{1{,}000{,}100}{1{,}000{,}200},$$

$$x_1 = \left(100 - \frac{100{,}010{,}000}{1{,}000{,}200} \right) 100 = \frac{1{,}000{,}000}{1{,}000{,}200}.$$

Thus, $x_1 \approx .9998$ and $x_2 \approx .9999$. On the other hand, our three-figure computer obtains

$$\begin{bmatrix} .01 & 100 & | & 100 \\ -100 & 200 & | & 100 \end{bmatrix} \sim \begin{bmatrix} .01 & 100 & | & 100 \\ 0 & 1{,}000{,}000 & | & 1{,}000{,}000 \end{bmatrix},$$

which leads to

$$x_2 = 1,$$
$$x_1 = (100 - 100)100 = 0.$$

The x_1-parts of the solution are very different. Our three-figure computer completely lost the second-row data entries 200 and 100 in the matrix when it added 10,000 times the first row to the second row.   ∎

EXAMPLE 4   Find the solution to system (1) that our three-figure computer would obtain using partial pivoting.

SOLUTION   Using partial pivoting, the three-figure computer obtains

$$\begin{bmatrix} .01 & 100 & | & 100 \\ -100 & 200 & | & 100 \end{bmatrix} \sim \begin{bmatrix} -100 & 200 & | & 100 \\ .01 & 100 & | & 100 \end{bmatrix}$$

$$\sim \begin{bmatrix} -100 & 200 & | & 100 \\ 0 & 100 & | & 100 \end{bmatrix},$$

which yields

$$x_2 = 1,$$
$$x_1 = \frac{100 - 200}{-100} = 1.$$

This is close to the solution $x_1 \approx .9998$ and $x_2 \approx .9999$ obtained in Example 3, and it is much better than the erroneous $x_1 = 0$, $x_2 = 1$ that the three-figure computer obtained in Example 3 without pivoting. The data entries 200 and 100 in the second row of the initial matrix were not lost in this computation.   ∎

To understand completely the reason for the difference between the solutions in Example 3 and in Example 4, consider again the linear system

$$.01x_1 + 100x_2 = 100$$
$$-100x_1 + 200x_2 = 100,$$

which was solved in Example 3. Multiplication of the first row by 10,000 produced a coefficient of x_2 of such magnitude compared with the second equation coefficient 200 that the coefficient 200 was totally destroyed in the ensuing addition. (Remember that we are using our three-figure computer. In a less dramatic example, some significant figures of the smaller coefficient might still contribute to the sum.) If we use partial pivoting, the linear system becomes

$$-100x_1 + 200x_2 = 100$$
$$.01x_1 + 100x_2 = 100$$

and multiplication of the first equation by a number less than 1 does not threaten the significant digits of the numbers in the second equation. This explains the success of partial pivoting in this situation.

Suppose now that we multiply the first equation of system (1) by 10,000, which of course does not alter the solution of the system. We then obtain the linear system

$$100x_1 + 1,000,000x_2 = 1,000,000 \tag{2}$$
$$-100x_1 + \qquad 200x_2 = \qquad 100.$$

If we solved system (2) using *partial pivoting*, we would not interchange the rows, because -100 is of no greater magnitude than 100. Addition of the first row to the second by our three-figure computer again totally destroys the coefficient 200 in the second row. Exercise 1 shows that the erroneous solution $x_1 = 0$, $x_2 = 1$ is again obtained. We could avoid this problem by **full pivoting**. In full pivoting, columns are also interchanged, if necessary, to make pivots as large as possible. That is, if a pivot is to be found in the row i and column j position of an intermediate matrix G, then not only rows but also columns are interchanged as needed to put the entry g_{rs} of greatest magnitude, where $r \geq i$ and $s \geq j$, in the pivot position. Thus, full pivoting for system (2) will lead to the matrix

$$\begin{array}{cc} x_2 & x_1 \end{array}$$
$$\begin{bmatrix} 1,000,000 & 100 \ \Big| & 1,000,000 \\ 200 & -100 \ \Big| & 100 \end{bmatrix}. \tag{3}$$

Exercise 2 illustrates that row reduction of matrix (3) by our three-figure computer gives a reasonable solution of system (2). Notice, however, that we now have to do some bookkeeping and must remember that the entries in column 1 of matrix (3) are really the coefficients of x_2, not of x_1. For a matrix of any size, the search for elements of maximum magnitude and the bookkeeping required in full pivoting take a lot of computer time. Partial pivoting is frequently used, representing a compromise between time and accuracy. The routine MATCOMP in LINTEK uses partial pivoting in its Gauss–Jordan reduction to reduced echelon form. Thus, if system (1) is modified so that the Gauss–Jordan method without pivoting fails to give a reasonable solution for a 20-figure computer, MATCOMP could probably handle it. However, one can no doubt create a similar modification of the 2×2 system (2) for which MATCOMP would give an erroneous solution.

Scaling

We display again system (2):

$$100x_1 + 1,000,000x_2 = 1,000,000$$
$$-100x_1 + 200x_2 = 100.$$

We might recognize that the number 1,000,000 dangerously dominates the data entries in the second row, at least as far as our three-figure computer is concerned. We might multiply the first equation in system (2) by .0001 to cut those two numbers down to size, essentially coming back to system (1). Partial pivoting handles system (1) well. Multiplication of an equation by a nonzero constant for such a purpose is known as **scaling**. Of course, one could equivalently scale by multiplying the second equation in system (2) by 10,000, to bring its coefficients into line with the large numbers in the first equation.

We box one other computer problem that can sometimes be addressed by scaling. In reducing a matrix to echelon form, we need to know whether the entry that appears in a pivot position as we start work on a column is truly nonzero, and indeed, whether there is any truly nonzero entry from that column to serve as pivot.

> Due to roundoff error, a computed number that should be zero is quite likely to be of small, nonzero magnitude.

Taking this into account, one usually programs row-echelon reduction so that entries of unexpectedly small size are changed to zero. MATCOMP finds the smallest nonzero magnitude m among 1 and all the coefficient data supplied for the linear system, and sets $E = rm$, where r is specified by the user. (Default r is .0001.) In reduction of the coefficient matrix, a computed entry of magnitude less than E is replaced by zero. The same procedure is followed in YUREDUCE. Whatever computed number we program a computer to choose for E in a program such as MATCOMP, we will be able to devise some linear system for which E is either too large or too small to give the correct result.

A procedure equivalent to the one in MATCOMP that we just outlined is to *scale* the original data for the linear system in such a way that the smallest nonzero entry is of magnitude roughly 1, and then always use the same value, perhaps 10^{-4}, for E.

When one of the authors first started experimenting with a computer, he was horrified to discover that a matrix inversion routine built into the mainframe BASIC program of a major computer company would refuse to invert a matrix if all the entries were small enough. An error message such as

"Nearly singular matrix. Inversion impossible."

would appear. He considers a 2×2 matrix

$$\begin{bmatrix} a & b \\ c & d \end{bmatrix}$$

to be nearly singular if and only if lines in the plane having equations of the form

$$ax + by = r$$
$$cx + dy = s$$

are nearly parallel. Now the lines

$$10^{-8}x + \quad 0y = 1$$
$$0x + 10^{-9}y = 1$$

are actually perpendicular, the first one being vertical and the second horizontal. This is as far away from parallel as one can get! It annoyed the author greatly to have the matrix

$$\begin{bmatrix} 10^{-8} & 0 \\ 0 & 10^{-9} \end{bmatrix}$$

called "nearly singular." The inverse is obviously

$$\begin{bmatrix} 10^{8} & 0 \\ 0 & 10^{9} \end{bmatrix}.$$

A scaling routine was promptly written to be executed before calling the inversion routine. The matrix was multiplied by a constant that would bring the smallest nonzero magnitude to at least 1, and then the inversion subroutine was used, and the result rescaled to provide the inverse of the original matrix. For example, applied to the matrix just discussed, this procedure becomes

$$\begin{bmatrix} 10^{-8} & 9 \\ 0 & 10^{-9} \end{bmatrix} \text{ multiplied by } 10^9 \text{ becomes } \begin{bmatrix} 10 & 0 \\ 0 & 1 \end{bmatrix},$$

$$\text{inverted becomes } \begin{bmatrix} \frac{1}{10} & 0 \\ 0 & 1 \end{bmatrix},$$

$$\text{multiplied by } 10^9 \text{ becomes } \begin{bmatrix} 10^{8} & 0 \\ 0 & 10^{9} \end{bmatrix}$$

Having more programming experience now, this author is much more charitable and understanding. The user may also find unsatisfactory things about MATCOMP. We hope that this little anecdote has helped explain the notion of scaling.

Ill-Conditioned Matrices

The line $x + y = 100$ in the plane has x-intercept 100 and y-intercept 100, as shown in Figure 10.1. The line $x + .9y = 100$ also has x-intercept 100, but it has y-intercept larger than 100. The two lines are almost parallel. The common x-intercept shows that the solution of the linear system

$$
\begin{aligned}
x + \;\;y &= 100 \\
x + .9y &= 100
\end{aligned}
\tag{4}
$$

is $x = 100$, $y = 0$, as illustrated in Figure 10.2. Now the line $.9x + y = 100$ has y-intercept 100 but x-intercept larger than 100, so the linear system

$$
\begin{aligned}
x + y &= 100 \\
.9x + y &= 100
\end{aligned}
\tag{5}
$$

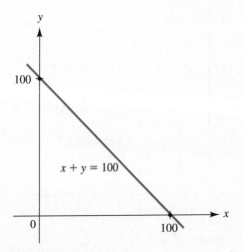

FIGURE 10.1
The line $x + y = 100$.

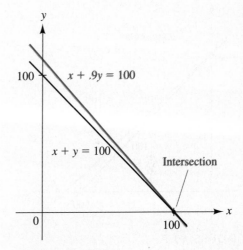

FIGURE 10.2
The lines are almost parallel.

has the very different solution $x = 0$, $y = 100$, as shown in Figure 10.3. Systems (4) and (5) are examples of **ill-conditioned** or **unstable** systems: small changes in the coefficients or in the constants on the right-hand sides can produce very great changes in the solutions. We say that a matrix A is **ill-conditioned** if a linear system $A\mathbf{x} = \mathbf{b}$ having A as coefficient matrix is ill-conditioned. For two equations in two unknowns, solving an ill-conditioned system corresponds to finding the intersection of two nearly parallel lines, as shown in Figure 10.4. Changing a coefficient of x or y slightly in one equation changes the slope of that line only slightly, but it may generate a big change in the location of the point of intersection of the two lines.

Computers have a lot of trouble finding accurate solutions of ill-conditioned systems such as systems (4) and (5), because the small roundoff errors created by the computer can produce large changes in the solution. Pivoting and scaling usually don't help the situation; the systems are basically unstable. Notice that the coefficients of x and y in systems (4) and (5) are of comparable magnitude.

Among the most famous ill-conditioned matrices are the Hilbert matrices. These are very bad matrices named after a very good mathematician, David Hilbert! (See the historical note on page 444.) The entry in the ith row and jth column of a Hilbert matrix is $1/(i + j - 1)$. Thus, if we let H_n be the $n \times n$ Hilbert matrix, we have

$$H_2 = \begin{bmatrix} 1 & \frac{1}{2} \\ \frac{1}{2} & \frac{1}{3} \end{bmatrix}, \qquad H_3 = \begin{bmatrix} 1 & \frac{1}{2} & \frac{1}{3} \\ \frac{1}{2} & \frac{1}{3} & \frac{1}{4} \\ \frac{1}{3} & \frac{1}{4} & \frac{1}{5} \end{bmatrix}, \qquad \text{and so on.}$$

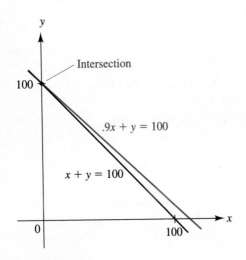

FIGURE 10.3
A very different solution from the one in Figure 10.2.

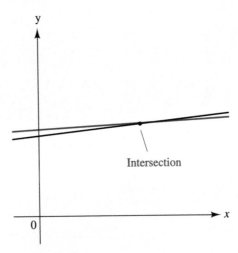

FIGURE 10.4
The intersection of two nearly parallel lines.

It can be shown that H_n is invertible for all n, so a square linear system $H_n\mathbf{x} = \mathbf{b}$ has a unique solution, but the solution may be very hard to find. When the matrix is reduced to echelon form, entries of surprisingly small magnitude appear. Scaling of a row in which all entries are close to zero may help a bit.

Bad as the Hilbert matrices are, powers of them are even worse. The software we make available includes a routine called HILBERT, which is modeled on YUREDUCE. The computer generates a Hilbert matrix of the size we specify, up to 10×10. It will then raise the matrix to the power 2, 4, 8, or 16 if we so request. We may then proceed roughly as in YUREDUCE. Routines such as YUREDUCE and HILBERT should help us understand this section, because we can watch and see just what is happening as we reduce a matrix. MATCOMP, which simply spits out answers, may produce an absurd result, but we have no way of knowing exactly where things went wrong.

SUMMARY

1. Addition of numbers of very different magnitudes by a computer can cause loss of some or all of the significant figures in the number of smaller magnitude.

2. Subtraction of nearly equal numbers by a computer can cause loss of significant figures.

3. Due to roundoff error, a computer may obtain a nonzero value for a number that should be zero. To attempt to handle this problem, a computer program might assign the value zero to certain computed numbers whenever the numbers have a magnitude less than some predetermined small positive number.

4. In partial pivoting, the pivot in each column is created by swapping rows, if necessary, so that the pivot has at least the maximum magnitude of any entry below it in that column. Partial pivoting may be helpful in avoiding the problem stated in summary item 1.

5. In full pivoting, columns are interchanged as well as rows, if necessary, to create pivots of maximum possible magnitude. Full pivoting requires much more computer time than does partial pivoting, and bookkeeping is necessary to keep track of the relationship between the columns and the variables. Partial pivoting is more commonly used.

6. *Scaling,* which is multiplication of a row by a nonzero constant, can be used to reduce the sizes of entries that threaten to dominate entries in lower rows, or to increase the sizes of entries in a row where some entries are very small.

7. A linear system $A\mathbf{x} = \mathbf{b}$ is ill-conditioned or unstable if small changes in the numbers can produce large changes in the solution. The matrix A is then also called *ill-conditioned.*

8. Hilbert matrices, which are square matrices with entry $1/(i + j - 1)$ in the ith row and jth column, are examples of ill-conditioned matrices.

9. With present technology, it appears hopeless to write a computer program that will successfully handle every linear system involving even very small coefficient matrices—say, of size at most 10×10.

EXERCISES

1. Find the solution by a three-figure computer of the system

$$100x_1 + 1,000,000x_2 = 1,000,000$$
$$-100x_1 + \qquad 200x_2 = \qquad 100$$

using just partial pivoting.

2. Repeat Exercise 1, but use full pivoting.

3. Find the solution, without pivoting, by a five-figure computer of the linear system in Exercise 1. Is the solution reasonably accurate? If so, modify the system a bit to obtain one for which a five-figure computer does not find a reasonable solution without pivoting.

4. Modify the linear system in Exercise 1 so that an eight-figure computer (roughly the usual single-precision computer) will not obtain a reasonable solution without partial pivoting.

5. Repeat Exercise 4 for an 18-figure computer (roughly the usual double-precision computer).

6. Find a linear system with two equations in x and y such that a change of .001 in the coefficient of x in the second equation produces a change of at least 1,000,000 in both of the values x and y in a solution.

7. Let A be an invertible square matrix. Show that the following scaling routine for finding A^{-1} using a computer is *mathematically* correct.
   a. Find the minimum absolute value m of all nonzero entries in A.
   b. Multiply A by an integer $n > 1/m$.
   c. Find the inverse of the resulting matrix.

d. Multiply the matrix obtained in part c by n to get A^{-1}.

Use the routines MATCOMP, YUREDUCE, and HILBERT in LINTEK for Exercises 8–13.

8. Use YUREDUCE to solve the system

$$1E{-}9 \; x_1 + 1E9 \; x_2 = 1$$
$$1E9 \; x_1 + 2E9 \; x_2 = 1,$$

without using any pivoting. Check the answer mentally to see if it is approximately correct. If it is, increase the exponent from 9 until a system is obtained in which the solution without pivoting is erroneous. [*Note:* $1E{-}9 = 10^{-9}$, $2E9 = 2(10^9)$]

9. Use YUREDUCE to solve the system in Exercise 8, which did not give a reasonable solution without pivoting, but this time use partial pivoting.

10. Repeat Exercise 9, but this time use full pivoting. Compare the answer with that in Exercise 9.

11. See how MATCOMP handles the linear system formed in Exercise 8.

12. Experiment with MATCOMP, and see if it gives a "nearly singular matrix" message when finding the inverse of a 2×2 diagonal matrix

$$rI = r\begin{bmatrix} 1 & 0 \\ 0 & 1 \end{bmatrix},$$

where r is a sufficiently small nonzero number. Use the default for roundoff control in MATCOMP.

13. Using MATCOMP, use the scaling routine suggested in Exercise 7 to find the inverse of the matrix

$$\begin{bmatrix} .000001 & -.000003 & .000011 & .000006 \\ -.000002 & .000013 & .000007 & .000010 \\ .000009 & -.000011 & 0 & -.000005 \\ .000014 & -.000008 & -.000002 & .000003 \end{bmatrix}.$$

Then see if the same answer is obtained without using the scaling routine. Check the inverse in each case by multiplying by the original matrix.

14. Mark each of the following True or False.

____ **a.** Addition and subtraction never cause any problem when executed by a computer.

____ **b.** Given any present-day computer, one can find two positive numbers whose sum the computer will represent as the larger of the two numbers.

____ **c.** A computer may have trouble representing as accurately as desired the sum of two numbers of extremely different magnitudes.

____ **d.** A computer may have trouble representing as accurately as desired the sum of two numbers of essentially the same magnitude.

____ **e.** A computer may have trouble representing as accurately as desired the sum of a and b, where a is approximately $2b$ or $-2b$.

____ **f.** Partial pivoting handles all problems resulting from roundoff error when a linear system is being solved by the Gauss method.

____ **g.** Full pivoting handles roundoff error problems better than partial pivoting, but it is generally not used because of the extra computer time required to implement it.

____ **h.** Given any present-day computer, one can find a system of two equations in two unknowns that the computer cannot solve accurately using the Gauss method with back substitution.

____ **i.** A linear system of two equations in two unknowns is unstable if the lines represented by the two equations are extremely close to being parallel.

____ **j.** The entry in the ith row and jth column of a Hilbert matrix is $1/(i + j)$.

Use the routine HILBERT in LINTEK in Exercises 15–22. Because the Hilbert matrices are nonsingular, diagonal entries computed during the elimination should never be zero. Except in Exercise 22, enter 0 for r when r is requested during a run of HILBERT.

15. Solve $H_2x = b, c$, where

$$b = \begin{bmatrix} 2 \\ 9 \end{bmatrix} \quad \text{and} \quad c = \begin{bmatrix} 3 \\ 8 \end{bmatrix}.$$

Use just one run of HILBERT; that is, solve both systems at once. Notice that the components of **b** and **c** differ by just 1. Find the difference in the components of the two solution vectors.

16. Repeat Exercise 15, changing the coefficient matrix to $H_2{}^4$.

17. Repeat Exercise 15, using as coefficient matrix H_4 with

$$b = \begin{bmatrix} 2 \\ 1 \\ 4 \\ 6 \end{bmatrix} \quad \text{and} \quad c = \begin{bmatrix} 3 \\ 0 \\ 5 \\ 7 \end{bmatrix}.$$

18. Repeat Exercise 15, using as coefficient matrix $H_4{}^4$.

19. Find the inverse of H_2, and use the I menu option to test whether the computed inverse is correct.

20. Continue Exercise 19 by trying to find the inverses of H_2 raised to the powers 2, 4, 8, and 16. Was it always possible to reduce the power of the Hilbert matrix to diagonal form? If not, what happened? Why did it happen? Was scaling (1 an entry) of any help? For those cases in which reduction to diagonal form was possible, did the computed inverse seem to be reasonably accurate when tested?

21. Repeat Exercises 19 and 20, using H_4 and the various powers of it. Are problems encountered for lower powers this time?

22. We have seen that, in reducing a matrix, we may wish to instruct the computer to assign the value zero to computed entries of sufficiently small magnitude, because entries that should be zero might otherwise be left nonzero because of roundoff error. Use HILBERT to try to solve the linear system

$$H_3^2 x = b, \qquad \text{where} \quad b = \begin{bmatrix} -3 \\ 1 \\ 4 \end{bmatrix},$$

entering as r the number .0001 when requested. Will the usual Gauss–Jordan elimination using HILBERT solve the system? Will HILBERT solve it using scaling (1 an entry)?

MATLAB

The command **hilb(n)** in MATLAB creates the $n \times n$ Hilbert matrix, and the command **invhilb(n)** attempts to create its inverse. For a matrix X in MATLAB, the command **max(X)** produces a row vector whose jth entry is the maximum of the entries in the jth column of X, whereas for a row vector v, the command **max(v)** returns the maximum of the components of v. Thus the command **max(max(X))** returns the maximum of the entries of X. The functions **min(X)** and **min(v)** return analogous minimum values. Thus the command line

$$n = 5; A = hilb(n)*invhilb(n); b = [max(max(A)) \quad min(min(A))]$$

will return the two-component row vector **b** whose first component is the maximum of the entries of A and whose second component is the minimum of those entries. Of course, if **invhilb(n)** is computed accurately, this vector should be [1, 0].

M1. Enter **format long** and then enter the line

$$n = 2; A = hilb(n)*invhilb(n); b = [max(max(A)) \quad min(min(A))]$$

in MATLAB. Using the up-arrow key to change the value of n, copy down the vector **b** with entries to three significant figures for the values 5, 10, 15, 20, 25, and 30 of n.

M2. Recall that the command $X = rand(n,n)$ generates a random $n \times n$ matrix with entries between 0 and 1; the Hilbert matrices also have entries between 0 and 1. Enter the command

$$X = rand(30,30); A = X*inv(X); b = [max(max(A)) \quad min(min(A))]$$

to see if MATLAB has difficulty inverting such a random 30×30 matrix. Using the up-arrow key, execute this command a total of seven times, and see if the results seem to be consistent.

A MATHEMATICAL INDUCTION

Sometimes we want to prove that a statement about positive integers is true for all positive integers or perhaps for some finite or infinite sequence of consecutive integers. Such proofs are accomplished using *mathematical induction*. The validity of the method rests on the following axiom of the positive integers. The set of all positive integers is denoted by Z^+.

Induction Axiom

Let S be a subset of Z^+ satisfying

1. $1 \in S$,
2. If $k \in S$, then $(k + 1) \in S$.

Then $S = Z^+$.

This axiom leads immediately to the method of mathematical induction.

Mathematical Induction

Let $P(n)$ be a statement concerning the positive integer n. Suppose that

1. $P(1)$ is true,
2. If $P(k)$ is true, then $P(k + 1)$ is true.

Then $P(n)$ is true for all $n \in Z^+$.

Most of the time, we want to show that $P(n)$ holds for all $n \in Z^+$. If we wish only to show that it holds for $r, r + 1, r + 2, \ldots, s - 1, s$, then we show that $P(r)$ is true and that $P(k)$ implies $P(k + 1)$ for $r \le k \le s - 1$. Notice that r may be any integer—positive, negative, or zero.

EXAMPLE A.1   Prove the formula

$$1 + 2 + \cdots + n = \frac{n(n + 1)}{2} \tag{A.1}$$

for the sum of the arithmetic progression, using mathematical induction.

SOLUTION   We let $P(n)$ be the statement that formula (A.1) is true. For $n = 1$, we obtain

$$\frac{n(n + 1)}{2} = \frac{1(2)}{2} = 1,$$

so $P(1)$ is true.

Suppose that $k \ge 1$ and $P(k)$ is true (our *induction hypothesis*), so

$$1 + 2 + \cdots + k = \frac{k(k + 1)}{2}.$$

To show that $P(k + 1)$ is true, we compute

$$1 + 2 + \cdots + (k + 1) = (1 + 2 + \cdots + k) + (k + 1)$$
$$= \frac{k(k + 1)}{2} + (k + 1) = \frac{k^2 + k + 2k + 2}{2}$$
$$= \frac{k^2 + 3k + 2}{2} = \frac{(k + 1)(k + 2)}{2}.$$

Thus, $P(k + 1)$ holds, and formula (A.1) is true for all $n \in Z^+$.   ∎

EXAMPLE A.2   Show that a set of n elements has exactly 2^n subsets for any nonnegative integer n.

SOLUTION   This time we start the induction with $n = 0$. Let S be a finite set having n elements. We wish to show

$$P(n): S \text{ has } 2^n \text{ subsets.} \tag{A.2}$$

If $n = 0$, then S is the empty set and has only one subset—namely, the empty set itself. Because $2^0 = 1$, we see that $P(0)$ is true.

Suppose that $P(k)$ is true. Let S have $k + 1$ elements, and let one element of S be c. Then $S - \{c\}$ has k elements, and hence 2^k subsets. Now every subset of S either contains c or does not contain c. Those not containing c are subsets of $S - \{c\}$, so there are 2^k of them by the induction hypothesis. Each subset containing c consists of one of the 2^k subsets not containing c, with c adjoined. There are 2^k such subsets also. The total number of subsets of S is then

$$2^k + 2^k = 2^k(2) = 2^{k+1},$$

so $P(k + 1)$ is true. Thus, $P(n)$ is true for all nonnegative integers n.   ∎

EXAMPLE A.3   Let $x \in \mathbb{R}$ with $x > -1$ and $x \neq 0$. Show that $(1 + x)^n > 1 + nx$ for every positive integer $n \geq 2$.

SOLUTION   We let $P(n)$ be the statement

$$(1 + x)^n > 1 + nx. \qquad \text{(A.3)}$$

(Notice that $P(1)$ is false.) Then $P(2)$ is the statement $(1 + x)^2 > 1 + 2x$. Now $(1 + x)^2 = 1 + 2x + x^2$, and $x^2 > 0$, because $x \neq 0$. Thus, $(1 + x)^2 > 1 + 2x$, so $P(2)$ is true.

Suppose that $P(k)$ is true, so

$$(1 + x)^k > 1 + kx. \qquad \text{(A.4)}$$

Now $1 + x > 0$, because $x > -1$. Multiplying both sides of inequality (A.4) by $1 + x$, we obtain

$$(1 + x)^{k+1} > (1 + kx)(1 + x) = 1 + (k + 1)x + kx^2.$$

Because $kx^2 > 0$, we see that $P(k + 1)$ is true. Thus $P(n)$ is true for every positive integer $n \geq 2$. ∎

In a frequently used form of induction known as *complete induction*, the statement

If $P(k)$ is true, then $P(k + 1)$ is true

in the second box on page A-1 is replaced by the statement

If $P(m)$ is true for $1 \leq m \leq k$, then $P(k + 1)$ is true.

Again, we are trying to show that $P(k + 1)$ is true, knowing that $P(k)$ is true. But if we have reached the stage of induction where $P(k)$ has been proved, we know that $P(m)$ is true for $1 \leq m \leq k$, so the strengthened hypothesis in the second statement is permissible.

EXAMPLE A.4   Recall that the set of all polynomials with real coefficients is denoted by P. Show that every polynomial in P of degree $n \in Z^+$ either is irreducible itself or is a product of irreducible polynomials in P. (An **irreducible polynomial** is one that cannot be factored into polynomials in P all of lower degree.)

SOLUTION   We will use complete induction. Let $P(n)$ be the statement that is to be proved. Clearly $P(1)$ is true, because a polynomial of degree 1 is already irreducible.

Let k be a positive integer. Our induction hypothesis is then: every polynomial in P of degree less than $k + 1$ either is irreducible or can be factored into irreducible polynomials. Let $f(x)$ be a polynomial of degree $k + 1$. If $f(x)$ is irreducible, we have nothing more to do. Otherwise, we may factor $f(x)$ into polynomials $g(x)$ and $h(x)$ of lower degree than $k + 1$, obtaining $f(x) = g(x)h(x)$. The induction hypothesis indicates that each of $g(x)$ and $h(x)$ can be factored into irreducible polynomials, thus providing such a factorization of $f(x)$. This proves $P(k + 1)$. It follows that $P(n)$ is true for all $n \in Z^+$. ∎

EXERCISES

1. Prove that

$$1^2 + 2^2 + 3^2 + \cdots + n^2 = \frac{n(n + 1)(2n + 1)}{6}$$

for $n \in Z^+$.

2. Prove that

$$1^3 + 2^3 + 3^3 + \cdots + n^3 = \frac{n^2(n + 1)^2}{4}$$

for $n \in Z^+$.

3. Prove that

$$1 + 3 + 5 + \cdots + (2n - 1) = n^2$$

for $n \in Z^+$.

4. Prove that

$$\frac{1}{1 \cdot 2} + \frac{1}{2 \cdot 3} + \frac{1}{3 \cdot 4} + \cdots + \frac{1}{n(n + 1)}$$

$$= \frac{n}{n + 1}$$

for $n \in Z^+$.

5. Prove by induction that if $a, r \in \mathbb{R}$ and $r \neq 1$, then

$$a + ar + ar^2 + \cdots + ar^n$$

$$= a\frac{1 - r^{n+1}}{1 - r}$$

for $n \in Z^+$.

6. Find the flaw in the following argument.

We prove that any two integers i and j in Z^+ are equal. Let

$$\max(i, j) = \begin{cases} i & \text{if } i \geq j, \\ j & \text{if } j > i. \end{cases}$$

Let $P(n)$ be the statement

$P(n)$: Whenever $\max(i, j) = n$, then $i = j$.

Notice that, if $P(n)$ is true for all positive integers n, then any two positive integers i and j are equal. We proceed to prove $P(n)$ for positive integers n by induction.

Clearly $P(1)$ is true, because, if $i, j \in Z^+$ and max $(i, j) = 1$, then $i = j = 1$.

Assume that $P(k)$ is true. Let i and j be such that max $(i, j) = k + 1$. Then

$\max(i - 1, j - 1) = k$, so $i - 1 = j - 1$ by the induction hypothesis. Therefore, $i = j$ and $P(k + 1)$ is true; so, $P(n)$ is true for all n.

7. Criticize the following argument.

Let us show that every positive integer has some interesting property. Let $P(n)$ be the statement that n has an interesting property. We use complete induction.

Of course $P(1)$ is true, because 1 is the only positive integer that equals its own square, which is surely an interesting property of 1.

Suppose that $P(m)$ is true for $1 \leq m \leq k$. If $P(k + 1)$ were not true, then $k + 1$ would be the smallest integer without an interesting property, which would, in itself, be an interesting property of $k + 1$. So $P(k + 1)$ must be true. Thus $P(n)$ is true for all $n \in Z^+$.

8. We have never been able to see any flaw in part a. Try your luck with it, and then answer part b.

a. A serial killer is sentenced to be executed. He asks the judge not to let him know the day of the execution. The judge says, "I sentence you to be executed at 10 A.M. some day of this coming January, but I promise that you will not be aware that you are being executed that day until they come to get you at 8 A.M." The killer goes to his cell and proceeds to prove, as follows, that he can't be executed in January.

Let $P(n)$ be the statement that I can't be executed on January $(31 - n)$. I want to prove $P(n)$ for $0 \leq n \leq 30$. Now I can't be executed on January 31, for since that is the last day of the month and I am to be executed that month, I would know that was the day before 8 A.M., contrary to the judge's sentence. Thus $P(0)$ is true. Suppose that $P(m)$ is true for $0 \leq m \leq k$, where $k \leq 29$. That is, suppose I can't be executed on January $(31 - k)$ through January 31. Then January $(31 - k - 1)$ must be the last possible day for execution, and I would be aware that was the day before 8 A.M., contrary to the judge's sentence. Thus I can't be executed on January $(31 - (k + 1))$, so $P(k + 1)$ is true. Therefore, I can't be executed in January. (Of course, the serial killer was executed on January 17.)

b. An instructor teaches a class five days a week, Monday through Friday. She tells her class that she will give one more quiz on one day during the final full week of classes, but that the students will not know for sure that the quiz will be that day until they come to the classroom. What is the last day of the week on which she can give the quiz in order to satisfy these conditions?

B TWO DEFERRED PROOFS

PROOF OF THEOREM 4.2 ON EXPANSION BY MINORS

Our demonstration of the various properties of determinants in Section 4.2 depended on our ability to compute a determinant by expanding it by minors on *any row or column*, as stated in Theorem 4.2. In order to prove Theorem 4.2, we will need to look more closely at the form of the terms that appear in an expanded determinant.

Determinants of orders 2 and 3 can be written as

$$\begin{vmatrix} a_{11} & a_{12} \\ a_{21} & a_{22} \end{vmatrix} = (1)(a_{11}a_{22}) + (-1)(a_{12}a_{21})$$

and

$$\begin{vmatrix} a_{11} & a_{12} & a_{13} \\ a_{21} & a_{22} & a_{23} \\ a_{31} & a_{32} & a_{33} \end{vmatrix} = (1)(a_{11}a_{22}a_{33}) + (-1)(a_{11}a_{23}a_{32}) + (1)(a_{12}a_{23}a_{31}) + (-1)(a_{12}a_{21}a_{33})$$
$$+ (1)(a_{13}a_{21}a_{32}) + (-1)(a_{13}a_{22}a_{31}).$$

Notice that each determinant appears as a sum of products, each with an associated sign given by (1) or (−1), which is determined by the formula $(-1)^{1+j}$ as we expand the determinant across the first row. Furthermore, each product contains exactly one factor from each row and exactly one factor from each column of the matrix. That is, the row indices in each product run through all row numbers, and the column indices run through all column numbers. This is an illustration of a general theorem, which we now prove by induction.

A-6

THEOREM B.1   Structure of an Expanded Determinant

The determinant of an $n \times n$ matrix $A = [a_{ij}]$ can be expressed as a sum of signed products, where each product contains exactly one factor from each row and exactly one factor from each column. The expansion of $\det(A)$ on any row or column also has this form.

PROOF   We consider the expansion of $\det(A)$ on the first row and give a proof by induction. We have just shown that our result is true for determinants of orders 2 and 3. Let $n > 3$, and assume that our result holds for all square matrices of size smaller than $n \times n$. Let A be an $n \times n$ matrix. When we expand $\det(A)$ by minors across the first row, the only expression involving a_{1j} is $(-1)^{1+j}a_{1j}|A_{1j}|$. We apply our induction hypothesis to the determinant $|A_{1j}|$ of order $n - 1$: it is a sum of signed products, each of which has one factor from each row and column of A except for row 1 and column j. As we multiply this sum term by term by a_{1j}, we obtain a sum of products having a_{1j} as the factor from row 1 and column j, and one factor from each other row and from each other column. Thus an expression of the stated form is indeed obtained as we expand $\det(A)$ by minors across the first row.

It is clear that essentially the same argument shows that expansion across any row or down any column yields the same type of sum of signed products.   ▲

Our illustration for 2×2 and 3×3 matrices indicates that we might always have the same number of products appearing in $\det(A)$ with a sign given by 1 as with a sign given by -1. This is indeed the case for determinants of order greater than 1, and the induction proof is left to the reader.

We now restate and prove Theorem 4.2.

Let $A = [a_{ij}]$ be an $n \times n$ matrix. Then

$$|A| = (-1)^{r+1}a_{r1}|A_{r1}| + (-1)^{r+2}a_{r2}|A_{r2}| + \cdots + (-1)^{r+n}a_{rn}|A_{rn}| \qquad \textbf{(B.1)}$$

for any r from 1 to n, and

$$|A| = (-1)^{1+s}a_{1s}|A_{1s}| + (-1)^{2+s}a_{2s}|A_{2s}| + \cdots + (-1)^{n+s}a_{ns}|A_{ns}| \qquad \textbf{(B.2)}$$

for any s from 1 to n.

PROOF OF THEOREM 4.2   We first prove Eq. (B.1) for any choice of r from 1 to n. Clearly, Eq. (B.1) holds for $n = 1$ and $n = 2$. Proceeding by induction, let $n > 2$ and assume that determinants of order less than n can be computed by using an expansion on *any row*. Let A be an $n \times n$ matrix. We show that expansion of $\det(A)$ by minors on row r is the same as expansion on row i for $i < r$. From Theorem B.1, we know that each of the expansions gives a sum of signed products, where each product contains a single factor from each row and from each column of A. We will compare the products containing as factors both a_{ij} and a_{rs} in each of the expansions. We consider two cases, as illustrated in Figures B.1 and B.2.

If det(A) is expanded on the ith row, the sum of signed products containing $a_{ij}a_{rs}$ is part of $(-1)^{i+j}a_{ij}|A_{ij}|$. In computing $|A_{ij}|$ we may, by our induction assumption, expand on the rth row. For $j < s$, terms of $|A_{ij}|$ involving a_{rs} are then $(-1)^{(r-1)+(s-1)}d$, where d is the determinant of the matrix obtained from A by crossing out rows i and r and columns j and s, as shown in Figure B.1. The exponent $(r - 1) + (s - 1)$ occurs because a_{rs} is in row $r - 1$ and column $s - 1$ of A_{ij}. Thus, the part of our expansion of det(A) across the ith row that contains $a_{ij}a_{rs}$ is equal to

$$(-1)^{i+j}(-1)^{(r-1)+(s-1)}a_{ij}a_{rs}d \quad \text{for} \quad j < s. \tag{B.3}$$

For $j > s$, we consult Figure B.2 and use similar reasoning to see that the part of our expansion of det(A) across the ith row, which contains $a_{ij}a_{rs}$, is equal to

$$(-1)^{i+j}(-1)^{(r-1)+s}a_{ij}a_{rs}d \quad \text{for} \quad j > s. \tag{B.4}$$

We now expand det(A) by minors on the rth row, obtaining $(-1)^{r+s}a_{rs}|A_{rs}|$ as the portion involving a_{rs}. Expanding $|A_{rs}|$ on the ith row, using our induction assumption, we obtain $(-1)^{i+j}a_{ij}d$ if $j < s$ and $(-1)^{i+(j-1)}a_{ij}d$ if $j > s$. Thus the part of the expansion of det(A) on the rth row, which contains $a_{ij}a_{rs}$, is equal to

$$(-1)^{r+s}(-1)^{i+j}a_{rs}a_{ij}d \quad \text{for} \quad j < s \tag{B.5}$$

or

$$(-1)^{r+s}(-1)^{i+(j-1)}a_{rs}a_{ij}d \quad \text{for} \quad j > s. \tag{B.6}$$

Expressions (B.3) and (B.5) are equal, because $(-1)^{r+s+i+j} = (-1)^{r+s+i+j-2}$, and expressions (B.4) and (B.6) are equal, because $(-1)^{r+s+i+j-1}$ is the algebraic sign of each. This concludes the proof that the expansions of det(A) by minors across rows i and r are equal.

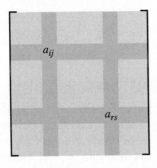

FIGURE B.1
The case $j < s$.

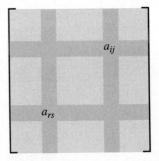

FIGURE B.2
The case $j > s$.

A similar argument shows that expansions of det(A) down columns j and s are the same.

Finally, we must show that an expansion of det(A) on a row is equal to expansion on a column. It is sufficient for us to prove that the expansion of det(A) on the first row is the same as the expansion on the first column, in view of what we have proved above. Again, we use induction and dispose of the cases $n = 1$ and $n = 2$ as trivial to check. Let $n > 2$, and assume that our result holds for matrices of size smaller than $n \times n$. Let A be an $n \times n$ matrix. Expanding det(A) on the first row yields

$$a_{11}|A_{11}| + \sum_{j=2}^{n} (-1)^{1+j}a_{1j}|A_{1j}|.$$

For $j > 1$, we expand $|A_{1j}|$ on the first column, using our induction assumption, and obtain $|A_{1j}| = \sum_{i=2}^{n} (-1)^{(i-1)+1}a_{i1}d$, where d is the determinant of the matrix obtained from A by crossing out rows 1 and i and columns 1 and j. Thus the terms in the expansion of det(A) containing $a_{1j}a_{i1}$ are

$$(-1)^{1+j+i}a_{1j}a_{i1}d. \tag{B.7}$$

On the other hand, if we expand det(A) on the first column, we obtain

$$a_{11}|A_{11}| + \sum_{i=2}^{n} (-1)^{i+1}a_{i1}|A_{i1}|.$$

For $i > 1$, expanding on the first row, using our induction assumption, shows that $|A_{i1}| = \sum_{j=2}^{n} (-1)^{1+(j-1)}a_{1j}d$. This results in

$$(-1)^{i+1+j}a_{i1}a_{1j}d$$

as the part of the expansion of det(A) containing the sum $a_{i1}a_{1j}$, and this agrees with the expression in formula (B.7). This concludes our proof.   ▲

PROOF OF THEOREM 4.7 ON THE VOLUME OF AN n-BOX IN $\mathbb{R}^m$

Theorem 4.7 in Section 4.4 asserts that the volume of an n-box in $\mathbb{R}^m$ determined by independent vectors $\mathbf{a}_1, \mathbf{a}_2, \ldots, \mathbf{a}_n$ can be computed as

$$\text{Volume} = \sqrt{\det(A^T A)}, \tag{B.8}$$

where A is the $m \times n$ matrix having $\mathbf{a}_1, \mathbf{a}_2, \ldots, \mathbf{a}_n$ as column vectors. Before proving this result, we must first give a proper definition of the *volume* of such a box. The definition proceeds inductively. That is, we define the volume of a 1-box directly, and then we define the volume of an n-box in terms of the volume of an $(n - 1)$-box. Our definition of volume is a natural one, essentially taking the product of the measure of the base of the box and the altitude of the box. We will think of the base of the n-box determined by $\mathbf{a}_1, \mathbf{a}_2, \ldots, \mathbf{a}_n$ as an

$(n - 1)$-box determined by some $n - 1$ of the vectors $\mathbf{a}_i$. In general, such a base can be selected in several different ways. We find it convenient to work with boxes determined by *ordered* sequences of vectors. We will choose one special base for the box and give a definition of its volume in terms of this order of the vectors. Once we obtain the expression $\det(A^T A)$ in Eq. (B.8) for the square of the volume, we can show that the volume does not change if the order of the vectors in the sequence is changed. We will show that $\det(A^T A)$ remains unchanged if the order of the columns of A is changed.

Observe that, if an n-box is determined by $\mathbf{a}_1, \mathbf{a}_2, \ldots, \mathbf{a}_n$, then $\mathbf{a}_1$ can be uniquely expressed in the form

$$\mathbf{a}_1 = \mathbf{b} + \mathbf{p}, \tag{B.9}$$

where $\mathbf{p}$ is the projection of $\mathbf{a}_1$ on $W = \mathrm{sp}(\mathbf{a}_2, \ldots, \mathbf{a}_n)$ and $\mathbf{b} = \mathbf{a}_1 - \mathbf{p}$ is orthogonal to W. This follows from Theorem 6.1 and is illustrated in Figure B.3.

DEFINITION B.1   Volume of an n-Box

> The **volume** of the 1-box determined by a nonzero vector $\mathbf{a}_1$ in $\mathbb{R}^m$ is $\|\mathbf{a}_1\|$. Let $\mathbf{a}_1, \mathbf{a}_2, \ldots, \mathbf{a}_n$ be an ordered sequence of n independent vectors, and suppose that the volume of an r-box determined by an ordered sequence of r independent vectors has been defined for $r < n$. The **volume of the n-box** in $\mathbb{R}^m$ determined by the ordered sequence of $\mathbf{a}_i$ is the product of the volume of the "base" determined by the ordered sequence $\mathbf{a}_2, \ldots, \mathbf{a}_n$ and the length of the vector $\mathbf{b}$ given in Eq. (B.9). That is,
>
> Volume = (Altitude $\|\mathbf{b}\|$)(Volume of the base).

As a first step in finding a formula for the volume of an n-box, we establish a preliminary result on determinants.

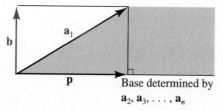

FIGURE B.3
The altitude vector b perpendicular to the base box.

THEOREM B.2 Property of $\det(A^T A)$

Let $\mathbf{a}_1, \mathbf{a}_2, \ldots, \mathbf{a}_n$ be vectors in $\mathbb{R}^m$, and let A be the $m \times n$ matrix with jth column vector $\mathbf{a}_j$. Let B be the $m \times n$ matrix obtained from A by replacing the first column of A by the vector

$$\mathbf{b} = \mathbf{a}_1 - r_2\mathbf{a}_2 - \cdots - r_n\mathbf{a}_n$$

for scalars $r_2, \ldots, r_n$. Then

$$\det(A^T A) = \det(B^T B). \tag{B.10}$$

PROOF The matrix B can be obtained from the matrix A by a sequence of $n - 1$ elementary *column-addition* operations. Each of the elementary column operations can be performed on A by multiplying A on the *right* by an elementary matrix formed by executing the same elementary column-addition operation on the $n \times n$ identity matrix I. Each elementary column-addition matrix therefore has the same determinant 1 as the identity matrix I. Their product is an $n \times n$ matrix E such that $B = AE$, and $\det(E) = 1$. Using properties of determinants and the transpose operation, we have

$$\det(B^T B) = \det((AE)^T(AE)) = \det(E^T(A^T A)E)$$
$$= 1 \cdot \det(A^T A) \cdot 1 = \det(A^T A). \quad \blacktriangle$$

We can now prove our volume formula in Theorem 4.7.

The volume of the n-box in $\mathbb{R}^m$ determined by the ordered sequence $\mathbf{a}_1, \mathbf{a}_2, \ldots, \mathbf{a}_n$ of n independent vectors is given by

$$\text{Volume} = \sqrt{\det(A^T A)},$$

where A is the $m \times n$ matrix with $\mathbf{a}_j$ as jth column vector.

PROOF OF THEOREM 4.7 Because our volume was defined inductively, we give an inductive proof. The theorem is valid if $n = 1$ or 2, by Eqs. (1) and (2), respectively, in Section 4.4. Let $n > 2$, and suppose that the theorem is proved for all k-boxes for $k \le n - 1$. If we write $\mathbf{a}_1 = \mathbf{b} + \mathbf{p}$, as in Eq. (B.9), then, because $\mathbf{p}$ lies in $\text{sp}(\mathbf{a}_2, \ldots, \mathbf{a}_n)$, we have

$$\mathbf{p} = r_2\mathbf{a}_2 + \cdots + r_n\mathbf{a}_n$$

for some scalars $r_2, \ldots, r_n$, so

$$\mathbf{b} = \mathbf{a}_1 - \mathbf{p} = \mathbf{a}_1 - r_2\mathbf{a}_2 - \cdots - r_n\mathbf{a}_n.$$

Let B be the matrix obtained from A by replacing the first column vector $\mathbf{a}_1$ of A by the vector $\mathbf{b}$, as in Theorem B.2. Because $\mathbf{b}$ is orthogonal to each of the vectors $\mathbf{a}_2, \ldots, \mathbf{a}_n$, which determine the base of our box, we obtain

$$B^T B = \begin{bmatrix} \mathbf{b} \cdot \mathbf{b} & 0 & \cdots & 0 \\ 0 & \mathbf{a}_2 \cdot \mathbf{a}_2 & \cdots & \mathbf{a}_2 \cdot \mathbf{a}_n \\ \vdots & \vdots & & \vdots \\ 0 & \mathbf{a}_n \cdot \mathbf{a}_2 & \cdots & \mathbf{a}_n \cdot \mathbf{a}_n \end{bmatrix}. \qquad \text{(B.11)}$$

From Eq. (B.11), we see that

$$\det(B^T B) = \|\mathbf{b}\|^2 \begin{vmatrix} \mathbf{a}_2 \cdot \mathbf{a}_2 & \cdots & \mathbf{a}_2 \cdot \mathbf{a}_n \\ \vdots & & \vdots \\ \mathbf{a}_n \cdot \mathbf{a}_2 & \cdots & \mathbf{a}_n \cdot \mathbf{a}_n \end{vmatrix}.$$

By our induction assumption, the square of the volume of the $(n-1)$-box in $\mathbb{R}^m$ determined by the ordered sequence $\mathbf{a}_2, \ldots, \mathbf{a}_n$ is

$$\begin{vmatrix} \mathbf{a}_2 \cdot \mathbf{a}_2 & \cdots & \mathbf{a}_2 \cdot \mathbf{a}_n \\ \vdots & & \vdots \\ \mathbf{a}_n \cdot \mathbf{a}_2 & \cdots & \mathbf{a}_n \cdot \mathbf{a}_n \end{vmatrix}.$$

Applying Eq. (B.10) in Theorem B.2, we obtain

$$\det(A^T A) = \det(B^T B) = \|\mathbf{b}\|^2 (\text{Volume of the base})^2 = (\text{Volume})^2.$$

This proves our theorem. ▲

COROLLARY   Independence of Order

The volume of a box determined by the independent vectors $\mathbf{a}_1$, $\mathbf{a}_2, \ldots, \mathbf{a}_n$ and defined in Definition B.1 is independent of the order of the vectors; in particular, the volume is independent of a choice of base for the box.

PROOF   A rearrangement of the sequence $\mathbf{a}_1, \mathbf{a}_2, \ldots, \mathbf{a}_n$ of vectors corresponds to the same rearrangement of the columns of matrix A. Such a rearrangement of the columns of A can be accomplished by multiplying A on the right by a product of $n \times n$ elementary *column-interchange* matrices, having determinant ± 1. As in the proof of Theorem B.2, we see that, for the resulting matrix $B = AE$, we have $\det(B^T B) = \det(A^T A)$ because $\det(E^T)\det(E) = 1$. ▲

C LINTEK ROUTINES

Below are listed the names and brief descriptions of the routines that make up the computer software LINTEK designed for this text.

VECTGRPH   Gives graded quizzes on vector geometry based on displayed graphics. Useful throughout the course.

MATCOMP   Performs matrix computations, solves linear systems, and finds real eigenvalues and eigenvectors. For use throughout the course.

YUREDUCE   Enables the user to select items from a menu for step-by-step row reduction of a matrix. For Sections 1.4–1.6, 9.2, and 10.3.

EBYMTIME   Gives a lower bound for the time required to find the determinant of a matrix using only repeated expansion by minors. For Section 4.2.

ALLROOTS   Provides step-by-step execution of Newton's method to find both real and complex roots of a polynomial with real or complex coefficients; may be used in conjunction with MAT-COMP to find complex as well as real eigenvalues of a small matrix. For Section 5.1.

QRFACTOR   Executes the Gram–Schmidt process, gives the QR-factorization of a suitable matrix A, and can be used to find least-squares solutions. For Section 6.2. Also, for Section 8.4, step-by-step computation by the QR-algorithm of both real and complex eigenvalues of a matrix. The user specifies shifts and when to decouple.

YOUFIT   The user can experiment with a graphic to try to find the least-squares linear, quadratic, or exponential fit of two-dimensional data. The computer can then be asked to find the best fit. For Section 6.5.

POWER   Executes, one step at a time, the power method (with deflation) for finding eigenvalues and eigenvectors. For Section 8.4.

JACOBI	Executes, one "rotation" at a time, the method of Jacobi for diagonalizing a symmetric matrix. For Section 8.4.
TIMING	Times algebraic operations and flops; also times various methods for solving linear systems. For Sections 10.1 and 10.2.
LUFACTOR	Gives the factorizations $A = LU$ or $PA = LU$, which can then be used to solve $A\mathbf{x} = \mathbf{b}$. For Section 10.2.
HILBERT	Enables the user to experiment with ill-conditioned Hilbert matrices. For Section 10.3.
LINPROG	A graphics program allowing the user to estimate solutions of two-variable linear programming problems. For use with the previous edition of our text.
SIMPLEX	Executes the simplex method with optional display of tableau to solve linear programming problems. For use with Chapter 10 of the previous edition of our text.

D

MATLAB PROCEDURES AND COMMANDS USED IN THE EXERCISES

The MATLAB exercises in our text illustrate and facilitate computations in linear algebra. Also, they have been carefully designed, starting with those in Section 1.1, to introduce students gradually to MATLAB, explaining procedures and commands as the occasion for them arises. It is not necessary to study this appendix before starting right in with the MATLAB exercises in Section 1.1. In case a student has forgotten a procedure or command, we summarize here for easy reference the ones needed for the exercises. The information given here is only a small fraction of that available from the MATLAB manual, which is of course the best reference.

GENERAL INFORMATION

MATLAB prints ≫ on the screen as its *prompt* when it is ready for your next command.

MATLAB is *case sensitive*—that is, if you have set $n = 3$, then N is undefined. Thus you can set X equal to a 3×2 matrix and x equal to a row vector.

To view on the screen the value, vector, or matrix currently assigned to a variable such as A or x, type the variable and press the Enter key.

For information on a command, enter **help** followed by a space and the name of the command or function. For example, entering **help *** will bring the response that $X * Y$ is the matrix product of X and Y, and entering **help eye** will inform us that eye(n) is the $n \times n$ identity matrix.

The *most recent* commands you have given MATLAB are kept in a *stack*, and you can move back and forth through them using the up-arrow and down-arrow keys. Thus if you make a mistake in typing a command and get an error message, you can correct the error by striking the up-arrow key to put the command by the cursor and then edit it to correct the error, avoiding retyping the entire command. The exercises frequently ask you to execute a command you recently gave, and you can simply use the up-arrow key until the command is at the cursor, and then press the Enter key to execute it again. This saves a lot of typing. If you know you will be using a command again, don't let it get too far back before asking to execute it again, or it may have been removed from the command stack. Executing it puts it at the top of the stack.

DATA ENTRY

Numerical, vector, and matrix data can be entered by the variable name for the data followed by an equal sign and then the data. For example, entering **x = 3** assigns the value 3 to the letter x, and then entering **y = sin(x)** will then assign to y the value sin(3). Entering **y = sin(x)** before a value has been assigned to x will produce an error message. Values must have been assigned to a variable before the variable can be used in a function or computation. If you do not assign a name to the result of a computation, then it is assigned the temporary name "ans." Thus this reserved name *ans* should not be used in your work because its value will be changed at the next computation having no assigned name.

The constant $\pi = 4 \tan^{-1}(1)$ can be entered as **pi**.

Data for a vector or a matrix should be entered between square brackets with a space between numbers and with rows separated by a semicolon. For example, the matrix

$$A = \begin{bmatrix} 2 & -1 & 3 & 6 \\ 0 & 5 & 12 & -7 \\ 4 & -2 & 9 & 11 \end{bmatrix}$$

can be entered as

$$A = [2\ -1\ 3\ 6;\ 0\ 5\ 12\ -7;\ 4\ -2\ 9\ 11],$$

which will produce the response

$A =$

$$\begin{matrix} 2 & -1 & 3 & 6 \\ 0 & 5 & 12 & -7 \\ 4 & -2 & 9 & 11 \end{matrix}$$

from MATLAB. If you do not wish your data entry to be *echoed* in this fashion for proofreading, put a semicolon after your data entry line. The semicolon can also be used to separate several data items or commands all entered on one line. For example, entering

$$x = 3;\ \sin(x);\ v = [1\ -3\ 4];\ C = [2\ -1;\ 3\ 5];$$

will result in the variable assignments

$$x = 3 \text{ and } ans = \sin(3) \text{ and } v = [1, -3, 4] \text{ and } C = \begin{bmatrix} 2 & -1 \\ 3 & 5 \end{bmatrix}$$

and no data will be echoed. If the final semicolon were omitted, the data for the matrix C would be echoed. If you run out of space on a line and need to continue data on the next line, type at least two periods .. and then immediately press Enter and continue entry on the next line.

Matrices can be glued together to form larger matrices using the same form of matrix entry, provided the gluing will still produce a rectangular array. For example, if A is a 3×4 matrix, B is 3×5, and C is 2×4, then entering **D = [A B]** will produce a 3×9 matrix D consisting of the four columns of A followed at the right by the five columns of B, whereas entering **E = [A ; C]** will produce a 5×4 matrix consisting of the first three rows of A followed below by the two rows of C.

NUMERIC OPERATIONS

Use + for addition, − for subtraction, * for multiplication, / for division, and ^ for exponentiation. Thus entering **6^3** produces the result ans = 216.

MATRIX OPERATIONS AND NOTATIONS

Addition, subtraction, multiplication, division, and exponentiation are denoted just as for numeric data, always assuming that they are defined for the matrices specified. For example, if A and B are square matrices of the same size, then we can form $A + B$, $A - B$, $A * B$, $2 * A$, and A^4. If A and B are invertible, then MATLAB interprets A/B as AB^{-1} and interprets $A \backslash B$ as $A^{-1}B$.

A period before an operation symbol, such as .* or .^, indicates that the operation is to be performed at each corresponding position i,j in a matrix rather than on the matrix as a whole. Thus entering **C = A .* B** produces the matrix C such that $c_{ij} = a_{ij}b_{ij}$, and entering **D = A.^2** produces the matrix D where $d_{ij} = a_{ij}^2$.

The transpose of A is denoted by A', so that entering **A = A'** will replace A by its transpose. Entering **b = [1; 2; 3]** or entering **b = [1 2 3]'** creates the same column vector **b**.

Let A be a 3×5 matrix in MATLAB. Entries are numbered, starting from the upper left corner and proceeding down each column in turn. Thus $A(6)$ is the last entry in the second column of A. This entry is also identified as $A(3,2)$; we think of the 3 and 2 as subscripts as in our text notation a_{32}. Use of a colon between two integers has the meaning of "through." Thus, entering **v = A(2,1:4)** will produce a row vector **v** consisting of entries 1 through 4 (the first four components) of the second row vector of A, whereas entering **B = A(1:2,2:4)** will produce the 2×3 submatrix of A consisting of columns 2, 3, and 4 of rows 1 and 2. The colon by itself in place of a subscript has the meaning of "all," so that entering **w = A(:,3)** sets **w** equal to the third column vector of A. For another example, if **x** is a row vector with five components, then entering **A(2,:) = x** will replace the second row vector of A by **x**.

The colon can be used in the "through" sense to enter vectors with components incremented by a specified amount. Entering **y = 4 : −.5 : 2** sets **y** equal to the vector [4, 3.5, 3, 2.5, 2] with increment −0.5, whereas entering **x = 1 : 6** will set **x** equal to the vector [1, 2, 3, 4, 5, 6] with default increment 1 between components.

MATLAB COMMANDS REFERRED TO IN THIS TEXT

Remember that all variables must have already been assigned numeric, vector, or matrix values. Following is a summary of the commands used in the MATLAB exercises of our text, together with a brief description of what each one does.

acos(x)	returns the arccosine of every entry in (the number, vector, or matrix) x.
bar(v)	draws a bar graph wherein the height of the ith bar is the ith component of the vector **v**.
clear	erases all variables; all data are lost.
clear A w x	erases the variables A, w, and x only.
clock	returns the vector [year month day hour minute second]
det(A)	returns the determinant of the square matrix A.
eig(A)	returns a column vector whose components are the eigenvalues of A. Entering **[V, D] = eig(A)** creates a matrix V whose column vectors are eigenvectors of A and a diagonal matrix D whose jth entry on the diagonal is the eigenvalue associated with the jth column of V, so that $AV = VD$.
etime(t1,t0)	returns the elapsed time between two clock output vectors **t0** and **t1**. If we set **t0 = clock** before executing a computation, then the command **etime(clock,t0)** immediately after the computation gives the time required for the computation.
exit	leaves MATLAB and returns us to DOS.
eye(n)	returns the $n \times n$ identity matrix.
for i = 1:n, . . . ; end	causes the routine placed in the . . . portion to be executed n times with the value of i incremented by 1 each time. The value of n must not exceed 1024 in the student version of MATLAB.
format long	causes data to be printed in scientific format with about 16 significant digits until we specify otherwise. (See **format short**.)
format short	causes data to be printed with about five significant digits until we specify otherwise. (See **format long**.) The default format when MATLAB is accessed is the short one.
help	produces a list of topics for which we can obtain an on-screen explanation. To obtain help on a specific command or function—say, the det function—enter **help det**.

hilb(n)	returns the $n \times n$ Hilbert matrix.
invhilb(n)	returns the result of an attempted computation of the inverse of the $n \times n$ Hilbert matrix.
inv(A)	returns the inverse of the matrix A.
log(x)	returns the natural logarithm of every entry in (the number, vector, or matrix) x.
lu	The command **[L, U, P] = lu(A)** returns a lower-triangular matrix L, an upper-triangular matrix U, and a permutation matrix P such that $PA = LU$.
max(x)	returns the maximum of the components of the vector **x**.
max(X)	returns a row vector whose jth component is the maximum of the components of the jth column vector of X.
mean(x)	returns the arithmetic mean of the components of the vector **x**. That is, if **x** has n components, then mean(**x**) $= (x_1 + x_2 + \cdots + x_n)/n$.
mean(A)	returns the row vector whose jth component is the arithmetic mean of the jth column of the matrix A. (See **mean(x)**.) If A is an $m \times n$ matrix, then mean(mean(A)) $=$ (the sum of all mn entries)/(mn).
mesh(A)	draws a three-dimensional perspective mesh surface whose height above the point (i,j) in a horizontal reference plane is a_{ij}. Let **x** be a vector of m components and **y** a vector of n components. The command **[X,Y] = meshdom(x, y)** creates matrices X and Y where matrix X is an array of x-coordinates of all the mn points (x_i, y_j) over which we want to plot heights for our mesh surface, and the matrix Y is a similar array of y-coordinates. See the explanation preceding Exercise M10 in Section 1.3 for a description of the use of **mesh** and **meshdom** to draw a surface graph of a function $z = f(x, y)$.
min(x)	returns the minimum of the components of the vector **x**.
min(X)	returns a row vector whose jth component is the minimum of the components of the jth column vector of X.
norm(x)	returns the usual norm $\|\mathbf{x}\|$ of the vector **x**.
ones(m,n)	returns the $m \times n$ matrix with all entries equal to 1.
ones(n)	returns the $n \times n$ matrix with all entries equal to 1.
plot(x)	draws the polygonal approximation graph of a function f such that $f(i) = x_i$ for a vector **x**.

plot(x, y)	draws the polygonal approximation graph of a function f such that $f(x_i) = y_i$ for vectors **x** and **y**.
poly(A)	returns a row vector whose components are the coefficients of the characteristic polynomial $\det(\lambda I - A)$ of the square matrix A. Entering **p = poly(A)** assigns this vector to the variable **p**. (See **roots(p)**.)
qr	The command **[Q, R] = qr(A)** returns an orthogonal matrix Q and an upper-triangular matrix R such that $A = QR$. If matrix A is an $m \times n$ matrix, then R is also $m \times n$ whereas Q is $m \times m$.
quit	leaves MATLAB and returns us to DOS.
rand(m,n)	returns an $m \times n$ matrix with random number entries between 0 and 1.
rand(n)	returns an $n \times n$ matrix with random number entries between 0 and 1.
rat(x, 's')	causes the vector (or matrix) **x** to be printed on the screen with entries replaced by rational (fractional) approximations. The entries of **x** are not changed. Hopefully, entries that should be equal to fractions with small denominators will be replaced by those fractions.
roots(p)	returns a column vector of roots of the polynomial whose coefficients are in the row vector **p**. (See **poly(A)**.)
rot90(A)	returns the matrix obtained by "rotating the matrix A" counterclockwise 90°. Entering **B = rot90(A)** produces this matrix B.
rref(A)	returns the reduced row-echelon form of the matrix A.
rrefmovie(A)	produces a "movie" of reduction of A to reduced row-echelon form. It may go by so fast that it is hard to study, even using the *Pause* key. See the comment following Exercise M9 in Section 1.4 about fixing this problem.
size(A)	returns the two-component row vector $[m, n]$, where A is an $m \times n$ matrix.
stairs(x)	draws a stairstep graph of the vector **x**, where the height of the ith step is x_i.
sum(x)	returns the sum of all the entries in the vector **x**.
triu(A)	returns the upper-triangular matrix obtained by replacing all entries a_{ij} for $j > i$ in A by zero.
who	creates a list on the screen of all variables that are currently assigned numeric, vector, or matrix values.
zeros(m,n)	returns the $m \times n$ matrix with all entries equal to 0.
zeros(n)	returns the $n \times n$ matrix with all entries equal to 0.

ANSWERS TO MOST ODD-NUMBERED EXERCISES

CHAPTER 1

Section 1.1

1. $\mathbf{v} + \mathbf{w} = [-1, -3]$
$\mathbf{v} - \mathbf{w} = [5, 1]$

3. $\mathbf{v} + \mathbf{w} = 2\mathbf{i} + 5\mathbf{j} + 6\mathbf{k}$
$\mathbf{v} - \mathbf{w} = \mathbf{j} - 2\mathbf{k}$

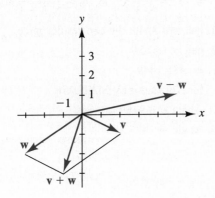

5. $[-11, 9, -4]$ **7.** $[6, 4, -5]$
9. $[17, -18, 3, -20]$ **11.** $[20, -12, 6, -20]$
13.

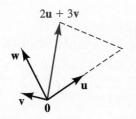

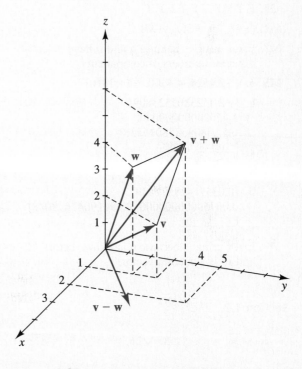

15.

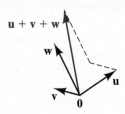

17. $r \approx 1.5$, $s \approx 1.8$ **19.** $r \approx .5$, $s \approx -1.2$

21. -1 **23.** $-\frac{2}{5}$ **25.** All $c \neq 15$

27. 0   **29.** -27 **31.** $5\mathbf{i} - \mathbf{j}$

33. $[1, -3, -4, 13]$

35. $x\begin{bmatrix} 3 \\ 1 \\ 2 \end{bmatrix} + y\begin{bmatrix} -2 \\ -1 \\ 1 \end{bmatrix} + z\begin{bmatrix} 4 \\ -3 \\ -5 \end{bmatrix} = \begin{bmatrix} 10 \\ 0 \\ -3 \end{bmatrix}$

37. a. $-3p \quad\ -4r + 6s = \ \ 8$
    $4p - 2q + 3r \qquad\ = -3$
    $6p + 5q - 2r + 7s = \ \ 1$

  b. $p\begin{bmatrix} -3 \\ 4 \\ 6 \end{bmatrix} + q\begin{bmatrix} 0 \\ -2 \\ 5 \end{bmatrix} + r\begin{bmatrix} -4 \\ 3 \\ -2 \end{bmatrix} + s\begin{bmatrix} 6 \\ 0 \\ 7 \end{bmatrix} = \begin{bmatrix} 8 \\ -3 \\ 1 \end{bmatrix}$

39. F T F F T F T F F T

M1. $[-58, 79, -36, -190]$

M3. Error using + because a and u have different numbers of components.

M5. a. $[-2.4524, 4.4500, -11.3810]$

  b. $[-2.45238095238095,$
    $4.45000000000000,$
    $-11.38095238095238]$

  c. $\left[-\frac{103}{42}, \frac{89}{20}, -\frac{293}{21}\right]$

M7. a. $[0.0357, 0.0075, 0.1962]$

  b. $[0.03571428571429,$
    $0.00750000000000, 0.19619047619048]$

  c. $\left[\frac{1}{28}, \frac{3}{400}, \frac{103}{525}\right]$

M9. Error using + because $\mathbf{u}$ is a row vector and $\mathbf{u}^T$ is a column vector.

Section 1.2

1. $\sqrt{26}$ **3.** $\sqrt{26}$ **5.** $\sqrt{478}$

7. $\frac{1}{\sqrt{26}}[-1, 3, 4]$ **9.** -3 **11.** 3

13. $\cos^{-1}\frac{11}{\sqrt{364}} \approx 54.8°$ **15.** $-\frac{43}{3}$

17. $[13, -5, 7]$ or any nonzero multiple of it

19. $15\sqrt{6}$ **21.** -540

25. Perpendicular

27. Parallel with opposite direction

29. Neither

31. The vector $\mathbf{v} - \mathbf{w}$, when translated to start at $(w_1, w_2, \ldots, w_n)$, reaches to the tip of the vector $\mathbf{v}$, so its length is the distance from $(w_1, w_2, \ldots, w_n)$ to $(v_1, v_2, \ldots, v_n)$.

33. $\sqrt{33}$ **35.** 10   **37.** $\frac{100}{\sqrt{3}}$ lb

39. b. $\mathbf{F}_2 = -T_2(\sin \theta_2)\mathbf{i} + T_2(\cos \theta_2)\mathbf{j}$

  c. $T_1(\sin \theta_1) - T_2(\sin \theta_2) = 0$, $T_1(\cos \theta_1) + T_2(\cos \theta_2) = 100$

  d. $T_1 = \frac{100\sqrt{2}}{\sqrt{3} + 1}$ lb, $T_2 = \frac{200}{\sqrt{3} + 1}$ lb

M1. $\|\mathbf{a}\| \approx 6.3246$; the two results agree.

M3. $\|\mathbf{u}\| \approx 1.8251$

M5. a. 485.1449

  b. Not found by MATLAB

M7. Angle ≈ 0.9499 radians

M9. Angle $\approx 147.4283°$

Section 1.3

1. $\begin{bmatrix} -6 & 3 & 9 \\ 12 & 0 & -3 \end{bmatrix}$ **3.** $\begin{bmatrix} 2 & 2 & 1 \\ 9 & -1 & 2 \end{bmatrix}$

5. $\begin{bmatrix} 6 & -3 \\ -3 & 1 \\ -2 & 5 \end{bmatrix}$ **7.** Impossible

9. $\begin{bmatrix} -130 & 140 \\ 110 & -60 \end{bmatrix}$ **11.** Impossible

13. $\begin{bmatrix} 27 & -35 \\ -4 & 26 \end{bmatrix}$ **15.** $\begin{bmatrix} 20 & -2 & -10 \\ -2 & 1 & 3 \\ -10 & 3 & 10 \end{bmatrix}$

17. a. $\begin{bmatrix} 4 & 0 & 0 \\ 0 & 1 & 0 \\ 0 & 0 & 1 \end{bmatrix}$ **b.** $\begin{bmatrix} 128 & 0 & 0 \\ 0 & -1 & 0 \\ 0 & 0 & 1 \end{bmatrix}$

19. $xy = [-14]$, $yx = \begin{bmatrix} -8 & 12 & -4 \\ 2 & -3 & 1 \\ -6 & 9 & -3 \end{bmatrix}$

21. T F F T F T T T F T

23. $(A\mathbf{b})^T(A\mathbf{c}) = \mathbf{b}^T A^T A \mathbf{c}$

27. The (i, j)th entry in $(r + s)A$ is $(r + s)a_{ij} = ra_{ij} + sa_{ij}$, which is the (i, j)th entry in $rA + sA$. Because $(r + s)A$ and $rA + sA$ have the same size, they are equal.

33. Let $A = [a_{ij}]$ be an $m \times n$ matrix, $B = [b_{jk}]$ be an $n \times r$ matrix, and $C = [c_{kq}]$ be an $r \times s$ matrix. Then the ith row of AB is

$$\left[\sum_{j=1}^{n} a_{ij}b_{j1}, \ \sum_{j=1}^{n} a_{ij}b_{j2}, \ \ldots, \ \sum_{j=1}^{n} a_{ij}b_{jr} \right],$$

so the (i, q)th entry in $(AB)C$ is

$$\left(\sum_{j=1}^{n} a_{ij}b_{j1} \right)c_{1q} + \left(\sum_{j=1}^{n} a_{ij}b_{j2} \right)c_{2q} + \cdots +$$

$$\left(\sum_{j=1}^{n} a_{ij}b_{jr} \right)c_{rq} = \sum_{k=1}^{r} \left(\sum_{j=1}^{n} (a_{ij}b_{jk}c_{kq}) \right).$$

Further, the qth column of BC has components

$$\sum_{k=1}^{r} b_{1k}c_{kq}, \ \sum_{k=1}^{r} b_{2k}c_{kq}, \ \cdots, \ \sum_{k=1}^{r} b_{nk}c_{kq},$$

so the (i, q)th entry in $A(BC)$ is

$$a_{i1}\left(\sum_{k=1}^{r} b_{1k}c_{kq} \right) + a_{i2}\left(\sum_{k=1}^{r} b_{2k}c_{kq} \right) + \cdots +$$

$$a_{in}\left(\sum_{k=1}^{r} b_{nk}c_{kq} \right) = \sum_{j=1}^{n} a_{ij}\left(\sum_{k=1}^{r} b_{jk}c_{kq} \right) =$$

$$\sum_{j=1}^{n} \left(\sum_{k=1}^{r} (a_{ij}b_{jk}c_{kq}) \right) = \sum_{k=1}^{r} \left(\sum_{j=1}^{n} (a_{ij}b_{jk}c_{kq}) \right).$$

Because $(AB)C$ and $A(BC)$ are both $m \times s$ matrices with the same (i, q)th entry, they must be equal.

35. a. $n \times m$ **b.** $n \times n$ **c.** $m \times m$

39. Because $(AA^T)^T = (A^T)^T A^T = AA^T$, we see that AA^T is symmetric.

41. a. The jth entry in column vector $A\mathbf{e}_j$ is $[a_{i1}, a_{i2}, \cdots, a_{in}] \cdot \mathbf{e}_j = a_{ij}$. Therefore, $A\mathbf{e}_j$ is the jth column vector of A.

b. (i) We have $A\mathbf{e}_j = \mathbf{0}$ for each $j = 1, 2, \ldots, n$; so, by part a, the jth column of A is the zero vector for each j. That is, $A = O$. **(ii)** $A\mathbf{x} = B\mathbf{x}$ for each $\mathbf{x}$

if and only if $(A - B)\mathbf{x} = \mathbf{0}$ for each $\mathbf{x}$,

if and only if $A - B = O$ by part (i),

if and only if $A = B$.

45. These matrices do not commute for any value of r.

47. 2989   **49.** 348650

51. 32   **53.** -41558

M1. a. 32   **b.** 14629   **c.** -41558

M3. $\begin{bmatrix} 141 & -30 & -107 \\ -30 & 50 & 18 \\ -107 & 18 & 189 \end{bmatrix}$ **M5.** -117

M7. The expected mean is approximately 0.5. Your experimental values will differ from ours, and so we don't give ours.

Section 1.4

1. a. $\begin{bmatrix} 1 & 3 & 2 \\ 0 & 5 & 0 \\ 0 & 0 & 0 \end{bmatrix}$;   **b.** $\begin{bmatrix} 1 & 0 & 2 \\ 0 & 1 & 0 \\ 0 & 0 & 0 \end{bmatrix}$

3. a. $\begin{bmatrix} -1 & 1 & 2 & 0 \\ 0 & 2 & -1 & 3 \\ 0 & 0 & 10 & 0 \\ 0 & 0 & 0 & 0 \end{bmatrix}$; **b.** $\begin{bmatrix} 1 & 0 & 0 & 1.5 \\ 0 & 1 & 0 & 1.5 \\ 0 & 0 & 1 & 0 \\ 0 & 0 & 0 & 0 \end{bmatrix}$

5. a. $\begin{bmatrix} 1 & -3 & 0 & 0 & -1 \\ 0 & 0 & 1 & 3 & -4 \\ 0 & 0 & 0 & 1 & 3 \\ 0 & 0 & 0 & 0 & 16 \end{bmatrix}$;

b. $\begin{bmatrix} 1 & -3 & 0 & 0 & 0 \\ 0 & 0 & 1 & 0 & 0 \\ 0 & 0 & 0 & 1 & 0 \\ 0 & 0 & 0 & 0 & 1 \end{bmatrix}$

7. $\mathbf{x} = \begin{bmatrix} 5 - 6r \\ 2 - 4r \\ r \end{bmatrix}$, $\begin{bmatrix} -7 \\ -6 \\ 2 \end{bmatrix}$

9. $\mathbf{x} = \begin{bmatrix} 1 - 2r \\ -2 - r - 3s \\ r \\ s \end{bmatrix}, \begin{bmatrix} -5 \\ 1 \\ 3 \\ -2 \end{bmatrix}$ 11. $\mathbf{x} = \begin{bmatrix} 0 \\ 2 \\ -5 \\ 2 \end{bmatrix}$

13. $x = 2, y = -4$

15. $x = -3, y = 2, z = 4$

17. $\mathbf{x} = \begin{bmatrix} 5 \\ 1 \end{bmatrix}$ 19. Inconsistent

21. $\mathbf{x} = \begin{bmatrix} -2 \\ 1 \end{bmatrix}$ 23. $\mathbf{x} = \begin{bmatrix} -8 \\ -23 - 5s \\ -7 + s \\ 2s \end{bmatrix}$

25. Yes   27. No

29. F F T T F T T T F T

31. $x_1 = -1, x_2 = 3$

33. $x_1 = 2, x_2 = -3$

35. $x_1 = 1, x_2 = -1, x_3 = 1, x_4 = 2$

37. $x_1 = -3, x_2 = 5, x_3 = 2, x_4 = -3$

39. All b_1 and b_2

41. All b_1, b_2, b_3 such that $b_3 + b_2 - b_1 = 0$

43. $\begin{bmatrix} 1 & 0 & 0 \\ 2 & 1 & 0 \\ 0 & 0 & 1 \end{bmatrix}$ 45. $\begin{bmatrix} 0 & 1 & 0 \\ 0 & 0 & 1 \\ 1 & 0 & 0 \end{bmatrix}$

47. $\begin{bmatrix} 0 & 0 & 1 & 0 \\ 0 & 1 & 0 & 0 \\ 4 & 0 & 0 & 0 \\ 0 & 0 & 0 & 1 \end{bmatrix}$ 49. $\begin{bmatrix} 1 & -60 & 0 & -15 \\ 0 & 1 & 0 & 0 \\ 0 & 0 & 1 & 0 \\ 0 & -12 & 0 & -3 \end{bmatrix}$

51. $\begin{bmatrix} 2 & -30 & 5 & -10 \\ -4 & 121 & -20 & 40 \\ 0 & -6 & 1 & -2 \\ 0 & 3 & 0 & 1 \end{bmatrix}$

57. $a = 1, b = -2, c = 1, d = 2$

59. $\mathbf{x} = \begin{bmatrix} -1 \\ 7 \end{bmatrix}$ 61. $\mathbf{x} \approx \begin{bmatrix} -1.2857 \\ 3.1429 \\ 1.2857 \end{bmatrix}$

63. $\mathbf{x} = \begin{bmatrix} 1 \\ 1 \\ -1 \\ 1 \\ 2 \end{bmatrix}$

65. $\begin{bmatrix} 1 & 2 & 0 & 0 & 3 & -6 \\ 0 & 0 & 1 & 0 & 1 & -2 \\ 0 & 0 & 0 & 1 & -1 & 3 \end{bmatrix}$ 67. $\mathbf{x} = \begin{bmatrix} 1 \\ 3 \\ 0 \end{bmatrix}$

M1. $\mathbf{x} = \begin{bmatrix} -2 \\ 1 \end{bmatrix}$ M3. Inconsistent

M5. $\mathbf{x} \approx \begin{bmatrix} 1 - 11s \\ 3 - 7s \\ s \end{bmatrix}$

M7. $\mathbf{x} = \begin{bmatrix} -13 - 2r + 14s \\ r \\ -5 + 5s \\ s \end{bmatrix}$

M9. $\mathbf{x} \approx \begin{bmatrix} 0.0345 \\ -0.5370 \\ -1.6240 \\ 0.1222 \\ 0.8188 \end{bmatrix}$

Section 1.5

1. **a.** $A^{-1} = \begin{bmatrix} 1 & -1 \\ 0 & 1 \end{bmatrix}$

b. The matrix A itself is an elementary matrix.

3. Not invertible

5. **a.** $A^{-1} = \begin{bmatrix} 1 & 0 & 1 \\ 0 & 1 & 1 \\ 0 & 0 & -1 \end{bmatrix}$

b. $A = \begin{bmatrix} 1 & 0 & 0 \\ 0 & 1 & 0 \\ 0 & 0 & -1 \end{bmatrix}\begin{bmatrix} 1 & 0 & 0 \\ 0 & 1 & 1 \\ 0 & 0 & 1 \end{bmatrix}\begin{bmatrix} 1 & 0 & 1 \\ 0 & 1 & 0 \\ 0 & 0 & 1 \end{bmatrix}$

(Other answers are possible.)

7. **a.** $A^{-1} = \begin{bmatrix} -7 & 5 & 3 \\ 3 & -2 & -2 \\ 3 & -2 & -1 \end{bmatrix}$;

b. $A = \begin{bmatrix} 2 & 0 & 0 \\ 0 & 1 & 0 \\ 0 & 0 & 1 \end{bmatrix}\begin{bmatrix} 1 & 0 & 0 \\ 3 & 1 & 0 \\ 0 & 0 & 1 \end{bmatrix}\begin{bmatrix} 1 & 0 & 0 \\ 0 & \frac{1}{2} & 0 \\ 0 & 0 & 1 \end{bmatrix}\begin{bmatrix} 1 & \frac{1}{2} & 0 \\ 0 & 1 & 0 \\ 0 & 0 & 1 \end{bmatrix} \cdot$

$\begin{bmatrix} 1 & 0 & 0 \\ 0 & 1 & 0 \\ 0 & -1 & 1 \end{bmatrix}\begin{bmatrix} 1 & 0 & 0 \\ 0 & 1 & 0 \\ 0 & 0 & -1 \end{bmatrix}\begin{bmatrix} 1 & 0 & 3 \\ 0 & 1 & 0 \\ 0 & 0 & 1 \end{bmatrix} \cdot$

$\begin{bmatrix} 1 & 0 & 0 \\ 0 & 1 & -2 \\ 0 & 0 & 1 \end{bmatrix}$

9.
$$\begin{bmatrix} 1 & 0 & 0 & 0 & 0 & 0 \\ 0 & -1 & 0 & 0 & 0 & 0 \\ 0 & 0 & \frac{1}{2} & 0 & 0 & 0 \\ 0 & 0 & 0 & \frac{1}{3} & 0 & 0 \\ 0 & 0 & 0 & 0 & \frac{1}{4} & 0 \\ 0 & 0 & 0 & 0 & 0 & \frac{1}{5} \end{bmatrix}$$

11. The span of the column vectors is $\mathbb{R}^4$.

13. a. $A^{-1} = \begin{bmatrix} -7 & 3 \\ -5 & 2 \end{bmatrix}$ **b.** $\begin{bmatrix} x_1 \\ x_2 \end{bmatrix} = \begin{bmatrix} -37 \\ -26 \end{bmatrix}$

15. $x = -7r + 5s + 3t,\ y = 3r - 2s - 2t,$
$z = 3r - 2s - t$

17. $\begin{bmatrix} 46 & 33 & 30 \\ 39 & 29 & 26 \\ 99 & 68 & 63 \end{bmatrix}$ **19.** $\begin{bmatrix} 3 & 2 & 1 \\ 0 & 3 & 2 \\ 1 & 3 & 4 \end{bmatrix}$

21. The matrix is invertible for any value of r except $r = 0$.

23. T T T F T T T F F F

25. a. No;   **b.** Yes

27. a. Notice that $A(A^{-1}B) = (AA^{-1})B = IB = B$, so $X = A^{-1}B$ is a solution. To show uniqueness, suppose that $AX = B$. Then $A^{-1}(AX) = A^{-1}B,\ (A^{-1}A)X = A^{-1}B,\ IX = A^{-1}B$, and $X = A^{-1}B$; therefore, this is the only solution.

b. Let $E_1, E_2, \ldots, E_k$ be elementary matrices that reduce $[A \mid B]$ to $[I \mid X]$, and let $C = E_k E_{k-1} \cdots E_2 E_1$. Then $CA = I$ and $CB = X$. Thus, $C = A^{-1}$ and so $X = A^{-1}B$.

29. a. $\begin{bmatrix} 0 & 0 \\ 1 & 0 \end{bmatrix}$

39. $\begin{bmatrix} 0.355 & -0.0645 & 0.161 \\ -0.129 & 0.387 & 0.0323 \\ -0.0968 & 0.290 & -0.226 \end{bmatrix}$

41. $\begin{bmatrix} 0.0275 & -0.0296 & -0.0269 & 0.0263 \\ 0.168 & -0.0947 & 0.0462 & -0.0757 \\ 0.395 & -0.138 & -0.00769 & -0.0771 \\ -0.0180 & 0.0947 & 0.00385 & 0.0257 \end{bmatrix}$

43. See answer to Exercise 9.

45. See answer to Exercise 41.

47. $\begin{bmatrix} 0.291 & 0.199 & 0.0419 & -0.00828 & -0.272 \\ -0.0564 & 0.159 & 0.148 & -0.0737 & -0.0699 \\ 0.0276 & 0.145 & -0.00841 & -0.0302 & -0.0250 \\ -0.0467 & 0.122 & -0.029 & 0.133 & -0.00841 \\ 0.0116 & -0.128 & -0.0470 & -0.0417 & 0.178 \end{bmatrix}$

M1. 0.001783   **M3.** 0.4397

M5. −418.07   **M7.** −0.001071

Section 1.6

1. A subspace   **3.** Not a subspace

5. Not a subspace   **7.** Not a subspace

9. A subspace

13. a. Every subspace of $\mathbb{R}^2$ is either the origin, a line through the origin, or all of $\mathbb{R}^2$.

b. Every subspace of $\mathbb{R}^3$ is either the origin, a line through the origin, a plane through the origin, or all of $\mathbb{R}^3$.

15. No, because $\mathbf{0} = r\mathbf{0}$ for all $r \in \mathbb{R}$, so $\mathbf{0}$ is not a *unique* linear combination of $\mathbf{0}$.

17. $\{[-1, 3, 0], [-1, 0, 3]\}$

19. $\{[-7, 1, 13, 0], [-6, -1, 0, 13]\}$

21. $\{[-60, 137, 33, 0, 1]\}$ **23.** Not a basis

25. A basis   **27.** A basis

29. Not a basis   **31.** $\{[-1, -3, 11]\}$

33. $\text{sp}(\mathbf{v}_1, \mathbf{v}_2, \ldots, \mathbf{v}_k) = \text{sp}(\mathbf{w}_1, \mathbf{w}_2, \ldots, \mathbf{w}_m)$ if and only if each $\mathbf{v}_i$ is a linear combination of the $\mathbf{w}_j$ and each $\mathbf{w}_j$ is a linear combination of the $\mathbf{v}_i$.

35. $\mathbf{x} = \begin{bmatrix} -\frac{5}{8} \\ \frac{7}{4} \\ 0 \\ 0 \end{bmatrix} + \begin{bmatrix} -3r/4 + s/8 \\ 3r/2 + s/4 \\ r \\ s \end{bmatrix}$

37. $\mathbf{x} = \begin{bmatrix} \frac{5}{3} \\ \frac{5}{3} \\ 0 \\ 0 \end{bmatrix} \begin{bmatrix} -(5r + s)/3 \\ (r + 2s)/3 \\ r \\ s \end{bmatrix}$

39. In this case, a solution is not uniquely determined. The system is viewed as underdetermined because it is insufficient to determine a unique solution.

41. $x - y = 1$
$2x - 2y = 2$
$3x - 3y = 3$

49. A basis   **51.** Not a basis

M1. Not a basis   **M3.** A basis

M5. The probability is actually 1.

M7. Yes, we obtained the expected result.

Section 1.7

1. Not a transition matrix

3. Not a transition matrix

5. A regular transition matrix

7. A transition matrix, but not regular

9. 0.28   **11.** 0.330   **13.** Not regular

15. Regular   **17.** Not regular

19. $\begin{bmatrix} \frac{3}{4} \\ \frac{1}{4} \end{bmatrix}$ **21.** $\begin{bmatrix} \frac{1}{4} \\ \frac{3}{8} \\ \frac{3}{8} \end{bmatrix}$ **23.** $\begin{bmatrix} \frac{12}{35} \\ \frac{8}{35} \\ \frac{15}{35} \end{bmatrix}$

25. T F F T T F T F F T

27. $A^{100} \approx \begin{bmatrix} \frac{12}{35} & \frac{12}{35} & \frac{12}{35} \\ \frac{8}{35} & \frac{8}{35} & \frac{8}{35} \\ \frac{15}{35} & \frac{15}{35} & \frac{15}{35} \end{bmatrix}$ **29.** 0.47   **31.** $\begin{bmatrix} .32 \\ .47 \\ .21 \end{bmatrix}$

33. The Markov chain is regular because T has no zero entry; $\mathbf{s} = \begin{bmatrix} \frac{11}{43} \\ \frac{17}{43} \\ \frac{15}{43} \end{bmatrix}$.

35. $\frac{1}{8}$ **37.** $\begin{bmatrix} \frac{1}{4} \\ \frac{1}{2} \\ \frac{1}{4} \end{bmatrix}$

39. The Markov chain is regular because T^2 has no zero entries; $\mathbf{s} = \begin{bmatrix} \frac{1}{4} \\ \frac{1}{2} \\ \frac{1}{4} \end{bmatrix}$.

41. $T = \begin{bmatrix} 0 & 0 & 0 \\ 1 & \frac{1}{2} & 0 \\ 0 & \frac{1}{2} & 1 \end{bmatrix} \begin{matrix} \text{D} \\ \text{H} \\ \text{R} \end{matrix}$

with column labels **D H R**.

The recessive state is absorbing, because there is an entry 1 in the row 3, column 3 position.

45. $\begin{bmatrix} \frac{1}{5} \\ \frac{4}{5} \end{bmatrix}$ **47.** $\begin{bmatrix} \frac{5}{9} \\ \frac{4}{9} \end{bmatrix}$ **49.** $\begin{bmatrix} .4037 \\ .3975 \\ .1988 \end{bmatrix}$

51. $\begin{bmatrix} .2115 \\ .3462 \\ .4423 \end{bmatrix}$ **53.** $\begin{bmatrix} .2682 \\ .2235 \\ .1916 \\ .1676 \\ .1490 \end{bmatrix}$

Section 1.8

1. 0001011 0000000 0111001 1111111 1111111 0100101 0000000 0100101 1111111 0111001

3. $x_4 = x_1 + x_2, x_5 = x_1 + x_3, x_6 = x_2 + x_3$

5. Note that each x_i appears in at least two parity-check equations and that, for each combination i, j of positions in a message word, some parity-check equation contains one of x_i, x_j but not the other. As explained following Eq. (2) in the text, this shows that the distance between code words is at least 3. Consequently, any two errors can be detected.

7. a. 110   **b.** 001   **c.** 110

d. 001 or 100 or 111   **e.** 101

9. We compute, using binary addition,

$$\begin{bmatrix} 1 & 1 & 0 & 1 & 0 & 0 \\ 1 & 0 & 1 & 0 & 1 & 0 \\ 0 & 1 & 1 & 0 & 0 & 1 \end{bmatrix} \begin{bmatrix} 1 & 0 & 1 & 1 & 1 \\ 1 & 0 & 1 & 0 & 0 \\ 0 & 1 & 1 & 1 & 0 \\ 1 & 0 & 0 & 0 & 1 \\ 1 & 1 & 1 & 1 & 0 \\ 1 & 1 & 1 & 0 & 1 \end{bmatrix} =$$

$$\begin{bmatrix} 1 & 0 & 0 & 1 & 0 \\ 0 & 0 & 1 & 1 & 1 \\ 0 & 0 & 1 & 1 & 1 \end{bmatrix},$$

where the first matrix is the parity-check matrix and the second matrix has as *columns* the received words.

a. Because the first column of the product is the fourth column of the parity-check matrix, we change the fourth position of the received word 110111 to get the code word 110011 and decode it as 110.

b. Because the second column of the product is the zero vector, the received word 001011 is a code word and we decode it as 001.

c. Because the third column of the product is the third column of the parity-check matrix, we change the third position of the received word 111011 to get the code word 110011 and decode it as 110.

d. Because the fourth column of the product is not the zero vector and not a column of the parity-check matrix, there are at least two errors, and we ask for retransmission.

e. Because the fifth column of the product is the third column of the parity-check matrix, we change the third position of the received word 100101 to get the code word 101101 and decode it as 101.

11. Because we add by components, this follows from the fact that, using binary addition, $1 + 1 = 1 - 1 = 0$, $1 + 0 = 1 - 0 = 1$, $0 + 1 = 0 - 1 = 1$, and $0 + 0 = 0 - 0 = 0$.

13. From the solutions of Exercises 11 and 12, we see that $u - v = u + v$ contains 1's in precisely the positions where the two words differ. The number of places where u and v differ is equal to the distance between them, and the number of 1's in $u - v$ is $\text{wt}(u - v)$, so these numbers are equal.

15. This follows immediately from the fact that $\mathbb{B}^n$ itself is closed under addition modulo 2.

17. Suppose that $d(u, v)$ is minimum in C. By Exercise 13, $d(u, v) = \text{wt}(u - v)$, showing that the minimum weight of nonzero code words is less than or equal to the minimum distance between two of them.

On the other hand, if w is a nonzero code word of minimum weight, then $\text{wt}(w)$ is the distance from w to the zero code word, showing the opposite inequality, so we have the equality stated in the exercise.

19. The triangle inequality in Exercise 14 shows that if the distance from received word v to code word u is at most m and the distance from v to code word w is at most m, then $d(u, v) \leq 2m$. Thus, if the distance between code words is at least $2m + 1$, a received word v with at most m incorrect components has a *unique* nearest-neighbor code word. This number $2m + 1$ can't be improved, because if $d(u, v) = 2m$, then a possible received word w at distance m from both u and v can be constructed by having its components agree with those of u and v where the components of u and v agree, and by making m of the $2m$ components of w in which u and v differ opposite from the components of u and the other m components of w opposite from those of v.

21. Let e_i be the word in $\mathbb{B}^n$ with 1 in the ith position and 0's elsewhere. Now e_i is not in C, because the distance from e_i to $000 \ldots 0$ is 1, and $000 \ldots 0 \in C$. Also, $v + e_i \neq w + e_j$ for any two distinct words v and w in C, because otherwise $v - w = e_j - e_i$ would be in C with $\text{wt}(e_j - e_i) = 2$. Let $e_i + C = \{e_i + v \mid v \in C\}$. Note that C and $e_i + C$ have the same number of words. The disjoint sets C and $e_i + C$ for $i = 1, 2, \ldots, k$ contain all words whose distances from some word in C are at most 1. Thus $\mathbb{B}^n$ must be large enough to contain all of these $(n + 1)2^k$ elements. That is, $2^n \geq (n + 1)2^k$. Dividing by 2^k gives the desired result.

23. a. 3   **b.** 3   **c.** 4   **d.** 5   **e.** 6   **f.** 7

25. $x_9 = x_1 + x_2 + x_3 + x_6 + x_7$, $x_{10} = x_5 + x_6 + x_7 + x_8$, $x_{11} = x_2 + x_3 + x_4 + x_6 + x_8$, $x_{12} = x_1 + x_3 + x_4 + x_5$ (Other answers are possible.)

CHAPTER 2

Section 2.1

1. Two nonzero vectors in $\mathbb{R}^2$ are dependent if and only if they are parallel.

3. Two nonzero vectors in $\mathbb{R}^3$ are dependent if and only if they are parallel.

5. Three vectors in $\mathbb{R}^3$ are dependent if and only if they all lie in one plane through the origin.

7. $\{[-3, 1], [6, 4]\}$　　9. $\{[2, 1], [1, 4]\}$

11. $\{[1, 2, 1, 2], [2, 1, 0, -1]\}$

13. $\{[1, 3, 5, 7], [2, 0, 4, 2]\}$

15. $\{[2, 3, 1], [5, 2, 1]\}$　　17. Independent

19. Dependent　　21. Independent

23. Dependent　　25. Dependent

27. $\{[2, 1, 1, 1], [1, 0, 1, 1], [1, 0, 0, 0], [0, 0, 1, 0]\}$

31. Suppose that $r_1\mathbf{w}_1 + r_2\mathbf{w}_2 + r_3\mathbf{w}_3 = \mathbf{0}$, so that $r_1(2\mathbf{v}_1 + 3\mathbf{v}_2) + r_2(\mathbf{v}_2 - 2\mathbf{v}_3) + r_3(-\mathbf{v}_1 - 3\mathbf{v}_3) = \mathbf{0}$. Then $(2r_1 - r_3)\mathbf{v}_1 + (3r_1 + r_2)\mathbf{v}_2 + (-2r_2 - 3r_3)\mathbf{v}_3 = \mathbf{0}$. We try setting these scalar coefficients to zero and solve the linear system

$$
\begin{aligned}
2r_1 \quad\quad - r_3 &= 0, \\
3r_1 + \quad r_2 \quad\quad &= 0, \\
-2r_2 - 3r_3 &= 0.
\end{aligned}
$$

$$
\begin{bmatrix} 2 & 0 & -1 & | & 0 \\ 3 & 1 & 0 & | & 0 \\ 0 & -2 & -3 & | & 0 \end{bmatrix} \sim \begin{bmatrix} 1 & 0 & -\frac{1}{2} & | & 0 \\ 0 & 1 & \frac{3}{2} & | & 0 \\ 0 & -2 & -3 & | & 0 \end{bmatrix} \sim
$$

$$
\begin{bmatrix} 1 & 0 & -\frac{1}{2} & | & 0 \\ 0 & 1 & \frac{3}{2} & | & 0 \\ 0 & 0 & 0 & | & 0 \end{bmatrix}.
$$

This system has the nontrivial solution $r_3 = 2$, $r_2 = -3$, and $r_1 = 1$. Thus, $(2\mathbf{v}_1 + 3\mathbf{v}_2) - 3(\mathbf{v}_2 - 2\mathbf{v}_3) + 2(-\mathbf{v}_1 - 3\mathbf{v}_3) = \mathbf{0}$ for all choices of vectors $\mathbf{v}_1, \mathbf{v}_2, \mathbf{v}_3 \in V$, so $\mathbf{w}_1, \mathbf{w}_2, \mathbf{w}_3$ are dependent. (Notice that, if the system had only the trivial solution, then any choice of independent vectors $\mathbf{v}_1, \mathbf{v}_2, \mathbf{v}_3$ would show this exercise to be false.)

33. No such scalars s exist.

35. Let $A = \begin{bmatrix} 1 & 1 & 0 \\ 0 & 0 & 0 \\ 0 & 0 & 1 \end{bmatrix}$, $\mathbf{v} = \begin{bmatrix} 1 \\ 0 \\ 0 \end{bmatrix}$, $\mathbf{w} = \begin{bmatrix} 0 \\ 1 \\ 0 \end{bmatrix}$. Then

$A\mathbf{v} = A\mathbf{w} = \begin{bmatrix} 1 \\ 0 \\ 0 \end{bmatrix}$. For the second part of the

problem, let $A = \begin{bmatrix} 1 & 0 & 0 \\ 0 & 1 & 0 \\ 0 & 0 & 0 \end{bmatrix}$, $\mathbf{v} = \begin{bmatrix} 1 \\ 0 \\ 0 \end{bmatrix}$,

$\mathbf{w} = \begin{bmatrix} 0 \\ 1 \\ 0 \end{bmatrix}$. Then $A\mathbf{v} = \mathbf{v}$ and $A\mathbf{w} = \mathbf{w}$, so $A\mathbf{v}$

and $A\mathbf{w}$ are still independent.

39. $\{\mathbf{v}_1, \mathbf{v}_2, \mathbf{v}_4\}$　　41. $\{\mathbf{u}_1, \mathbf{u}_2, \mathbf{u}_4, \mathbf{u}_6\}$

M1. $\{\mathbf{v}_1, \mathbf{v}_2, \mathbf{v}_4\}$　　M3. $\{\mathbf{u}_1, \mathbf{u}_2, \mathbf{u}_4, \mathbf{u}_6\}$

Section 2.2

1. a. 2

   b. $\{[2, 0, -3, 1], [3, 4, 2, 2]\}$ by inspection

   c. The set consisting of $\begin{bmatrix} 2 \\ 3 \end{bmatrix}$, $\begin{bmatrix} 0 \\ 4 \end{bmatrix}$

   or of $\begin{bmatrix} 1 \\ 0 \end{bmatrix}$, $\begin{bmatrix} 0 \\ 1 \end{bmatrix}$

   d. The set consisting of $\begin{bmatrix} 12 \\ -13 \\ 8 \\ 0 \end{bmatrix}$, $\begin{bmatrix} -4 \\ -1 \\ 0 \\ 8 \end{bmatrix}$

3. a. 3

   b. $\{[1, 0, -1, 0], [0, 1, 1, 0], [0, 0, 0, 1]\}$

   c. The set consisting of $\begin{bmatrix} 0 \\ 1 \\ 4 \\ 1 \end{bmatrix}$, $\begin{bmatrix} 6 \\ 2 \\ 1 \\ 3 \end{bmatrix}$, $\begin{bmatrix} 3 \\ 1 \\ 4 \\ 0 \end{bmatrix}$

   d. The set consisting of the vector $\begin{bmatrix} 1 \\ -1 \\ 1 \\ 0 \end{bmatrix}$

5. a. 3

   b. $\{[6, 0, 0, 5], [0, 3, 0, 1], [0, 0, 3, 1]\}$

   c. The set consisting of $\begin{bmatrix} 0 \\ 2 \\ 0 \end{bmatrix}$, $\begin{bmatrix} 1 \\ 1 \\ 2 \end{bmatrix}$, $\begin{bmatrix} 2 \\ 0 \\ 1 \end{bmatrix}$ or the

   set of column vectors $\{\mathbf{e}_1, \mathbf{e}_2, \mathbf{e}_3\}$

d. The set consisting of $\begin{bmatrix} -5 \\ -2 \\ -2 \\ 6 \end{bmatrix}$

7. The rank is 3; therefore, the matrix is not invertible.

9. The rank is 3; therefore, the matrix is invertible.

11. T F T T F F T F F T

15. No. If A is $m \times n$ and C is $n \times s$, then AC is $m \times s$. The column space of AC lies in $\mathbb{R}^m$ whereas the column space of C lies in $\mathbb{R}^n$.

17. $\begin{bmatrix} 0 & 1 & 0 \\ 0 & 0 & 1 \\ 0 & 0 & 0 \end{bmatrix}$

19. Let $A = I$ and $C = \begin{bmatrix} 1 & 1 \\ 1 & 1 \end{bmatrix}$. Then

rank $(AC) = 1 < 2 = $ rank (A).

25. a. 3   **b.** Rows 1, 2, and 4

Section 2.3

1. Yes. We have $T([x_1, x_2, x_3]) = \begin{bmatrix} 1 & 1 & 0 \\ 1 & -3 & 0 \end{bmatrix} \begin{bmatrix} x_1 \\ x_2 \\ x_3 \end{bmatrix}$.

Example 3 then shows that T is a linear transformation.

3. No. $T(0[1, 2, 3]) = T([0, 0, 0]) = [1, 1, 1, 1]$, whereas $0T([1, 2, 3]) = 0[1, 1, 1, 1] = [0, 0, 0, 0]$.

5. $[24, -34]$ **7.** $[2, 7, -15]$

9. $[4, 2, 4]$

11. $[802, -477, 398, 57]$ **13.** $\begin{bmatrix} 1 & 1 \\ 1 & -3 \end{bmatrix}$

15. $\begin{bmatrix} 1 & 1 & 1 \\ 1 & 1 & 0 \\ 1 & 0 & 0 \end{bmatrix}$ **17.** $\begin{bmatrix} 1 & -1 & 3 \\ 1 & 1 & 1 \\ 1 & 0 & 0 \end{bmatrix}$

19. The matrix associated with $T' \circ T$ is

$\begin{bmatrix} 2 & 0 \\ 3 & 1 \end{bmatrix}$; $(T' \circ T)([x_1, x_2]) = [2x_1, 3x_1 + x_2]$.

21. Yes; $T^{-1}([x_1, x_2]) = \left[\dfrac{3x_1 + x_2}{4}, \dfrac{x_1 - x_2}{4} \right]$.

23. Yes; $T^{-1}([x_1, x_2, x_3]) = [x_3, x_2 - x_3, x_1 - x_2]$.

25. Yes; $T^{-1}([x_1, x_2, x_3]) = $
$\left[x_3, \dfrac{-x_1 + 3x_2 - 2x_3}{4}, \dfrac{x_1 + x_2 - 2x_3}{4} \right]$.

27. Yes; $(T' \circ T)^{-1}([x_1, x_2]) = \left[\dfrac{x_1}{2}, \dfrac{-3x_1 + 2x_2}{2} \right]$.

29. T F T T T F T T F T

33. a. Because

$$T_k(\mathbf{u} + \mathbf{v}) = [u_1 + v_1, u_2 + v_2, \ldots,$$
$$u_k + v_k, 0, \ldots, 0]$$
$$= [u_1, u_2, \ldots, u_k, 0, \ldots, 0] +$$
$$[v_1, v_2, \ldots, v_k, 0, \ldots, 0]$$
$$= T_k(\mathbf{u}) + T_k(\mathbf{v})$$

and

$$T_k(r\mathbf{u}) = [ru_1, ru_2, \ldots, ru_k, 0, \ldots, 0]$$
$$= r[u_1, u_2, \ldots, u_k, 0, \ldots, 0]$$
$$= rT_k(\mathbf{u}),$$

we know that T_k is a linear transformation. We see that $T_k[V] \subseteq W_k$, which means that it is a subspace of W_k, by the box on page 143.

b. Because $T_k[V]$ is a subspace of $T_{k+1}[V]$, we have

$$d_1 \le d_2 \le d_3 \le \cdots \le d_n \le n.$$

If the jth column of H has a pivot but the $(j + 1)$st column has no pivot, then $d_{j+1} = d_j$. However, if the jth and $(j + 1)$st columns both have pivots, then $d_{j+1} = d_j + 1$. See parts c and d for illustrations.

c. If $d_1 = d_2 = 1$ and $d_3 = d_4 = 2$, then H must have the form

$$H = \begin{bmatrix} p & X & X & X \\ 0 & 0 & p & X \\ 0 & 0 & 0 & 0 \\ 0 & 0 & 0 & 0 \\ \cdot & \cdot & \cdot & \cdot \\ \cdot & \cdot & \cdot & \cdot \\ \cdot & \cdot & \cdot & \cdot \\ 0 & 0 & 0 & 0 \end{bmatrix},$$

where a p denotes a pivot and an X denotes a possibly nonzero entry.

d. If $d_1 = 1$, $d_2 = d_3 = d_4 = 2$, and $d_5 = d_6 = 3$, then H must have the form

$$H = \begin{bmatrix} p & \times & \times & \times & \times & \times \\ 0 & p & \times & \times & \times & \times \\ 0 & 0 & 0 & 0 & p & \times \\ 0 & 0 & 0 & 0 & 0 & 0 \\ 0 & 0 & 0 & 0 & 0 & 0 \\ \cdot & \cdot & \cdot & \cdot & \cdot & \cdot \\ \cdot & \cdot & \cdot & \cdot & \cdot & \cdot \\ \cdot & \cdot & \cdot & \cdot & \cdot & \cdot \\ 0 & 0 & 0 & 0 & 0 & 0 \end{bmatrix},$$

where a p denotes a pivot and an $\times$ denotes a possibly nonzero entry.

e. The number of pivots in H is the number of distinct dimension numbers in the list $d_1, d_2, d_3, \ldots, d_n$. Moreover, a pivot occurs in the $(j + 1)$st column if and only if $d_{j+1} = d_j + 1$; the row and column positions of pivots are completely determined by the list $d_1, d_2, d_3, \ldots, d_n$. The pivot in the kth row occurs in the jth column if and only if d_j is the kth *distinct* positive integer in the list $d_1, d_2, d_3, \ldots, d_n$. Because the numbers d_i depend only on A (they are defined just in terms of the row space of A), so do the number and locations of the pivots; the number and locations are the same for all echelon forms of A.

f. Let H be a *reduced* echelon form of A. Part e shows that the number of pivots and their locations depend only on A, and consequently the number of zero rows depends only on A and is the same for all choices of H. Suppose that the pivot in the kth row of H is in column j. Consider now a nonzero row vector in $\mathbb{R}^n$ that has entries zero in all components corresponding to columns of H containing pivots except for the jth component, where the entry is 1; entries in components not corresponding to pivot column locations in H may be arbitrary. We claim that there is a *unique* such vector in the row space of A—namely, the kth row vector of H. Such a vector must be a linear combination of the nonzero rows of H, which span the row space of A. The fact that there are zeros above

as well as below pivots in the *reduced* row-echelon form H shows that the only possible such linear combination of nonzero rows of H is 1 times the kth row plus 0 times the other rows. This gives a characterization of the kth nonzero row of H in terms of the row space of A and completes the demonstration that the reduced echelon form of A is unique.

35. [588, 160, 8]

37. Undefined because $T_2 \circ T_3$ is not invertible.

Section 2.4

1. Because $\begin{bmatrix} 1 & -3 \\ 2 & -6 \end{bmatrix} \begin{bmatrix} x \\ y \end{bmatrix} = \begin{bmatrix} x - 3y \\ 2x - 6y \end{bmatrix}$, which

has the form $\begin{bmatrix} t \\ 2t \end{bmatrix}$, we see that the range of T_A consists of vectors along the line $y = 2x$. Now $T([1, 2]) = [-5, -10] \neq [1, 2]$, and projection onto a line must leave fixed the points that lie on the line, so T_A is not a projection on the line.

3. a. $\begin{bmatrix} 1/\sqrt{2} & 1/\sqrt{2} \\ -1/\sqrt{2} & 1/\sqrt{2} \end{bmatrix}$ b. $\begin{bmatrix} \frac{1}{2} & \sqrt{3}/2 \\ -\sqrt{3}/2 & \frac{1}{2} \end{bmatrix}$

c. $\begin{bmatrix} -\sqrt{3}/2 & \frac{1}{2} \\ -\frac{1}{2} & -\sqrt{3}/2 \end{bmatrix}$

7. $\begin{bmatrix} \dfrac{1 - m^2}{1 + m^2} & \dfrac{2m}{1 + m^2} \\ \dfrac{2m}{1 + m^2} & \dfrac{m^2 - 1}{1 + m^2} \end{bmatrix}$

11. In column-vector notation, we have
$$T\left(\begin{bmatrix} x \\ y \end{bmatrix}\right) = \begin{bmatrix} -y \\ x \end{bmatrix} = \begin{bmatrix} 0 & -1 \\ 1 & 0 \end{bmatrix} \begin{bmatrix} x \\ y \end{bmatrix} =$$
$$\begin{bmatrix} -1 & 0 \\ 0 & 1 \end{bmatrix} \begin{bmatrix} 0 & 1 \\ 1 & 0 \end{bmatrix} \begin{bmatrix} x \\ y \end{bmatrix}, \text{ which represents a}$$
reflection in the line $y = x$ followed by a reflection in the y-axis.

13. In column-vector notation, we have
$$T\left(\begin{bmatrix} x \\ y \end{bmatrix}\right) = \begin{bmatrix} -x \\ -y \end{bmatrix} = \begin{bmatrix} -1 & 0 \\ 0 & -1 \end{bmatrix} \begin{bmatrix} x \\ y \end{bmatrix} =$$
$$\begin{bmatrix} -1 & 0 \\ 0 & 1 \end{bmatrix} \begin{bmatrix} 1 & 0 \\ 0 & -1 \end{bmatrix} \begin{bmatrix} x \\ y \end{bmatrix}, \text{ which represents a}$$
reflection in the x-axis followed by a reflection in the y-axis.

15. In column-vector notation, we have

$$T\left(\begin{bmatrix} x \\ y \end{bmatrix}\right) = A\begin{bmatrix} x \\ y \end{bmatrix} = \begin{bmatrix} 1 & 0 \\ 3 & 1 \end{bmatrix}\begin{bmatrix} 1 & 0 \\ 0 & 2 \end{bmatrix}\begin{bmatrix} 1 & 1 \\ 0 & 1 \end{bmatrix}\begin{bmatrix} x \\ y \end{bmatrix},$$

which represents a horizontal shear followed by a vertical expansion followed by a vertical shear.

19. $\|\mathbf{v}\| = \sqrt{\mathbf{v} \cdot \mathbf{v}}$ and $\theta =$
$\cos^{-1} \dfrac{\mathbf{u} \cdot \mathbf{v}}{\sqrt{\mathbf{u} \cdot \mathbf{u}} \sqrt{\mathbf{v} \cdot \mathbf{v}}}.$

Section 2.5

1.

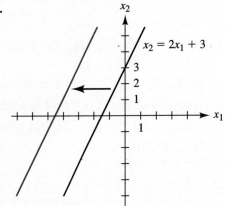

3.

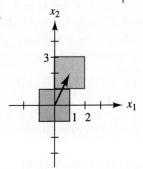

5.

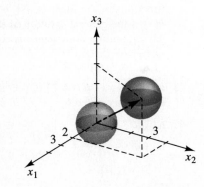

7. $x_1 = 3 - 2t$
  $x_2 = -3 + t$

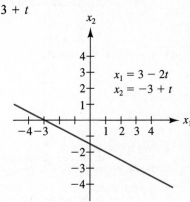

9. $x_1 = \dfrac{d_1 c}{d_1^2 + d_2^2} + d_2 t,\ x_2 = \dfrac{d_2 c}{d_1^2 + d_2^2} - d_1 t$

11. a. $x_1 = -2 + t,\ x_2 = 4 - t;$
  b. $x_1 = 3 - 3t,\ x_2 = -1 - 2t,\ x_3 = 6 - 7t;$
  c. $x_1 = 2 - 3t,\ x_2 = 5t,\ x_3 = 4 - 12t$

13. The lines contain exactly the same set $\{(5 - 3t, -1 + t) \mid t \in \mathbb{R}\}$ of points.

15. a. $\left(\dfrac{1}{2}, \dfrac{3}{2}\right)$ **b.** $\left(\dfrac{3}{2}, -2, \dfrac{5}{2}\right)$ **c.** $\left(-2, \dfrac{9}{2}, \dfrac{17}{2}\right)$

17. $\left(-\dfrac{3}{2}, -\dfrac{1}{2}, \dfrac{15}{4}\right)$ **19.** $\left(\dfrac{3}{2}, \dfrac{3}{2}, 1, \dfrac{7}{2}, -\dfrac{1}{2}\right)$

21. $(2, 3, 4)$ **23.** $x_1 + x_2 + x_3 = 1$

25. $2x_1 + x_2 - x_3 = 2$

27. $x_1 + 4x_2 + x_3 = 10$

29. $x_2 + x_3 = 3,\ -3x_1 - 7x_2 + 8x_3 + 3x_4 = 0$
  (There are infinitely many other correct linear systems.)

31. $9x_1 - x_2 + 2x_3 - 6x_4 + 3x_5 = 6$

33. $x_1 + x_2 + x_3 + x_4 + x_5 + x_6 = 1$

35. The 0-flat $\mathbf{x} = \begin{bmatrix} -1 \\ 0 \\ 2 \end{bmatrix}$

37. The 1-flat $\mathbf{x} = \begin{bmatrix} -1 \\ -1 \\ 0 \end{bmatrix} + t\begin{bmatrix} 2 \\ 1 \\ 1 \end{bmatrix}$

39. The 1-flat $\mathbf{x} = \begin{bmatrix} -8 \\ -23 \\ -7 \\ 0 \end{bmatrix} + t\begin{bmatrix} 0 \\ -5 \\ 1 \\ 2 \end{bmatrix}$

41. The 0-flat $\mathbf{x} = \begin{bmatrix} -43 \\ -12 \\ 7 \\ 1 \end{bmatrix}$

43. T F F T F T F T F T

CHAPTER 3

Section 3.1

1. Not a vector space **3.** A vector space

5. Not a vector space **7.** Not a vector space

9. A vector space   **11.** A vector space

13. A vector space   **15.** A vector space

17. a. $[-1, 0]$ is the "zero vector."
  b. Part 5 of Theorem 3.1 in this vector space becomes $r[-1, 0] = [-1, 0]$, for all $r \in \mathbb{R}$. That is, $[0, 0]$ is not the zero vector $\mathbf{0}$ in this vector space.

27. Both 2×6 matrices and 3×4 matrices contain 12 entries. If we number entries in some fashion from 1 to 12, say starting at the upper left-hand corner and proceeding down each column in turn, then each matrix can be viewed as defining a function from the set $S = \{1, 2, \ldots, 12\}$ into the set $\mathbb{R}$. The rules for adding matrices and multiplying by a scalar correspond in each case to our definitions of addition and scalar multiplication in the space of functions mapping S into $\mathbb{R}$. Thus we can identify both $M_{2,6}$ and $M_{3,4}$ with this function space, and hence with each other.

29. $\mathbb{R}^{24}$ $\mathbb{R}^{25}$ $\mathbb{R}^{26}$ P_{26} $M_{4,7}$
$M_{2,12}$ P_{24} P_{25} $M_{3,9}$
$M_{4,6}$ $M_{5,5}$ $M_{2,13}$
$M_{3,8}$

Section 3.2

1. Not a subspace   **3.** Not a subspace

5. A subspace

7. a. Because $1 = \sin^2 x + \cos^2 x$, we have $c = c(\sin^2 x) + c(\cos^2 x)$, which shows that $c \in \text{sp}(\sin^2 x, \cos^2 x)$.

b. Now $\cos 2x = \cos^2 x - \sin^2 x = (-1) \sin^2 x + (1) \cos^2 x$, which shows that $\cos 2x \in \text{sp}(\sin^2 x, \cos^2 x)$

c. Now $\cos 4x = \cos^2 2x - \sin^2 2x = (1 - \sin^2 2x) - \sin^2 2x = \frac{1}{7}(7) + (-2)\sin^2 2x$, which shows that $\cos 4x$, and thus $8 \cos 4x$, is in $\text{sp}(7, \sin^2 2x)$.

9. a. We see that $\mathbf{v}_1, 2\mathbf{v}_1 + \mathbf{v}_2 \in \text{sp}(\mathbf{v}_1, \mathbf{v}_2)$; and therefore,

$$\text{sp}(\mathbf{v}_1, 2\mathbf{v}_1 + \mathbf{v}_2) \subseteq \text{sp}(\mathbf{v}_1, \mathbf{v}_2).$$

Furthermore, $\mathbf{v}_1 = 1\mathbf{v}_1 + 0(2\mathbf{v}_1 + \mathbf{v}_2)$ and $\mathbf{v}_2 = (-2)\mathbf{v}_1 + 1(2\mathbf{v}_1 + \mathbf{v}_2)$, showing that $\mathbf{v}_1, \mathbf{v}_2 \in \text{sp}(\mathbf{v}_1, 2\mathbf{v}_1 + \mathbf{v}_2)$; and therefore,

$$\text{sp}(\mathbf{v}_1, \mathbf{v}_2) \subseteq \text{sp}(\mathbf{v}_1, 2\mathbf{v}_1 + \mathbf{v}_2).$$

Thus, $\text{sp}(\mathbf{v}_1, \mathbf{v}_2) = \text{sp}(\mathbf{v}_1, 2\mathbf{v}_1 + \mathbf{v}_2)$.

11. Dependent   **13.** Dependent

15. Independent   **17.** Dependent

19. Independent   **21.** Not a basis

23. $\{1, 4x + 3, x^2 + 2\}$

25. T F T T T F F F T T

35. Let $W = \text{sp}(\mathbf{e}_1, \mathbf{e}_2)$ and $U = \text{sp}(\mathbf{e}_3, \mathbf{e}_4, \mathbf{e}_5)$ in $\mathbb{R}^5$. Then $W \cap U = \{\mathbf{0}\}$ and each $\mathbf{x} \in \mathbb{R}^5$ has the form $\mathbf{x} = \mathbf{w} + \mathbf{u}$, where $\mathbf{w} = x_1\mathbf{e}_1 + x_2\mathbf{e}_2$ and $\mathbf{u} = x_3\mathbf{e}_3 + x_4\mathbf{e}_4 + x_5\mathbf{e}_5$.

39. In deciding whether $\sin x$ and $\cos x$ are independent, we consider linear combinations of them *with scalar coefficients*. The given coefficients $f(x)$ and $g(x)$ are not scalars. For a counterexample, consider $f(x) = -\cos x$ and $g(x) = \sin x$. We have $(-\cos x)(\sin x) + (\sin x)(\cos x) = 0$.

41. The set of solutions consists of all functions of the form $h(x) + p(x)$, where $h(x)$ is the general solution of the corresponding homogeneous equation $f_n(x)y^{(n)} + f_{n-1}(x)y^{(n-1)} + \cdots + f_2(x)y'' + f_1(x)y' + f_0(x)y = 0$.

43. a. $\{a \sin x + b \cos x \mid a, b \in \mathbb{R}\}$
  b. $\{a \sin x + b \cos x + x \mid a, b \in \mathbb{R}\}$

45. a. $\{ae^{3x} + be^{-3x} + c \mid a, b, c \in \mathbb{R}\}$
  b. $\{ae^{3x} + be^{-3x} + c - \dfrac{x^3}{27} - \dfrac{1}{9}x^2 - \dfrac{2}{81}x \mid a, b, c \in \mathbb{R}\}$

47. One basis B for W consists of those f_a for a in S defined by $f_a(a) = 1$ and $f_a(s) = 0$ for $s \neq a$ in S. If $f \in W$ and $f(s) \neq 0$ only for $s \in \{a_1, a_2, \ldots, a_n\}$, then $f = f(a_1)f_{a_1} + f(a_2)f_{a_2} + \cdots + f(a_n)f_{a_n}$. Now the linear combination $g = c_1 f_{b_1} + c_2 f_{b_2} + \cdots + c_m f_{b_m}$ is a function satisfying $g(b_j) = c_j$ for $j = 1, 2, \ldots, m$ and $g(s) = 0$ for all other $s \in S$. Thus $g \in W$, so B spans only W. (The crucial thing is that all linear combinations are sums of only *finite* numbers of vectors.)

Section 3.3

1. $[1, -1]$ **3.** $[2, 6, -4]$ **5.** $[2, 1, 3]$

7. $[4, 1, -2, 1]$ **9.** $\left[\frac{1}{2}, \frac{1}{2}, 1, -\frac{1}{2}, 0\right]$

11. $[1, 2, -1, 5]$

13. $p(x) = [(x + 1) - 1]^3 + [(x + 1) - 1]^2$
$\qquad - [(x + 1) - 1] - 1$
$\qquad = (x + 1)^3 - 3(x + 1)^2 + 3(x + 1) - 1$
$\qquad\quad + (x + 1)^2 - 2(x + 1) + 1$
$\qquad\qquad - (x + 1) + 1$
$\qquad\qquad\quad - 1$
$\qquad = (x + 1)^3 - 2(x + 1)^2 + 0(x + 1) + 0$

so the coordinate vector is $[1, -2, 0, 0]$.

15. $[4, 3, -5, -4]$

17. Let $x^3 - 4x^2 + 3x + 7 = b_3(x - 2)^3 + b_2(x - 2)^2 + b_1(x - 2) + b_0$. Setting $x = 2$, we find that $8 - 16 + 6 + 7 = b_0$, so $b_0 = 5$. Differentiating and setting $x = 2$, we obtain $3x^2 - 8x + 3 = 3b_3(x - 2)^2 + 2b_2(x - 2) + b_1$; $12 - 16 + 3 = b_1$; $b_1 = -1$. Differentiating again and setting $x = 2$, we have $6x - 8 = 6b_3(x - 2) + 2b_2$; $12 - 8 = 2b_2$; $b_2 = 2$. Differentiating again, we obtain $6 = 6b_3$, so $b_3 = 1$. The coordinate vector is $[1, 2, -1, 5]$.

19. b. $\{f_1(x), f_2(x)\}$ **21.** $2x^2 + 6x + 2$

Section 3.4

1. A linear transformation, $\ker(T) = \{f \in F \mid f(-4) = 0\}$, not invertible

3. A linear transformation, $\ker(T)$ is the zero function, invertible

5. A linear transformation, $\ker(T)$ is the zero function, invertible

7. The zero function is the only function in $\ker(T)$.

9. $\{c_1 e^{2x} + c_2 e^{-2x} - \frac{1}{5} \sin x \mid c_1, c_2 \in \mathbb{R}\}$

11. $\{c_1 x + c_2 + \cos x \mid c_1, c_2 \in \mathbb{R}\}$

13. $\{c_1 \sin 2x + c_2 \cos 2x + \frac{1}{4}x^2 - \frac{1}{8} \mid c_1, c_2 \in \mathbb{R}\}$

15. $\{c_1 e^{2x} + c_2 x + c_3 - \frac{1}{12}x^3 - \frac{1}{8}x^2 \mid c_1, c_2, c_3 \in \mathbb{R}\}$

17. $T(v) = 4b_1' + 4b_2' + 7b_3' + 6b_4'$

19. $T(v) = -17b_1' + 4b_2' + 13b_3' - 3b_4'$

21. a. $A = \begin{bmatrix} 0 & 0 & 0 & 0 \\ 3 & 0 & 0 & 0 \\ 0 & 2 & 0 & 0 \\ 0 & 0 & 1 & 0 \end{bmatrix}$

b. $A \begin{bmatrix} 4 \\ -5 \\ 10 \\ -13 \end{bmatrix} = \begin{bmatrix} 0 \\ 12 \\ -10 \\ 10 \end{bmatrix}$; $12x^2 - 10x + 10$

c. The second derivative is given by
$A^2 \begin{bmatrix} -5 \\ 8 \\ -3 \\ 4 \end{bmatrix} = \begin{bmatrix} 0 \\ 0 \\ -30 \\ 16 \end{bmatrix}$; $-30x + 16$

23. a. $D(x^2 e^x) = x^2 e^x + 2x e^x$; $D^2(x^2 e^x) = x^2 e^x + 4x e^x + 2e^x$; $D(x e^x) = x e^x + e^x$; $D^2(x e^x) = x e^x + 2e^x$; $D(e^x) = e^x$; $D^2(e^x) = e^x$

$A = \begin{bmatrix} 1 & 0 & 0 \\ 4 & 1 & 0 \\ 2 & 2 & 1 \end{bmatrix}$

b. From the computations in part a, we have $A_1 = \begin{bmatrix} 1 & 0 & 0 \\ 2 & 1 & 0 \\ 0 & 1 & 1 \end{bmatrix}$, and computation shows that $A_1^2 = A$.

25. We obtain $A = \begin{bmatrix} -2 & 2 \\ 2 & -2 \end{bmatrix}$ in both part a and part b.

27. $\begin{bmatrix} 9 & 0 & 0 \\ 0 & 25 & 0 \\ 0 & 0 & 81 \end{bmatrix}$

29. $(a + c)e^{2x} + b e^{4x} + (a + c)e^{8x}$ **31.** $\begin{bmatrix} -3 & -3 \\ 4 & -4 \end{bmatrix}$

33. $-2b \sin 2x + 2a \cos 2x$

45. If A_1 is the representation of T_1 and A_2 is the representation of T_2 relative to B,B', then $A_1 + A_2$ is the representation of $T_1 + T_2$ relative to B,B'. If A is the representation of T relative to B,B', then rA is the representation of rT relative to B,B'

51. Let $V = D_\infty$, the space of infinitely differentiable functions mapping $\mathbb{R}$ into $\mathbb{R}$. Let $T(f) = f'$ for $f \in D_\infty$. Then range$(T) = D_\infty$ because every infinitely differentiable function is continuous and thus has an antiderivative, but $T(x) = T(x + 1)$ shows that T is not one-to-one.

Section 3.5

1. Not an inner product

3. Not an inner product

5. Not an inner product

7. An inner product

9. Not an inner product

11. **a.** $\dfrac{5}{6}$ **b.** $\dfrac{1}{\sqrt{3}}$ **c.** $\dfrac{1}{\sqrt{30}}$ **d.** $\dfrac{1}{\sqrt{2}}$

13. $f(x) = -x + \frac{1}{2}$ and $g(x) = \cos \pi x$. Other answers are possible.

15. 77

CHAPTER 4

Section 4.1

1. -15 **3.** 15

5. $\mathbf{b} \cdot (\mathbf{b} \times \mathbf{c}) = \begin{vmatrix} b_1 & b_2 & b_3 \\ b_1 & b_2 & b_3 \\ c_1 & c_2 & c_3 \end{vmatrix}$
$= b_1(b_2c_3 - b_3c_2) - b_2(b_1c_3 - b_3c_1) + b_3(b_1c_2 - b_2c_1)$
$= 0,$

$\mathbf{c} \cdot (\mathbf{b} \times \mathbf{c}) = \begin{vmatrix} c_1 & c_2 & c_3 \\ b_1 & b_2 & b_3 \\ c_1 & c_2 & c_3 \end{vmatrix}$
$= c_1(b_2c_3 - b_3c_2) - c_2(b_1c_3 - b_3c_1) + c_3(b_1c_2 - b_2c_1)$
$= 0$

7. 120   **9.** -9

11. $\begin{vmatrix} a_1 & a_2 \\ b_1 & b_2 \end{vmatrix} = a_1b_2 - a_2b_1 = -(b_1a_2 - b_2a_1)$

$\qquad = -\begin{vmatrix} b_1 & b_2 \\ a_1 & a_2 \end{vmatrix}$

13. $-6\mathbf{i} + 3\mathbf{j} + 5\mathbf{k}$ **15.** $0\mathbf{i} + 0\mathbf{j} + 0\mathbf{k}$

17. $22\mathbf{i} + 18\mathbf{j} + 2\mathbf{k}$

19. F T T F F T F T T F

21. 38   **23.** $\sqrt{62}$ **25.** $\dfrac{19}{2}$

27. $\dfrac{\sqrt{230}}{2}$ **29.** 16   **31.** $\sqrt{390}$

33. $\mathbf{a} \cdot (\mathbf{b} \times \mathbf{c}) = -6,$
$\mathbf{a} \times (\mathbf{b} \times \mathbf{c}) = 12\mathbf{i} + 4\mathbf{k}$

35. $\mathbf{a} \cdot (\mathbf{b} \times \mathbf{c}) = 19,$
$\mathbf{a} \times (\mathbf{b} \times \mathbf{c}) = 3\mathbf{i} - 7\mathbf{j} + \mathbf{k}$

37. 20   **39.** 9   **41.** 1   **43.** $\dfrac{7}{3}$

45. Not collinear   **47.** Collinear

49. Not coplanar   **51.** Not coplanar

53. 0   **55.** $\|\mathbf{a}\|^2\|\mathbf{b}\|^2$

57. $\mathbf{i} \times (\mathbf{i} \times \mathbf{j}) = \mathbf{i} \times \mathbf{k} = -\mathbf{j}$, but $(\mathbf{i} \times \mathbf{i}) \times \mathbf{j} = \mathbf{0} \times \mathbf{j} = \mathbf{0}$.

59. $\mathbf{b} \times \mathbf{c} = \begin{vmatrix} \mathbf{i} & \mathbf{j} & \mathbf{k} \\ b_1 & b_2 & b_3 \\ c_1 & c_2 & c_3 \end{vmatrix} = (b_2c_3 - b_3c_2)\mathbf{i} -$

$(b_1c_3 - b_3c_1)\mathbf{j} + (b_1c_2 - b_2c_1)\mathbf{k}$. Thus,

$\mathbf{a} \cdot (\mathbf{b} \times \mathbf{c}) = a_1(b_2c_3 - b_3c_2) -$
$\qquad a_2(b_1c_3 - b_3c_1) +$
$\qquad a_3(b_1c_2 - b_2c_1)$

$= \begin{vmatrix} a_1 & a_2 & a_3 \\ b_1 & b_2 & b_3 \\ c_1 & c_2 & c_3 \end{vmatrix}.$

Equation (4) in the text shows that this determinant is $\pm$(Volume of the box determined by $\mathbf{a}$, $\mathbf{b}$, and $\mathbf{c}$). Similarly,

$(\mathbf{a} \times \mathbf{b}) \cdot \mathbf{c} = \mathbf{c} \cdot (\mathbf{a} \times \mathbf{b}) = \begin{vmatrix} c_1 & c_2 & c_3 \\ a_1 & a_2 & a_3 \\ b_1 & b_2 & b_3 \end{vmatrix}$

$= c_1(a_2b_3 - a_3b_2) -$
$\qquad c_2(a_1b_3 - a_3b_1) +$
$\qquad c_3(a_1b_2 - a_2b_1),$

which is the same number.

61. 322   **63.** 130.8097

M3. a. $67.38\mathbf{i} - 88.76\mathbf{j} - 8.11\mathbf{k}$

    b. 111.7326

Section 4.2

1. 13   **3.** -21 **5.** 19   **7.** 320

9. 0   **11.** -6 **13.** 2   **15.** 4

17. 54   **19.** $\frac{1}{2}$

21. F T T F T T F T F

23. 0   **25.** 6   **27.** 5, -2 **29.** 1, 6, -2

33. Let A have column vectors $\mathbf{a}_1, \mathbf{a}_2, \ldots, \mathbf{a}_n$. Let

$$B = \begin{bmatrix} b_1 & & & \\ & b_2 & & \mathbf{0} \\ & & \ddots & \\ \mathbf{0} & & & b_n \end{bmatrix}.$$

Then $AB = \begin{bmatrix} | & | & & | \\ b_1\mathbf{a}_1 & b_2\mathbf{a}_2 & \cdots & b_n\mathbf{a}_n \\ | & | & & | \end{bmatrix}$.

Applying the column scaling property n times, we have

$$\det(AB) = b_1 \begin{vmatrix} | & | & & | \\ \mathbf{a}_1 & b_2\mathbf{a}_2 & \cdots & b_n\mathbf{a}_n \\ | & | & & | \end{vmatrix}$$

$$= b_1 b_2 \begin{vmatrix} | & | & | & & | \\ \mathbf{a}_1 & \mathbf{a}_2 & b_3\mathbf{a}_3 & \cdots & b_n\mathbf{a}_n \\ | & | & | & & | \end{vmatrix}$$

$$= \cdots = (b_1 b_2 \cdots b_n)\det(A)$$

$$= \det(B)\det(A).$$

Section 4.3

1. 27   **3.** -8 **5.** -195

7. -207 **9.** 496

13. Let $R_1, R_2, \ldots, R_k$ be square submatrices positioned corner-to-corner along the diagonal of a matrix A, where A has zero entries except possibly for those in the submatrices R_i. Then $\det(A) = \det(R_1)\det(R_2) \cdots \det(R_k)$.

15. $A^{-1} = \begin{bmatrix} \frac{1}{2} & -\frac{1}{2} \\ -1 & 2 \end{bmatrix}$

17. $A^{-1} = -\dfrac{1}{17} \begin{bmatrix} 1 & 4 & -4 \\ 7 & -6 & -11 \\ -5 & -3 & 3 \end{bmatrix}$

19. $A^{-1} = -\dfrac{1}{13} \begin{bmatrix} -7 & 5 & 1 \\ 4 & -1 & -8 \\ -1 & -3 & 2 \end{bmatrix}$

21. $\text{adj}(A) = \begin{bmatrix} -7 & -1 & 4 \\ -3 & 2 & -8 \\ 6 & -4 & -1 \end{bmatrix}$

25. The coefficient matrix has determinant 0, so the system cannot be solved using Cramer's rule.

27. $x_1 = 0$, $x_2 = 1$

29. $x_1 = 3$, $x_2 = -\frac{11}{3}$, $x_3 = -\frac{7}{3}$

31. $x_1 = 0$, $x_2 = \frac{1}{2}$, $x_3 = 0$

33. $x_2 = \frac{41}{59}$

35. F T T F T T F T T F

41. 19095. The answers are the same.

43. With default roundoff, the smallest such value of m is 6, which gave 0 rather than 1 as the value of the determinant. Although this wrong result would lead us to believe that A^6 is singular, the error is actually quite small compared with the size of the entries in A^6. We obtained 0 as $\det(A^m)$ for $6 \leq m \leq 17$, and we obtained -1.470838×10^{10} as $\det(A^{18})$. MATCOMP gave us -1.598896×10^{13} as $\det(A^{20})$.

    With roundoff control 0, we obtained $\det(A^6) = 1$ and $\det(A^7) = 0.999998$, which has an error of only 0.000002. The error became significantly larger with $m = 12$, where we obtained 11.45003 for the determinant. We obtained the same result with $m = 20$ as we did with the default roundoff.

With the default roundoff, computed entries of magnitude less than (0.0001)(Smallest nonzero magnitude in A^m) are set equal to zero when A^m is reduced to echelon form. As soon as m is large enough (so that the entries of A^m are large enough), this creates a false zero entry on the diagonal, which produces a false calculation of 0 for the determinant. With roundoff zero, this does not happen. But when m get sufficiently large, roundoff error in the computation of A^m and in its reduction to echelon form creates even greater error in the calculation of the determinant, no matter what roundoff control number is taken. Notice, however, that the value given for the determinant is always small compared with the size of the entries in A^m.

45. $\begin{bmatrix} 12 & -8 & -7 \\ -11 & 13 & 5 \\ 9 & -6 & -1 \end{bmatrix}$

47. $\begin{bmatrix} -4827 & 114 & -2211 & 2409 \\ -3218 & 76 & -1474 & 1606 \\ 8045 & -190 & 3685 & -4015 \\ 1609 & -38 & 737 & -803 \end{bmatrix}$

Section 4.4

1. $\sqrt{213}$ **3.** $\sqrt{30}$ **5.** 11   **7.** 38

9. 2   **11.** $\dfrac{\sqrt{6}}{2}$ **13.** $\dfrac{2}{3}$

15. Let **a**, **b**, **c**, and **d** be the vectors from the origin to the four points. Let A be the $n \times 3$ matrix with column vectors $\mathbf{b} - \mathbf{a}$, $\mathbf{c} - \mathbf{a}$, and $\mathbf{d} - \mathbf{a}$. The points are coplanar if and only if $\det(A^T A) = 0$.

17. The points are not coplanar.   **19.** 32

21. 144π **23.** 264   **25.** $\dfrac{2816\pi}{3}$

27. $5\sqrt{3}$ **29.** $25\pi\sqrt{3}$ **31.** $16\sqrt{17}$

33. a. 0   **b.** 0

35. T F T F T F T T F T

CHAPTER 5

Section 5.1

1. $\mathbf{v}_1$, $\mathbf{v}_3$, and $\mathbf{v}_5$ are eigenvectors of A_1, with corresponding eigenvalues -1, 2, and 2, respectively. $\mathbf{v}_2$, $\mathbf{v}_3$, and $\mathbf{v}_4$ are eigenvectors of A_2, with corresponding eigenvalues 5, 5, and 1, respectively.

3. *Characteristic polynomial*: $\lambda^2 - 4\lambda + 3$
Eigenvalues: $\lambda_1 = 1$, $\lambda_2 = 3$
Eigenvectors: for $\lambda_1 = 1$: $\mathbf{v}_1 = \begin{bmatrix} -r \\ r \end{bmatrix}$, $r \neq 0$,

for $\lambda_2 = 3$: $\mathbf{v}_2 = \begin{bmatrix} -s \\ 2s \end{bmatrix}$, $s \neq 0$

5. *Characteristic polynomial*: $\lambda^2 + 1$
Eigenvalues: There are no real eigenvalues.

7. *Characteristic polynomial*: $-\lambda^3 + 2\lambda^2 + \lambda - 2$
Eigenvalues: $\lambda_1 = -1$, $\lambda_2 = 1$, $\lambda_3 = 2$
Eigenvectors: for $\lambda_1 = -1$: $\mathbf{v}_1 = \begin{bmatrix} 0 \\ r \\ 0 \end{bmatrix}$, $r \neq 0$,

for $\lambda_2 = 1$: $\mathbf{v}_2 = \begin{bmatrix} 0 \\ -s \\ s \end{bmatrix}$, $s \neq 0$,

for $\lambda_3 = 2$: $\mathbf{v}_3 = \begin{bmatrix} -t \\ -t \\ t \end{bmatrix}$, $t \neq 0$

9. *Characteristic polynomial*: $-\lambda^3 + 8\lambda^2 + \lambda - 8$
Eigenvalues: $\lambda_1 = -1$, $\lambda_2 = 1$, $\lambda_3 = 8$
Eigenvectors: for $\lambda_1 = -1$: $\mathbf{v}_1 = \begin{bmatrix} 0 \\ r \\ 0 \end{bmatrix}$, $r \neq 0$,

for $\lambda_2 = 1$: $\mathbf{v}_2 = \begin{bmatrix} 0 \\ -s \\ s \end{bmatrix}$, $s \neq 0$

for $\lambda_3 = 8$: $\mathbf{v}_3 = \begin{bmatrix} -t \\ -t \\ t \end{bmatrix}$, $t \neq 0$

11. *Characteristic polynomial*: $-\lambda^3 - \lambda^2 + 8\lambda + 12$
Eigenvalues: $\lambda_1 = \lambda_2 - 2$, $\lambda_3 = 3$
Eigenvectors: for λ_1, $\lambda_2 = -2$: $\mathbf{v}_1 = \begin{bmatrix} -r \\ s \\ r \end{bmatrix}$,

r and s not both 0,

for $\lambda_3 = 3$: $\mathbf{v}_3 = \begin{bmatrix} 0 \\ -t \\ t \end{bmatrix}, t \neq 0$

13. Characteristic polynomial: $-\lambda^3 + 2\lambda^2 + 4\lambda - 8$
Eigenvalues: $\lambda_1 = -2, \lambda_2 = \lambda_3 = 2$
Eigenvectors: for $\lambda_1 = -2$: $\mathbf{v}_1 = \begin{bmatrix} -r \\ -3r \\ r \end{bmatrix}, r \neq 0,$

for $\lambda_2, \lambda_3 = 2$: $\mathbf{v}_2 = \begin{bmatrix} 0 \\ s \\ 0 \end{bmatrix}, s \neq 0$

15. Characteristic polynomial: $-\lambda^3 - 3\lambda^2 + 4$
Eigenvalues: $\lambda_1 = \lambda_2 = -2, \lambda_3 = 1$
Eigenvectors: for $\lambda_1, \lambda_2 = -2$: $\mathbf{v}_1 = \begin{bmatrix} 0 \\ r \\ 0 \end{bmatrix}, r \neq 0,$

for $\lambda_3 = 1$: $\mathbf{v}_3 = \begin{bmatrix} 3s \\ -s \\ 3s \end{bmatrix}, s \neq 0$

17. Eigenvalues: $\lambda_1 = -1, \lambda_2 = 5$
Eigenvectors: for $\lambda_1 = -1$: $\mathbf{v}_1 = \begin{bmatrix} r \\ r \end{bmatrix}, r \neq 0,$

for $\lambda_2 = 5$: $\mathbf{v}_2 = \begin{bmatrix} -s \\ s \end{bmatrix}, s \neq 0$

19. Eigenvalues: $\lambda_1 = 0, \lambda_2 = 1, \lambda_3 = 2$
Eigenvectors: for $\lambda_1 = 0$: $\mathbf{v}_1 = \begin{bmatrix} -r \\ 0 \\ r \end{bmatrix}, r \neq 0,$

for $\lambda_2 = 1$: $\mathbf{v}_2 = \begin{bmatrix} 0 \\ s \\ 0 \end{bmatrix}, s \neq 0,$

for $\lambda_3 = 2$: $\mathbf{v}_3 = \begin{bmatrix} t \\ 0 \\ t \end{bmatrix}, t \neq 0$

21. Eigenvalues: $\lambda_1 = -2, \lambda_2 = 1, \lambda_3 = 5$
Eigenvectors: for $\lambda_1 = -2$: $\mathbf{v}_1 = \begin{bmatrix} 0 \\ r \\ r \end{bmatrix}, r \neq 0,$

for $\lambda_2 = 1$: $\mathbf{v}_2 = \begin{bmatrix} -6s \\ 5s \\ 8s \end{bmatrix}, s \neq 0,$

for $\lambda_3 = 5$: $\mathbf{v}_3 = \begin{bmatrix} 0 \\ -5t \\ 2t \end{bmatrix}, t \neq 0$

23. F F T T F T F T F T

31. Eigenvalues: $\lambda_1 = \dfrac{1 + \sqrt{5}}{2}, \lambda_2 = \dfrac{1 - \sqrt{5}}{2}$

Eigenvectors: for $\lambda_1 = \dfrac{1 + \sqrt{5}}{2}$:

$\mathbf{v}_1 = r \begin{bmatrix} (1 + \sqrt{5})/2 \\ 1 \end{bmatrix}, r \neq 0,$

for $\lambda_2 = \dfrac{1 - \sqrt{5}}{2}$:

$\mathbf{v}_2 = s \begin{bmatrix} (1 - \sqrt{5})/2 \\ 1 \end{bmatrix}, s \neq 0$

33. a. Work with the matrix $A + 10I$, whose eigenvalues are approximately 30, 12, 7, and -9.5. (Other answers are possible.);

b. Work with the matrix $A - 10I$, whose eigenvalues are approximately -9.5, -8, -13, and -30. (Other answers are possible.)

37. When the determinant of an $n \times n$ matrix A is expanded according to the definition using expansion by minors, a sum of $n!$ (signed) terms is obtained, each containing a product of one element from each row and from each column. (This is readily proved by induction on n.) One of the $n!$ terms obtained by expanding $|A - \lambda I|$ to obtain $p(\lambda)$ is

$(a_{11} - \lambda)(a_{22} - \lambda)(a_{33} - \lambda) \cdots (a_{nn} - \lambda).$

We claim that this is the only one of the $n!$ terms that can contribute to the coefficient of λ^{n-1} in $p(\lambda)$. Any term contributing to the coefficient of λ^{n-1} must contain at least $n - 1$ of the factors in the preceding product; the other factor must be from the remaining row and column, and hence it must be the remaining factor from the diagonal of $A - \lambda I$. Computing the coefficient of λ^{n-1} in the preceding product, we find that, when $-\lambda$ is chosen from all but the factor $a_{ii} - \lambda$ in expanding the product, the resulting contribution to the coefficient of λ^{n-1} in $p(\lambda)$ is $(-1)^{n-1}a_{ii}$. Thus, the coefficient of λ^{n-1} in $p(\lambda)$ is

$(-1)^{n-1}(a_{11} + a_{22} + a_{33} + \cdots + a_{nn}).$

Now $p(\lambda) = (-1)^n(\lambda - \lambda_1)(\lambda - \lambda_2)(\lambda - \lambda_3)$ $\cdots (\lambda - \lambda_n)$, and computation of this product shows in the same way that the coefficient of λ^{n-1} is also

$(-1)^{n-1}(\lambda_1 + \lambda_2 + \lambda_3 + \cdots + \lambda_n)$.

Thus, $\text{tr}(A) = \lambda_1 + \lambda_2 + \lambda_3 + \cdots + \lambda_n$.

39. Because $\begin{vmatrix} 2 - \lambda & -1 \\ 1 & 3 - \lambda \end{vmatrix} = \lambda^2 - 5\lambda + 7$, we compute

$$A^2 - 5A + 7I = \begin{bmatrix} 2 & -1 \\ 1 & 3 \end{bmatrix}^2 - 5\begin{bmatrix} 2 & -1 \\ 1 & 3 \end{bmatrix} +$$

$$7\begin{bmatrix} 1 & 0 \\ 0 & 1 \end{bmatrix} = \begin{bmatrix} 3 & -5 \\ 5 & 8 \end{bmatrix} + \begin{bmatrix} -10 & 5 \\ -5 & -15 \end{bmatrix} + \begin{bmatrix} 7 & 0 \\ 0 & 7 \end{bmatrix}$$

$$= \begin{bmatrix} 0 & 0 \\ 0 & 0 \end{bmatrix},$$

illustrating the Cayley–Hamilton theorem.

43. The desired result is stated as Theorem 5.3 in Section 5.2. Because an $n \times n$ matrix is the standard matrix representation of the linear transformation T: $\mathbb{R}^n \to \mathbb{R}^n$ and the eigenvalues and eigenvectors of the transformation are the same as those of the matrix, proof of Exercise 42 for transformations implies the result for matrices, and vice versa.

45. a. $F_8 = 21$;

b. $F_{30} = 832040$;

c. $F_{50} = 12586269025$;

d. $F_{77} = 5527939700884757$;

e. $F_{150} \approx 9.969216677189304 \times 10^{30}$

47. a. 0, 1, 2, 1, −3, −7, −4, 10, 25;

b. $\begin{bmatrix} 2 & -3 & 1 \\ 1 & 0 & 0 \\ 0 & 1 & 0 \end{bmatrix}$;

c. $a_{30} = -191694$

49. $\lambda_1 = -1, \mathbf{v}_1 = \begin{bmatrix} r \\ -r \\ r \\ r \end{bmatrix}, r \neq 0; \lambda_2 = 3, \mathbf{v}_2 = \begin{bmatrix} 0 \\ 0 \\ 0 \\ s \end{bmatrix},$

$s \neq 0; \lambda_3 = \lambda_4 = 2, \mathbf{v}_3 = \begin{bmatrix} 0 \\ -t \\ u \\ t \end{bmatrix}, t$ and u not

both zero

51. $\lambda_1 = 16, \mathbf{v}_1 = \begin{bmatrix} 0 \\ r \\ 0 \\ 0 \end{bmatrix}, r \neq 0; \lambda_2 = 36, \mathbf{v}_2 = \begin{bmatrix} 0 \\ 0 \\ s \\ s \end{bmatrix},$

$s \neq 0; \lambda_3 = \lambda_4 = 4, \mathbf{v}_3 = \begin{bmatrix} t + u \\ t \\ t \\ u \end{bmatrix}, t$ and u

not both zero

53. *Characteristic polynomial:* $-\lambda^3 - 16\lambda^2 + 48\lambda - 261$

Eigenvalues: $\lambda_1 \approx 1.6033 + 3.3194i$,
$\lambda_2 \approx 1.6033 - 3.3194i, \lambda_3 \approx -19.2067$

55. *Characteristic polynomial:* $\lambda^4 - 56\lambda^3 + 210\lambda^2 + 22879\lambda + 678658$

Eigenvalues: $\lambda_1 \approx -12.8959 + 13.3087i$,
$\lambda_2 \approx -12.8959 - 13.3087i$,
$\lambda_3 \approx 40.8959 + 17.4259i$,
$\lambda_4 \approx 40.8959 - 17.4259i$

M1. a. $\lambda^3 + 16\lambda^2 - 48\lambda + 261$;
$\lambda_1 \approx -19.2067$,
$\lambda_2 \approx 1.6033 + 3.3194i$,
$\lambda_3 \approx 1.6033 - 3.3194i$

b. $\lambda^4 + 14\lambda^3 - 131\lambda^2 + 739\lambda - 21533$;
$\lambda_1 \approx -22.9142$,
$\lambda_2 \approx 10.4799$,
$\lambda_3 \approx -0.7828 + 9.4370i$,
$\lambda_4 \approx -0.7828 - 9.4370i$

M3. λ of minimum magnitude $\approx -2.68877026277667$

M7. λ of maximum magnitude ≈ 19.2077
λ of minimum magnitude ≈ 8.0147

Section 5.2

1. *Eigenvalues:* $\lambda_1 = -5, \lambda_2 = 5$

Eigenvectors: for $\lambda_1 = -5$: $\mathbf{v}_1 = \begin{bmatrix} -2r \\ r \end{bmatrix}, r \neq 0$,

for $\lambda_2 = 5$: $\mathbf{v}_2 = \begin{bmatrix} s \\ 2s \end{bmatrix}, s \neq 0$

$C = \begin{bmatrix} -2 & 1 \\ 1 & 2 \end{bmatrix}, D = \begin{bmatrix} -5 & 0 \\ 0 & 5 \end{bmatrix}$

3. *Eigenvalues:* $\lambda_1 = -1,\ \lambda_2 = 3$

Eigenvectors: for $\lambda_1 = -1$: $\mathbf{v}_1 = \begin{bmatrix} -r \\ r \end{bmatrix},\ r \neq 0,$

for $\lambda_2 = 3$: $\mathbf{v}_2 = \begin{bmatrix} -2s \\ s \end{bmatrix},\ s \neq 0$

$C = \begin{bmatrix} -1 & -2 \\ 1 & 1 \end{bmatrix},\ D = \begin{bmatrix} -1 & 0 \\ 0 & 3 \end{bmatrix}$

5. *Eigenvalues:* $\lambda_1 = -3,\ \lambda_2 = 1,\ \lambda_3 = 7$

Eigenvectors: for $\lambda_1 = -3$: $\mathbf{v}_1 = \begin{bmatrix} r \\ 0 \\ 0 \end{bmatrix},\ r \neq 0,$

for $\lambda_2 = 1$: $\mathbf{v}_2 = \begin{bmatrix} s \\ s \\ s \end{bmatrix},\ s \neq 0,$

for $\lambda_3 = 7$: $\mathbf{v}_3 = \begin{bmatrix} t \\ t \\ 0 \end{bmatrix},\ t \neq 0$

$C = \begin{bmatrix} 1 & 1 & 1 \\ 0 & 1 & 1 \\ 0 & 1 & 0 \end{bmatrix},\ D = \begin{bmatrix} -3 & 0 & 0 \\ 0 & 1 & 0 \\ 0 & 0 & 7 \end{bmatrix}$

7. *Eigenvalues:* $\lambda_1 = -1,\ \lambda_2 = 1,\ \lambda_3 = 2$

Eigenvectors: for $\lambda_1 = -1$: $\mathbf{v}_1 = \begin{bmatrix} -r \\ 0 \\ r \end{bmatrix},\ r \neq 0,$

for $\lambda_2 = 1$: $\mathbf{v}_2 = \begin{bmatrix} -s \\ 0 \\ 3s \end{bmatrix},\ s \neq 0,$

for $\lambda_3 = 2$: $\mathbf{v}_3 = \begin{bmatrix} 0 \\ t \\ 0 \end{bmatrix},\ t \neq 0$

$C = \begin{bmatrix} -1 & -1 & 0 \\ 0 & 0 & 1 \\ 1 & 3 & 0 \end{bmatrix},\ D = \begin{bmatrix} -1 & 0 & 0 \\ 0 & 1 & 0 \\ 0 & 0 & 2 \end{bmatrix}$

9. Yes, the matrix is symmetric.

11. Yes, the eigenvalues are distinct.

13. F T T F T F F T F T

15. Assume that A is a square matrix such that $D = C^{-1}AC$ is a diagonal matrix for some invertible matrix C. Because $CC^{-1} = I$, we have $(CC^{-1})^T = (C^{-1})^TC^T = I^T = I$, so $(C^T)^{-1} = (C^{-1})^T$. Then $D^T = (C^{-1}AC)^T = C^TA^T(C^T)^{-1}$, so A^T is similar to the diagonal matrix D^T.

19. b. $A = O$ **c.** $\begin{bmatrix} 0 & 1 \\ 0 & 0 \end{bmatrix}$

21. $T([x, y]) = \dfrac{1}{1+m^2}[(1 - m^2)x + 2my,$
$2mx + (m^2 - 1)y]$

23. If A and B are similar square matrices, then $B = C^{-1}AC$ for some invertible matrix C. Let λ be an eigenvalue of A, and let $\{\mathbf{v}_1, \mathbf{v}_2, \ldots, \mathbf{v}_k\}$ be a basis for E_λ. Then Exercise 22 shows that $C^{-1}\mathbf{v}_1, C^{-1}\mathbf{v}_2, \ldots,$ $C^{-1}\mathbf{v}_k$ are eigenvectors of B corresponding to λ, and Exercise 37 in Section 2.1 shows that these vectors are independent. Thus the eigenspace of λ relative to B has dimension at least as great as the eigenspace of λ relative to A. By the symmetry of the similarity relation (see Exercise 16), the eigenspace of λ relative to A has dimension at least as great as the eigenspace of λ relative to B. Thus the dimensions of these eigenspaces are equal—that is, the geometric multiplicity of λ is the same relative to A as relative to B.

31. Diagonalizable

33. Complex (but not real) diagonalizable

35. Complex (but not real) diagonalizable

37. Not diagonalizable

39. Real diagonalizable

41. Complex (but not real) diagonalizable

Section 5.3

1. a. $A = \begin{bmatrix} \frac{1}{2} & \frac{1}{2} \\ 1 & 0 \end{bmatrix}$;

b. Neutrally stable;

c. $\begin{bmatrix} a_{k+1} \\ a_k \end{bmatrix} = A^k \begin{bmatrix} 1 \\ 0 \end{bmatrix} = \dfrac{2}{3}(1)^k \begin{bmatrix} 1 \\ 1 \end{bmatrix} - \dfrac{1}{3}\left(-\dfrac{1}{2}\right)^k \begin{bmatrix} -1 \\ 2 \end{bmatrix}$. The sequence starts $0,\ 1,\ \dfrac{1}{2},\ \dfrac{3}{4},\ \dfrac{5}{8}$

and $A^3 \begin{bmatrix} 1 \\ 0 \end{bmatrix} = \dfrac{2}{3} \begin{bmatrix} 1 \\ 1 \end{bmatrix} + \dfrac{1}{24} \begin{bmatrix} -1 \\ 2 \end{bmatrix} = \begin{bmatrix} \frac{5}{8} \\ \frac{3}{4} \end{bmatrix}$,

which checks.

d. For large k, we have $\begin{bmatrix} a_{k+1} \\ a_k \end{bmatrix} \approx \dfrac{2}{3} \begin{bmatrix} 1 \\ 1 \end{bmatrix} = \begin{bmatrix} \frac{2}{3} \\ \frac{2}{3} \end{bmatrix}$, so $a_k \approx \dfrac{2}{3}$.

3. a. $A = \begin{bmatrix} \frac{1}{2} & \frac{1}{2} \\ 1 & 0 \end{bmatrix}$;   **b.** Neutrally stable;

c. $\begin{bmatrix} a_{k+1} \\ a_k \end{bmatrix} = A^k \begin{bmatrix} 0 \\ 1 \end{bmatrix} = \frac{1}{3}(1)^k \begin{bmatrix} 1 \\ 1 \end{bmatrix} + \frac{1}{3}\left(-\frac{1}{2}\right)^k \begin{bmatrix} -1 \\ 2 \end{bmatrix}$. The sequence starts $1, 0, \frac{1}{2}, \frac{1}{4}, \frac{3}{8}$

and $A^3 \begin{bmatrix} 1 \\ 1 \end{bmatrix} = \frac{1}{3}\begin{bmatrix} 1 \\ 1 \end{bmatrix} - \frac{1}{24}\begin{bmatrix} -1 \\ 2 \end{bmatrix} = \begin{bmatrix} \frac{3}{8} \\ \frac{1}{4} \end{bmatrix}$,

which checks.

d. For large k, we have $\begin{bmatrix} a_{k+1} \\ a_k \end{bmatrix} \approx \frac{1}{3}\begin{bmatrix} 1 \\ 1 \end{bmatrix} = \begin{bmatrix} \frac{1}{3} \\ \frac{1}{3} \end{bmatrix}$,

so $a_k \approx \frac{1}{3}$.

5. a. $A = \begin{bmatrix} 1 & \frac{3}{4} \\ 1 & 0 \end{bmatrix}$;   **b.** Unstable;

c. $\begin{bmatrix} a_{k+1} \\ a_k \end{bmatrix} = A^k \begin{bmatrix} 1 \\ 0 \end{bmatrix} = \frac{1}{4}\left(\frac{3}{2}\right)^k \begin{bmatrix} 3 \\ 2 \end{bmatrix} - \frac{1}{4}\left(-\frac{1}{2}\right)^k \begin{bmatrix} -1 \\ 2 \end{bmatrix}$. The sequence starts $0, 1, 1, \frac{7}{4}, \frac{5}{2}$

and $A^3 \begin{bmatrix} 1 \\ 0 \end{bmatrix} = \frac{27}{32}\begin{bmatrix} 3 \\ 2 \end{bmatrix} + \frac{1}{32}\begin{bmatrix} -1 \\ 2 \end{bmatrix} = \begin{bmatrix} \frac{80}{32} \\ \frac{56}{32} \end{bmatrix}$

$= \begin{bmatrix} \frac{5}{2} \\ \frac{7}{4} \end{bmatrix}$, which checks.

d. For large k, we have $\begin{bmatrix} a_{k+1} \\ a_k \end{bmatrix} \approx \frac{1}{4}\left(\frac{3}{2}\right)^k \begin{bmatrix} 3 \\ 2 \end{bmatrix}$,

so $a_k \approx \frac{3^k}{2^{k+1}}$; and a_k approaches ∞ as k approaches ∞.

7. $\begin{bmatrix} x_1 \\ x_2 \end{bmatrix} = \begin{bmatrix} -k_1 e^{-3t} + 4k_2 e^{4t} \\ k_1 e^{-3t} + 3k_2 e^{4t} \end{bmatrix}$

9. $\begin{bmatrix} x_1 \\ x_2 \end{bmatrix} = \begin{bmatrix} -2k_1 e^t + k_2 e^{4t} \\ k_1 e^t + k_2 e^{4t} \end{bmatrix}$

11. $\begin{bmatrix} x_1 \\ x_2 \\ x_3 \end{bmatrix} = \begin{bmatrix} k_1 e^{-3t} + k_2 e^t + k_3 e^{7t} \\ k_2 e^t + k_3 e^{7t} \\ k_2 e^t \end{bmatrix}$

13. $\begin{bmatrix} x_1 \\ x_2 \\ x_3 \end{bmatrix} = \begin{bmatrix} -k_1 e^{-t} - k_2 e^t \\ k_3 e^{2t} \\ k_1 e^{-t} + 3k_2 e^t \end{bmatrix}$

CHAPTER 6

Section 6.1

1. $\frac{2}{5}[3, 4]$

3. $\mathbf{p}_1 = [1, 0, 0]$, $\mathbf{p}_2 = [0, 2, 0]$, $\mathbf{p}_3 = [0, 0, 1]$

5. $\mathbf{p} = -\frac{1}{3}[2, -3, 1, 2]$

7. $\text{sp}([1, 0, 1], [-2, 1, 0])$

9. $\text{sp}([-12, 4, 5])$

11. $\text{sp}([2, -7, 1, 0], [-1, -2, 0, 1])$

13. a. $-5\mathbf{i} + 3\mathbf{j} + \mathbf{k}$ **b.** $-5\mathbf{i} + 3\mathbf{j} + \mathbf{k}$

15. $\frac{1}{3}[5, 4, 1]$ **17.** $\frac{1}{7}[5, 3, 1]$

19. $\frac{1}{6}[2, -1, 5]$ **21.** $\frac{1}{3}[3, -2, -1, 1]$

23. F T T T F T T F F T   **29.** $\frac{4}{5}\sqrt{5}$

31. $\sqrt{\frac{14}{5}}$ **33.** $\frac{\sqrt{161}}{3\sqrt{3}}$ **35.** $\sqrt{10}$ **37.** $\frac{3}{2}x$

Section 6.2

1. $[2, 3, 1] \cdot [-1, 1, -1] = -2 + 3 - 1 = 0$, so the generating set is orthogonal.
$\mathbf{b}_W = \frac{1}{42}[136, 29, 103]$.

3. $[1, -1, -1, 1] \cdot [1, 1, 1, 1] = 1 - 1 - 1 + 1 = 0$,

$[1, -1, -1, 1] \cdot [-1, 0, 0, 1] = -1 + 0 + 0 + 1 = 0$, and

$[1, 1, 1, 1] \cdot [-1, 0, 0, 1] = -1 + 0 + 0 + 1 = 0$,

so the generating set is orthogonal; $\mathbf{b}_W = [2, 2, 2, 1]$.

5. $\left\{ \frac{1}{\sqrt{5}}[1, 0, -2], \frac{1}{\sqrt{70}}[6, -5, 3] \right\}$

7. $\left\{ [0, 1, 0], \frac{1}{\sqrt{2}}[1, 0, 1] \right\}$

9. $\left\{ \frac{1}{\sqrt{2}}[1, 0, 1], \frac{1}{\sqrt{3}}[-1, 1, 1], \frac{1}{\sqrt{6}}[1, 2, -1] \right\}$

11. $\left\{ \frac{1}{\sqrt{2}}[1, 0, 1, 0], [0, 1, 0, 0], \frac{1}{\sqrt{6}}[1, 0, -1, 2] \right\}$

13. $\left[\dfrac{9}{2}, -3, \dfrac{9}{2}\right]$ **15.** $\left[\dfrac{4}{3}, 0, -\dfrac{1}{3}, \dfrac{5}{3}\right]$

17. $\left\{\dfrac{1}{\sqrt{2}}[1, 0, 1, 0], \dfrac{1}{\sqrt{6}}[-1, 2, 1, 0],\right.$

$\left.\dfrac{1}{\sqrt{3}}[1, 1, -1, 0], [0, 0, 0, 1]\right\}$

19. $\{[3, -2, 0, 1], [-9, -8, 14, 11]\}$

21. $\left\{\dfrac{1}{\sqrt{6}}[2, 1, 1], \dfrac{1}{\sqrt{2}}[0, -1, 1]\right\}$

23. $\{[2, 1, -1, 1], [1, 1, 3, 0], [-24, 9, 5, 44]\}$

25. F T T F F T F T T T

27. $Q = \begin{bmatrix} 1/\sqrt{2} & -1/\sqrt{3} & 1/\sqrt{6} \\ 0 & 1/\sqrt{3} & 2/\sqrt{6} \\ 1/\sqrt{2} & 1/\sqrt{3} & -1/\sqrt{6} \end{bmatrix}$,

$R = \begin{bmatrix} \sqrt{2} & \sqrt{2} & \sqrt{2} \\ 0 & \sqrt{3} & -1/\sqrt{3} \\ 0 & 0 & 4/\sqrt{6} \end{bmatrix}$

33. $\left\{\sqrt{\dfrac{2}{\pi}} \sin x, \sqrt{\dfrac{2}{\pi}} \cos x\right\}$

35. $\left\{1, \sqrt{\dfrac{2}{4e - e^2 - 3}} (e^x - e + 1)\right\}$

Section 6.3

1. Let A be the given matrix. Then

$A^T A = \dfrac{1}{\sqrt{2}} \begin{bmatrix} 1 & -1 \\ 1 & 1 \end{bmatrix} \dfrac{1}{\sqrt{2}} \begin{bmatrix} 1 & 1 \\ -1 & 1 \end{bmatrix} = \dfrac{1}{2} \begin{bmatrix} 2 & 0 \\ 0 & 2 \end{bmatrix}$

$= \begin{bmatrix} 1 & 0 \\ 0 & 1 \end{bmatrix}$,

so A is orthogonal and $A^{-1} = A^T = \dfrac{1}{\sqrt{2}} \begin{bmatrix} 1 & -1 \\ 1 & 1 \end{bmatrix}$.

3. Let A be the given matrix. Then

$A^T A = \dfrac{1}{7} \begin{bmatrix} 2 & 3 & -6 \\ -3 & 6 & 2 \\ 6 & 2 & 3 \end{bmatrix} \dfrac{1}{7} \begin{bmatrix} 2 & -3 & 6 \\ 3 & 6 & 2 \\ -6 & 2 & 3 \end{bmatrix} = $

$\dfrac{1}{49} \begin{bmatrix} 49 & 0 & 0 \\ 0 & 49 & 0 \\ 0 & 0 & 49 \end{bmatrix} = \begin{bmatrix} 1 & 0 & 0 \\ 0 & 1 & 0 \\ 0 & 0 & 1 \end{bmatrix}$,

so A is orthogonal and $A^{-1} = A^T = \dfrac{1}{7} \begin{bmatrix} 2 & 3 & -6 \\ -3 & 6 & 2 \\ 6 & 2 & 3 \end{bmatrix}$.

5. $\begin{bmatrix} \frac{1}{2} & -\frac{1}{2} \\ \frac{1}{2} & \frac{1}{2} \\ \frac{1}{6} & \frac{1}{6} \end{bmatrix}$ **7.** $\dfrac{1}{49} \begin{bmatrix} 1 & \frac{3}{2} & -3 \\ -3 & 6 & 2 \\ 6 & 2 & 3 \end{bmatrix}$

9. $\pm\dfrac{1}{\sqrt{6}} \begin{bmatrix} -1 \\ 2 \\ -1 \end{bmatrix}$ **11.** $\pm\dfrac{\sqrt{23}}{6}$

13. $\dfrac{1}{\sqrt{2}} \begin{bmatrix} -1 & 1 \\ 1 & 1 \end{bmatrix}$ **15.** $\dfrac{1}{2} \begin{bmatrix} 1 & -2 & 1 \\ -\sqrt{2} & 0 & \sqrt{2} \\ 1 & 2 & 1 \end{bmatrix}$

17. $\begin{bmatrix} 0 & \frac{1}{2} & -1\sqrt{2} & \frac{1}{2} \\ -1/\sqrt{2} & -\frac{1}{2} & 0 & \frac{1}{2} \\ 1/\sqrt{2} & -\frac{1}{2} & 0 & \frac{1}{2} \\ 0 & \frac{1}{2} & 1/\sqrt{2} & \frac{1}{2} \end{bmatrix}$

19. F T T T T T T T F T   **23.** $\begin{bmatrix} 2 & 1 \\ 1 & 1 \end{bmatrix}$

27. An orthogonal matrix A gives rise to an orthogonal linear transformation $T(\mathbf{x}) = A\mathbf{x}$ that preserves the magnitude of vectors. Thus, if $A\mathbf{v} = \lambda\mathbf{v}$, so that $T(\mathbf{v}) = \lambda\mathbf{v}$, we must have $\|\mathbf{v}\| = \|\lambda\mathbf{v}\| = |\lambda| \|\mathbf{v}\|$. If $\mathbf{v}$ is an eigenvector, so that $\mathbf{v} \neq \mathbf{0}$, it follows that $|\lambda| = 1$; so $\lambda = \pm 1$.

33. No   **35.** Yes   **37.** Yes

Section 6.4

1. $P = \dfrac{1}{6} \begin{bmatrix} 4 & 2 & -2 \\ 2 & 1 & -1 \\ -2 & -1 & 1 \end{bmatrix}$, projection $= \dfrac{1}{2} \begin{bmatrix} 2 \\ 1 \\ -1 \end{bmatrix}$

3. $P = \dfrac{1}{35} \begin{bmatrix} 34 & -3 & 5 \\ -3 & 26 & 15 \\ 5 & 15 & 10 \end{bmatrix}$, projection $= \dfrac{1}{35} \begin{bmatrix} 86 \\ 13 \\ 25 \end{bmatrix}$

5. $P = \dfrac{1}{6} \begin{bmatrix} 5 & -1 & 2 \\ -1 & 5 & 2 \\ 2 & 2 & 2 \end{bmatrix}$, projection $= \dfrac{1}{3} \begin{bmatrix} 2 \\ 8 \\ 5 \end{bmatrix}$

7. $P = \dfrac{1}{21} \begin{bmatrix} 10 & -1 & 3 & 10 \\ -1 & 19 & 6 & -1 \\ 3 & 6 & 3 & 3 \\ 10 & -1 & 3 & 10 \end{bmatrix}$, projection $=$

$\dfrac{1}{21} \begin{bmatrix} 41 \\ 40 \\ 27 \\ 41 \end{bmatrix}$

9. $\begin{bmatrix} 1 & 0 & 0 \\ 0 & 1 & 0 \\ 0 & 0 & 0 \end{bmatrix}$ **11.** $\begin{bmatrix} 1 & 0 & 0 & 0 \\ 0 & 1 & 0 & 0 \\ 0 & 0 & 0 & 0 \\ 0 & 0 & 0 & 1 \end{bmatrix}$

13. If P is the projection matrix for a subspace W of $\mathbb{R}^n$ and if $\mathbf{b} \in \mathbb{R}^n$ is a column vector, then the projection of $\mathbf{b}$ on W is $P\mathbf{b}$. Because $P\mathbf{b}$ is in W, geometry indicates that the projection of $P\mathbf{b}$ on W is again $P\mathbf{b}$. Thus, $P(P\mathbf{b}) = P\mathbf{b}$, so $P^2\mathbf{b} = P\mathbf{b}$ and $(P^2 - P)\mathbf{b} = \mathbf{0}$ for all $\mathbf{b} \in \mathbb{R}^n$. It follows from Exercise 41 in Section 1.3 that $P^2 - P = O$, so $P^2 = P$.

15. F T T F F T F F T T   **17.** I

19. a. 0, 1

   b. 0 has geometric and algebraic multiplicity $n - k$,

   1 has geometric and algebraic multiplicity k.

   c. Because the algebraic and geometric multiplicities of each eigenvalue are equal, P is a diagonalizable matrix.

21. The $n \times n$ identity matrix I for each positive integer n.

23. $\begin{bmatrix} \frac{9}{25} & \frac{12}{25} & 0 \\ \frac{12}{25} & \frac{16}{25} & 0 \\ 0 & 0 & 1 \end{bmatrix}$

25. $\dfrac{1}{49}\begin{bmatrix} 13 & -18 & 0 & -12 \\ -18 & 36 & 12 & 0 \\ 0 & 12 & 13 & -18 \\ -12 & 0 & -18 & 36 \end{bmatrix}$

27. $\dfrac{1}{3}\begin{bmatrix} 3 & 0 & 0 & 0 \\ 0 & 2 & -1 & 1 \\ 0 & -1 & 2 & 1 \\ 0 & 1 & 1 & 2 \end{bmatrix}$ **29.** $\begin{bmatrix} 10 \\ -4 \\ -2 \end{bmatrix}$ **31.** $\begin{bmatrix} 14 \\ 0 \\ -7 \\ 14 \end{bmatrix}$

33. Referring to Figure 6.11, we see that, for $\mathbf{p} = P\mathbf{b}$, the vector from the tip of $\mathbf{b}$ to the tip of $\mathbf{p}$ is $\mathbf{p} - \mathbf{b}$, which is also the vector from the tip of $\mathbf{p}$ to the tip of $\mathbf{b}_r$. Thus, the vector $\mathbf{b}_r = \mathbf{b} + 2(\mathbf{p} - \mathbf{b}) = 2\mathbf{p} - \mathbf{b} = 2(P\mathbf{b}) - \mathbf{b} = (2P - I)\mathbf{b}$.

35. The projections are approximately
$\begin{bmatrix} 1.151261 \\ -1.184874 \\ 3.89916 \end{bmatrix}$, $\begin{bmatrix} 3 \\ 3 \\ -1 \end{bmatrix}$, and $\begin{bmatrix} 1.932773 \\ -.806723 \\ 4.378151 \end{bmatrix}$,
respectively.

37. The projections are approximately
$\begin{bmatrix} 1.864516 \\ 1.496774 \\ -.135484 \\ 2.819355 \end{bmatrix}$, $\begin{bmatrix} 1.058064 \\ .787097 \\ -.941936 \\ 2.077419 \end{bmatrix}$, and $\begin{bmatrix} 4.116129 \\ 2.574194 \\ 1.116129 \\ 3.154839 \end{bmatrix}$,
respectively.

Section 6.5

1. a. $y = \dfrac{116.4}{59} + \dfrac{60.4}{59}x$

   b. ≈ 7.092 inches

3. a. $y = .528e^{.274x}$ **b.** $\approx \$27,300$

5. $y = \dfrac{33}{35} + \dfrac{102}{35}x$

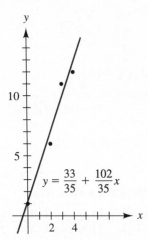

7. $y = -0.9 + 2.6x$

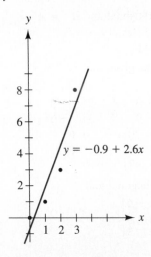

9. $y = 0.1 - 0.4x + x^2$

11. $y = 1.6 + 2x$ **13.** 4.5 min

15. Let $t = x - c$, where $c = (\sum_{i=1}^{m} a_i)/m$. The data points $(a_1 - c, b_1)$, $(a_2 - c, b_2)$, ..., $(a_m - c, b_m)$ have the property that $\sum_{i=1}^{m} (a_i - c) = 0$. Exercise 14 then shows that these data points have least-squares linear fit given by $y = r_0 + r_1 t$, where r_0 and r_1 have the values given in Exercise 14. Making the substitution $t = x - c$, we see that the data points (a_1, b_1), (a_2, b_2), ..., (a_m, b_m) have the least-squares linear fit given by $y = r_0 + r_1(x - c)$.

17. $\bar{x} = \begin{bmatrix} -\frac{1}{5} \\ \frac{3}{5} \end{bmatrix}$ **19.** $\bar{x} = \begin{bmatrix} 0 \\ 2 \\ -\frac{1}{4} \end{bmatrix}$

21. F F T T F F T T F F

23. See answer to Exercise 17.

25. See answer to Exercise 19.

27. The computer gave the fit $y = 0.7587548 + 1.311284x$ with a least-squares sum of 0.03891051.

29. We achieved a least-squares sum of 5.838961 with the expoential fit $y = 0.8e^{0.2x}$. The computer achieved a least-squares sum of 6.34004 with the exponential fit $y = 0.8874836e^{0.1960377x}$. The fit using logarithms tries to fit the smaller y-value data accurately at the expense of the larger y-value data, so that the *percent* accuracy of fit to the y-coordinates is as good as possible.

31. The computer gave the fit $y = 12.03846 - 1.526374x$ with a least-squares sum of 0.204176.

33. $y \approx 5.476 - 0.75x + 0.2738x^2$

35. $y \approx 5.632 - 1.139x + 0.1288x^2 + 0.05556x^3 + 0.01512x^4$

37. $y = -5 - 8x + 9x^2 - x^3$

CHAPTER 7

Section 7.1

1. $[-1, 1]$ **3.** $[-4, -2, 1, 5]$

5. $[3, 5, 1, 1]$ **7.** $2x^2 + 6x + 2$ **9.** $\begin{bmatrix} -1 \\ -4 \\ -2 \end{bmatrix}$

11. a. $C_{B,B'} = \begin{bmatrix} 2 & 1 & 2 \\ 3 & 2 & 0 \\ 1 & 0 & 3 \end{bmatrix}$;

b. $C_{B',B} = \begin{bmatrix} -6 & 3 & 4 \\ 9 & -4 & -6 \\ 2 & -1 & -1 \end{bmatrix}$

13. a. $C_{B,B'} = \begin{bmatrix} 1 & 1 & 1 & 1 \\ 1 & 1 & 1 & 0 \\ 1 & 1 & 0 & 0 \\ 1 & 0 & 0 & 0 \end{bmatrix}$;

b. $C_{B',B} = \begin{bmatrix} 0 & 0 & 0 & 1 \\ 0 & 0 & 1 & -1 \\ 0 & 1 & -1 & 0 \\ 1 & -1 & 0 & 0 \end{bmatrix}$

15. $C_{B',B} = \begin{bmatrix} 1 & 2 & 1 \\ -1 & -2 & -2 \\ 0 & 1 & 0 \end{bmatrix}$

17. $C_{B',B} = \begin{bmatrix} 1 & 0 & 0 & 1 \\ -1 & 1 & 0 & 0 \\ 0 & -1 & 1 & 0 \\ 0 & 0 & -1 & 1 \end{bmatrix}$

19. $C_{B,B'} = \begin{bmatrix} 1 & 2 & 1 & 0 \\ 1 & -1 & 0 & 1 \\ 0 & 1 & -1 & 1 \\ 1 & 1 & 1 & 1 \end{bmatrix}$

21. $C_{B,B'} = \begin{bmatrix} -1 & 1 \\ 1 & 1 \end{bmatrix}$

23. T F T F T F T F T T

25. $C_{B,B''} = C_{B',B''} C_{B,B'}$

Section 7.2

1. $R_B = \begin{bmatrix} 6 & 7 \\ -3 & -3 \end{bmatrix}$, $R_{B'} = \begin{bmatrix} 1 & -1 \\ 1 & 2 \end{bmatrix}$,

$C = \begin{bmatrix} -1 & 2 \\ 1 & -1 \end{bmatrix}$

3. $R_B = \begin{bmatrix} 0 & 1 & 0 \\ 2 & 0 & 1 \\ 0 & 1 & 0 \end{bmatrix}$, $R_{B'} = \begin{bmatrix} 1 & 1 & 0 \\ 1 & 0 & 1 \\ 0 & 1 & -1 \end{bmatrix}$,

$C = \begin{bmatrix} 0 & 0 & 1 \\ 0 & 1 & -1 \\ 1 & -1 & 0 \end{bmatrix}$

5. $R_B = \begin{bmatrix} \frac{13}{5} & \frac{4}{5} & 2 \\ -\frac{11}{5} & -\frac{3}{5} & -2 \\ -\frac{9}{5} & -\frac{2}{5} & -2 \end{bmatrix}$, $R_{B'} = \begin{bmatrix} -\frac{4}{3} & -\frac{1}{6} & -\frac{10}{3} \\ -\frac{4}{3} & -\frac{5}{3} & -\frac{16}{3} \\ 1 & \frac{1}{2} & 3 \end{bmatrix}$,

$C = \begin{bmatrix} 0 & -\frac{9}{5} & -\frac{8}{5} \\ 1 & \frac{13}{5} & \frac{21}{5} \\ 0 & \frac{12}{5} & \frac{14}{5} \end{bmatrix}$

7. $R_B = \begin{bmatrix} 1 & 0 & 0 \\ 0 & 1 & 0 \\ 0 & 0 & -1 \end{bmatrix}$, $R_{B'} = \frac{1}{3}\begin{bmatrix} 1 & -2 & -2 \\ -2 & 1 & -2 \\ -2 & -2 & 1 \end{bmatrix}$,

$C = \frac{1}{3}\begin{bmatrix} 1 & 1 & -2 \\ 1 & -2 & 1 \\ 1 & 1 & 1 \end{bmatrix}$

9. $R_B = \begin{bmatrix} 1 & 0 & 0 \\ 0 & 0 & 0 \\ 0 & 0 & 1 \end{bmatrix}$, $R_{B'} = \begin{bmatrix} 1 & 0 & 0 \\ 0 & 1 & 0 \\ 0 & 0 & 0 \end{bmatrix}$,

$C = \frac{1}{2}\begin{bmatrix} 1 & 0 & 1 \\ 1 & 0 & -1 \\ 0 & 2 & 0 \end{bmatrix}$

11. $R_B = \begin{bmatrix} 2 & 0 & 0 \\ 2 & 2 & 0 \\ 1 & 1 & 2 \end{bmatrix}$, $R_{B'} = \begin{bmatrix} 2 & 1 & 1 \\ 0 & 2 & 2 \\ 0 & 0 & 2 \end{bmatrix}$,

$C = \begin{bmatrix} 0 & 0 & 1 \\ 0 & 1 & 0 \\ 1 & 0 & 0 \end{bmatrix}$

13. $R_B = \begin{bmatrix} 0 & 0 & 0 & 0 \\ 3 & 0 & 0 & 0 \\ 0 & 2 & 0 & 0 \\ 0 & 0 & 1 & 0 \end{bmatrix}$, $R_{B'} = \begin{bmatrix} 0 & 1 & -2 & -3 \\ 0 & 0 & 2 & 0 \\ 0 & 0 & 0 & 3 \\ 0 & 0 & 0 & 0 \end{bmatrix}$,

$C = \begin{bmatrix} 0 & 0 & 0 & 1 \\ 0 & 0 & 1 & 0 \\ 0 & 1 & 0 & 0 \\ 1 & 1 & 1 & 1 \end{bmatrix}$

15. $\begin{bmatrix} 2 & 1 & -3 \\ -1 & 0 & 1 \\ 0 & 0 & 1 \end{bmatrix}$

17. $\lambda_1 = -1$, $\lambda_2 = 5$; $E_{-1} = \text{sp}\left(\begin{bmatrix} 1 \\ 1 \end{bmatrix}\right)$,

$E_5 = \text{sp}\left(\begin{bmatrix} -1 \\ 1 \end{bmatrix}\right)$; diagonalizable

19. $\lambda_1 = 0$, $\lambda_2 = 1$, $\lambda_3 = 2$; $E_0 = \text{sp}\left(\begin{bmatrix} -1 \\ 0 \\ 1 \end{bmatrix}\right)$,

$E_1 = \text{sp}\left(\begin{bmatrix} 0 \\ 1 \\ 0 \end{bmatrix}\right)$, $E_2 = \text{sp}\left(\begin{bmatrix} 1 \\ 0 \\ 1 \end{bmatrix}\right)$; diagonalizable

21. $\lambda_1 = -2$, $\lambda_2 = \lambda_3 = 5$; $E_{-2} = \text{sp}\left(\begin{bmatrix} 0 \\ 1 \\ 1 \end{bmatrix}\right)$,

$E_5 = \text{sp}\left(\begin{bmatrix} 0 \\ -5 \\ 2 \end{bmatrix}\right)$; not diagonalizable

23. F T T F T T T F F T

CHAPTER 8

Section 8.1

1. $U = \begin{bmatrix} 3 & -6 \\ 0 & 1 \end{bmatrix}$, $A = \begin{bmatrix} 3 & -3 \\ -3 & 1 \end{bmatrix}$

3. $U = \begin{bmatrix} 1 & -4 & 3 \\ 0 & -1 & -8 \\ 0 & 0 & 0 \end{bmatrix}$, $A = \begin{bmatrix} 1 & -2 & \frac{3}{2} \\ -2 & -1 & -4 \\ \frac{3}{2} & -4 & 0 \end{bmatrix}$

5. $U = \begin{bmatrix} -2 & 8 \\ 0 & 3 \end{bmatrix}$, $A = \begin{bmatrix} -2 & 4 \\ 4 & 3 \end{bmatrix}$

7. $U = \begin{bmatrix} 8 & 5 & -4 \\ 0 & 1 & -2 \\ 0 & 0 & 10 \end{bmatrix}$, $A = \begin{bmatrix} 8 & \frac{5}{2} & -2 \\ \frac{5}{2} & 1 & -1 \\ -2 & -1 & 10 \end{bmatrix}$

9. $\begin{bmatrix} x \\ y \end{bmatrix} = \begin{bmatrix} -1/\sqrt{2} & 1/\sqrt{2} \\ 1/\sqrt{2} & 1/\sqrt{2} \end{bmatrix}\begin{bmatrix} t_1 \\ t_2 \end{bmatrix}$, $-t_1^2 + t_2^2$

11. $\begin{bmatrix} x \\ y \end{bmatrix} = \begin{bmatrix} 3/\sqrt{10} & -1/\sqrt{10} \\ 1/\sqrt{10} & 3/\sqrt{10} \end{bmatrix}\begin{bmatrix} t_1 \\ t_2 \end{bmatrix}$, $-t_1^2 + 9t_2^2$

13. $\begin{bmatrix} x \\ y \end{bmatrix} = \begin{bmatrix} 1/\sqrt{2} & -1/\sqrt{2} \\ 1/\sqrt{2} & 1/\sqrt{2} \end{bmatrix}\begin{bmatrix} t_1 \\ t_2 \end{bmatrix}$, $t_1^2 + 5t_2^2$

15. $\begin{bmatrix} x_1 \\ x_2 \\ x_3 \end{bmatrix} = \begin{bmatrix} 1/\sqrt{3} & -1/\sqrt{2} & -1/\sqrt{6} \\ 1/\sqrt{3} & 1/\sqrt{2} & -1/\sqrt{6} \\ 1/\sqrt{3} & 0 & 2/\sqrt{6} \end{bmatrix}\begin{bmatrix} t_1 \\ t_2 \\ t_3 \end{bmatrix}$,

$-t_1^2 + 2t_2^2 + 2t_3^2$

17. $a + c = k$, $ac = b^2$

19. $-5.472136t_1^2 + 3.472136t_2^2$

21. $-4.021597t_1^2 + 1.323057t_2^2 + 4.69854t_3^2$

23. $-4t_1^2 + \frac{1}{2}t_2^2 + 4t_3^2 + \frac{11}{2}t_4^2$

Section 8.2

1. $C = \begin{bmatrix} 1/\sqrt{2} & -1/\sqrt{2} \\ 1/\sqrt{2} & 1/\sqrt{2} \end{bmatrix}$

$t_1^2 - t_2^2 = 1$

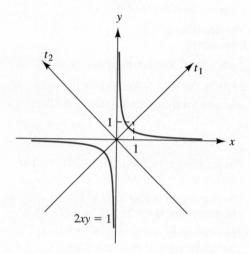

$2xy = 1$

3. $C = \begin{bmatrix} 1/\sqrt{2} & -1/\sqrt{2} \\ 1/\sqrt{2} & 1/\sqrt{2} \end{bmatrix}$

$2t_1^2 + 0t_2^2 = 4$

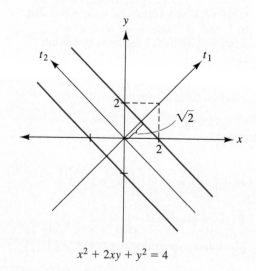

$x^2 + 2xy + y^2 = 4$

5. $C = \begin{bmatrix} 3/\sqrt{10} & -1/\sqrt{10} \\ 1/\sqrt{10} & 3/\sqrt{10} \end{bmatrix}$

$11t_1^2 + t_2^2 = 4$

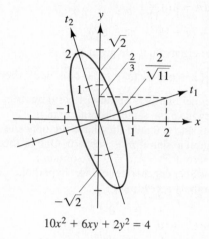

$10x^2 + 6xy + 2y^2 = 4$

7. $C = \begin{bmatrix} 1/\sqrt{5} & -2/\sqrt{5} \\ 2/\sqrt{5} & 1/\sqrt{5} \end{bmatrix}$

$7t_1^2 + 2t_2^2 = 8$

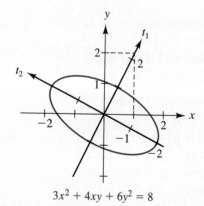

$3x^2 + 4xy + 6y^2 = 8$

9. The symmetric matrix of the quadratic-form portion is $\begin{bmatrix} a & b/2 \\ b/2 & c \end{bmatrix}$. Thus,

$$\det(A - \lambda I) = \begin{vmatrix} a - \lambda & b/2 \\ b/2 & c - \lambda \end{vmatrix} =$$

$\lambda^2 - (a + c)\lambda + ac - b^2/4$.

The eigenvalues are given by $\lambda = \left(\frac{1}{2}\right)\left(a + c \pm \sqrt{(a + c)^2 + b^2 - 4ac}\right)$; they are real numbers because A is symmetric; they have the same algebraic sign if $b^2 - 4ac < 0$, and they have the opposite algebraic sign if $b^2 - 4ac > 0$. One of them is zero if $b^2 - 4ac = 0$. We obtain a (possibly degenerate) ellipse, hyperbola, or parabola accordingly.

11. Let $C^T = \begin{bmatrix} 1 & 0 & 0 \\ 0 & b/r & c/r \\ 0 & -c/r & b/r \end{bmatrix}$, where

$r = \sqrt{b^2 + c^2}$. Then C is an orthogonal matrix such that $\det(C) = 1$ and

$$\begin{bmatrix} t_1 \\ t_2 \\ t_3 \end{bmatrix} = \mathbf{t} = C^T \begin{bmatrix} x \\ y \\ z \end{bmatrix} = \begin{bmatrix} x \\ (by + cz)/r \\ (-cy + bz)/r \end{bmatrix}.$$

Thus,

$$\begin{bmatrix} x \\ y \\ z \end{bmatrix} = C\mathbf{t} = \begin{bmatrix} t_1 \\ (bt_2 - ct_3)/r \\ (ct_2 + bt_3)/r \end{bmatrix}$$

represents a rotation of axes that transforms the equation $ax^2 + by + cz = d$ into the form $at_1^2 + rt_2 = d$.

13. Hyperboloid of two sheets

15. Hyperboloid of one sheet

17. Hyperbolic cylinder

19. Hyperboloid of one sheet

21. Circular cone or hyperboloid of one or two sheets

23. Elliptic cone or hyperboloid of one or two sheets

25. Parabolic cylinder or two parallel planes or one plane or empty

27. Hyperbolic paraboloid or hyperbolic cylinder or two intersecting planes

Section 8.3

1. g has a local minimum of -7 at $(0, 0)$.

3. g has no local extremum at $(0, 0)$.

5. g has a local maximum of 3 at $(-5, 0)$.

7. g has no local extremum at $(0, 0)$.

9. The behavior of g at $(3, 1)$ is not determined.

11. g has a local maximum of 4 at $(0, 0, 0)$.

13. g has no local extremum at $(0, 0, 0)$.

15. g has no local extremum at $(7, -6, 0)$.

17. $g(x, y) = y^2 + 10$ **19.** $g(x, y) = y^2 + x^3$

21. $g(x, y) = x^4 + y^4 + 40$

23. The maximum is .5 at $\pm(1/\sqrt{2})(1, 1)$. The minimum is $-.5$ at $\pm(1/\sqrt{2})(1, -1)$.

25. The maximum is 9 at $\pm(1/\sqrt{10})(-1, 3)$. The minimum is -1 at $\pm(1/\sqrt{10})(3, 1)$.

27. The maximum is 6 at $\pm(1/\sqrt{2})(1, -1)$. The minimum is 0 at $\pm(1/\sqrt{2})(1, 1)$.

29. The maximum is 3 at $\pm(1/\sqrt{3})(1, -1, 1)$. The minimum is 0 at $\pm(1/\sqrt{2a^2 + 2b^2 - 2ab})(a - b, a, b)$.

31. The maximum is 2 at $(\pm 1/\sqrt{2a^2 + 2b^2})(a, b, b, -a)$. The minimum is -2 at $\pm(1/\sqrt{2})(0, -1, 1, 0)$.

33. The local maximum of f is $\lambda_1 a^2$, where λ_1 is the maximum eigenvalue of the symmetric coefficient matrix of the form f; it is assumed at any eigenvector corresponding to λ_1 and of length a. An analogous statement holds for the local minimum of f.

35. g has a local maximum of 5 at $(0, 0, 0)$.

37. g has no local extremum at $(0, 0, 0)$.

Section 8.4

1. $\mathbf{w}_2 = \begin{bmatrix} 0 \\ -1 \end{bmatrix}$, $\mathbf{w}_3 = \begin{bmatrix} 1 \\ -\frac{1}{3} \end{bmatrix}$, $\mathbf{w}_4 = \begin{bmatrix} .75 \\ -1 \end{bmatrix}$

Rayleigh quotients: -2, 1, 5.2

Maximum eigenvalue 6, eigenvector $\begin{bmatrix} 1 \\ -1 \end{bmatrix}$

3. $\mathbf{w}_2 = \begin{bmatrix} 1 \\ \frac{5}{7} \end{bmatrix}$, $\mathbf{w}_3 = \begin{bmatrix} 1 \\ \frac{19}{29} \end{bmatrix}$, $\mathbf{w}_4 = \begin{bmatrix} 1 \\ \frac{65}{103} \end{bmatrix}$

Rayleigh quotients: $6, \frac{298}{74} \approx 4, \frac{4222}{1202} \approx 3.5$

Maximum eigenvalue 3, eigenvector $\begin{bmatrix} 1 \\ \frac{3}{5} \end{bmatrix}$

5. $5 \begin{bmatrix} \frac{1}{2} & \frac{1}{2} \\ \frac{1}{2} & \frac{1}{2} \end{bmatrix} - \begin{bmatrix} \frac{1}{2} & -\frac{1}{2} \\ -\frac{1}{2} & \frac{1}{2} \end{bmatrix}$

7. $2 \begin{bmatrix} \frac{2}{3} & \frac{1}{3} & \frac{1}{3} \\ \frac{1}{3} & \frac{1}{6} & \frac{1}{6} \\ \frac{1}{3} & \frac{1}{6} & \frac{1}{6} \end{bmatrix} - \begin{bmatrix} \frac{1}{3} & -\frac{1}{3} & -\frac{1}{3} \\ -\frac{1}{3} & \frac{1}{3} & \frac{1}{3} \\ -\frac{1}{3} & \frac{1}{3} & \frac{1}{3} \end{bmatrix} + 0\mathbf{b}_3\mathbf{b}_3{}^T$

9. $\begin{bmatrix} -\frac{1}{2} & \frac{1}{2} \\ \frac{1}{2} & -\frac{1}{2} \end{bmatrix}$

11. $\begin{bmatrix} -\frac{1}{3} & \frac{1}{3} & \frac{1}{3} \\ \frac{1}{3} & -\frac{1}{3} & -\frac{1}{3} \\ \frac{1}{3} & -\frac{1}{3} & -\frac{1}{3} \end{bmatrix}$

13. $\lambda = 12$, $\mathbf{v} = [-.7059, 1, -.4118]$

15. $\lambda = 6$, $\mathbf{v} = [-.9032, 1, -.4194]$

17. $\lambda = .1883$, $\mathbf{v} = [1, .2893, .3204]$

19. $\lambda_1 = 4.732050807568877$,
    $\mathbf{v}_1 = r[1, -.7320508, 1]$

    $\lambda_2 = 1.267949192431123$,
    $\mathbf{v}_2 = s[.3660254, 1, .3660254]$

    $\lambda_3 = -4$, $\mathbf{v}_3 = t[-1, 0, 1]$,
    for nonzero r, s, t

21. $\lambda_1 = 16.87586339619508$,
    $\mathbf{v}_1 = r[.9245289, 1, .3678643, .7858846]$

    $\lambda_2 = -15.93189429348535$,
    $\mathbf{v}_2 = s[-.3162426, .6635827, -1, -.004253739]$

    $\lambda_3 = 6.347821447472841$,
    $\mathbf{v}_3 = t[-.5527083, .9894762, .8356429, -1]$

    $\lambda_4 = -.291790550182573$,
    $\mathbf{v}_4 = u[-1, .06924058, .3582734, .9206089]$

    for nonzero r, s, t, and u

23. **a.** The characteristic polynomial $|A - \lambda|$
    $= \begin{vmatrix} a - \lambda & b \\ b & c - \lambda \end{vmatrix} = \lambda^2 + (-a - c)\lambda + (ac - b^2)$ has roots

$\lambda = \frac{1}{2} \left(a + c \pm \sqrt{(a + c)^2 - 4(ac - b^2)} \right)$
$= \frac{1}{2} \left(a + c \pm \sqrt{(a - c)^2 + 4b^2} \right)$.

b. If we use part a, the first row vector of $A - \lambda I$ is

$[a - \lambda, b] =$
$\left[\frac{1}{2} \left(a - c \mp \sqrt{(a - c)^2 + 4b^2} \right), b \right]$
$= [g \mp \sqrt{g^2 + b^2}, b]$.

c. From part a, eigenvectors for the matrix are $[-b, g \pm \sqrt{g^2 + b^2}] = [-b, g \pm h]$. Normalizing, we obtain $\frac{(-b, g \pm h)}{\sqrt{b^2 + (g \pm h)^2}}$. Using the upper choice of sign and setting $r = \sqrt{b^2 + (g + h)^2}$, we obtain $[-b/r, (g + h)/r]$ as the first column of C. Using the lower choice of sign and setting $s = \sqrt{b^2 + (g - h)^2}$, we obtain $[-b/s, (g - h)/s]$ as the second column of C.

d. $\det(C) = \frac{-b(g - h)}{rs} + \frac{b(g + h)}{rs} = \frac{2bh}{rs}$; because h, r, $s \geq 0$, we see that the algebraic sign of $\det(C)$ is the same as that of b.

25. $\lambda_1 = -12.00517907692924$,
    $\lambda_2 = 7.906602974286551$,
    $\lambda_3 = 17.09857610264269$

27. $\lambda_1 = -5.210618568922174$,
    $\lambda_2 = 2.856693936892428$,
    $\lambda_3 = 3.528363748899602$,
    $\lambda_4 = 7.825560883130143$

29. 5.823349919059785,
    $-11.91167495952989 \pm 1.357830063519836i$

31. 57.22941613544168,
    -92.88108454947197,
    -54.25594801085533,
    $47.45380821244281 \pm 44.48897425527453i$

CHAPTER 9

Section 9.1

1. **a.** $z + \mathbf{w} = 4 + i$, $zw = 5 + 5i$
   b. $z + \mathbf{w} = 3 + 2i$, $zw = -1 + 3i$

3. a. $|z| = \sqrt{13}$, $\bar{z} = (3 - 2i)$, $z\bar{z} =$
$(3 + 2i)(3 - 2i) = 13 = |z|^2$

   b. $|z| = \sqrt{17}$, $\bar{z} = 4 + i$, $z\bar{z} =$
$(4 - i)(4 + i) = 17 = |z|^2$

7. a. $\dfrac{3}{2} + \dfrac{1}{2}i$ **b.** $\dfrac{13}{25} + \left(-\dfrac{9}{25}\right)i$

9. a. Modulus $2\sqrt{2}$, principal argument $3\pi/4$

11. 16   **17.** F T F F T F F T F T

19. $\sqrt{2} + \sqrt{2}i$, $-\sqrt{2} + \sqrt{2}i$, $-\sqrt{2} - \sqrt{2}i$,
$\sqrt{2} - \sqrt{2}i$

21. $1, i, -1, -i$

23. $2, \sqrt{2} + \sqrt{2}i, 2i, -\sqrt{2} + \sqrt{2}i, -2,$
$-\sqrt{2} - \sqrt{2}i, -2i, \sqrt{2} - \sqrt{2}i$

Section 9.2

3. $AB = \begin{bmatrix} -3 + 2i & 2i & 2i \\ 2 & 2i & 1 \\ 2 + 3i & -1 + i & 2 + i \end{bmatrix}$,

$BA = \begin{bmatrix} -2 + 2i & i & 2 - i \\ 2 + 3i & 1 + 3i & 0 \\ 2i & -1 + i & 0 \end{bmatrix}$

5. $\dfrac{1}{3}\begin{bmatrix} 2 + i & -i \\ -1 - i & 1 \end{bmatrix}$

7. $\dfrac{1}{10}\begin{bmatrix} 9 - 3i & 1 + 3i & -4 + 8i \\ -3 + i & 3 - i & -2 - 6i \\ -2 + 4i & 2 - 4i & 2 - 4i \end{bmatrix}$

9. $z = \dfrac{1}{10}\begin{bmatrix} -7 + 9i \\ 9 - 3i \\ 6 - 2i \end{bmatrix}$ **11.** $\text{sp}\left(\begin{bmatrix} 1 + i \\ 1 + 3i \\ 2 \end{bmatrix}\right)$

13. 3   **15. a.** $\langle u, v \rangle = 0$, $\langle v, u \rangle = 0$

          b. $\langle u, v \rangle = 5 - 3i$, $\langle v, u \rangle = 5 + 3i$

21. a. Perpendicular   **d.** Parallel

     b. Parallel   **e.** Perpendicular

     c. Neither

23. $\dfrac{2}{\sqrt{7}}[i, 1 - i, 1 + i, 1 - i]$

25. $[-3i, 1, 2 + 2i]$

27. $\{[2 + i, 1 + i], [1 - i, -2 + i]\}$

29. $\{[1, i, i], [1 + 3i, 3 - 2i, i],$
$[1 + i, i, 1 - 2i]\}$

31. a. Both

     b. Hermitian but not unitary

     c. Not Hermitian but unitary

     d. Neither

33. T T F F T T T T F F

41. Diagonal matrices with entries of modulus 1 on the diagonal.

M1. See answer to Exercise 3.

M3. $\begin{bmatrix} -i & 1 + i & 0 \\ 1 + i & -1 + i & 1 \\ -1 + i & -1 - 2i & i \end{bmatrix}$ **M5.** $\begin{bmatrix} 2 + 4i \\ -4 + i \\ -6i \end{bmatrix}$

M7. 2

M9. Entering **[Q, R] = qr(A)**, where A is the matrix having the given vectors $\mathbf{a}_1, \mathbf{a}_2, \mathbf{a}_3$ as column vectors, returns a matrix Q having as column vectors an *orthonormal* basis $\{\mathbf{q}_1, \mathbf{q}_2, \mathbf{q}_3\}$, where

$\mathbf{q}_1 \approx [-0.5774, -0.5774i, -0.5774i]$,

$\mathbf{q}_2 \approx [-0.4695 - 0.1719i,$
$-0.4695 - 0.1719i, 0.2977 + 0.6414i]$,

$\mathbf{q}_3 \approx [0.4695 + 0.429i, 0.0726 - 0.6414i,$
$\qquad\qquad\qquad 0.3703 + 0.1719i]$.

To check, using MATLAB, the Student's Solutions Manual's answers

$\mathbf{v}_1 = \mathbf{a}_1$, $\mathbf{v}_2 = \mathbf{a}_2$, $\mathbf{v}_3 = [1 - 3i, -3 + i, -2i]$

for an *orthogonal* basis, enter

((1−i)/Q(1,1))∗Q(:,1) to check $\mathbf{v}_1$, enter
(1/Q(1,2))∗Q(:,2) to check $\mathbf{v}_2$, and enter
((1−3∗i)/Q(1,3))∗Q(:,3) to check $\mathbf{v}_3$.

M11. a. $\sqrt{274}, \sqrt{476}, \sqrt{458}$, and $\sqrt{353}$ for rows 1, 2, 3, and 4, respectively.

     b. $\sqrt{277}, \sqrt{192}, \sqrt{529}, \sqrt{124}$, and $\sqrt{439}$ for columns 1, 2, 3, 4, and 5, respectively.

     c. $-45 - 146i$

     d. $31 + 14i$

Section 9.3

1. $U = \dfrac{1}{\sqrt{2}}\begin{bmatrix} -i & i \\ 1 & 1 \end{bmatrix}$, $D = \begin{bmatrix} 0 & 0 \\ 0 & 2 \end{bmatrix}$

3. $U = \begin{bmatrix} (1+i)/\sqrt{3} & (1+i)/\sqrt{6} \\ -1/\sqrt{3} & 2/\sqrt{6} \end{bmatrix}$, $D = \begin{bmatrix} 0 & 0 \\ 0 & 3 \end{bmatrix}$

5. $U = \dfrac{1}{\sqrt{2}} \begin{bmatrix} -i & 0 & i \\ 1 & 0 & 1 \\ 0 & \sqrt{2} & 0 \end{bmatrix}$, $D = \begin{bmatrix} -1 & 0 & 0 \\ 0 & 1 & 0 \\ 0 & 0 & 1 \end{bmatrix}$

7. $U = \begin{bmatrix} (1-i)/\sqrt{6} & 0 & (1-i)/\sqrt{3} \\ -2\sqrt{6} & 0 & 1 \\ 0 & 1 & 0 \end{bmatrix}$,

$D = \begin{bmatrix} -3 & 0 & 0 \\ 0 & 3 & 0 \\ 0 & 0 & 3 \end{bmatrix}$

9. $U = \begin{bmatrix} (1+i)/\sqrt{6} & 0 & (1+i)/\sqrt{3} \\ 0 & 1 & 0 \\ -2/\sqrt{6} & 0 & 1/\sqrt{3} \end{bmatrix}$,

$D = \begin{bmatrix} -3 & 0 & 0 \\ 0 & -3 & 0 \\ 0 & 0 & 3 \end{bmatrix}$

11. $U = \begin{bmatrix} (-1-i)/\sqrt{8} & 0 & (3+3i)/\sqrt{24} \\ (1-i)/\sqrt{8} & (1+i)/\sqrt{3} & (1-i)/\sqrt{24} \\ 2/\sqrt{8} & -i/\sqrt{3} & 2/\sqrt{24} \end{bmatrix}$,

$D = \begin{bmatrix} 0 & 0 & 0 \\ 0 & 1 & 0 \\ 0 & 0 & 4 \end{bmatrix}$

13. $\{a \in \mathbb{C} \mid |a| = 4\}$ **15.** $a = -1$

19. F T T F T F T T F F

Section 9.4

1. Yes **3.** No **5.** No

7. a. $\lambda_1 = \lambda_2 = \lambda_3 = \lambda_4 = -2$.

b. $J + 2I$ has rank 3 and nullity 1,

$(J + 2I)^2$ has rank 2 and nullity 2,

$(J + 2I)^3$ has rank 1 and nullity 3,

$(J + 2I)^k$ has rank 0 and nullity 4 for $k \geq 4$.

c. $J + 2I$: $\mathbf{e}_4 \to \mathbf{e}_3 \to \mathbf{e}_2 \to \mathbf{e}_1 \to \mathbf{0}$.

d. $J\mathbf{e}_1 = -2\mathbf{e}_1$, $J\mathbf{e}_2 = \mathbf{e}_1 - 2\mathbf{e}_2$,
$J\mathbf{e}_3 = \mathbf{e}_2 - 2\mathbf{e}_3$, $J\mathbf{e}_4 = \mathbf{e}_3 - 2\mathbf{e}_4$.

9. a. $\lambda_1 = -1$, $\lambda_2 = \lambda_3 = \lambda_4 = \lambda_5 = 2$.

b. $(J + I)^k$ has rank 4 and nullity 1 for $k \geq 1$,

$(J - 2I)$ has rank 3 and nullity 2,

$(J - 2I)^k$ has rank 1 and nullity 4 for $k \geq 2$.

c. $J + I$: $\mathbf{e}_1 \to \mathbf{0}$,

$J - 2I$: $\mathbf{e}_3 \to \mathbf{e}_2 \to \mathbf{0}$, $\mathbf{e}_5 \to \mathbf{e}_4 \to \mathbf{0}$.

d. $J\mathbf{e}_1 = -\mathbf{e}_1$, $J\mathbf{e}_2 = 2\mathbf{e}_2$, $J\mathbf{e}_3 = \mathbf{e}_2 + 2\mathbf{e}_3$,
$J\mathbf{e}_4 = 2\mathbf{e}_4$, $J\mathbf{e}_5 = \mathbf{e}_4 + 2\mathbf{e}_5$.

11. $\begin{bmatrix} 3 & 0 & 0 & 0 & 0 \\ 0 & 3 & 1 & 0 & 0 \\ 0 & 0 & 3 & 1 & 0 \\ 0 & 0 & 0 & 3 & 1 \\ 0 & 0 & 0 & 0 & 3 \end{bmatrix}$

13. $\begin{bmatrix} 1 & 1 & 0 & 0 & 0 & 0 & 0 & 0 \\ 0 & 1 & 0 & 0 & 0 & 0 & 0 & 0 \\ 0 & 0 & 1 & 1 & 0 & 0 & 0 & 0 \\ 0 & 0 & 0 & 1 & 1 & 0 & 0 & 0 \\ 0 & 0 & 0 & 0 & 1 & 0 & 0 & 0 \\ 0 & 0 & 0 & 0 & 0 & -2 & 0 & 0 \\ 0 & 0 & 0 & 0 & 0 & 0 & -2 & 0 \\ 0 & 0 & 0 & 0 & 0 & 0 & 0 & -2 \end{bmatrix}$

15. $\begin{bmatrix} 0 & 1 \\ 0 & 0 \end{bmatrix}$, $\left\{ \begin{bmatrix} 2 \\ 5 \end{bmatrix}, \begin{bmatrix} 1 \\ 3 \end{bmatrix} \right\}$ (Other bases are possible.)

17. $\begin{bmatrix} 1 & 0 & 0 \\ 0 & 4 & 1 \\ 0 & 0 & 4 \end{bmatrix}$, $\left\{ \begin{bmatrix} 0 \\ 1 \\ 0 \end{bmatrix}, \begin{bmatrix} 0 \\ 5 \\ 5 \end{bmatrix}, \begin{bmatrix} 1 \\ 0 \\ 1 \end{bmatrix} \right\}$ (Other answers are possible.)

19. $\begin{bmatrix} 2 & 1 & 0 & 0 & 0 \\ 0 & 2 & 0 & 0 & 0 \\ 0 & 0 & -1 & 1 & 0 \\ 0 & 0 & 0 & -1 & 0 \\ 0 & 0 & 0 & 0 & -1 \end{bmatrix}$, $\left\{ \begin{bmatrix} 0 \\ 1 \\ 0 \\ 0 \\ 0 \end{bmatrix}, \begin{bmatrix} 5 \\ 0 \\ 0 \\ 0 \\ 0 \end{bmatrix}, \begin{bmatrix} 0 \\ 0 \\ -1 \\ 0 \\ 0 \end{bmatrix}, \begin{bmatrix} 0 \\ 0 \\ 0 \\ 0 \\ 1 \end{bmatrix}, \begin{bmatrix} 0 \\ 0 \\ 1 \\ 1 \\ 0 \end{bmatrix} \right\}$

(Other answers are possible.)

21. $\begin{bmatrix} 2 & 1 & 0 & 0 & 0 \\ 0 & 2 & 0 & 0 & 0 \\ 0 & 0 & 2 & 0 & 0 \\ 0 & 0 & 0 & 2 & 0 \\ 0 & 0 & 0 & 0 & 2 \end{bmatrix}$, $\{\mathbf{e}_1 + \mathbf{e}_3, \mathbf{e}_5, \mathbf{e}_2, \mathbf{e}_4, \mathbf{e}_1 - \mathbf{e}_3\}$

(Other answers are possible.)

23. T F T T F T T F F **25.** O **27.** O

29. $A^4 + (3 - i)A^3 + (3 - 3i)A^2 + (1 - 3i)A - iI$

CHAPTER 10

Section 10.1

1. There are $n - 1$ flops performed on **b** while the first column of A is being fixed up, $n - 2$ flops while the second column of A is being fixed up, and so on. The total number is $(n - 1) + (n - 2) + \cdots + 2 + 1 = n(n - 1)/2$, which has order of magnitude $n^2/2$ for large n.

3. mn, if we call each indexed addition a flop

5. mn^2 7. $2n^3$ 9. $3n^3$ 11. $6n^3$

13. $3n^3/2$ 15. n^3 17. $2n$

19. w^2n, counting each final division as a flop

21. F T F F T T F F T F

23. (No text answer data are possible for this problem, since different computers run at different speeds. However, for n large enough to require 6 seconds or more to solve an $n \times n$ system, our computer did require roughly 50% more time when using the Gauss–Jordan method than when using the Gauss method with back substitution.)

25. See answer to Exercise 23.

Section 10.2

1. It is not significant; no arithmetic operations are involved, just storing of indexed values.

3. $\begin{bmatrix} -18 \\ 5 \end{bmatrix}$ 5. $\begin{bmatrix} -27 \\ -11 \\ -1 \end{bmatrix}$

7. $\begin{bmatrix} 1 \\ -2 \\ 1 \\ 3 \end{bmatrix}$

9. $P = \begin{bmatrix} 1 & 0 & 0 \\ 0 & 0 & 1 \\ 0 & 1 & 0 \end{bmatrix}$, $L = \begin{bmatrix} 1 & 0 & 0 \\ 1 & 1 & 0 \\ 3 & 0 & 1 \end{bmatrix}$,

$U = \begin{bmatrix} 2 & 1 & -3 \\ 0 & -2 & 8 \\ 0 & 0 & 1 \end{bmatrix}$, $\mathbf{x} = \begin{bmatrix} 2 \\ -1 \\ -1 \end{bmatrix}$

11. $L = \begin{bmatrix} 1 & 0 & 0 \\ 3 & 1 & 0 \\ 4 & -10 & 1 \end{bmatrix}$, $U = \begin{bmatrix} 1 & 2 & -1 \\ 0 & 1 & 5 \\ 0 & 0 & 55 \end{bmatrix}$,

$\mathbf{x} = \begin{bmatrix} -1 \\ 0 \\ 2 \end{bmatrix}$

13. $L = \begin{bmatrix} 1 & 0 & 0 & 0 \\ 2 & 1 & 0 & 0 \\ -1 & 0 & 1 & 0 \\ 4 & 2 & -\frac{3}{2} & 1 \end{bmatrix}$, $U = \begin{bmatrix} 1 & 2 & -3 & 0 \\ 0 & 1 & 0 & 0 \\ 0 & 0 & -2 & -1 \\ 0 & 0 & 0 & -\frac{1}{2} \end{bmatrix}$,

$\mathbf{x} = \begin{bmatrix} 3 \\ 1 \\ -1 \\ 2 \end{bmatrix}$

15. $\begin{bmatrix} 0 \\ -1 \end{bmatrix}$ 17. $\begin{bmatrix} -1 \\ 2 \\ 2 \end{bmatrix}$

19. $LDU = \begin{bmatrix} 1 & 0 & 0 \\ 0 & 1 & 0 \\ 3 & -4 & 1 \end{bmatrix} \begin{bmatrix} 1 & 0 & 0 \\ 0 & 1 & 0 \\ 0 & 0 & 14 \end{bmatrix} \begin{bmatrix} 1 & 1 & -3 \\ 0 & 1 & 1 \\ 0 & 0 & 1 \end{bmatrix}$

21. F F T F T F T F F T

23. Display:

$$\begin{bmatrix} 1 & 3 & -5 & 2 & 1 \\ 4 & -18 & 30 & 0 & -1 \\ 3 & .1666667 & 9 & -2 & 4.166667 \\ 2 & .2777778 & 1.407407 & -4.185185 & 5.413581 \\ -6 & -1.166667 & .6666667 & -4.141593 & 45.4764 \end{bmatrix}.$$

For $\mathbf{b}_1$, $\mathbf{x} = \begin{bmatrix} 1 \\ -1 \\ 0 \\ 2 \\ -2 \end{bmatrix}$; for $\mathbf{b}_2$, $\mathbf{x} = \begin{bmatrix} -5 \\ -8 \\ 9 \\ 11 \\ 4 \end{bmatrix}$; for $\mathbf{b}_3$, $\mathbf{x} = \begin{bmatrix} -8 \\ 9 \\ -11 \\ 3 \\ 7 \end{bmatrix}$

25. $\begin{bmatrix} .4814815 \\ -1.592593 \end{bmatrix}$ 27. $\begin{bmatrix} .8125 \\ 1.875 \\ -.3125 \end{bmatrix}$

29. $\begin{bmatrix} -.09131459 \\ -.08786787 \\ -.4790014 \\ -.01573263 \\ .08439226 \\ .2712737 \end{bmatrix}$

31. The ratios roughly conform on our computer. One solution that requires 100 flops took about $\frac{1}{3}$ of the time to form the L/U display, which requires about $\frac{1000}{3}$ flops, etc.

33. The ratios roughly conform on our computer, in the sense explained in the answer to Exercise 31.

Section 10.3

1. $x_1 = 0$, $x_2 = 1$

3. $x_1 = 1$, $x_2 = .9999$. Yes, it is reasonably accurate. $10x_1 + 1000000x_2 = 1000000$, $-10x_1 + 20x_2 = 10$ is a system that can't be solved without pivoting by a five-figure computer.

5. $x_1 + 10^{19}x_2 = 10^{19}$, $-x_1 + 2x_2 = 1$

7. We need to show that $(n(nA)^{-1})A = I$. It is easy to see that $(rA)B = A(rB) = r(AB)$ for any scalar r and matrices A and B such that AB is defined. Thus we have $(n(nA)^{-1})A = (nA)^{-1}(nA) = I$.

9. On our computer, $[x_1, x_2] = [-10^{-9}, 10^{-9}]$, which is approximately correct.

11. MATCOMP yielded the solution $x_1 = -10^{-9}$, $x_2 = 10^{-9}$, which is approximately correct for the system

$$10^{-9}x_1 + \quad 10^9x_2 = 1$$
$$10^9x_1 + 2(10^9)x_2 = 1.$$

13. Both with and without the scaling routine, MATCOMP successfully inverted the given matrix on our computer.

15. For **b**, we have $\mathbf{x} = \begin{bmatrix} -46 \\ 96 \end{bmatrix}$, and for **c**, we have $\mathbf{x} = \begin{bmatrix} -36 \\ 78 \end{bmatrix}$.

The two components of **b** differ from the corresponding ones of **c** by 10 and by 18.

17. For **b**, we have $\mathbf{x} = \begin{bmatrix} 32 \\ 240 \\ -1500 \\ 1400 \end{bmatrix}$, and for **c**, we

have $\mathbf{x} = \begin{bmatrix} 268 \\ -2100 \\ 3720 \\ -1820 \end{bmatrix}$. Corresponding

components of the solutions of $H_4\mathbf{x} = \mathbf{b}$ and of $H_4\mathbf{x} = \mathbf{c}$ differ by as much as 5220.

19. We find that $H_2^{-1} = \begin{bmatrix} 4 & -6 \\ -6 & 12 \end{bmatrix}$.

21. We were able to complete a reduction for inverses of the 1st, 2nd, 4th, 8th, and 16th powers of H_4. The inverse checks for H_4 and H_4^2 were pretty good, and we expect that

$$H_4^{-1} = \begin{bmatrix} 16 & -120 & 240 & -140 \\ -120 & 1200 & -2700 & 1680 \\ 240 & -2700 & 6480 & -4200 \\ -140 & 1680 & -4200 & 2800 \end{bmatrix}$$

and

$$(H_4^2)^{-1} \approx \begin{bmatrix} 91856 & -1029120 & 2471040 & -1603840 \\ -1029120 & 11566800 & -27820800 & 18076800 \\ 2471040 & -27820800 & 66978000 & -43545600 \\ -1603840 & 18076800 & -43545600 & 28322000 \end{bmatrix}.$$

The entries in our "inverse" of H_4^4 were of the right order of magnitude; if the entries of H_4^m are of order of magnitude roughly 10^s, then we expect the entries of H_4^{2m} to be of order roughly 10^{2s}. Also, our inverse matrix for H_4^4 was symmetric, but the inverse check was not very good. For H_4^8 and H_4^{16}, our "inverses" contained entries of completely wrong orders of magnitude, and the inverse matrices were not symmetric.

M1. For $n = 5$, $\mathbf{b} \approx [1, 0]$;

for $n = 10$, $\mathbf{b} \approx [1, 0]$;

for $n = 15$, $\mathbf{b} \approx 10^3[1.68, -2.90]$;

for $n = 20$, $\mathbf{b} \approx 10^{10}[6.265, -6.688]$;

for $n = 25$, $\mathbf{b} \approx 10^{18}[2.09, -3.09]$;

for $n = 30$, $\mathbf{b} \approx 10^{25}[-6.56, 6.97]$.

APPENDIX A

1. Let $P(n)$ be the equation to be proved. Clearly $P(1)$ is true, because

$$1 = \frac{1(1 + 1)(2 + 1)}{6}.$$ Assume that $P(k)$ is true.

Then

$$1^2 + 2^2 + 3^2 + \cdots + k^2 + (k + 1)^2$$

$$= \frac{k(k + 1)(2k + 1)}{6} + (k + 1)^2$$

$$= \frac{(k + 1)(2k^2 + k + 6k + 6)}{6}$$

$$= \frac{(k + 1)(k + 2)(2k + 3)}{6}.$$

Thus, $P(k + 1)$ is true, and $P(n)$ holds for all $n \in Z^+$.

3. Let $P(n)$ be the equation to be proved. We see that $P(1)$ is true. Assume that $P(k)$ is true. Then

$$1 + 3 + 5 + \cdots + (2k - 1) + (2k + 1)$$
$$= k^2 + 2k + 1 = (k + 1)^2,$$

as required. Therefore, $P(n)$ holds for all $n \in Z^+$.

5. Let $P(n)$ be the equation to be proved. We see that $P(1)$ is true, because
$a(1 - r^2)/(1 - r) = a(1 + r) = a + ar$.
Assume that $P(k)$ is true. Then

$$a + ar + ar^2 + \cdots + ar^k + ar^{k+1}$$

$$= \frac{a(1 - r^{k+1})}{1 - r} + ar^{k+1}$$

$$= \frac{a(1 - r^{k+1} + r^{k+1}(1 - r))}{1 - r}$$

$$= \frac{a(1 - r^{k+2})}{1 - r},$$

which establishes $P(k + 1)$. Therefore, $P(n)$ is true for all $n \in Z^+$.

7. The notion of an "interesting property" has not been made precise; it is not well defined. Moreover, we work in mathematics with two-valued logic: a statement is either true or false, *but not both*. The assertion that not having an interesting property would be an interesting property seems to contradict this two-valued logic. We would be saying that the integer both has and does not have an interesting property.

INDEX